World *of* Chemistry

FOURTH EDITION

ESSENTIALS

Melvin Joesten
Vanderbilt University

Mary E. Castellion

John L. Hogg
Texas A&M University

THOMSON
™
BROOKS/COLE

Australia • Brazil • Canada • Mexico • Singapore • Spain
United Kingdom • United States

THOMSON

BROOKS/COLE

World of Chemistry: Essentials,
Fourth Edition
Melvin Joesten, Mary E. Castellion, John L. Hogg

Publisher: DAVID HARRIS
Acquisitions Editor: LISA LOCKWOOD
Development Editor: ELLEN BITTER
Editorial Assistant: LAUREN OLIVEIRA
Technology Project Manager: ERICKA YEOMAN
Marketing Manager: AMEE MOSLEY
Marketing Assistant: MICHELE COLELLA
Advertising Project Manager: BRYAN VANN
Project Manager, Editorial Production: SHELLEY RYAN
Creative Director: ROB HUGEL
Art Director: LEE FRIEDMAN
Print Buyer: DOREEN SURUKI
Permissions Editor: BOB KAUSER

Production Service: PRE-PRESS COMPANY, INC.
Text Designer: JOHN WALKER
Photo Researcher: DENA DIGILIO BETZ
Copy Editor: LAURENE SORESEN
Illustrator: PRE-PRESS COMPANY, INC.
Cover Designer: JEREMY MENDE
Cover Images (left to right): © GUY GRENIER/MASTERFILE;
ALBERT NORMANDIN/MASTERFILE: PHOTO RESEARCHERS:
GARRY BLACK/MASTERFILE; MASTERFILE; MAIN IMAGE:
COURTESY OF NASA
Cover Printer: QUEBECOR WORLD-DUBUQUE
Compositor: PRE-PRESS COMPANY, INC.
Printer: QUEBECOR WORLD-DUBUQUE

Printed in the United States of America
1 2 3 4 5 6 7 10 09 08 07 06

ISBN: 0-495-01213-0

Library of Congress Control Number: 2005937827

Thomson Higher Education
10 Davis Drive
Belmont, CA 94002-3098
USA

For more information about our products, contact us at:

Thomson Learning Academic Resource Center
1-800-423-0563

For permission to use material from this text or product, submit a request online at **http://www.thomsonrights.com.** Any additional questions about permissions can be submitted by e-mail to **thomsonrights@thomson.com.**

Dedication

I would like to dedicate this book to the people I love most in the world:
Janet, Ryan, Krista, John, my parents, and my brother.
–John Hogg

We dedicate this book to the memory of our longtime editor and friend,
John Vondeling (1993–2001).

Brief Contents

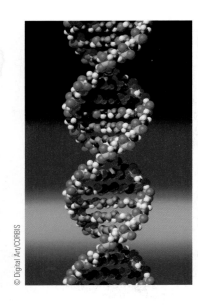

© Digital Art/CORBIS

Contents

Page 126

Hematite, Fe_2O_3

Calcite, $CaCO_3$

Gypsum, $CaSO_4 \cdot 2 H_2O$

Fluorite, CaF_2

Orpiment, As_2S_3

Charles D. Winters

Page 93

Charles D. Winters

Page 171

Charles D. Winters

Page 339

Charles D. Winters

Page 60

© CHAIWAT SUBRASOM/Reuters/CORBIS

Page 279

Index to Boxed Features

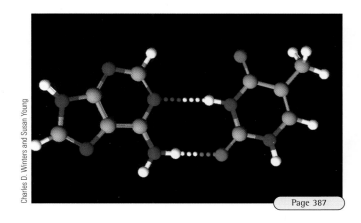

Charles D. Winters and Susan Young

Page 387

Preface

World of Chemistry: Essentials is designed for a one-semester chemistry course for non-science majors. The fundamental approach of the text is to teach the basic chemical principles within a framework of real world applications. Our goal is to teach students the chemical concepts they need in order to make appropriate consumer decisions, to read about and understand chemically-related issues in the media, to make informed decisions as voters, and to appreciate the significance of chemistry's impact on their daily lives.

Charles D. Winters

Changes to the Fourth Edition

In revising this edition, we have remained focused on our goal of teaching chemical concepts in the context of real world applications. With this goal and reviewer feedback in mind, we have reorganized and added content, including unique features.

First and foremost in our list of changes is the addition of new co-author John L. Hogg, professor of chemistry and Thamann Professor of Undergraduate Teaching Excellence at Texas A&M University. He received his BS degree in chemistry at Southwestern State College in Oklahoma (1970) and his PhD from the University of Kansas (1974). Hogg joined the faculty at Texas A&M University in 1975. He has nearly 40 publications in the general areas of physical organic and bioorganic chemistry. He has taught the Liberal Arts Chemistry class and sophomore Organic Chemistry at Texas A&M nearly every semester since 1993 and taught a mixture of graduate and undergraduate organic chemistry courses prior to that time. He is the chief advisor for all undergraduate chemistry majors at Texas A&M and has written a monthly newsletter for them during the academic year since 1985. He has received the Texas A&M University Association of Former Students Faculty Distinguished Achievement Award in Teaching (1982 and 2005) and in Student Relationships (1989). He is one of only three faculty members to have received three such awards in the more than 50-year history of the awards. He also received a College of Science teaching award (2002) and the Mervin and Annette Peters Advising Award as the outstanding undergraduate faculty advisor at Texas A&M in 2004. Dr. Hogg was one of the first four faculty members at Texas A&M to receive the title of University Professor of Undergraduate Teaching Excellence (1996), and in 2002 he was honored with the named professorship he now holds. He has performed an estimated 250 chemistry demonstration shows for over 40,000 people.

ORGANIZATION: A major organizational change from the third to fourth edition has been the conversion and expansion of the topics previously found in Chapter 16 (Air: The Precious Canopy) into three separate chapters. This material, now Chapter 4 (The Air We Breathe), Chapter 6 (Carbon Dioxide and the Greenhouse Effect), and Chapter 7 (Chlorofluorocarbons and the Ozone Layer), has been moved to a more prominent place early in the text. The very limited treatment of acid rain in Chapter 16 of the third edition has been expanded and combined with Chapter 15 (Water, Water Everywhere, But Not a Drop to Drink?) to create the significantly modified Chapter 11 (Water and Acid Rain) which comes earlier in the text. Most of the discussion of the chemistry of water in Chapter 5 of the third edition has been moved to this new Chapter 11. The major discussion of water now occurs after the students are more familiar with chemical bonding, states of matter, chemical reactions, and, more importantly, acid-base chemistry. This allows a better appreciation of the special chemistry and properties of water and its critical importance to life, and promotes discussion of environmental issues related to water. These changes reflect a desire to encourage students to get excited about the course material by presenting relevant environmental issues earlier in the text. These environmental issues are in the news on an almost daily basis, and therefore are easily used to illustrate the importance of fundamental chemistry knowledge for the non-science major.

In addition, Nuclear Changes and Nuclear Power (Chapter 4 in the third edition) has been moved to follow Chapter 12 (Energy and Hydrocarbons) since the major focus of the chapter is on nuclear power and its beneficial uses, as opposed to history and the atomic bomb. This placement logically follows a discussion of the use of fossil fuels and the problems associated with their diminishing supply. It also reinforces atomic concepts and encourages students to think about atoms in some detail again later in the semester as they prepare for a final examination.

Throughout the text, new artwork, photography, examples, and problems have been added. Several additions have been made to The Personal Side of Chemistry features.

CHEMISTRY IN THE MOVIES BOXES: This new feature, authored by Mark Griep, associate professor of chemistry at the University of Nebraska-Lincoln, uses innovative boxed material to relate in-chapter chemistry concepts to popular and classic film content. Problems are included in each movie box for possible student assignments. Answers are provided on the instructor website, along with video clip information for each film. Comments and suggestions on the movie boxes can be directed to Mark Griep at mgriep1@unl.edu.

CHEMISTRY ANIMATIONS: Margin annotations throughout the text direct students to the Joesten website where they can work through animations and simulations related to specific chapter content.

CHEMISTRY ON THE WEB: This end of chapter section lists websites related to material referenced in the chapter. These websites allow further exploration of important topics discussed within the chapter and often provide unique features not easily included within a textbook. For instance, one link allows real-time monitoring of ozone concentrations in major cities and regions of the United States. Website URLs are provided on the Joesten student website.

Chapter-by-Chapter Changes and Additions

Chapter 1 continues to set the philosophical tone for the book. Three major and current societal issues involving chemistry are introduced. An introduction to fossil fuel use and its relationship to global warming has been added. There is an expanded discussion of risk/benefit analysis.

Chapters 2 and 3 introduce students to the fundamental properties of the atom, the electromagnetic spectrum, chemical compounds, heterogeneous and homogeneous mixtures, the quantitative side of science, and the symbolism and importance of the periodic table and periodic trends. Content and application updates in the new edition include a new section (Beyond the Bohr Atom) dealing with orbital shapes and energies.

Chapter 4 is a mostly new chapter dealing with air and pollution. It contains a significantly expanded treatment of this material in comparison to the previous edition. There is considerably more discussion of the relationships between percent, ppm, and ppb and the use of these concentration units in everyday situations. There is also a new discussion on automobile pollution regulations.

Chapter 5 introduces bonding, both ionic and covalent, and includes a new discussion of electronegativity and its utility in predicting bond types and the shapes of molecules. There is added emphasis on Lewis dot structures, and new material dealing with resonance structures and the use of VSEPR theory to predict molecular shapes. A discussion of molecular geometry versus electron pair geometry has been added. There is a new section on reversible changes of state and use of the concepts of endothermic and exothermic heat changes. The states of matter are introduced. The majority of the discussion of water has been moved to a later chapter.

Chapter 6 is an entirely new chapter focusing on carbon dioxide, the greenhouse effect, and global warming. This has been expanded from the two pages devoted to this issue in the previous edition. The global warming discussion focuses on a comparison of greenhouse gases, the scientific status of the phenomenon, and political responses to it. Pardon the pun, but this is a hot topic these days!

Chapter 7 is a new chapter on the relationship between CFCs and the ozone hole, expanded from the four pages in the previous edition. The mechanisms by which oxygen and ozone protect us are discussed in the context of the Chapman Cycle. Scientific responses to the issue and the replacement of CFCs are considered. There is also discussion addressing the relationship between UV exposure and skin cancer.

Chapter 8 continues to focus on chemical equations and what they mean as well as on the concept of the mole. The discussion of reaction rates and pathways now includes an expanded section on energy profiles and an introduction to enthalpy of reaction. The importance of recycling remains a feature of this chapter.

Chapter 9 focuses on acids and bases and their chemistry. This chapter now explains how to quickly estimate pH from hydronium ion or hydroxide ion concentrations to a reasonable degree of accuracy without the use of calculators. The non-linearity of the pH scale is emphasized and the blood buffering system is introduced. The practical acid-base properties of many common household chemicals are discussed.

Chapter 10 explores oxidation and reduction chemistry. There is a new treatment of fuel cells and hybrid vehicles to illustrate the relevance of this chemistry topic.

Chapter 11 is a new chapter including material from Chapters 5, 15, and 16 of the previous edition. The focus is on water, its unique properties and its importance for life, acid rainfall and water pollution, and water purification. There is a significantly expanded discussion illustrating the uniqueness of water by comparing it with several compounds of comparable size and molecular weight. The issue of water shortage problems worldwide is discussed.

Chapter 12 introduces organic chemistry by focusing on the energy content of hydrocarbons and their use as fuels. The chapter includes an expanded section on the use of bond energies to calculate heat changes associated with the combustion of various fuels, allowing a more complete comparison of various fuels. Wedge-dash or sawhorse projections of hydrocarbons have been introduced. The chair conformation of cyclohexane has been added to allow comparison to aromatic compounds such as benzene. Alcohols are also introduced in this chapter as their potential as fuels is considered.

Chapter 13 is Chapter 4 from the previous edition and focuses on fundamental nuclear chemistry. Useful applications of radioactivity are considered. In this context, a discussion of the use of radiocarbon dating has been added. A major focus of the chapter is the use of energy from nuclear reactions and this discussion has driven the chapter placement after the chapter on hydrocarbon fuels. Comparisons between low-level radioactive waste and high-level radioactive waste have been added.

Chapter 14 expands the coverage of organic chemistry beyond that introduced in Chapter 12 to include the major functional groups. Ethylene glycol poisoning is used to illustrate some functional group chemistry of alcohols and carboxylic acids and reinforces the importance of the roles of acid-base chemistry and ionic bonding in this phenomenon. There is a new discussion of waterlock polymers and their applications, used as an example of the importance of polymers, are treated in some detail in this chapter. There is an expanded discussion of plastics recycling.

Chapter 15 presents the fundamental chemical principles of some important biochemical phenomena in a straightforward manner. This edition of the text has an expanded discussion of optical activity and stereoisomers that includes more details of the thalidomide tragedy and current use of drugs. Chair conformations of simple sugars are introduced. There is a discussion of the use of Splenda as an artificial sweetener. Steroids and steroid abuse are discussed in more detail. The discussion of soaps and how they work has been improved. The discussion of proteins has been expanded to include prosthetic groups, cofactors, and coenzymes, and the discussion of sickle cell anemia has been enlarged. The treatment of enzyme action using lysozyme, as an example, has been improved.

Chapter 16 has been expanded with the addition of the Dietary Guidelines for Americans 2005 and the structures of some important molecules such as beta-carotene and vitamin A.

Chapter 17 has had significant updating and expansion of the lists of important prescription drugs and the statistics related to the leading causes of death. There is a new section on antibiotic resistant infections and even a discussion of a *Sports Illustrated* article about methicillin-resistant infections picked up in locker rooms. The treatment of progesterone, oral contraceptives, RU-486, and "morning after pills" has been significantly increased. There is a new discussion of Cox-2 inhibitors and the side effects of aspirin. An important addition to this chapter is the topic of methamphetamine abuse and the relationship between the availability of pseudoephedrine in over-the-counter medications and meth labs in the United States. Cancer statistics and the list of carcinogens have been updated. Another new topic tackled is homeopathic medicine and the issues surrounding its validity.

Chapter 18 covers the chemistry of useful materials such as glass, steel, and semiconductors, and has had only minor content changes and updates when compared with the previous edition.

Chapter 19 has been updated to include additional emphasis on global population growth and its problems compared with the previous edition and includes the introduction of the "ecological footprints" of humans as an important concept. The discussion of pesticides and fungicides has been expanded. All statistics have been updated.

Special Features

Chapter opening questions highlight the most important aspects of the chapter topics and stimulate student interest.

Key terms are boldfaced upon their first use and marginal definitions are provided for each.

Worked Examples demonstrate how to utilize the concepts in this course in problem solving. Each example is accompanied by one or more exercises (**Try It**) that are answered in Appendix F, thereby helping students to check their understanding. The examples and exercises, combined with the **Concept Checks**, are designed to help students gain confidence about smaller segments of material before they try to answer larger questions and problems at the end of each chapter.

Key Terms at the end of each chapter summarize the essential terms listed in the margins.

Applying Your Knowledge sections include end-of-chapter questions that review chapter material and encourage thought about the meaning and application of chemical concepts. Quantitative questions are included where appropriate.

Personal Side boxes highlight the achievements of important contributors to the science of chemistry.

World of Chemistry boxes describe new developments related to chapter topics, or focus on science-related societal issues that often raise difficult questions that do not always have clear-cut answers. In the previous edition, these boxes were called Science and Society and Frontiers in the World of Chemistry.

Accompanying Materials to World of Chemistry: Essentials, Fourth Edition

Supporting instructor materials are available to qualified adopters. Please consult your local Thomson Brooks/Cole representative for details. Visit http:// chemistry.brookscole.com to:

   locate your local representative.

   download electronic files of text art and ancillaries, or

   request a desk copy.

Online Instructor's Resources Manual, Test Bank Updated by Chris Truitt, Texas Tech University, the test bank includes:

   detailed solutions to the end-of-chapter Apply Your Knowledge questions and problems that do not appear in the text, and

   a test bank of multiple choice questions and problems.

iLrn Computerized Test Bank This easy-to-use software, containing questions and problems authored specifically for the text, allows professors to create, deliver, and customize tests in minutes.

Online Electronic Images Artwork from the textbook is available at the Instructor's website for download and use in PowerPoint presentations.

NEW! JoinIn on TurningPoint® CD-ROM JoinIn content for Response Systems, authored by Chris Truitt, allows you to transform your classroom and assess your students' progress with instant in-class quizzes and polls. Our exclusive agreement to offer TurningPoint® software lets you pose book-specific questions and display students' answers seamlessly within the Microsoft PowerPoint® slides of your own lecture, in conjunction with the "clicker" hardware of your choice. Enhance how your students interact with you, your lecture, and each other. Please consult your Brooks/Cole representative for further details.

NEW! Student Companion Website Organized by chapter, this outstanding site features chapter-by-chapter online quizzes, weblinks from the Chemistry on the Web sections in the textbook, and animations related to specific textbook content.

NEW! OWL: Online Web-based Learning System Developed over the last several years at the University of Massachusetts, Amherst, and class-tested by thousands of students, OWL is a fully customizable and flexible web-based homework system and assessment tool with course management. With both numerical and chemical parameterization and useful, specific feedback built in, OWL produces thousands of chemistry questions correlated to the text. The OWL system also features a database of simulations, guided tutorials, and problems associated with the textbook content. Instructors are able to customize the OWL program, use the grade book feature and generate multiple reports to measure each student's performance

NEW! Inquiry-based Laboratories for Liberal Arts Chemistry by Vickie Williamson and Larry Peck, Texas A&M University. This guidebook offers 19 experiments focusing on conceptual learning of the chemical phenomena in our everyday lives. The manual employs the learning cycle approach, used as the underlying model for the guided and open inquiry/application laboratories. An on-line instructor's guide is also available.

Acknowledgments

The publication of any textbook requires the efforts of many people beyond the authors. We would like to thank Ragu Raghavan, sales representative for Thomson-Brooks/Cole and Michelle Julet, vice-president and editor-in-chief at Thomson-Brooks/Cole, for helping us to bring John Hogg to this project as a co-author.

The efforts of Ellen Bitter, assistant editor of physical sciences at Thomson, are also greatly appreciated. Dena Diglio Betz did a wonderful job with the photo research and made valuable suggestions to improve the book. Our association with Gordon Laws, development manager at Pre-Press Company, Inc. in Massachusetts, has been a wonderful experience. Shelley Ryan at Thomson skillfully guided the overall production process. Lisa Lockwood, acquisitions editor, has been a guiding force throughout the development of this edition.

Reviews always contribute significantly to any book and that is certainly true in this case. In particular, we would like to acknowledge the contributions of John Vincent, University of Alabama. His comments and suggestions have been quite valuable in shaping this new edition and have made the book much better than it would have been without his help.

We are grateful to Mark Greip, University of Nebraska-Lincoln, for his contribution of the innovative Chemistry in the Movies boxes. Larry Peck and Vickie Williamson, two Texas A&M colleagues, agreed to develop an inquiry-based laboratory manual to accompany the book. Senior undergraduate chemistry major Jessica Biles did some library research for us and made valuable suggestions for material in Chapter 17. John would like to specifically thank his administrative assistant, Marylin Warren, and his wife, Janet Hogg, for their support throughout the development of this edition.

Reviewers for the Fourth Edition

SIMON BOTT	*University of Houston*
KEENAN DUNGEY	*University of Illinois at Springfield*
DAN FREEMAN	*University of South Carolina*
MARCIA GILLETTE	*Indiana University, Kokomo*
CLIFFORD GOTLIEB	*Shasta College*
WILLIAM GREEN	*Evergreen State College*
STEVEN GWALTNEY	*Mississippi State University*
ALTON HASSELL	*Baylor University*
STEVEN HOLMGREN	*Montana State University, Bozeman*
BRYAN LEES	*Kean University*
ISAM MARAWI	*Xavier University*
SUE MCCLURE	*Cal Poly, San Luis Obispo*
KATHLEEN MURPHY	*Daemen College*
SUN O. JANE MYONG	*Sinclair Community College*
LAURA PENCE	*University of Hartford*
PHIL REID	*University of Washington*
JAN SIMEK	*Cal Poly, San Luis Obispo*
ERIC TAYLOR	*University of Louisiana*
CHRISTOPHER TRUITT	*Texas Tech University*
JOHN VINCENT	*University of Alabama*
FRED WALTERS	*University of Louisiana*

LIVING IN A WORLD OF CHEMISTRY

© Scientific American, Inc./George V. Kelvin

1.1   THE WORLD OF CHEMISTRY

On the first day of class each semester, professors often distribute a questionnaire to the students to gather some general impressions about their backgrounds, aspirations, and career goals. Typical reasons for enrolling in a chemistry course designed for liberal arts students range from "I'm just taking this course to fulfill a graduation requirement" to "I really liked chemistry in high school." Most do not plan to pursue careers that they perceive to be heavily dependent on a knowledge of chemistry. The professor then sets out on a semester-long journey to increase students' awareness of the impact of chemistry on their daily life. He or she wants them to develop an appreciation of the fact that some level of understanding of chemistry is critical to an understanding of life and its associated activities.

An informed citizen of the world makes use, consciously or not, of a knowledge of chemical phenomena daily. Try to imagine an activity that does not directly or indirectly depend on chemistry. Cooking? Playing golf? Word processing on a computer? Brushing your teeth? Farming? Banking? Teaching third-grade students? Voting? Sunbathing? Reading a book? Watching a movie? If you can readily see how chemistry, or some product or device dependent on chemistry, influences each of these activities, then you may already have an appreciation of the importance of chemistry. Clearly, every activity in which we are involved on a daily basis depends on chemicals in some way. However, you may still question how much fundamental chemistry you actually need to know.

Many argue that chemistry is uniquely positioned at the crossroad between the biological and physical sciences. By exploiting their understanding of both realms, chemists and other professionals with strong backgrounds in chemistry have made, and continue to make, valuable contributions to improve the human condition. Major technological and biological discoveries almost always depend on a fundamental understanding of chemistry. The pursuit of these discoveries, as a way to improve the world in which we live, drives those who seek to be a part of the process. While you probably aspire to improve the world, at least to some extent, you probably don't imagine your effort to revolve around chemistry. A little analysis of even the most mundane of your daily activities, however, reveals that chemistry's influence is pervasive in modern society.

Consider the following. Why are some consumer products touted as being "chemical-free," and why are so many people drinking bottled water these days? Does that mean that chemicals are bad? Should you be concerned about this as you shop for consumer goods and food products? We'll answer those questions many times as we proceed through this course.

Many of the issues on which a responsible citizen votes depend heavily on the ability to evaluate scientific information. For instance, should you vote to elect a congressional representative who is convinced that global warming is a myth or one who is convinced that it is a serious issue about which something needs to be done? Should you object to legislation to allow irradiation of foods because you think this will render the foods radioactive? Should you really understand something about carbon dioxide and water, the combustion products formed by burning gasoline, the next time you decide to purchase the latest SUV? These are just a few of the social issues facing us today that are heavily dependent on some knowledge of chemistry.

Such issues often make it seem as if chemistry only causes problems. We also want to focus on the contributions that chemistry has made to the development of the modern conveniences, wonderful medicines, and consumer goods we enjoy. These contributions have certainly led to an increased life span and a higher quality of life for many of the people of the world. However, not everyone yet enjoys these benefits. Could it possibly be that there are problems that may arise, or have already arisen, as a result of our quest for this wonderful lifestyle? There is an old saying that "every cloud has a silver lining." Could it be that almost every silver lining has a little tarnish on it? As wonderful as some of our chemical and technological advances have been, they all carry with them some not-so-desirable aspects. A study of chemistry will leave you better prepared to understand that benefit must be balanced against risk many times daily. Often you do this analysis without even thinking, as you've already established that the benefits of a certain activity or use of a certain product outweigh any risks. New information often calls these decisions into question, however, so you must be capable of making informed decisions as you encounter these dilemmas.

Charles D. Winters

Are these foods chemical free?

Let's briefly consider how you might encounter "chemistry" in the future. You may be surprised to know that third-graders are required to know the difference between chemical and physical changes and between atoms and molecules. Would you, as a future third-grade teacher, feel comfortable teaching these concepts to your students? As a future farmer, would you understand enough chemistry to utilize the chemicals necessary for modern farming, or would you feel that "organic farming without chemicals" is the way to go? Is such farming even possible? Would you, as a stockbroker, feel comfortable recommending that your clients invest in a company to develop automobile engines that run on water instead of gasoline? Would you, as a parent, feel comfortable when considering unconventional medicines or medical treatments recommended by a friend, or would you feel more comfortable with advice from your physician? We could discuss hypothetical situations like these forever but, hopefully, it is apparent by now that maybe, just maybe, you do need to know a little more chemistry than you do now.

Making you a little more comfortable with chemistry is a major goal of this course. Let's begin.

Maybe *chemistry* means *chemicals* to you. And perhaps you think this word should be used only with the adjective *toxic*. That belief wouldn't be surprising, because you have probably heard of "toxic chemical spills" or warnings about "toxics" in the environment. Indeed, some chemicals are toxic—very toxic: the arsenic of mystery stories, the poisonous gases of World War I, the chemicals released by the microorganism that grows in badly canned food and causes severe food poisoning.

Learning about chemistry doesn't, however, mean learning only about harmful substances. Nor does it mean learning only about the wonders of our modern world provided by chemistry, such as the medications that cure once-incurable diseases; the synthetic fabrics that are beautiful, durable, and inexpensive; or the colors on the television screen.

Instead, we suggest you come to the subject with a "What's in it for me?" attitude, not from a selfish viewpoint, but as a citizen of the world of chemistry. The first photographs of planet Earth taken from the Moon provided a forceful reminder to many of us that this planet and the materials on it are finite. The world of chemistry really has but one concern: the materials provided by our planet and what we do with them.

In a global view, some understanding of chemistry will be helpful in dealing with major social issues that lie ahead in the 21st century. The present world population of over 6 billion is expanding at an ever-increasing rate and is expected to double in the next 50 years. How can everyone be fed, clothed, and housed? How can we prevent spoiling what planet Earth has provided us? How can we reverse some of the damage done when we knew less than we do now about the consequences of our activities? How can we prevent doing new damage to our environment? None of these questions can be fully addressed without some serious applied chemistry.

Knowing something about chemistry adds a new dimension to everyday life, too. If an advertisement proclaims that a product "contains no chemicals," what does that tell you about the producers of that product? Is it worth paying twice the price for something that proclaims itself to be "all natural"? (Sometimes it is and sometimes it isn't.) Do you know that you shouldn't mix household ammonia and chlorine bleach because they react to form a very toxic gas?

Lying ahead in the chapters of this book is a look into the chemical view of the world. It is based on observations and facts. A *scientific fact* results from repeated observations that produce the same result every time. (Water boils at 100°C at sea level—that's a scientific fact.) The chemical world is also based on models, theories, and experiments. Scientists use models and theories to organize knowledge and make predictions. The predictions must

The real meaning of *chemical* is discussed in Chapter 2.

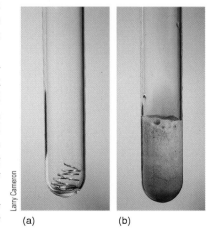

Larry Cameron

(a)   (b)

The amazing effect of nitric acid (b) on copper (a). As a young man, Ira Remsen (1864–1927) dropped a copper penny into nitric acid to see what would happen. Remsen, who became an outstanding chemist and teacher, had this reaction: "It resulted in a desire on my part to learn more about that remarkable kind of action. Plainly the only way to learn about it was to see its results, to experiment, to work in a laboratory."

then be tested by experiments. If experimental results disagree with the predictions, the reason must be explored. Possibly the theory is wrong.

You will also see that the chemical view of the world requires learning to look at materials in two ways. The first is what direct observation provides; the second is the mental image of what scientists have learned about the world that cannot be observed directly.

To continue this introduction to the world of chemistry, we offer a glimpse of three chemistry-related topics, which are sufficiently newsworthy to be mentioned often in your daily newspaper—the DNA that governs heredity for all plants and animals, the disappearance of ozone in our atmosphere, and the connection between fossil fuel use and global warming. As you progress through this and subsequent chapters of this book, you will learn more about the connection between chemistry-related topics that make the news and the chemists' view of matter.

1.2   DNA, BIOCHEMISTRY, AND SCIENCE

DNA (an abbreviation for the chemical name *deoxyribonucleic acid*) is present in cells throughout our bodies. The composition of an individual's DNA provides the equivalent of a fingerprint in identifying that individual. Because methods have now become available for analyzing DNA, such analysis has played a more and more important role in forensic science.

In response to arguments about the validity of courtroom use of DNA identification, the National Academy of Sciences conducted a review. They concluded that there is no reason to question the reliability of properly analyzed DNA evidence.

Comparison of the DNA analysis of a sample of blood or other biological material found at the scene of a crime with that of a sample from a suspect can weigh heavily as evidence for guilt or innocence. In the early days of the courtroom presentation of DNA evidence, however, there was concern over the use of competing scientific "experts" by opposing sides. Judges were often put in the position of ruling on scientific matters about which they knew very little. In 1995, the U.S. Supreme Court decided that federal judges, and presumably also state judges, have the power to hold pretrial hearings to validate the competence of experts before allowing them to testify. One goal is to prevent "junk science"—science not widely accepted and not developed with the necessary care—from reaching the courtroom. As a result, an effort was begun to educate judges about DNA. In 1995 the first group of 35 judges spent six days studying the subject; they even sequenced their own DNA in the laboratory.

The results of establishing the validity of DNA testing have been widespread. By 2003, nationwide more than 130 individuals, including 12 people awaiting execution, had been proven innocent by DNA tests. In New York City, the police department had solved 150 cold cases by examining DNA evidence. Proof of innocence or guilt by DNA testing is expected to expand rapidly, thanks to the DNA Index System established by the FBI. This system links the DNA evidence collected in almost every state and makes it available for searching by law enforcement officials in all states.

Chapter 15 is devoted to biochemistry.

The study of DNA is one aspect of *biochemistry*, the natural science that unites chemistry, a physical science, with the biological sciences. **Chemistry** is often defined as the study of matter and the changes it can undergo. A moment's thought should convince you that this makes an understanding of chemistry fundamental to an understanding of everything in nature. And, of course, with an understanding of natural matter, chemists are able to modify matter and synthesize new kinds of matter.

Historically, the natural sciences have been associated with observations of nature—our physical and biological environment. A traditional classification

Chemistry The study of matter and the changes it can undergo.

THE CHEMISTRY OF

HARRY POTTER AND THE CHAMBER OF SECRETS (2002)

by **Mark A. Griep**, *University of Nebraska-Lincoln*

The second movie based on J. K. Rowling's Harry Potter series has two main themes. The non-chemical theme is the main theme: Harry Potter is a parselmouth (he communicates with snakes using parseltongue language). The other theme is that someone or something at the school is petrifying muggle-born students and the cat Mrs. Norris. In the selected clip, studious Hermione Granger explains that mandrake root (or mandragora) is an antidote for petrification. Then, Professor Sprout teaches the students how to re-pot their young mandrakes. The message is that if you are going to use herbal remedies, then you need to maintain a garden.

Of course, everyone knows that Harry Potter is fictional, that real magic doesn't exist, and that J. K. Rowling's series is a complete fantasy. But, wait a minute. Mandrake tuber actually does exist (botanical name: *Mandragora officinarum*) and it does have a humanoid shape. In fact, there is even a Western myth that its screams will kill any person who uproots it (Flavius Josephus in 93 AD). In contradiction to Hermione's description, though, mandrake root has been used since Greek and Roman times as a "sleeping draft" (called an anesthetic today) at low doses and as a deadly poison at high doses.

In the late 1800s, Scottish surgeon Benjamin Richardson was searching for anesthetics. Even though ether anesthesia had revolutionized dentistry and surgery in the 1840s, it had a number of undesirable side effects. In 1874, Richardson published his experiments describing how he prepared an alcoholic

Atropine (mandragorine)

Scopolamine

extract of the mandrake root, tested its effects on lab animals, and then self-experimented. He decided that it also had too many side effects to be useful.

For many herbal remedies, chemists were able to find a single active ingredient that was responsible for the physiological activity. The French chemist Crouzel thought it might be possible to separate the desired effects from the side effects by isolating the active ingredient from mandrake root. He isolated the major active molecule and named it mandragorine (mandragora + amine, for the nitrogen). Later chemists showed that it was identical to the molecule named atropine isolated from the deadly belladonna plant (*Atropa belladonna*). Another active molecule isolated from mandrake root and belladonna was named scopolamine. Atropine and scopolamine cause similar but not identical effects. Today, they are both used medicinally although not as anesthetics. For instance, atropine is used as an antidote for certain types of nerve gases (see *The Rock* (1996)).

Questions to Consider

1. Use http://www.botanical.com to find other toxic and nontoxic members of the nightshade family of plants (Solanaceae). Do the toxic members produce the same set of toxic molecules?

2. Use the Internet to discover the other medicinal uses of atropine and scopolamine.

© Stapleton Collection / CORBIS

DIOSCURIDES: Umzeichnung eines Bildes des Wiener Anicia-Juliana-Codex nach Lambecius (Lambeck), Comm. de Bibl.Caes.Vindob., vol.II. D. empfängt von der „Heuresis" eine → Mandragora-Wurzel

Image from a thirteenth-century medical manuscript. Plant screams were reported to kill humans.

DNA was identified as the basic genetic material in chromosomes in 1944 by O. T. Avery, C. M. MacLeod, and M. McCarty. Nine years later James D. Watson and Francis Crick solved the difficult problem of how DNA is constructed (Section 15.11).

of the natural sciences is represented in Figure 1.1, with an emphasis on the relationship of chemistry to other sciences. As each science grows more sophisticated, the boundaries become more blurred, however. The dynamic character of science is illustrated by the emergence of new disciplines. There are now individuals who refer to themselves as biophysicists, bioinorganic chemists, geochemists, chemical physicists, and even molecular paleontologists (they use modern chemical methods to analyze ancient artifacts).

The study of DNA is one of the frontier areas of science. Hardly a week passes without the report of some new development, which might be the unraveling of the cause of an inherited disease or the introduction of a new product of the technology known as *genetic engineering*. Because these developments often relate to human health or behavior, they make the newspapers.

About 55 years ago, DNA was identified as the carrier of genetic information from one generation to the next, determining how we will be the same or different from our ancestors. In the intervening years, the chemical composition of DNA and the shape of the molecule came to be understood.

In 2000, the door opened on a revolution in our knowledge of DNA. Two research endeavors, one in a for-profit company (Celera Genomics) and one in a not-for-profit consortium (the Human Genome Project) jointly announced their first results in mapping human DNA. The map of human DNA was completed in April 2003. The location of every segment of human DNA had been identified.

In their original publication the leaders of the Human Genome Project made the following prophetic statement: "It has not escaped our notice that

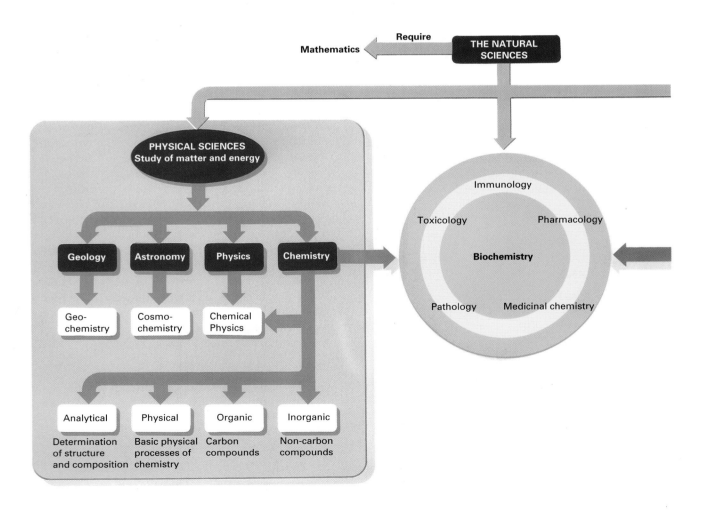

the more we learn about the human genome, the more there is to explore." Among these ongoing explorations are the following:

- Development of products that improve the health of humans, other animals, and plants
- Understanding of hereditary diseases
- Development of drugs to cure hereditary diseases
- Alteration of an individual's genetic makeup to prevent or cure a disease
- The study of why a drug can be effective in some individuals but not in others
- Exact matching of drugs to an individual's genetic makeup
- Development of improved agricultural crops and animals
- Creation of genetically modified bacteria that will mass-produce desirable chemical products
- Use of genetic information for a better understanding of evolution

1.3 AIR-CONDITIONING, THE OZONE HOLE, AND TECHNOLOGY

Next, we consider something that directly affects more of us than DNA testing—the chemicals used as refrigerants. The development of refrigerants is rooted in some interesting history of chemistry. Refrigeration as we know it

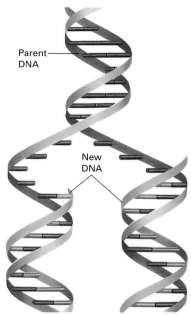

An abstract picture of how DNA unravels and is reproduced. In this way, a parent's DNA is passed to the off-spring. (More information about how this happens is given in Section 15.11.)

FIGURE 1.1 The natural sciences, with some of their subdivisions. The number of subdivisions continues to increase as the boundaries between the sciences disappear.

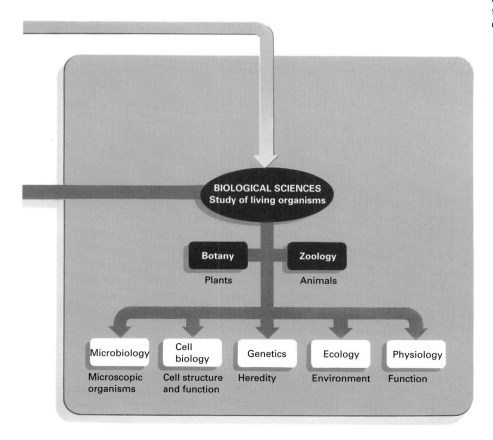

has been in use since the late 1800s. The process requires a fluid that absorbs heat as it evaporates, releases heat when it condenses, and can be continuously cycled through evaporation and condensation without breaking down.

In the early 1900s, the fluids used as refrigerants were mostly flammable or toxic. One day, Charles Kettering, then director of research at General Motors, passed along a challenge to Thomas Midgley, a young engineer: "The refrigeration industry needs a new refrigerant if they ever expect to get anywhere." Midgley and a colleague vigorously dug into the problem. They hunted for candidates in the extensive tables of data kept at hand in every chemistry laboratory and surveyed the systematic variations in the properties of the chemical elements. Although fluorine had a reputation for being toxic, they decided that in some cases it might not be. To begin with, they decided to make a new fluorine-containing substance in the laboratory. Because one of the necessary raw materials was a fluorine chemical that was scarce, they quickly purchased the only available supply of five small bottles. One of the bottles was used to make the first sample of the new chemical. In Midgley's own words:

> A guinea pig was placed under a bell jar with it and, much to the surprise of the physician in charge, didn't suddenly gasp and die. In fact, it wasn't even irritated. Our predictions were fulfilled.

> A second sample made from the starting material in a different bottle did, however, kill the guinea pig. Here was a problem to be solved. Some investigation showed that only the first bottle contained pure material. It was a contaminant that had killed the guinea pig, not Midgley's newly prepared chemical.

> Of five bottles . . . one had really contained good material. We had chosen that one by accident for our first trial. Had we chosen any one of the other four, the animal would have died as expected by everyone else in the world except ourselves. I believe we would have given up what would then have seemed like a "bum hunch."

> And the moral of this last little story is simply this: You must be lucky as well as have good associates and assistants to succeed in this world of applied chemistry.

The outcome of this experiment was the proliferation of refrigerators in our homes. Food spoilage diminished, and after World War II the food processing industry expanded to produce frozen foods.

So far, this is a success story—the development of chlorofluorocarbons, often referred to as CFCs, as much less hazardous refrigerants. Through the 1950s and 1960s, their use expanded, as all kinds of new consumer products were brought to market. The properties of CFCs made them ideal for propellants in aerosol cans and for blowing the tiny holes into materials such as the polyurethane foam used in pillows and furniture cushions. The problem being created did not surface until the 1970s because no one was aware of the fate of these very stable compounds as they were released into the environment.

By the 1980s, it had become clear that CFCs drift unchanged into the stratosphere, where they interact with ozone and destroy it. The result is the "ozone hole" and the potential for damage from the resulting increase in solar radiation that reaches Earth's surface.

Ozone in the stratosphere absorbs ultraviolet radiation. With ozone depletion comes a corresponding increase in the radiation that reaches us and an increase in its harmful effects, which include human skin cancer and cataracts that can lead to blindness. Increased ultraviolet exposure is also harmful throughout the environment. In the oceans, the phytoplankton population is decreased, resulting in a loss of the fish that feed on

Charles D. Winters

A supermarket refrigerator, a product of modern technology.

To a chemist, a stable substance is one that is not easily changed into something else. Water is a stable substance.

Ozone is a gaseous pollutant at low altitudes in our atmosphere (Section 4.7).

phytoplankton. There is also speculation that ultraviolet radiation is killing coral and causing an increase in the number of physically deformed frogs.

In 1987 nations that produce CFCs drafted a plan of action designed to reduce CFC use. Later amendments called for complete phaseout of all chemicals that harm the ozone layer and accelerated the schedule for that phaseout. (These chemicals and the ozone layer are further discussed in Chapter 7.)

It is helpful in reading about refrigerants and similar topics to recognize the distinctions among basic science, applied science, and technology. **Basic science,** or basic research, is the pursuit of knowledge about the universe with no short-term practical objectives for application in mind. The biochemists who struggled for years to understand exactly how DNA functions within cells were doing basic science.

Applied science has the well-defined, short-term goal of solving a specific problem. The search for a better refrigerant by Midgley and his colleagues is an excellent example of applied science: They had a clear-cut, practical goal. To reach the goal, as is done in either basic or applied science, they utilized the recorded observations of earlier studies to make a prediction and then performed experiments to test their prediction.

Technology, also an application of scientific knowledge, is a bit more difficult to define. In essence, it is the sum of the way we apply science in the context of our society, our economic system, and our industry. The first refrigerators and automobile air conditioners designed to use CFCs were the products of a new technology. The rapidly expanding number of ways to manipulate DNA to make new medicines or other marketable products is referred to as *biotechnology.*

Regardless of the type of scientific discovery, there is a delay between the discovery and its technological application. The incubation intervals for a number of practical applications of various types are given in Table 1.1.

The important point is that technology, like science, is a human activity. Decisions about the uses of technology and priorities for technological developments are made by men and women. How scientific knowledge is used to promote technology depends on those persons who have the authority to make the decisions. Sometimes those persons are all of us—when we go to the polls in a democratic society, we can influence decisions about technology. When we have this chance, it is important to be well-enough informed to critically evaluate the societal issues related to the technology.

As the history of CFCs illustrates, science and technology, like social conditions, are constantly evolving. When CFCs were introduced as refrigerants in 1930, they were a great advance for the economy and replaced hazardous

Basic science Pursuit of scientific knowledge without the goal of a practical application

Applied science Science with a well-defined, short-term goal of solving a specific problem

Technology Application of scientific knowledge in the context of industrial production, our economic system, and our societal goals

---

TABLE 1.1   Time Needed to Develop Technology for Some Fruitful Ideas

INNOVATION	CONCEPTION	REALIZATION	INCUBATION INTERVAL (YEARS)
Antibiotics	1910	1940	30
Nylon	1927	1939	12
Photocopying	1935	1950	15
Photography	1782	1838	56
Roll-on deodorant	1948	1955	7
Videotape recorder	1950	1956	6
X rays in medicine	Dec. 1895	Jan. 1896	0.08
Zipper	1883	1913	30

materials. At that time, the number of technologically advanced nations and the world population were smaller. People were in the habit of assuming that natural processes would keep the environment healthy, and to a greater extent than today that was true. Moreover, some of the sophisticated instruments that have since revealed ozone depletion were not available then. Only by staying informed can we be ready to adjust to changing times.

1.4 FOSSIL FUEL USE AND GLOBAL WARMING

The world has become increasingly dependent on the use of *fossil fuels* (coal, oil, and natural gas) to support consumer-driven societies and their associated lifestyles. Enormous quantities of these fuels are used daily. The fuels may be burned in a variety of vehicles to propel them down the road, through the air, or across the oceans. The fuels may be utilized to drive industrial manufacturing processes directly, such as by powering an engine, or they may be used indirectly to generate electricity to power the economy by producing the consumer products upon which we've become so dependent. Our ability to heat and cool our homes; grow, harvest, transport, and cook our food; provide health care and medicines; and entertain ourselves during our free time all depend on fossil fuels in one way or another. Some activities depend on burning the fuels whereas others depend on utilizing the fossil fuels as raw materials in the chemical industry. Fossil fuels are nonrenewable resources, however, and the supply will eventually run out.

It is becoming increasingly apparent that the products of fossil fuel combustion may be causing Earth to warm beyond what is natural and desirable. It is a known fact that both carbon dioxide and water vapor absorb and trap heat radiation. It is also known that the levels of carbon dioxide and some other heat-trapping substances have increased significantly in the atmosphere over the past hundred or so years. The debate among scientists, politicians, and concerned citizens centers on whether the warming of Earth is due to the increase in concentration of these substances. If so, what will be the consequences of this warming? And is there anything we can or should do about it?

Some people believe this *global warming* will be good for Earth, some think it will be inconsequential, some think we will adjust to it, and some think it threatens life as we know it. Some people think there is nothing that can or should be done about this warming, whereas others feel it is imperative that we do everything within our power to halt this phenomenon. The average citizen must become knowledgeable enough about this issue to make informed decisions concerning factors affecting this global warming. Such knowledge could influence the type of automobile you purchase, the size of the house you live in, and your lifestyle in general. It could eventually influence where and how you live or, possibly, even if you live. Knowledge is power and only by obtaining, properly assessing, and utilizing this knowledge can we adjust to changing times.

Charles D. Winters

Even the combustion of charcoal, a product produced by heating wood to high temperatures in the absence of oxygen, produces carbon dioxide, a compound implicated in global warming.

Global warming is discussed in Chapter 5.

1.5 BENEFITS/RISKS AND THE LAW

A glance through the stories in any daily newspaper or weekly news magazine or a half hour spent watching almost any television newscast is bound to demonstrate that the potential for risk to public health and safety is newsworthy.

It is important to consider how society weighs the benefit of some activity, the use of some chemical, or the use of some new technology against the potential risks. For example, automobiles offer many transportation advantages over horses, yet automobiles are responsible for thousands of deaths and considerable pollution each year. Clearly, we've accepted these risks while, at the same time, making efforts to minimize or eliminate them.

The Boston-based Harvard Center for Risk Analysis (HCRA) points out that in the days when measles and polio were prevalent, most people agreed that the benefits of vaccination against these conditions far outweighed the risk associated with the vaccines and opted for immunization. However, now these diseases are quite rare in most parts of the world. Many parents of small children have decided that the benefit to be gained from vaccinations does not outweigh the risks from side effects and are opting out of voluntary vaccination programs. As concern has grown that terrorists might release the smallpox virus into a world in which smallpox has essentially been eliminated, some have called for renewed vaccination against this deadly disease. However, many physicians and health care providers who would be "first responders" in such a terrorist attack have been reluctant to voluntarily participate in renewed vaccination programs, due to the small risk of side effects. The current perception is frequently that the benefit (protection against an unlikely smallpox attack) does not outweigh the risks. On the other hand, when one of the authors of this book was a young boy, people were routinely vaccinated against smallpox because the risk of side effects was small compared to the benefit—protection against a deadly killer. Less clear are situations in which there are small benefits and small risks or large benefits with large risks.

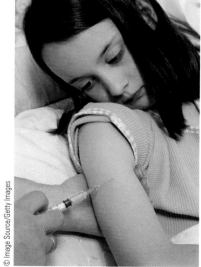

What is the risk-benefit tradeoff for vaccination?

In any benefit versus risk analysis, several factors come into play. An excellent discussion of these factors is presented in the June 2003 HCRA publication *Risk in Perspective* (http://www.hcra.harvard.edu). The authors identify the following major risk perception factors to consider:

- Level of dread: Which do you dread more, for example, heart attack or cancer?

- Control: We feel safer when we are in control.

- Natural or human-made risk: Natural risks seem less worrisome.

- Choice: Do you have a choice in the risk activity?

- Involvement of children: Risks that are acceptable for adults may not be seen in the same light when children are involved.

- Newness of risk: Newly perceived risks seem worse that "old" risks.

- Awareness: Simply becoming more aware of a risk makes it seem more serious than it may really be. The popular press may contribute to this problem in its effort to keep us well-informed.

- Can it happen to you? Not long ago American citizens didn't worry much about terrorism because "it doesn't happen here." September 11, 2001, certainly changed that.

- Trust: We tend to be more willing to accept risks imposed on us by those we trust than by those we do not.

- Risk-benefit tradeoff: Benefits tend to make risks more acceptable.

Consider the following risk comparisons of "natural" versus "human-made" risks. Many people are opposed to the use of "toxic chemicals" around food products, yet toxicologists estimate that the typical American diet contains ten thousand times more naturally occurring cancer-causing chemicals than those of the synthetic variety. The Centers for Disease Control estimate that about nine thousand Americans die each year from natural food poisoning, whereas worst-case estimates of deaths due to pesticide residues in foods are generally less than a few thousand. Suppose legislation was proposed that banned the sale of any foodstuff containing a human-made chemical for which a reasonable rate of consumption would lead to one additional cancer death in ten thousand people. Most people probably would be in favor of banning the use of the foodstuff containing this chemical. However, the following activities are estimated by experts to carry this same risk of cancer due to naturally occurring carcinogens, shown in parentheses, in the foodstuff: drinking a glass of orange juice (*d*-limonene) every other week; consuming a head of lettuce (caffeic acid) every other year or one carrot (caffeic acid) per week; or drinking one glass of wine (ethyl alcohol) every three years. Does this place a different perspective on such legislation? See Table 1.2 for some activities estimated to increase the probability of one additional death per one million people involved in the activity.

It is rather ironic that as life expectancies have increased dramatically because of chemical and technological progress, our level of acceptance of risk is declining in areas associated with chemicals. Can there be a rational and logical explanation for this? Almost every chemical, natural or human-made, has some risk associated with it at some level of exposure. The risk associated with a chemical is a function of its toxicity and a person's level of exposure to it. The toxicity of a given chemical is an inherent property of the particular chemical, but exposure to that toxicity is something we may have some control over. For instance, water is inherently nontoxic but overexposure to it under certain circumstances may lead to death by drowning. Peanut butter contains a potent, naturally occurring toxic compound known as aflatoxin, yet most of us are not concerned about its presence because its concentration, and hence our exposure to it, is very low. As our ability grows to detect increasingly lower concentrations of such chemicals in our foods and environments, the level of anxiety over their presence appears to have risen.

Over time our ability to detect the presence of toxic substances has changed from levels in the parts per hundred (%) to levels below parts per million (ppm), parts per billion (ppb), and now parts per trillion (ppt). It is not necessarily true that the concentrations of these toxic substances have increased over the years. Surely there must be some balance between a certain

TABLE 1.2 Estimates of Risk: Activities with a Probability of One Additional Death per One Million People Exposed to the Risk

ACTIVITY	CAUSE OF DEATH
Smoking 1.4 cigarettes	Cancer, lung disease
Living 2 months with a cigarette smoker	Cancer, lung disease
Eating 100 charcoal-broiled steaks	Cancer
Traveling 6 minutes by canoe	Accident
Traveling 10 minutes by bicycle	Accident
Traveling 300 miles by car	Accident
Traveling 1000 miles by jet aircraft	Accident
Drinking Miami tap water for 1 year	Cancer from chloroform
Living 2 months in Denver	Cancer caused by cosmic radiation
Living 5 years at the boundary of a typical nuclear power plant	Cancer

level of informed, serious concern and chemical or technical paranoia. We must assess the risk from the use of a certain chemical or the pursuit of a certain technology or activity and then balance that against the benefit to be obtained. Some individuals seem to believe that a certain amount of public activism and government regulation can guarantee risk-free living. Is that possible?

No absolute answer can be provided to the question, "How safe is safe enough?" The determination of acceptable levels of risk requires value judgments that are difficult and complex, involving the consideration of scientific, social, economic, and political factors. Over the years a number of laws designed to protect human health and the environment have been enacted.

The Safe Drinking Water Act, the Toxic Substances Control Act, and the Clean Air Act are laws of this type. The federal Environmental Protection Agency (EPA) is required to balance regulatory costs and benefits in its decision-making activities. Risk assessments are used here. Chemicals are regulated or banned when they pose "unreasonable risks" to or have "adverse effects" on human health or the environment. Parts of the Clean Air Act and the Clean Water Act impose pollution controls based on the best economically available technology or the best practical technology. Such laws

Cottonwood Lake near Buena Vista, Colorado. It's up to us to keep such places beautiful.

© Grant Heilman/Grant Heilman Photography

assume that complete elimination of the discharge of human and industrial wastes into water or air is not feasible. Controls are imposed to reduce exposure, but true balancing is not attempted. The goal is to provide an "ample margin of safety" to protect the public. As technology advances and the cost of the technology decreases, the margin of safety can be adjusted.

Risk management requires value judgments that integrate social, economic, and political issues with risk assessment. Risk assessment is the province of scientists, but determination of the acceptability of the risk is a societal issue. It is up to all of us to weigh the benefits against the risks in an intelligent and competent manner.

1.6 WHAT IS YOUR ATTITUDE TOWARD CHEMISTRY?

Before proceeding with this study of chemistry and its relationship to our society, you might want to examine your prejudices (if any) and attitudes about chemistry, science, and technology. Many nonscientists regard science and its various branches as a mystery and think that they cannot possibly comprehend the basic concepts and their relation to societal issues. Many also have what has been christened *chemophobia* (an unreasonable fear of chemicals) and a feeling of hopelessness about the environment. These attitudes are most likely the result of news stories about the harmful effects of technology. Some of these harmful effects are indeed tragic.

What is needed, however, is a full realization of both the benefits and the harmful effects of science and technology. In the analysis of these pluses and minuses, we need to determine why the harmful effects occurred and if the risk can be reduced for future generations. This book will give you the basics in chemistry along with some insight into the role of chemistry in both positive and negative aspects of the world today. We hope this knowledge will afford you a healthier and more satisfying life by allowing you to make wise decisions about personal problems and problems of concern to society as a whole.

W Assess your understanding of this chapter's topics with an online chapter quiz at **www.brookscole.com/chemistry/joesten4**

■ KEY TERMS

chemistry

basic science

applied science

technology

■ APPLYING YOUR KNOWLEDGE

1. Which of the following is not one of the physical sciences?
   (a) Physics (b) Chemistry
   (c) Zoology (d) Geology

2. A scientist who is trying to make an insecticide that is more toxic to mosquitoes than to humans would be involved in _____.
   (a) basic research (b) applied research
   (c) technology (d) serendipity

3. An appropriate risk assessment regarding the use of a toxic chemical for control of household pests would include a consideration of the inherent toxicity of the chemical and what other factor?

4. Define chemistry.

5. In which branch of the biological sciences would the study of DNA fall?

6. Is it possible that scientists in other branches of biological and physical sciences might also study DNA? If so, explain what they might be looking for.

7. Dallas, Houston, and San Antonio are three of the ten largest cities in the United States. They have certainly benefitted from the technology associated with air-conditioning and the internal combustion engine. Have there been risks or downsides associated with the application of this technology? Do you think the benefits have outweighed the risks?

8. List as many risk-perception factors as you can that you believe most people consider, consciously or subconsciously, when evaluating a new activity, product, medicine, or technology. Refer to pages 10–14.

9. How would you assess the risks and benefits of eating only organically grown fruits and vegetables (that is, those grown without the use of human-made chemicals) compared to the risks and benefits of eating fruits and vegetables grown using human-made chemicals allowed by federal and state laws? Does your level of *chemophobia* (fear of chemicals) enter into this?

Ⓦ CHEMISTRY ON THE WEB

For up-to-date URLs, visit the text website at **www.brookscole.com/ chemistry/joesten4**

- DNA Fingerprinting
- Ozone Depletion
- Global Warming
- Risk Versus Benefit

THE CHEMICAL VIEW OF MATTER

Native copper

Charles D. Winters

Practically everything we use has been changed from its natural state of little or no utility to one of very different appearance and much greater utility. Some of these changes are mechanical, such as turning trees into the lumber used to build houses. Many others are chemical, such as heating sand to make glass. Exploring, understanding, and managing the processes by which natural materials can be changed are basic to the science of chemistry. These activities require close examination of the composition and structure of matter, the stuff of which everything is made, which we begin to do in this chapter.

• What are elements and chemical compounds made of?

• What is the difference between a mixture and a pure substance?

FIGURE 2.1 A scanning tunneling microscope photo of 48 iron atoms in the shape of a corral on a copper surface.

- What is the difference between a chemical and a physical process?
- What is the basic theme of chemistry?
- How are symbols for the elements used in formulas and equations to communicate chemical information?
- Why are standards and measurements essential to chemistry?
- What are the metric units most commonly used in chemistry?

Many times a day you expect things to behave in certain ways and you act on these expectations. If you are cooking spaghetti, you know that you must first boil some water and that after the spaghetti has been in the boiling water for a while it will soften. If you are pouring gasoline into the fuel tank of your lawn mower, you know that you must protect the gasoline from sparks or it will explode into fire. To cook the spaghetti or safely fill the gas tank, you don't need to think about why these expectations are correct.

Chemistry, however, has taken on the job of understanding why matter behaves as it does. Answering questions such as, "Why does water boil at 100°C?" or "Why does gasoline burn, but water doesn't?" or "Why does spaghetti get soft in boiling water?" requires a different approach than direct observation. We have to look more deeply into the nature of matter than we can actually see, at least under everyday conditions.

Samples of matter large enough to be seen, felt, and handled—and thus large enough for ordinary laboratory experiments—are called **macroscopic** samples. In contrast, **microscopic** samples are so small that they have to be viewed with the aid of a microscope. The structure of matter that really interests chemists, however, is at the **nanoscopic** level. Our senses have limited access into this small world of structure, although new kinds of instruments are beginning to change this condition (Figure 2.1).

In the chemical model of matter, everything around us, matter of every possible kind, is pictured as collections of very small particles. The picture begins with elements and their atoms—*all matter is composed of atoms*. In fact, techniques have been developed that allow the manipulation of individual atoms.

2.1 ELEMENTS—THE SIMPLEST KIND OF MATTER

The historical development of the chemical model of matter is discussed in Chapter 3.

Solids, liquids, and gases are the three common **states of matter.** The fourth state, **plasmas**, occurs in flames, stars, and the outer atmosphere of Earth.

Macroscopic Large enough to be seen, felt, and handled

Microscopic Visible only with the aid of a microscope

Nanoscopic In the range of the nanometer (0.000000001 m); atoms and molecules are in the nanoscopic size range

Most of the materials you handle every day are pretty complex in their structure, and many are mixtures. Only occasionally do you encounter elements not combined with something else. One example is the helium used to fill a birthday balloon. The gas is helium, an element. From the chemical point of view, what does this mean? First, it means that helium cannot be separated or broken down into any other kind of matter that can still be recognized as helium. Second, it means that the balloon contains the very, very small particles known as helium atoms. Third, it means that helium has a unique and consistent set of properties by which it can be identified.

Another element you may have seen around your house is mercury. It is used to conduct electricity in switches like those in some thermostats and as the liquid in some thermometers. Mercury is certainly very different from helium in its properties. It is a liquid, not a gas, and it conducts electricity, which helium cannot do. What mercury and helium have in common is that they are both elements and each is composed of a single kind of atom. The liquid in the mercury thermometer is made of mercury atoms. Because the properties of helium and mercury are so different, we can safely conclude that helium and mercury atoms are not the same (Figure 2.2).

The chemical model of matter, therefore, starts with the following:

A **pure substance** is something with a uniform and fixed composition at the nanoscopic level. As you will see, pure substances can be recognized by the unchanging nature of their properties.

An **element** is a pure substance composed of only one kind of atom.

An **atom** is the smallest particle of an element, and the atoms of different elements are different.

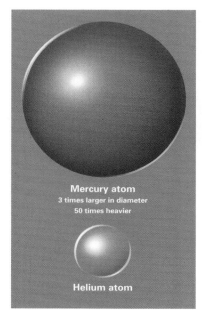

Mercury atom
3 times larger in diameter
50 times heavier

Helium atom

FIGURE 2.2 The relative sizes of helium and mercury atoms. As you might guess, mercury atoms not only are larger than helium atoms, but also are much heavier (about 50 times heavier).

2.2 CHEMICAL COMPOUNDS— ATOMS IN COMBINATION

If a pure substance is not an element, it is a chemical compound. To take our most common example, what does it mean that water is a chemical compound? You probably know that water is referred to as "h-2-oh" even if you have never studied chemistry. The "h-2-oh" is a way of reading the notation that chemists use to represent water: H_2O. In this kind of notation, H represents the element hydrogen, and O represents the element oxygen.

Chemical compounds are pure substances made of atoms of different elements combined in definite ways. We know that water is a pure substance because it has the same properties no matter where it comes from. We know that water is a chemical compound because passing energy through it in the form of an electric current causes decomposition to the elements hydrogen and oxygen. Also, under the right conditions, hydrogen and oxygen combine to form water.

Remember that from a chemical point of view, all matter consists of tiny particles. In water the smallest particle that can still be recognized as water is a water **molecule**. Each water molecule is composed of two hydrogen atoms (shown by the subscript 2 in H_2O) and only one oxygen atom (shown by the absence of a subscript after the O in H_2O). Experimentally then, pure substances are classified into two categories: (1) *chemical compounds*, which can be broken down into simpler substances called elements, and (2) *elements*, which cannot be broken down into any smaller particle still recognizable as that element.

Once elements are combined in compounds, the original, characteristic properties of the elements are replaced by the characteristic properties of the compounds. Consider the difference between table sugar, a white crystalline substance that is soluble in water, and its constituent elements: carbon, which is usually a black powder and is not water soluble; hydrogen, the lightest gas known; and oxygen, the atmospheric gas needed for respiration.

The symbolism used to represent elements and compounds is explained further in Sections 2.6 and 2.7.

2.3 MIXTURES AND PURE SUBSTANCES

Most samples of matter as they occur in nature are mixtures. Some are obviously **heterogeneous mixtures**—their nonuniform composition is clearly visible, as in chocolate chip cookies or the different kinds of crystals in many rocks (Figure 2.3). **Homogeneous mixtures**, which are uniform in composition, are referred to as **solutions**. No amount of optical magnification will reveal a solution to be heterogeneous, because it is a uniform mixture of particles too small to be seen with ordinary light. In addition, different samples of the same homogeneous mixture have the same composition. Examples of solutions are *clean* air, which consists mostly of nitrogen and oxygen; freshly brewed tea; and some brass alloys, which are homogeneous

Chemical compounds Pure substances composed of atoms of different elements combined in definite, fixed ratios

Molecule The smallest chemical unit of a compound that retains the composition and properties of the compound and can exist independently

Homogeneous mixtures are uniform in composition and **heterogeneous mixtures** are not.

Solutions Homogeneous mixtures, which may be in the solid, liquid, or gaseous state

(a)

(b)

FIGURE 2.3 A natural heterogeneous mixture—in rocks—and a human-made one—in chocolate chip cookies.

As science advances, smaller and smaller concentrations of impurities can be detected.

mixtures of the elements copper and zinc. As these examples show, *solutions can be found in the gaseous, liquid, and solid states.*

Other mixtures may appear to be homogeneous even if they are not. For example, the air in your classroom appears homogeneous until a beam of light enters the room and reveals floating dust particles. Blood appears homogeneous as it gushes from a cut in your finger, but a microscope shows that it is quite a heterogeneous mixture (Figure 2.4).

When a mixture is separated into its components, the components are said to be *purified*. Efforts at separation are usually incomplete in a single step, and repetition of the process is necessary to produce a purer substance. Ultimately, the goal is to arrive at substances that cannot be purified further.

Separation of mixtures is done by taking advantage of the different properties of substances in the mixture. For example, many people want to separate certain substances from their tap water before they drink it. One way to do this is based on the forces of attraction between particles at the nanoscopic level. The water is passed through a material that attracts the undesirable substances and holds them back.

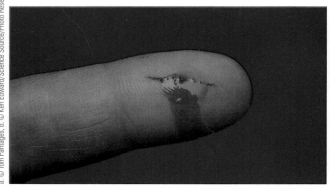

(a)

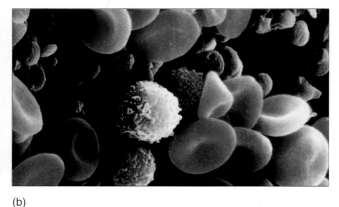

(b)

FIGURE 2.4 Blood, another heterogeneous mixture. (a) Blood looks homogeneous as it flows from a cut. (b) A microscope shows that it is quite a mixture. Color has been added to this electron microscope picture to show the different kinds of blood cells. The doughnut-shaped ones are red blood cells; the fuzzy ones are immune system cells; the small tan ones are platelets.

2.4 CHANGES IN MATTER: IS IT PHYSICAL OR CHEMICAL?

Sulfur is yellow; iron is magnetic; water boils at 100°C. These are different properties of matter, but with something in common. We can observe the color of a substance, pick it up with a magnet, measure its boiling point or determine its **mass** without changing its chemical identity. Such properties are classified as **physical properties**. In chemistry, the word *physical* is used to refer to processes that do not change chemical identities. Separating sulfur and iron with a magnet is a *physical* separation. Boiling water is a *physical* change—the steam that forms is still water (H_2O).

Whenever possible, the physical properties of substances are pinpointed by making numerical measurements. You might describe lead as heavier than aluminum—if you pick up same-sized pieces of these metals, lead will definitely feel heavier. The difference is better represented though by giving the numerical value of the **density** of each metal, which takes into account the need to compare pieces of equal size, that is, of equal volume. One cubic centimeter (cm^3) of lead weighs 11.3 grams (g) and 1 cubic centimeter of aluminum weighs 2.7 g. Lead is *denser* than aluminum. Stated in the usual way:

	DENSITIES (AT 20°C)
Lead	11.3 g/cm^3
Aluminum	2.7 g/cm^3

By contrast to physical changes such as boiling, there are processes that do result in changes of identity. When gasoline burns, it is converted to a mixture of carbon dioxide, carbon monoxide, and water. Burning in air, a property that gasoline, kerosene, and similar substances have in common, is classified as a **chemical property**.

The word *chemical* is used to describe processes that result in a change in identity. The combination of hydrogen and oxygen to form water is a *chemical change* or **chemical reaction**—a process in which one or more substances (the **reactants**, which can be elements or compounds, or both) are converted to one or more different substances (the **products**, which can also be elements or compounds, or both).

$$\text{Reactants} \longrightarrow \text{products}$$
$$\text{Hydrogen + oxygen} \longrightarrow \text{water}$$

A chemical reaction produces a new arrangement of atoms. The number and kinds of atoms in the reactants and products remain the same, but the reactants and products are different substances that can be recognized by their different properties. To distinguish between a hydrogen–oxygen mixture and a compound composed of hydrogen and oxygen, we might say that in water, hydrogen and oxygen are *chemically* combined. The meaning is that the hydrogen and oxygen atoms are held together strongly enough to form the individual units we call water molecules.

Some easily observed results of chemical reactions are the rusting of iron, the change of leaf color in the fall, and the formation of carbon dioxide bubbles by an antacid tablet. Often, though not always, the occurrence of a chemical reaction can be detected because there is some observable change (Figure 2.5).

The word *chemical* is often used as a noun. In this sense, every substance in the universe is a chemical. You might read of polluted waters that contain

---

See an animation of **Water Vaporization and Air Pressure** at http://brookscole.com/chemistry/joesten4

Some physical properties are color, odor, melting point, boiling point, solubility, hardness, density, and state (solid, liquid, or gas).

Don't confuse density and mass. A large piece of aluminum could be heavier than a small piece of lead, but the lead is still denser.

See an animation of **A Chemical Reaction Between Hydrogen and Oxygen** at http://brookscole.com/chemistry/joesten4

See an animation of **A Chemical Reaction Between Bromine and Aluminum** at http://brookscole.com/chemistry/joesten4

Compounds composed of molecules at the nanoscopic level are referred to as **molecular compounds**.

Mass A measure of the quantity of matter in an object

Physical properties Properties that can be observed without changing the identity of a substance

Density is mass per unit volume. The customary measurement units (Section 2.8) for density are grams (g) for mass (weight) and cubic centimeters (cm^3) for the volume of solids or milliliters (mL) for the volume of liquids.

Chemical properties Properties that result in chemical reactions and changes in the identity of one or more reactants

Chemical reaction Process in which one or more pure substances are converted to one or more different pure substances

Reactants Substances that undergo chemical change in a chemical reaction

Products Substances formed as the result of a chemical reaction

The diver has potential energy because of his position above the surface of the water.

As the diver falls, potential energy is converted to kinetic energy.

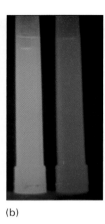

Charles D. Winters

(a)   (b)   (c)   (d)

FIGURE 2.5 Observable indications of chemical reactions. Chances are that a chemical reaction has occurred when (a) a flame is visible, (b) mixing substances produces a glow, (c) a solid appears when two solutions are mixed or (d) there is a color change when two substances are mixed.

 See an animation of **Energy Transfer** at http://brookscole.com/chemistry/joesten4

 See an animation of **Separation of a Mixture** at http://brookscole.com/chemistry/joesten4

Energy The capacity for doing work or causing change

Potential energy Energy in storage by virtue of position or arrangement

Kinetic energy The energy of objects in motion

chemicals. Used this way, the word really stands for *chemical compounds*. The chemical compounds in polluted water may be unhealthful or harmful. Anyone who has studied even a little chemistry understands that useful substances—penicillin, table salt, nylon fabrics, laundry detergents, and all natural herbs and spices—are also made of chemical compounds. So are all plants and animals. Pure water is also a chemical compound.

Some physical changes and almost all chemical reactions are accompanied by changes in energy. Frequently, the energy is taken up or released in the form of heat. Heat must be added for the physical changes of melting or boiling to occur. Heat is released when the chemical reactions of combustion, or burning, take place. But many chemical reactions, such as those of photosynthesis, require energy from their surroundings in order to take place.

Energy is the ability to cause change or, in the formal terms of physics, to do work. Energy in storage, waiting to be used, is **potential energy**. There is potential energy in gasoline, known as *chemical energy*—it is released as heat and light when gasoline burns. Chemical energy can also be released in the form of electrical energy or mechanical energy (Table 2.1). Energy in use, rather than in storage, is **kinetic energy**, the energy associated with motion.

2.5   CLASSIFICATION OF MATTER

The kinds of matter we have described—elements, compounds, and mixtures—can be classified according to their composition and how they can be separated into other substances, as shown in Figure 2.6.

Heterogeneous samples of matter are all mixtures and can be physically separated into various kinds of homogeneous matter. Homogeneous matter can be a pure substance or a mixture. If it is a mixture, it is described as a

TABLE 2.1   Some Examples of Conversion of Chemical Potential Energy to Other Forms of Energy

CONVERSION TO—	ELECTRICAL ENERGY	HEAT	LIGHT	MECHANICAL ENERGY

Done by—	Batteries
	Fuel cells
		Combustion
		Digestion of food
		Many kinds of
		   chemical reactions
			Burning candle, logs in
			   fireplace
			Luminescence in firefly
			Luminescence from
			   chemical reaction
				Rocket
				Animal muscles
				Dynamite explosion

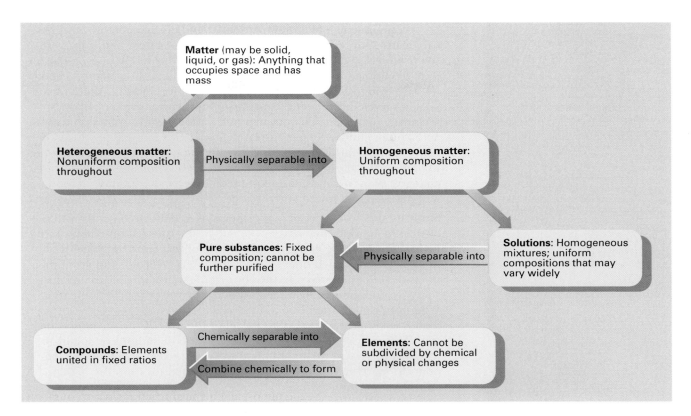

FIGURE 2.6 A classification of matter. Once elements have combined into compounds, only chemical reactions can separate them.

solution and has the same composition throughout. Different solutions of the same substances, however, can vary in composition, sometimes over a very wide range. A teacup full of water, for example, might dissolve anywhere from a few grains of sugar to more than one measuring cup of sugar. The water and sugar can be separated by the physical process of evaporating the water.

If the homogeneous matter is a pure substance, it has a fixed composition and must be either an element or a compound. While elements contain only atoms of that element, compounds contain atoms of different elements combined in distinctive ways. At the nanoscopic level, each sugar molecule contains 12 carbon (C) atoms, 22 hydrogen (H) atoms, and 11 oxygen (O) atoms, represented by $C_{12}H_{22}O_{11}$. At the macroscopic level, every 100 g of table sugar contains 42.1 g of carbon, 6.5 g of hydrogen, and 51.4 g of oxygen in chemical combination. Only chemical reactions could be used to produce pure carbon, hydrogen, and oxygen from sugar.

Why Study Pure Substances?

Perhaps by now you are wondering why we should be interested in elements and compounds and their properties. There are three basic reasons.

1. Only by studying pure matter can we understand how to utilize its properties, design new kinds of matter, and make desirable changes in the nature of everyday life. A long time ago, for example, it was known that swallowing extracts from the willow tree could relieve pain, although there were some unpleasant side effects. Once chemists identified the pain-relieving substance in the mixture of extracts, they were able to create a molecule that was similar enough to relieve pain, but different enough to have fewer side effects. We know this pure substance as aspirin, which can be represented by $C_9H_8O_4$.

At one time most natural materials could be changed only by physical means. Only a few useful materials, such as pottery and iron, were the products of chemical change. Otherwise, the design of everyday objects was limited to utilizing the properties of natural substances. Today we have evidence all around us that the chemical modification of materials has indeed changed the quality of life. Synthetic fibers, plastics, life-saving drugs, latex paints, new and better fuels, photographic films, audiotapes, and videotapes are but a few of the materials produced by controlled chemical change. It is equally important that understanding the properties of substances allows us to prevent *undesirable* chemical changes, such as the corrosion of metal objects, the spoiling of food, and the effects of hereditary diseases.

We shall return to examine the chemistry of many of these modern materials in later chapters.

2. Studying the properties of matter helps us to deal intelligently with our environment, both in everyday life and in the social and political arena. Knowing something about the properties of fertilizers or pesticides helps us decide which ones to use, which ones to avoid, and when their use is or is not necessary. In the context of community life, many decisions are enlightened by some knowledge of the properties of matter. When you pour waste down a sewer drain, what happens to it? Does it matter whether the waste is motor oil or the water from washing your car? Does your town or city really need to spend millions of dollars to upgrade its sewage treatment plant to keep nitrogen-containing compounds out of natural waters or to remove asbestos from the high school auditorium ceiling?

3. The third reason for studying the properties of elements and compounds is simple curiosity. Chemicals and chemical change are a part of nature that is open to investigation. For some people, this is as big a challenge as a high mountain is to a climber. If we hope to understand matter, the first steps are to discover the simplest forms of matter and study their interactions. Curiosity draws many chemists to basic research.

The Structure of Matter Explains Chemical and Physical Properties

An essential part of basic research is investigation into the *structure of matter*—how atoms of the elements are connected in larger units of matter and how these units are arranged in samples of matter large enough to be seen (Figure 2.7).

(a)

(b)

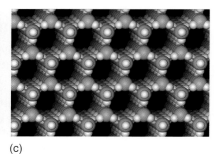

(c)

FIGURE 2.7 Three ways to look at snow. (a) A snowy landscape. (b) The beautiful patterns of snowflakes as they can be seen under a microscope. (c) The orderly arrangement, in ice and snow, of water molecules represented by molecular models.

THE PERSONAL SIDE

Alfred Bernhard Nobel (1833–1896)

To receive a Nobel Prize is looked on by most scientists as the highest possible honor, and the accomplishments recognized by the prize are always among the most significant advances in science. Textbooks usually make note of achievements and individuals worthy of Nobel Prizes as a way of highlighting their importance. It was breakthroughs in the chemistry of explosives that made the Nobel Prizes possible.

Alfred Nobel was a Swedish chemist and engineer, the son of an inventor and industrialist. Together, Nobel and his father established a factory to manufacture nitroglycerin, a liquid explosive that is extremely sensitive to light and heat. After two years in operation, there was an explosion at the factory that killed five people, including Alfred Nobel's younger brother. Perhaps motivated by this event, Alfred searched for a way to handle nitroglycerin more safely. He found the answer and in 1867 patented dynamite. By mixing nitroglycerin with a soft, absorbent, nonflammable natural material

© The Granger Collection

ALFRED NOBEL

known as *kieselguhr*, or diatomaceous earth, Nobel had stabilized the explosive power of nitroglycerin. His dynamite could be shipped safely and would explode only when this was desirable. He also invented the blasting caps (containing a mercury compound) needed to set off the dynamite explosion.

Nobel's talent as an entrepreneur combined with his many inventions (he held 355 patents) made him a very rich man. He never married and left his fortune to establish the Nobel Prizes, awarded annually to individuals who "have conferred the greatest benefits on mankind in the fields of physics, chemistry, physiology or medicine, literature and peace."

The properties of a sample of matter are determined by the nature of its parts, just as the capabilities of a computer are determined by the parts that have been assembled. If we hope to understand the nature of matter, it is absolutely necessary that we understand the minute parts and how they are related to each other. Indeed, *the basic theme of this text and of chemistry itself is the relationship between the structure of matter and its properties.* Armed with an understanding of this relationship, chemists are developing an ever-increasing ability to create new substances and predict the properties of these substances.

CONCEPT CHECK 2A

1. Which of the following questions would a chemist study?
   (**a**) Does a grapefruit contain vitamin E?
   (**b**) What gives a grapefruit its yellow color?
   (**c**) How many grapefruits does the average tree produce in a season?
2. If a sample of matter contains only one kind of atom it is a(an) _____ and _____ be broken down to simpler substances.
3. The two kinds of pure substances are _____ and _____.
4. A solution is a(an) _____ mixture.
5. Solutions may exist in solid, liquid, or gaseous states. (**a**) True (**b**) False
6. If you burn butter in a frying pan, you have carried out a(an) _____ change because the butter changed in _____.
7. Which of the following contain chemicals? (**a**) Aspirin (**b**) Baking soda (**c**) Rubbing alcohol (**d**) Vanilla (**e**) Soap (**f**) Thoughts (**g**) Electricity

8. Which of the following mixtures is homogeneous? (**a**) Sugar dissolved in water (**b**) Oil and water (**c**) Cement, sand, and gravel

9. Identify by name the reactant(s) shown in the following two equations:

(**a**) $\underset{\text{Methane}}{CH_4} + \underset{\text{Oxygen}}{2\,O_2} \longrightarrow \underset{\text{Carbon dioxide}}{CO_2} + \underset{\text{Water}}{2\,H_2O}$

(**b**) $\underset{\text{Hydrogen peroxide}}{2\,H_2O_2} \longrightarrow \underset{\text{Water}}{2\,H_2O} + \underset{\text{Oxygen}}{O_2}$

10. Which of the following are physical properties and which are chemical properties?
    (**a**) Flammability
    (**b**) Density
    (**c**) Souring of milk

11. Chemical potential energy is converted into light and heat in combustion.
    (**a**) True
    (**b**) False

Charles D. Winters

Two nonmetals: bromine, a liquid, and iodine, a solid

By "everyday conditions" chemists refer to usual temperatures and pressures at Earth's surface. Changing these conditions can change the state of a substance, for example, the freezing or boiling of water.

(2.6) THE CHEMICAL ELEMENTS

Iron, gold, sulfur, tin, and a few other substances that occur in Earth's crust were recognized as elements long before the modern meaning of the term developed. Other naturally occurring elements were identified only after long and laborious efforts to separate them from the compounds in which they are found. Although it isn't always easy, chemists can now separate and identify the elements present in the most complex mixtures and compounds.

The names of the elements are listed inside the front cover of this book. About 18 of these elements are not found anywhere in nature, but have only been produced artificially in laboratories, usually in extremely small quantities.

The elements range widely in their properties. As is discussed in Chapter 3 and represented by the arrangement of the elements in the periodic table (inside the front cover of this book), the properties of the elements are largely determined by the composition of their atoms.

Under everyday conditions some elements, including helium, hydrogen, nitrogen, oxygen, and chlorine, are gases. Only two elements—mercury and bromine—are ordinarily liquids. Most of the elements are solids, and of these most are silvery metals. Two distinguishing properties of metals are that they conduct electric current and have a shiny appearance. A second major class of elements, the nonmetals, do not conduct electricity and are not lustrous. The familiar nonmetals include all the elemental gases, and carbon, sulfur, phosphorus, and iodine, which are solids.

Symbols for the Elements

Sometimes people are discouraged about understanding chemistry when they encounter chemical formulas and names such as CH_3CH_2OH, H_2SO_4, acetylsalicylic acid, nitric oxide. . . . There is nothing unique about chemistry, however, in needing its own vocabulary. Surely the symbols and language in the margin are equally mysterious to anyone who has never been a musician or played football. Chemistry, just like music or football, needs special symbols and language. Without them, communication would be impossible.

Since elements are the building blocks of all matter, symbols for the elements are fundamental to communicating about chemistry. The symbols for the elements are listed next to their names inside the front cover of this book. Some symbols are single letters that are the first letter of the name. Others are two letters from the name, always with the first letter capitalized and the second lowercase. For example, *io*dine is represented by I and *in*dium is represented by In; *ni*trogen is represented by N and *ni*ckel, by Ni; *ma*gnesium is represented by Mg and *ma*nganese, by Mn. You can see that it is important to recognize the symbols and know how they differ for different elements.

Eleven metals are represented by symbols not derived from their modern names, but instead from older names in other languages. Most of these metals, listed in Table 2.2, were known in ancient times because they are found free in nature or are easily obtained from their naturally occurring compounds.

Symbols for the elements are the alphabet for the language of chemistry.

Of the known elements, only 18 are nonmetals. Most of the common nonmetals have single letter symbols

H	C	N	O
Hydrogen	Carbon	Nitrogen	Oxygen
P	S	F	I
Phosphorus	Sulfur	Fluorine	Iodine

Consult the alphabetical listing inside the front cover to find the symbol from the name of any element, or vice versa.

Earlier we noted that a sample of gaseous helium (also a nonmetal) contains helium atoms. The symbol He is therefore used to represent a single He atom or a large collection of helium atoms. One of the first milestones in understanding the structure of matter was the discovery that a number of nonmetallic elements are composed not of atoms, but of molecules. A sample of pure hydrogen gas does contain only hydrogen atoms, as expected for an element, but the atoms are joined together in two-atom hydrogen molecules, represented by H_2. Seven nonmetals exist under everyday conditions as two-atom molecules, known as **diatomic molecules**:

ELEMENT	SYMBOL	PROPERTIES
Hydrogen	H_2	Colorless, odorless, occurs as a very light gas, burns in air
Oxygen	O_2	Colorless, odorless gas, reactive, constituent of air
Nitrogen	N_2	Colorless, odorless gas, rather unreactive
Chlorine	Cl_2	Greenish-yellow gas, very sharp choking odor, poisonous
Fluorine	F_2	Pale yellow, highly reactive gas
Bromine	Br_2	Dark red liquid, vaporizes readily, very corrosive
Iodine	I_2	Dark purple solid that changes directly to gas when heated gently

Chemical formulas are the words in the language of chemistry.

Diatomic molecules Molecules composed of two atoms

TABLE 2.2 Elements with Symbols Not Based on Their Modern Names

MODERN NAME	SYMBOL	ORIGIN OF SYMBOL
Antimony	Sb	*stibium:* Latin name for *antimony*, which comes from the Greek *anti + monos*, or element not found alone
Copper	Cu	*cuprum:* Latin name, meaning from the island of Cyprus
Gold	Au	*aurum:* Latin name, meaning shining dawn
Iron	Fe	*ferrum:* Latin name for the element
Lead	Pb	*plumbum:* Latin name for the element
Mercury	Hg	*hydrargyrum:* Greek name, meaning liquid silver or quick silver
Potassium	K	*kalium:* Latin name, meaning alkali
Silver	Ag	*argentum:* Latin name for the element
Sodium	Na	*natrium:* Latin name for the element
Tin	Sn	*stannum:* Latin name for the element
Tungsten	W	*wolfram:* Swedish name, meaning devourer of tin (because it interferes with purifying tin)

2.7 USING CHEMICAL SYMBOLS

Formulas for Chemical Compounds

We have already used a few **chemical formulas**, combinations of the symbols for the elements that represent the stable combinations of atoms in molecules: H_2 for elemental hydrogen, H_2O for water, and $C_{12}H_{22}O_{11}$ for table sugar. The principle is the same no matter how many atoms are combined. The symbols represent the elements and the **subscripts** (for example, the $_2$ in H_2O) indicate the relative numbers of atoms of each kind. The formulas and properties of a few simple molecular compounds are given in Table 2.3. Such formulas can represent one molecule or a large sample of the compound.

Sometimes a line or lines are drawn between symbols to indicate which atoms are connected in molecules:

$$H—H \qquad H—\overset{\displaystyle H}{\underset{}{N}}—H \qquad H—\overset{\displaystyle H}{\underset{\displaystyle H}{C}}—H \qquad O=C=O$$

A hydrogen molecule   An ammonia molecule   A methane molecule   A carbon dioxide molecule

Table sugar, $C_{12}H_{22}O_{11}$
A pure substance
An organic compound
Contains carbon, hydrogen, and oxygen

Charles D. Winters

Formulas that show the connections in this way are known as **structural formulas**, whereas formulas that give just one symbol for each element present are called **molecular formulas**. The molecular formula for methane is CH_4, for example. Methane and the many millions of compounds that contain carbon combined with hydrogen—and often also with nitrogen, oxygen, phosphorus, or sulfur—are known as **organic compounds**.

Compounds not based on carbon are referred to as **inorganic compounds**. A few that are in everyday use are

	NaCl	NH_4NO_3	H_2SO_4	$Mg(OH)_2$
Chemical name:	Sodium chloride	Ammonium nitrate	Sulfuric acid	Magnesium hydroxide
Use:	Table salt	Fertilizer	Battery acid	A laxative

When it is necessary to explore the shapes of molecules beyond what can be shown on a flat piece of paper, chemists resort to physical models or computer-drawn pictures, such as those shown in Figure 2.8.

The lines in structural formulas represent chemical bonds, which we'll have much more to say about in Chapter 5.

Organic compounds are so numerous and so important that they are the basis for an entire field of chemistry—**organic chemistry** (Chapters 12 and 14). Most compounds produced by living things, most medicines, and the components of most plastics are organic compounds.

Chemical formula Written combination of element symbols that represents the atoms combined in a chemical compound

Subscripts In chemical formulas, numbers written below the line (for example, $_2$ in H_2O) to show numbers or ratios of atoms in a compound

Structural formulas Chemical formulas that show the connections between atoms in molecules as straight lines

Molecular formulas Chemical formulas that represent molecules with atom symbols and subscripts

Organic compounds Compounds of carbon with hydrogen, and derivatives of these compounds

Inorganic compounds Compounds of all elements other than the organic compounds of carbon

TABLE 2.3	Some Simple Molecular Compounds	
COMPOUND	**FORMULA**	**PROPERTIES**
Water	H_2O	Odorless liquid
Carbon monoxide	CO	Odorless, flammable, toxic gas
Carbon dioxide	CO_2	Odorless, nonflammable, suffocating gas
Sulfur dioxide	SO_2	Nonflammable gas, suffocating odor
Ammonia	NH_3	Colorless, nonflammable gas, pungent odor
Methane	CH_4	Odorless, flammable gas
Carbon tetrachloride	CCl_4	Nonflammable, dense liquid
Nitrogen dioxide	NO_2	Reddish-brown gas, irritant

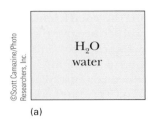
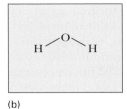

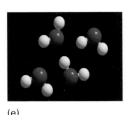

(a)   (b)   (c)   (d)   (e)

FIGURE 2.8 The water molecule, as represented in print (a and b), in physical models that can be handled (c and d), or on a computer screen (e).

EXAMPLE   2.1   Chemical Symbols

Which of the following represent elements and which represent compounds?
(**a**) KI   (**b**) Co   (**c**) Ag   (**d**) NO   (**e**) Cl_2

SOLUTION
The chemical symbols in (**a**) and (**d**) represent compounds—the second letter in an element symbol is never a capital letter; (**b**) and (**c**) represent the elements cobalt and silver; and (**e**) represents the element chlorine, which exists as diatomic molecules (p. 27).

TRY IT 2.1
Write the symbols or formulas for (**a**) lead, (**b**) phosphorus, (**c**) a molecule containing one hydrogen atom and one chlorine atom, (**d**) a molecule containing one aluminum atom and three bromine atoms, and (**e**) elemental fluorine.

Chemical Equations

To concisely represent chemical reactions, symbols and formulas are arranged in **chemical equations**. For example, in words, a chemical reaction could be expressed as "carbon reacts with oxygen to form carbon monoxide," a reaction that happens whenever something containing carbon burns incompletely. Like most solid elements, carbon is represented just by its symbol, C; oxygen must be represented by the molecular formula for its diatomic molecules, O_2; and carbon monoxide molecules, which contain two atoms, one each of carbon and oxygen, are represented by their molecular formula, CO.

Earlier, we pointed out that in chemical reactions atoms are rearranged into different substances, but the number of atoms stays the same. To represent this correctly, chemical equations must be **balanced**—the number of atoms of each kind in the reactants and products must be the same. In this case, one oxygen molecule will combine with two carbon atoms to form two carbon monoxide molecules. All this information is contained in the equation

$$2\,C(s) + O_2(g) \longrightarrow 2\,CO(g)$$

We have also included here the information that carbon is a solid (*s*) and that oxygen and carbon monoxide are gases (*g*). The arrow ($\longrightarrow$) is often read as "yields." The equation then states the following:

At the macro level: Carbon, a solid, plus oxygen gas yields carbon monoxide gas.

At the nanoscopic level: Two atoms of carbon plus one diatomic molecule of oxygen yield two molecules of carbon monoxide.

Chemical equations are the sentences in the language of chemistry.

States of matter are shown in equations as (*g*) for a gas, (ℓ) for a liquid, and (*s*) for a solid; (*aq*) is used for a substance dissolved in water (an **aqueous solution**).

Chemical equations Representations of chemical reactions by the formulas of reactants and products

Balanced chemical equations Chemical equations in which the total number of atoms of each kind is the same in reactants and products

The number written before a formula in an equation, the **coefficient**, gives the relative amount of the substance involved. The subscripts give the composition of the pure substances. Changing the coefficient changes only the amount of the element or compound involved, whereas changing a subscript would change the identity of the reactant or product. For example, 2 CO represents two molecules of carbon monoxide, whereas CO_2 represents a molecule of carbon dioxide, a very different substance formed in the complete combustion of carbon-containing materials:

$$C(s) + O_2(g) \longrightarrow CO_2(g)$$

EXAMPLE 2.2 **Chemical Equations**

Nitrogen dioxide (NO_2) is a red-brown gas often visible in the haze during a period of air pollution over a city. (**a**) Interpret in words the information given in the equation for formation of nitrogen dioxide from nitrogen monoxide (NO):

$$2\,NO(g) + O_2(g) \longrightarrow 2\,NO_2(g)$$

(**b**) Prove that the equation is balanced.

SOLUTION
(**a**) Nitrogen monoxide, a gas, reacts with oxygen gas to give nitrogen dioxide, also a gas. The coefficients show that two molecules of NO react with one molecule of O_2 to give two molecules of NO_2.
(**b**) *Reactants:* The coefficient 2 in 2 NO shows the presence of two molecules of NO, which means two N atoms and two O atoms. The O_2 has no coefficient; thus, there is one O_2 molecule, but the subscript shows that the single molecule has two O atoms. Therefore, there are two N atoms and four O atoms on the reactants side. *Products:* The coefficient in 2 NO_2 shows two molecules of NO_2, which gives a total of two N atoms (N has no subscript) and four O atoms (two in each of the two molecules, as shown by the subscript on O). The equation is balanced.

TRY IT 2.2
Write a balanced equation, including the state designations, for the reaction of hydrogen gas with oxygen gas to give water.

TRY IT 2.3
Repeat the question of Example 2.2 for the equation for the formation of hydrogen chloride (HCl):

$$H_2(g) + Cl_2(g) \longrightarrow 2\,HCl(g)$$

CONCEPT CHECK 2B

1. Chemists strive to understand the _____ of matter in order to create new materials.
2. Name two elements that are solids, two that are liquids, and two that are gases.
3. Write the symbols for the six elements you named in question 2.
4. Name the elements combined in the compound K_2HPO_4. How many atoms in total are represented in this chemical formula?
5. Are the following equations balanced? If not, balance them.
   (**a**) $2\,K(s) + Cl_2(g) \longrightarrow KCl(s)$
   (**b**) $2\,Mg(s) + O_2(g) \longrightarrow 2\,MgO(s)$

NO is commonly known as nitric oxide.

Coefficients In a chemical equation, numbers written before formulas to balance the equation

(c) $SO_3(g) + H_2O(\ell) \longrightarrow 2\,H_2SO_4(aq)$

(d) $NaCl(aq) + AgNO_3(aq) \longrightarrow NaNO_3(aq) + AgCl(s)$

6. Consider the chemical equation $CH_4 + 2\,O_2 \longrightarrow CO_2 + 2\,H_2O$. Explain what is meant by the symbols:

   (a) O _____

   (b) $2\,O_2$ _____

   (c) CH_4 _____

   (d) $\longrightarrow$ _____

   (e) H_2O _____

   (f) $2\,H_2O$ _____

7. A solid element does not conduct electricity. Is it a metal or a nonmetal?

8. Which of the following elements have symbols not based on their modern names? **(a)** Lead **(b)** Carbon **(c)** Oxygen **(d)** Potassium

2.8 THE QUANTITATIVE SIDE OF SCIENCE

Several times in the preceding sections we used the numerical results of measurements of the boiling points, masses, or densities of pure substances. These and hundreds of other kinds of measurements are fundamental to chemistry and every other science. The result of a measurement is what we refer to as **quantitative** information—it uses numbers. Weighing yourself is a quantitative experiment. By contrast, there is **qualitative** information that does not deal with numbers. For example, sticking your hand into a tub of water and determining it to be "hot" is a qualitative observation.

The result of a measurement is recorded as a number plus a unit. "A boiling point of 52" doesn't mean anything without the unit. Was the temperature measured in Fahrenheit degrees, as is done by the usual household thermometer in the United States, or in Celsius degrees, as would be done in the rest of the world?

The establishment of scientific facts and laws is obviously dependent on accurate observations and measurements. Although measurements can be reported as precisely in one system of measurement as another, there has been an effort since the time of the French Revolution in the late 1700s to have all scientists embrace the same simple system. The hope was and is to facilitate communication in science. The metric system, which was born of this effort, has two advantages. First, it is easy to convert from one unit to another, since smaller and larger units for the same physical quantity differ only by multiples of ten. Consequently, to change millimeters to meters, the decimal point need only be moved three places to the left

$$1 \text{ millimeter} = 0.001 \text{ meter} \quad \text{therefore,} \quad 5.0 \text{ millimeters} = 0.0050 \text{ meter}$$

Compare the decimal shift to the problem of changing inches to yards. Knowing

$$1 \text{ ft} = 12 \text{ inches} \quad 1 \text{ yard} = 3 \text{ ft}$$

we can calculate

$$5.0 \text{ inches} \times \frac{1 \text{ ft}}{12 \text{ inches}} \times \frac{1 \text{ yard}}{3 \text{ ft}} = 0.14 \text{ yard}$$

Take a moment to be sure you understand how units are included and canceled in this unit conversion. An excellent way to keep track of what you are

Everyone making scientific measurements must understand the *significance* of the digits when they are using the results of measurements. This becomes especially important in using electronic calculators. If a 3-cm³ piece of something weighs 4.52 g, is its density 1.5 g/cm³, 1.50 g/cm³, or—as given by a calculator—1.506666667 g/cm³? Appendix A explains how questions like this are dealt with.

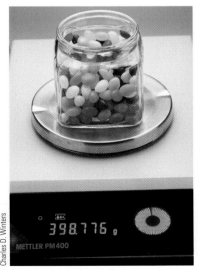

Charles D. Winters

METTLER PM400

A metric laboratory balance.

Quantitative Describes information or experiments that are numerical

Qualitative Describes information or experiments that are not numerical

For more information and practice with units and problem solving, consult Appendix C at the end of this book.

doing is to include units in setting up the problem and then to cancel them like numbers. You will find that this method applies to far more than unit conversion or the mathematics of chemistry. Often by including the units with numbers you can figure out what is needed to find a mathematical answer; and when the units cancel to give the desired unit with the answer, chances are that the answer will be the right one (if, of course, your arithmetic is correct).

The second advantage of the modern metric system is that standards for most fundamental units are defined by reproducible phenomena of nature. For example, the metric unit for time—the second—is now defined in terms of a specific number of cycles of radiation from a radioactive cesium atom, a time period believed never to vary.

The units still in everyday use in the United States have evolved from the English system of measurement, which began with the decrees of various English monarchs. The *yard* started out as the length of the waist sash worn by Saxon kings and obviously varied with the girth of the king. A step toward standardization came when King Henry I decreed that the yard should be the distance from the tip of his nose to the end of his thumb. Standards for the English units in use in the United States today are designated by the National Institute of Standards and Technology (NIST).

Since its introduction in 1790, the metric system has been continually modified and improved. The current version—the International System of Units, abbreviated SI (from the French, Système International d'Unités)— was adopted by the International Bureau of Weights and Measures in 1960. The SI system defines seven fundamental units from which other units are derived. For example, the fundamental SI unit for length is the meter (abbreviated m), which then gives the SI unit for area as the square meter (m^2) and the SI unit for volume as the cubic meter (m^3). Multiples and fractions of SI units are designated by adding prefixes (Table 2.4). Therefore, when a length unit smaller than the meter is needed, we can choose the *centi*meter (cm), the *milli*meter (mm), or the *nano*meter (nm); and when a longer length unit is needed, we can use the *kilo*meter (km). For mass units, prefixes are added to gram, such as the kilogram (kg), the milligram (mg), or the microgram (μg, where μ is the Greek letter mu), or the nanogram (ng).

Because of their convenient sizes, the five units listed in Table 2.5 are commonly used in chemistry and indeed are in everyday use in the rest of the world. We will use the unit symbols listed in Table 2.5 in this text. The volume

The prefixes in Table 2.4 are represented by the powers of 10 used in scientific, or exponential, notation for writing large and small numbers. For example, $10^3 = 10 \times 10 \times 10 = 1000$. Appendix B reviews this notation.

TABLE 2.4 Common Prefixes for Multiples and Fractions of SI Units

PREFIX	ABBREVIATION	MEANING	EXAMPLE
mega-	M	10^6 (1 million)	1 megaton = 1×10^6 tons
kilo-	k	10^3 (1 thousand)	1 kilogram (kg) = 1×10^3 g
deci-	d	10^{-1} (1 tenth)	1 decimeter (dm) = 0.1 m
centi-	c	10^{-2} (1 one-hundredth)	1 centimeter (cm) = 0.01 m
milli-	m	10^{-3} (1 one-thousandth)	1 millimeter (mm) = 0.001 m
micro-	μ*	10^{-6} (1 one-millionth)	1 micrometer (μm) = 1×10^{-6} m
nano-	n	10^{-9} (1 one-billionth)	1 nanometer (nm) = 1×10^{-9} m
pico-†	p	10^{-12} (1 one-trillionth)	1 picometer (pm) = 1×10^{-12} m

*This is the Greek letter mu (pronounced "mew").
†This prefix is pronounced "peako."

TABLE 2.5		Some Common Units in Chemistry
NAME OF UNIT	**SYMBOL**	**COMMON EQUIVALENT**
meter	m	39.4 inches
liter	L	1.06 quarts
gram	g	0.0352 ounce
degrees Celsius	°C	Water boils at 100°C and freezes at 0°C
calorie	cal	Energy required to raise the temperature of 1 g of water by 1°C

NET WT 10 OZ (283g)

6-16 OZ BOTTLES (96 FL OZ)
6-473 mL BOTTLES (2.84 L)

Charles D. Winters

Two labels that are in compliance with the changeover from English units to metric units.

EXAMPLE 2.3   Metric Unit Prefixes

How many meters are in 2 km?

SOLUTION
The prefix *kilo-* (k) means 1000 times, meaning that 1000 m = 1 km. It is hardly worth the trouble to write anything down in this solution. One just thinks 1000 m for every kilometer as one thinks ten dimes for every one dollar bill. The best way to write this problem, or any more complicated problem, however, is to include the units and cancel them out (like numbers in algebra). This way, if you are left with the correct unit, you have probably done the problem correctly.

$$2 \text{ km} \times \frac{1000 \text{ m}}{1 \text{ km}} = 2000 \text{ m}$$

TRY IT 2.4
How much in dollars is 20 *mega*bucks?

---

unit of liters is preferable to the SI-derived unit of cubic meters, which is much too large (1 cubic meter = 1000 liters). (Figure 2.9 gives the relative sizes of some other length units.)

We tend to use the words *mass* and *weight* interchangeably, and that's all right because we understand what we mean. Strictly speaking though, *mass* measures the amount of matter in an object and is the same everywhere in the universe, whereas *weight* measures the force of gravity on that object and is different where the force of gravity is different (for example, on the Moon).

CONCEPT CHECK   2C

1. The result of a measurement must include both a(an) _____ and a(an) _____.
2. The metric system uses _____ to indicate multiples of _____.
3. The volume unit derived from the centimeter is a(an) _____.
4. The prefix meaning 1000 times bigger is _____, and the prefix meaning 0.001, or 1000 times smaller, is _____.
5. What are the units for the answer to the following calculation?

$$0.500 \text{ ft} \times \frac{12 \text{ inches}}{1 \text{ ft}} \times \frac{1 \text{ m}}{39.4 \text{ inches}} \times \frac{1000 \text{ mm}}{1 \text{ m}} = 152 \text{ _____}$$

6. Which conversion factor should you use to convert miles to kilometers (km)?

   (a) $\dfrac{1 \text{ mile}}{1.61 \text{ km}}$ (b) $\dfrac{1.61 \text{ km}}{1 \text{ mile}}$

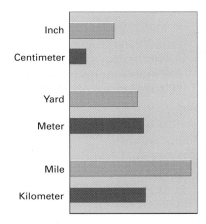

Inch
Centimeter
Yard
Meter
Mile
Kilometer

FIGURE 2.9 Relative sizes of some English and metric length units: 1 inch = 2.54 centimeters; 1 yard = 0.91 meter; 1 mile = 1.61 kilometers.

7. Which of the following experiments is qualitative and which is quantitative?
   (a) Determination of the distance between two atoms in a molecule.
   (b) Determination of the identity of the metal in a piece of wire.

 Assess your understanding of this chapter's topics with an online chapter quiz at **www.brookscole.com/chemistry/joesten4**

■ KEY TERMS

macroscopic	physical properties	subscripts
microscopic	density	structural formulas
nanoscopic	chemical property	molecular formulas
pure substance	chemical reaction	organic compounds
element	reactants	inorganic compounds
atom	products	organic chemistry
chemical compounds	molecular compounds	chemical equations
molecule	energy	balanced chemical equations
homogeneous mixtures	potential energy	aqueous solution
heterogeneous mixtures	kinetic energy	coefficient
solutions	diatomic molecules	quantitative
mass	chemical formula	qualitative

■ LANGUAGE OF CHEMISTRY

1. Produces new substance(s)
2. Air
3. Unchanged by further purification
4. Cannot be reduced to simpler substances
5. SI system
6. Symbol for iron
7. Volume unit
8. Molecule containing three oxygen atoms
9. Carbon monoxide
10. 100 cm
11. SI fundamental unit
12. Deci-
13. Centi-
14. Structural formula
15. Solution
16. Molecular formula

a. Chemical change
b. Element
c. Properties of pure substance
d. Mixture
e. Uses multiples of ten
f. O_3
g. CO
h. Fe
i. 1 m
j. H—S—H
k. Liter
l. One-tenth
m. Meter
n. C_4H_{10}
o. Homogeneous mixture
p. One-hundredth

■ APPLYING YOUR KNOWLEDGE

1. Name as many materials as you can that you have used during the past day and that were not chemically changed from their natural states.

2. Identify the following as physical or chemical changes. Explain your choices.
   (**a**) Formation of snowflakes
   (**b**) Rusting of a piece of iron
   (**c**) Ripening of fruit
   (**d**) Fashioning a table leg from a piece of wood
   (**e**) Fermenting grapes
   (**f**) Boiling a potato

3. Would it be possible for two pure substances to have exactly the same set of properties? Give reasons for your answer.

4. List three physical properties that can be used to identify pure substances. Give a specific example of each property.

5. What was the family business of Alfred Nobel, the man for whom the Nobel Prizes are named?

6. Nitroglycerine, the chemical compound that led to Alfred Nobel's fortune, is also used for the treatment of heart conditions. What does this illustrate about the versatility of many chemical compounds? Are the risks and benefits associated with the use of nitroglycerin as an explosive the same as for using it to treat angina, a heart condition?

7. Classify each of the following as a physical property or a chemical property. Explain your answers.
   (**a**) Density
   (**b**) Melting temperature
   (**c**) Decomposition of a substance into two elements on heating
   (**d**) Electrical conductivity
   (**e**) The failure of a substance to react with sulfur
   (**f**) The ignition temperature of a piece of paper

8. Chemical changes can be both useful and destructive to humanity's purposes. Cite a few examples of each kind of change from your own experience. Also give evidence from observation that each is indeed a chemical change and not a physical change.

9. Classify each of the following as an element, a compound, or a mixture. Justify each answer.
   (**a**) Mercury
   (**b**) Milk
   (**c**) Pure water
   (**d**) A piece of lumber
   (**e**) Ink
   (**f**) Iced tea

   (**g**) Pure ice
   (**h**) Carbon
   (**i**) Antimony

10. Write your last name. How many element symbols can you produce from the letters in your name?

11. Is it possible for the properties of iron to change? What about the properties of steel, which is an alloy and a homogeneous mixture? Explain your answer.

12. An element is the smallest part of a compound that may still be identified as that compound.
    (**a**) True
    (**b**) False

13. An atom is the smallest part of an element that may still be identified as the element.
    (**a**) True
    (**b**) False

14. An atom is the smallest part of a compound that may still be identified as the compound.
    (**a**) True
    (**b**) False

15. A pure compound may be broken down into two or more different kinds of pure elements.
    (**a**) True
    (**b**) False

16. You have a mixture of sand (SiO_2) and salt (NaCl). How would you separate these two substances? When you have them as separate substances, how can you prove which is which?

17. Consider the following five elements: nitrogen, sulfur, chlorine, magnesium, and cobalt. By using this text or any other source available at the library, find the major source for these elements and at least one compound that uses the element in combined form.

18. Atrazine is a selective herbicide that has the molecular formula $C_8H_{14}N_5Cl$. This compound is used for season-long weed control in corn, sorghum, and certain other crops. What elements are present in atrazine?

19. Cytoxan, also known as cyclophosphamide, is widely used alone or in combination in the treatment of certain kinds of cancer. It interferes with protein synthesis and in the process kills rapidly replicating cells, particularly malignant ones. Cytoxan has the molecular formula $C_7H_{15}O_2N_2PCl_2$.
    (**a**) How many atoms are in one molecule of Cytoxan?
    (**b**) What elements are present in Cytoxan?

(c) What is the ratio of hydrogen atoms to nitrogen atoms in Cytoxan?

(d) Would Cytoxan be classified as an organic compound?

20. By reading labels, identify a commercial product that contains each of the following compounds:

(a) Calcium carbonate

(b) Phosphoric acid

(c) Water

(d) Fructose

(e) Sodium chloride

(f) Potassium sorbate

(g) Potassium iodide

(h) Glycerol

(i) Aluminum

(j) Butylated hydroxytoluene (BHT)

21. There are three states of matter—gas, liquid, and solid. Name a material that is a pure substance for each state of matter. Do not use water, oxygen, or salt. Name a material that is a mixture for each state of matter. Do not use air, gasoline, or brass.

22. Which of the following statements is true of a balanced chemical equation?

(a) The total number of molecules of reactants has to be the same as the total number of product molecules.

(b) The total number of atoms in the reactant molecules doesn't have to equal the total number of atoms in the product molecules.

(c) The identity of the atoms in the reactants has to be the same as the identity of the atoms in the products.

(d) The physical state of the reactants has to be the same as the physical state of the products.

23. Is it possible to have a mixture of two elements and also to have a compound of the same two elements? Explain. Can you think of an example?

24. Given the following sentence, write a chemical equation using chemical symbols that convey the same information: "One nitrogen molecule reacts with three hydrogen molecules to produce two ammonia molecules, each containing one nitrogen and three hydrogen atoms."

25. Chemical equations are a shorthand way of describing how species known as _____ are converted into _____.

(a) reactants; solvents

(b) products; solutes

(c) solutes; solutions

(d) reactants; products

26. Name four kinds of energy.

27. Describe in words the chemical processes represented by the following equations:

(a) $2\ Na(s) + Cl_2(g) \longrightarrow 2\ NaCl(s)$

(b) $N_2(g) + 3\ Cl_2(g) \longrightarrow 2\ NCl_3(g)$ [named as nitrogen trichloride]

(c) $CO_2(g) + H_2O(\ell) \longrightarrow H_2CO_3(aq)$ [carbonic acid]

(d) $2\ H_2O_2(aq)$ [hydrogen peroxide] $\longrightarrow O_2(g) + 2\ H_2O(\ell)$

28. Prove that each of the equations in question 27 is balanced.

29. For equations b and d in question 27, identify the reactant(s) and the product(s).

30. Are the following equations balanced?

(a) $AgNO_3\ (aq) + Na_2SO_4\ (aq) \longrightarrow Ag_2SO_4(s) + NaNO_3(aq)$

(b) $AgNO_3(aq) + HCl(aq) \longrightarrow AgCl(s) + HNO_3(aq)$

31. Is the tea in tea bags a pure substance? Use the process of making tea to make an argument for your answer. How would your argument apply to instant tea?

32. Find and list as many pure substances as you can in a kitchen.

33. (a) How many milligrams are there in 1 g?

(b) How many meters are there in 1 km?

(c) How many centigrams are there in 1 g?

34. What are the most common units in chemistry for mass, length, and volume?

35. Which of the following quantities is a density?

(a) 9 cal/g   (b) 100 cm/m

(c) 1.5 g/mL   (d) 454 g/lb

36. A cook wants to pour 1.5 L of batter into a 2-quart bowl. Will it fit?

37. A rancher needs one acre of grazing land for ten cows. How many acres are needed for 55 cows? Solve this problem (and others where appropriate) by including units. In this case, the answer will have the "units" of acres per cow.

38. There are 200 mg of ibuprofen in an Advil tablet. How many grams is this? How many micrograms?

39. How many meters does a runner cover in a 10-km race?

40. If a 1-ounce portion of cereal contains 3.0 g of protein, how many milligrams of protein does it contain?

41. Complete the following:

(a) 4 cm = _____ m

(b) 0.043 g = _____ mg

(c) 15.5 m = _____ mm

(d) 328 mL = _____ L

(e) 0.98 kg = _____ g

42. An average African gorilla has a mass of 163 kg. What is this mass in grams?

43. An average adult man has a mass of 70 kg. Convert this mass to milligrams.

44. An aspirin tablet usually contains 325 mg of aspirin. What is this in grams?

45. What are the results for the following operations in exponential notation? You may need to refer to Appendix B on scientific notation for questions 46 and 47.
    (a) (4.0×10^5) (2.0×10^2)
    (b) $(6.0 \times 10^7)/(2.0 \times 10^2)$
    (c) $(7.4 \times 10^4)/(4.6 \times 10^9)$
    (d) (2.3×10^{-3}) (4.2×10^4)

46. Express the following in exponential (scientific) notation:
    (a) 8,000,000   (b) 0.000075
    (c) 23,600,000,000   (d) 37,000
    (e) 6492   (f) 0.000000028

47. An electric power plant produces 450 megawatts.
    (a) What is this expressed in watts?
    (b) How many 100-watt lightbulbs will this light?

48. A computer has a 60 gigabyte hard drive. How many bytes is this?

49. What is the mass in grams for a 1000-mL volume (1 liter) of iridium if it has a density of 22.42 g/mL?

50. Lead has a density of 11.4 g/cm³. What is the mass in grams of a cube of lead that has the dimensions 10 cm × 10 cm × 10 cm?

51. The densities for aluminum and iron are 2.7 g/cm³ and 7.86 g/cm³, respectively. What is the ratio of the mass of an object made of aluminum to that of the same object made of iron?

Ⓦ CHEMISTRY ON THE WEB

For up-to-date URLs, visit the text website at **www.brookscole.com/ chemistry/joesten4**

- Properties of Matter
- Scientific Notation
- Dimensional Analysis
- Significant Figures

ATOMS AND THE PERIODIC TABLE

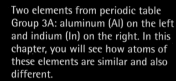

Two elements from periodic table Group 3A: aluminum (Al) on the left and indium (In) on the right. In this chapter, you will see how atoms of these elements are similar and also different.

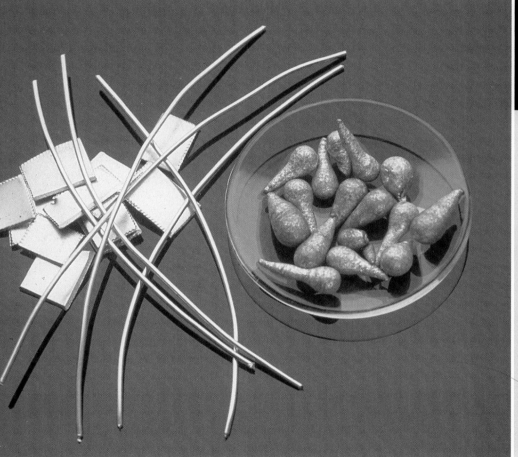

Charles D. Winters

Why does an element or compound have the properties it has? Why does one element or compound undergo a change that another element or compound will not undergo? Inanimate matter behaves the way it does because of the nature of its parts. The use of atoms to represent these parts dates back to about 400 B.C. when the Greek philosopher Leucippus and his student, Democritus (460–370 B.C.), argued for a limit to the divisibility of matter, which was counter to the prevailing view of Greek philosophers that matter is endlessly divisible. Democritus used the Greek word *atomos*, which literally means "uncuttable," to describe the ultimate particles of matter, particles that could not be divided further. However, it wasn't until John Dalton (1766–1844) introduced his atomic theory in 1803 that the importance of using atoms to explain properties of matter was recognized.

Dalton's atomic theory and the development of the periodic table by Mendeleev in 1869 led to the rapid growth of chemistry as a science. In particular, the influence of the location and number of electrons in atoms on the properties of elements has become one of the essential ideas of chemistry. In this chapter we will use current knowledge about the atom together with the periodic table as the basis for our understanding of the chemical view of matter.

- What are the milestones in the development of atomic theory?
- What is the experimental evidence for the existence of subatomic particles within atoms?
- What are the three basic subatomic particles of the atom?
- What are isotopes?
- Where are electrons in atoms, and how are they arranged?
- How was the periodic table developed?
- Why do elements in the same group in the periodic table have similar chemical properties?
- How does the activity of an element relate to its position in the periodic table?

3.1 JOHN DALTON'S ATOMIC THEORY

John Dalton, drawing from his own quantitative experiments and those of earlier scientists, proposed in 1803 that

1. All matter is made up of indivisible and indestructible particles called atoms.
2. All atoms of a given element are identical, both in mass and in properties. By contrast, atoms of different elements have different masses and different properties.
3. Compounds form when atoms of different elements combine in ratios of small whole numbers.
4. Elements and compounds are composed of definite arrangements of atoms. Chemical change occurs when the atomic arrays are rearranged.

John Dalton's atomic theory was accepted because it could be used to explain several scientific laws that Dalton and other scientists of the time had established. Two of these laws are (1) the law of conservation of matter, and (2) the law of definite proportions.

Some years before Dalton proposed his atomic theory, Antoine Lavoisier (1743–1794) had carried out a series of experiments in which the reactants were carefully weighed before a chemical reaction and the products were carefully weighed afterward. He found no change in mass when a reaction occurred, proposed that this was true for every reaction, and called his proposal the **law of conservation of matter**: *Matter is neither lost nor gained during a chemical reaction.* Others verified his results, and the law became accepted. Points (2) and (4) in Dalton's theory imply the same thing: If each kind of atom has a particular characteristic mass, and if there are exactly the same numbers of each kind of atom before and after a reaction, the masses before and after must also be the same.

Another chemical law known in Dalton's time had been proposed by Joseph Louis Proust (1754–1826) as a result of his analyses of minerals. Proust found that a particular compound, once purified, always contained the same elements in the same ratio by mass. One such study, which Proust

W See an animation of **Conservation of Matter** at http://brookscole.com/chemistry/joesten4

Law of conservation of matter Matter is neither lost nor gained during a chemical reaction

John Dalton (1766–1844)

John Dalton, a gentle man and a devout Quaker, gained acclaim because of his work. He made careful measurements, kept detailed records of his research, and expressed them convincingly in his writings. However, he was a poor speaker and was not well received as a lecturer. When Dalton was 66 years old, some of his admirers sought to present him to King William IV. Dalton resisted because he would not wear the court dress. Because he had a graduate degree from Oxford University, the scarlet robes of Oxford were deemed suitable, but a Quaker could not wear scarlet. Dalton, being color blind, saw scarlet as gray; thus, he was presented in scarlet to the court, but in gray to himself. This

JOHN DALTON

He kept more than 200,000 notes on meteorology. Despite his accomplishments he shunned glory and maintained he could never find time for marriage.

remarkable man was, in fact, the first to describe color blindness. He began teaching in a Quaker school when only 12 years old, discovered a basic law of physics (the law of partial pressure of gases), and helped found the British Association for the Advancement of Science.

made in 1799, involved copper carbonate. Proust discovered that, regardless of how copper carbonate was prepared in the laboratory or how it was isolated from nature, it always contained the same proportions by mass—five parts copper, four parts oxygen, and one part carbon. Careful analyses of this and other compounds led Proust to propose the **law of definite proportions**: *In a compound, the constituent elements are always present in a definite proportion by mass.* For example, pure water, a compound, is always made up of 11.2% hydrogen and 88.8% oxygen by weight. Pure table sugar, another compound, always contains 42.1% carbon, 6.5% hydrogen, and 51.4% oxygen by weight. The source of the pure substance is irrelevant.

3.2   STRUCTURE OF THE ATOM

Scientists now have experimental evidence for the existence of more than 60 subatomic particles. However, only three are important to our understanding of the chemical view of matter: **protons** and **neutrons**, found in the **nucleus** of the atom, and **electrons**, found outside the nucleus (Table 3.1). The mass and charge of electrons and protons were determined with experiments that used electric and magnetic fields. Electrons and protons have the same quantity of charge but different signs. The mass of the positively charged proton is about the same as the mass of the neutron. Both have masses 1800 times the mass of the negatively charged electron. Although studies in the early 1900s indicated the presence of a third type of particle with no charge but a mass similar to that of the proton, the lack of charge meant that electric and magnetic fields could not be used to detect this type of particle. As a result, the neutron was not discovered until 1932 when James Chadwick (1891–1974) detected neutrons with experiments using emissions from radioactive elements.

Even though protons and electrons had been identified as subatomic particles, the arrangement of these particles within the atom was not known. An atom was thought to be a uniform sphere of protons within which electrons circulated in rings. Only after the discovery of natural radioactivity did Ernest Rutherford (1871–1937) carry out experiments that led to the idea of a nucleus as a tiny core of the atom.

Law of definite proportions In a compound, the constituent elements are always present in a definite proportion by weight

Protons Positively charged subatomic particles found in the nucleus

Neutrons Electrically neutral subatomic particles found in the nucleus

Nucleus Small central core of the atom; contains the protons and neutrons

Electrons Negatively charged subatomic particles found in the space around the nucleus

TABLE 3.1	Summary of Properties of Electrons, Protons, and Neutrons				
	RELATIVE CHARGE	**MASS (G)**	**MASS (AMU)***	**APPROXIMATE RELATIVE MASS (AMU)***	**LOCATION**
Electron	−1	0.000911×10^{-24}	0.00055	~0	Outside the nucleus
Proton	+1	1.672623×10^{-24}	1.00727	~1	Nucleus
Neutron	0	1.674929×10^{-24}	1.00867	~1	Nucleus

*amu is the abbreviation for *atomic mass* unit (page 48) 1 amu = 1.6605×10^{-24} g

Natural Radioactivity

Henri Becquerel (1852–1908) discovered natural radioactivity in natural uranium and radium ores in 1896 (Section 13.1). In 1898, Marie Sklodowska Curie (1867–1934)—a student of Becquerel—and her husband Pierre discovered two radioactive elements, *radium* and *polonium*. In 1899, Marie Curie suggested that atoms of radioactive substances disintegrate when they emit these unusual rays. She named this phenomenon **radioactivity**. A given radioactive element gives off exactly the same type of radioactive particles or rays regardless of whether it is found in its pure state or combined with other elements. About 25 elements exist only in radioactive forms. Marie Curie's suggestion that atoms disintegrate contradicted Dalton's idea that atoms are indivisible.

Ernest Rutherford (1871–1937) began studying the radiation emitted from radioactive elements soon after experiments by the Curies and others had shown that three types of radiation may be spontaneously emitted by radioactive elements (Section 13.1). These are referred to as **alpha (α) particles**, **beta (β) particles**, and **gamma (γ) rays**. They behave differently when they pass between electrically charged plates, as shown in Figure 3.1. Alpha and beta rays are deflected, while gamma rays pass straight through undeflected. This implies that alpha and beta rays are electrically charged particles, because particles with a charge would be attracted or repelled by the charged plates. Even though an alpha particle has an electrical charge (+2) twice as large as that of a beta particle (−1), alpha particles are deflected less; hence, alpha particles must be heavier than beta particles. Careful studies by Rutherford showed that alpha particles are helium atoms that have lost two

Radioactivity Spontaneous decomposition of unstable atomic nuclei; produces alpha, beta, and gamma radiation

Alpha particles Positively charged particles identical to helium-4 nuclei; emitted by certain radioactive isotopes

Beta particles Electrons ejected at high speeds from certain radioactive isotopes

Gamma rays High-energy electromagnetic radiation emitted from radioactive isotopes

FIGURE 3.1 Separation of alpha and beta particles and gamma rays by an electric field. Alpha particles are deflected toward the negative plate; beta particles are attracted toward the positive plate; and gamma rays are not deflected. Additional studies showed that alpha particles are high-energy helium nuclei, beta particles are high-energy electrons, and gamma rays are high-energy electromagnetic radiation.

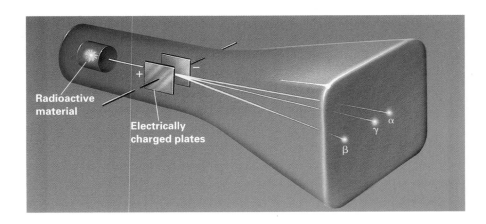

electrons and, thus have a +2 charge (He^{2+}). Beta particles are negatively charged particles identical to electrons. Gamma rays have no detectable charge or mass—they behave like rays of light.

The Nucleus of the Atom

Rutherford's experiments with alpha particles led him to consider using them in experiments on the structure of the atom. In 1909, he suggested to two of his co-workers that they bombard a piece of gold foil with alpha particles. Hans Geiger (1882–1945), a German physicist, and Ernest Marsden (1889–1970), an undergraduate student, set up the apparatus diagrammed in Figure 3.2 and observed what happened when alpha particles hit the thin gold foil. Most passed straight through, but Geiger and Marsden were amazed to find that a few alpha particles were deflected through large angles, and some came almost straight back. Rutherford later described this unexpected result by saying, "It was about as incredible as if you had fired a 15-inch [artillery] shell at a piece of paper and it came back and hit you."

What allowed most of the alpha particles to pass through the gold foil in a rather straight path? According to Rutherford's interpretation, the atom is mostly empty space and therefore offers little resistance to the alpha particles (Figure 3.3).

What caused a few alpha particles to be deflected? According to Rutherford's model of the atom, all of the positive charge and most of the mass of the atom must be concentrated in a very small volume at the center of the atom. He named this part of the atom, which contained most of the mass of the atom and all of the positive charge, the nucleus. When an alpha particle passes near the nucleus, the positive charge of the nucleus repels the positive charge of the alpha particle; the path of the smaller alpha particle is consequently deflected. The closer an alpha particle comes to a target nucleus, the more it is deflected. Those alpha particles that meet a nucleus head-on bounce back toward the source as a result of the strong positive–positive repulsion, since the alpha particles do not have enough energy to penetrate the nucleus.

Rutherford's calculations, based on the observed deflections, indicated that the nucleus is a very small part of an atom. *The diameter of an atom is about 100,000 times greater than the diameter of its nucleus.*

Truly, Rutherford's model of the atom was one of the most dramatic interpretations of experimental evidence to come out of this period of significant discoveries.

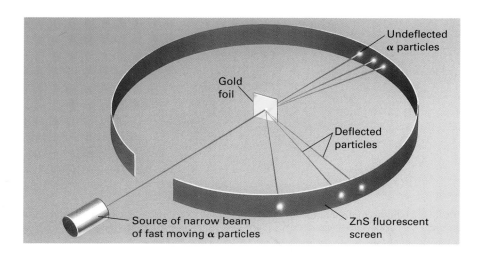

FIGURE 3.2   Rutherford gold foil experiment. A beam of positively charged alpha particles was directed at a very thin piece of gold foil. A luminescent screen was used to detect particles passing through or deflected by the foil. Most particles passed straight through. Some were deflected slightly, and a few were deflected back toward the source. (Rutherford actually used a movable luminescent screen instead of the circular screen shown.)

Undeflected α particles

Gold foil

Deflected particles

Source of narrow beam of fast moving α particles

ZnS fluorescent screen

THE PERSONAL SIDE

Ernest Rutherford (1871–1937)

ERNEST RUTHERFORD (RIGHT) WITH A COLLEAGUE IN THE CAVENDISH LABORATORY AT CAMBRIDGE UNIVERSITY.

C. E. Wynn-Williams. Courtesy AIP/Emilio Segre Visual Archives

Lord Ernest Rutherford was born in New Zealand in 1871, but went to Cambridge University in England to pursue his Ph.D. in physics in 1895. His original interest was in a phenomenon that we now call radio waves, and he apparently hoped to make his fortune in the field, largely so he could marry his fiancée back in New Zealand. However, his professor at Cambridge, J. J. Thomson, convinced him to work on the newly discovered phenomenon of radioactivity. Rutherford discovered alpha and beta radiation while at Cambridge. In 1899, he moved to McGill University in Canada where he did further experiments to prove that alpha radiation is actually composed of helium nuclei and that beta radiation consists of electrons. For this work he received the Nobel Prize in chemistry in 1908.

In 1903, Rutherford and his young wife visited Pierre and Marie Curie in Paris on the very day that Madame Curie received her doctorate in physics. That evening, during a party in the garden of the Curies' home, Pierre Curie brought out a tube coated with a phosphor and containing a large quantity of radioactive radium in solution.

The phosphor glowed brilliantly from the radiation given off by the radium. Rutherford later said that the light was so bright that he could clearly see that Pierre Curie's hands were "in a very inflamed and painful state due to exposure to radium rays."

In 1907, Rutherford moved from Canada to Manchester University in England, and there he performed the experiments that gave us the modern view of the atom. In 1919, he moved back to Cambridge and assumed the position formerly held by J. J. Thomson. Not only was Rutherford responsible for very important work in physics and chemistry, but he also guided the work of no fewer than ten future recipients of the Nobel Prize.

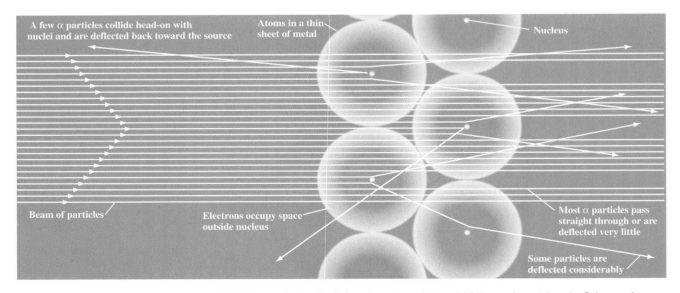

FIGURE 3.3 Rutherford's interpretation of the gold foil experiment done by Geiger and Marsden. Each circle represents an atom and the dots represent their nuclei. The gold foil was about 1000 atoms thick.

CONCEPT CHECK **3A**

1. The law of conservation of matter states that matter is neither lost nor _____ in a(an) _____ reaction.
2. A 64 g sample of pure SO_2 was found to be composed of 32 g of sulfur combined with 32 g of oxygen. What is the percentage of sulfur in SO_2?
3. Hydrogen peroxide, H_2O_2, is always 94.12% O. What is the percentage of H in hydrogen peroxide?
4. According to Dalton's atomic theory, a compound has a definite percentage by mass of each element because (**a**) all atoms of a given element weigh _the same_, and (**b**) all molecules of a given compound contain a definite number and kind of _atoms_.
5. If the law of definite proportion is true, will the percent mass of silver, Ag, in silver sulfide be the same for all lumps or pieces of Ag_2S? Explain your answer.
6. Natural gas is essentially methane (CH_4). Will methane produced from Texas gas fields have the same composition as methane produced from gas fields in China? Explain your answer.
7. What is the charge on the proton? (**a**) −1, (**b**) +1, (**c**) 3.
8. A beta particle is a high-energy (**a**) neutron, (**b**) electron, (**c**) proton, (**d**) helium nucleus.
9. Ernest Rutherford proposed the modern nuclear model of the atom. (**a**) True, (**b**) False.
10. Most of the mass of an atom is concentrated in its (**a**) nucleus, (**b**) electrons, (**c**) protons.
11. The mass of the proton is _____ times the mass of the electron.

(3.3) MODERN VIEW OF THE ATOM

Early experiments on the structure of the atom clearly showed that the three primary constituents of atoms are electrons, protons, and neutrons. The nucleus or core of the atom is made up of protons with a positive electrical charge and neutrons with no charge. The electrons, with a negative electrical charge, are found in the space around the nucleus (Figure 3.4). For an atom, which always has no net electrical charge, *the number of negatively charged electrons around the nucleus equals the number of positively charged protons in the nucleus.*

Atoms are extremely small, far too small to be seen with even the most powerful optical microscopes. The diameters of most atoms range from 1×10^{-8} cm to 5×10^{-8} cm. For example, the diameter of a gold atom is 3×10^{-8} cm. To visualize how small this is, consider that it would take approximately 517 million gold atoms to run the length (15.5 cm) of a dollar bill. If this weren't hard enough to imagine, remember that Rutherford's experiments provided evidence that the diameter of the nucleus is 100,000 times smaller than the diameter of the atom. For example, if an atom were scaled upward in size so that the nucleus were the size of a small marble, the atom would be the size of the Houston Astrodome, and most of the space in between would be empty. Because the nucleus carries most of the mass of the atom (the neutrons and protons) in such a small volume, a small matchbox full of atomic nuclei would weigh more than 2.5 billion tons. The interior of a collapsed star is made up of nuclear material estimated to be nearly this dense.

Scientists have been able to obtain computer-enhanced images of the outer surface of atoms (Figure 3.5) using the scanning tunneling microscope (STM) and the atomic force microscope (AFM).

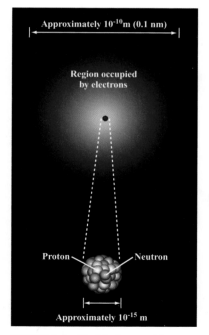

FIGURE 3.4   Model of atom. All atoms consist of one or more protons (positively charged) and usually at least as many neutrons (no charge) packed into an extremely small nucleus. Electrons (negatively charged) are arranged in a cloud around the nucleus.

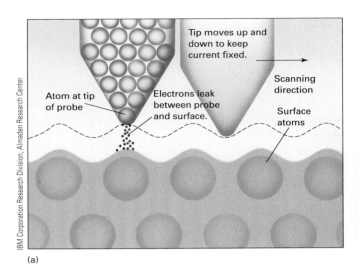

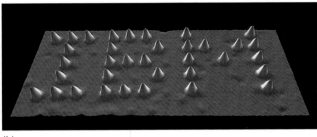

FIGURE 3.5 **Scanning tunneling microscope (STM).** (a) When an electric current passes through a tungsten needle with a narrow tip (atom's width) into the atoms on the surface of the sample being examined, the electron flow between the tip and the surface changes in relation to the electron clouds around the atoms. By adjusting the position of the needle to maintain a constant current, the positions of the atoms are measured. (b) IBM spelled out in xenon (Xe) atoms. A few years after invention of the scanning tunneling microscope, scientists discovered that not only could they take pictures of atoms on a surface, but they could also push them around with the tip of the microscope. They created this picture to demonstrate what they could do. The "IBM" is 600 billionths of an inch (16.8 nm) wide.

Atomic Number

The **atomic number** of an element indicates the number of protons in the nucleus of the atom. All atoms of the same element have the same number of protons in the nucleus. In the periodic table (inside this book's front cover) the atomic number for each element is given above the element's symbol. There is a different atomic number for each element, beginning with the atomic number 1 for hydrogen. Phosphorus, for example, has an atomic number of 15; thus, the nucleus of a phosphorus atom contains 15 protons.

It is the number of protons, and nothing more, that determines the identity of an atom. In a neutral atom the number of protons is equal to the number of electrons; therefore, the atomic number also gives the number of electrons, all of which exist outside the nucleus in an atom of an element.

Mass Number

The **mass number** of a particular atom is the total number of neutrons and protons present in the nucleus of an atom. Because the atomic number gives the number of protons in the nucleus, the difference between the mass number and the atomic number equals the number of neutrons in the nucleus.

A notation frequently used for showing the mass number and atomic number of an atom places subscripts and superscripts to the left of the symbol.

$$\text{Mass number} \longrightarrow {}^{19}_{9}\text{F} \longleftarrow \text{Symbol of element}$$
$$\text{Atomic number} \longrightarrow$$

Atomic number The number of protons in the nuclei of the atoms of an element

Mass number Number of neutrons plus number of protons in the nucleus of an atom

The subscript giving the atomic number is optional because the element symbol tells you what the atomic number must be. For example, the fluorine atom can have the notation ${}^{19}_{9}\text{F}$ or simply ${}^{19}\text{F}$. For an atom of fluorine, ${}^{19}_{9}\text{F}$, the number of protons is 9, the number of electrons is also 9, and the number of neutrons is $19 - 9 = 10$.

EXAMPLE   3.1   Atomic Composition

How many protons, neutrons, and electrons are in an atom of gold (Au) with a mass number of 197?

SOLUTION

The atomic number of an element gives the number of protons and electrons. If the element is known, its atomic number can be obtained from the periodic table (or the alphabetical list inside the front cover of this book). The atomic number of gold is 79. Gold has 79 protons and 79 electrons. The number of neutrons is obtained by subtracting the atomic number from the mass number:

$$197 - 79 = 118 \text{ neutrons}$$

TRY IT 3.1

How many protons, neutrons, and electrons are there in a neutral $^{59}_{28}\text{Ni}$ atom?

Isotopes

When a natural sample of almost any element is analyzed, the element is found to be composed of atoms with different mass numbers. Atoms of the same element having different mass numbers are called **isotopes** of that element.

The element neon is a good example to consider. A natural sample of neon gas is found to be a mixture of three isotopes of neon:

$$^{20}_{10}\text{Ne} \qquad ^{21}_{10}\text{Ne} \qquad ^{22}_{10}\text{Ne}$$

The fundamental difference between isotopes is the different number of neutrons per atom. All atoms of neon have 10 electrons and 10 protons. About 90.92% of the atoms have 10 neutrons, 0.26% have 11, and 8.82% have 12. Because they have different numbers of neutrons, it follows that they must have different masses. Note that all the isotopes have the same atomic number because they are all neon. An analogy may be helpful here. Consider three brand new, identical automobiles that have had different numbers of 50-lb. weights hidden in their trunks. The automobiles are identical in every way except for their different masses.

Over 100 elements are known, and more than 1000 isotopes have been identified, many of them produced artificially (see Section 13.5). Some elements have many isotopes; tin, for example, has 10 natural isotopes. Hydrogen has three isotopes, which are the only known isotopes generally referred to by different names: $^{1}_{1}\text{H}$ is commonly called hydrogen or protium, $^{2}_{1}\text{H}$ is called deuterium, and $^{3}_{1}\text{H}$ is called tritium. Tritium is radioactive.

Relative abundances of neon isotopes:

$^{20}_{10}\text{Ne}$	90.92%
$^{21}_{10}\text{Ne}$	0.26%
$^{22}_{10}\text{Ne}$	8.82%

To represent isotopes with words instead of symbols, the mass number is added to the name, for example, neon-20, neon-21, and neon-22.

EXAMPLE   3.2   Isotopes

Carbon has seven known isotopes. Three of these have six, seven, and eight neutrons, respectively. Write the complete chemical notation for these three isotopes, giving mass number, atomic number, and symbol.

SOLUTION

The atomic number of carbon is 6. The mass number for the three isotopes is 6 plus the number of neutrons, which equal 12, 13, and 14, respectively.

$$^{12}_{6}\text{C} \qquad ^{13}_{6}\text{C} \qquad ^{14}_{6}\text{C}$$

Isotopes Atoms of the same element having different mass numbers

Charles D. Winters

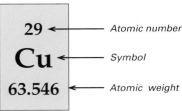

29 ←———— *Atomic number*

Cu ←———— *Symbol*

63.546 ←———— *Atomic weight*

The periodic table entry for copper. (The complete periodic table and its significance are discussed in Section 3.6.)

Chemists usually use the term *atomic weight* of an element instead of *atomic mass* when they are referring to a naturally occurring sample of an element. Although the quantity is more properly called a mass than a weight, the term "atomic weight" is so commonly used that it has become accepted. Atomic weights are established by exacting experiments. The same values are used by all scientists because the relative abundances of the elements are essentially the same everywhere on our planet.

Atomic mass unit (amu) The unit for relative atomic masses of the elements; 1 amu = 1/12 the mass of the carbon-12 isotope. 1 amu = 1.6605×10^{-24} g

Atomic weight The number that represents the average atomic mass of the element's isotopes weighted by percentage abundance

TRY IT 3.2

Silver has two isotopes, one with 60 neutrons and the other with 62 neutrons. Give the complete chemical notation for these isotopes.

Atomic Mass and Atomic Weight

Although Dalton knew nothing about subatomic particles, he proposed that atoms of different elements have different masses. Eventually, it was found that an oxygen atom is about 16 times heavier than a hydrogen atom. This fact, however, does not tell us the mass of either atom. These are relative masses in the same way that a grapefruit may weigh twice as much as an orange. This information gives neither the mass of the grapefruit nor that of the orange. Nevertheless, if a specific number is assigned as the mass of any particular atom, this fixes the numbers assigned to the masses of all other atoms. The present atomic mass scale, adopted by scientists worldwide in 1961, is based on assigning the mass of a particular isotope of the carbon atom, the carbon-12 isotope, as exactly 12 **atomic mass units (amu)**.

The atomic masses are given in the periodic table (inside front cover) below the symbol for the element, as illustrated in the margin for copper. These values are average masses, which take into account the relative abundances of the different isotopes as found in nature. This average is often referred to as the **atomic weight** of the element. For example, boron has two naturally occurring isotopes, $^{10}_5$B and $^{11}_5$B, with natural percent abundances of 19.91% and 80.09%. The masses in amu are 10.0129 and 11.0093, respectively. The atomic weight listed in the periodic table is the average mass of a natural sample of atoms, expressed in atomic mass units. The average mass of 10.81 for B is closer to the mass of $^{11}_5$B because of its higher abundance.

> **CONCEPT CHECK**   **3B**

1. If an atom has an atomic number of 10, then it has _____ protons and _____ electrons. If its mass number is 21, then it has _____ neutrons.
2. In the symbol $^{80}_{35}$Br, the number 35 is the _____, and the number 80 is the _____.
3. Isotopes of an element are atoms that have nuclei with the same number of _____ but different numbers of _____.
4. The negatively charged particles in an atom are _____; the positively charged particles are _____; the neutral particles are _____.
5. In a neutral atom there are equal numbers of _____ and _____.
6. The number of protons per atom is called the _____ number of the element.
7. An atom of arsenic, $^{75}_{33}$As, has _____ electrons, _____ protons, and _____ neutrons.
8. The diameter of the nucleus is _____ smaller than the diameter of the atom.
9. Draw the nuclei of $^1_1$H, $^2_1$H, and $^3_1$H, representing protons and neutrons with circles.
10. The nuclei of all helium atoms contain exactly 2 protons. (**a**) True, (**b**) False.
11. All the isotopes of an element have the same relative abundance. (**a**) True, (**b**) False.

3.4 WHERE ARE THE ELECTRONS IN ATOMS?

Experiments on the interactions of light with matter provide important information about the energy and location of electrons in atoms.

Continuous and Line Spectra

We are familiar with the spectrum of colors that makes up visible light. The spectrum of white light is a display of separated colors. This type of spectrum is referred to as a **continuous spectrum** and is obtained by passing sunlight or light from an incandescent lightbulb through a glass prism. When we see a rainbow, we are looking at the continuous visible spectrum that forms when raindrops act as prisms and disperse the sunlight. The different colors of light have different wavelengths. Red light has longer wavelengths than does blue light, but in the continuous spectrum the colors merge from one to another with no break in the spectrum. White light is a combination of all the colors of different wavelengths.

If a high voltage is applied to an element in the gas phase in a partially evacuated tube, the atoms absorb energy and are said to be "excited." The excited atoms emit light. An example of this is a neon advertising sign in which excited neon atoms emit orange-red light (Figure 3.6a). When light from such a source passes through a prism, a different type of spectrum is obtained, one that is not continuous but has characteristic lines at specific wavelengths (Figure 3.6b). This type of spectrum is called a *line emission spectrum*.

Visible light is only a small portion of the electromagnetic spectrum (Figure 3.7). Ultraviolet radiation, the type that leads to sunburn and some forms of skin cancer, has wavelengths shorter than those of visible light; X rays and gamma rays (the latter emitted from radioactive atoms) have even shorter wavelengths. Infrared radiation, the type that is sensed as heat from a fire, has longer wavelengths than visible light. Longer still are the wavelengths of the types of radiation in a microwave oven and in television and radio transmissions. Although the example of a line spectrum shown in Figure 3.6b is only for the visible region, excited atoms of elements also emit characteristic wavelengths in other regions of the electromagnetic spectrum, as demonstrated by the experiments described in the next section.

Bohr Model of the Atom

In 1913, Niels Bohr introduced his model of the hydrogen atom. He proposed that the single electron of the hydrogen atom could occupy only certain energy levels. He referred to these energy levels as orbits and represented the energy difference between any two adjacent orbits as a single

See an animation of **Atomic Line Spectra** at http://brookscole.com/chemistry/joesten4

Bohr assumed that atoms can exist only in certain energy states.

In 1900 Max Planck proposed that energy is not continuous, but comes in discrete "bundles" or "packets" called *photons* or *quanta*. Something that can have only certain values, with none in between, is referred to as *quantized*.

Continuous spectrum A spectrum that contains radiation distributed over all wavelengths

(a)   (b)

FIGURE 3.6   Neon. (a) A partially evacuated tube that contains neon gas gives a reddish–orange glow when high voltage is applied. (b) The line emission spectrum of neon is obtained when light from a neon source passes through a prism.

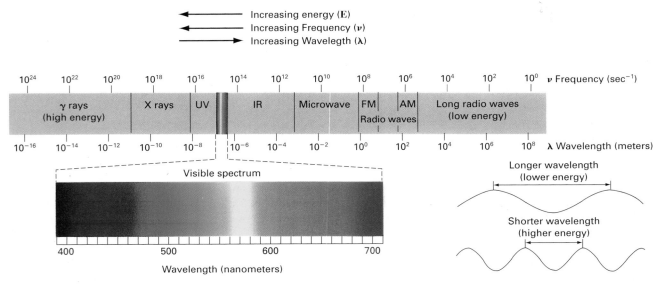

FIGURE 3.7 **The electromagnetic spectrum.** Visible light (enlarged section) is but a small part of the entire spectrum. The energy of electromagnetic radiation increases from the radio wave end to the gamma ray end. The frequency of electromagnetic radiation is related to the wavelength by $\nu\lambda = c$ where ν = frequency; λ = wavelength; and c = speed of light, 3.00×10^8 meters (m)/second (s). The higher the frequency, the lower the wavelength and the larger the energy. The energy (E) of a *photon* (or *quantum*) of light is given by the expression $E_{photon} = h\nu$ where h is Planck's constant (6.6262×10^{-34} J · s).

It is important to understand the relationships among λ, ν, and E

As $\lambda\downarrow$, $\nu\uparrow$ E$\uparrow$

As $\lambda\uparrow$, $\nu\downarrow$ E$\downarrow$

Wavelength and frequency (or energy) are inversely related.

Bohr used the term *orbits*, but the modern equivalents of his orbits are called *shells* or *energy levels*.

Quantum The smallest increment of energy, for example, in an atom emitting or absorbing radiation

Ground state The condition of an atom in which all electrons are in their normal, lowest energy levels

Excited state An unstable, higher energy state of an atom

quantum of energy. When the hydrogen electron absorbs a quantum of energy, it moves to a higher energy level. When this electron returns to the lower, more stable energy level, the quantum of energy is emitted as a specific wavelength of light.

In Bohr's model, each allowed orbit is assigned an integer, n, known as the principal quantum number. The values of n for the orbits range from 1 to infinity. Any atom with its electrons in their normal, lowest energy levels is said to be in the **ground state**. Energy must be supplied to move the electron farther away from the nucleus because the positive nucleus and the negative electron attract each other. When the electron of a hydrogen atom occupies a higher energy level with n greater than 1, the atom has more energy than in its ground state and is said to be in an **excited state**. The excited state is an unstable state, and the extra energy is emitted when the electron returns to the ground state. According to Bohr, the light forming the lines in the bright-line emission spectrum of hydrogen comes from electrons moving toward the nucleus after having first been excited to energy levels farther from the nucleus. Because the energy levels have only certain energies, the emitted light has only certain wavelengths.

With brilliant imagination, Bohr applied a little algebra and some classic mathematical equations of physics to his tiny solar system model of the hydrogen atom and was able to calculate the wavelengths of the lines in the hydrogen spectrum. By 1900, scientists had measured the wavelengths of lines for hydrogen in the ultraviolet, visible, and infrared regions, and Bohr's calculated values agreed with the measured values. Niels Bohr had tied the unseen (the interior of the atom) with the seen (the observable lines in the hydrogen spectrum)—a fantastic achievement. The concepts of quantum number and energy level are valid for all atoms and molecules.

THE PERSONAL SIDE

Niels Bohr (1885–1962)

Niels Bohr was born in Copenhagen, Denmark. He earned a Ph.D. in physics in Copenhagen in 1911. He worked first with J. J. Thomson in Cambridge, England, and later with Ernest Rutherford in Manchester, England. It was there that he began to develop the ideas that a few years later led to the publication of his theory of atomic structure and his explanation of atomic spectra. He received the Nobel Prize in 1922 for this work. After working with Rutherford for a very short time, Bohr returned to Copenhagen, where he eventually became the director of the Institute of Theoretical Physics.

American Institute of Physics

NIELS BOHR

Bohr was still in Denmark when Hitler's army suddenly invaded the country in 1940. In 1943, to avoid imprisonment, he escaped to Sweden. There he helped to arrange the escape of nearly every Danish Jew from Hitler's gas chambers. He was later flown to England in a tiny plane, in which he passed into a coma and nearly died from lack of oxygen.

He went on to the United States, where until 1945 he worked with other physicists on the atomic bomb development at Los Alamos, New Mexico. His insistence on sharing the secret of the atomic bomb with other allies, to permit international control over nuclear energy, so angered Winston Churchill that he had to be restrained from ordering Bohr's arrest. Bohr worked hard and long on behalf of the development and use of atomic energy for peaceful purposes. For his efforts, he was awarded the first Atoms for Peace Prize in 1957. He died in Copenhagen on November 18, 1962.

The Bohr model was accepted almost immediately after its presentation, and Bohr was awarded the Nobel Prize in physics in 1922 for his contribution to the understanding of the hydrogen atom. However, his model gave only approximate agreement with line spectra of atoms having more than one electron. Later models of the atom have been more successful by considering electrons as having both particle and wave characteristics. This led to mathematical treatment of the locations of electrons as probabilities instead of as the precise locations envisioned by Bohr. Thus, in the modern view of the atom, we can picture a space around the nucleus occupied by each electron, somewhat like a cloud of electrical charge of a particular energy. We just don't know where within the cloud a particular electron is at any given instant. Bohr's concept of the main energy levels represented by the quantum number n remains valid, however; and for our purposes, this concept is all that we need to discuss the location of electrons in atoms.

Atom Building Using the Bohr Model

Recall that the atomic number is the number of electrons (or protons) per atom of an element. Imagine building atoms by adding one electron to the appropriate energy level as another proton is added to the nucleus. As part of his theory, Bohr proposed that only a fixed number of electrons could be accommodated in any one level; and he calculated that this number was given by the formula $2n^2$, where n equals the number of the energy level. For the lowest energy level (first level), n equals 1, and the maximum number of electrons allowed is $2(1)^2$, or 2. For the second energy level, the maximum number of electrons is $2(2)^2$, or 8. By using $2n^2$, the maximum number of electrons allowed

ENERGY LEVEL (n)	$2n^2$	MAXIMUM NUMBER OF ELECTRONS
1	$2(1)^2$	2
2	$2(2)^2$	8
3	$2(3)^2$	18
4	$2(4)^2$	32
5	$2(5)^2$	50
6	$2(6)^2$	72

TABLE 3.2	Electron Arrangements of the First 20 Elements*			
ELEMENT	**ATOMIC NUMBER**	**NUMBER OF ELECTRONS IN EACH ENERGY LEVEL**		
		1ST	**2ND**	**3RD**	**4TH**
Hydrogen (H)	1	1 e			
Helium (He)	2	2 e			
Lithium (Li)	3	2 e	1 e		
Beryllium (Be)	4	2 e	2 e		
Boron (B)	5	2 e	3 e		
Carbon (C)	6	2 e	4 e		
Nitrogen (N)	7	2 e	5 e		
Oxygen (O)	8	2 e	6 e		
Fluorine (F)	9	2 e	7 e		
Neon (Ne)	10	2 e	8 e		
Sodium (Na)	11	2 e	8 e	1 e	
Magnesium (Mg)	12	2 e	8 e	2 e	
Aluminum (Al)	13	2 e	8 e	3 e	
Silicon (Si)	14	2 e	8 e	4 e	
Phosphorus (P)	15	2 e	8 e	5 e	
Sulfur (S)	16	2 e	8 e	6 e	
Chlorine (Cl)	17	2 e	8 e	7 e	
Argon (Ar)	18	2 e	8 e	8 e	
Potassium (K)	19	2 e	8 e	8 e	1 e†
Calcium (Ca)	20	2 e	8 e	8 e	2 e

*Valence electrons are shown in color.
†The discussion which follows, Beyond the Bohr Atom (p. 53), will more clearly explain why the nineteenth and twentieth electrons do not add to the third energy level as might have been predicted using the $2n^2$ guide.

for levels 3, 4, and 5 are 18, 32, and 50, respectively. A general overriding rule to the preceding numbers is that the highest energy level can have no more than eight electrons, except for transition elements, for a stable atom. Table 3.2 lists the Bohr electron arrangements for the first 20 elements.

Electrons in the highest occupied energy level listed for the elements in Table 3.2 are at the greatest stable distance from the nucleus. These are the most important electrons in the study of chemistry because they are the ones that interact when atoms react with each other. G. N. Lewis first proposed that each energy level could hold only a characteristic number of electrons and that only those electrons in the highest occupied level were involved when one atom combined with another. These outermost electrons came to be known as **valence electrons**. The number of valence electrons an atom has is important in determining how the atom combines with other atoms.

For example, look at phosphorus (P). The Bohr arrangement of electrons is 2–8–5. This means that the stable state of the phosphorus atom has electrons in three energy levels. The one closest to the nucleus has two electrons; the second energy level has eight electrons; and the highest energy level has five electrons. The energy level with five electrons is farthest from the nucleus; thus, P has *five valence electrons,* and these are the electrons that are most available for interactions with valence electrons of other atoms in chemical reactions. This interaction results in the formation of bonds, which will be discussed in Chapter 5.

Valence electrons Outermost electrons in an atom

1. Look at Figure 3.7 to predict which of the following has the highest energy? (**a**) blue light, (**b**) red light, (**c**) ultraviolet light, (**d**) infrared light.
2. Which color of light has the shortest wavelength? (**a**) blue, (**b**) green, (**c**) orange, (**d**) red.
3. According to Bohr's theory, light of characteristic wavelength is emitted as an electron drops from an energy level (closer to/farther from) the nucleus to an energy level (closer to/farther from) the nucleus.
4. The maximum number of the electrons in $n = 3$ energy level is _____.
5. The ground-state Bohr representation for electrons in an atom of K is _____. K has _____ valence electron(s).
6. The ground-state Bohr representation for electrons in an atom of Cl is _____. Cl has _____ valence electron(s).

Beyond the Bohr Atom

The simple Bohr model of the atom, in which each **shell** is capable of holding $2n^2$ electrons, can be used to explain many properties of atoms and their associated electrons. However, more sophisticated treatments of atoms depend on the fact that electrons moving at very high speeds, up to 2 million meters per second, exhibit properties of waves. Although it is appropriate at some level of understanding to envision electrons as particles moving around the nucleus like planets moving around the sun, electron locations around the nucleus can be more accurately described by using wave theory. This theory defines electron locations in terms of regions of probability known as **orbitals**. A full treatment of highly mathematical wave theory, developed by an Austrian scientist named Erwin Schrodinger, is well beyond the scope of this text, but his results are nonetheless useful.

In essence, Schrodinger concluded that the volume of space in which each pair of electrons could most likely be found defines an orbital and that each orbital can accommodate a maximum of two electrons. Furthermore, he concluded that Bohr's planetary shells can contain more than one such orbital and the orbitals within a given shell may be of different types. Quantum theory shows that the two electrons residing in the first Bohr shell, occupy a single, spherical orbital called a *1s orbital* (Figure 3.8).

The second Bohr shell can accommodate a total of eight electrons (2×2^2) but quantum theory shows that these eight electrons occupy two different types of orbitals. This second energy shell, thus, contains two **subshells** comprised of a single spherical *2s orbital* and three separate, dumbbell-shaped orbitals oriented at right angles with respect to each other and known as *2p orbitals*. All of these orbitals, like all atomic orbitals regardless of type, are centered about the nucleus of the atom. The 2s orbital is higher in energy than the 1s orbital but it is slightly lower in energy than the three 2p orbitals. The 2s orbital holds two electrons and the three 2p orbitals each hold two electrons for a total of eight electrons in the second energy shell.

A continuation of this treatment showed the third energy level, which can accommodate a maximum of 2×3^2 or 18 electrons, is actually composed of three subshells. The lowest energy orbital in this shell is the spherical *3s orbital*, capable of holding two electrons. The next-lowest energy orbitals in this shell are three dumbbell-shaped *3p orbitals* (Figure 3.8) capable of holding a total of six electrons. Lastly, the third energy shell contains five orbitals known as *3d orbitals*. These five 3d orbitals can

Shell A principal energy level defined by a given value of *n*

Orbital A region of three-dimensional space around an atom within which there is a significant probability (usually shown as 90%) that a given electron will be found

Subshells Different energy levels (orbitals) within a given shell

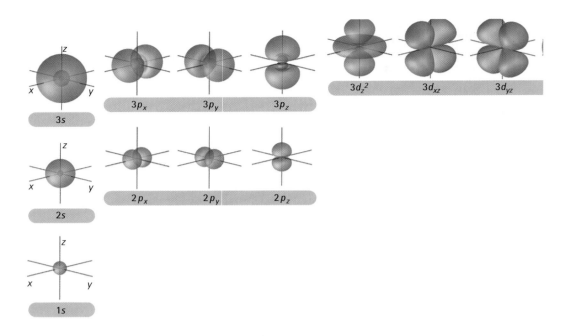

FIGURE 3.8 Atomic orbitals. Boundary surface diagrams for electron densities of 1s, 2s, 2p, 3s, 3p, and 3d orbitals. For the p orbitals, the subscript letter on the orbital notation (x, y, z) indicates the cartesian axis along which the orbital lies.

accommodate a total of 10 electrons. Summarizing, we see that the third shell contains a single 3s orbital (2-electron capacity), three 3p orbitals (6-electron capacity), and five 3d orbitals (10-electron capacity) for a maximum of 18 electrons in the third shell.

The fourth energy shell contains, after the 4s orbital, three 4p orbitals, five 4d orbitals, and seven 4f orbitals (not pictured) that can accommodate a maximum of 14 electrons. Thus, the fourth shell or energy level can accommodate a total of 2×4^2 (or 32) electrons. The trend continues in this fashion until no additional orbitals are required to accommodate the electrons.

Although the Bohr model correctly predicts the number of electrons in each energy shell according to the $2n^2$ calculation with n = the number of the energy shell, the quantum wave model more accurately describes the location probability of the electrons in three-dimensional space.

As one progresses from one atom to the atom of next higher atomic number in the periodic table, electrons are added singly to the lowest energy orbital in a given subshell that does not already have an electron. Once each orbital in a subshell has a single electron, additional electrons are then added to the singly-occupied orbitals in that subshell until they are each doubly occupied. At that point, additional electrons are then added to the orbitals with the next highest energy until one exhausts the number of electrons associated with the atom.

Figure 3.9 provides an order-of-filling chart that allows the exact specification of orbitals occupied by electrons in a given atom. Simply follow the arrowed pathway shown in Figure 3.9 and place the appropriate number of electrons in each orbital or set of orbitals in the order specified. Remember that there are 2 electrons in the 2s orbitals, 6 electrons in the 2p orbitals, 10 electrons in the 3d orbitals and 14 electrons in the 4f orbitals.

Using this method, we can write the electron configuration of each atom as shown in Table 3.3 where superscripts are used to indicate the number of electrons in the orbitals. All electrons that occupy a subshell with the same

 See an animation of **Atomic Subshell Energies** at http://brookscole.com/chemistry/joesten4

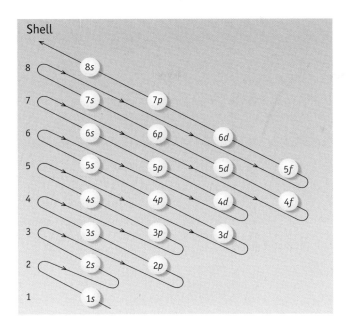

FIGURE 3.9 **Subshell filling order.** Subshells in atoms are filled in order of increasing energy, as this diagram shows. The order of filling is $1s \Rightarrow 2s \Rightarrow 2p \Rightarrow 3s \Rightarrow 3p \Rightarrow 4s \Rightarrow 3d$ and so on.

main shell number (for example, where the second shell is signified by using a 2 in front of s and p orbitals) are valence electrons.

This discussion helps explain why the last electron added when going from $_{18}$Ar to $_{19}$K goes in the fourth energy level (as shown in Table 3.2) and not in the third energy level, as one might have predicted using the simple Bohr model. This happens because, according to the filling model, electrons are added to the 4s orbital before being added to the 3d orbital. Following this model, the next electron added to arrive at $_{20}$Ca goes into the 4s orbital; the 3d orbitals are filled next, before the 4p orbitals. If one were to follow the filling method across the entire periodic table the transition metals (Section 3.6) would occupy the region of the periodic table where the d and f orbitals come into play. Figure 3.10 presents a graphic, "building up" version of the periodic table.

TABLE 3.3 Electron Configurations of the First 20 Elements*

ELEMENT	ATOMIC NUMBER		ELEMENT	ATOMIC NUMBER	
Hydrogen (H)	1	$1s^1$	Sodium (Na)	11	$1s^2 2s^2 2p^6\ 3s^1$
Helium (He)	2	$1s^2$	Magnesium (Mg)	12	$1s^2 2s^2 2p^6\ 3s^2$
Lithium (Li)	3	$1s^2\ 2s^1$	Aluminum (Al)	13	$1s^2 2s^2 2p^6\ 3s^2 3p^1$
Beryllium (Be)	4	$1s^2\ 2s^2$	Silicon (Si)	14	$1s^2 2s^2 2p^6\ 3s^2 3p^2$
Boron (B)	5	$1s^2\ 2s^2 2p^1$	Phosphorus (P)	15	$1s^2 2s^2 2p^6\ 3s^2 3p^3$
Carbon (C)	6	$1s^2\ 2s^2 2p^2$	Sulfur (S)	16	$1s^2 2s^2 2p^6\ 3s^2 3p^4$
Nitrogen (N)	7	$1s^2\ 2s^2 2p^3$	Chlorine (Cl)	17	$1s^2 2s^2 2p^6\ 3s^2 3p^5$
Oxygen (O)	8	$1s^2\ 2s^2 2p^4$	Argon (Ar)	18	$1s^2 2s^2 2p^6\ 3s^2 3p^6$
Fluorine (F)	9	$1s^2\ 2s^2 2p^5$	Potassium (K)	19	$1s^2 2s^2 2p^6 3s^2 3p^6\ 4s^1$
Neon (Ne)	10	$1s^2\ 2s^2 2p^6$	Calcium (Ca)	20	$1s^2 2s^2 2p^6 3s^2 3p^6\ 4s^2$

*Valence electrons are shown in color.

Number of Electrons

Subshells being filled

Subshell	1	2	3	4	5	6	7	8	9	10	11	12	13	14	
7p	113 Uut	114 Uuq	115 Uup	116 Uuh	117	(118)									End of period 7
6d	103 Lr	104 Rf	105 Db	106 Sg	107 Bh	108 Hs	109 Mt	110 Ds	111 Rg	112 Uub					
5f	89 Ac $6d^1 7s^2$	90 Th $6d^2 7s^2$	91 Pa $5f^2 6d^1 7s^2$	92 U $5f^3 6d^1 7s^2$	93 Np $5f^4 6d^1 7s^2$	94 Pu	95 Am	96 Cm $5f^7 6d^1 7s^2$	97 Bk	98 Cf	99 Es	100 Fm	101 Md	102 No	
7s	87 Fr	88 Ra													
6p	81 Tl	82 Pb	83 Bi	84 Po	85 At	(86 Rn)									End of period 6
5d	71 Lu	72 Hf	73 Ta	74 W	75 Re	76 Os	77 Ir	78 Pt $5d^9 6s^1$	79 Au $5d^{10} 6s^1$	80 Hg					
4f	57 La $5d^1 6s^2$	58 Ce $4f^1 5d^1 6s^2$	59 Pr	60 Nd	61 Pm	62 Sm	63 Eu	64 Gd $4f^7 5d^1 6s^2$	65 Tb	66 Dy	67 Ho	68 Er	69 Tm	70 Yb	
6s	55 Cs	56 Ba													
5p	49 In	50 Sn	51 Sb	52 Te	53 I	(54 Xe)									End of period 5
4d	39 Y	40 Zr	41 Nb $4d^4 5s^1$	42 Mo $4d^5 5s^1$	43 Tc	44 Ru $4d^7 5s^1$	45 Rh $4d^8 5s^1$	46 Pd $4d^{10}$	47 Ag $4d^{10} 5s^1$	48 Cd					
5s	37 Rb	38 Sr													
4p	31 Ga	32 Ge	33 As	34 Se	35 Br	(36 Kr)									End of period 4
3d	21 Sc	22 Ti	23 V	24 Cr $3d^5 4s^1$	25 Mn	26 Fe	27 Co	28 Ni	29 Cu $3d^{10} 4s^1$	30 Zn					
4s	19 K	20 Ca													
3p	13 Al	14 Si	15 P	16 S	17 Cl	(18 Ar)									End of period 3
3s	11 Na	12 Mg													
2p	5 B	6 C	7 N	8 O	9 F	(10 Ne)									End of period 2
2s	3 Li	4 Be													
1s	1 H	(2 He)													End of period 1

FIGURE 3.10 In this "building-up" version of the periodic table, the lightest elements are at the bottom. Electrons fill subshells from bottom to top in order of energy as the atomic number of the atom increases. The numbers across the top give the number of electrons in each subshell. The ground-state electron configurations of most elements are apparent from their positions in the table. Those that are known to differ from expectation are indicated explicitly.

In summary, quantum theory's more sophisticated treatment of electrons allows us to more accurately describe the positions of electrons around a given nucleus. This can help us understand chemical bonding in a more detailed way.

3.5 DEVELOPMENT OF THE PERIODIC TABLE

So far, in exploring the chemical view of matter, you have seen that everything is made of atoms. Atoms of different elements combined in specific ratios are present in chemical compounds. All matter consists of elements, compounds, or mixtures of elements or compounds. You have also been introduced to the subatomic composition of atoms—the electrons, protons, and neutrons. The next part of the story is to see how atomic structure and the properties of elements and compounds are related to each other. Fortunately, the chemistry of the elements can be organized and classified in a way that helps both chemists and nonchemists. The periodic table is the single most important classification system in chemistry because it summarizes, correlates, and predicts a wealth of chemical information. Chemists consult it every day during every possible kind of work. It can simply be a reminder of the symbols and names of the elements, of which elements have similar properties, and of where each element lies on the continuum of atomic numbers. It can also be an inspiration in the search for new compounds or mixtures that will fulfill a specific need. Memorizing the periodic table is no more necessary than memorizing the map of your home state. Nevertheless, in both cases, it's very helpful to have a general idea of the major features.

On the evening of February 17, 1869, at the University of St. Petersburg in Russia, a 35-year-old professor of general chemistry—Dmitri Ivanovich Mendeleev (1834–1907)—was writing a chapter of his soon-to-be-famous textbook on chemistry. He had the properties of each element written on cards, with a separate card for each element. While he was shuffling the cards trying to gather his thoughts before writing his manuscript, Mendeleev realized that if the elements were arranged in the order of their atomic weights, there was a trend in properties that repeated itself several times. He arranged the elements into groups that had similar properties and used the resulting periodic chart to predict the properties and places in the chart of as yet undiscovered elements.

Oesper Collection in the History of Chemistry, University of Cincinnati

Dmitri Mendeleev

Thus, the periodic law and the periodic table were born, although only 63 elements had been discovered by 1869 (for example, the noble gases were not discovered until after 1893), and the clarifying concept of the atomic number was not known until 1913. Mendeleev's idea and textbook achieved great success, and he rose to a position of prestige and fame while he continued to teach at the University of St. Petersburg.

Mendeleev aided the discovery of new elements by predicting their properties with remarkable accuracy, and he even suggested the geographical regions in which minerals containing the elements could be found. The properties of a missing element were predicted by consideration of the properties of its neighboring elements in the table. For example, for the element we now know as germanium, which falls below silicon in the modern periodic table (Figure 3.11), Mendeleev predicted a gray element of atomic weight 72 with a density of 5.5 g/cm^3. Germanium, once discovered, proved to be a gray element of atomic weight 72.59 with a density of 5.36 g/cm^3.

The empty spaces in the table and Mendeleev's predictions of the properties of missing elements stimulated a flurry of prospecting for elements in the 1870s and 1880s and eight more were discovered by 1886.

Mendeleev found that a few elements did not fit under other elements with similar chemical properties when arranged according to increasing atomic

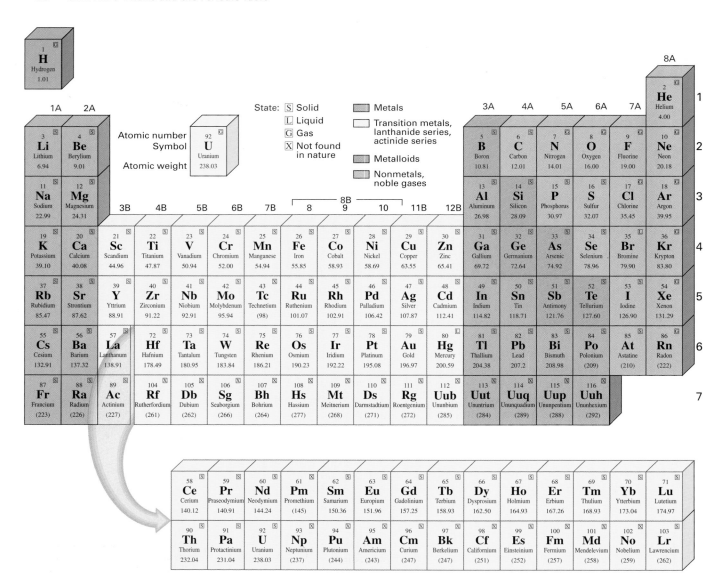

FIGURE 3.11 Modern periodic table of elements.

weight. Eventually, it was found that the atomic weight is *not* the property that governs the similarities and differences among the elements. This was discovered in 1913 by H. G. J. Moseley (1888–1915), a young scientist working with Ernest Rutherford. Moseley found that the wavelengths of X-rays emitted by a particular element are related in a precise way to the atomic number of that element. He quickly realized that other atomic properties may be similarly related to atomic number and not, as Mendeleev had believed, to atomic weight.

By building on the work of Mendeleev and others, and by using the concept of the atomic number, we are now able to state the modern **periodic law**: *When elements are arranged in the order of their atomic numbers, their chemical and physical properties show repeatable, or periodic, trends.* Other familiar periodic phenomena include the changing of the seasons and the orbits of the planets, which are periodic with time. To illustrate this idea on a simple level, a shingle roof, which has the same pattern over and over, is periodic.

Thus, to build up a periodic table according to the periodic law, the elements are lined up in a horizontal row in the order of their atomic numbers, as in Figure 3.11. At an element with similar properties to one already previously placed in the row, a new row is started. *Each column then contains elements*

TABLE 3.4   Some Properties of the First 20 Elements

| ELEMENT | ATOMIC NUMBER | DESCRIPTION | COMPOUND FORMATION* | |
			WITH CL (OR NA)	WITH O (OR MG)
Hydrogen (H)	1	Colorless gas, reactive	HCl	H_2O
Helium (He)	2	Colorless gas, unreactive	None	None
Lithium (Li)	3	Soft metal, low density, very reactive	LiCl	Li_2O
Beryllium (Be)	4	Harder metal than Li, low density, less reactive than Li	$BeCl_2$	BeO
Boron (B)	5	Both metallic and nonmetallic, very hard, not very reactive	BCl_3	B_2O_3
Carbon (C)	6	Brittle nonmetal, unreactive at room temperature	CCl_4	CO_2
Nitrogen (N)	7	Colorless gas, nonmetallic, not very reactive	NCl_3	N_2O_5
Oxygen (O)	8	Colorless gas, nonmetallic, reactive	Na_2O, Cl_2O	MgO
Fluorine (F)	9	Greenish-yellow gas, nonmetallic, extremely reactive	NaF, ClF	MgF_2, OF_2
Neon (Ne)	10	Colorless gas, unreactive	None	None
Sodium (Na)	11	Soft metal, low density, very reactive	NaCl	Na_2O
Magnesium (Mg)	12	Harder metal than Na, low density, less reactive than Na	$MgCl_2$	MgO
Aluminum (Al)	13	Metal as hard as Mg, less reactive than Mg	$AlCl_3$	Al_2O_3
Silicon (Si)	14	Brittle nonmetal, not very reactive	$SiCl_4$	SiO_2
Phosphorus (P)	15	Nonmetal, low melting point, white solid, reactive	PCl_3	P_2O_5
Sulfur (S)	16	Yellow solid, nonmetallic, low melting point, moderately reactive	Na_2S, SCl_2	MgS
Chlorine (Cl)	17	Yellow-green gas, nonmetallic, extremely reactive	NaCl	$MgCl_2$, Cl_2O
Argon (Ar)	18	Colorless gas, unreactive	None	None
Potassium (K)	19	Soft metal, low density, very reactive	KCl	K_2O
Calcium (Ca)	20	Harder metal than K, low density, less reactive than K	$CaCl_2$	CaO

*The chemical formulas shown are lowest ratios. The molecular formula for $AlCl_3$ is Al_2Cl_6, and that for P_2O_5 is P_4O_{10}.

with similar properties. Some chemical and physical properties of the first 20 elements are summarized in Table 3.4. Do you see any trends and similarities among the elements in Table 3.4? For example, lithium (Li) is a soft metal with low density that is very reactive. It combines with chlorine gas to form lithium chloride with the formula LiCl. The other elements in Table 3.4 that have properties similar to those of lithium are sodium (Na) and potassium (K). According to the periodic law, lithium, sodium, and potassium should be in the same group; and they are. Look for similarities among other elements listed in Table 3.4 and check your grouping with that shown in the periodic table in Figure 3.11.

3.6 THE MODERN PERIODIC TABLE

In the modern **periodic table**, Figure 3.11, the elements are arranged in order of their atomic numbers so that elements with similar chemical and physical properties fall together in vertical columns. These vertical columns are called **groups**. The periodic table commonly used in the United States has groups numbered 1 through 8, with each numeral followed by a letter A or B. The A groups are the **representative** or **main-group** elements. The B groups are the **transition elements** that link the two areas of representative elements. The **inner transition elements** are the *lanthanide series* and the *actinide series.* They are placed at the bottom of the periodic table because the similarity of properties within the two series would require their placement between

 See an animation of an **Exploration of the Periodic Table** at http://brookscole.com/chemistry/joesten4

Periodic table Arrangement of elements in atomic number order in rows so that elements with similar properties fall together in vertical columns

Groups Vertical columns of elements with similar properties in the periodic table

Representative or main–group elements Elements in the A groups of the periodic table

Transition elements Elements in periodic table rows 4–7 in which *d* or *f* orbitals are being filled; they lie between main-group elements

Inner transition elements Elements 58–71 (lanthanides) and 90–103 (actinides), which lie between Groups 2A and 3A of the main group elements and are usually placed at the bottom of the periodic table

Chemists from all over the world belong to the International Union of Pure and Applied Chemistry (IUPAC). IUPAC has recommended that groups be labeled 1 through 18 consecutively from left to right.

Three elements: sulfur *(bottom)* and diamond (which is carbon, *top left*), which are main-group nonmetals; and gold *(top right)*, a transition metal.

In our periodic table, hydrogen is separated from the elements of Group 1 because it is a nonmetal, not an alkali metal.

lanthanum and hafnium, and between actinium and rutherfordium, respectively, which would make the table inconveniently wide.

The horizontal rows are called **periods**. These periods or rows are related to energy levels for electrons in atoms (Figures 3.10 and 3.11). The length of a row is linked to the maximum number of electrons, $2n^2$, that can fit into an energy level. The periods are not equal in size because the maximum number of electrons per energy level increases as the distance of the energy level from the nucleus increases. Periods one through seven have 2, 8, 8, 18, 18, 32, and 23 (incomplete) elements, respectively. Larger periods, seen as the atoms of elements get larger, are similar to the longer rows and more seats per row in a stadium as you proceed up from field level in the stands (Figure 3.12).

Most of the elements are **metals** and are found in Groups 1A, 2A, parts of Groups 3A to 6A (red in Figure 3.11), and the B groups (yellow) in Figure 3.11. Characteristic physical properties of metals include malleability (ability to be beaten into thin sheets such as aluminum foil), ductility (ability to be stretched or drawn into wire such as copper), and good conduction of heat and electricity.

Eighteen elements are **nonmetals** (in green in Figure 3.11), and except for hydrogen they are found on the right side of the periodic table. Hydrogen is shown above Group 1A because its atoms have one electron. However, hydrogen is a nonmetal and probably should be in a group by itself, although you may see H in both Group 1A and Group 7A in some periodic tables. Hydrogen forms compounds with formulas similar to those of the Group 1A elements, but with vastly different properties. For example, compare NaCl (table salt) with HCl (a strong acid), or compare Na_2O (an active metal oxide) with H_2O (water). Hydrogen also forms compounds similar to those of the Group 7A elements: NaCl and NaH (sodium hydride), and $CaBr_2$ and CaH_2 (calcium hydride).

The physical and chemical properties of nonmetals are opposite those of metals. For example, nonmetals are **insulators**; that is, they are extremely poor conductors of heat and electricity.

Elements that border the staircase in Figure 3.11—those between metals and nonmetals—are six **metalloids** (in blue). Their properties are intermediate between those of metals and nonmetals. For example, silicon (Si), germanium (Ge), and arsenic (As) are **semiconductors** and are the elements that form the basic components of computer chips. Semiconductors conduct electricity less well than metals, such as silver and copper, but better than

Periods Horizontal rows in the periodic table

Metals Elements that conduct electric current; most are malleable and ductile

Nonmetals Elements that do not conduct electrical current

Insulators Poor conductors of heat and electricity

Metalloids Elements with properties intermediate between those of metals and nonmetals, and falling between them in the periodic table

Semiconductors Metalloid elements with electrical conductivity intermediate between that of metals and nonmetals

FIGURE 3.12 **Analogy of periods in periodic table to rows in a football stadium.** Larger periods in the periodic table as atoms of elements get larger are similar to longer rows and more seats per row in a stadium in the rows that are farther from the playing field.

insulators, such as sulfur. The six **noble gases** in Group 8A have little tendency to undergo chemical reactions and at one time were considered inert (totally unreactive). The classifications of metals, nonmetals, and metalloids will enable you to predict the kinds of compounds formed between elements.

EXAMPLE 3.3 | **Periodic Table**

For the elements with atomic numbers 17, 33, and 82, give the names and symbols and identify the elements as metals, metalloids, or nonmetals.

SOLUTION

Chlorine (Cl) is the element with atomic number 17. It is in Group 7A. Chlorine and all the other elements in Group 7A are nonmetals.

Arsenic (As) is the element with atomic number 33. It is in Group 5A. Because it lies along the line between metals and nonmetals, arsenic is a metalloid.

Lead (Pb) is the element with atomic number 82. It is in Group 4A. Like other elements at the bottom of Groups 3A to 6A, lead is a metal.

TRY IT 3.3

List the main groups (the A groups) in the periodic table that (**a**) consist entirely of metals, (**b**) consist entirely of nonmetals, and (**c**) include metalloids. Identify the numbers of valence electrons in atoms from groups listed under (**a**), (**b**), and (**c**).

3.7 PERIODIC TRENDS

Why do elements in the same group in the periodic table have similar chemical behavior? Why do metals and nonmetals have different properties? G. N. Lewis was seeking answers to these questions during his development of the concept of valence electrons. He assumed that each noble gas atom had a completely filled outermost shell, which he regarded as a stable configuration because of the lack of reactivity of noble gases. He also assumed that the reactivity of other elements was influenced by their numbers of valence electrons.

Lewis Dot Symbols

Lewis used the element's symbol to represent the atomic nucleus together with all but the outermost electrons. *It is important to note that the number of valence electrons for main group elements (Groups 1A–8A) equals the group number of the element.* The valence electrons which are the outermost electrons of an atom, are represented by dots. The dots are placed around the symbol one at a time until they are used up or until all four sides are occupied; any remaining electron dots are paired with the ones already there. Lewis dot symbols for atoms of the first 20 elements are shown in Table 3.5. Notice that all atoms of elements in a given A group have the same number of valence electrons and that the number of valence electrons equals the group number. All atoms of Group 1A elements have one valence electron; Group 2A atoms have two valence electrons, and so forth. *The importance of valence electrons in the study of chemistry cannot be overemphasized.* The identical number of valence electrons primarily accounts for the similar properties of elements in the same group. *The chemical view of matter is primarily concerned with what valence electrons are doing in the course of chemical reactions.*

Lewis dot symbols will be used extensively in Chapter 5 in the discussions of bonding.

See Section 3.4 for a discussion of valence electrons.

The *reactivity* of an element or a compound is its tendency to undergo chemical reactions. Some chemical pollutants, such as DDT, have low reactivity and therefore remain in the environment unchanged for a long time. An element such as potassium that on exposure reacts immediately with water or oxygen in the air is highly reactive.

Noble gases Group 8A elements in the periodic table

TABLE 3.5		Lewis Dot Symbols for Atoms					
1A	2A	3A	4A	5A	6A	7A	8A
H·							He:
Li·	·Be·	·Ḃ·	·Ċ·	·N̈·	:Ö·	:F̈·	:N̈e:
Na·	·Mg·	·Al̇·	·Ṡi·	·P̈·	·S̈·	:C̈l·	:Är:
K·	·Ca·						

Atomic Properties

Metallic character means having properties of metals, such as ductility, malleability, and conductivity.

From left to right across each period, metallic character gives way to nonmetallic character (Figure 3.11). The elements with the most metallic character are at the lower-left part of the periodic table near cesium (Cs). The elements with the most nonmetallic character are at the upper-right portion of the periodic table near fluorine. The six metalloid elements (in blue) that begin with boron and move down like a staircase to astatine (At) roughly separate the metals and the nonmetals.

Atomic radius is one-half the distance between the nuclei centers of two like atoms in a molecule.

Atomic radii show periodicity by increasing with atomic number down the main groups of the periodic table (Figure 3.13). Why do atoms get larger from the top to the bottom of a group? The larger atoms simply have more energy levels occupied by electrons than do the smaller atoms.

Atomic radii decrease across a period from left to right (Figure 3.13). You may see a paradox in adding electrons and getting smaller atoms, but protons also are being added. The increasing positive charge of the nucleus pulls electrons closer to the nucleus and causes contraction of the atom.

We can also use the trends in size of atomic radii to predict trends in reactivity. The valence electrons in larger atoms are farther from the nucleus. The larger the atom, the easier it is to remove the valence electrons because the attractive forces between protons in the nucleus and valence electrons decrease with increasing size of the atom.

A general equation for the reaction of water and a metal from Group 1A, using M to represent any of the 1A metals, is

$$2 M(s) + 2 H_2O(\ell) \longrightarrow$$
$$2 MOH(aq) + H_2(g)$$

In ionic compounds, one class of chemical compounds (Section 5.2), atoms have gained and lost electrons to form **ions**, which have positive or negative charges. Metal atoms lose valence electrons to form positive ions. The larger the metal atom, the greater the tendency to lose valence electrons and the more reactive the metal. Therefore, we would predict that the most reactive metal in Figure 3.13 is cesium (Cs), the metal with the largest radius, and this is correct. The resulting *increase* in reactivity of metals down a given group in the periodic table is dramatically illustrated by lithium, sodium, and potassium—the first three metals in Group 1A. Their atoms increase in size in this order down the group. Each element reacts with water—lithium quietly and smoothly, sodium more vigorously, and potassium much more quickly. The reactions of both sodium and potassium give off enough heat to ignite the hydrogen gas produced by the reaction, but as shown in Figure 3.14, potassium reacts with explosive violence. For elements at the bottom of Group 1A, exposure to moist air produces a vigorous explosion.

A general equation for the reaction of hydrogen with a nonmetal from Group 7A, using X to represent any of the 7A nonmetals, is

$$H_2(g) + X_2(g) \longrightarrow 2 HX(g)$$

Ions Atoms or groups of atoms with a positive or negative charge, for example, sodium ion (Na^+), chloride ion (Cl^-), and sulfate ion (SO_4^{2-})

Nonmetal atoms gain electrons from metals to form negative ions. The smaller the nonmetal atom, the higher the reactivity of the nonmetal. For example, fluorine atoms are the smallest of Group 7A elements, and fluorine is the most reactive nonmetal. It reacts with all other elements except three

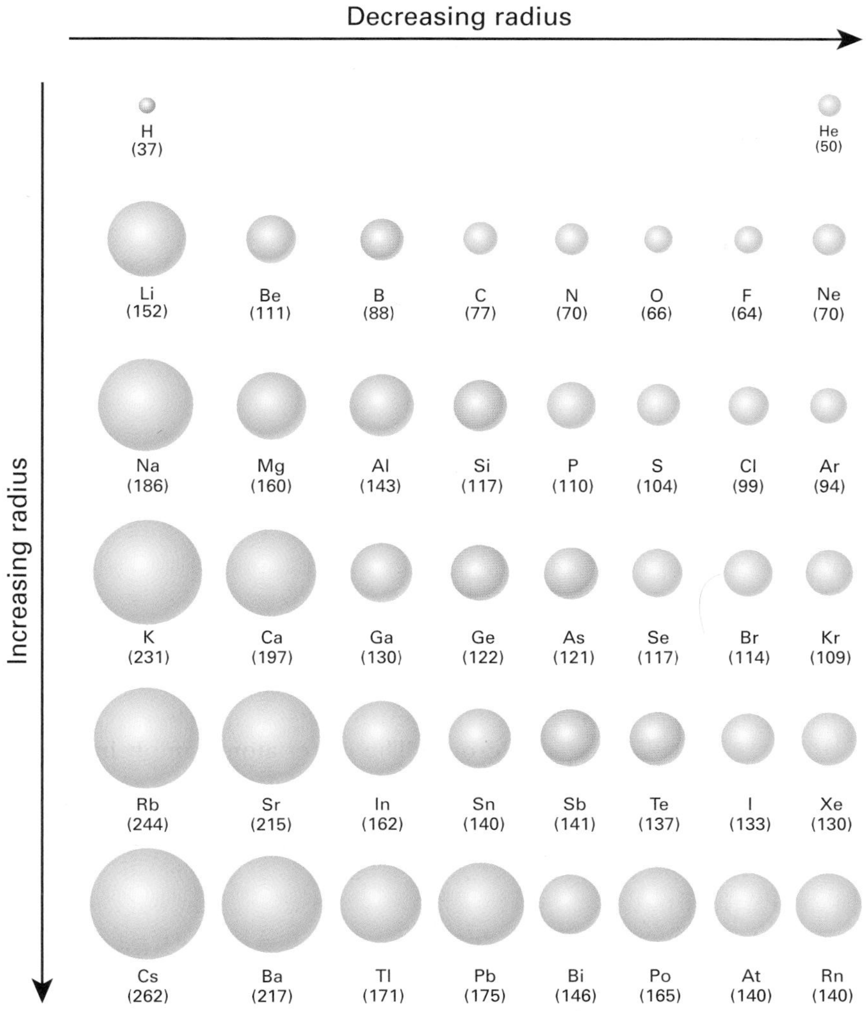

Decreasing radius →

Increasing radius ↓

H
(37)

He
(50)

Li
(152)

Be
(111)

B
(88)

C
(77)

N
(70)

O
(66)

F
(64)

Ne
(70)

Na
(186)

Mg
(160)

Al
(143)

Si
(117)

P
(110)

S
(104)

Cl
(99)

Ar
(94)

K
(231)

Ca
(197)

Ga
(130)

Ge
(122)

As
(121)

Se
(117)

Br
(114)

Kr
(109)

Rb
(244)

Sr
(215)

In
(162)

Sn
(140)

Sb
(141)

Te
(137)

I
(133)

Xe
(130)

Cs
(262)

Ba
(217)

Tl
(171)

Pb
(175)

Bi
(146)

Po
(165)

At
(140)

Rn
(140)

See an animation of **Atomic Size** at http://brookscole.com/chemistry/joesten4

FIGURE 3.13 **Atomic radii of the A Group elements (in picometers).** A picometer (pm) is 1×10^{-12} m (100 nm). Remember that the radius is one-half of the diameter.

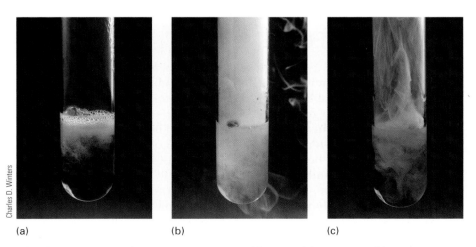

See an animation of **Periodic Reactivity Trends** at http://brookscole.com/chemistry/joesten4

Charles D. Winters

(a)   (b)   (c)

FIGURE 3.14 Reaction of alkali metals with water: (a) lithium, (b) sodium, and (c) potassium.

noble gases—helium, neon, and argon. The reaction of Group 7A elements with hydrogen illustrates how the reactivity of nonmetals *decreases* down the group. Fluorine reacts explosively with hydrogen, but the reaction with hydrogen is less violent for chlorine and is very slow for iodine.

Why are there repeatable patterns of properties across the periods in the periodic table? Again, it is because there is a repeatable pattern in atomic structure, and properties depend on atomic structure. *Each period begins with one valence electron for atoms of the elements in Group 1A. Each period builds up to eight valence electrons, and the period ends. This pattern repeats across periods two through six.* As more elements are made by nuclear accelerators (Section 13.6), it is possible that period seven will be completed someday.

EXAMPLE 3.4 **Atomic Radii and Reactivity**

Which element in each pair is more reactive: (**a**) O or S, (**b**) Be or Ca, or (**c**) P or As?

SOLUTION
(**a**) Oxygen and sulfur are Group 6A nonmetals. Generally, the smaller the atomic radius of a nonmetal, the more reactive the nonmetal is. Because oxygen atoms are smaller than sulfur atoms, oxygen should be more reactive than sulfur.
(**b**) Beryllium (Be) and calcium (Ca) are Group 2A metals. Reactivity of metals in a given group increases down the group as the atomic radius increases. Therefore, calcium should be more reactive than beryllium.
(**c**) Phosphorus and arsenic are Group 5A nonmetals; thus, phosphorus, which has a smaller atomic radius, is predicted to be more reactive than arsenic.

TRY IT 3.4A
Which element in each pair has the larger radius, that is, which is the larger atom in each pair: (**a**) Ca or Ba, (**b**) S or Se, (**c**) Si or S, or (**d**) Ga or Br?

TRY IT 3.4B
Which element in each pair is more reactive: (**a**) Mg or Sr, (**b**) Cl or Br, or (**c**) Rb or Cs?

Reaction of calcium with water. It is easy to see by comparison with Figure 3.14 that calcium is less reactive than the alkali metals.

Charles D. Winters

3.8 PROPERTIES OF MAIN-GROUP ELEMENTS

Elements in a periodic table group have similar properties, but not the same properties. Some properties, as already illustrated for atomic radii and reactivity, increase or decrease in a predictable fashion from top to bottom of a periodic group.

Elements in a group generally react with other elements to form similar compounds, a fact accounted for by their identical numbers of valence electrons. For example, the formula for the compound of Li and Cl is LiCl; thus, you can expect there to be a compound of Rb and Cl with the formula RbCl and a compound of Cs and Cl with the formula of CsCl. Likewise, the formula Na_2O is known; therefore, a compound with the formula Na_2S predictably exists, because oxygen and sulfur are in the same group. In general, elements in the same group of the periodic table form some of the same types of compounds. In fact, several main groups have group names because of the distinctive and similar properties of the group elements.

The Group 1A elements (Li, Na, K, Rb, Cs, and Fr) are called the **alkali metals**. The name *alkali* derives from an old word meaning "ashes of burned plants." All the alkali metals are soft enough to be cut with a knife. None are found in nature as free elements, because all combine rapidly and completely with virtually all the nonmetals and, as illustrated in Figure 3.14, with water. These elements form positive ions (**cations**) with a +1 charge. Francium, the last member of Group 1A, is found only in trace amounts in nature, and all its 21 isotopes are naturally radioactive.

The Group 2A elements (Be, Mg, Ca, Sr, Ba, and Ra) are called the **alkaline earth metals**. Compared with the alkali metals, the alkaline earth metals are harder, are more dense, and melt at higher temperatures. These elements form cations with a +2 charge. Because the valence electrons of alkaline earths are held more tightly, they are less reactive than their alkali metal neighbors. All the alkaline earth metals react with oxygen to form an oxide MO, where M is the alkaline earth.

$$2\,\mathrm{M}(s) + \mathrm{O}_2(g) \longrightarrow 2\,\mathrm{MO}(s)$$

The **halogens** (F, Cl, Br, I, and At) are Group 7A elements. In the elemental state, each of these elements exists as diatomic molecules (X_2). Fluorine (F_2) and chlorine (Cl_2) are gases at room temperature, whereas bromine (Br_2) is a liquid and iodine (I_2) is a solid. This illustrates a typical trend within a group—an increase in melting point and boiling point in going down a group. All isotopes of astatine (At) are naturally radioactive and disintegrate quickly. The name *halogen* comes from a Greek word and means "salt producing." The best-known salt containing a halogen is sodium chloride (NaCl), table salt. However, there are many other halogen salts, such as calcium fluoride (CaF_2), a natural source of fluorine; potassium iodide (KI), an additive to table salt that prevents goiter; and silver bromide (AgBr), the active photosensitive component of photographic film. The halogen salts contain negatively charged ions known as **anions**.

The *noble gases* (He, Ne, Ar, Kr, Xe, and Rn) are Group 8A elements. They are all colorless gases composed of single atoms at room temperature. They are referred to as "noble" because they lack chemical reactivity and generally do not react with "common" elements. Neon is the gas that glows orange-red in tubes of neon lights. Other gases and color-tinted tubes are used to give different colors. Radon (Rn) is naturally radioactive. Problems associated with indoor radon pollution are discussed in Section 13.7.

See an animation of **Ion Formation** at http://brookscole.com/chemistry/joesten4

Recall that the only other diatomic elements are H_2, N_2, and O_2.

CONCEPT CHECK **3D**

1. How many valence electrons are in each of the following atoms?
   - (**a**) Sodium (Na), Group 1A
   - (**b**) Calcium (Ca), Group 2A
   - (**c**) Boron (B), Group 3A
   - (**d**) Aluminum (Al), Group 3A
   - (**e**) Neon (Ne), Group 8A
   - (**f**) Iodine (I), Group 7A
2. The periodic table organized by Mendeleev placed elements in order of increasing atomic weight. (**a**) True (**b**) False
3. The modern periodic table has the elements placed in order of increasing atomic weight. (**a**) True (**b**) False
4. Metals typically have low numbers of valence electrons. (**a**) True (**b**) False
5. Nonmetals are typically found in Groups 5A through 8A. (**a**) True (**b**) False

Alkali metals Elements in Group 1A in the periodic table

Cation An ion with an overall positive charge; for example, Li^+, Ca^{2+}, NH_4^+

Alkaline earth metals Elements in Group 2A in the periodic table

Halogens Elements in Group 7A in the periodic table

Anion An ion with an overall negative charge; for example, Cl^-, NO_3^-

6. Which of the following are true about metals?
   (a) They are found only in Group 1A and 2A.
   (b) They are good conductors of heat.
   (c) They are good insulators.
   (d) They usually react to form positive ions.

7. Group 2A includes Be, Mg, Ca, Sr, Ba, and Ra. What is the predicted formula for the compound formed between Sr and Cl, if Be and Ba form $BeCl_2$ and $BaCl_2$?

8. Which of the elements in Group 1A is the most reactive: Li, Na, K, Rb, or Cs?

9. Which of the following pairs of atoms has the greater number of valence electrons?
   (a) Lithium (Li) or oxygen (O)   (b) Calcium (Ca) or sodium (Na)
   (c) Oxygen (O) or fluorine (F)   (d) Boron (B) or nitrogen (N)
   (e) Sulfur (S) or arsenic (As)   (f) Neon (Ne) or carbon (C)

10. Group 1A elements are all metals except for hydrogen. (a) True (b) False

11. Isotopes have different numbers of valence electrons. (a) True (b) False

12. In each of the following atoms, which atom will be more reactive?
   (a) Lithium (Li) or cesium (Cs)
   (b) Fluorine (F) or bromine (Br)
   (c) Beryllium (Be) or calcium (Ca)
   (d) Oxygen (O) or sulfur (S)
   (e) Lithium (Li) or sodium (Na)
   (f) Neon (Ne) or xenon (Xe)

 Assess your understanding of this chapter's topics with an online chapter quiz at **www.brookscole.com/chemistry/joesten4**

■ KEY TERMS

law of conservation of matter	continuous spectrum	metals
law of definite proportions	quantum	nonmetals
nucleus	ground state	insulators
electrons	excited state	metalloids
protons	valence electrons	semiconductors
neutrons	shell	reactivity
radioactivity	orbital	noble gases
alpha (α) particles	subshell	ions
beta (β) particles	periodic table	alkali metals
gamma (γ) rays	groups	alkali earth metals
atomic number	representative or main-group elements	cation
mass number		halogens
isotopes	transition elements	anion
atomic mass unit (amu)	inner transition elements	
atomic weight	periods	

■ LANGUAGE OF CHEMISTRY I

1. Mass number
2. Unlike electric charges
3. $2n^2$
4. Nucleus
5. Electron
6. ^{22}Ne and ^{20}Ne
7. Atomic number
8. Particles in an H atom
9. Neutron
10. Rutherford
11. Bohr

a. Attract
b. Equal to number of protons in the nucleus
c. Negative particle found outside the nucleus
d. Neutrons plus protons
e. Discovered the nucleus
f. Proton and an electron
g. Developed a theory for representing electrons in energy levels
h. Isotopes
i. Maximum number of electrons in an energy level
j. Uncharged elementary particle
k. Central part of an atom consisting of protons and neutrons

■ LANGUAGE OF CHEMISTRY II

1. Periodic
2. Larger atoms
3. Two valence electrons
4. A noble gas
5. A transition metal
6. A main-group metal
7. Decrease in atomic radius
8. A halogen
9. An inner transition element
10. Valence electrons

a. Electron arrangement 2-8-2
b. Outermost occupied shell
c. Chromium (Cr)
d. Repeated pattern
e. At the bottom of a group
f. Praseodymium (Pr)
g. Eight valence electrons
h. Seven valence electrons
i. Across a period
j. Cesium

■ APPLYING YOUR KNOWLEDGE

1. What is the law of conservation of matter? Give an example of the law in action.
2. State the law of definite proportions. Give an example to illustrate what it means.
3. Which Quaker scientist is given credit for much of the development of the atomic theory of atoms, molecules, elements, and compounds?
4. What kinds of evidence did Dalton have for atoms that the early Greeks (Democritus, Leucippus) did not have?
5. How does Dalton's atomic theory explain these?
   (a) The law of conservation of matter
   (b) The law of definite proportions
6. Describe the most significant observation to be derived from Lord Ernest Rutherford's experiments in which he bombarded gold foil with alpha particles.

7. The fact that carbon dioxide always contains 27.27% carbon and 72.73% oxygen by mass is a manifestation of the:
   (a) Law of mass action
   (b) Law of definite proportions
   (c) Law of conservation of matter
   (d) Periodic law
8. What is meant by the term *subatomic particles*? Give two examples.
9. Give short definitions for the following terms:
   (a) Atomic number   (b) Mass number
   (c) Atomic weight   (d) Isotope
   (e) Natural abundance   (f) Atomic mass unit
10. There are more than 1000 kinds of atoms, each with a different weight, yet only 116 elements were known as this book was being written. How does one explain this in terms of subatomic particles?

11. Dmitri Mendeleev was the last born of 17 children in his family. Can you write the atomic symbol and give the total number of electrons and the number of valence electrons for each of the first 17 elements in his periodic table?

12. Which of the following pairs are isotopes? Explain your answers.
    (a) ^{50}Ti and ^{50}V   (b) ^{12}C and ^{14}C
    (c) ^{40}Ar and ^{40}K

13. What do all the atoms of an element have in common?

14. If $^{37}_{17}$Cl is used to designate a certain isotope of chlorine, what would $^{35}_{17}$Cl represent?
    (a) an isotope with two fewer neutrons
    (b) an isotope with two fewer protons
    (c) a different element
    (d) an isotope with two more electrons
    (e) an isotope with one less proton and one less neutron

15. A common isotope of Li has a mass of 7. The atomic number of Li is 3. How can this information be used to determine the number of protons and neutrons in the nucleus?

16. The element iodine (I) occurs naturally as a single isotope of atomic mass 127; its atomic number is 53. How many protons and how many neutrons does it have in its nucleus?

17. Which of the following elements would need only to gain two electrons to have a full outer shell (energy level) of electrons equivalent to the noble gas of next highest atomic number?
    (a) $_7$N (b) $_{20}$Ca (c) $_{50}$Sn (d) $_{35}$Br (e) $_{34}$Se

18. Complete the following table so that it contains information about five different atoms.

	Number of Protons	Number of Neutrons	Number of Electrons	Atomic Number	Mass Number
(a)	32	—	—	—	73
(b)	—	14	14	—	—
(c)	—	—	—	28	59
(d)	48	64	—	—	—
(e)	—	115	—	77	—

19. Identify the elements in question 18.

20. What number is most important in identifying an atom?

21. The atomic weight listed in the periodic table for magnesium is 24.305 amu. Someone said that there wasn't a single magnesium atom on the entire earth with a weight of 24.305 amu. Is this statement correct? Why or why not?

22. Complete the following table:

Isotope	Atomic No.	Mass No.	No. of Protons	No. of Neutrons	No. of Electrons
Bromine-81	—	81	—	—	—
Boron-11	5	—	—	—	—
^{35}Cl	17	—	—	—	—
^{52}Cr	—	52	—	—	—
Ni-60	—	—	—	—	—
Sr-90	—	—	—	—	—
Lead-206	—	—	—	—	—

23. The longest wavelength for visible light is approximately 700 nm. What is the frequency for this type of light? Recall: $c = \lambda\nu$; $c = 3.00 \times 10^8$ m/s.

24. Krypton is the name of Superman's home planet and also that of an element. Look up the element krypton and list its symbol, atomic number, atomic weight, and electron arrangement.

25. If the frequency of a given type of electromagnetic radiation increases, the:
    (a) speed of light increases
    (b) energy decreases
    (c) wavelength increases
    (d) energy increases

26. Some ultraviolet (UV) radiation has a wavelength of 350 nm. What is this equivalent to according to the following analysis?

$$350 \text{ nm} \times \frac{10^{-9} \text{ m}}{1 \text{ nm}} \times \frac{10^2 \text{ cm}}{1 \text{ m}} = \underline{\qquad}$$

    (a) 350×10^{-7} cm   (b) 350×10^{9} m
    (c) 35 cm   (d) 350×10^{-11} mm

27. Infrared (IR) radiation (wavelengths in the range of 10^{-4} cm) is significantly less energetic than UV radiation. This means that:
    (a) Infrared radiation has a higher frequency than UV radiation
    (b) Ultraviolet radiation has a longer wavelength than IR radiation
    (c) Infrared radiation has a lower frequency than UV radiation
    (d) Infrared radiation has a shorter wavelength than UV radiation

28. Which equation represents the maximum number of electrons that can occupy a given energy level (shell) in an element if n is an integer representing the energy level?
    (a) n^2 (b) $2n^2$ (c) $2n$ (d) n (e) $n/2$

29. Write the placement of electrons in their ground-state energy levels according to the Bohr theory for atoms having 6, 10, 13, and 20 electrons.

30. Write the Bohr electron notation for atoms of the elements sodium through argon.

31. How many valence electrons do atoms of each of the elements in question 30 contain?

32. Ultraviolet light has a wavelength of approximately 200 nm. What is the frequency for this form of UV light? Recall: $c = \lambda\nu; c = 3.00 \times 10^8$ m/s.

33. State the periodic law.

34. Give definitions for the following terms:
    (**a**) Group   (**b**) Period
    (**c**) Chemical properties   (**d**) Transition element
    (**e**) Inner transition element
    (**f**) Representative element

35. How did the discovery of the periodic law lead to the discovery of elements?

36. Describe the relative locations of metals, nonmetals, and metalloids in the periodic table.

37. How do metals differ from nonmetals?

38. Identify each of the following elements as either a metal, nonmetal, or metalloid.
    (**a**) Nitrogen   (**b**) Arsenic
    (**c**) Argon   (**d**) Calcium
    (**e**) Uranium

39. Answer the following questions about the periodic table.
    (**a**) How many periods are there?
    (**b**) How many representative groups or families are there?
    (**c**) How many groups consist of all metals?
    (**d**) How many groups consist of all nonmetals?
    (**e**) Is there a period that consists of all metals?

40. What do the electron structures of alkali metals have in common?

41. The diameter of an atom is approximately 10^{-10} m. The diameter of an atomic nucleus is about 10^{-15} m. What is the ratio of the atomic diameter to the nuclear diameter?

42. Give the number of valence electrons for each of the following:
    (**a**) Ba   (**b**) Al
    (**c**) P   (**d**) Se
    (**e**) Br   (**f**) K

43. Give the symbol for an element that has
    (**a**) 3 valence electrons   (**b**) 4 valence electrons
    (**c**) 7 valence electrons   (**d**) 1 valence electron

44. Draw the Lewis dot symbols for Be, Cl, K, As, and Kr.

45. From their position in the periodic table, predict which will be more metallic.
    (**a**) Be or B   (**b**) Be or Ca
    (**c**) As or Ge   (**d**) As or Bi

46. Which atom in the following pairs is more metallic?
    (**a**) Li or F   (**b**) Li or Cs
    (**c**) Be or Ba   (**d**) C or Pb
    (**e**) B or Al   (**f**) Na or Ar

47. What general electron arrangement is conducive to chemical inactivity?

48. Use the information in the periodic table to locate the following details.
    (**a**) The nuclear charge on cadmium (Cd)
    (**b**) The atomic number of As
    (**c**) The atomic mass (or mass number) of an isotope of Br having 46 neutrons
    (**d**) The number of electrons in an atom of Ba
    (**e**) The number of protons in an isotope of Zn
    (**f**) The number of protons and neutrons in an isotope of Sr, atomic mass (or mass number) of 88
    (**g**) An element forming compounds similar to those of Ga

49. Complete the following table.

Atomic Number	Name of Element	Number of Valence Electrons	Period	Metal or Nonmetal
6	_____	_____	_____	_____
12	_____	_____	_____	_____
17	_____	_____	_____	_____
37	_____	_____	_____	_____
42	_____	_____	_____	_____
54	_____	_____	_____	_____

50. Write the symbols of the halogen family in the order of increasing size of their atoms.

51. Why does Cs have larger atoms than Li?

52. How does the atomic radius for a metal atom relate to the reactivity of the metal?

53. How does the atomic radius for a nonmetal atom relate to reactivity?

54. Which atom in the following pairs is more reactive?
    (**a**) Li or Rb   (**b**) Mg or Ba
    (**c**) Na or Ar   (**d**) Ne or O
    (**e**) He or Xe   (**f**) Br or F

55. What are the approximate wavelengths for the following colors of visible light?
    (**a**) Red (**b**) Green (**c**) Yellow (**d**) Blue

56. Describe the variation in atomic size
    (**a**) across a period
    (**b**) down a group

57. Rank the following atoms by size, with the largest on the left and the smallest on the right: K, S, Al, P, and Cl.

58. The elements at the bottom of Groups 1A, 2A, 6A, 7A, and 8A are all radioactive; thus, less is known about their physical and chemical properties. By using group and period trends described in this chapter, predict which of these five elements—Fr, Ra, Po, At, and Rn—would:
    (a) be the most metallic
    (b) be the most nonmetallic
    (c) have the largest atomic radius
    (d) be the most unreactive
    (e) react most readily with water

59. If element 36 is a noble gas, in what groups would you expect elements 35 and 37 to occur?

60. Oxygen and sulfur are very different elements, in that one is a colorless gas and the other is a yellow crystalline solid. Why then are they both in Group 6A?

61. How many seconds are needed for a photon to travel through 3200 miles of fiber optic cable from New York to Los Angeles? Recall: $c = 3.00 \times 10^8$ m/s.

62. How many seconds are needed for a microwave photon to travel through 24,000 miles of space to and from a television relay satellite? Recall: $c = 3.00 \times 10^8$ m/s.

63. Most α-particles, which are helium atoms without their electrons, when fired at a thin gold (Au) foil would:
    (a) pass right through the foil or be deflected very little, because atoms of Au are mostly empty space
    (b) collide head on with the nuclei of the Au atoms, which occupy most of the space of the foil
    (c) collide with the very massive electrons associated with the Au atoms of the foil
    (d) make a right-angle turn when confronted with the foil's huge number of atoms

64. In a chemical reaction, _____ must be conserved but the identity of the _____ has to change.
    (a) mass and matter; elements
    (b) mass and matter; molecules
    (c) physical state and color; isotopes
    (d) heat; elements
    (e) electrons; protons

 # CHEMISTRY ON THE WEB

For up-to-date URLs, visit the text website at **www.brookscole.com/ chemistry/joesten4**

- How to Use the Periodic Table and Examples of Periodic Tables
- Novelty Periodic Tables of Science Fiction and of Comics
- The History of the Atom (John Dalton's Model)
- A Look Inside the Atom
- The Discovery of Radioactivity
- Scanning Tunneling Microscope and Images
- The Electromagnetic Spectrum
- History of Atomic Structure and Some More Periodic Tables
- Video of Reactions of Sodium and Potassium with Water

THE AIR WE BREATHE

There is much to be learned about the composition of our air whether over a city street or over a lake like this one.

Charles D. Winters

In the beginning, there was no molecular oxygen (O_2). It is generally agreed that until about 4.5–5 billion years ago, Earth's atmosphere was quite different than it is today. There was gaseous nitrogen (N_2), water vapor (H_2O), carbon dioxide (CO_2), methane (CH_4), and ammonia (NH_3). In the beginning, not all life required oxygen and life was limited to deep in the oceans.

Scientists believe that by about 2.5 billion years ago, cyanobacteria capable of photosynthesis had evolved and that these bacteria began a process of changing our atmosphere. They produced oxygen as a waste product of their metabolism! The oxygen built up very slowly in our atmosphere because Earth contained abundant amounts of materials that are very reactive with oxygen. Iron and other materials found both in the oceans and on land, and reactive toward oxygen, consumed most of the waste oxygen at first. Slowly, however, these materials exhausted their capacity to consume more oxygen and the concentration of oxygen in our atmosphere began to increase. By about 1.75 billion years ago, the oxygen content of our atmosphere had increased to near its present value of about 21%. The increase in oxygen concentration played two key roles in the development of oxygen-dependent life. First, oxygen was needed for respiration. Second, the increased oxygen concentration allowed the ozone layer to form in our stratosphere. Without the protection from high-energy ultraviolet radiation afforded by this ozone layer, oxygen-dependent organisms, of which humans are but one example, would not have survived or, if they had, would have been dramatically different than they are today.

Although we've just begun our exploration of chemistry, our goal here is to illustrate that one can use a relatively limited knowledge of elements, atoms, molecules, and chemical reactions to understand some pretty important phenomena. In this chapter we will focus on air, a precious commodity that, as has become increasingly apparent, is at considerable risk. Have we begun to pollute our atmosphere to such an extent that our health and way of life are endangered? We will look at our air, that life-sustaining chemical canopy that surrounds Earth, and address some of the issues related to its composition and our influence on it.

- What are the major components of Earth's atmosphere?
- What are the major regions of Earth's atmosphere?
- What is a pollutant and what are its characteristics?
- What are the major pollutants of our atmosphere and where do they come from?
- How is the concentration of a pollutant reported?
- Are all pollutants equally dangerous?
- What can we do to minimize air pollution?
- Is legislation necessary to control pollution?

(4.1) THE LOWER ATMOSPHERIC REGIONS AND THEIR COMPOSITION

Earth is enveloped by a few vertical miles of chemicals that compose the gaseous medium in which we exist. We refer to that region of space as our *atmosphere*. The gases in Earth's atmosphere are held in place by gravitational force. The thermal energy that keeps the molecules in motion ensures that these gases do not simply sink to Earth's surface, while gravity ensures that they do not simply escape into outer space. The percentage composition of our atmosphere remains essentially constant with increasing altitude but the density of the atmosphere drops significantly as the altitude increases and the temperature decreases. The **atmospheric pressure** decreases from $14.7 \text{ lbs/in}^2 = 1 \text{ kg/cm}^2$ (1 atm) at sea level, to about 12 lbs/in^2 (~0.8 atm) in Denver, the "mile-high city," and finally to about 4.4 lbs/in^2 (~0.3 atm) at the top of Mt. Everest.

Even though the air at the top of Mt. Everest is still about 21% oxygen, the fact that the air there is only about one-third as dense as that at sea level means that a mountain climber would have to take in three times the normal

Atmospheric pressure The pressure exerted by the weight of the atmosphere at a given altitude

volume of air to take in the same amount of oxygen. It is little wonder that most such climbers use supplemental oxygen at high altitudes where even slight exertion can leave one gasping for breath.

The region of Earth's atmosphere closest to Earth, which extends to an altitude of roughly 10 km (~6 mi), is known as the **troposphere** (Figure 4.1). This is the region in which our weather occurs. Propeller and jet aircraft and balloons routinely travel in the troposphere. Denver sits at about 1.6 km above sea level, while Mt. Everest reaches almost 9 km into the troposphere. At the upper level of the troposphere the temperature of the air is about $-60°C$ ($-76°F$), much colder than Earth's average surface temperature of $15°C$ ($59°F$). Most of the pollution discussed in this chapter occurs near the surface of Earth. We will refer to this as *ground-level air pollution* even though it may extend for several kilometers into our atmosphere.

Above the troposphere is the **stratosphere**, which extends to about 50 km (~30 mi). The temperature of the stratosphere rises gradually to about $0°C$ ($32°F$) at 50 km. This temperature increase is primarily due to the absorption of ultraviolet light (UV light) by the protective **ozone layer**, which is a band of higher-level ozone concentrations at about 15–30 km. Ozone, O_3, although a pollutant in the troposphere, plays a protective role in the stratosphere. Only some supersonic planes and some high-flying weather balloons venture into this region, although spacecraft travel through this region on their voyages beyond Earth's atmosphere.

Before we consider atmospheric pollutants, let's examine the normal composition of dry air in more detail. As shown in Table 4.1, nitrogen and oxygen make up about 99% of Earth's atmosphere. There are lesser amounts of many other gaseous substances.

The concentration values in Table 4.1 are listed as percentages or as parts per million (ppm). All of us are familiar with percentages. For instance, in

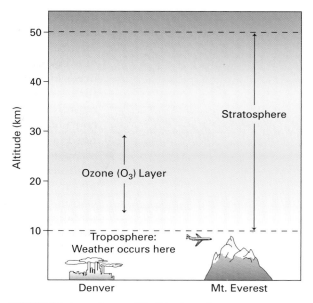

FIGURE 4.1 Lower layers of Earth's atmosphere.

Troposphere The region of Earth's atmosphere that extends from sea level up to about 10 km (1 km = 0.62 mi)

Stratosphere The region of Earth's atmosphere that extends from the troposphere to an altitude of about 50 km (~31 mi)

Ozone layer The stratospheric layer of gaseous ozone (O_3) that protects life on Earth by filtering out most of the harmful ultraviolet radiation emitted by the Sun

TABLE 4.1	Composition of Dry Air at Sea Level	
COMPOSITION (BY VOLUME)		
GAS	**PERCENTAGE**	**PPM**
Nitrogen (N_2)	78.084	780,840
Oxygen (O_2)	20.948	209,480
Argon (Ar)	0.934	9,340
Carbon dioxide (CO_2)	0.037	370
Neon (Ne)	0.00182	18.2
Helium (He)	0.00052	5.2
Methane (CH_4)	0.0002	2
Hydrogen (H_2)	0.0001	1
Krypton (Kr)	0.0001	1
Carbon monoxide (CO)	0.00001	0.1
Xenon (Xe)	0.000008	0.08
Ozone (O_3)	0.000002	0.02
Ammonia (NH_3)	0.000001	0.01
Nitrogen dioxide (NO_2)	0.0000001	0.001
Sulfur dioxide (SO_2)	0.00000002	0.0002

a class composed of 136 females and 64 males, we can easily determine the percentage of females by dividing the number of females by the total class membership (200) and multiplying by 100 to get 68%. This means that out of every 100 students, there are 68 females. Thus, 68% is 68 parts per hundred. Many of the constituents of air are present at concentrations far less than 1%. It gets quite awkward to deal with such small numbers; for example, neon is present at a level of 0.00182%. It is easier to express such low-level concentrations in **parts per million (ppm)** or **parts per billion (ppb)**. You will need to become comfortable with these units in your role as a chemically informed citizen.

To convert 0.00182% to ppm we are really converting parts per hundred (%) to ppm. In other words, we are asking the following question: If 0.00182 out of every 100 particles in the air are Ne atoms, what number would that be out of 1,000,000 particles? A simple proportional equation can be set up as shown here to solve for this value (represented by x).

$$\frac{0.00182}{100} = \frac{x}{1,000,000}$$

By cross multiplying, we get $100x = 1820$ or $x = 18.2$ ppm. This is the same as saying that if we could reach into the air and grab 1,000,000 particles, 18.2 would, on average, be Ne atoms. Of course, we can't actually have two-tenths of a neon atom, but this is a statistical average. With a little practice you should be easily able to convert among percent, ppm, and ppb in either direction.

EXAMPLE 4.1 **Conversions between percent, ppm, and ppb**

(a) On a particularly humid day, the water vapor concentration in the air might be 0.85%. How would this concentration of water vapor be expressed in ppm and ppb?

SOLUTION

$$\frac{0.85}{100} = \frac{x}{1,000,000} \quad \text{so} \quad 100x = 850,000 \text{ and } x = 8,500 \text{ or } 8,500 \text{ ppm}$$

and

$$\frac{0.85}{100} = \frac{x}{1,000,000,000} \quad \text{so} \quad 100x = 850,000,000 \text{ and } x = 8,500,000 \text{ ppb}$$

By solving the proportional equations above you can see that 0.85% is the same as 8,500 ppm or 8,5000,000 ppb.

(b) Breathing air that contains 1,000 ppm of carbon monoxide for about four hours will probably cause death in most adults. To what percentage of carbon monoxide does 1,000 ppm correspond?

SOLUTION

The following equation is another way of stating that, if 1,000 out of every 1,000,000 particles in the air are CO, (x) of every 100 particles are CO.

$$\frac{1,000}{1,000,000} = \frac{x}{100} \quad \text{so} \quad 1,000,000 x = 100,000 \text{ and } x = 0.1 \text{ or } 0.1\%$$

In this example, in every 100 particles in the air, an average of 0.1 is CO. Of course, we can't have 0.1 molecules. To make this statistical average more realistic, we might choose to convert 0.1% to parts per thousand to see that this corresponds to 1 molecule per thousand (ppt).

If you are familiar with scientific notation, you will see that the following statements are true. To convert from % to ppm or ppb, simply multiply by 10^4 or 10^7, respectively. To convert from ppb to ppm or %, simply divide by 10^4 or 10^7, respectively. To review the use of scientific notation, see Appendix B.

Parts per million 1 part out of every million parts (1 ppm)

Parts per billion 1 part out of every billion parts (1 ppb)

4.2   AIR: A SOURCE OF PURE GASES

Before air can be separated into pure nitrogen, oxygen, and other gases by the process known as *fractionation of air*, water vapor and carbon dioxide must be removed. This is usually done by precooling the air to separate ice and frozen carbon dioxide. Afterward, the air is compressed to more than 100 times normal atmospheric pressure, cooled to room temperature, and allowed to expand into a chamber. Overcoming the forces that attract molecules to each other requires energy (Section 5.7), so as the gas expands, the molecules lose energy, they slow down, and the gas gets cooler. If the compression and expansion are repeated many times and controlled properly, the expanding air cools to the point of liquefaction. Once air has been liquefied, its pure gases can be separated by taking advantage of their different boiling points.

When liquid air is allowed to warm up, nitrogen (bp $-196°C$) vaporizes first, and the liquid becomes more concentrated in oxygen and argon. Further processing allows separation of high-purity oxygen (bp $-183°C$) and argon (bp $-189°C$). The less-abundant gases (neon, bp $-246°C$; xenon, bp $-108°C$; and krypton, bp $-153°C$) are also separated from liquid air.

Helium (bp $-271°C$) is not commercially recovered from air because it is cheaper to isolate it from natural gas, where it is sometimes present in as much as 7% by volume in the gas fields of West Texas.

> Air, like most gases, heats up when it is compressed and cools down when it expands.

> Because helium atoms are very light, once released they can achieve velocities great enough to escape from Earth's gravity. The decay of radioactive elements resupplies helium, creating a roughly constant amount in the atmosphere.

Oxygen

Most oxygen produced by the fractionation of liquid air is used in steelmaking. Some is also used in rocket propulsion and in other controlled reactions with oxygen.

Liquid oxygen (LOX) can be stored and shipped at its boiling temperature of $-183°C$ under atmospheric pressure. Substances this cold are called *cryogens* (from the Greek *kryos,* meaning "icy cold"). Cryogens represent special hazards, since contact produces instantaneous frostbite and structural materials such as plastics, rubber gaskets, and some metals become brittle and fracture easily at these low temperatures. Because liquid oxygen can react explosively with some substances, contact between it and substances that can ignite and burn in air must be prevented.

> A small but vital use of oxygen is respiratory therapy.

Nitrogen

Because nitrogen gas is so chemically unreactive, it is used as an inert atmosphere for applications such as welding and other high-temperature metallurgical processes. If air is not excluded from these processes, unwanted reactions of hot metals with oxygen would occur.

Liquid nitrogen is used in medicine *(cryosurgery),* for example, in cooling a localized area of skin prior to removal of a wart or other unsightly or pathogenic tissue. Because of its low temperature and inertness, liquid nitrogen is also widely used in frozen food preparation and preservation during transit. Trucks or railroad boxcars with nitrogen atmospheres present health hazards, since they contain little (if any) oxygen to support life. Workers must either enter such areas with breathing apparatus or first allow fresh air to enter.

Nitrogen is an essential element for plants, but they cannot derive it from gaseous nitrogen. It must first be "fixed." **Nitrogen fixation** is the process of changing atmospheric nitrogen into compounds that dissolve in water and can be absorbed through plant roots (see Section 19.3).

Charles D. Winters

Liquid nitrogen, which boils at $-196°C$.

Nitrogen fixation Conversion of atmospheric nitrogen into soluble nitrogen compounds available as plant nutrients

A neon sign outside a boutique in Taos, New Mexico.

Noble Gases

Approximately 250,000 tons of argon, the most abundant of the **noble gases** in the air, are recovered each year in the United States. Most of the argon is used to provide inert atmospheres for metallurgical processes. It is also used as a filler gas in incandescent lightbulbs to prolong the life of the hot filament. Neon is used in many "neon" signs, but argon, krypton, and xenon are also used for this purpose.

4.3 NATURAL VERSUS ANTHROPOGENIC AIR POLLUTION

Nature pollutes the air on a massive scale with ash, mercury vapor, hydrogen chloride, and hydrogen sulfide, which are released during volcanic eruptions. Carbon dioxide and chlorinated organic compounds are released during forest and grassland fires. Many reactive organic compounds are naturally released into the atmosphere from coniferous and deciduous plants. We can do little to control these natural events so our main concern in this chapter is pollution that results from human activity, known as **anthropogenic pollution**.

Human activity, especially in heavily populated (urban) areas, seems to have the most noticeable effects on the quality of the air we breathe. Automobiles, fossil fuel–burning power plants, smelting plants, other metallurgical plants, and petroleum refineries add significant quantities of polluting chemicals to the atmosphere. These atmospheric pollutants, especially in the concentrations found in urban areas, and increasingly in rural areas, can cause people to have burning eyes, to cough, to have breathing difficulties, and even to die. Air pollution is nothing new; Shakespeare wrote about it in the 17th century.

Prior to 1960, there was little concern about air pollution and little effort toward its control in the United States, in spite of some dramatic episodes in which many people suffered as a direct result of polluted air. For example, in October 1948, the town of Donora, Pennsylvania, was overcome by five days of air pollution that caused almost 6000 residents to become ill and 18 to die. In the past, smoke, carbon monoxide, sulfur dioxide, nitrogen oxides, and organic vapors were emitted into the air from industrial facilities with little apparent thought about their harmful nature as long as they were scattered into the atmosphere and away from human smell and sight.

In the 1960s, people in the United States began to realize that air pollution was a growing problem. The result, in 1970, was passage of the first Clean Air Act. This law helped to control air pollution from sources such as industry and automobiles, but it was not very comprehensive. The Clean Air Act was amended in 1977 to add stricter requirements.

In June 1989, President George Bush Sr. signed into law the 1990 Clean Air Act (CAA) amendments, a major overhaul of the earlier Clean Air Act. These amendments affect almost everything that is manufactured in this country. The substances regulated by the 1990 amendments include particulates, ozone, carbon monoxide, oxides of nitrogen and sulfur, **hydrocarbons**, volatile toxic substances, carbon dioxide, and stratospheric-ozone-depleting chemicals. In the next section, we look first at the particles that obscure our vision, aggravate respiratory illnesses, and cause regional and global cooling by scattering sunlight. Subsequent sections will cover the other major pollutants.

(a)

(b)

Air pollution by nature and industry. (a) Volcanoes, like the one pictured in Mauna Loa, Hawaii, emit ash and a variety of inorganic chemicals. (b) Industrial smokestacks, like these at a steel plant in Pennsylvania, also emit pollutants, but are subject to enforcement of the Clean Air Act.

Noble gases The gases of periodic table group 8A; helium, neon, argon, krypton, xenon, and radon; long thought to be completely unreactive. They are colorless and exist as single atoms at room temperature.

Anthropogenic pollution Pollution that is attributable to human activity

Hydrocarbons Compounds that contain only carbon and hydrogen atoms, such as methane (CH_4).

4.4 AIR POLLUTANTS: PARTICLE SIZE MAKES A DIFFERENCE

Air pollutant particles range in size from fly ash particles, which are big enough to see, down to individual molecules, **ions**, or atoms. Many pollutants are attracted into the water droplets of fog. Solids and liquid droplets suspended in the atmosphere are collectively known as **particulates**. The solids may be metal oxides, soil particles, sea salt, fly ash from electric generating plants and incinerators, elemental carbon, or even small metal particles. **Aerosol** particles range upward from a diameter of 1 nanometer (nm) to about 10,000 nm and may contain as many as a trillion atoms, ions, or small molecules. Particles in the 2000-nm range are largely responsible for the deterioration of visibility.

Aerosol particles are small enough to remain suspended in the atmosphere for long periods. Such particles are easily breathable and can cause lung diseases. They may also contain mutagenic or carcinogenic compounds. Because of their relatively large surface area, aerosol particles have great capacities to *adsorb* and concentrate chemicals on their surfaces. Liquid aerosols or particles covered with a thin coating of water may *absorb* air pollutants into the water, thereby concentrating them and providing a medium in which reactions may occur.

Millions of tons of soot, dust, and smoke particles are emitted into the atmosphere of the United States each year. The average suspended particulate concentrations in the United States vary from about 0.00001 g/m^3 of air in rural areas to about six times as much in urban locations. In heavily polluted areas, concentrations of particulates may increase to 0.002 g/m^3.

Particulates in the atmosphere can cool Earth by scattering and partially reflecting light from the Sun. Large volcanic eruptions such as those from Mt. St. Helens in 1980 and Mt. Pinatubo in 1991 had measurable cooling effects on Earth.

Particulates and aerosols are removed naturally from the atmosphere by gravitational settling and by rain and snow. Industrial emissions of particulates can be prevented by treating the emissions with one or more of a variety of physical methods such as filtration, centrifugal separation, and scrubbing. Another method often used is electrostatic precipitation, which is more than 98% effective in removing aerosols and dust particulates even smaller than 1 μm from exhaust gases. The effects of an efficient electrostatic precipitator can be quite dramatic, as Figure 4.2 shows.

---

1 μm = 10^{-6} m, or 1000 nm

1 nm = 10^{-9} m

Major contributors to the amount of atmospheric particulates were volcanic eruptions by Krakatoa, Indonesia, 1883; Mt. Katmai, Alaska, 1912; Hekla, Iceland, 1947; Mt. Spurr, Alaska, 1953; Bezymyannaya, U.S.S.R., 1956; Mt. St. Helens, Washington, 1980; and Mt. Pinatubo, 1991.

4.5 SMOG

The poisonous mixture of smoke, fog, air, and other chemicals was first called *smog* in 1911 by Dr. Harold de Voeux in his report on a London air pollution disaster that caused the deaths of 1150 people. Through the years, smog has been a technological plague in many communities and industrial regions.

What general conditions are necessary to produce smog? Although the chemical ingredients of smogs vary depending on the sources of the pollutants, certain geographical and meteorological conditions exist in nearly every instance of smog.

First, there must be a period of windlessness so that pollutants can collect without being dispersed vertically or horizontally. This sets the conditions for a **thermal inversion**, which is an abnormal temperature arrangement for air masses (Figure 4.3). Normally, warmer air is on the bottom, nearer the

Ion An atom or group of atoms that is positively (a cation) or negatively (an anion) charged

Particulates Solid particles and liquid droplets suspended in the atmosphere

Aerosols Tiny pollutant particles, so small that they remain suspended in the atmosphere for long periods

Thermal inversion An atmospheric condition in which warmer air is on top of cooler air

FIGURE 4.2 Electrostatic precipitation and its effectiveness. (a) An electrostatic precipitator. The central electrode is negatively charged and imparts a negative charge to particles in smoke that passes over it. The particles are attracted to the positively charged walls and fall into the collector. (b) Smokestacks at a steel mill with the electrostatic precipitator turned off. (c) The same smokestacks with the electrostatic precipitator turned on.

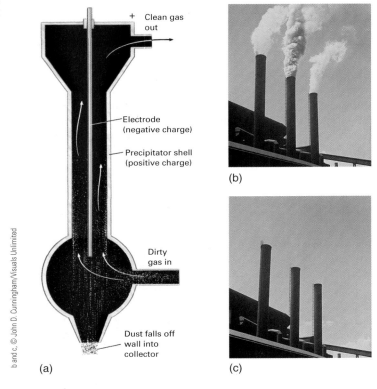

b and c, © John D. Cunningham/Visuals Unlimited

(a)

(b)

(c)

warm earth and this warmer, less dense air rises and transports most of the pollutants to the upper troposphere, where they are dispersed. In a thermal inversion the warmer air is on top, and the cooler, denser air retains its position nearer Earth. The air becomes stagnated. If the land is bowl shaped (surrounded by mountains, cliffs, or the like), a stagnant air mass can remain in place for quite some time. When these atmospheric conditions exist, the pollutants supplied by combustion in automobiles, electric power plants, and industrial plants accumulate to form smog.

Two general kinds of smog have been identified. One is derived largely from the combustion of coal and oil, and contains sulfur dioxide mixed with soot, fly ash, smoke, and partially oxidized organic compounds. Because it was first characterized in and around the city of London, it is sometimes called "London" smog. Smogs of this kind, also called **industrial smogs** because of their association with industrial activity, generally diminish in

Industrial smog Smog generally caused by industrial activity such as coal burning; also known as London smog

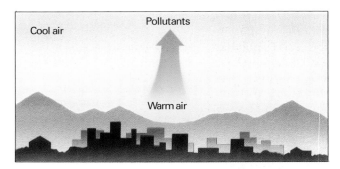

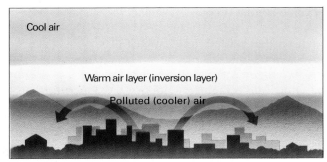

FIGURE 4.3 A thermal inversion. Normally, air that is warmed near the surface rises, carrying pollutants with it. During a thermal inversion, a blanket of warm air becomes stationary over a layer of cooler, denser air. The result is that pollutants are trapped near the surface.

intensity and frequency as less coal is burned and more controls are installed on industrial emissions.

The main ingredient in industrial smog is sulfur dioxide. Laboratory experiments have shown that sulfur dioxide increases aerosol formation, particularly in the presence of mixtures of hydrocarbons and nitrogen oxides. For example, mixtures of 3 parts per million (ppm) hydrocarbons, 1 ppm NO_2, and 0.5 ppm SO_2 at 50% relative humidity form aerosols that have sulfuric acid (H_2SO_4) as a major product. Breathing a sulfuric acid aerosol is very harmful, especially to people suffering from respiratory diseases such as asthma or emphysema. At a concentration of 5 ppm for 1 hour, this kind of aerosol can cause constriction of bronchial tubes. A level of 10 ppm for 1 hour can cause severe breathing distress.

A notorious case of London smog occurred in 1952. On Thursday, December 4 of that year, a thermal inversion, a particularly foggy night, and freezing temperatures along the Thames River valley prompted Londoners to stoke their house fires with large amounts of coal. As was later revealed, low-quality coal had been supplied to the citizenry so that the higher quality coal could be exported. This low-quality coal had a high sulfur content, which contributed to the resultant problems.

In response to the continuing cold, fires continued to burn throughout the weekend, adding their smoke and chemical fumes to an increasingly polluted atmosphere. Accounts of the incident indicated that no sunlight shone through the smog for miles around London as early as Saturday. As the weekend wore on and weather conditions persisted things only got worse. Respiratory problems mounted and people resorted to sleeping in upright positions to ease their breathing. Hospitals were flooded with people gasping for breath and more than 100 people died from respiration-related heart conditions on Monday.

Weather conditions began to improve on Tuesday, December 9, but much damage had been done by that time. An estimated 4,000 deaths were initially attributed to the smog but some recent (2001) estimates, based on a new assessment of the data, put the total death count, including belated deaths directly related to the smog, at 11,000–12,000 people. There have been other deadly London smogs since then but none has ever matched the magnitude of the 1952 disaster. For comparison, consider that the London smog death total from this single incident exceeded the toll of either the Japanese attack on Pearl Harbor in 1941 or the collapse of New York City's World Trade Center twin towers following the terrorist attacks in 2001. And yet, this is an event which most people have never even heard of. Pollution may not be as glamorous or newsworthy as war or terrorism but it can be more deadly.

The second type of smog is typical of Los Angeles and other urban centers where exhaust fumes from internal combustion engines are highly concentrated in the atmosphere. This predominantly urban smog is called **photochemical smog** because light—in this instance sunlight—is important in initiating several chemical reactions that together make the smog harmful. Photochemical smog is practically free of sulfur dioxide but contains substantial amounts of nitrogen oxides, ozone, oxygenated and ozonated hydrocarbons, and organic peroxide compounds, together with unreacted hydrocarbons of varying complexity. The automobile is a direct or indirect source of many of the components of photochemical smog. Consider Figure 4.4. It shows how several components of photochemical smog increase during the rush hour in Los Angeles.

Many of the chemical reactions that create photochemical smog take place in aerosol particles. These reactions produce **secondary pollutants**—pollutants that are not directly released from some source but are formed by reactions with other components in the air.

Organic peroxides contain the R—O—O—R′ structure and are produced by ozone reacting with organic molecules. Hydrogen peroxide is H—O—O—H. R and R′ represent hydrocarbon groups.

Photochemical smog Smog formed by the action of sunlight on photoreactive pollutants in the air

Secondary pollutants Pollutants formed by chemical reactions in the atmosphere

FIGURE 4.4 The average concentrations of pollutants NO, NO₂, and O₃ on a smoggy day in Los Angeles. The NO concentration builds up during the morning rush hour. Later in the day, the concentrations of NO₂ and O₃ build up.

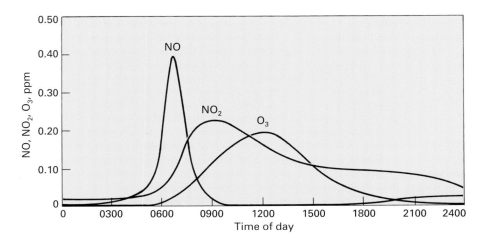

Both NO and oxygen atoms are free radicals. They are usually very reactive.

Photochemical smog over San Diego.

One process by which **primary pollutants** form the secondary pollutants of photochemical smog begins with the absorption of light energy by a molecule of nitrogen dioxide. Nitrogen dioxide reacts with light with a wavelength between 280 nm and 430 nm. This **photodissociation** reaction (*photo*, light; *dissociation*, breaking apart) produces nitric oxide and free oxygen atoms (O, oxygen **free radicals**) that can react with a molecule of oxygen to produce a molecule of ozone (O_3), which is an important secondary pollutant.

$$NO_2(g) \xrightarrow{\text{light}} NO(g) + O(g)$$
$$O_2(g) + O(g) \longrightarrow O_3(g)$$

Atomic oxygen can also react with hydrocarbons to form other chemicals that are toxic and also impart an odor to the air. On a sunny day only about 0.2 ppm of nitrogen oxides and 1 ppm of reactive hydrocarbons are sufficient to initiate these photochemical smog reactions. The hydrocarbons involved in these reactions come mostly from unburned petroleum products such as gasoline, and the nitrogen oxides come from the exhausts of internal combustion engines.

In the following sections we shall look at the major ingredients of photochemical smog—the oxides of nitrogen and hydrocarbons, and the secondary pollutant *ozone*.

4.6 NITROGEN OXIDES

There are eight known oxides of nitrogen, three of which are recognized as important components of the atmosphere: dinitrogen oxide (N_2O), nitric oxide (NO), and nitrogen dioxide (NO_2). These oxides of nitrogen are collectively known as "NO_x." About 97% of the nitrogen oxides in the atmosphere are naturally produced; only 3% result from human activity. Certain bacteria can produce N_2O, so this oxide of nitrogen is also commonly found in the atmosphere in trace amounts.

One of the key culprits in the formation of photochemical smog is NO (nitric oxide), a colorless but reactive gas. Its nominal concentration in our atmosphere is quite low, and there are natural sources of nitric oxide about which we can do little. It is the anthropogenic nitric oxide about which we are most concerned, as it is a pollutant over which we may have some control. Where does this nitric oxide come from? You are probably aware that an

Primary pollutants Pollutants emitted directly into the atmosphere

Photodissociation Decomposition of a reactant, caused by the energy of light

Free radical A species that has one or more unpaired electrons

internal combustion engine takes in huge amounts of air as it runs. This air comprises about 78% nitrogen gas, which is unreactive under normal circumstances. About 21% of the air sucked into the engine is oxygen gas. Nitrogen and oxygen do not react with each other at normal air temperatures, but the combustion chamber of an engine may reach temperatures as high as 2,500°C (the average is probably around 400–600°C). Under these conditions the reaction that forms nitric oxide, which requires heat, takes place quite readily. The same reaction is triggered in our atmosphere by lightning strikes.

$$N_2(g) + O_2(g) \xrightarrow{\text{energy}} 2\,NO(g)$$

Nitric oxide itself is a toxic pollutant and the U.S. Environmental Protection Agency (the EPA) includes it among the primary toxic pollutants it attempts to regulate. If this nitric oxide is able to escape the exhaust system of the engine, it reacts quite rapidly with atmospheric oxygen to produce nitrogen dioxide, a reddish-brown, highly toxic gas.

$$2\,NO(g) + O_2(g) \xrightarrow{\text{energy}} \underset{\text{nitrogen dioxide}}{2\,NO_2(g)}$$

Almost anyone who has ever observed the skyline of a major city such as Houston, Los Angeles, or Mexico City has seen a reddish-brown haze hanging over the city at times. This haze is composed largely of nitrogen dioxide, an ocular and respiratory irritant. The EPA has set the allowable yearly average exposure to all nitrogen oxides at 53 ppb. A city or state whose level of nitrogen oxides exceeds this number is subject to fine and government intervention. In this chapter, you will learn what actions may be taken to lower these emission levels.

Normally the atmospheric concentration of NO_2 is a few parts per billion (ppb) or less; most of the nitrogen oxides formed during lightning storms are washed out by rain. This is one of the ways nitrogen is made available to plants. Looking at all the sources of oxides of nitrogen (Table 4.2), it is apparent that combustion processes are their primary sources. In the United States, most oxides of nitrogen from sources other than nature are produced from fossil fuel combustion in vehicles, industry, and power plants.

A primary role of nitrogen dioxide as a pollutant is in the formation of the secondary pollutant ozone (Section 4.7) and photochemical smog. Nitrogen dioxide can also react with water to form nitric acid and nitrous acid.

Charles D. Winters

Nitrogen dioxide.

Another name for nitric oxide is *nitrogen monoxide*.

TABLE 4.2	Emissions of NO_x	
SOURCE	**EMISSIONS (MILLIONS OF TONS)** *United States*
Fossil fuel combustion	66
Biomass burning	1.1
Lightning	3.3
Microbial activity in soil	3.3
Input from the stratosphere	0.3
Total (uncertainty in estimates)	74 (±1)

Source: Stanford Research Institute.

This reaction takes place readily in aqueous aerosols, producing acids that help to stabilize the droplet.

$$2\,NO_2(g) + H_2O(\ell) \longrightarrow \underset{\substack{\text{Nitric}\\\text{acid}}}{HNO_3(aq)} + \underset{\substack{\text{Nitrous}\\\text{acid}}}{HNO_2(aq)}$$

Of course, breathing air containing these aerosol droplets is harmful because of the corrosive nature of the acids. The acids in turn can react with ammonia or metallic particles in the atmosphere to produce nitrate or nitrite salts. For example,

$$\underset{\text{Ammonia}}{NH_3(g)} + HNO_3\,(aq) \longrightarrow \underset{\substack{\text{Ammonium nitrate}\\\text{(a salt)}}}{NH_4NO_3(aq)}$$

Both the acids and the salts stabilize the aerosol particles, which eventually settle from the air or dissolve in larger raindrops. Nitrogen dioxide, besides causing the formation of ozone, is a primary cause of haze in urban or industrial atmospheres because of its participation in the process of aerosol formation.

CONCEPT CHECK) **4A**

1. What were the major components of Earth's atmosphere 4.5 billion years ago?
2. What gas, critical to life as we know it, was absent from Earth's early atmosphere?
3. How can the fact that Earth's atmosphere is approximately 21% oxygen at all altitudes be reconciled with the fact that it becomes increasingly difficult to breathe at higher altitudes?
4. What regions of Earth's atmosphere comprise the troposphere and the stratosphere?
5. In what region of Earth's atmosphere does the ozone layer lie?
6. An average breath of inhaled air is about 0.03% carbon dioxide. Convert this to ppm and ppb.
7. An average breath of exhaled air is about 4% carbon dioxide. Convert this to ppm and ppb.
8. What is anthropogenic pollution?
9. Distinguish between primary and secondary pollutants in our atmosphere.
10. Name a chemical that is considered both an air pollutant and a beneficial chemical.
11. Because of their large surface area, aerosol particles can (absorb/adsorb) chemicals onto their surfaces.
12. A liquid aerosol particle will probably (absorb/adsorb) certain chemicals.
13. A thermal inversion occurs when (warm/cool) air is above (warm/cool) air.
14. Industrial-type smog is often associated with (coal burning/sunlight).
15. In all combustion processes in air, some nitrogen _____ are formed.
16. Explain the time course of pollution formation shown in Figure 4.4 and write the chemical equations associated with each process.

4.7 OZONE AS A POLLUTANT

Ozone consists of three oxygen atoms bound together in a molecule with the formula O_3. It is an **allotrope** of molecular oxygen. It has a pungent odor that we often smell near sparking electrical appliances or after a thunderstorm when rainfall washes lightning-produced ozone out of the troposphere.

As you've seen, any nitric oxide that escapes into our atmosphere can react with oxygen to form nitrogen dioxide. Nitrogen dioxide can, under the intensity of a bright summer sun, be photochemically degraded into nitric oxide and very reactive oxygen atoms (O). Reaction between these oxygen atoms and oxygen molecules leads to the formation of ozone.

Because it is generated by chemical reactions in the atmosphere, ozone is one of the most difficult pollutants to control. One can observe the daily course of ozone formation in cities throughout the United States by accessing the EPA's ozone monitoring web site, http://www.epa.gov/airnow/, and clicking on ozone now. Pick a city such as Dallas-Fort Worth or Los Angeles and view the variation in ozone concentrations over the course of several days. Try to find the day on which the ozone pollution was worst. (Check August 2, 2005, for Dallas-Fort Worth and July 17 or August 28, 2005, for Los Angeles for some particularly interesting data.) You may even find that your local newspaper includes color-coded ozone alerts on its weather page.

Table 4.3 shows the relationship between the local ozone concentration in the atmosphere, air quality indices, and air quality descriptors in common use. A period of unhealthy air may be quite short or quite lengthy depending on weather conditions, intensity of sunlight, prevailing winds, topography, and the level of nitrogen dioxides emitted into the air in a given area by automobiles, buses, trucks, trains, ships, industry, and even lawnmowers.

Although there has been an overall improvement in air quality in the United States in recent years, there are still problems with ozone in many urban areas and even in some remote areas such as Big Bend National Park deep in southwest Texas. Federal, state, and local efforts to control emissions of nitrogen oxides are constantly discussed and debated in the local newspapers and by political action groups in many areas of the United States. The popularity of the automobile in American culture has had a dramatic impact on air quality. In an approach not yet tried in the United States, there have been efforts in some major cities in other parts of the world to limit the number of automobiles allowed into those cities on a given day or during a given time period through the use of permits or taxes. There have been noticeable

Allotropes Different molecular forms of the same element; in this case the two molecular forms are diatomic O_2 and triatomic O_3

TABLE 4.3 Air Quality Index and Air Quality Descriptors for Ozone

Ozone Concentration, ppm (8-hr average unless noted)	Air Quality Index Values	Air Quality Descriptor
0.00 to 0.064	0 to 50	Good
0.065 to 0.084	51 to 100	Moderate
0.085 to 0.104	101 to 150	Unhealthy for sensitive groups
0.105 to 0.124	151 to 200	Unhealthy
0.125 (8 hr) to 0.404 (1 hr)	201 to 300	Very unhealthy

improvements in the air quality of these cities during the times these programs have been in effect but, as you can imagine, the programs have met some opposition.

In the United States, the EPA's efforts to control ozone concentrations in our ground-level air have had some success. In 2004 the EPA reported that, thanks to their emission control programs, ozone levels had continuously decreased during the period 1980–2003. The result was the lowest national ozone level since 1980, making 2003 one of the cleanest years on record. However, concerns about the prospect that the downward trend in ozone levels may be slowing, caused regulators to propose additional national rules designed to decrease vehicle exhaust and industrial emissions.

(4.8) SULFUR DIOXIDE: A MAJOR PRIMARY POLLUTANT

Sulfur dioxide is produced when sulfur or sulfur-containing compounds are burned in air.

$$S(s) + O_2(g) \longrightarrow SO_2(g)$$

Although volcanoes put large amounts of SO_2 into the atmosphere annually, human activities probably account for up to 70% of all emissions on a global basis. Once formed, SO_2 generally becomes distributed in aerosol droplets, which are numerous enough to contribute to significantly reduced visibility and can affect both global and regional climate by causing the scattering of sunlight that would otherwise warm Earth. Emissions of SO_2 cause the mean temperature in the United States to be about 1°C cooler than it would be otherwise.

Most of the coal burned in the United States contains sulfur in the form of the mineral pyrite (FeS_2). The percentage (by weight) of sulfur in this coal ranges from 1 to 4%. The pyrite is oxidized as the coal burns, forming SO_2.

$$4 FeS_2(s) + 11 O_2(g) \longrightarrow 2 Fe_2O_3(s) + 8 SO_2(g)$$

Large amounts of coal are burned in this country to generate electricity. A 1000-megawatt (MW) coal-fired generating plant can burn about 700 tons of coal an hour. If the coal contains 4% sulfur, that equals 56 tons of SO_2 an hour, or 490,560 tons of SO_2 every year. About 800 million tons of coal are burned each year to produce electricity.

Oil-burning electric generating plants can also produce comparable amounts of SO_2 because some fuel oils can contain up to 4% sulfur. The sulfur in the oil is in the form of compounds in which sulfur atoms are bound to carbon and hydrogen atoms. Gasoline contains relatively low concentrations of sulfur-containing compounds. Even so, the EPA has mandated that the maximum concentration of sulfur-containing compounds be reduced from 120 ppm in 2004 to 90 ppm in 2005 and to 30 ppm by 2006. Further mandated reductions may be expected.

States that rely mainly on coal for their electricity production and industrial furnaces have the highest SO_2 emissions in the United States. Operators of all coal-fired burners are under EPA orders to eliminate most of the SO_2 before it reaches the exhaust stack.

The removal of sulfur from high-sulfur coal is costly and incomplete. One method is to pulverize the coal to the consistency of talcum powder and remove the pyrite (FeS_2) by magnetic separation. Reducing the sulfur content of fuel oil is also costly. It involves the formation of hydrogen sulfide (H_2S) by bubbling hydrogen through the oil in the presence of metal catalysts.

At present, most sulfur-containing coal is burned without prior treatment, and SO_2 is removed from the exhaust gases. In one method, lime (calcium oxide) reacts with SO_2 to form calcium sulfite, a solid particulate, which can be removed from an exhaust stack by an electrostatic precipitator. Lime is prepared by heating limestone, which is mainly calcium carbonate.

$$CaCO_3(s) \xrightarrow{\text{Heat}} \underset{\text{Lime}}{CaO(s)} + CO_2(g)$$
$$\underset{\text{Limestone}}{} $$

$$CaO(s) + SO_2(g) \longrightarrow \underset{\text{Calcium sulfite}}{CaSO_3(s)}$$

In another method, exhaust gases containing SO_2 are passed through molten sodium carbonate and solid sodium sulfite is formed (Figure 4.5).

Notice how both of these methods of SO_2 removal emit additional CO_2, a major contributor to global warming (Section 6.3), into the atmosphere. Newer technology to address this problem continues to be explored.

A less desirable method of lowering the effects of SO_2 emissions, but one still being used, is sending the smoke up very tall exhaust stacks. Tall stacks emit SO_2 at a high elevation and away from the immediate vicinity, which allows the SO_2 to be diluted before it forms aerosol particles. The fact remains, however, that the longer it stays in the air, the greater chance SO_2 has to become sulfuric acid. A 10-year study in Great Britain showed that although SO_2 emissions from power plants increased by 35%, the construction of tall stacks decreased the ground-level concentrations of SO_2 by as much as 30%. The question is, who got the SO_2? In this case, Britain's solution was others' pollution. In the United States, the EPA may have added to a pollution problem unwittingly by implementing rules in 1970 that caused plants to increase the height of smokestacks and caused pollutants to be carried longer distances by winds.

Most of the SO_2 that does get into the atmosphere reacts with oxygen to form sulfur trioxide (SO_3). This compound has a strong affinity for water and dissolves in aqueous aerosol particles, forming sulfuric acid, which in turn contributes to acid rain (Section 11.2).

$$SO_3(g) + H_2O(\ell) \longrightarrow H_2SO_4(aq)$$

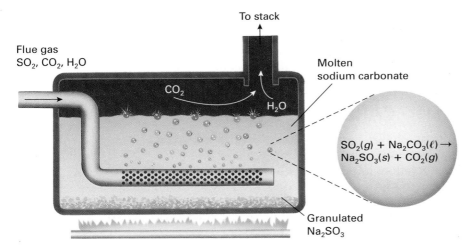

FIGURE 4.5 Removal of SO_2 from flue gas by reaction with molten sodium carbonate.

TABLE 4.4	Emission Rates for Hydrocarbons (HC), Carbon Monoxide (CO), and NO$_x$	
1960 (Precontrol—no catalytic mufflers installed)	*1993 (Catalytic mufflers required on all automobiles)*	*2004 Standards [a,b]*
HC 10.6 g/mi	HC 0.41 g/mi	NMHC[c] 0.075 g/mi
CO 84.0 g/mi	CO 3.4 g/mi	CO 3.4 g/mi
NO$_x$ 4.1 g/mi	NO$_x$ 1.0 g/mi	NO$_x$ 0.05 g/mi

[a] Source: EPA Office of Transportation and Air Quality.
[b] Today's car and truck fleets offered for sale must meet minimum Corporate Average Fuel Economy (CAFE) standards. Cars must achieve an average fuel economy of 27.5 miles per gallon (mpg); light trucks and SUVs must average 20.7 mpg today and 22.7 mpg by 2007. A new set of federal emission standards known as Tier 2 will hold all passenger cars and trucks to the same set of standards beginning in 2009. Tier 2 criteria currently encompass 11 emission categories (known as "bins") ranging from Bin 1 (cleanest) to Bin 11 (dirtiest). Automakers can sell vehicles in different bins as long as they meet the fleetwide average of Bin 5. Under Tier 2, Bin 5 standards, every model certified as Bin 5 produces less than 0.16 of smog-forming pollutants per mile at 5 years or 50,000 miles. By comparison, Tier 2, Bin 10 standards allow no more than 0.76 g of smog-forming pollutants per mile. Some states require higher emission standards. Source: "Update from the Union of Concerned Scientists." *Earthwise*, vol. 6 No. 3 (Summer 2004).
[c] NMHC stands for non-methane hydrocarbons.

In 1988, William Chameides, of Georgia Tech in Atlanta, published a report in *Science* magazine in which he stated that in some cities trees may account for more hydrocarbons in the atmosphere than those produced from human activities. The EPA has since found this to be true.

Charles D. Winters

A pine tree, which can release hydrocarbons into the air.

4.9 HYDROCARBONS AND AIR POLLUTION

Hydrocarbons enter the atmosphere from both natural sources and human activities. Certain natural hydrocarbons are produced in large quantities by both coniferous and deciduous trees. Methane gas (CH$_4$) is produced by such diverse sources as rice growing, ruminant animals such as cows, termites, ants, and decay-causing bacteria acting on dead plants and animals. Human activities such as the use of industrial solvents, petroleum refining and distribution, and the release of unburned gasoline and diesel fuel components account for a large amount of hydrocarbons in the atmosphere.

In addition to simpler hydrocarbons, some larger hydrocarbon molecules are released into the atmosphere, primarily from motor vehicle exhaust. The greatest danger of some of these pollutants and the organic derivatives formed from them is their toxicity.

Although it is practically impossible to control hydrocarbon emissions from living plants and other natural sources, hydrocarbon emissions from automobiles can be controlled. Two means of control are in use at present. First, the spouts and hoses on gasoline pumps have been redesigned to prevent gasoline from entering the air. Second, catalytic converters that reduce emissions of hydrocarbons, CO, and NO$_x$ are now part of every automobile's exhaust system.

The effectiveness of these catalytic converters, which have been on automobiles sold in the United States since the mid-1970s, can be seen by comparing the emissions of hydrocarbons, CO, and NO$_x$ in grams per vehicle-mile in 1960, before there were controls, to the same values now (Table 4.4).

4.10 CARBON MONOXIDE

At least ten times more carbon monoxide enters the atmosphere from natural sources than from all industrial and automotive sources combined. Of the 3.8 billion tons of carbon monoxide emitted every year, about 3 billion tons are emitted by the oxidation of decaying organic matter in the topsoil. In spite of this fact, carbon monoxide is considered an air pollutant, primarily because so much of it is produced by human activities in the urban environment.

Like ozone, carbon monoxide is one of the most difficult pollutants to control. Cities such as Los Angeles, Houston, and other highly populated urban centers with high densities of automobiles tend to be repeatedly cited by the EPA for not attaining the required ambient air quality for carbon monoxide.

Carbon monoxide is always produced when carbon or carbon-containing compounds are burned using an insufficient quantity of oxygen.

$$2\,C(s) + O_2(g) \longrightarrow 2\,CO(g)$$

Gasoline engines are notorious sources of CO. This is because the rapid combustion inside the combustion chamber does not burn all the carbon to CO_2 before the exhaust gases are swept out. Modern catalytic converters convert much of this carbon monoxide to carbon dioxide, but the amounts that are not converted make being near a heavily traveled street dangerous because of the carbon monoxide concentrations. At peak traffic times, concentrations as high as 50 ppm can occur. In the countryside, carbon monoxide levels are closer to the global average of 0.1 ppm. The only effective means of controlling carbon monoxide concentrations in urban air is to control the major emitters—automobiles. Of course, transportation that does not depend on burning hydrocarbon fuels does not emit *any* CO.

Carbon monoxide is a colorless, odorless gas at room temperature. This means carbon monoxide has no warning properties. Because it is a gas, it mixes with air, is inhaled, and comes in contact with the blood while in the lungs. The interference of carbon monoxide with oxygen transport in the blood is one of the best-understood kinds of metabolic poisoning. Carbon monoxide, like oxygen, combines with the hemoglobin in red blood cells.

In some occupations, it is difficult to avoid long-term exposure to carbon monoxide.

$$O_2(g) + \text{hemoglobin}(aq) \rightleftharpoons \text{oxyhemoglobin}(aq)$$
$$CO(g) + \text{hemoglobin}(aq) \rightleftharpoons \text{carboxyhemoglobin}(aq)$$

Both of these reactions are *reversible,* as indicated by the double arrow ($\rightleftharpoons$). That is, oxyhemoglobin can give up its oxygen (that's what it does when it transports oxygen to a cell in the body), and carboxyhemoglobin can give up its carbon monoxide molecule, although not as easily. Laboratory tests show that carboxyhemoglobin is 140 times more stable than oxyhemoglobin.

Both O_2 and CO bind to the iron atom in the hemoglobin molecule. The greater stability of carboxyhemoglobin occurs because the Fe—CO attraction is stronger than the Fe—O_2 attraction. Since hemoglobin is so effectively tied up by carbon monoxide, those hemoglobin molecules that contain a CO molecule cannot perform their vital function of transporting oxygen. Breathing air with a concentration of 30 ppm of CO for 8 hours is sufficient to cause headache and nausea for most people. Breathing air that is 0.1% (1000 ppm) carbon monoxide for 4 hours converts approximately 60% of the hemoglobin of an average adult to carboxyhemoglobin, and death is likely to result unless the carboxyhemoglobin molecules can be freed of the attached CO molecules.

Carbon monoxide exposures are quite common. In fact, low exposures are almost impossible to avoid. This means that some of the hemoglobin molecules in your blood are always bound to carbon monoxide. Any organic material that undergoes **incomplete combustion** will liberate carbon monoxide. Sources include auto exhausts, smoldering leaves, lighted cigars or cigarettes, and charcoal burners. In the United States alone, **combustion** sources of all types dump about 200 million tons of carbon monoxide per year into the atmosphere, where it mixes with the other molecules in the air.

Reversible reactions may proceed in both directions.

Carbon monoxide poisoning can occur when kerosene heaters or charcoal burners are used indoors without proper ventilation.

For incomplete combustion of carbon, the ratio is 2 C to 1 O_2: $2\,C(s) + O_2(g) \longrightarrow 2\,CO(g)$. For complete combustion of carbon, the ratio is 1 C to 1 O_2: $C(s) + O_2(g) \longrightarrow CO_2(g)$.

High O_2 concentrations favor oxyhemoglobin, while high CO concentrations favor carboxyhemoglobin.

Incomplete combustion Combustion in which there is insufficient oxygen for complete combustion

Combustion The rapid combination of a fuel with oxygen; the process of burning

Since the reactions of both carbon monoxide and oxygen with hemoglobin are reversible, the concentrations of the two gases, in addition to the relative strengths of their binding to hemoglobin, affect how much hemoglobin will be combined with either molecule.

In air that contains 0.1% or less CO, oxygen molecules outnumber the CO molecules by at least 200 to 1. Under these conditions, there is little danger from carbon monoxide poisoning. However, breathing air that contains higher concentrations of CO allows CO to effectively compete with the oxygen in binding to hemoglobin and this leads to problems. It's as if a bunch of hungry college students all enter a cafeteria and are randomly given trays of food containing only salad (O_2) or pizza (CO) as they pass through the line. If most of the trays contain salad, most of the students will receive what their body really needs. However, the students who receive a pizza grab it and won't let go. If more pizza is served, a larger percentage of students will receive an unhealthy meal. A school nutritionist, seeing this, grabs the trays of pizza and replaces them with salads. Hemoglobin responds in much the same way. It has a strong affinity for CO but is normally fed a diet high in oxygen. Once exposed to a diet high in CO, it is necessary to overwhelm the system with excess oxygen to restore a healthy situation. A victim of carbon monoxide poisoning can be saved if he or she is quickly exposed to fresh air, or, still better, pure oxygen.

(4.11)　A LOOK AHEAD

The quality of the air we breathe can be dramatically affected by the presence of the air pollutants we've discussed in this chapter. Both legislative and technological efforts to reduce the concentrations of these pollutants have shown positive results but there is clearly still much to be done. As we progress through the text we will frequently revisit some of the chemistry involved in these efforts. For instance, in Chapter 6 we will focus on the combustion of fossil fuels and the production of, among other things, carbon dioxide. We don't normally think of carbon dioxide as a pollutant, but it is having a major impact on our environment in the form of **global warming**. Although CO_2's concentration in our atmosphere is low, as is its toxicity, its ability to trap heat near the surface of Earth may be of even more concern than the pollutants discussed in this chapter.

In Chapter 12 we will discuss, in more detail, some of the technology associated with reducing air pollution from internal combustion engines.

You will also see that some pollutants (sulfur oxides and nitrogen oxides) have a major impact on our environment as primary contributors to acid rainfall. You will need some knowledge of water chemistry and of the pH scale to fully appreciate this problem, so we will defer that discussion until Chapter 11.

Lastly, some chemicals known as chlorofluorocarbons (CFCs) are pollutants of major concern in an entirely different region of our atmosphere. These chemicals, whose effect at tropospheric levels is minimal, play havoc with the protective ozone layer of our stratosphere.

Air pollution can take many forms, come from many sources, have its impact in different regions of our atmosphere, and require different forms of personal, legislative, and technological action to control. You will decide what contribution, if any, you as an individual can make to these efforts. Alternatively, you may decide that no action is required. Only by being informed will you be able to make these decisions.

Global warming Warming of Earth above a level that is desirable from a human viewpoint as a result of increased concentrations of one or more greenhouse gases.

CONCEPT CHECK) **4B**

1. What chemical species reacts with molecular oxygen to form ozone?
2. What is the main source of anthropogenic sulfur dioxide in Earth's atmosphere?
3. Write a balanced chemical equation for the reaction of sulfur dioxide with molecular oxygen to produce sulfur trioxide. What compound is formed when sulfur trioxide reacts with water?
4. What two primary pollutants are formed when coal or fuel oil is burned?
5. Which chemical would be most likely to react with sulfur dioxide to remove it from combustion gases—sodium chloride (NaCl), lime (CaO), or nitric oxide (NO)?
6. List three natural processes that release hydrocarbons into our atmosphere.
7. Using the data in Table 4.4, determine how many grams of hydrocarbons would be released in an automobile trip from Dallas, Texas, to Walt Disney World in Orlando, Florida, a distance of about 1,000 miles, by a car that exactly meets the 2004 standards for emissions.
9. With what biological molecule does carbon monoxide combine to inhibit the distribution of oxygen in the body?
10. You discover a still-breathing victim of carbon monoxide poisoning lying in a room with a malfunctioning gas heating system. You've just called 911. What should you do next?

 Assess your understanding of this chapter's topics with an online chapter quiz at **www.brookscole.com/chemistry/joesten4**

■ KEY TERMS

atmospheric pressure	anthropogenic pollution	secondary pollutants
troposphere	hydrocarbons	primary pollutants
stratosphere	ion	photodissociation
ozone layer	particulates	free radical
parts per million (ppm)	aerosols	allotropes
parts per billion (ppb)	thermal inversion	incomplete combustion
nitrogen fixation	industrial smog	combustion
noble gases	photochemical smog	global warming

■ THE LANGUAGE OF CHEMISTRY

1. Gas with the highest percentage by volume in the atmosphere
2. Consisting mostly of very small water droplets
3. Abnormal temperature arrangement for air masses
4. A mixture of smoke, fog, air, and various other chemicals
5. A pollutant caused by reactions of other chemicals
6. Second most abundant gas in the atmosphere
7. A pollutant that is directly discharged into the atmosphere
8. A chemical species with an unpaired valence electron
9. Main ingredient in industrial smog
10. The product of a reaction between an oxygen atom and an oxygen molecule
11. An oxide of nitrogen produced by certain bacteria
12. The oxide of nitrogen produced by lightning, forest fires, and internal combustion engines
13. "Bad" ozone
14. An air pollutant that bonds to hemoglobin

a. Oxygen
b. NO_2 and SO_2
c. Nitrogen
d. Aerosol
e. CO_2
f. Secondary pollutant
g. SO_2
h. Smog
i. NO
j. Fossil fuel electricity generators
k. Primary pollutant
l. Thermal inversion
m. CO
n. Ozone in the air we breathe
o. N_2O
p. Free radical
q. Ozone molecule

■ APPLYING YOUR KNOWLEDGE

1. Define the following terms:
   (a) Polluted air (b) Aerosol
   (c) Photochemical smog (d) Secondary pollutant
2. Define the following terms:
   (a) Photodissociation (b) Particulate
   (c) Thermal inversion (d) Free radical
3. Name three major air pollutants and give a source for each.
4. Which of these would not be considered a primary pollutant in our atmosphere—nitric oxide (NO), carbon monoxide (CO), ozone (O_3), hydrocarbons, or sulfur dioxide (SO_2)?
5. What federal legislation has the abbreviation CAA? Describe how this legislation has changed over the years.
6. Describe what a thermal inversion is and explain how it can aggravate the problems caused by smog.
7. How is an industrial-type smog different from a photochemical smog?
8. Explain how ozone is formed in the troposphere.
9. Place the four reactions below in a sequence that explains the production of ozone in a typical urban environment
   (a) $NO_2 + \text{sunlight} \longrightarrow NO + O$
   (b) $N_2 + O_2 \longrightarrow 2\,NO$
   (c) $2\,NO + O_2 \longrightarrow 2\,NO_2$
   (d) $O + O_2 \longrightarrow O_3$

10. Which reaction in Question #9 involves an atom combining with a diatomic molecule to produce a triatomic molecule?
11. What are the main sources of nitrogen oxides in the atmosphere?
12. What is an aerosol and how does it play a role in air pollution?
13. Name two sources of hydrocarbons in the atmosphere. Which one is more readily controlled?
14. Write the equation for the reaction that occurs when lightning causes nitrogen to react with oxygen.
15. Nitrogen dioxide plays a role in the formation of what secondary pollutant?
16. What are the names of the two regions of the atmosphere discussed in this chapter? Which one is closer to the surface of Earth?
17. What happens when a photon of light strikes a molecule of NO_2? Write the reaction.
18. Describe how ozone can be harmful when it is present in the air we breathe.
19. How are the harmful effects of SO_2-containing aerosols, NO_2, and ozone similar?
20. Describe two sources of carbon monoxide.
21. Describe how pure oxygen can be obtained from air.
22. How is the control of ozone in the lower atmosphere connected to the control of NO_x emissions?

23. What health risks are posed by breathing air contaminated with particulates?

24. What is the product of the combustion of sulfur or sulfur-containing compounds in air? Write the reaction.

25. Pick a source of SO_2 emissions and describe two ways SO_2 emissions from that source can be controlled.

26. Write the chemical equation for the reaction between carbon and oxygen when an insufficient amount of oxygen is present. What is the name of the product of this reaction?

27. A human forced to breathe air that is 7% carbon dioxide would probably become unconscious, suffer brain damage, and die within a few minutes. However, we are worried that Earth's atmosphere may approach 400 ppm of carbon dioxide within the next few years. Which of the following statements is correct?
    (a) 400 ppm of carbon dioxide is greater than 7% carbon dioxide.
    (b) 7% is equal to 7000 ppm
    (c) 400 ppm is equal to 0.04%
    (d) 400 ppm of carbon dioxide in our air means the percentage of oxygen will drop by 4%, making it harder to breathe.
    (e) 400 ppm is equal to 40,000 ppb

28. A December 2002 ice storm in the southeastern United States knocked out electricity for millions of people. About 150 people were hospitalized and treated in hyperbaric chambers where they could breathe much higher than normal concentrations of oxygen; several people died. Use your chemical literacy to decide which of the following scenarios probably led to this.
    (a) People purchased large blocks of dry ice (solid carbon dioxide) to keep the items in their refrigerators from spoiling and the passage of the dry ice from the solid state to the gaseous state caused carbon dioxide poisoning. The increased amounts of carbon dioxide also made it more difficult to keep fires burning so many people experienced hypothermia.
    (b) Massive quantities of broken tree limbs released huge amounts of carbon dioxide into the atmosphere, leading to severe breathing difficulties for those with respiratory problems.
    (c) Downed electrical wires sparked and this led to the production of significant amounts of ozone from atmospheric oxygen in nearby areas.
    (d) Trying to keep warm without electricity, people resorted to poorly ventilated indoor fires that produced significant amounts of carbon monoxide due to incomplete combustion. This led to many cases of CO poisoning.

W CHEMISTRY ON THE WEB

For up-to-date URLs, visit the text website at **www.brookscole.com/chemistry/joesten4**

- Evolution of Earth's Atmosphere
- Regions of the Atmosphere
- Noble Gases
- The Chemistry of Atmospheric Pollutants
- Air Pollution Simulation
- Photochemical Smog
- Ground Level Ozone
- Real Time Animations and Archival Animations of Ozone Pollution Throughout the United States
- All Types of Air Pollution
- London Smog Disaster and Donora Smog Disaster
- Nitrogen Oxides
- Sulfur Oxides
- Carbon Monoxide
- Plain English Guide to the Clean Air Act
- Determine How Much Your Car Pollutes by Model, Make, and Year

CHEMICAL BONDING AND STATES OF MATTER

Hematite, Fe₂O₃

Calcite, CaCO₃

osum, CaSO₄ · 2 H₂O

Fluorite, CaF₂

Orpiment, As₂S₃

Charles D. Winters

The majority of chemical compounds have either ionic bonds or covalent bonds. Ionic compounds, like those in this photo, are usually crystalline solids.

Atoms are the fundamental building blocks of all matter. Everything that exists is made up of atoms held together in innumerable combinations that range from the simplest molecules, composed of only two atoms, to the most complex molecules, composed of thousands of atoms. Atoms of the 90 naturally occurring elements, and occasionally the 25 or so non-natural elements, combine to make everything that we can see, touch, and smell and lots of things that we can't so easily observe.

What makes atoms stick together? Valence electrons form the glue, but how? Atoms form a few different types of bonds when they combine with other atoms. In this chapter we describe two of them—*ionic bonds* and *covalent bonds*. Transfer of valence electrons from an atom of a metal to an atom of a nonmetal produces an ionic bond. Sharing of electrons between atoms of nonmetals produces covalent bonds. These simple ideas about valence electrons are the basis for understanding the bonding in two major classes of chemical compounds, *ionic compounds* and *molecular compounds*. In addition, once you understand how valence electrons function, you can better understand how the states of matter—gas, liquid, and solid—are formed.

In this chapter we will address the following questions:

- How can the periodic table be used to predict the tendency of an atom to form ionic bonds or covalent bonds?
- How are ionic bonds formed?
- How can the periodic table be used to predict the formulas of ionic compounds?
- How are covalent bonds formed?
- Why are the electrons often unequally shared by the atoms in a covalent bond?
- How are the states of matter related to the attractive forces between molecules?
- What is hydrogen bonding and why is it so important?
- How are the properties of solids, liquids, and gases related to chemical bonding?

The concept of valence electrons developed by G. N. Lewis is useful for understanding how atoms of different elements interact and why elements in the same group have similar properties (Section 3.7). Lewis assumed that each noble gas atom had a completely filled outermost shell, which he regarded as a stable configuration because of the lack of reactivity of noble gases. Since all noble gases (except He) have eight valence electrons, the observation came to be known as the **octet rule**: *When atoms of elements react, they tend to lose, gain, or share electrons to achieve the same electron arrangement as the noble gas nearest them in the periodic table.* Metals can achieve a noble gas electron arrangement by giving up electrons, and nonmetals can achieve a noble gas electron arrangement by adding or sharing electrons. This is the basis for our discussion of the two major classes of bonding—ionic bonding and covalent bonding.

⟨ 5.1 ⟩ ELECTRONEGATIVITY AND BONDING

The tendency of an atom to participate in bonding with another atom is related directly to a property of the atom known as its **electronegativity**. The electronegativity of the elements is shown in the periodic table in Figure 5.1. The electronegativity values range from a low of 0.8 for Cs to a high of 4.0 for F. Atoms with high values are said to be electronegative and attract the valence electrons of other atoms quite strongly. Atoms with lower electronegativity values are said to be electropositive, and they donate their valence electrons to a significantly more electronegative atom.

Inspection of the electronegativity values in Figure 5.1 leads to the observation that nonmetals, the elements in the upper right corner of the periodic table, are the most electronegative, while the metals in the lower left corner are the least electronegative. The general trend is that electronegativity increases across a period in the periodic table and decreases down a group of the periodic table. A key to predicting the type of bonding interactions between atoms is the difference in the electronegativity of the bonding atoms. As you will see, bonding interactions between atoms on opposite sides of the

Octet rule In forming bonds, main-group elements gain, lose, or share electrons to achieve a stable electron configuration with eight valence electrons

Electronegativity A measure of the relative tendency of an atom to attract electrons to itself

1A	2A										1B	2B	3A	4A	5A	6A	7A
															H 2.2		
Li 1.0	Be 1.6												B 2.0	C 2.5	N 3.0	O 3.5	F 4.0
Na 0.9	Mg 1.3	3B	4B	5B	6B	7B	8B			1B	2B	Al 1.6	Si 1.9	P 2.2	S 2.6	Cl 3.2
K 0.8	Ca 1.0	Sc 1.4	Ti 1.5	V 1.6	Cr 1.7	Mn 1.5	Fe 1.8	Co 1.9	Ni 1.9	Cu 1.9	Zn 1.6	Ga 1.8	Ge 2.0	As 2.2	Se 2.6	Br 3.0
Rb 0.8	Sr 1.0	Y 1.2	Zr 1.3	Nb 1.6	Mo 2.2	Tc 1.9	Ru 2.2	Rh 2.3	Pd 2.2	Ag 1.9	Cd 1.7	In 1.8	Sn 2.0	Sb 1.9	Te 2.1	I 2.7
Cs 0.8	Ba 0.9	La 1.1	Hf 1.3	Ta 1.5	W 2.4	Re 1.9	Os 2.2	Ir 2.2	Pt 2.3	Au 2.5	Hg 2.0	Tl 1.6	Pb 2.3	Bi 2.0	Po 2.0	At 2.2

▨ <1.0   ☐ 1.5–1.9   ▦ 2.5–2.9
▨ 1.0–1.4   ▩ 2.0–2.4   ▨ 3.0–4.0

FIGURE 5.1   Electronegativity values. Nonmetals have high electronegativity values (e.g., O, 3.5), metalloids have intermediate values (e.g., Ge, 2.0), and metals have low values (e.g., Mg,1.3).

periodic table, where the difference in electronegativity values is greatest, result from the tendency of the more electropositive metal atom to transfer its valence electrons to the more electronegative nonmetal atom. This type of bonding interaction is referred to as an *ionic bond* (Section 5.2). Interactions between atoms of similar or less dramatically different electronegativities are significantly different in nature. In these cases, the valence electrons are shared between two atoms, resulting in the formation of a bond referred to as a *covalent bond* (Section 5.3). The following analysis of both types of bonding interactions illustrates the magnitude of electronegativity differences that leads to ionic or covalent bonding and helps us predict bonding interactions in other compounds.

5.2   IONIC BONDS AND IONIC COMPOUNDS

The type of bonding between sodium (Group 1A) and chlorine (Group 7A) atoms can be predicted using the octet rule and the difference in electronegativities of the two atoms. Sodium, with a single valence electron, has an electronegativity of 1.0, and chlorine, with seven valence electrons, has an electronegativity of 3.0. The difference in these values (ΔEN), 2.0, indicates a situation in which one atom (chlorine, in this case) is capable of taking a valence electron away from the other atom (sodium, in this case) to complete its octet of outer shell electrons. As a general rule of thumb, ΔEN values of 2.0 or greater indicate the likelihood of transfer of electrons from the less electronegative atom to the more electronegative atom. In this way, both atoms attain full outer shells of valence electrons, one by gaining electrons and the other by losing electrons. Atoms become **anions** when they gain electron(s) to become negatively charged; atoms become **cations** when they lose electron(s) to become positively charged. The chlorine atom becomes a chloride anion (Cl^-) and the sodium atom becomes a sodium cation (Na^+).

Let's look at this situation more closely. A sodium atom (atomic number 11) has a total of 11 electrons and 11 protons. Only one of these electrons is a valence electron. By giving its single valence electron away, the sodium atom becomes a sodium cation with a net $+1$ charge because it now has 11 protons but only 10 electrons. Furthermore, the sodium cation now has exactly the same number of outer shell electrons (valence electrons) as the nearest noble gas, Ne.

The chlorine atom (atomic number 17) has a total of 17 electrons and 17 protons. Seven of these electrons are outer shell or valence electrons. By gaining the electron from a sodium atom, the chlorine atom becomes a chloride

See an animation of **Formation of Sodium Chloride** at http:// brookscole.com/chemistry/joesten4

Anion An atom or group of bonded atoms that has a negative charge

Cation An atom or group of bonded atoms that has a positive charge

To represent the reaction of sodium with chlorine as it actually occurs requires writing Cl_2 for elemental chlorine:

$$2\ Na + Cl_2 \longrightarrow 2\ NaCl$$

The positive ion is written before the negative ion, i.e., NaCl not ClNa.

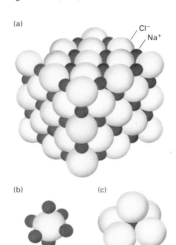

FIGURE 5.2 Structure of sodium chloride. (a) Model of the three-dimensional sodium chloride crystal lattice. (b) Each Cl^- ion is surrounded by 6 Na^+ ions. (c) Each Na^+ ion (in red) is surrounded by 6 Cl^- ions.

See an animation of **Sodium Metal Reacting with Chlorine Gas** at http://brookscole.com/chemistry/joesten4

FIGURE 5.3 Sodium (stored under mineral oil to protect it from oxygen in the air), chlorine, and sodium chloride (which we know as table salt).

Lewis dot symbol A representation of a molecule, ion, or formula unit by atomic symbols with the valence electrons shown as dots

Ionic bond The attraction between positive and negative ions

Ionic compounds Compounds composed of positive and negative ions

Formula unit In ionic compounds, the simplest ratio of oppositively charged ions that gives an electrically neutral unit

anion with a net −1 charge since it now has 17 protons and 18 electrons. Furthermore, the chloride anion now has exactly the same number of outer shell (valence electrons) as the nearest noble gas, Ar. The reaction between sodium and chlorine can be illustrated using the **Lewis dot symbols** as shown.

$$Na\cdot + \cdot\ddot{\underset{\cdot\cdot}{Cl}}: \longrightarrow Na^+ + :\ddot{\underset{\cdot\cdot}{Cl}}:^-$$

The electropositive atom transfers its valence electron(s) to the more electronegative atom when the difference in electronegativity is large. By this process, both atoms obtain the same number of outer shell electrons as the nearest noble gas and satisfy the octet rule. This is a generalization that can be successfully used to predict the nature of bonding between atoms of many different elements.

You have already learned that the strong electrostatic attraction between the positive and negative ions is known as the **ionic bond**. Compounds that are held together by ionic bonds are known as **ionic compounds**. Because chemical compounds are overall electrically neutral, sodium chloride must be composed of one sodium ion for every chloride ion. To show this composition, the formula of the compound is written as NaCl. (Note that the ionic charges are not indicated in the formula.) In ionic compounds the simplest ratio of oppositely charged ions that gives an electrically neutral unit is represented in the formula and is called a **formula unit**. The formula unit for sodium chloride is NaCl, or one sodium ion and one chloride ion.

Ionic crystals are made up of large numbers of formula units that form a regular three-dimensional crystalline lattice. A model of the sodium chloride crystalline lattice is shown in Figure 5.2. Note that each Na^+ ion has six Cl^- ions around it. Similarly, each Cl^- ion has six Na^+ ions around it. In this way, the one-to-one ratio of the singly charged ions is preserved. There is no *unique molecule* in ionic structures; no particular ion is attached exclusively to another ion, but each ion is attracted to all the oppositely charged ions surrounding it.

When atoms become ions, properties are drastically altered. For example, a collection of Br_2 molecules is red, but bromide ions (Br^-) contribute no color to a crystal of a compound such as NaBr. A chunk of sodium atoms (Figure 5.3, *left*) is soft, metallic, and violently reactive with water, but Na^+ ions are stable in water. A large collection of Cl_2 molecules (Figure 5.3, *center*) constitutes a greenish-yellow poisonous gas, but chloride ions (Cl^-) produce no color in compounds and are not poisonous. We have everyday evidence of this because we use NaCl (Figure 5.3, *right*) to season food. When atoms become ions, atoms change their nature.

Metals in Groups 1A, 2A, and 3A are capable of donating one, two, or three valence electrons, respectively, to the significantly more electronegative nonmetals in Groups 5A, 6A, and 7A. These nonmetal atoms can accept three, two, or one electron, respectively. Thus, cations of +1, +2, or +3 charge are formed from Group 1A, 2A, and 3A metals, respectively, and anions of −3, −2, or −1 charge are formed from Group 5A, 6A, and 7A nonmetals, respectively. Carbon, in Group 4A, represents a sort of crossover point from elements that tend to form cations to ones more likely to form anions. Carbon itself rarely forms the predicted +4 or −4 ions since losing four electrons to another atom is simply too energetically unfavorable, as is gaining four electrons from another atom. Thus, carbon tends to be found mostly in molecules and not in ionic compounds.

Main group elements form ions of a single type as shown in Figure 5.4. Transition metal elements form multiple ions in many cases. For instance, iron (atomic number 26) readily forms both +2 and +3 ions whereas silver (atomic number 47) forms only the +1 ion. Therefore, it is necessary to memorize some of the most important transition metal ions shown in Figure 5.4.

1A	2A	3B	4B	5B	6B	7B	8B	8B	8B	1B	2B	3A	4A	5A	6A	7A	8A
H^+																H^-	N
Li^+														N^{3-}	O^{2-}	F^-	O, B, L, E
Na^+	Mg^{2+}											Al^{3+}		P^{3-}	S^{2-}	Cl^-	E
K^+	Ca^{2+}		Ti^{2+}		Cr^{2+} Cr^{3+}	Mn^{2+}	Fe^{2+} Fe^{3+}	Co^{2+} Co^{3+}	Ni^{2+}	Cu^+ Cu^{2+}	Zn^{2+}				Se^{2-}	Br^-	G
Rb^+	Sr^{2+}									Ag^+	Cd^{2+}		Sn^{2+}		Te^{2-}	I^-	A, S
Cs^+	Ba^{2+}									Hg_2^{2+} Hg^{2+}			Pb^{2+}	Bi^{3+}			E, S

FIGURE 5.4 Common ions. Main-group metals usually form positive ions with a charge given by the group number (blue). For transition metals, the positive charge is variable (red), and ions other than those shown are possible. Nonmetals generally form negative ions with a charge equal to 8 minus the group number (yellow). The Group 8A elements, known as the *noble gases,* form very few compounds.

Predicting Formulas of Ionic Compounds

Because opposite charges are attracted to each other, it stands to reason that cations of the metals of Groups 1A, 2A, and 3A tend to form ionic compounds with the anions of the nonmetals of Groups 5A, 6A, and 7A. There are, of course, ionic compounds formed by combination of the cations of the transition metal elements with various nonmetal anions. The formula units for ionic compounds are always represented by using *the simplest ratio of cation and anion necessary to give an electrically neutral formula unit.* The transfer of two valence electrons from a calcium atom to an oxygen atom gives Ca^{+2} and O^{-2} ions. Calcium oxide, formed by the combination of these two ions is represented as CaO and not Ca_2O_2. Similarly, magnesium and bromine atoms would combine to give the formula unit $MgBr_2$, since magnesium forms a $+2$ ion and bromine forms a -1 ion.

See an animation of **Ion Formation** at http://brookscole.com/chemistry/joesten4

An uncut sapphire. Blue sapphire is aluminum oxide with a small proportion of its aluminum ions replaced by iron(II) and titanium(IV) ions.

Charles D. Winters

EXAMPLE 5.1   Predicting Formulas of Ionic Compounds

What is the formula of aluminum oxide, the ionic compound formed by the combination of aluminum and oxygen?

SOLUTION
1. Aluminum (Al) is in Group 3A, so remove three electrons from an Al atom to give the Al^{3+} ion.
2. Oxygen (O) is in Group 6A, so add two electrons to give eight. The resulting ion is O^{2-}.
3. To make this formula electrically neutral: Al^{3+} ion is $+3$, and two ions would give a total charge of $+6$; O^{2-} ion is -2, and three ions would give a total charge of -6; $+6$ and $-6 = 0$; the formula is Al_2O_3.

TRY IT 5.1
What is the formula of calcium fluoride, the ionic compound formed by the combination of calcium (Ca) and fluorine (F)?

| EXAMPLE 5.2 | Predicting Formulas of Ionic Compounds |

What is the formula of lithium nitride, the ionic compound formed by the combination of lithium and nitrogen?

SOLUTION

1. Lithium (Li) is in Group 1A; so remove one electron from the Li atom to give the Li^+ ion.
2. Nitrogen (N) is in Group 5A; so add three electrons to the N atom to give eight outer shell electrons and form an N^{3-} ion.
3. The electrically neutral compound would be formed from the combination of three Li^+ ions with one N^{3-} ion to give Li_3N.

TRY IT 5.2

Write the formula units for the ionic compounds formed by the combination of: (**a**) cesium (Cs) and iodine (I); (**b**) strontium (Sr) and chlorine (Cl); and (**c**) barium (Ba) and sulfur (S).

Naming Binary Ionic Compounds

By starting once again with table salt, we can use the formula and chemical name of this compound, NaCl (sodium chloride), to help remember the rules for naming ionic compounds. The positive ion is named first, followed by the name of the negative ion. The element name is used for the positive ion. If the compound is made up of only one metal and one nonmetal (a **binary compound**), the name of the negative ion ends in *-ide*. Examples we have already used besides sodium chloride include calcium oxide (CaO), aluminum oxide (Al_2O_3), and lithium nitride (Li_3N). (Note that *binary* means only two elements are present, but the number of atoms in the formula can be more than two.) For transition metals that form ions with different charges, Roman numerals are used with the names to indicate the charge. For example, iron(II) ion refers to Fe^{2+}, and iron(III) ion means Fe^{3+}. The name of $FeCl_2$ is iron(II) chloride; and the name of $FeCl_3$ is iron(III) chloride.

| EXAMPLE 5.3 | Naming Binary Ionic Compounds |

Name the following ionic compounds: (**a**) K_2S, (**b**) $BaBr_2$, (**c**) Li_2O, (**d**) Fe_2O_3.

SOLUTION

All these compounds are binary compounds—made up of ions of two elements. The positive ion is named first, followed by the negative ion. The positive ion is named as the element. The negative ion is named by adding *-ide* to the stem of the name of the element. The correct names of the first three are (**a**) potassium sulfide, (**b**) barium bromide, and (**c**) lithium oxide. Compounds of iron require the use of a Roman numeral after *iron* to identify its charge. The charge on the iron ion in Fe_2O_3 is determined by calculating the total negative charge (-6 from $3 O^{2-}$) and dividing by 2 (because there are two iron ions). This gives a charge of $+3$; thus, the correct name of Fe_2O_3 is iron(III) oxide.

TRY IT 5.3

Name the following compounds: (**a**) RbCl, (**b**) Ga_2O_3, (**c**) $CaBr_2$, and (**d**) Fe_3N_2.

Binary compound Chemical compound composed of two elements

<table>
<tr><td>EXAMPLE</td><td>5.4</td></tr>
</table>

EXAMPLE 5.4 Writing Formulas for Binary Ionic Compounds

Write formulas for the following compounds: (**a**) cesium bromide, (**b**) cobalt(III) chloride, (**c**) barium oxide.

SOLUTION

(**a**) The correct formula is CsBr since Cs (Group 1A) forms Cs^+ and Br (Group 7A) forms Br^-. (**b**) The Roman numeral III indicates the Co^{3+} ion, and Cl (Group 7A) forms Cl^-; thus, the correct formula is $CoCl_3$. (**c**) The correct formula is BaO since Ba (Group 2A) forms Ba^{2+} and O (Group 6A) forms O^{2-}.

TRY IT 5.4

What are the formulas of (**a**) cobalt(II) sulfide, (**b**) magnesium fluoride, and (**c**) potassium iodide?

Ionic Compounds with Polyatomic Ions

Atoms of two or more elements can also combine to form a **polyatomic ion**, a chemically distinct species with an electric charge. Communication in the world of chemistry and understanding many applications require one to know the names, formulas, and charges of the common polyatomic ions listed in Table 5.1.

The bonding between the atoms within polyatomic ions is just like the bonding within molecular compounds (Section 5.3), but the group of atoms has either more or fewer electrons than protons and therefore has an overall charge. Compounds that contain polyatomic ions are ionic, and their formulas are written by the same procedure described for binary ionic compounds. The only difference is that the polyatomic ion formula is enclosed in parentheses when more than one such ion is present. For example, the formula of aluminum nitrate is $Al(NO_3)_3$. The compounds are also named in the same manner as binary ionic compounds, with the

> Most common polyatomic ions are negatively charged; ammonium ion (NH_4^+) is the major exception.

> **Polyatomic ion** A positive or negative ion composed of two or more atoms

---

TABLE 5.1 Names and Composition of Some Common Polyatomic Ions

Cation (Positive Ion)

NH_4^+	Ammonium ion

Anions (Negative Ions)

OH^-	Hydroxide ion	CO_3^{2-}	Carbonate ion
$CH_3CO_2^-$	Acetate ion	HCO_3^-	Hydrogen carbonate ion
NO_2^-	Nitrite ion		(or bicarbonate ion)
NO_3^-	Nitrate ion	PO_4^{3-}	Phosphate ion
SO_3^{2-}	Sulfite ion	HPO_4^{2-}	Hydrogen phosphate ion
HSO_3^-	Hydrogen sulfite ion	$H_2PO_4^-$	Dihydrogen phosphate ion
SO_4^{2-}	Sulfate ion	ClO^-	Hypochlorite ion
HSO_4^-	Hydrogen sulfate ion	ClO_3^-	Chlorate ion
	(or bisulfate ion)	ClO_4^-	Perchlorate ion
CN^-	Cyanide ion		

TABLE 5.2	Some Commercially Important Ionic Compounds with Polyatomic Ions	
FORMULA	**NAME (COMMON NAME)**	**USES**
NH_4NO_3	Ammonium nitrate	Fertilizers and explosives
KNO_3	Potassium nitrate	Gunpowder and matches
$NaOH$	Sodium hydroxide (lye)	Extracting Al from ore; preparing soaps, detergents, and rayon; pulp and paper industry
$Mg(OH)_2$	Magnesium hydroxide	Milk of magnesia
Na_2CO_3	Sodium carbonate (washing soda, soda ash)	Water softening, detergents and cleansers; pulp and paper industry; glass and ceramics
$NaHCO_3$	Sodium bicarbonate (baking soda)	Household use, food industry; fire extinguishers
Na_3PO_4	Sodium phosphate	Food additive
$Ca(H_2PO_4)_2$	Calcium dihydrogen phosphate	Fertilizer
$CaSO_4$	Calcium sulfate	Gypsum, drywall (wallboard)
$Al_2(SO_4)_3$	Aluminum sulfate	Water purification

name of the positive ion followed by the name of the negative ion. Examples of some important ionic compounds with polyatomic ions are given in Table 5.2.

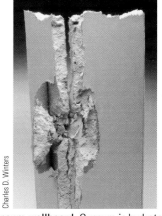

Charles D. Winters

Gypsum wallboard. Gypsum is hydrated calcium sulfate, $CaSO_4 \cdot 2\,H_2O$.

EXAMPLE 5.5   **Writing Formulas for Compounds of Polyatomic Ions**

Write the formulas of (**a**) magnesium sulfate and (**b**) calcium hydrogen sulfite.

SOLUTION
(**a**) The sulfate ion is SO_4^{2-} and the charge on the magnesium ion is $+2$; so the formula is $MgSO_4$. No parentheses are needed around the sulfate ion because only one SO_4^{2-} is present.
(**b**) The hydrogen sulfite ion is HSO_3^- and the charge on the calcium ion is $+2$; thus, the formula is $Ca(HSO_3)_2$.

TRY IT 5.5
Write the formulas of (**a**) magnesium carbonate and (**b**) sodium dihydrogen phosphate.

CONCEPT CHECK   **5A**

1. The attractive forces between positive and negative ions in a crystal lattice are called _____ bonds.
2. Positive ions are formed from neutral atoms by (**a**) losing electrons (**b**) gaining electrons.

3. What charge is expected when the following atoms form ions?
   (**a**) Lithium (Li)   (**b**) Aluminum (Al)
   (**c**) Sulfur (S)   (**d**) Bromine (Br)
4. Which of the following atoms form positive ions?
   (**a**) Potassium (K)   (**b**) Iodine (I)   (**c**) Nitrogen (N)
   (**d**) Sodium (Na)   (**e**) Strontium (Sr)   (**f**) Silver (Ag)
5. Negative ions are formed from neutral atoms by (**a**) losing electrons (**b**) gaining electrons.
6. When nutritionists refer to the importance of low salt intake, they are referring to the compound with the formula _____ and the name _____.
7. Which is the more electronegative atom in each set?
   (**a**) Ca or Cl   (**b**) F or O   (**c**) Na or S
   (**d**) K or N   (**e**) Fr or Br

5.3 COVALENT BONDS

Most chemical compounds are not ionic. The atoms in most chemical compounds are held together in units known as *molecules* (Section 2.2). In the chemical bonds in molecules, one or more pairs of electrons are shared between two atoms in what are known as **covalent bonds**. The shared electrons feel the positive attraction of the protons in the nucleus of each of the two atoms connected by the bond. Most molecules are composed of atoms of nonmetals, whereas ionic compounds are composed of metal cations and nonmetal anions.

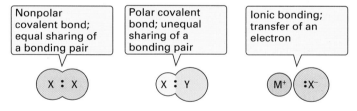

Nonpolar covalent, polar covalent, and ionic bonding.

Some covalent bonds connect two identical atoms of the elements that exist naturally as diatomic molecules (i.e., H_2, F_2, Cl_2, Br_2, I_2, O_2, and N_2). Lewis dot structures of the type discussed earlier can also be used to represent covalent bonding.

In drawing Lewis structures for molecules, you first need to determine the total number of valence electrons the collected atoms will contribute to the structure. You should then connect the individual atoms with pairs of electrons to form the basic skeleton. In doing this, some generalizations will be useful:

- Avoid H—H, H—X, and X—X bonds (where X is a halogen atom) because these bonds exist only in H—H, H—F, H—Cl, H—Br, H—I, and the diatomic halogens.
- Avoid O—O bonds because these are relatively rare except in molecular oxygen (O_2), ozone (O_3), H—O—O—H (hydrogen peroxide), and a few related compounds.
- Carbon–carbon, carbon–hydrogen, and carbon–oxygen bonds are quite common, so maximizing the number of these in a structure is a good way to start.

Covalent bond A bond in which two atoms share electrons

Once you establish what the skeletal structure of the molecule will be, start adding electrons in pairs to the outermost, more electronegative atoms of the structure until you satisfy the octet rule for those atoms or until you run out of electrons. At this point, stop and see if each atom has an octet. If not, you must share pairs of electrons between an atom with an octet and an atom to which it is bonded but is short of electrons for the octet. This will give rise to multiple bonds (double bonds or triple bonds) as illustrated under "Multiple Covalent Bonds" later in this chapter.

After you've arrived at a structure you will see that shared electrons are involved in covalent bonds but there may also sometimes be unshared electrons associated with an individual atom in the structure. The electrons in these Lewis structures are classified as **bonding pairs** or **lone pairs**.

$$\text{bonding pair} \quad H - \overset{..}{\underset{..}{O}} - H \quad \text{lone pair}$$

For many molecules or ions, the actual connectivity of the atoms may not be obvious and more information may be needed to choose the best among several possible Lewis structures. Compounds containing single, double, and triple bonds are described in the rest of this chapter, with consideration given to choosing among several possible Lewis structures.

Single Covalent Bonds

A single covalent bond is formed when two atoms share a single pair of electrons. The simplest case in which this can occur is the hydrogen molecule (H_2). Since each hydrogen atom (Group 1A) has a single valence electron, the two hydrogen atoms are connected by a single pair of electrons that you can represent in shorthand fashion with a line, as shown here.

$$H\cdot + H\cdot \longrightarrow H:H \quad \text{or} \quad H-H$$
Lewis structure

In a molecule such as F_2, each fluorine atom has seven valence electrons because fluorine is in Group 7A. Thus, the Lewis structure will contain 14 electrons. Connect the two F atoms with a pair of electrons and then add the remaining 12 electrons in pairs until each F atom has an octet and you arrive at the structure below. Note that you do not necessarily have to consider the dot structure for the individual atoms to arrive at the structure of the molecule as long as you use only valence electrons in pairs to connect atoms.

$$\cdot\overset{..}{\underset{..}{F}}: + \cdot\overset{..}{\underset{..}{F}}: \longrightarrow :\overset{..}{\underset{..}{F}}:\overset{..}{\underset{..}{F}}: \quad \text{or} \quad :\overset{..}{\underset{..}{F}}-\overset{..}{\underset{..}{F}}:$$

You can use this process to arrive at Lewis structures for many molecules. For instance, a Lewis structure for water (H_2O) should contain a total of eight valence electrons (two from the two hydrogen atoms and six from the oxygen atom). After connecting the two hydrogens to the oxygen atom by pairs of electrons, we are left with four electrons which we must add (as two non-bonding pairs) to the oxygen atom to arrive at the structure shown. Note that we can arrive at the correct structure without any consideration of the fact that the actual bonds are formed by each hydrogen atom sharing its valence electron with one of the unpaired valence electrons from the oxygen atom, although this is what happens.

$$2 H\cdot + \cdot\overset{..}{\underset{..}{O}}: \longrightarrow H:\overset{..}{\underset{..}{O}}:H \quad \text{or} \quad H-\overset{..}{\underset{..}{O}}-H$$

A single covalent bond is formed when two atoms share a single pair of electrons.

 See an animation of **Bond Length and Energy** at http://brookscole.com/chemistry/joesten4

 See an animation of **Electrostatic Interactions Between Hydrogen Atoms** at http://brookscole.com/chemistry/joesten4

Bonding pairs Pairs of electrons shared between two atoms, frequently represented by lines (instead of dots) for convenience in Lewis structures

Lone pairs (or *non-bonding* or *unshared pairs*) Pairs of valence electrons that are not shared between two atoms in the Lewis structure but instead are associated with a single atom in the structure; should always be shown as dots in Lewis structures

EXAMPLE	5.6	**Writing Lewis Dot Structures**

Use your knowledge of valence electrons to draw a Lewis dot structure for ammonia (NH_3).

SOLUTION

The total number of valence electrons to be used in the structure will be five from the nitrogen atom and one from each of the three hydrogen atoms for a total of eight valence electrons.

1. Recalling that you should avoid H—H bonds, you should connect the atoms together using pairs of electrons to arrive at the following structure. This uses six of the eight valence electrons.

   H:N:H
   ··
   H

2. Add the remaining two electrons as a lone pair to the nitrogen atom to obtain the complete Lewis structure shown here.

   ··
   H:N:H
   ··
   H

3. For ease of representation, you may replace the bonding pairs of electrons with lines to arrive at the following structure.

   ··
   H—N—H
   |
   H

TRY IT 5.6

Use the process just discussed to draw Lewis dot structures for the following molecules: HI, Br_2, CH_4, and HOCl.

Single Bonds in Hydrocarbons

Hydrocarbons, compounds that contain only carbon and hydrogen atoms, represent a large class of compounds that contain two different types of atoms. They are extremely important molecules in the fuel industry, as will be seen later. Hydrocarbons that contain only C—H and C—C single bonds are

Hydrocarbons Compounds containing only carbon and hydrogen

THE PERSONAL SIDE

Gilbert Newton Lewis (1875–1946)

G. N. Lewis was born in Massachusetts but raised in Nebraska. After earning his B.A. and Ph.D. degrees at Harvard University, he began his academic career. In 1912, he was appointed chairman of the Chemistry Department at the University of California at Berkeley, and he remained there for the rest of his life. Lewis felt that a chemistry department should both teach and advance fundamental chemistry, and he was not only a productive researcher but also a teacher who profoundly affected his students. He developed his concepts about valence electrons and the stability of the noble gas electron

GILBERT NEWTON LEWIS

Frances Simon, Courtesy AIP Emilio Segrè Visual Archives

configuration (octet rule) to explain periodic trends to students in his introductory chemistry course. Although he was teaching these concepts as early as 1902, he didn't publish them until 1916. Lewis also made major contributions to other areas of chemistry such as isotope studies and acid–base theory.

All organic compounds can be pictured as derived from hydrocarbons by the addition of various other kinds of atoms or groups of atoms. Organic compounds are so numerous and so important that they are the basis for an entire field of chemistry—organic chemistry.

known as **alkanes**. In the previous exercise you drew a Lewis dot structure for the simplest alkane, known as methane (CH_4). Methane is the major component of natural gas. One could imagine bringing four hydrogen atoms, each with a single valence electron, together with a carbon atom with four unpaired valence electrons to form the methane molecule.

$$\text{H : }\overset{\overset{\displaystyle H}{\cdot}}{\underset{\underset{\displaystyle H}{\cdot \cdot}}{C}}\text{ : H} \quad \text{or} \quad \text{H}-\overset{\overset{\displaystyle H}{|}}{\underset{\underset{\displaystyle H}{|}}{C}}-\text{H} \quad \text{or}$$

Three ways to represent methane

As will be seen over and over again, carbon has a tendency to form four bonds. In alkanes these four bonds will be either single bonds to H or single bonds to C. Another simple alkane is propane, C_3H_8, which is represented by the following Lewis structure.

$$\text{H}-\overset{\overset{\displaystyle H}{|}}{\underset{\underset{\displaystyle H}{|}}{C}}-\overset{\overset{\displaystyle H}{|}}{\underset{\underset{\displaystyle H}{|}}{C}}-\overset{\overset{\displaystyle H}{|}}{\underset{\underset{\displaystyle H}{|}}{C}}-\text{H}$$

Structural formulas like the one shown here for propane quickly get cumbersome as the number of atoms increases. One way to handle this is to write all of the symbols of the atoms on one line, with the atoms connected together written together. Propane would therefore be $CH_3CH_2CH_3$. Formulas written this way are known as *condensed formulas*.

TRY IT 5.7

(**a**) Draw a Lewis dot structure for the alkane intermediate between methane and propane. It is known as ethane and has the formula C_2H_6.
(**b**) There are two different ways to connect the fourteen atoms of butane (C_4H_{10}). The two different molecules are known as **isomers**. See if you can draw Lewis dot structures for both of these. This should lead you to see that, as the number of carbon atoms increases, the number of ways to connect the atoms together increases as well.

Alkanes are often referred to as **saturated hydrocarbons** because they have the highest possible ratio of hydrogen to carbon atoms that can be bonded in a molecule. The general formula for saturated hydrocarbons is C_nH_{2n+2} where n is the number of carbon atoms.

The simplest hydrocarbon molecule, methane, has a tetrahedral shape, with the carbon atom at the center of the tetrahedron and the four hydrogen atoms at the corners. Every carbon atom in a saturated hydrocarbon molecule has its bonds tetrahedrally arranged. This arrangement allows the hydrogen atoms bonded to carbon atoms to be located as far as possible from one another.

Although one cannot be absolutely certain of the geometry of a molecule just by looking at a Lewis structure, some generalizations are useful in predicting the geometrical arrangement of atoms and lone pairs around a central atom. The first of these is: *If the central atom under consideration is bonded to four other atoms or has a total of four other atoms and/or lone pairs attached to it, the arrangement of the four bonded atoms and/or lone pairs will be approximately tetrahedral.*

Tetrahedral methane. The four faces of a tetrahedron are equilateral triangles, and the angle between any two lines drawn from the center to two corners is 109.5°.

Multiple Covalent Bonds

Frequently one encounters molecules in which more than one pair of electrons is shared between two atoms. An atom with fewer than seven valence electrons can form covalent bonds in two ways. The atom can contribute one electron to each of several single covalent bonds, or it can share two (or three) pairs of electrons with a single other atom in forming multiple bonds to that atom.

Alkanes Hydrocarbons with carbon–carbon single bonds

Isomers Different compounds with the same formula but different structures

Saturated hydrocarbons Hydrocarbons that are alkanes

For instance, we can arrive at the Lewis dot structure for carbon dioxide (CO_2) in the following manner. First, determine that there are 16 total valence electrons (4 from carbon and 6 each from the 2 oxygen atoms) and connect the carbon to the two oxygen atoms with pairs of electrons (structure *a*); remember to avoid oxygen–oxygen bonds. This uses 4 of the 16 electrons, leaving 12 electrons. Next, add electrons in pairs to the outermost electronegative oxygen atoms until you satisfy the octet rule for the oxygen atoms (structure *b*). At this point the carbon only shares four electrons and does not satisfy the octet rule. Share a pair of electrons from each oxygen atom with the central carbon atom to arrive at structure *c,* in which the octet rule is satisfied for all atoms. The carbon dioxide molecule has two **double bonds**.

$$\text{O:C:O} \qquad \text{:Ö:C:Ö:} \qquad \text{:Ö::C::Ö:} \qquad \text{or} \qquad \text{:O}=\text{C}=\text{O:}$$

   a   b   c

We could now ask about the geometry around the central carbon of carbon dioxide. This allows us to consider our second generalization about the geometry around an atom. *If the cental atom under consideration is bonded to only two other atoms and has no lone pairs associated with it or has a total of two other atoms and/or lone pairs attached to it, the arrangement of the two bonded atoms or equivalents will be approximately linear.* To illustrate, carbon dioxide is a linear molecule with an O—C—O bond angle of 180°.

In the N_2 molecule, each nitrogen atom contributes five valence electrons for a total of ten valence electrons. Using the method illustrated above, connect the two nitrogen atoms with a single pair of electrons to give *a*. Next add the remaining eight electrons in pairs to each nitrogen until the octet rule is satisfied for that nitrogen or until there are no more electrons. This gives *b*. From structure *b* we will share two lone pairs from the nitrogen atom with three lone pairs with the other nitrogen, resulting in structure *c,* in which the octet rule is satisfied by both atoms. These atoms are connected by a **triple bond** and the nitrogen molecule is linear, as any two-atom molecule must be.

$$\text{N:N} \qquad \text{:N̈:N:} \qquad \text{:N:::N:} \qquad \text{or} \qquad \text{:N}\equiv\text{N:}$$

   a   b   c

(EXAMPLE) 5.7)   **Molecules with Triple Bonds**

Show that the best Lewis dot structure for acetylene (C_2H_2) contains a triple bond between the carbon atoms and a single bond between the carbon and hydrogen atoms.

SOLUTION:

There are 10 valence electrons associated with the 2 carbon atoms (4 each) and the 2 hydrogen atoms (1 each). Connect the basic skeleton of the molecule using pairs of electrons to arrive at structure *a*. This leaves 4 electrons which can be added in two pairs to one of the two carbon atoms to give structure *b*. This leaves the remaining carbon without an octet of electrons so two electron pairs from the first carbon must be shared with the other carbon to give the final structure *c*.

$$\text{H:C:C:H} \qquad \text{H:C̈:C:H} \qquad \text{H:C:::C:H} \qquad \text{or} \qquad \text{H}-\text{C}\equiv\text{C}-\text{H}$$

   a   b   c

Double bond A bond in which two pairs of electrons are shared between two atoms

Triple bond A bond in which three pairs of electrons are shared between two atoms

Ozone (O_3) is an interesting molecule for which to consider Lewis dot structures as it illustrates another important feature of such structures. Three oxygen atoms contain a total of 18 valence electrons. This is clearly a case where one cannot avoid oxygen–oxygen bonds, so the three oxygens must be connected by pairs of electrons to give structure *a*. This leaves 14 valence electrons that are added in pairs to the outermost oxygen atoms to give structure *b* and leaves two electrons, which are then placed on the central oxygen atom to give structure *c*. The central oxygen atom in *c* only shares six electrons so a pair of electrons may be shared from an outer oxygen atom to either side of the central oxygen to give structures *d* and *e*. Structures *d* and *e* differ only in the position of electrons and not in the position of atoms and are known as **resonance structures**. The real structure of ozone is neither *d* or *e*. It is a hybrid structure of the two in which each oxygen–oxygen bond is about halfway between a single bond and a double bond. The two resonance structures are the limiting Lewis dot structures that can be drawn to represent the actual structure of ozone, a compound for which the bonding cannot accurately be represented by a single Lewis dot structure.

$$O:O:O \qquad :\ddot{O}:O:\ddot{O}: \qquad :\ddot{O}:\ddot{O}:\ddot{O}:$$
$$\text{a} \qquad\qquad \text{b} \qquad\qquad \text{c}$$

$$:\ddot{O}::\ddot{O}:\ddot{O}: \quad \text{or} \quad :\ddot{O}=\ddot{O}-\ddot{O}: \qquad :\ddot{O}:\ddot{O}=\ddot{O}: \quad \text{or} \quad :\ddot{O}-\ddot{O}::\ddot{O}:$$
$$\text{d} \qquad\qquad\qquad\qquad\qquad \text{e}$$

The geometry around the central oxygen atom of this molecule reflects the O—O—O bond angle and allows us to consider our third and final generalization about geometry around an atom. *If the central atom under consideration has a **total** of three other atoms and/or lone pairs attached to it, the three bonded atoms or equivalents will have a bond angle of ~120°.* Note that the central atom in ozone has two bonded atoms and one lone pair. In fact, the O—O—O bond angle in ozone is about 117°, so the prediction here is fairly accurate, as it will be in general.

Single, double, and triple bonds differ in length and strength. Triple bonds are shorter than double bonds, which in turn are shorter than single bonds. Bond energies normally increase with decreasing bond length. **Bond energy** is defined as the amount of energy required to break a defined number of the bonds (this number, a *mole*, is discussed in Section 8.2). Some typical bond lengths and energies are listed in Table 5.3.

Multiple Covalent Bonds in Hydrocarbons

Ethylene (Figure 5.5) contains a double bond between the carbon atoms and single bonds between the hydrogen atoms and the carbon atoms. Ethylene is the first member of the **alkene** series of hydrocarbons, compounds that have

Resonance structures Two or more possible Lewis dot representations of a molecule or ion in which the only difference is the position of the valence electrons. The positions of the nuclei remain constant. The actual structure is a hybrid of the structures

Bond energy Amount of energy required to break one mole of bonds between a specified pair of atoms

Alkenes Hydrocarbons with one or more carbon-carbon double bonds

FIGURE 5.5   Structure of an ethylene molecule. All the atoms in this molecule lie in the same plane.

TABLE 5.3	Some Bond Lengths and Bond Energies					
Bond type	C—C	C=C	C≡C	N—N	N=N	N≡N
Bond length (nm)	0.154	0.134	0.120	0.140	0.124	0.109
Bond energy (kcal/mol)*	83	146	200	40	100	225

*kcal/mol (kilocalories per mole) = thousands of calories necessary to break 6.02×10^{23} bonds (see Section 8.2).

one or more C=C bonds, that is, carbon–carbon double bonds and the general formula C_nH_{2n}.

$$H \diagdown \atop H \diagup C=C \diagup H \atop \diagdown H$$

Ethylene (C_2H_4)

The structural formula of ethylene illustrates why alkenes are said to be **unsaturated hydrocarbons**: they contain fewer hydrogen atoms than the corresponding alkanes and react with hydrogen to form alkanes.

$$H_2 + \begin{matrix} H \\ \diagdown \\ \end{matrix} C=C \begin{matrix} H \\ \diagup \\ \end{matrix} \longrightarrow \begin{matrix} H & H \\ | & | \\ H-C-C-H \\ | & | \\ H & H \end{matrix}$$

Acetylene, the gas that produces a flame hot enough to cut steel when it is mixed with oxygen and burned, has a carbon–carbon triple bond, C≡C.

$$H-C\equiv C-H$$

5.4 SHAPES OF MOLECULES

Lewis dot structures can only be drawn in two dimensions. Since the shapes of very few molecules can be adequately represented using two dimensions, it is necessary to go beyond the Lewis structure to infer the shapes of molecules.

The *valence shell electron pair repulsion (VSEPR) model* is quite useful in this regard. According to this model, the valence electrons represented in a Lewis dot structure are key to predicting the geometry of the molecule. These valence electrons repel each other. This repulsion leads to geometries in which repulsion is minimized. The result is a geometry in which the electrons around a central atom are as far away from each other as possible while maintaining their association with the central atom. More than one atom may be considered as the central atom as one focuses in turn on the geometry of different parts of a molecule.

Imagine differing numbers of identical balloons attached to a central point. If the balloons each have identical electrostatic charges, they repel each other. The final geometry adopted by the balloons will be one of the following: the balloons will either be 180° apart (two balloons) to define a *linear geometry*, 120° apart (three balloons) to define a *trigonal planar geometry*, or 109.5° apart (four balloons) to define a *tetrahedral geometry* (Figure 5.6). These are some of the more common situations found in chemical bonding. Five- and six-electron pairs assume trigonal bipyramidal and octahedral geometries.

Polyethylene (Section 14.4) is the most widely used polymer. Examples of plastics made from polyethylene include milk bottles, sandwich bags, garbage bags, toys, and molded objects.

Unsaturated hydrocarbons Hydrocarbons that contain fewer than the maximum number of hydrogen atoms; alkenes and alkynes

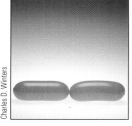
Linear

Trigonal planar

Tetrahedral

Trigonal bipyramidal

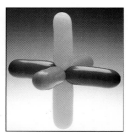
Octahedral

FIGURE 5.6 Balloon models of electron-pair geometries for two to six electron pairs. Balloons of similar size and shape, when tied together, naturally assume the arrangements shown.

Charles D. Winters

If we now extend this idea to the regions of high electron density around a central atom, we can predict both the *molecular geometry* and the *electron pair geometry* of a molecule. Electron pair geometry refers to the geometrical relationships between regions of high electron density; molecular geometry refers to the geometrical relationships between atoms. Regions of high electron density can be defined by looking at the Lewis dot structure. These regions are either bonds to another atom (single, double, or triple bonds all count as one region) or unshared pairs of electrons. The geometry that is defined when these regions of high electron density are as far apart as possible is called *electron pair geometry*.

If all of the regions of high electron density are bonds, then the electron pair geometry is the same as the molecular geometry. This is what we've seen, thus far, when looking at methane and carbon dioxide. Methane has four regions of electron density, and its electron pair and molecular geometry are both tetrahedral with 109.5° bond angles. Carbon dioxide and acetylene each have two regions of electron density about the central atom, and their electron pair and molecular geometries are linear with a 180° bond angle. These two examples illustrate the four-balloon and two-balloon analogies, respectively.

Let's look at another simple molecule, formaldehyde, to illustrate the three-balloon analogy. Formaldehyde has the formula H_2CO. The central carbon atom has three regions of electron density (comprised of two C—H bonds and one C=O bond). The VSEPR model suggests that these three groups assume a trigonal planar geometry with approximately 120° bond angles. That is what is found. The four atoms lie in a plane, and the bond angles are very close to the predicted angle. This is another case in which the electron pair geometry and the molecular geometry are the same. Note that the geometry at the carbon atoms of ethylene ($H_2C=CH_2$), Figure 5.5, is the same as that about the central carbon atom in formaldehyde.

If a central atom has unshared pairs, the *electron pair geometry* will not be the same as the *molecular geometry*. This situation can be illustrated using water and ammonia. A Lewis dot structure for water reveals that there are two bonds to hydrogen atoms and two unshared pairs on the central oxygen atom. Thus, if we take into account the actual geometric relationship between the atoms and unshared pairs, we can predict that water has a tetrahedral electron pair geometry with an H—O—H bond angle of about 109.5°. The actual bond angle is slightly compressed to 104.5° due to excess repulsion by the two unshared pairs. The lone pairs exert a larger influence than do bonded pairs in determining repulsive interactions. However, *the molecular geometry of water is said to be "bent"* because the three atoms must all lie in a plane. Thus, the electron pair geometry of water is tetrahedral but the molecular geometry is bent.

A similar analysis of ammonia (NH_3) reveals that there are four regions of electron density around the central nitrogen atom. Thus, we predict the *electron pair geometry of ammonia to be tetrahedral* with an H—N—H bond angle of about 109.5°. The actual bond angle is again slightly compressed to 107° due to excess repulsion by the single unshared pair. (Compare the amount of repulsion and angle compression from one unshared pair in ammonia to two unshared pairs in water.) The *molecular geometry of ammonia is said to be trigonal pyramidal* since the nitrogen atom is at the apex of a shallow trigonal pyramid with the three hydrogen atoms forming the base of the pyramid.

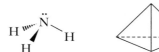

So, let us compare methane, ammonia, and water.

MOLECULE	BONDS	LONE PAIRS	BOND ANGLE (DEGREES)	MOLECULAR GEOMETRY	ELECTRON PAIR GEOMETRY
CH_4	4	0	109.5	tetrahedral	tetrahedral
NH_3	3	1	107	trigonal pyramidal	tetrahedral
H_2O	2	2	104.5	bent	tetrahedral

Names for Binary Molecular Compounds

Hydrogen forms binary compounds with all the nonmetals except the noble gases. For compounds of oxygen, sulfur, and the halogens, the H atom is generally written first in the formula and is named first using the element name. The other nonmetal is named as if it were a negative ion. For example, HF is hydrogen fluoride and H_2S is hydrogen sulfide.

When there is more than one possible combination of two elements, the number of atoms of a given type in the compound is designated with a prefix such as *mono-, di-, tri-, tetra-,* and so on. Table 5.4 lists the prefixes for up to ten atoms, and some common molecular compounds and their names are given in Table 5.5.

Many molecular compounds were discovered years ago and have names so common they continue to be used. Examples include water (H_2O), ammonia (NH_3), nitric oxide (NO), and nitrous oxide (N_2O).

EXAMPLE 5.8 **Naming Molecular Compounds**

What is the name of N_2O_4?

SOLUTION
The prefix for 2 N atoms is *di-* and the prefix for 4 O atoms is *tetra-*; thus, the name of N_2O_4 is dinitrogen tetraoxide.

TRY IT 5.8
What is the name of P_4S_3?

TABLE 5.4 Prefixes for Number of Atoms in a Compound

NUMBER OF ATOMS	PREFIX	NUMBER OF ATOMS	PREFIX
1	mono-	6	hexa-
2	di-	7	hepta-
3	tri-	8	octa-
4	tetra-	9	nona-
5	penta-	10	deca-

TABLE 5.5 Common Molecular Compounds

Compound	Name	Use
CO	Carbon monoxide	Preparation of methanol and other organic chemicals
CO_2	Carbon dioxide	Carbonated beverages; fire extinguisher, inert atmosphere; dry ice
NO	Nitrogen monoxide (nitric oxide)	Preparation of nitric acid
NO_2	Nitrogen dioxide	Preparation of nitric acid
N_2O	Dinitrogen oxide (nitrous oxide)	Spray-can propellant; anesthetic
SO_2	Sulfur dioxide	Preparation of sulfuric acid; food preservative; metal refining
SO_3	Sulfur trioxide	Preparation of sulfuric acid
CCl_4	Carbon tetrachloride	Solvent
SF_6	Sulfur hexafluoride	Insulator in electric transformers
P_4O_{10}	Tetraphosphorus decaoxide	Preparation of phosphoric acid

CONCEPT CHECK 5B

1. The name of SO_2 is _____.
2. The formula of sulfur trioxide is _____.
3. The name of HBr is _____.
4. The formula of dichlorine monoxide is _____.
5. The formula of sulfur dichloride is _____.
6. The name of $SiCl_4$ is _____.
7. How many electrons are shared in **(a)** a double bond? **(b)** a triple bond?
8. Which is the strongest bond? **(a)** C—C, **(b)** C=C, **(c)** C≡C.
9. Draw Lewis dot structures for the following molecules. Show all lone pairs.
   (a) HCN **(b)** $HONO_2$
   (c) CH_4 **(d)** CO
   (e) CO_2 **(f)** O_3
   (g) N_2H_4 **(h)** H_2O_2
   (i) HONO **(j)** C_2H_4
   (k) CH_3OH
10. Compare the Lewis structures of NH_3 and PBr_3. Note that N and P are in the same group of the periodic table, and comment on this fact as it relates to the Lewis structures of these two compounds.
11. Predict the structure of SiH_4. Hint: Look at your answer to question 9c and remember that C and Si are in the same group in the periodic table.
12. Which molecules in question 9 have double bonds? triple bonds?

5.5 POLAR AND NONPOLAR BONDING

In a molecule like H_2 or F_2, where both atoms are alike, there is equal sharing of the electron pair. Where two unlike atoms are bonded, however, the sharing of the electron pair is unequal and results in a shift of electric charge toward one partner. The more electronegative an element is, the more that element attracts electrons.

When two atoms are bonded covalently and their abilities to attract electrons are the same, there is an equal sharing of the bonding electrons, and the bond is a **nonpolar** covalent bond. The bonds in H_2, F_2, and NCl_3 (N and Cl have equal abilities to attract electrons) are nonpolar.

Two atoms with different abilities to attract electrons bonded covalently form a **polar** covalent bond. The bonds in HF, NO, SO_2, H_2O, CCl_4, and BeF_2 are polar. In a molecule of HF, for example, the bonding pair of electrons is drawn more toward the fluorine atom and away from the hydrogen atom (Figure 5.7). The unequal sharing of electrons makes the fluorine end of the molecule more negative than the hydrogen end.

Polar bonds fall between the extremes of nonpolar covalent bonds and ionic bonds. In a covalent bond between two atoms of the same element, there is no charge separation; that is, the negative charge of the electrons is evenly distributed over the bond. In ionic bonds there is complete separation of the charges, and in polar bonds the separation falls somewhere in between.

Polar bonds in molecules can result in the molecule itself being polar when the shape of the molecule allows a permanent separation of charge. A linear diatomic molecule containing a polar bond, such as HF, is polar because there is a separation between the positive end of the molecule and the negative end (Figure 5.7b). Some linear molecules can be nonpolar because the separated charges cancel out each other. The carbon dioxide (CO_2) molecule is an example. The oxygen atoms attract electrons to themselves, leaving the carbon atom with a partial positive charge, but because the oxygen atoms are opposite one another, the molecule itself is not polar.

$$\overset{\longleftrightarrow \quad \longleftrightarrow}{O=C=O}$$

The water molecule, on the other hand, is an example of a polar molecule. The shape of the water molecule is bent; thus, the partial charges of the polar bonds between the hydrogen atoms and the oxygen atom do not cancel out one another.

The $\longleftrightarrow$ indicates bond polarity. The arrow points in the direction where the partial negative charge is found.

Nonpolar Describes a bond or molecule which is not polar because it has no polar bonds or because its polar bonds are oriented symmetrically so that they cancel each other.

Polar Describes a bond or molecule in which charge is unevenly distributed, creating positive and negative regions

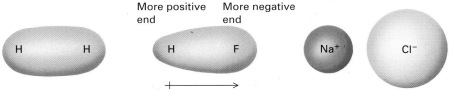

More positive end   More negative end

(a) Nonpolar covalent bond   (b) Polar covalent bond   (c) Ionic bond

FIGURE 5.7 **Nonpolar, polar, and ionic bonds.** (a) In a nonpolar molecule such as H_2, the valence electron density is equally shared by both atoms. (b) In a polar molecule like HF, the valence electron density is shifted toward the fluorine atom. An arrow is used to show the direction of molecular polarity, with the arrowhead pointing toward the negative end of the molecule and the plus sign at the positive end of the molecule. (c) In ionic compounds such as NaCl, the valence electron or electrons of the metal are transferred completely to the nonmetal to give ions.

5.6 PROPERTIES OF MOLECULAR AND IONIC COMPOUNDS COMPARED

The general properties of ionic and molecular compounds are summarized in Table 5.6. All these properties can be interpreted using the chemical view of matter. Ionic compounds form hard, brittle, crystalline solids with high melting points. Their crystalline nature is explained by the regular arrangement of positive and negative ions needed to place each ion in contact with those of the opposite charge. Their hardness is accounted for by the strong electrostatic forces of attraction that hold the ions in place. The differences in melting and boiling points (high for ionic compounds and low for molecular compounds) are largely accounted for by the strength of the electrostatic attraction between ions, which is greater than that between molecules. Melting requires adding enough energy to overcome these forces.

Whether a molecular compound is a gas, liquid, or solid is a function of the strength of the attractions between molecules. Only when there are huge molecules, with attractions between many polar regions in the molecule, as in polymers (Figure 14.11), do we find very hard and strong molecular substances.

Another major difference between ionic and molecular compounds is in their ability to conduct electricity, an ability that depends on the presence of mobile carriers of positive or negative charge. A solid ionic compound cannot conduct electricity because the ions are held in fixed positions. When melted or dissolved in water, the situation is different—the ions are free to move and current can flow. A compound that conducts electricity under these conditions is referred to as an **electrolyte**. Ionic compounds, to whatever extent they dissolve, are electrolytes in water solution because the ions separate from the crystal and can move about in the solution. Molecules, whether in pure compounds or dissolved in water or any other liquid, have no overall charge and therefore do not carry current; they are referred to as **nonelectrolytes**.

 See an animation of **Electric Current Conduction by Molten Salts** at http://brookscole.com/chemistry/joesten4

 See an animation of **An Ionic Substance Dissolving in Water** at http://brookscole.com/chemistry/joesten4

Electrolyte A compound that conducts electricity when melted or dissolved in water

Nonelectrolyte A compound that does not conduct electricity when melted or dissolved in water

---

TABLE 5.6 Properties of Ionic and Molecular Compounds

IONIC COMPOUNDS	MOLECULAR COMPOUNDS
Examples: NaCl, CaF$_2$	*Examples: CH$_4$, CO$_2$, NH$_3$, CH$_3$CH$_2$CH$_3$*
Many are formed by combination of reactive metals with reactive nonmetals	Many are formed by combination of nonmetals with other nonmetals or with less reactive metals
Crystalline solids	Gases, liquids, and solids
Hard and brittle	Solids are brittle and weak, or soft and waxy
High melting points	Low melting points
High boiling points (700°C to 3500°C)	Low boiling points (−250°C to 600°C)
Good conductors of electricity when molten; poor conductors of heat and electricity when solid	Poor conductors of heat and electricity
Many are soluble in water	Many are insoluble in water but soluble in organic solvents

CONCEPT CHECK **5C**

1. (**a**) An example of a molecule with covalent bonds in which the electrons are equally shared between the atoms is _____.
   (**b**) An example of a molecule in which electrons are unequally shared is _____.
2. What are the bond angles in CH_4?
3. Which is a more polar bond?
   (**a**) H—H or H—F
   (**b**) C—H or C—O
   (**c**) H—N or C—N
4. Which is a polar molecule? (**a**) H_2O, (**b**) H_2, (**c**) O_2, (**d**) CCl_4
5. Which is a nonpolar molecule with polar covalent bonds?
   (**a**) H_2O, (**b**) H_2, (**c**) O_2, (**d**) CCl_4

5.7 INTERMOLECULAR FORCES

Although atoms within molecules are held together by strong chemical bonds, the attractive forces between two separate molecules are much weaker. For example, only about 1% of the energy required to break one of the C—H bonds in a methane (CH_4) molecule is required to pull two methane molecules away from one another. These small forces between molecules are called **intermolecular forces**. Even though these attractions are small, they give samples of matter exceedingly important properties. For example, intermolecular attractions are responsible for the fact that all gases condense to form liquids and solids. Both pressure and temperature play a role in the liquefaction of a gas. When the pressure is high, the volume decreases and the molecules of a gas are close together. This condition allows the effect of attractive forces to be appreciable (see Section 5.8).

For polar molecules (Section 5.5), intermolecular forces act between the positive end of one polar molecule and the negative end of an adjacent polar molecule. Molecules of SO_2 (Figure 5.8) are polar, with the partially negative region of one molecule (represented by δ^-) being attracted to the partially positive region (represented by δ^+) of an adjacent molecule.

Even nonpolar molecules can have momentary unequal distribution of their electrons, which results in weak intermolecular forces. For example, why would nitrogen—composed of nonpolar molecules—liquefy? Why would carbon dioxide—also composed of nonpolar molecules—form a solid? When the molecules get close enough, one molecule will cause an uneven distribution of charge in its neighbor. The two molecules become momentarily polar and are attracted to each other. These induced attractive forces are weak but become more pronounced when the molecules are larger and contain more electrons.

Hydrogen Bonding

An especially strong intermolecular force called **hydrogen bonding** acts between molecules in which a hydrogen atom is covalently bonded to a strongly **electronegative** atom with nonbonding electron pairs. The electronegative atom may be fluorine, oxygen, or nitrogen. The hydrogen bond is the attraction between such a hydrogen atom with a partial positive charge on one molecule and an electronegative atom (F, O, or N) of another molecule that has a partial negative charge and one or more unshared electron pairs. The

Methane (CH_4) is the major component of natural gas.

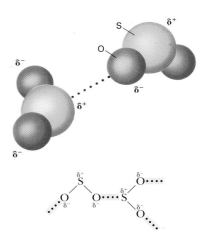

FIGURE 5.8 Polar sulfur dioxide molecules attracted to one another (oxygen atoms attract electrons more than do sulfur atoms).

Intermolecular forces Attractive forces that act between molecules; weaker than covalent bonds

Hydrogen bonding Attraction between a hydrogen atom bonded to a highly electronegative atom (O, N, F) and the lone pair on an electronegative atom in another or the same molecule

FIGURE 5.9 Boiling points of simple hydrogen-containing compounds. Lines connect molecules in which hydrogen combines with atoms from the same periodic table group. As shown for water, the temperature at which the compounds would boil if there were no hydrogen bonding is found by following the lines on the right down to the left.

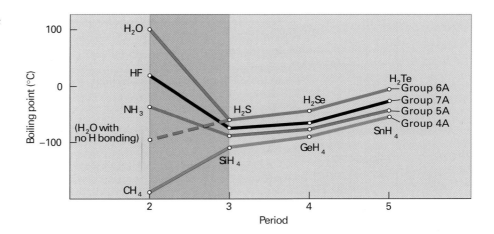

 See an animation of **Atomic Motion and Thermal Energy** at http://brookscole.com/chemistry/joesten4

 See an animation of **Atomic Properties of Solids, Liquids and Gases** at http://brookscole.com/chemistry/joesten4

greater the electron attraction of the atom connected to H, the greater the partial positive charge on the H, hence, the stronger the hydrogen bond is between it and a partially negative atom on another molecule. Hydrogen bonds are typically shown as dotted lines between atoms.

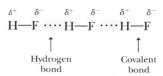

Water provides the most common example of hydrogen bonding. Hydrogen compounds of oxygen's neighbors and family members in the periodic table are all gases at room temperature. However, water is a liquid at room temperature, and this indicates a strong degree of intermolecular attraction. Figure 5.9 illustrates that the boiling point of H_2O is about 200°C higher than would be predicted if hydrogen bonding were not present.

Because each hydrogen atom can form a hydrogen bond to an oxygen atom in another water molecule and each oxygen atom has two nonbonding electron pairs, each water molecule can form a maximum of four hydrogen bonds to four other water molecules (Figure 5.10). The result is a tetrahedral cluster of water molecules around the central water molecule.

Although hydrogen bonds are much weaker (with a bond energy of only ~3–5 kcal/mol) than ordinary covalent bonds (O—H bond energies are ~111 kcal/mol), they play a key role in the chemistry of life (Chapter 15). For instance, all of the water in your body would simply boil away if the hydrogen bonds were all suddenly broken.

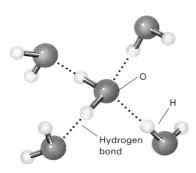

FIGURE 5.10 The four hydrogen bonds between one water molecule and its neighbors.

5.8 THE STATES OF MATTER

By now you know enough about bonding to understand the states of matter: gas, liquid, and solid. Matter in any state is composed of atoms, molecules, or ions in constant motion. In gases (Figure 5.11a), the particles are in rapid motion relative to one another. In addition, they are, on average, far apart from one another. In liquids (Figure 5.11b), the particles are in motion, but they are so close together that they touch one another. In solids (Figure 5.11c), the particles are also touching one another and are in motion. They do not, however, move appreciable distances, but instead vibrate in fixed positions.

The particles in a sample of gas move at high speeds when they are at or near room temperature. For example, the molecules found in the air in

Nitrogen molecules move at about 1230 mph and oxygen molecules, which are slightly more massive, move at about 1150 mph at 77°F. Hydrogen, the lightest gas, moves at about 3400 mph.

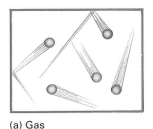
(a) Gas

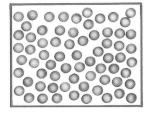
(b) Liquid

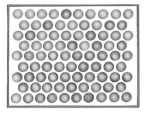
(c) Solid

FIGURE 5.11   The three states of matter. The particles represented by the circles can be atoms, molecules, or ions (in liquids and solids). (a) In a gas, particles are very far apart and move rapidly in straight lines. (b) In a liquid, the particles move about at random, alone or in clusters. (c) In a solid, the particles or ions are in fixed positions and can only vibrate in place.

your room are moving at over 1000 miles per hour. They only travel about 400 times their own diameter before colliding with another gas molecule, a short distance. If you cool a gas to a very low temperature, the molecules move much less rapidly. On further cooling, the molecules are moving so slowly that they attract one another and begin to form tiny droplets of liquid, a phenomenon called **condensation**. If that liquefied sample of gas is cooled even further, it will become a solid. All gases behave this way. For those gases consisting of nonpolar molecules, the condensation and solidification temperatures are quite low. However, for those gases whose molecules are polar, the temperatures are much higher. In the table in the margin, the condensation and solidification temperatures for SO_2, which is a polar molecule, and H_2O, which can engage in hydrogen bonding, are given for comparison.

In contrast to solids, whose particles are in rigid arrangements, liquids and gases have the property of being **fluid**—that is, they flow—because their atoms, ions, or molecules are not as strongly attracted to each other as they are in solids. Not being confined to specific locations, the particles in a liquid can move past one another. For most substances, the particles are a little farther apart in a liquid than in the corresponding solid, so that the volume occupied by a given mass of the liquid is a little larger than the volume occupied by the same mass of the solid. This means the liquid is less dense than the solid, and the solid form of a sample of matter sinks in its liquid (Figure 5.12). *There is a rather important exception to this rule: solid water*

Condensation and solidification temperatures for some gases at 1 atm pressure:

	Condenses (°C)	Solidifies (°C)
He	−269	−
Ar	−185.9	−189
O_2	−183.0	−218.4
N_2	−195.8	−209.8
Cl_2	−34.6	−101.0
SO_2	−10.1	−72.7
H_2O	100	0

Condensation The change of molecules from the gaseous state to the liquid state

Fluid A state of matter that is capable of flowing; a gas or a liquid

(a)

(b)

FIGURE 5.12   Water (H_2O) and benzene (C_6H_6). (a) Ice floats on water. (b) Solid benzene (melting point 5.5°C) sinks in liquid benzene.

Charles D. Winters

floats on liquid water. The importance of this property of water is discussed in Section 11.1.

For solids, liquids, or gases, the higher the temperature, the faster the particles move. A solid melts when its temperature is raised to the point at which the particles vibrate fast enough and far enough to get away from the attraction of their neighbors and move out of their regularly spaced positions. As the temperature rises, the particles move even faster, until finally they escape their neighbors and become independent; the substance becomes a gas.

5.9 GASES

Gases surround us in our atmosphere. We breathe a mixture of nitrogen, oxygen, and other gases. Every breath (~0.5 liter in volume) we take carries with it the oxygen gas we need to burn the foods we eat. When we exhale, the carbon dioxide gas that is produced by the food-burning processes in our cells leaves our bodies, as does all of the inhaled nitrogen. This is one of the ways we rid ourselves of waste materials.

Interestingly, all gases possess a set of common physical properties. At constant temperature, all gases expand when the surrounding pressure decreases and contract when the pressure increases. At constant pressure, all gases expand with increasing temperature and contract with decreasing temperature. All gases have the ability to mix in any proportion with other gases. These properties are explained by the fact that the gas particles are far apart, move fast, and have very little chance to interact. Their molecular properties, such as size, number of electrons, and shape, have virtually no effect on the properties of the gas as a whole.

A sample of gas confined inside a container exerts a **pressure**, which is caused by the individual particles of the gas sample striking the walls of the container. Earth's atmosphere, a mixture of gases, exhibits a pressure that is dependent on the altitude relative to sea level and temperature.

Another general property of gases is *compressibility*. All gases can be compressed by applying a pressure on a confined sample (Figure 5.13). This property of all gases, known as *Boyle's law* after Robert Boyle who discovered it in 1661, is explained by the great distances between gas particles—applying pressure only confines the particles in a smaller space. Of course, a gas will always expand if the pressure is reduced. If a sample of gas is released into deep space, where the pressure is effectively zero, the molecules would begin randomly moving away from one another and eventually would become attracted to some nearby star system.

Perhaps one of the more interesting properties of all gases is that of **miscibility**—the ability to mix in all proportions with other gases. The miscibility of gases is explained by the great distances between gas molecules. In effect, there is always room for more molecules.

To illustrate how gases mix with one another, consider what happens when a person wearing a strong perfume enters a room. The smell of the perfume is noticed immediately by those close by and eventually by everyone in the room. The perfume contains **volatile**, meaning easily vaporized, compounds that gradually mix with the other gases in the room's atmosphere. Even if there were no apparent movement of the air in the room, the aroma of perfume would eventually reach everywhere in the room. This mixing of two or more gases due to random molecular motion is called *gaseous diffusion*. Given time, the molecules of one component in a gas mixture will thoroughly and completely mix with all the other components to form a homogeneous mixture.

An average exhaled breath is only about 4% carbon dioxide. It is mostly nitrogen (~75%) and unused oxygen (~16%) with some water vapor.

Charles D. Winters

FIGURE 5.13 **A bicycle pump.** The effects described by Boyle's Law (Section 5.9) in action. The pump compresses air into a smaller volume. You experience Boyle's law because you can feel the increasing pressure of the gas as you press down on the plunger.

Molecules of a gaseous compound must physically enter your nose for you to be able to smell them.

Pressure Force per unit area

Miscibility Ability to mix in all proportions

Volatile Easily vaporized

CONCEPT CHECK   5D

1. Arrange the states of matter (liquid, gas, solid) by increasing order among the particles.
2. As temperature increases, molecular motion (decreases/increases).
3. Name the two states of matter in which the particles are very close to each other, on average.
4. Which two states of matter can be described as fluid?
5. Which state of matter is compressible? Which two states of matter are noncompressible?
6. As the temperature of a sample of gas decreases, the volume _____ at constant pressure.
7. All gases mix with one another in all proportions. (**a**) True, (**b**) False.
8. Which kind of intermolecular force is stronger—hydrogen bonding or attractions caused by shifting electrons in molecules?

5.10   LIQUIDS

In a liquid the molecules are close enough together that, unlike a gas, a liquid is only slightly compressible. However, the molecules remain mobile enough that the liquid flows. Because they are difficult to compress and their molecules are moving in all directions, confined liquids can transmit applied pressure equally in all directions. This property is used in the *hydraulic fluids* that operate automotive brakes and airplane wing surfaces, tail flaps, and rudders.

Every liquid has a **vapor pressure**, which is the pressure in the gaseous state of those molecules that have escaped from the liquid at a given temperature. As you would expect, the vapor pressure of a liquid increases with increasing temperature because more molecules escape the liquid (vaporize) and enter the gaseous state. The higher the temperature, the greater the volatility, because a larger fraction of the molecules have sufficient energy to overcome the attractive forces at the liquid's surface. Everyday experiences, such as heating water on a stove or spilling a liquid on a hot pavement in the summer, tell us that raising the temperature of the liquid makes vaporization take place more readily. Conversely, at lower temperatures, the volatility of a liquid will be lower. The same amount of water spilled on the pavement on a winter day will remain there much longer.

The temperature at which the vapor pressure equals the atmospheric pressure is the **boiling point** of the liquid. If the atmospheric pressure is 1 atm, the boiling temperature is called the **normal boiling point**. The normal boiling point of a compound is a reflection of the extent of intermolecular interactions (e.g., hydrogen bonding) and the molecular weight of the compound. In general, compounds with the ability to form hydrogen bonds have higher boiling points than those that do not. Higher molecular weight compounds also have higher boiling points. Water and ethanol, both capable of hydrogen bonding, boil at 100°C and 78.5°C, respectively. Ethane, which is incapable of hydrogen bonding, boils at −88°C, and octane, a higher molecular weight hydrocarbon also incapable of hydrogen bonding, boils at 125.6°C.

5.11   SOLUTIONS

Solutions, as explained in Section 2.3, are homogeneous mixtures that can be in the gaseous, liquid, or solid state. Most commonly, however, we encounter liquid solutions. How many liquid solutions are familiar to you? How about

Vapor pressure The pressure above a liquid caused by molecules that have escaped from the liquid surface

Boiling point The temperature at which the vapor pressure of a liquid equals the atmospheric pressure and boiling occurs. If the atmospheric pressure is 1 atm, it is the **normal boiling point.**

FIGURE 5.14 Solubilities. (a) Polar water, with a bit of nonpolar iodine (I_2) dissolved in it, floats on top of nonpolar carbon tetrachloride (CCl_4), with which it is immiscible. (b) Nonpolar iodine is much more soluble in nonpolar carbon tetrachloride than in water. Therefore, shaking the mixture in (a) causes the iodine molecules to migrate into the carbon tetrachloride, where they produce a purple color.

(a)   (b)

Charles D. Winters

 See an animation of **Like Dissolves Like Solubility** at http://brookscole.com/chemistry/joesten4

Solubility of Oxygen in Water at Various Temperatures*

TEMPERATURE (°C)	SOLUBILITY OF O_2 (g O_2/L H_2O)
0	0.0141
10	0.0109
20	0.0092
25	0.0083
30	0.0077
35	0.0070
40	0.0065

*These data are for water in contact with air at 1 atm pressure.

Solvent In a solution, the substance present in the greater amount

Solute In a solution, the substance dissolved in the solvent

Solubility The maximum quantity of a solute that will dissolve in a given amount of a solvent at a specified temperature

Saturated solution Solution that has dissolved all the solute that it can dissolve at a given temperature

Unsaturated solution Solution in which less than the maximum amount of solute is dissolved at a given temperature

Insoluble Describes a substance that will not dissolve in a given solvent

sugar or salt dissolved in water, oil paints dissolved in turpentine, or grease dissolved in gasoline? In each of these solutions, the substance present in the greater amount, the liquid, is the **solvent**; and the substance dissolved in the liquid, the one present in a smaller amount, is the **solute**. For example, in a glass of iced tea, water is the solvent; and sugar, lemon juice, and the components extracted from the tea leaves that impart taste and color are solutes.

A substance's **solubility** is defined as the quantity of solute that will dissolve in a given amount of solvent at a given temperature. Solubility is determined by the strengths of the forces of attraction between solvent molecules, between solute atoms, molecules, or ions, and between solute and solvent. The forces acting between the solvent and solute particles must be greater than those within the solute for the solute to dissolve. In other words, the solute and solvent must like each other more than they like themselves. If the forces between a liquid solvent and a solute are strong enough, then the solute will dissolve (Figure 5.14).

Although some solutes are so much like the solvent they are dissolving in that they dissolve in all proportions (a property called miscibility, like that of gases), there are limits to the solubility of most solutes in a given solvent. When a quantity of solvent has dissolved all the solute it can, the solution is said to be **saturated**. If more solute can be dissolved in the solution, the solution is said to be **unsaturated**. Our everyday experiences illustrate this kind of behavior in solutions. For example, if a spoonful of sugar is added to a glass of iced tea, it quickly dissolves with a little stirring, which hastens the mixing process. The resulting solution is unsaturated (it can dissolve more sugar); if you like your iced tea a little sweeter, you just add another spoonful of sugar and stir. However, if sugar is continually added to the solution, a point is reached when no more appears to dissolve and the added sugar simply sinks to the bottom of the glass. At this point the solution has become saturated in the dissolved sugar.

If almost none of the solute dissolves in a solvent, then it is said to be **insoluble** in that solvent. Metals, except those that react with water, are insoluble in water. Oily substances are also insoluble in water, as is water in oily substances. The low solubility of oily substances in water is caused by a lack of attraction between the highly polar water molecules and the nonpolar molecules of an oily substance. The water solubility of a nonpolar molecule can be increased if a polar part can be added. Conversely, the water solubility of a molecule decreases as the nonpolar portion of the molecule increases. The simpler alcohols are very water soluble because the —OH group is polar and forms strong hydrogen bonds with water. As the hydrocarbon chain lengthens, the influence of the

—OH group decreases, and with it, the water solubility of the molecule. A general rule of solubility is that "like dissolves like," which means that structurally similar compounds are more likely to be soluble in each other.

The effect of temperature on the solubility of solutes in various solvents is difficult to predict. Experience with everyday solutions such as sugar in water would lead us to predict that increasing the temperature of the solvent will cause more solute to dissolve. Although this is true for sugar and most other nonionic solutes, it is not true for all ionic compounds. In fact, the solubility of table salt in water is about the same at all temperatures.

Gases dissolve in liquids to an extent dependent on the similarity of the gas molecules and the solvent molecules. Polar gas molecules dissolve to a greater extent in polar solvents than nonpolar molecules do. Pressure also affects gas solubility. At higher pressure more gas dissolves in a given volume of liquid than at lower pressure. When the pressure is lowered, gas will be evolved from a gas-in-liquid solution. The behavior of a carbonated beverage when the cap is removed is a common illustration of this principle.

Temperature has a greater effect on gas solubility in liquids than it has on solids or liquids dissolving in liquids. Without exception, lower temperatures cause more gas to dissolve in a given volume of liquid. Higher temperatures cause less gas to dissolve. As every fisherman knows, fish prefer deeper water in the summer months. This is because more oxygen dissolves in colder water, and the water temperature is generally cooler at greater depths.

Air, a mixture of gases, is a solution. Sterling silver, a mixture of silver and copper metals, is also a solution.

5.12 SOLIDS

In contrast to gases and liquids, in which molecules are in continual random motion, the movement of atoms, molecules, or ions in solids is restricted to vibration and sometimes rotation around an average position. This leads to an orderly array of particles and to properties different from those of liquids or gases.

Because the particles that make up a solid are very close together, solids are difficult to compress. In this respect solids are like liquids, the particles of which are also very close together. Unlike liquids, however, solids are rigid, so they cannot transmit pressure in all directions. Solids have definite shapes, occupy fixed volumes, and have varying degrees of hardness. Hardness depends on the kinds of bonds that hold the particles of the solid together. Graphite is one of the softest solids known; diamond is one of the hardest. Graphite is used as a lubricant. At the atomic level, graphite consists of layered sheets that contain carbon atoms. The attractive forces between the sheets are very weak; as a result one sheet can slide along another and be removed easily from the rest. In diamond, on the other hand, each carbon atom is strongly bonded to four neighbors, and each of those neighbors is strongly bonded to three more carbon atoms, and so on throughout the solid. Because of the number and strength of the bonds holding each carbon atom to its neighbors, diamond is so hard that it can scratch or cut almost any other solid. The cutting and abrasive uses of diamonds are far more important commercially than their gemstone uses.

Properties of Solids

When a solid is heated to a temperature at which molecular motions are violent enough to partially overcome the interparticle forces, the orderliness of the solid's structure collapses, and the solid melts. The temperature at which melting occurs is the **melting point** of the solid. Melting requires energy to overcome the attractions between the particles in a solid lattice. The energy required to melt a compound is called the *heat of fusion*. The heat of fusion of ice (solid water) is 80 cal/g (334.7 J/g). In the reverse of melting, **solidification** or

Melting point The temperature at which a solid substance turns into a liquid

Crystallization (solidification) Formation of a crystalline solid from a melt or a solution

FIGURE 5.15 Ice cubes shrink in the freezer because of sublimation. Even a solid like ice has a vapor pressure caused by its molecules in the vapor state. Dry air will sweep these molecules away.

Dry air

Air containing water molecules from ice

Ice at 0°C, v.p. = 4.60 mm Hg

TABLE 5.7	Melting Points of Some Solids
SOLID	**MELTING POINT (°C)**
Molecular Solids: Nonpolar Molecules	
H_2	−259
O_2	−218.4
N_2	−209.8
F_2	−219.6
Cl_2	−101.0
Molecular Solids: Polar Molecules	
HCl	−115
HBr	−88
H_2S	−86
HF*	−83.1
NH_3	−77.7
SO_2	−72.7
HI	−51
H_2O	0
Ionic Solids	
NaI	662
NaBr	747
$CaCl_2$	782
NaCl	800
Al_2O_3	>2072
MgO	2852

*Strongly hydrogen-bonded molecules shown in color.

Basic types of solids are
(1) Molecular (most molecules)
(2) Metallic Fe, Cu, etc.
(3) Ionic NaCl, KNO_3
(4) Network (SiN or diamond)

Sublimation The process whereby molecules escape from the surface of a solid into the gas phase

crystallization, energy is evolved. Melting points of solids depend on the kinds of forces holding the particles together in the solid, because these forces must be overcome for a solid to become a liquid. Table 5.7 gives some melting points for several types of solids.

Molecules can also escape directly from the solid to enter the gaseous state by a process known as **sublimation**. A common substance that sublimes at normal atmospheric pressure is solid carbon dioxide, which has a vapor pressure of 1 atm at −78°C. Solid CO_2 is known as dry ice, because it is cold like ice (although much colder at −78°C) and produces no liquid residue because it does not melt. Ice can sublime or melt. Have you ever noticed that ice and snow slowly disappear even if the temperature never gets above freezing? The reason is that ice sublimes readily in dry air (Figure 5.15). Given enough air passing over it, a sample of ice will sublime away, even at temperatures well below its melting point, leaving no trace behind. This is what happens in a frost-free refrigerator. A current of dry air periodically blows across any ice formed in the freezer compartment, taking away water vapor (and hence the ice) without having to warm the freezer compartment to melt the ice.

5.13 REVERSIBLE STATE CHANGES

You are certainly aware that ice turns into liquid water upon melting and that liquid water turns into water vapor upon evaporation, but did you ever stop to consider the microscopic changes associated with these changes of state? At 0°C the water molecules in ice are tightly held in a crystalline lattice in which there is little molecular motion. As the ice absorbs heat and starts to melt, the lattice structure is disrupted, and the molecular motion increases. However, the temperature of the liquid water immediately upon melting is still 0°C. If the water is heated even more, the liquid-state molecular motion becomes greater and greater while the water temperature increases from 0°C to 100°C. At 100°C, there is enough thermal energy to overcome all intermolecular interactions between water molecules. The water begins to boil and turn into the vapor state. The temperature of the water molecules in the vapor state is exactly 100°C. It is possible, however, to superheat steam above this point, which is why burns from steam sometimes are worse than those from boiling water.

This process of changing a solid to a liquid and then changing the liquid to a gas involves melting to the liquid state, heating of the liquid state to the boiling point, and vaporization of the liquid state to the gaseous state. Note that in this direction the processes are all *endothermic* (heat input is required). The reverse process (condensing water vapor to liquid water, cooling the water to its freezing point, and allowing the water to freeze) is *exothermic*, meaning that heat is released. By using the heat of fusion, the specific heat, and the heat of vaporization for water, one can calculate the amount of heat required to, for instance, convert 5 grams of ice at 0°C into 5 grams of steam at 100°C. Melting requires 400 cal (5 g × 80 cal/g), increasing the temperature of the water from 0°C to 100°C requires 500 cal (5 g × 1 cal/g-°C × 100°C), and converting the water to steam requires 2700 cal (5 g × 540 cal/g). Thus, the entire process is endothermic by 3600 cal. The reverse of this process would be exothermic by 3600 cal. Assuming we know the appropriate numbers, we can perform these calculations for any substance that undergoes these phase changes in a regular manner.

The fact that liquid water releases heat as it cools down and then freezes is sometimes used to advantage by citrus and vegetable farmers who anticipate that crops may experience only a very light freeze. They spray crops liberally with water, which releases heat as it freezes to form an insulating ice blanket around the crop. If the temperature does not drop much below freezing, the crop may be saved. You might use this to your advantage by making sure your outdoor plants are well watered on a night when the temperature is expected to hover near freezing.

CONCEPT CHECK   5E

1. Most solids will not float in their liquids. (**a**) True, (**b**) False.
2. The density of solid water is less than that of liquid water. (**a**) True, (**b**) False.
3. In a solution, the substance dissolved is called the _____, and the substance doing the dissolving is called the _____.
4. Carbon tetrachloride is a nonpolar compound. In which solvent would you expect it to dissolve? (**a**) Water (a polar solvent), (**b**) Corn oil (a nonpolar solvent).
5. More of a gas will dissolve in warm water than in cold water. (**a**) True, (**b**) False.
6. Increasing the pressure of a gas over a liquid (increases/decreases) its solubility in the liquid.
7. Ionic solids generally have melting points higher than those of nonpolar molecular solids. (**a**) True, (**b**) False.
8. The opposite of crystallization is called _____.
9. The process of molecules escaping the surface of a solid and going into the vapor state is called _____.

 Assess your understanding of this chapter's topic with an online chapter quiz at **www.brookscole.com/chemistry/joesten4**

■ KEY TERMS

octet rule	ionic compound	lone pairs
electronegativity	formula unit	hydrocarbons
anion	binary compound	alkanes
cation	polyatomic ion	isomers
Lewis dot symbol	covalent bond	saturated hydrocarbons
ionic bond	bonding pairs	double bond

triple bond	hydrogen bonding	solvent
resonance structures	electronegative	solute
bond energy	condensation	solubility
alkenes	fluid	miscibility
unsaturated hydrocarbons	pressure	saturated solution
nonpolar	miscibility	unsaturated solution
polar	volatile	insoluble
electrolyte	vapor pressure	melting point
nonelectrolyte	boiling point	crystallization (solidification)
intermolecular forces	normal boiling point	sublimation

■ THE LANGUAGE OF CHEMISTRY

1. Ionic bonds
2. Covalent bonds
3. Single covalent bond
4. Double covalent bond
5. Triple covalent bond
6. NaCl
7. Molecule
8. NH_3
9. Mixable in all proportions
10. Made of carbon atoms all covalently bonded to four other carbon atoms
11. Causes water's high boiling point
12. Solvent that will dissolve most salts
13. Causes more gas to dissolve in a liquid
14. Number of neighbors around a water molecule in ice
15. Going from solid to gas
16. Going from gas to liquid
17. Phosphate ion

a. An electrically neutral arrangement of covalently bonded atoms
b. Positive ions attracted to negative ions
c. Polyatomic ion
d. Bond with four shared electrons
e. Bond with two shared electrons
f. Molecule
g. Ionic compound
h. Bond with six shared electrons
i. Shared electrons
j. Hydrogen bonding
k. Any two gases
l. Condensation
m. Sublimation
n. Four
o. Water
p. Pressure
q. Diamond

■ APPLYING YOUR KNOWLEDGE

1. Give definitions for the following terms:
   (a) Cation   (b) Anion
   (c) Octet rule   (d) Formula unit
2. Give definitions for the following terms:
   (a) Nonmetal   (b) Binary compound
3. Give definitions for the following terms:
   (a) Shared pair   (b) Double bond
   (c) Triple bond   (d) Unshared pair
   (e) Single bond   (f) Multiple bond
4. Give definitions for the following terms:
   (a) Covalent bond   (b) Polyatomic ion
   (c) Ionic bond   (d) Binary compound
5. Give definitions for the following terms:
   (a) Nonpolar bond   (b) Polar bond
6. What is the octet rule?

7. Describe what each of the following terms means:
   (a) Hydrocarbon
   (b) Saturated hydrocarbon
   (c) Unsaturated hydrocarbon
   (d) Alkenes
8. Is Ca^{3+} a possible ion under normal chemical conditions? Why or why not?
9. Predict the ions that would be formed by:
   (a) Br   (b) Al
   (c) Na   (d) Ba
   (e) Ca   (f) Ga
   (g) I   (h) S
   (i) All Group 1A metals
   (j) All Group 7A nonmetals

10. An ion has 12 protons, 13 neutrons, and 10 electrons. What is its charge? Consult the periodic table and write the symbol of the ion.

11. Write the formula and name of the ionic compounds formed from atoms of each of the following pairs of elements:
    (a) Al and I   (b) Sr and Cl
    (c) Ca and N   (d) K and S
    (e) Al and S   (f) Li and N

12. What holds ionic solids together?

13. Name the following compounds:
    (a) $CaSO_4$ (b) Na_3PO_4
    (c) $NaHCO_3$ (d) K_2HPO_4
    (e) $NaNO_2$ (f) $Cu(NO_3)_2$

14. The following are known mostly by the names that follow the formula for naming such compounds. Give another acceptable name for these molecules.
    (a) H_2O water or _____
    (b) O_2 oxygen or _____
    (c) H_2O_2 hydrogen peroxide or _____

15. Write correct formulas for the ionic compounds you expect to be formed when the following pairs of elements react:
    (a) Li and Te   (b) Mg and Br
    (c) Ga and S

16. Describe the difference between an ionic bond and a covalent bond.

17. Predict the type of bond formed between each of the following pairs of elements:
    (a) Sodium and sulfur
    (b) Nitrogen and bromine
    (c) Calcium and oxygen
    (d) Phosphorus and iodine
    (e) Carbon and oxygen

18. Complete the following table by writing the predicted formulas for each pair of elements:

	F	O	Cl	S	Br	Se
Na	NaF					
K				K_2S		
B		B_2O_3				
Al						
Ga			$GaCl_3$			
C					CBr_4	
Si						$SiSe_2$

19. Name the following compounds:
    (a) NO   (b) SO_3
    (c) N_2O (d) NO_2

20. Draw Lewis structures for the following:
    (a) CO   (b) SiF_4
    (c) C_2H_4 (d) H_2S
    (e) C_2H_2 (f) C_2H_6
    (g) OH^- (h) NF_3

21. Summarize the differences between ionic, polar covalent, and nonpolar covalent bonding.

22. Draw Lewis dot structures for the following molecules and then use your knowledge of the VSEPR model to predict the shapes of the molecules.
    (a) dichlorodifluoromethane
    (b) nitrogen trichloride
    (c) oxygen dichloride

23. Which are the more polar bonds in the following molecules?

    (a) Chloroethane   (b) Freon 12
       (CH_3CH_2Cl)   (CCl_2F_2)

24. A compound that has an odd number of total valence electrons can never satisfy the octet rule. Why?

25. Nitric oxide (NO) is a molecule that does not satisfy the octet rule. How many total valence electrons are present in the molecule? Draw the best Lewis structure you can for this molecule.

26. Which of the following compounds has the most polar bonds? (a) H—F, (b) H—Cl, (c) H—Br.

27. Which of the following molecules is (are) not polar? For each polar molecule, which is the negative and which is the positive end of the molecule?
    (a) CO, (b) GeH_4, (c) BCl_3, (d) HF.

28. Which of the following molecules is polar and which is nonpolar? Explain.
    (a) Acetone (CH_3COCH_3), a common solvent

    (b) Butane ($CH_3CH_2CH_2CH_3$), a common fuel

    (c) Ammonia (NH_3)

29. Describe the following states of matter: (a) gas, (b) liquid, (c) solid.

30. The structural formula for ethanol, the alcohol in alcoholic beverages, is

$$
\begin{array}{ccc}
& H & H \\
& | & | \\
H- & C- & C- O- H \\
& | & | \\
& H & H
\end{array}
$$

   Give the total number of: (a) valence electrons, (b) single bonds, and (c) bonding pairs of electrons. How many extra pairs of electrons are left? What are these called, and where should they be placed in the structural formula?

31. In which state of matter are the particles in fixed positions? (a) liquid, (b) solid, (c) gas.

32. In which state of matter are the particles the greatest average distance apart? (a) liquid, (b) solid, (c) gas.

33. Explain why the pressure of the atmosphere decreases with increasing altitude.

34. Which state of matter can be described as "particles close together and in constant, random motion"? (a) liquid, (b) solid, (c) gas.

35. Explain why molecules of a perfume can be detected a few feet away from the person wearing the perfume.

36. What happens to the pressure in an automobile tire in cold weather? Explain.

37. Name two properties of water that are unusual because of the presence of hydrogen bonding between adjacent water molecules.

38. Draw a structure showing four water molecules bonded to a central water molecule by means of hydrogen bonding. Indicate all the hydrogen bonds by drawing arrows to them.

39. Whenever a liquid evaporates, heat is required. Use this statement to explain why you get chilled when you emerge from a swimming pool on a windy day.

40. Which of the following compounds would you expect to exhibit hydrogen bonding? Explain your answer. (a) CH_3OH, (b) NH_3, (c) SO_2, (d) CO_2, (e) CH_4, (f) HF, (g) CH_3OCH_3.

41. Based on Figure 5.9, approximately what would be the boiling point of ammonia if there were no hydrogen bonding between the molecules?

42. Why are gases compressible, whereas liquids and solids are not?

43. Why is a gas more soluble in a solvent when a higher pressure is applied? Give an example of where you see this behavior of gases dissolving in liquids.

44. What causes surface tension in liquids? Name a compound that has a high surface tension.

45. Victims of carbon monoxide poisoning are placed in a hyperbaric chamber where the pressure is raised above 1 atm and a richer oxygen environment exists. If the pressure is raised from 1 atm to 3 atm, what will be the proportional change in oxygen solubility in blood? Use Henry's law and assume blood behaves like water.

46. The solubility of gases in liquids depends on temperature and gas pressure. Henry's law says the concentration of dissolved gas, C, is equal to the pressure of the gas multiplied by a constant, $C = kP$. The solubility of O_2 in water at 10°C is 0.0109 g O_2/L H_2O when the pressure is 1 atm. What would be the solubility if the pressure is decreased to 0.5 atm?

47. Gold has a normal melting point of 1063°C and a normal boiling point of 2600°C when the pressure is 1 atm. What state (solid, liquid, gas) will gold be in at 800°C and 1 atm?

48. A common solvent, acetone, $(CH_3)_2CO$, has a normal melting point of −95°C and a normal boiling point of 56.5°C when the pressure is 1 atm. What state (solid, liquid, gas) will acetone be in at −50°C and 1 atm?

49. Francium is radioactive; therefore, physical data for francium compounds are not readily available. What melting point do you expect for francium chloride, FrCl, based on the normal melting points for other Group 1A ionic chlorides? Justify your estimate. NaCl, 801°C; KCl, 776°C; CsCl, 646°C; FrCl, _____.

50. Predict the boiling point for octane, C_8H_{18}, using the observed boiling points for the similar hydrocarbons hexane, C_6H_{14}; 68.7°C; heptane, C_7H_{16}; 98.4°C; and nonane, C_9H_{20}; 150.8°C.

W CHEMISTRY ON THE WEB

For up-to-date URLs, visit the text website at **www.brookscole.com/ chemistry/joesten4**

- Covalent Bonding
- Solids and Liquids and Gases, Oh Why?
- States of Matter
- Gases, Liquids and Solids

CARBON DIOXIDE AND THE GREENHOUSE EFFECT

The Manshuk Mametova glacier is melting down to a lake in the northern Tien Shan mountains.

© Alexei Kalmykov/Reuters/CORBIS

As pollutants go, carbon dioxide might be considered a relatively innocuous component of our atmosphere. After all, every breath we exhale contains about 4% carbon dioxide even though the inhaled air is only about 0.04% carbon dioxide. Fossil fuel combustion and natural decay processes combined with forest and grassland fires release billions of metric tons of carbon dioxide into the atmosphere annually. At the same time, trees, grasses, and other plants remove equivalent quantities of carbon dioxide from our atmosphere each year. Carbon dioxide is constantly being dissolved in and released from the ocean and other bodies of water as their temperature fluctuates. In other words, carbon dioxide is constantly being added to and removed from our atmosphere by a variety of processes, some natural and some of human origin.

Carbon dioxide is not considered a pollutant in the same sense that nitrogen oxides, sulfur oxides, ozone, carbon monoxide, and other gaseous pollutants are. However, there is much concern that the increasing concentration of carbon dioxide in our lower atmosphere is having a major impact on our climate. In this chapter we will consider the following issues:

- What is the normal concentration of carbon dioxide in our atmosphere?
- What factors are causing the concentration of carbon dioxide to increase?
- What are the implications of rising carbon dioxide concentrations?
- What are greenhouse gases?
- Why is carbon dioxide the greenhouse gas about which we are most concerned?
- What can be done to control greenhouse gas emissions?

6.1 ATMOSPHERIC CARBON DIOXIDE CONCENTRATION OVER TIME

Although the concentration of carbon dioxide in our atmosphere is relatively low (about 0.04% by volume), evidence indicates that this concentration has fluctuated quite significantly over the past 200,000 years. This is not a long time, considering Earth has existed for about 4.5 billion years, but this is the time frame for which the concentration can be reliably inferred using available data. Some have suggested that significantly higher concentrations of carbon dioxide may have been, at least in part, responsible for the 10–20°C higher terrestrial temperatures about 100 million years ago when dinosaurs roamed Earth.

Data from ice cores taken from deep below the surface of Antarctica have been used to estimate Earth's temperature in the past. The exact manner in which these determinations have been made depends on the fact that water molecules containing different isotopes of hydrogen (protium, the most common isotope, or deuterium) have slightly different boiling points. By looking at the relative amounts of these isotopes in the ice, one can determine the atmospheric temperature at the time the ice was laid down. Thus, one can compare the temperature data with the carbon dioxide data to see if there is a correlation. This correlation is shown in Figure 6.1 on page 128. It reveals that as the amount of carbon dioxide trapped in these ice cores fluctuated, the temperature fluctuated in the same direction. Higher carbon dioxide concentrations correlate with higher than average temperatures in a strikingly similar pattern. This has led scientists to believe that the amount of carbon dioxide in our atmosphere has a profound impact on Earth's temperature.

This correlation suggests, but does not prove, that the high concentration of carbon dioxide is the cause of the temperature increase. It is theoretically possible, but thought to be less likely, that other phenomena led to higher temperatures which, in turn, caused the carbon dioxide concentration to rise. However, carbon dioxide is known to trap heat so it is incumbent on us to see if the increasing concentration of carbon dioxide in our atmosphere and the resultant temperature increase in the past two hundred years, a mere blip in time on Figure 6.1, are related to human activity.

Ice core samples from the Greenland Ice Sheet. The samples, stored at −33°F, are examined for evidence of global warming.

© Roger Ressmeyer/CORBIS

THE CHEMISTRY OF

APOLLO 13 (1995)

© 2005 by Mark A. Griep, *University of Nebraska–Lincoln*

The movie *Apollo 13* (released in 1995) is based on the real story of the ill-fated Apollo 13 moon mission of 1970. The most dramatic scene in the movie *Apollo 13* combines carbon dioxide, inorganic chemistry, and human ingenuity. The three astronauts nearly died due to toxic levels of the carbon dioxide that they exhaled into the confined space of their lunar module.

After two uneventful days on its way to the moon, an electrical short caused two of the oxygen tanks mounted on the outside of the command and space module to explode. In response, pilot Jack Swigert radioed Ground Control: "Houston, we've had a problem here."

The oxygen in the tanks was used for several purposes. It was inhaled by the astronauts and it was combined with hydrogen to make water and energy. The released energy was used to generate electrical power for the spacecraft in conjunction with fuel cells. The loss of two oxygen tanks meant that the moon-landing mission had to be aborted. Bringing the astronauts back safely became the primary objective.

To conserve energy, Ground Control ordered the three-man crew to shut down all electrical instruments, to move into the smaller Lunar Module, and to breathe as little as possible. Soon, however, Ground Control realized that the "CO_2 scrubbers" were also turned off and that CO_2 levels would build up unless something was done. During the process of respiration, CO_2 and water are exhaled.

$$O_2 + [CH_2O] \longrightarrow CO_2 + H_2O + \text{energy}$$

Respiration is a passive process, which means that ambient gas levels determine the levels in the body. High carbon dioxide in the blood is toxic because it prevents oxygen from reaching the cells. The astronauts had to devise a way to keep the CO_2 scrubbers working.

© NASA

Lithium hydroxide canister rigged up during the Apollo 13 flight.

Questions to Consider

1. The Apollo 13 mission used lithium hydroxide to remove the exhaled CO_2 from their atmosphere. The lunar module was equipped with enough LiOH for two astronauts during their two-day mission to the moon. How many days would the LiOH in the lunar module last for three men exhaling for four days, if they didn't change the canister?

   $$2\,LiOH_{(s)} + CO_2 \longrightarrow Li_2CO_{3(s)} + H_2O$$

2. When the engineers were designing Apollo 13, they had to estimate the amount of CO_2 that each astronaut would exhale. Unlike the Lunar Module, the Command and Space Module had enough LiOH to scrub all of the CO_2 exhaled by the three men. If each astronaut exhales 18.2 mol of CO_2 every day and the mission is eight days, what mass of LiOH would exactly meet the requirement?

3. On spaceships, it is important to use lightweight, non-bulky materials. The engineers chose LiOH over $Ba(OH)_2$. What mass of $Ba(OH)_2$ would handle the CO_2 exhaled by the three astronauts during their eight-day mission? How does this mass compare to LiOH?

   $$Ba(OH)_{2(s)} + CO_2 \longrightarrow BaCO_{3(s)} + H_2O$$

FIGURE 6.1 Estimated long-term variations in mean global surface temperature and average tropospheric carbon dioxide levels of the past 160,000 years. These carbon dioxide levels were obtained by inserting metal tubes deep into Antarctic glaciers, removing the ice, and analyzing bubbles of ancient air trapped in ice at various depths throughout the past. Such analyses reveal that, since the last great ice age ended about 10,000 years ago, we have enjoyed a warm interglacial period. The rough correlation between tropospheric carbon dioxide level and temperatures suggests, but does not prove, a connection between the two.

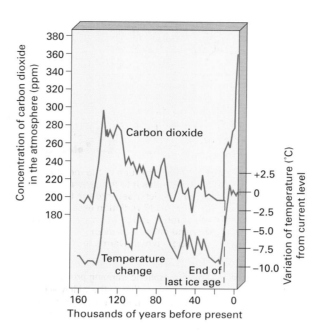

6.2 WHAT IS THE GREENHOUSE EFFECT?

The **greenhouse effect** is a well-known phenomenon that results in warming of Earth's atmosphere by absorption of **infrared (IR) radiation** by molecules in the atmosphere such as carbon dioxide, water vapor, and methane. Figure 6.2 shows that infrared radiation accounts for about 53% of the radiation coming from our Sun. About 8% is higher-energy ultraviolet (UV) radiation and 39% is visible (Vis) radiation. What is the impact of this radiation as it travels through our atmosphere and then strikes Earth?

Radiation making its way from the Sun to Earth encounters many types of molecules once it reaches Earth's stratosphere. Certain molecules are more efficient than others at absorbing some types of radiation and different types of radiation bring about different responses in the molecules they encounter. Both features are critical to understanding the greenhouse effect.

Figure 6.3 illustrates the fate of solar radiation. Much of the radiation coming from the sun is simply reflected back into space once it encounters our atmosphere. This fact is of little practical consequence to us except as a

Greenhouse effect A phenomenon of Earth's atmosphere by which solar radiation, trapped by Earth and re-emitted from the surface as infrared radiation, is prevented from escaping by various gases in the atmosphere. This leads to warming of Earth.

Infrared (IR) radiation Electromagnetic radiation whose wavelength is in the range between 2.5×10^{-5} m and 2.5×10^{-6} m (see Figure 3.6); it is often referred to as heat radiation

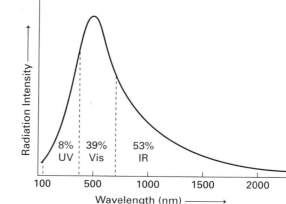

FIGURE 6.2 A comparison of the intensity of solar radiation and the wavelength of the radiation.

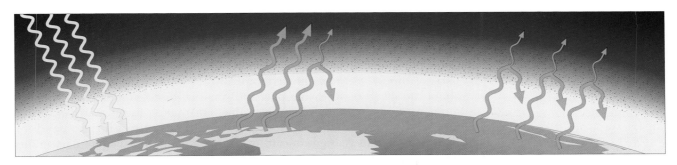

(a) Rays of sunlight penetrate the lower atmosphere and warm the earth's surface.

(b) The earth's surface absorbs much of the incoming solar radiation and degrades it to longer-wavelength infrared (IR) radiation, which rises into the lower atmosphere. Some of this IR radiation escapes into space as heat and some is absorbed by molecules of greenhouse gases and emitted as even longer wavelength IR radiation, which warms the lower atmosphere.

(c) As concentrations of greenhouse gases rise, their molecules absorb and emit more infrared radiation, which adds more heat to the lower atmosphere.

FIGURE 6.3 **The natural greenhouse effect.** Without the atmospheric warming provided by this natural effect, Earth would be a cold and mostly lifeless planet. According to the widely accepted greenhouse theory, when concentrations of greenhouse gases in the atmosphere rise, the average temperature of the troposphere rises. (Modified by permission from Cecie Starr, *Biology Concepts and Applications*, 4th ed, Brooks/Cole [Wadsworth] 2000)

reminder that Earth would be too hot to support life if this did not happen. Fortunately for us, much of the higher-energy ultraviolet radiation is absorbed by molecules in our atmosphere (primarily oxygen and ozone) so that only a small portion of this potentially damaging radiation reaches the surface of Earth.

Most of the solar radiation that hits Earth is in the visible and infrared range. At Earth's surface, this radiation is absorbed by all manner of objects including water, dirt, buildings, sidewalks, ice, snow, grass, and, of course, humans. By and large, the absorbed radiation increases molecular vibrations in whatever object it strikes. Sunlight striking your skin causes the molecules that comprise the skin to vibrate, making you feel warm. The same is true of sunlight striking Earth, a building, or a body of water. The energy of this absorbed radiation increases the temperature of the body that has absorbed the radiation. An object, once warmed, immediately begins to reradiate this energy away to cooler objects with which it is in contact, including the atmosphere. This energy is lost from the warmer body in the form of infrared radiation.

As infrared radiation heads away from Earth back toward space, about 84% of it is reabsorbed by molecules (water vapor, carbon dioxide, and methane) in our atmosphere and then reradiated back to Earth. This "greenhouse blanket" is critical to the balance of incoming and outgoing radiation that keeps our Earth's temperature tolerable for life. Thankfully, this phenomenon warms the surface of Earth to a comfortable average temperature of 15°C or about 59°F (compared to about −270°C in outer space) and allows us to flourish here. Without this warming, our oceans would be frozen solid!

The greenhouse effect is desirable up to a certain point. However, imagine what would happen if Earth were encased in a glass sphere capable of letting in UV and visible light but incapable of letting any of the reradiated IR radiation pass back out to space. The result would be catastrophic, as everyone knows who has ever entered a greenhouse at a garden store or at a botanical garden on a sunny day. Lots of visible light passes through the glass, is absorbed by objects, and is converted to infrared radiation which is then trapped in the greenhouse. It is not uncommon for a greenhouse to be uncomfortably warm

We will look at the importance of the oxygen-ozone screen as a mechanism for protecting us from dangerous UV radiation in Chapter 7.

A row of greenhouses.

for humans even when the outside temperature is below freezing. Greenhouses often require supplemental cooling and shading during summer months to offset this effect. Vehicles sitting in the hot sun in a parking lot all day serve as excellent examples of the greenhouse phenomenon. In many parts of the world the summer months bring with them constant warnings about the dangers of leaving pets and small children in locked cars, as the temperature increase may be unbearable. Fortunately for us, Earth's control mechanisms have regulated the natural greenhouse effect for thousands of years. However, increasing concern exists that humans have begun to upset this fine balance, and that is why we now speak of global warming (Section 4.11) as a potential environmental disaster whose time may have come.

TRY IT 6.1
Convert the 0.03%–0.04% average concentration of carbon dioxide in our atmosphere to ppm and ppb.

TRY IT 6.2

Convert the range of wavelengths for infrared radiation (2.5×10^{-5} m to 2.5×10^{-6} m) to nm and compare these wavelengths to those for ultraviolet radiation (200–400 nm). Which is longer? Which is more energetic?

6.3 WHY WORRY ABOUT CARBON DIOXIDE?

By using the relationships among energy, wavelength, and frequency (Figure 3.7), it is possible to calculate the energies associated with the types of electromagnetic radiation discussed above. For now we will just list the relative energies of infrared (IR), visible (Vis), and ultraviolet (UV) radiation and note, once again, that longer wavelength radiation corresponds to lower energy radiation. Later, after additional concepts have been introduced, we will be able to compute the exact energies in more meaningful terms (Section 8.2).

ENERGY TYPE	WAVELENGTH RANGE	RELATIVE ENERGY
Infrared	2.5×10^{-5} to 2.5×10^{-6} m	Low
Visible	700×10^{-9} to 400×10^{-9} m	Medium
Ultraviolet	400×10^{-9} to 200×10^{-9} m	High

Each type of radiation elicits a certain response when it impinges on a molecule or an object. An inspection of the relative energies associated with IR, Vis, and UV radiation reveals that IR radiation has the lowest energy of the three. IR radiation is sufficiently energetic to cause certain chemical bonds to bend or stretch but not to break. Infrared radiation is sometimes referred to as heat radiation because it causes things to warm up as the bonds comprising the molecules stretch and bend.

Only certain types of bending and stretching vibrations are allowed in molecules. Furthermore, an exact match must exist between the energy of the radiation and the energy necessary for this molecular motion to occur. Most of the molecules that comprise the major portion of our atmosphere do not absorb IR radiation. The ability of a molecule to absorb IR radiation is related to its shape and to the changes in electron distribution in the molecule that result when a given bond stretches or bends. The two major atmospheric gases (N_2 and O_2) do not absorb IR radiation and, thus, play no part in greenhouse warming.

W You can see amazing infrared photos of some common everyday objects at the following website: http://coolcosmos.ipac.caltech.edu/image_galleries/our_ir_world_gallery.html This site also has some intriguing visuals of the electromagnetic spectrum: http://coolcosmos.ipac.caltech.edu/cosmic_classroom/ir_tutorial/

That brings us to carbon dioxide, methane, and water. These gases do absorb infrared radiation because certain stretches and bends of the bonds in these molecules disrupt the electron distribution in ways necessary for IR absorption. Because carbon dioxide, water vapor, and methane are three molecules that are capable of absorbing IR radiation, these molecules play a major role in keeping Earth warm.

You may observe an animation of the motions of SO_2 and CO_2 molecules at: **http://jchemed.chem.wisc.edu/ JCEWWW/Articles/WWW0001/**

Carbon Dioxide in the Air

Carbon dioxide's effect on temperature can be illustrated by comparing the temperature on Venus (the second planet from the Sun) to that on Earth. Both planets are warmer than can be explained simply by looking at their distance from the Sun. Venus, whose average temperature is 450°C, is estimated to be 300°C warmer than might be predicted based on its distance from the Sun. Earth, which is almost 26 million miles farther from the Sun, is still about 33°C higher in average temperature (15°C) than the −18°C predicted. Why is that?

These elevated temperatures are largely attributable to the composition of the atmosphere of the two planets. Venus's atmosphere is 96% CO_2 with H_2SO_4 clouds and an atmospheric pressure of 90 atm, whereas Earth's atmosphere is 78% N_2, 21% O_2, 0.03–0.04% CO_2, and 0.1–1.0% H_2O at 1 atm of pressure at sea level. The percentage of water in Earth's atmosphere and, thus, the other percentages fluctuate over relatively small ranges depending on humidity. The average CO_2 concentration in Earth's atmosphere has increased from about 280 ppm in 1860 to 380 ppm in 2004, and it continues to rise (Figure 6.4). It took 150 years for the concentration to increase by 50 ppm from 280 ppm to 330 ppm but it has only taken 30 years for the most recent 30 ppm increase from 330 to 380 ppm. Earth has experienced an increase in CO_2 concentration of over 25% in the last century or so. During this time Earth's average temperature has increased by about 0.5–0.7°C, as shown in Figure 6.5. The big question is, has the increase in carbon dioxide concentration in our atmosphere been responsible for the increase in temperature? Furthermore, if there is a cause and effect relationship here, what will be the consequences of further increases in atmospheric carbon dioxide concentrations?

TRY IT 6.3
Use relationships between the Fahrenheit and Centigrade temperature scales to convince yourself that 59°F, Earth's average temperature, corresponds to 15°C. Do the same to show that Venus's average temperature of 450°C is 840°F. [Relationships: °F = (9/5)°C + 32 and °C = (5/9)(°F − 32)]

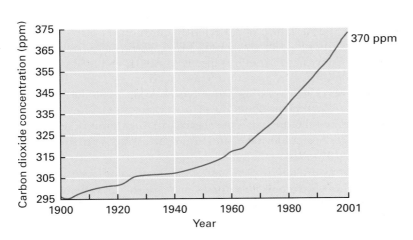

FIGURE 6.4 Atmospheric carbon dioxide concentrations have been rising steadily. Source: http://www.mlo.noaa .gov/Projects/GASES/co2graph.htm.

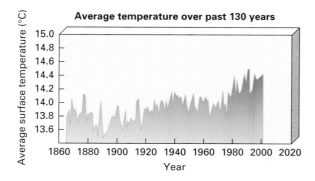

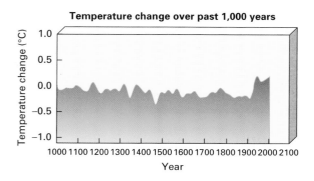

FIGURE 6.5 Trends in Earth's surface temperature over the past 1000 years and a more detailed analysis of the temperature trend near Earth's surface over the past 140 years. A more detailed presentation of this data may be found at http://www.ipcc.ch/present/graphics/2001syr/large.0516jpg.

TRY IT 6.4

The Kelvin scale of absolute temperature is often used by scientists. The relationship between the Celsius and Kelvin scales (abbreviated as K only) is: K = °C − 273.15. What is Earth's average temperature in kelvins?

Although unanimous agreement does not exist regarding the answer to the first question, more and more scientists are becoming convinced that the recent increases in Earth's temperature are largely, or entirely, due to increasing atmospheric concentrations of carbon dioxide. Although difficult to predict exactly, the general estimate is that, if the concentration of carbon dioxide reaches around 600 ppm, Earth's temperature will increase by between 1.5 and 6°C (2.4–10.5°F). This could come as early as 2050. While the magnitude of this predicted temperature change may not seem significant, temperature decreases within this range resulted in the last ice age about 20,000 years ago. (We will return to the possible consequences of an increase in Earth's temperature in Section 6.6.)

CONCEPT CHECK **6A**

1. Compare the percentage of carbon dioxide in the air to the percentage of carbon dioxide in an exhaled breath.
2. Compare the percentages of ultraviolet, visible, and infrared radiation emitted by the Sun.
3. Define "greenhouse effect."
4. What type of radiation is responsible for the greenhouse effect?
5. What percentage of Earth's radiated energy is absorbed and reradiated back to Earth?
6. What is Earth's average temperature? What would Earth's average temperature be without the greenhouse effect?
7. List the following types of solar radiation in terms of decreasing wavelength: infrared, visible, and ultraviolet. What is the trend in energy for these forms of radiation?
8. Identify two atmospheric gases that do not absorb infrared radiation.
9. Identify two atmospheric gases that absorb infrared radiation.

6.4 OTHER GREENHOUSE GASES

Some atmospheric gases other than carbon dioxide also absorb IR radiation. Some, such as **chlorofluorocarbons** (also known as **CFCs**), owe their atmospheric presence to human activities while others, such as water vapor, methane, and nitrous oxides, do not. Most, with the exception of water vapor, are present in relatively low concentrations, as is shown in Table 6.1. Although water vapor is a major greenhouse gas, we clearly can do little about the amount of water vapor in the air. However, one can observe the effect of water vapor on **radiational cooling** on a clear winter night versus a cloudy winter night. The temperature drops more quickly after sunset on clear nights due to the absence of heat-trapping clouds in the lower atmosphere. For this same reason, desert nights are often very cool due to low humidity. Is there anything we can do about potential greenhouse gases other than water? If so, on which gases should we focus?

The efficiency with which these gases absorb IR radiation is not equal, as shown by the greenhouse factors in Table 6.2. Some gases are orders of magnitude more efficient at absorbing IR radiation. We are fortunate that the concentration of these efficient absorbers in our atmosphere (the *tropospheric abundance*) is quite low. Since the seriousness of the problem is related both to the efficiency with which the gas absorbs IR and its concentration, we will focus on CO_2 and methane, both present in relatively high and increasing concentrations.

TRY IT 6.5
Convert the tropospheric abundance values listed in Table 6.2 into ppm and ppb.

6.5 SOURCES OF GREENHOUSE GASES

Prior to the Industrial Revolution, which began in the late 18th century in Europe, the average citizen of the world was responsible for the release of relatively small amounts of carbon dioxide gas into the atmosphere. This release

Chlorofluorocarbons (CFCs) A term used to refer collectively to compounds containing only carbon, chlorine, fluorine, and sometimes hydrogen. CF_2Cl_2, developed initially for use as a refrigerant in air conditioning systems, is an example.

Radiational cooling The phenomenon of enhanced loss of heat from Earth on a clear, cloudless night due to the absence of significant water vapor in the atmosphere in the form of clouds

---

TABLE 6.1 Atmospheric Concentration and Lifetimes of Some Key Greenhouse Gases

TIME/GAS	CO_2	CH_4	N_2O	CFC-11 (CCl_3F)	CFC-12 (CCl_2F_2)	HCFC-22 ($CHClF_2$)
Pre-1750 (preindustrial) Concentration[a]	~280 ppm	~700 ppb	270 ppb	0[b]	0[b]	0[b]
Concentration in 2001/2002	372 ppm	~1800 ppb	318 ppb	258 ppt[c]	546 ppt	146 ppt
Atmospheric lifetime (years)	Variable[d] (50–200)	12	114	45	100	11.9

All data taken from a tabulation put together by T. J. Blasing and Sonja Jones at http://cdiac.esd.ornl.gov/pns/current_ghg.html, which contains extensive documentation of the original sources.
[a] Pre-1750 concentrations are assumed to be practically uninfluenced by human activities associated with significant fossil fuel combustion.
[b] Not yet discovered at this time.
[c] This stands for parts per trillion (not parts per thousand).
[d] The mechanisms by which carbon dioxide is removed have quite different lifetimes for removal, so this is just an estimate.

TABLE 6.2	Relative Efficiencies of IR Absorption by Some Key Greenhouse Gases Compared with Their Tropospheric Abundance[a]	
Substance	**Greenhouse Factor[b]**	**Tropospheric Abundance (%)**
CO_2	1 (assigned value for comparison)	3.7×10^{-2}
CH_4	30	1.8×10^{-4}
N_2O	160	3.1×10^{-5}
O_3	2000	4.0×10^{-6}
CCl_3F	21,000	2.6×10^{-8}
CCl_2F_2	25,000	5.2×10^{-8}

[a] Intergovernmental Panel on Climate Change, 1996.
[b] The greenhouse factor is a measure of the effectiveness of infrared absorption by various atmospheric gases when compared with carbon dioxide, which is arbitrarily assigned a greenhouse factor of 1.

Burning fossil fuels produces carbon dioxide and water vapor, both greenhouse gases.

occurred through the burning of coal and wood for cooking and personal heating. However, the internal combustion engine brought with it an enormous appetite for fossil fuel. This fuel was coal in the early days of the revolution. Oil, gasoline, diesel fuel, and natural gas became major fuels as time progressed and our dependence on factories, electricity, and motorized transportation increased to its present level.

No way exists to get around the fact that burning fossils fuels produces carbon dioxide and water vapor, both greenhouse gases. A gallon of gasoline produces about 18 lbs. of carbon dioxide and about 1 gallon of water upon combustion (Section 12.1). The increased release of these gases into our atmosphere brings up the question of the rate of change of the concentration of these gases in our atmosphere. In other words, are we putting more carbon dioxide into our atmosphere than is or can be taken out of the atmosphere by natural mechanisms? The answer to this question appears to be "yes," as evidenced by the increasing carbon dioxide concentration.

Spewing volcanoes, natural vents at the bottom of some lakes and the oceans, naturally decaying plant and vegetable materials, forest fires (natural and those of human origin) and a few other mechanisms contribute significantly to the estimated 25 billion tons of carbon dioxide added to the atmosphere each year. Of this amount, 22 billion tons is generated by burning fossil fuels! It is this **anthropogenic** contribution of carbon dioxide to our atmosphere about which we are most concerned.

The two main mechanisms for removing carbon dioxide from the air are photosynthesis and dissolution into water (primarily the oceans).

Plants, trees, and grasses take enormous amounts of carbon dioxide from the air as part of their natural growth cycle. In photosynthesis, carbon dioxide and water are combined in the presence of sunlight with the chlorophyll in plants to produce sugars and oxygen gas. This is one of the main reasons there is so much concern about deforestation and loss of grasslands as the world's population increases.

Anthropogenic Originating from human activity

$$6\,CO_2 + 6\,H_2O \xrightarrow[\text{Chlorophyll}]{\text{Sunlight}} \underset{\text{Sugars}}{C_6H_{12}O_6} + 6\,O_2$$

Carbon dioxide readily dissolves in water and reacts with it to produce carbonic acid (H_2CO_3). The acid can stay dissolved or form bicarbonates and carbonate minerals that precipitate out as solids in the form of limestone and other materials. The temperature of the water greatly affects its ability to dissolve carbon dioxide and other gases. As anyone knows who has ever allowed a cold carbonated beverage to warm up and go flat, warm water dissolves less carbon dioxide. Warming of the world's oceans could lead to release of more carbon dioxide into the atmosphere, which, in turn, could lead to even more warming and greater release of carbon dioxide. This is not a problem to be taken lightly.

It is estimated that natural mechanisms remove 10 billion tons of carbon dioxide per year from our atmosphere. Thus, at present levels of fossil fuel combustion, we put about 25 billion tons of carbon dioxide into the atmosphere each year. Natural mechanisms remove only 10 billion tons, so we are producing a net increase of about 15 billion tons of carbon dioxide in our atmosphere each year! That explains the trend in carbon dioxide concentrations. The carbon dioxide concentration is increasing and we are almost certainly responsible for that increase.

Compare the lifestyle of a typical U.S. citizen to that of a peasant farmer in India and imagine the difference in the amounts of carbon dioxide released into the atmosphere as a function of their differing lifestyles. Consider, for example, transportation, housing, food production, electricity use, consumer products, and the energy required for their production. Would it surprise you to learn that the average U.S. citizen is responsible for releasing about 6 metric tons of carbon in the form of carbon dioxide into the atmosphere each year, compared to less than 0.5 metric ton for the Indian farmer? Since carbon dioxide is only 27.3% carbon, that equates to almost 22 metric tons of CO_2 per person in the United States versus 1.8 metric tons per person for the people of India.

TRY IT 6.6

Show that 6 metric tons of carbon in the form of carbon dioxide equates to 22 metric tons of carbon dioxide.

Farmers winnow grain near the settlement of Colcabamba in the Cordillera Blanca mountain range (Peruvian Andes).

Methane, another problematic greenhouse gas, is primarily released into the atmosphere from natural sources such as decaying vegetation in swamps (marsh gas); decaying organic materials in landfills and rice paddies; cows; and leaking natural gas lines and wells.

The digestive system of cows and other ruminant animals contain bacteria that break down cellulose and produce methane gas. Some large ruminants produce hundreds of liters of methane per day. The large increase in the world population of cattle and sheep has given rise to a large increase in methane emissions from this source.

Scientists agree that energy production, energy distribution, and livestock contribute most to methane emissions, followed by landfill emissions and biomass burning. Thus, even though the concentration of methane in our atmosphere is far less than that of carbon dioxide, its role in global warming cannot be totally discounted, because methane is an efficient absorber of IR radiation. However, the major focus is on carbon dioxide due to the enormous quantities of this gas released into the atmosphere by our carbon-based economy.

TRY IT 6.7

A gallon is 3.785 liters. How many gallons of methane would an average cow produce in one day if the cow produces 500 liters of methane per day?

6.6 WHAT DO WE KNOW FOR SURE ABOUT GLOBAL WARMING?

There are some things about global warming that can be stated with increasing confidence.

- Earth is definitely warming, and most scientists agree that humans are responsible for the warming.

- The warming is almost certainly due, at least in part, to increased anthropogenic emissions of greenhouse gases (primarily carbon dioxide). The Intergovernmental Panel on Climate Change (IPCC) sponsored by the United Nations recently predicted a potentially catastrophic warming of 1.4–5.8°C (2.5–10.4°F) by 2100 if nothing is done to curb these emissions.

- A temperature increase of only 2°C would mean Earth would be warmer than it has been for the past 2 million years.

- As the world's population increases, the rate and amount of greenhouse gases released will increase unless some significant changes are made. The population exceeds 6 billion at present.

6.7 CONSEQUENCES OF GLOBAL WARMING—GOOD OR BAD?

W A website that may give you an indication of some of the possible early warning signs and harbingers of things to come if global warming continues to increase is: http://www.climatehotmap.org

If global warming is a reality, as most now agree it is, then its consequences must be considered. The possible consequences of global warming are controversial and difficult to predict. However, most of the scientists who study the atmosphere agree that the following are likely results of increased global warming, although the extent to which each of these phenomena may manifest itself is debatable.

- Oceans may warm and release more carbon dioxide due to the reduced solubility of gases in warm water. This would increase warming even more and exacerbate the problem. However, warmer oceans might allow more phytoplankton to grow, which could consume more carbon dioxide.

- Solid, frozen methane deposits at the bottom of the oceans may become gaseous as the ocean warms, and the gaseous methane would be released into air. This would increase warming even more.

- Ice and snow cover may decrease significantly in parts of the world. This would cut down on reflected light so Earth would absorb even more radiation and warm up at an accelerated rate.

- Ocean levels may rise as ice and snow at the polar caps melt. The consequences for islands and low-lying areas along coasts could be disastrous. There is already evidence of this occurring. Ice shelves that have been frozen for thousands of years are breaking away. The Arctic Climate Impact Assessment, a study by over 300 scientists released in late 2004, suggested that, among other consequences of such warming, polar bears could become extinct by 2099.

- More water might evaporate from the oceans as they warm. This could increase global warming because water is a greenhouse gas. However, this could be offset by the reflective effect that additional cloud cover might have on incoming solar radiation.

- Warming of the oceans could cause major shifts in ocean currents, leading to unpredictable influences on weather, fisheries, etc.

© Royalty-Free/CORBIS

- Rainfalls would become much heavier in some areas as a result of increased evaporation. This would lead to increased flooding.

- The tree line would move farther north, causing trees to grow over a broader region. This could consume more carbon dioxide.

- Deserts and arable regions of the world could shift location and change size as a function of shifts in areas of rainfall. The consequences are difficult to predict.

- Tropical diseases might become more prevalent and widespread as Earth warms.

- The intensity of storms and weather systems would probably increase.

- Increased heat may mean worsened air pollution, damaged crops, and depleted resources in some areas.

- Some scientists have speculated that when a certain atmospheric concentration of greenhouse gases is reached, it may trigger a relatively rapid and cataclysmic climate system reorganization.

- Possibly, nothing of consequence, that is, nothing we can't handle, will happen.

CONCEPT CHECK   6B

1. What is radiational cooling? On what type of night might this phenomenon be least noticeable?
2. List the following greenhouse gases in order of decreasing concentration in the atmosphere: CCl_2F_2, carbon dioxide, methane. Compare the efficiencies with which these three gases absorb IR radiation.
3. What greenhouse gas is present in the highest concentration in our atmosphere?
4. About how much carbon dioxide and how much water would burning a gallon of gasoline produce?
5. List some natural mechanisms by which greenhouse gases are added to the atmosphere.
6. What are the two principal natural mechanisms for removing greenhouse gases from the atmosphere?
7. List three possible benefits of global warming and three possible negative consequences of global warming.
8. How does global warming differ from the daily heating that might occur in your state during the daylight hours, especially the summer hours?
9. Write and explain the equation for photosynthesis.

6.8   THE KYOTO CONFERENCE ADDRESSES GLOBAL WARMING

More and more people from less-developed countries are adopting the lifestyle of the United States. Because of this trend, the problems associated with global warming are expected to be dramatically exacerbated. Right now the United States is first worldwide in per capita carbon dioxide emissions, and China is eleventh. However, China is second in total emissions, and the United States is first. What will the world be like if China, India, or another highly populated country forges to the lead to become the largest source of per capita emissions as its population adopts those lifestyles of the modern world that are characterized by prodigious production of carbon dioxide?

Cognizant of the potential for a global warming disaster and unwilling to wait for utter certainty that global warming is a real problem, many of the world's nations sent representatives to Kyoto, Japan, in 1997 to begin to address the issue. At this conference, 161 countries set greenhouse gas emissions goals for "developed" countries but chose not to set them for "developing" countries. This was a contentious issue and was partly responsible for the fact that, at least as of this writing, the United States, alone among major world powers, has chosen not to support the goals established at Kyoto. Of course, economic issues were addressed, and the United States took the position that it would suffer more economic hardships than the other countries if these goals were enforced. There was also concern that the United States was being asked to bear an unfair share of the burden relative to less-developed countries. Neither Presidents Bill Clinton nor George Bush, Jr., nor the Democrats or Republicans have generally supported these goals, although some individual American politicians have come out in favor of the Kyoto Protocol. Russia ratified the Kyoto Protocol in October 2004. This ratification made carbon dioxide reductions mandatory among all 124 countries that had accepted the accord as of the date when it went into effect as law, February 16, 2005. On that date, carbon emissions became a commodity that can be traded by the signatory countries. The 30 industrialized countries that have signed the protocol will have to meet their protocol emissions targets by 2012. Some countries have vowed to cut emissions even more dramatically than is called for by the protocol. Great Britain plans to cut emissions by 60% by 2050. Even though the United States government has not signed the Kyoto Protocol, many state governments, organizations, and industries have taken action to address the issue. The attorneys general of eight states sued the nation's largest utility companies in July 2004, demanding that they reduce emissions of greenhouse gases. California has proposed a 30% reduction in car emissions by 2015 and other states are considering similar actions. Some power companies and other industries responsible for the production of large amounts of greenhouse gases are implementing voluntary reduction plans even in the absence of government dictates. *Business Week* magazine, in its August 16, 2004, cover story, reviewed the status of global warming from a business perspective and concluded that "remarkably, business is far ahead of Congress and the White House" in recognizing and addressing the issue of global warming. However, as pointed out in the October 9, 2004, issue of *New Scientist* magazine, even the most ardent supporters of the Kyoto Protocol recognize that it alone will not come close to solving the problem of global climate change. It is, according to UN Environment Program director Klaus Toepfner, "only the first step in a long journey."

TRY IT 6.8

Comment on the following quotation as it might apply to the United States versus some other countries. "One part of the world cannot live in an orgy of unrestrained consumption while the rest destroys its environment just to survive." *New Scientist* 17 August 2002, 31.

6.9 POSSIBLE RESPONSES TO GLOBAL WARMING

Assuming that the problem of global warming is real, scientists have begun to make suggestions about what can be done to either slow its progress or allow us to live with its consequences. Some of the actions suggested are obvious and probably practical. Others are obvious but probably impractical. Still other suggested actions are quite unusual, and their degree of practicality is debatable. However, with a problem as serious as global warming, maybe even the most outrageous of responses deserves some consideration. Engineering feats that, at one time, would have been considered folly are now routine. Who is to say how the future will judge our response to global warming? For instance, one or more of the following actions might prove effective:

- Decreasing the consumption of fossil fuels and developing and using alternative fuel sources that are not carbon based.

- Developing more efficient internal combustion engines or cars that do not depend so heavily on fossil fuels.

- Developing and requiring the use of more mass transit systems.

- Imposing taxes on goods and services that take into account the amount of greenhouse gases released in their production (a "carbon tax").

- Developing a worldwide emissions trading program wherein countries or other entities could work together to collectively exploit effective reduction programs.

- Developing and using new technologies that do not generate greenhouse gases. For instance, nuclear power plants and wind power generate no greenhouse gases. It is interesting to note that the Freedom Tower in New York City, to be built at the site of the former World Trade Center towers destroyed in the September 11, 2001, attack was originally designed to obtain 20% of its power from wind turbines at the top of the tower.

- Using the ocean as a waste dump for carbon dioxide by pumping carbon dioxide from factory stacks to the bottom of the ocean, where it will become liquid at the high pressure there. This might change the chemistry of the ocean, however. Also, problems might occur if the ocean suddenly belches to release large amounts of carbon dioxide.

- Fertilizing the oceans to increase phytoplankton growth. The idea is that this would increase carbon dioxide consumption. Even if this works, what happens when the phytoplankton die and decompose?

- Putting a giant parasol between the Sun and Earth to block or reflect part of the Sun's light.

- Transporting large amounts of dust into the stratosphere to reflect sunlight and counteract warming. What happens if we get too much sunlight? Some have suggested an alternative that would involve releasing large numbers of small, reflective spheres into the stratosphere.

- Changing our lifestyles to produce less carbon dioxide per capita.

- Reducing the world's population or at least slowing its growth.

- Banning the use of internal combustion engines and the burning of fossil fuels.

- Doing nothing and trying to adapt to a warmer world.

Although some of the actions described above seem sort of like science fiction, all of them have been seriously considered or have been tested on a small scale. Some combination of several of these actions will probably be necessary, although some of the suggestions will never get off the ground.

Finally, even though the evidence for a major anthropogenic contribution to global warming appears to be stronger and stronger with each passing day, we must recognize there are still different opinions about global warming. Most people will associate themselves with one of the following or a

Governor Arnold Schwarzenegger drives a hydrogen fuel cell vehicle during his announcement of the California Hydrogen Highways Network. The Governor also signed an executive order to build hydrogen-filling pumps every 20 miles by 2010.

Look at the following website to see what happened due to a spontaneous release of carbon dioxide by a belching lake. **http://www.geology.sdsu.edu/how_volcanoes_work/Nyos.html**

Clean coal-burning plant in Florida.

minor variation thereof. Which is most consistent with your opinion about this issue?

- *Wait*—Wait until we're 100% certain that this problem is for real, that it is serious, that we are causing it, and only then take action.

- *Act now*—Make our best scientific analysis of the problem based on the current evidence and take appropriate action immediately to try and buy time before it's too late to do anything. Continue to collect evidence and work on solutions to the problem, which may require new technology.

- *Ignore it*—If it's natural, just ignore it because humans cannot control nature.

Only time will tell if we respond appropriately.

 Assess your understanding of this chapter's topics with an online chapter quiz at **www.brookscole.com/chemistry/joesten4**

■ KEY TERMS

greenhouse effect	chlorofluorocarbons (CFCs)	anthropogenic
infrared (IR) radiation	radiational cooling	

■ APPLYING YOUR KNOWLEDGE

1. Define the following terms:
   (a) Greenhouse effect
   (b) Global warming
   (c) Anthropogenic
   (d) CFC
   (e) Kyoto Protocol
   (f) Infrared radiation

2. Most of the radiation coming from the sun is ultraviolet (UV) radiation. (a) True (b) False

3. Carbon dioxide is less efficient at absorbing infrared radiation than methane, CFCs, and some other molecules, but its higher atmospheric concentration makes it the global warming gas with which we are most concerned. (a) True (b) False

4. Photosynthesis removes huge amounts of carbon dioxide from our atmosphere. (a) True (b) False

5. Which of the following actions has not been proposed as a possible solution or partial solution to global warming?
   (a) Fertilizing the oceans with iron to increase phytoplankton growth
   (b) Only burning gasoline that contains no carbon compounds
   (c) Sequestering CO_2 by pumping it into the ground or the bottom of the oceans
   (d) Putting a giant parasol (a screen) between Earth and the Sun to cut down on the amount of radiation hitting Earth
   (e) Decreasing the rate of deforestation and increasing the planting of trees

6. Visible light coming from the sun can be absorbed and re-emitted as infrared radiation since visible light is of a higher energy than infrared radiation. (a) True (b) False

7. The temperature of Earth would be about $-18°C$ ($0°F$) instead of $+15°C$ ($+59°F$) if it were not for the greenhouse effect and the oceans would be frozen. (a) True (b) False

8. The concentration of CO_2 in our atmosphere has increased from around 315 ppm to more than 370 ppm in the last 40 or so years, and this appears to be related to global warming. (a) True (b) False

9. IR radiation is sometimes referred to as heat radiation, and all molecules absorb IR radiation. (a) True (b) False

10. A reasonable political agenda for an environmental group might be to propose legislation aimed at reducing the concentration of all greenhouse gases in our atmosphere to zero. (a) True (b) False

11. Some of the UV and visible radiation striking Earth's surface is re-emitted from Earth as IR radiation. (**a**) True (**b**) False

12. Which of the following is not a greenhouse gas?
    (**a**) Methane
    (**b**) Water
    (**c**) Carbon dioxide
    (**d**) Nitrogen
    (**e**) CFC-12

13. The oceans will be able to dissolve more carbon dioxide when they become warmer. (**a**) True (**b**) False

14. The United States is number 1 in per capita production of carbon dioxide emissions. (**a**) True (**b**) False

15. The United States government has not been very supportive of the terms of the Kyoto Protocol. (**a**) True (**b**) False

16. Most greenhouse gases are produced by photochemical reactions in much the same way as photochemical smog is produced. (**a**) True (**b**) False

17. Global warming is mainly associated with which of the following?
    (**a**) Molecules that absorb UV radiation in the stratosphere
    (**b**) Molecules that decompose the ozone layer
    (**c**) Molecules that readily react with ultraviolet light
    (**d**) The type of electromagnetic radiation that is responsible for stretching and bending chemical bonds
    (**e**) The huge concentrations of CFCs that have built up in our atmosphere

18. The single major reason that Earth does not continuously lose energy to the surrounding space is
    (**a**) Earth is close enough to the Sun in the summer to absorb enough extra energy to compensate for that lost during the winter
    (**b**) Oxygen and nitrogen in our atmosphere absorb enough reflected IR energy to maintain the present energy balance
    (**c**) The "heavy" CFCs added to our atmosphere have trapped a carbon dioxide layer below them in the atmosphere
    (**d**) Greenhouse gases like carbon dioxide and water vapor in the atmosphere absorb reflected energy and radiate it back to Earth's surface
    (**e**) The ozone layer

19. Every major industrialized country has ratified the Kyoto Protocol. (**a**) True (**b**) False

20. All greenhouse gas molecules are equally efficient at absorbing infrared radiation. (**a**) True (**b**) False

21. The average concentration of carbon dioxide in our atmosphere has increased to around 380 ppm from a preindustrial concentration of around 280 ppm. This is most likely due to which of the following?
    (**a**) An increase in the animal population of the world
    (**b**) Increased volcanic activity during the past 150 years
    (**c**) Increased gasoline efficiency of the internal combustion engine
    (**d**) A thinning of the ozone layer that allowed more CO to be converted to CO_2 by UV radiation
    (**e**) Increased use of fossil fuels by an increasingly populous world

22. Some scientists have proposed pumping carbon dioxide to the bottom of the oceans as a partial solution to global warming. (**a**) True (**b**) False

23. Discuss the pros and cons of the various methods that have been proposed to regulate or control global warming.

24. Should industrial processes that raise our standard of living but that produce greenhouse gases be exempt from regulation of greenhouse gas emissions?

25. Should developing countries be held to the same standards of greenhouse gas emissions as developed countries if doing so can be shown to slow their rate of development?

26. Would you support a carbon tax added to the cost of consumer goods based on a calculation of how much their production contributed to the greenhouse gas problem?

27. Should people who insist on driving vehicles that get below a certain gas mileage be assessed an additional tax since they contribute more carbon dioxide to the atmosphere?

28. Why do you think it has been so hard to convince the average person that global warming is a real problem?

29. Calculate the energy associated with a photon of infrared radiation that has a wavelength of 2.5×10^{-3} cm and that with a wavelength of 2.5×10^{-4} cm. Use the expressions $\lambda \nu = c$ and $E_{photon} = h\nu$ where $h = 1.58 \times 10^{-34}$ cal.

30. Repeat the calculation of problem 31 using appropriate wavelengths (Section 6.3) for ultraviolet and visible radiation and check to see if the order of these energies when compared with that for infrared radiation agree with the order listed in Section 6.3.

CHEMISTRY ON THE WEB

For up-to-date URLs, visit the text website at **www.brookscole.com/ chemistry/joesten4**

- Frequently Asked Questions about Global Warming
- Pictorial Evidence of Global Warming
- Global Warming
- Climate Change Stories

CHLOROFLUOROCARBONS AND THE OZONE LAYER

Before it was known that chloro-fluorocarbons (CFCs) could damage the stratospheric ozone layer, they were widely used as the propellant gas in aerosol spray cans, among other things.

Eyebyte/Alamy

The major effects of many air pollutants such as the nitrogen oxides, sulfur oxides, carbon monoxide, and ozone are felt at altitudes low enough that they are considered "ground-level" pollutants. The greenhouse gases that have been implicated in global warming also trap infrared radiation at low altitudes. These pollutants are so chemically reactive and/or water-soluble that they never make it out of our troposphere, the atmospheric region extending from sea level to about 10 km. Many of them react there in the gas phase with other chemicals, including water, to form compounds that are soluble in water and are removed by natural precipitation in a process called **rainout**. When sulfur oxides, nitrogen oxides, and carbon dioxide react with water they form, respectively, sulfuric acid, nitric acid, and carbonic acid.

Humans had done little, if anything, to affect the chemistry of the stratospheric ozone layer until we began to make significant use of a class of chemicals known as *chlorofluorocarbons*. Chlorofluorocarbons were first synthesized by chemists at DuPont in the 1930s. Prior to that time the little refrigeration available depended on the use of ammonia and sulfur dioxide as heat exchangers. Because of the potential danger associated with these corrosive, noxious, and toxic gases, scientists had been searching for replacement chemicals that would mitigate these problems. As was pointed out in Section 1.3, the development of CFC-12 (dichlorodifluoromethane) presented scientists with a chemical that was nontoxic, extremely stable, and nonflammable and had the right physical properties for use as a refrigerant. Little did anyone realize at the time that some of these properties would, one day, lead to their widespread use in products including aerosol spray cans and trigger an environmental problem of global proportions.

In this chapter we'll address the following issues:

- Where is the ozone layer and why is it so important?
- How does the ozone layer protect us from solar ultraviolet radiation?
- What other atmospheric chemical protects us from high-energy solar radiation?
- How is ozone produced in the stratosphere?
- How have humans interfered with this protective layer?
- Why is human interference manifesting itself at the South Pole?
- How do chlorofluorocarbons and other chemicals destroy ozone?
- Can anything be done to address the destruction of the ozone layer?

(7.1) THE OXYGEN–OZONE SCREEN

As you saw in Chapter 6, our Sun bombards Earth with electromagnetic radiation, of which about 53% is infrared (IR) radiation, about 39% is visible radiation, and about 8% is ultraviolet (UV) radiation. IR radiation is the lowest in energy of these three, and ultraviolet radiation is the highest in energy. We have focused on the effect that these forms of radiation can have upon reaching Earth and have briefly mentioned that some of this electromagnetic radiation never reaches Earth because it is either absorbed or reflected.

Ultraviolet radiation has wavelengths in the 200–400 nm range. This energy is sufficient to break some chemical bonds and to create havoc with many biological systems. Fortunately for us, much of the UV radiation that streams from the Sun towards Earth never reaches Earth. It is absorbed primarily by two important screening molecules, oxygen and ozone. Let's examine how this happens.

First we must look at the bonding in oxygen (O_2) and ozone (O_3) molecules. A classical Lewis dot structure of molecular oxygen indicates that it contains an oxygen–oxygen double bond. Ozone can be represented by two different, but equal, structures in which there is one oxygen–oxygen double bond and one oxygen–oxygen single bond. The actual structure is one in which both oxygen–oxygen bonds are essentially intermediate between single bonds and double bonds, but that is difficult to represent pictorially. We, therefore, resort to writing the two separate but equal resonance structures (page 106), which differ only in the location of lone pair electrons and multiple bonds. The real structure is a hybrid of the two resonance structures.

Rainout The process by which pollutants in the air are removed by natural precipitation such as rain

$$\ddot{O} = \ddot{O} - \ddot{O}: \longleftrightarrow :\ddot{O} - \ddot{O} = \ddot{O}$$

Ozone Resonance Structures

| TABLE 7.1 | Comparison of Types of Ultraviolet Radiation |

RADIATION TYPE	WAVELENGTH RANGE	RELATIVE ENERGY	COMMENT
UV-A	320–400 nm	Least energetic	Reaches Earth in great amounts but has less potential than UV-B or UV-C to cause biological damage due to its lower energy. Even so, UV-A is suspected of being responsible for premature aging and wrinkling of the skin. It also causes some types of skin cancers that can be fatal.
UV-B	280–320 nm	Intermediate	Most of the UV-B is absorbed by ozone in the atmosphere; some reaches Earth, however. UV-B radiation is very effective at damaging DNA, weakening the human immune system, causing some skin cancers and eye damage, and causing significant crop and marine organism damage.
UV-C	200–280 nm	Most energetic	Completely absorbed by oxygen and/or ozone in the atmosphere; may cause significant biological damage.

We can imagine that oxygen–oxygen bonds in ozone are 1.5 bonds. Thus, the oxygen–oxygen double bond in O_2 is stronger than the oxygen–oxygen 1.5 bond in ozone (O_3) and requires more energy to break.

It turns out that the energy required to break these oxygen–oxygen bonds is supplied by UV radiation of slightly different wavelengths. UV radiation in the 200–400 nm range is generally divided into three separate ranges referred to as *UV-A* (320–400 nm), *UV-B* (280–320 nm), and *UV-C* (200–280 nm) as shown in Table 7.1.

Because a double bond is stronger than a 1.5 bond and requires more energy to break, oxygen molecules absorb higher-energy UV radiation (UV-C) with wavelengths less than about 242 nm. The 1.5 bond of an ozone molecule is somewhat weaker than the double bond of oxygen and absorbs lower energy (UV-B) with wavelengths less than about 320 nm. Therefore, ozone is more reactive toward UV light than is oxygen.

7.2 WHERE IS THE OZONE?

The ozone layer is a band of higher than average ozone concentration within the stratosphere (Section 4.1). The maximum concentration of ozone in the densest region of the stratospheric ozone layer is slightly less than 10^{19} molecules of ozone per cubic meter. This is a big number. However, it has been estimated that an average human breath (about 0.5 liter) contains 2×10^{22} particles. That equates to 4×10^{22} particles per liter or 4×10^{25} particles per cubic meter (a cubic meter is 1000 liters). In other words, the density of ozone molecules at the densest part of the ozone layer is only 0.000001 times that of the air we breathe. If all the ozone in a column of air in the atmosphere were compressed to 1 atmosphere pressure at 0°C, the layer would be only 3 mm thick (Figure 7.1). This is about the thickness of two nickels stacked together. However, we must remember that this amount of ozone is spread out over many kilometers of space because the region of maximum ozone concentration is at least 15 km thick.

In the absence of anthropogenic influences, the concentration of ozone in the stratospheric ozone layer had maintained a **steady state** (a state in which the net rate of destruction equals the net rate of formation) for millions of years. On the average, ozone molecules were destroyed by a variety

Steady state A state in which the concentration of a chemical remains essentially constant even though the chemical may be undergoing numerous chemical reactions

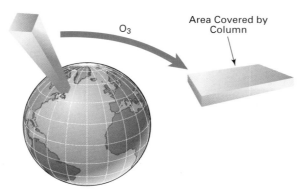

O₃

Area Covered by Column

FIGURE 7.1 If all the ozone over a certain area was compressed down to 0°C and 1 atm pressure, it would form a slab 3 mm thick, corresponding to 300 Dobson units (Section 7.3). Figure located at http://www.atm.ch.cam.ac.uk/tour/dobson.html. Used by permission of the Centre for Atmospheric Science, Cambridge University.

An animation of the Chapman cycle is found at a NASA website. Click on *Atmospheric Chemistry* under the *Atmosphere* heading. Then click on *ozone* and follow by clicking on the panel showing ozone creation to watch a movie of ozone formation. **http://visibleearth.nasa.gov/**

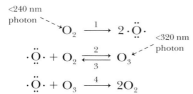

The Chapman cycle for the formation and destruction of stratospheric ozone.

of chemical reactions and UV-B absorption at the same rate that they were reformed by other processes. As a result the net concentration of ozone in the stratosphere remained constant. In other words, a finely balanced system was in place, capable of protecting any life on Earth from an over-abundance of high energy ultraviolet radiation.

We can illustrate a steady state condition by using a bathtub analogy. A bathtub is filled to a certain level with water, and then the drain is opened. One can maintain a constant level of water in the tub, even with an open drain, by carefully adding water to the tub with the faucet at exactly the rate at which it is draining away.

A slightly simplified version of the steady-state process for ozone formation (reactions 1 and 2) and destruction (reactions 3 and 4) is known as the **Chapman cycle** (shown in left margin) after the scientist who proposed it.

An inspection of the Chapman cycle reveals that the following chemical reactions are occurring:

- Reaction (1) indicates that molecular oxygen can absorb high energy UV radiation with wavelengths less than 240 nm to dissociate into two reactive oxygen atoms.

- Reaction (2) shows that an oxygen atom can combine with an oxygen molecule to form an ozone molecule. This is the process by which ozone is generated in the stratosphere.

- Reaction (3) shows that ozone molecules will react with ultraviolet radiation streaming in from the Sun if the radiation has wavelengths less than about 320 nm. In doing so, the ozone is reconverted to an oxygen atom and an oxygen molecule. Oxygen and ozone protect us from nearly all ultraviolet radiation with wavelengths less than 320 nm. The ozone concentration is maintained by this steady state process.

- Reaction (4) illustrates that it is possible for an oxygen atom to combine with an ozone molecule to generate two oxygen molecules. This reaction leads to net loss of ozone and is, therefore, not desirable. This is a relatively rare process under natural conditions, due to the exceedingly low concentrations of the species involved. Any chemical reactions whose net result is the promotion of reaction (4) are deleterious to the steady state concentration of the ozone layer.

Because the ozone layer developed naturally in the presence of many other perturbing influences such as destruction of ozone by water vapor, these reactions have no real effect on the steady state concentration of ozone. However, as you will soon see, anthropogenic influences have begun to perturb the steady state concentration of ozone.

TRY IT 7.1
For the average person, it is probably more important to be able to use words to describe the process by which ozone is formed in the stratosphere than it is to write out chemical equations describing the process. Can you do this?

TRY IT 7.2
Following the logic of Try It 7.1, use words to describe the reactions by which ozone and oxygen protect us from different wavelengths of ultraviolet radiation. Why do oxygen and ozone protect us from different wavelengths of ultraviolet radiation?

Chapman Cycle The set of four reactions that represents the steady-state formation and destruction of ozone in the stratosphere.

TRY IT 7.3

Use the bathtub analogy to describe how the steady state concentration of stratospheric ozone could be maintained. Also use the bathtub analogy to describe how anthropogenic influences might perturb the steady state concentration of ozone in the stratosphere.

It is known that measured ozone concentrations are lower than can be accounted for by the simple Chapman cycle. This has led scientists to look for other influences on the concentration of ozone. First, let's briefly consider one of the natural reactions that destroys ozone. UV radiation can break the oxygen–hydrogen bond of a water molecule in the stratosphere to generate hydrogen atoms and hydroxyl radicals ($\cdot OH$). These two species are involved in many reactions, some of which actually convert O_3 to O_2. However, this process, which scientists now believe is an efficient process above 50 km, has been occurring since the ozone layer developed, and there is little, if anything, that humans can do about it. The system has obviously attained a steady state that includes this perturbation.

Another natural mechanism for ozone destruction, shown in the margin, involves nitric oxide (NO). The NO can be generated when nitrous oxide (N_2O), produced naturally by microorganisms in soil and water, is released into the lower atmosphere and slowly makes its way into the stratosphere. Once in the stratosphere, one molecule of nitrous oxide can react with oxygen atoms to form two molecules of NO (reaction 1). The NO can then react with ozone to produce NO_2 and O_2 (reaction 2). The NO_2 can then react further with an oxygen atom to regenerate NO and O_2 (reaction 3). The net result of these last two reactions is exactly the same as the last reaction of the Chapman cycle and leads to destruction of ozone. The most significant sources of N_2O are natural biological and oceanic processes, so, once again, the system attains a steady state including this perturbation.

$$N_2O + \cdot \ddot{O} \cdot \xrightarrow{\ 1\ } 2 \cdot NO$$

$$\cdot NO + O_3 \xrightarrow{\ 2\ } \cdot NO_2 + O_2$$

$$\cdot NO_2 + \cdot \ddot{O} \cdot \xrightarrow{\ 3\ } \cdot NO + O_2$$

$$\cdot \ddot{O} \cdot + O_3 \xrightarrow{\ 4\ } 2O_2 \qquad \text{sum of } 2 + 3$$

Reaction sequence for ozone destruction promoted by nitric oxide.

CONCEPT CHECK) **7A**

1. What percentage of radiation coming from the Sun is ultraviolet radiation?
2. Draw a complete Lewis dot structure for ozone.
3. What wavelengths are considered to fall in the UV range of the spectrum?
4. Which is higher in energy, UV-A, UV-B, or UV-C?
5. Which form of radiation has the longer wavelength, UV-A, UV-B, or UV-C?
6. Write out the reactions of the Chapman cycle and discuss the meaning of each.
7. Name two chemicals that are involved in the natural destruction of ozone in the stratosphere.

(7.3) THE OZONE LAYER IS DISAPPEARING

Scientists, who had no inkling of what they were about to discover, began measuring stratospheric ozone concentrations over 80 years ago using relatively unsophisticated instrumentation. Eventually, as a credit to G. M. B. Dobson, one of the scientists who invented an early instrument used for these measurements, the unit of ozone concentration in the atmosphere became

THE PERSONAL SIDE

Susan Solomon (1956–   )

In 1985, a British team at Halley Bay Station, Antarctica, discovered the existence of a hole in the ozone layer above that continent. This totally unexpected phenomenon needed an explanation, and Susan Solomon—a young National Oceanic and Atmospheric Administration (NOAA) scientist—first proposed a good theory for it. While attending a lecture on polar stratospheric clouds, she realized that ice crystals in the clouds might do more than just scatter light over the Antarctic. Her chemist's intuition told her that the ice crystals could provide a surface on which chemical reactions of CFC compounds could take place.

In 1986, the National Aeronautics and Space Administration (NASA) chose Solomon (then 30 years old) to lead a team to Antarctica to sort out the right explanation for the ozone hole. Experiments during that visit to Antarctica showed that her cloud theory was correct, and a second

Courtesy of S. Solomon

SUSAN SOLOMON IN THE DRY VALLEYS.

expedition the next year added further evidence of its validity. Solomon's team and their experiments led to the first solid proof that there is a connection between CFCs and ozone depletion.

Susan Solomon became one of the youngest members of the National Academy of Sciences. She decided to become a scientist at age 10, having been influenced by watching Jacques Cousteau on TV. At age 16 she won first place in the Chicago Science Fair for a project called "Using Light to Determine Percentage of Oxygen," and went on to place third in the national science fair that year. She said that her winters as a young girl in Chicago prepared her for her visits to Antarctica.

Dobson Unit A measure of stratospheric ozone concentration; 1 Dobson unit is 1 ppb of ozone in air.

known as a **Dobson unit** (DU). This unit equates to about 1 ozone molecule for every billion molecules of air, so 1 DU is 1 ppb of ozone. Satellite measurements began to be used as time progressed. Measurements showed that near the equator an average value of about 250 DU of ozone was found, while the value was around 320 in the Northern Hemisphere. The concentration of ozone is actually higher at the North and South Poles.

Measurements taken at the South Pole, beginning in 1979, began to reveal a significant decline in stratospheric ozone, one that was especially prevalent during the months of September to November of each year (Figure 7.2). This

FIGURE 7.2 Seasonal ozone deviations from pre-ozone–hole (1957–1978) averages over Antarctica show a rapid decline during the southern spring and a lesser decline during the summer. From *The Changing Ozone Layer* by R.D. Bojkov, World Meteorological Association, 1995. Used by permission.

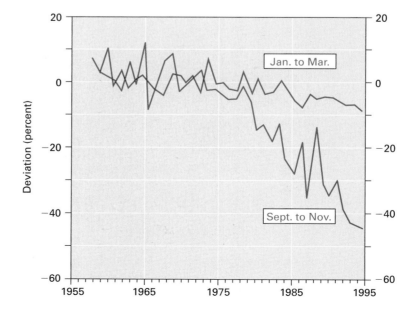

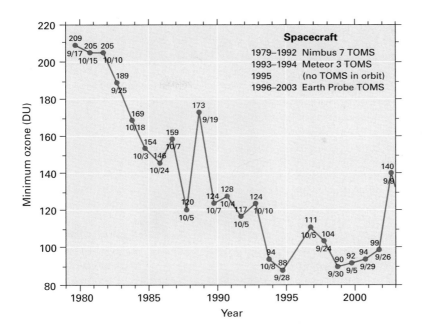

FIGURE 7.3 Antarctic ozone minima, 1979–2003. The ozone minima are given above the data points and the dates of recorded minima are given below the data points. The data, in Dobson units (DU), were measured by the Total Ozone Mapping Spectrometer (TOMS) spacecraft. (Source: http://jwocky.gsfc.nasa.gov/multi/min_ozone.gif)

was so unexpected that, early on, scientists thought there was something wrong with the data being collected. In fact, computers had been programmed to discard some of the data collected because it was so far outside the expected range. By 1985 the results were unmistakable. Although the expected predictable seasonal fluctuations in ozone concentration over Antarctica were present due to weather conditions and available sunlight, overall decline was obvious (Figure 7.3). This observed decline, which manifests itself as the annual ozone hole (Figure 7.4), could not simply be explained by the previously observed annual drop in late September or early October (the Antarctic spring) of each year. Clearly something else was going on. But what?

You may view animations of seasonal ozone variation anywhere in the world by going to the Earth Observatory website. Go to **http://earthobservatory.nasa.gov/**, then click on ozone and follow the instructions.

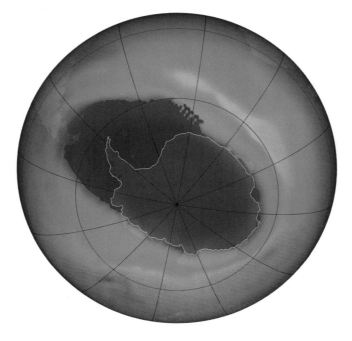

FIGURE 7.4 Ozone hole over Antarctica on September 22, 2004. The hole in darker blue can be seen to cover the entire South Pole (outlined in white) and extend over southern South America in the upper left quadrant of the image. Higher concentrations of ozone are represented by green, yellow, and red. (Source: http://earthobservatory.nasa.gov/Newsroom/NewImages/images.php3?img_id=16763)

Ozone Concentration (Dobson units)

125   280   435

THE PERSONAL SIDE

Paul J. Crutzen (1933–   ), Mario J. Molina (1943–   ), and F. Sherwood Rowland (1927–   )

Paul Crutzen, Mario Molina, and Sherwood Rowland shared the 1995 Nobel Prize in Chemistry for their work in atmospheric chemistry, particularly concerning the formation and decomposition of ozone. Born in Amsterdam, Mexico City, and Delaware, Ohio, respectively, in 1974 these three scientists alerted the world to an impending environmental disaster, possible destruction of the protective ozone layer by manmade chemicals. And yet nothing in their early backgrounds would have predicted this.

Crutzen started his career when he applied for, and got, a job as a computer programmer in 1961 even though he had not the slightest experience in the field. By 1965 he found himself employed at a meteorological institute in Sweden helping an American scientist to study the distribution of oxygen allotropes in the atmosphere. By 1972, he had decided to do pure science related to natural processes in the atmosphere and picked stratospheric ozone as his project. By his own admission, he had not the slightest anticipation of what lay ahead.

In similar fashion, Rowland and Molina came to the study of the influence of CFCs on stratospheric ozone almost by chance. In 1972, Rowland, then a professor of chemistry at the University of California at Irvine, heard a presentation about some measurements made by English scientist James Lovelock. On the cruise of the *Shackleton* to Antarctica, Lovelock had detected the presence of the manmade chlorofluorocarbon CCl_3F in both the northern and southern hemispheres. Rowland knew that this compound, inert at lower altitudes, would likely not remain so once it was exposed to the solar photochemistry of the stratosphere and decided this might make an interesting research project.

In the fall of 1973, Molina, who had just completed his Ph.D. at the University of California at Berkeley in 1972, moved to the University of California at Irvine to do postdoctoral research with Rowland. Among the choices of projects to work on

PAUL J. CRUTZEN

MARIO J. MOLINA

F. SHERWOOD ROWLAND

presented to Molina by Rowland was the question of the fate of the inert CFCs in the upper atmosphere. According to Rowland, Molina chose the project furthest from his previous experience in an effort to enter a field about which he knew very little.

Within three months of his arrival at Irvine, Molina and Rowland had developed the rudiments of their CFC–ozone depletion theory. Both became aware that Crutzen's earlier work on the influence of nitrogen oxides on ozone levels provided a model for how chlorine atoms could influence the concentration of ozone as well. By 1985, Molina, working at the Jet Propulsion Laboratory in California, was able to show that the chlorine activation reactions took place effectively in the presence of ice crystals in the polar stratosphere and, thus, account for the seasonal fluctuation of the ozone discovered earlier by Joseph Farman and his coworkers. Molina has stated that he chose the initial research project with Rowland simply out of scientific curiosity, never dreaming of the impact their work would have on the world's environment. You may read about their exploits, and those of other Nobel laureates, at **http://www.nobel.se/chemistry/laureates/**.

Sherwood Roland, Mario Molina, and Paul Crutzen made the most dramatic contribution to the puzzle of the missing ozone when they proposed that manmade chemicals called *chlorofluorocarbons* were the key ingredient. They subsequently were awarded the 1995 Nobel Prize in Chemistry for their work.

Scientific observations over the past 45 years or so reveal that the ozone concentration over Antarctica shows some seasonal variation (Figure 7.2) with the greatest and most rapid decline occurring from September to November. The interval corresponds to spring in the Southern Hemisphere. Furthermore, detailed observations from 1980 to the present show that the value of the minimum ozone concentration observed annually has dropped significantly (Figure 7.3). A close inspection of the data in Figure 7.3 suggests that this trend may be reversing itself. However, data collected in the fall of 2005 as this book was being written indicated the size of the hole at ~28 million sq km approached the record size of 29 million sq km observed in 2003.

(7.4) WHY CFCs AND WHY AT THE POLES?

After the discovery of the refrigerant gas properties of CFCs, their use became widespread in applications such as automobile and home air-conditioning. Soon the inertness, low boiling points, and insolubility in water of the CFCs made them popular as propellants in aerosol cans and as agents used to blow the holes into foam plastics. Because of their excellent solvent properties for greases and oils, many of the CFCs were used as degreasers during the manufacture of printed computer circuit boards, television sets, and other kinds of appliances. In effect, the increased use of CFCs paralleled the growth of modern, urbanized industrial society.

During the time CFCs were in common use, little effort was made to prevent them from escaping into the atmosphere. For example, if your automobile air conditioner needed repair, the common practice was to vent all of the CFC refrigerant into the atmosphere before any work was done. In fact, most of the CFCs ever manufactured have probably been released into the atmosphere.

It is obvious that the chlorofluorocarbons used as refrigerants, propellants, and solvents for the past half century were not released into the atmosphere at the South Pole, so why did the ozone hole first show up there? CFCs were, for years, released into our lower atmosphere by a variety of mechanisms, including venting of air conditioning systems when recharging, spraying of aerosol sprays, evaporation of solvents, and decomposition of blown foams in which CFCs were trapped. At lower altitudes, these chemicals were essentially unreactive and hung around for days, months, or years until the natural air currents and weather patterns transported them into the stratosphere. Although many have argued that these heavier-than-air molecules could never make their way

Recycling CFCs during repair of an automotive air conditioner. At many service centers, CFCs were simply vented into the atmosphere.

Courtesy of Robinair

into the stratosphere, that is simply not true. Vertical mixing of such chemicals within the troposphere has been shown to occur within weeks, and such chemicals released into one hemisphere can mix between hemispheres within about a year. Although it may take years for vertical mixing of such chemicals between the troposphere and the stratosphere, it does occur. Most chemicals are removed from the atmosphere by rainout or chemical reactions before they have a chance for this to occur, but CFCs are not water soluble and they are very unreactive until they reach the stratosphere.

Once in the stratosphere, these CFCs are subject to bombardment by intense UV radiation because they no longer are protected by the oxygen–ozone screen. High energy (<220 nm) radiation is sufficient to break the relatively weak carbon–chlorine bond of CFC molecules, as shown in Figure 7.5, for the common CFC known as trichlorofluoromethane or CFC-11. This generates a very reactive chlorine atom with an odd number of electrons that can then collide with ozone molecules to generate chlorine monoxide (ClO) and an oxygen molecule (O_2) (reaction 2). Reaction 3 shows how the chlorine monoxide can react with an oxygen atom to regenerate the destructive chlorine atom. Reaction 4, which is the sum of reactions 2 and 3, is exactly the same net reaction as the last reaction of the Chapman cycle (p. 146).

An additional destruction pathway in which two chlorine monoxide species combine (reaction 1) can occur is shown in the margin on page 153. The resultant ClOOCl undergoes two subsequent reactions (2 and 3), triggered by

 To see an animation of the sequence of reactions for ozone destruction, click on Atmosphere and then click on the panel showing ozone destruction to watch a movie of the process. **http://visibleearth.nasa.gov/**

FIGURE 7.5 A simplified summary of how chlorofluorocarbons and other chlorine-containing compounds can destroy ozone in the stratosphere faster than it is formed. Note that chlorine atoms are continuously regenerated as they react with ozone. Thus, they act as *catalysts*, chemicals that speed up chemical reactions without being used up by the reaction. Bromine atoms released from bromine-containing compounds that reach the stratosphere destroy ozone by a similar mechanism.

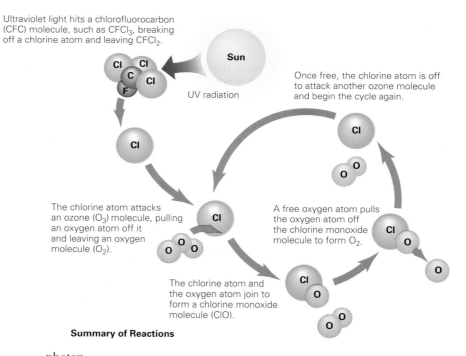

the UV light from the Sun, to generate two chlorine atoms which may then combine with ozone (reaction 4) in exactly the same process as shown in reaction 2 on page 152. This entire sequence converts two ozone molecules to three oxygen molecules, a reaction that consumes ozone.

Although the overall process is certainly more complicated, these simplified schemes provide an understanding of the role of Cl atoms in triggering the ozone depletion that leads to the seasonal ozone hole over Antarctica. A single Cl atom is thought to be capable of destroying 100,000 ozone molecules before it is rendered unreactive by some other chemical process. Collected data (Figure 7.6) also show that there is an anticorrelation between the concentration of ClO and the concentration of ozone in the stratosphere.

It is also likely that the chlorine atoms undergo other reactions to generate somewhat stable compounds that render them temporarily unreactive in the stratosphere where they are not going to be removed by rainout. In effect, they are "stored" in less reactive forms such as chlorine nitrate ($ClONO_2$), which can be produced when ClO reacts with NO_2. Another chlorine reservoir or storage molecule is HCl, which can form from the reaction between a chlorine atom and naturally occurring methane (CH_4) in the stratosphere.

The lower stratosphere above Antarctica is the coldest spot in Earth's atmosphere, and that, coupled with the relative lack of dispersive wind currents above the pole during the long dark periods of Antarctic winter, means that the conditions in the stratosphere above the South Pole are ripe for ice crystal formation. During the dark, cold ($-80°C$) polar winter, polar stratospheric ice clouds comprised of nitric acid trihydrate ($HNO_3 \cdot 3\ H_2O$) form. These ice clouds trigger the breakdown of the normally stable molecules ($ClONO_2$ and HCl) to generate HOCl and Cl_2, which photodissociate when sunlight returns to the South Pole in the spring (September–November). This photodissociation produces Cl atoms, and the destruction of ozone thus triggered is highly effective during this time of the year.

Scientists have also implicated some bromine-containing chemicals in the destruction of the ozone layer. The chief culprit has been CH_3Br, methyl bromide, a commonly used agricultural fumigant. Here again, it is the relatively weak C—Br bond that is broken by UV radiation to produce Br atoms, which act in the same role as Cl atoms to destroy ozone.

TRY IT 7.4
Explain why the ozone hole has occurred over the South Pole even though the molecules that triggered the hole formation were not released at the Pole.

$$ClO \cdot + ClO \cdot \xrightarrow{1} ClOOCl$$

$$ClOOCl \xrightarrow[sunlight]{2} ClOO \cdot + \dot{Cl}$$

$$ClOO \cdot \xrightarrow{3} \dot{Cl} + O_2$$

$$\dot{Cl} + O_3 \xrightarrow{4} ClO \cdot + O_2$$

Ozone destruction triggered by ClOOCl.

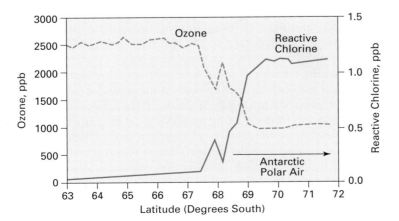

FIGURE 7.6 The anticorrelation between stratospheric ozone and chlorine monoxide concentration versus latitude near the South Pole in September 1987. (Source: http://www .elmhurst.edu/~chm/onlcourse/ chm110/assess/images/ozoant.gif)

TRY IT 7.5

What chemical bond in chlorofluorocarbons is most critical to their ability to destroy the ozone layer?

TRY IT 7.6

Why are CFCs not removed by rainout like many other chemicals released into the atmosphere?

7.5 WHAT ARE THE IMPLICATIONS OF INCREASES IN THE SIZE OF THE OZONE HOLE?

Although the decrease in ozone concentrations manifests itself most readily over Antarctica, this is simply a reflection of worldwide decreasing ozone concentrations throughout the ozone layer. There is evidence that the lowered ozone concentrations are already starting to have a negative effect on crop yields in the Southern Hemisphere. There is much concern about increases in skin cancers as the lowered ozone concentrations and ozone hole start to spread over populated areas. Australian officials have started to warn citizens about the enhanced danger associated with these lowered levels. Australian schoolchildren are not allowed out at recess without long sleeves and hats with visors and neck flaps to protect them from the sun. In the fall of 2000, the ozone hole extended over a population center for the first time and exposed thousands of Chileans in the city of Punta Arenas to extremely high levels of UV radiation. Citizens in Punta Arenas are now warned not to go out into the sunlight between the hours of 11 A.M. and 3 P.M. during the height of the seasonal reduction. If the size of the hole expands, the number of people living under the hole and subjected to such restrictions will increase.

Even people living in regions of the world not affected by the ozone hole are increasingly warned about the dangers of UV radiation exposure. Most daily newspapers and TV weather reports routinely report a UV index indicating the expected danger associated with outdoor activities. Skin cancer (melanoma) rates are on the rise. A 1% decrease in ozone concentration has been predicted to cause a 2% increase in skin cancers. As early as 1999, the amount of UV radiation reaching New Zealand had increased by over 12% compared to the previous decade.

Not all increases in skin cancer can be attributed directly to the thinning ozone layer, however. Our obsession with bronzed bodies and the increased amount of tanning, whether by natural sunlight or in tanning salons, has also contributed to the dramatic increase in skin cancer. A 2002 report in the *Journal of the National Cancer Institute* reported that people who used tanning devices were anywhere from 1.5–2.5 times more likely to develop skin cancer! An earlier study indicated that people under the age of 30 who had used tanning lamps more than 10 times a year faced an eightfold likelihood of developing melanoma (skin cancer). Although many tanning salons argue that they have switched from UV-B lamps to lower energy UV-A lamps, studies have indicated that this has not reduced the problem because longer tanning periods are required to get the same effect. A golden tan may look great, but is the increased risk of skin cancers really worth it? California enacted legislation in 2005 to prohibit the use of tanning booths by anyone under age 14.

TRY IT 7.7

List three possible consequences of a significant decrease in stratospheric ozone concentration.

TRY IT 7.8

Is tanning associated with UV radiation in tanning salons any less dangerous than tanning associated with sunlight? Why or why not?

7.6 CAN WE DO ANYTHING ABOUT THE OZONE HOLE?

Since most (>80%) of the chlorine and bromine in the stratosphere comes from manmade chemicals, it stands to reason that this would be the place to start addressing the ozone hole problem. As early as 1978, the use of chlorofluorocarbons as propellants in aerosol spray cans was legislatively banned in the United States. By 1990, their use as foaming agents in blown foams was also discontinued in the United States when scientists began to realize the seriousness of the problems posed due to industrial release of these chemicals. The subsequent release of these chemicals from foam products so prevalently used in our society was also recognized as a major problem. The CFC gases trapped in the foam products ultimately find their way into the atmosphere as these foam products, found in everything from couches, to car seats, to dashboards, to carpet backing, to Styrofoam products of every sort, begin to decompose with age.

The first major international response to this problem came with the Montreal Protocol on Substances that Deplete the Ozone Layer (1987). At that time a number of countries agreed to guidelines that called for the reduction of CFC production to half of 1986 levels by 1998. In the intervening years, increased awareness of the seriousness of the problem led nations around the world to agree to accelerated schedules for the elimination of these chemicals. In 1990, one hundred countries agreed to ban the production of CFCs by 2000. Subsequent meetings led to the addition of methyl bromide and some other related compounds to the list, although some developing countries have been given permission to use methyl bromide until 2015. The United States and 140 other countries agreed to completely halt production of CFCs after December 31, 1995. However, many of these compounds have lifetimes of 100 or more years in the atmosphere so they'll be around for a while even if no more are being produced. Those already produced will almost certainly find their way into the atmosphere.

These and continuing actions on a worldwide scale have had a significant impact on the amount of chlorine in the atmosphere. Its concentration in the atmosphere is known to have peaked in the mid-1990s and is now declining. There is hope that the ozone layer may recover within the next 50 years or so due to this worldwide effort to address a very serious environmental problem. It is hoped that during this time the Cl concentration will drop to the pre-ozone–hole level of around 2 ppb.

The demand for CFCs did not suddenly disappear once the problem was recognized. International agreements allowed the sale and use of existing stockpiles of these CFCs. Some countries have imposed taxes on CFCs to encourage the switch to alternatives. This has led to price increases and black-market smuggling of CFCs into the United States. Some reports have suggested it is more profitable to smuggle CFCs into the United States from Mexico, Russia, or China than it is to smuggle cocaine. There is also the concern that some countries may allow or even encourage continued production of banned CFCs to take advantage of the demand for them. One thing is certain, no longer are CFCs being vented into the atmosphere in most places as a result of increased concern about the consequences.

CONCEPT CHECK 7B

1. What is a Dobson unit (DU) used for?
2. What concentration unit does 1 DU denote?
3. Which three scientists received the Nobel Prize in chemistry in 1995 for their work related to the destruction of ozone in the stratosphere?
4. Explain how weather conditions at the South Pole contribute to the observed decline in ozone concentrations during the period September–November.
5. The absorption of UV radiation with wavelengths less than 220 nm is sufficient to cleave certain bonds in a molecule of CFC. What bonds are most susceptible to cleavage? What other halogen–carbon bond is thought to be problematic as well in this regard?
6. What reactive species is formed when a CFC molecule absorbs high-energy UV radiation?
7. What two products are formed in the reaction between a chlorine atom and an ozone molecule?
8. What is the estimated number of ozone molecules that may be destroyed by a single chlorine atom?
9. List at least three uses of CFCs.
10. List at least two results that might occur if the ozone concentration in the stratosphere continues to decline.

7.7 WHAT WILL REPLACE CFCs?

The demand for air conditioning, aerosol sprays, refrigerants, blown foams, industrial solvents, and agricultural fumigants has not disappeared, so scientists have been working hard to find replacements for CFCs. Since it is the relatively weak C—Cl bonds and C—Br bonds in these molecules that are problematic, a major effort has focused on replacing as many as possible of the Cl or Br atoms with H atoms. Compounds in which some, but not all, of the Cl atoms have been replaced by H atoms are referred to as *hydrochlorofluorocarbons* or HCFCs. The HCFCs have a tendency to break down before they reach the stratosphere and would appear to be acceptable substitutes for CFCs at first glance. Compounds in which all Cl atoms have been replaced by H are *hydrofluorocarbons* or HFCs. HFCs are very unreactive and are not susceptible to UV-induced breakdown in the stratosphere. These compounds would also seem to be acceptable substitutes for the CFCs. However, one must remember that the replacement compounds must still have the necessary physical properties (correct boiling point, low toxicity, high stability, low flammability) that would make them appropriate as refrigerants, if that is the intended use of the compounds. Thus, this has not been a simple task. One cannot simply replace most or all of the Cl atoms with H or F atoms and be guaranteed of having an acceptable replacement compound. Even if one finds such a candidate among the HCFCs and HFCs, one must remember that these compounds are still potential greenhouse gases with varying abilities to absorb infrared radiation. For every solution, we appear to encounter a new problem.

TABLE 7.2	Alternatives to CFCs and Halons for Common Applications[a]	
APPLICATION	**CFC OR HALON USED PREVIOUSLY**	**ALTERNATIVES**
Aerosols	CFC-11, CFC-12	hydrocarbons, HFC-152a, dimethyl ether and water, pumps, roll-ons
Refrigerators		
home	CFC-12	HFC-134a
commercial	CFC-11	HCFC-123*
Air conditioners		
home	HCFC-22*	HFC blends
automobiles	CFC-12	HFC-134a
commercial	CFC-11, CFC-12	HCFC-123*, HFC-134a
(large buildings)		
Foam blowing		
polystyrene	CFC-12	hydrocarbons, HCFC-22*, HFC-152a
(food containers)		
polyurethane	CFC-11	HCFC-141b*, HFC-134a, vacuum panels
(refrigerator insulation)		
rigid foams	CFC-11, CFC-12	HCFC-142b*
(construction materials)		
Cleaning solvents	CFC-113, CH_3CCl_3	water-based processes; other solvents; supercritical CO_2[‡]
(electronics industry)		
Fire extinguishers	Halon-1301, Halon-1211	CF_3I and other possibilities are being tested
(aircraft, military, office)		

* Alternatives denoted by an asterisk (*) are interim solutions only, as they are gradually being phased out.

‡ Supercritical CO_2 is above the critical temperature (T_c), where gas and liquid coexist with no visible interface.

[a] Source: Institute for Chemical Education

Table 7.2 shows some of the replacement compounds and their properties. Many of these compounds are in current use and have dramatically reduced impacts on the ozone layer. Even though these HCFCs and HFCs also absorb infrared radiation, a report from the United Nations Intergovernmental Panel on Climate Change in the spring of 2005 reported that these replacements generally, with some significant exceptions, have much lower global warming potential than CFCs. The global warming potential of emissions of CFCs, HCFCs, and HFCs was equivalent to 7.5 billion metric tons of carbon dioxide in 1990, compared with the 22 billion metric tons of carbon dioxide released due to fossil fuel combustion. This equivalence had dropped to 2.5 billion metric tons of carbon dioxide by 2000 and the expectation is that the equivalence will drop to 2.3 billion metric tons of carbon dioxide by 2015.

CFCs in aerosol cans and blown foams have been replaced by a variety of compounds such as propane, butane, HCFCs, and HFCs. Most commercial products now routinely advertise that they contain no CFCs, but most fail to acknowledge that the replacement compounds still may cause ozone depletion and global warming. The search for safe and effective alternatives to CFCs continues. Maybe new technology will come to the rescue soon.

Assess your understanding of this chapter's topics with an online chapter quiz at www.brookscole.com/chemistry/joesten4

■ KEY TERMS

rainout   Chapman cycle

resonance structures   Dobson unit

steady state

■ APPLYING YOUR KNOWLEDGE

1. Define the following terms:
   - (**a**) Ozone hole
   - (**b**) CFCs
   - (**c**) HCFCs
   - (**d**) Montreal Protocol
   - (**e**) Resonance structures
   - (**f**) UV-A
   - (**g**) UV-B
   - (**h**) UV-C
   - (**i**) Chapman cycle
   - (**j**) HFCs
   - (**k**) Stratosphere
   - (**l**) Steady state
   - (**m**) Dobson unit

2. Explain the different mechanisms by which ozone is formed in the troposphere as a pollutant and in the stratosphere as a protective shield.

3. Classify the following statements as true or false.
   - (**a**) Increased destruction of the ozone layer would probably lead to an increase in the incidence of skin cancers.
   - (**b**) Volcanic eruptions produce CFCs, which have been a major factor in the destruction of the ozone layer.
   - (**c**) The ozone hole that appears in the Southern Hemisphere each fall has now started to spread over some populated areas.
   - (**d**) Ozone in our stratosphere absorbs primarily UV-B radiation in its protective role, while oxygen absorbs UV-C radiation.
   - (**e**) UV-A radiation is not absorbed by stratospheric ozone.
   - (**f**) The ozone molecules in the protective ozone layer of the stratosphere are quite different than the harmful ozone molecules found in ground-level smog and air pollution.
   - (**g**) The replacement of CFCs by HCFCs has been totally effective at reversing the damage done to the ozone layer since HCFCs do not destroy the ozone layer.
   - (**h**) The concentration of CFCs in the stratosphere is relatively low. However, the fact that a single chlorine atom generated by the action of UV radiation on the CFC may cause the destruction of as many as 100,000 ozone molecules is cause for much concern.
   - (**i**) Less than 50% of the chlorine and bromine in the stratosphere comes from manmade chemicals.
   - (**j**) Oxygen molecules absorb lower energy ultraviolet radiation than ozone molecules absorb because the oxygen–oxygen double bond is shorter and stronger than the 1.5 bonds of ozone.

4. What is the difference, if any, between "good ozone" and "bad ozone"?

5. Discuss the mechanism by which CFCs and related compounds destroy the ozone layer. What chemical bond in the CFC is critical in this mechanism?

6. What were CFCs primarily used for before their potential as ozone-destroying chemicals was recognized? What compounds have been used as replacements?

7. How thick would the stratospheric ozone layer be if it were compressed to atmospheric pressure at Earth's average temperature?

8. List some of the characteristics that were sought in the chemicals used as replacements for early refrigerants.

9. What type of radiation does ozone shield us from? What wavelengths of this radiation are absorbed by ozone?

10. Oxygen protects us from ultraviolet radiation as does ozone, but there are differences in the wavelength of absorption. What are they?

11. Why do oxygen and ozone absorb UV radiation of different wavelengths? Which absorbs UV radiation of higher energy?

12. Explain why the "ozone hole" has occurred primarily over Antarctica.

13. Give an example of a bromine-containing compound that has been implicated in destroying ozone.

14. Write out the sequence of reactions of the Chapman cycle and explain what is happening in each reaction.

15. Give examples of two common CFCs and draw Lewis structures for them.

16. Discuss the bathtub analogy as it relates to the steady state concentration of ozone in the stratosphere.

17. Why are there seasonal fluctuations in ozone concentration above the South Pole?

18. CFC vapors are fairly heavy, and many people have suggested that this makes it impossible for them to make their way into the stratosphere. Explain how it is possible for that to happen and what factors allow it.

19. Many CFC replacements are greenhouse gases that are worse than CFC-11. How do CFCs and CFC replacements act as greenhouse gases?

20. What might be some of the effects of a significant reduction in stratospheric ozone?

21. Which molecule is primarily responsible for absorbing the UV-C (high energy UV) emitted by the Sun?
    - (a) nitrogen
    - (b) oxygen
    - (c) carbon dioxide
    - (d) carbon monoxide
    - (e) ozone

22. The ozone hole has developed primarily over the South Pole because:
    - (a) The heavy CFCs sink to the bottom of the world.
    - (b) Volcanoes in South America release significant amounts of CFCs into the atmosphere.
    - (c) The weather conditions at the South Pole are critical for the chemical reactions that destroy the ozone layer.
    - (d) The South Pole is closer to the Sun so there is more UV radiation there.
    - (e) Excess CFCs, now banned in the industrialized world, were shipped there for storage and leaked into the atmosphere.

23. Which of the following is a likely explanation for the fact that, even though volcanoes produce HCl gas, the HCl thus produced does not seem to be able to generate chlorine atoms to destroy the ozone layer?
    - (a) The chlorine atoms that would be produced by dissociation of HCl are different than the chlorine atoms produced by interaction of CFCs with ultraviolet radiation.
    - (b) HCl is a liquid that runs down the sides of the volcano during an eruption.
    - (c) HCl is a solid and simply falls out of the sky.
    - (d) HCl is probably removed from the troposphere by being rained out because it reacts readily with water.

24. The Chapman cycle is a series of chemical reactions related to the oxygen–ozone screening process. These reactions indicate that, in the absence of human interference, oxygen and ozone concentrations in our stratosphere would:
    - (a) continually increase
    - (b) maintain a steady state
    - (c) continually decrease
    - (d) change with the seasons
    - (e) drop dramatically as global warming increases

25. As we have progressed from using CFCs to HCFCs to HFCs we have been:
    - (a) gradually eliminating the chlorine atoms from these refrigerant molecules
    - (b) replacing many of the carbon atoms with less-harmful chlorine atoms
    - (c) trying to prepare refrigerants that will destroy tropospheric ozone before it contributes too much to global warming
    - (d) trying to produce a refrigerant that is so heavy that it cannot possibly ever make its way into the stratosphere
    - (e) come up with molecules that produce ozone by reaction with oxygen and UV light when they arrive in the stratosphere

26. The _____ deals with global warming and the _____ deals with the hole in the ozone layer and the phase-out of _____.
    - (a) Montreal Protocol; Kyoto Conference; greenhouse gases
    - (b) Kyoto Conference; Montreal Protocol; substances that deplete the ozone layer
    - (c) Chapman cycle; law of conservation of mass; greenhouse gases
    - (d) Kyoto Conference; Montreal Protocol; CFCs
    - (e) Kyoto Conference; Montreal Protocol; photons

27. UV-C radiation has wavelengths less than 280 nm while UV-B radiation has wavelengths between 280–320 nm. Which is higher in energy?

28. If the ozone layer in the stratosphere were to be brought to atmospheric pressure and the average surface temperature of Earth (15°C), the ozone would:
    - (a) be about an inch thick
    - (b) be about the thickness of two nickels stacked on top of each other
    - (c) totally block all UV radiation from reaching Earth
    - (d) totally block all UV and IR radiation from reaching Earth
    - (e) would be impervious to damage by Cl and Br atoms

29. Why do you think that ozone produced at ground level in large urban environments doesn't simply make its way to the stratosphere to replace the ozone that has been destroyed by CFCs?

30. In a previous chapter, we looked at some of the unusual ideas that had been proposed to address global warming. The ozone hole problem does not seem to have generated the same variety of unusual ideas. For instance, no one appears to have suggested putting "sunglasses" between Earth and the Sun. Why do you think that is? Can you suggest differences between global warming and the ozone hole that suggest we may have turned the corner in addressing the ozone hole but have not done so with global warming?

31. Compare and contrast global warming and the ozone hole problem in terms of:
    (a) atmospheric region involved
    (b) type of radiation involved
    (c) molecules involved
    (d) public recognition of the problem
    (e) legislative action and other action taken
    (f) public recognition of the problem
    (g) results of actions taken

32. Spend some time investigating one or more of the following websites related to the ozone hole problem.
    (a) The official site of the Environmental Protection Agency related to The Science of Ozone Depletion: **http://www.epa.gov/ozone/science/**
    (b) A comic book version about the ozone hole called "On The Trail of the Missing Ozone": **http://www.epa.gov/ozone/science/missoz/index.html**
    (c) A historical account of the development of the ozone hole: **http://www.atm.ch.cam.ac.uk/tour/index.html**
    (d) The National Academy of Science's website related to the ozone hole: **http://www.beyonddiscovery.org/content/view.article.asp?a=73**

 ## CHEMISTRY ON THE WEB

For up-to-date URLs, visit the text website at **www.brookscole.com/chemistry/joesten4**

- CFCs and Their Regulation
- The Ozone Hole
- Stratospheric Ozone Depletion
- On The Trail of the Missing Ozone (Cartoon Version)
- The UV Index
- Skin Cancer and Ozone Loss
- Protecting the Ozone Layer: A Checklist for Citizen Action

CHEMICAL REACTIVITY: CHEMICALS IN ACTION

© Brand X Pictures/Getty Images

Fireworks are an extraordinary display of chemical reactions.

Biologists, physicians, chemists, psychologists, and sometimes even sociologists and economics professors depend on experiments to test their theories. Chemists have one advantage in such experiments. Under *identical conditions,* pure chemicals always react with each other in the same way. Sometimes it is difficult, but with effort identical conditions can be achieved. Those who study living things or social interactions can never be sure. Are two plants, two laboratory rats, or two social groups ever identical?

In this chapter the conditions that influence the outcome of chemical reactions are explored.

- What is the meaning of a balanced equation?
- Why are the mole and the molar mass essential concepts?

- In what ways can reaction rates be influenced?
- What is happening in a chemical reaction that has come to equilibrium?
- What two kinds of changes influence the favorability of chemical reactions?
- What are the first and second laws of thermodynamics?
- What are the major issues in recycling metals and other materials?

When journalists investigate a story for page one in the newspaper, they want to answer six questions:

Who?   What?   When?   Where?   How?   Why?

To fully investigate a chemical reaction requires answering a similar list of questions:

What?   How much?   How fast?   How far?   Why?

Some scientists spend a lifetime seeking answers to these questions about a single complex reaction. Others devote themselves to answering one of these questions about many chemical reactions.

In Chapter 2 we introduced chemical reactions and the information needed to answer the "What?" question. What are the reactants and products? Hydrogen and oxygen, the reactants, combine to form water, the product in the following chemical reaction:

$$2 H_2(g) + O_2(g) \longrightarrow 2 H_2O(\ell)$$
Hydrogen   Oxygen   Water

Now we're going to pick up the story of chemical reactions and pursue the meaning of the other questions listed here.

 See an animation of **Hydrogen and Oxygen Combustion Use for Powering the Space Shuttle** at http://brookscole.com/chemistry/joesten4

8.1 BALANCED CHEMICAL EQUATIONS AND WHAT THEY TELL US

Chemical equations are the best way we have to represent what happens in chemical reactions at the nanoscopic level that we cannot see. An equation will not be faithful to reality if the chemical formulas are wrong or if the equation is not balanced.

In balancing chemical equations, we are applying the law of conservation of matter—the atoms in the reactants must all be there in the products. To be sure an equation is balanced requires counting up the atoms of each kind in the reactants and products (Section 2.7). Doing this, of course, requires knowing the identity and correct formulas of the reactants and products. An equation can *never* be balanced by changing the subscript in a chemical formula. This changes the identity of the compound. Only coefficients can be changed to achieve balance.

The products of the complete burning, or complete combustion, of any hydrocarbon are always carbon dioxide and water. Thus, for the burning of methane (CH_4), the major ingredient in natural gas, the *unbalanced* equation is

$$CH_4(g) + O_2(g) \longrightarrow CO_2(g) + H_2O(g)$$
1 C atom   1 C atom

The law of conservation of matter: *Matter is neither lost nor gained in chemical reactions* (Section 3.1). The only known exception to this law is in nuclear reactions, which occur only with radioactive isotopes or under the special conditions of artificial nuclear reactions (Section 13.5). The conservation law (so far) has always been reliable for chemical changes other than nuclear reactions.

The states of the reactants and products are indicated by
(*g*) for a gas
(*s*) for a solid
(*ℓ*) for a liquid
(*aq*) for something dissolved in water

The C atoms are balanced here—one C on each side of the arrow. But the O and H atoms are not balanced. Often, it is easiest to first balance atoms that appear in only one formula on each side. Balancing H requires a coefficient of 2 in front of H_2O:

$$CH_4(g) + O_2(g) \longrightarrow CO_2(g) + 2\ H_2O(g)$$

4 H atoms   4 H atoms

Now the O atoms must be balanced. With four of them on the right, providing two O_2 molecules on the left finishes the job:

$$CH_4(g) + 2\ O_2(g) \longrightarrow CO_2(g) + 2\ H_2O(g)$$

4 O atoms   4 O atoms

See an animation of **Balancing Chemical Equations** at http://brookscole.com/chemistry/joesten4

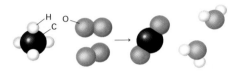

Note in these equations the difference between *subscripts* (e.g., the subscript 4 in CH_4), which relate to the need for correct formulas, and *coefficients* (e.g., the 2 with O_2 as a reactant), which relate to the need for a balanced equation.

One cannot predict the product(s) of a chemical reaction without prior knowledge of the reaction or knowledge of similar reactions. This built-up knowledge serves as a database that will allow you to use previous examples and chemical intuition to predict the products of many, but not all, chemical reactions.

EXAMPLE 8.1   Equation Balancing

Balance the following equation for the reaction of hydrofluoric acid [$HF(aq)$] with glass, which can be represented as calcium silicate ($CaSiO_3$). Decorative glass is etched using this reaction:

$$CaSiO_3(s) + HF(aq) \longrightarrow CaF_2(s) + SiF_4(g) + H_2O(\ell)$$

SOLUTION
The Ca and Si atoms are balanced. To balance the three O atoms on the left requires three H_2O molecules on the right. There must then be six H atoms on the left. Putting in both of these coefficients gives

$$CaSiO_3(s) + 6\ HF(aq) \longrightarrow CaF_2(s) + SiF_4(g) + 3\ H_2O(\ell)$$

There are now six F atoms on each side of the equation, and it is fully balanced.

On the left: 1 Ca   1 Si   6 H   6 F   3 O
On the right: 1 Ca   1 Si   6 H   6 F   3 O

TRY IT 8.1A
Balance the following equation for the preparation of aluminum trichloride, which is an ingredient in some antiperspirants.

$$Al(s) + Cl_2(g) \longrightarrow AlCl_3(s)$$

TRY IT 8.1B
Are the following equations balanced? If not, balance them.
(a) $CaO(s) + H_2O(\ell) \longrightarrow Ca(OH)_2$
(b) $SiO_2(s) + C(s) \longrightarrow Si(s) + CO(g)$

8.2 # THE MIGHTY MOLE AND THE "HOW MUCH?" QUESTION

Moles and Molar Masses

You might want to look back at the description of atomic weights in Section 3.3.

Eventually, anyone who wants to carry out a chemical reaction must figure out how much of the reactants must be combined to make the desired amount of product. Somehow a connection must be made between atoms, molecules, and ions at the nanoscopic scale and weighable amounts of chemicals. Here is where balanced equations are essential. The relative atomic weight scale and balanced chemical equations together make it possible to answer "How much?" questions.

To understand the situation, consider the balanced equation for hydrogen burning in chlorine to form hydrogen chloride:

$$H_2(g) + Cl_2(g) \longrightarrow 2\,HCl(g)$$

The equation shows that one molecule of hydrogen and one molecule of chlorine combine to form two molecules of hydrogen chloride. Is this information of any help in figuring out, for instance, how much hydrogen and chlorine would be needed to make 100 g of hydrogen chloride? The essential problem is that molecules are very small, so small that it is impossible to count them one by one.

The solution to the problem is counting by weighing. The atomic weights given for each element in the periodic table are relative. The mass of one average neon atom is 20 atomic mass units (amu), and the mass of one calcium atom is 40 amu, both relative to an atomic weight of exactly 12 amu for carbon-12. Translating these numbers to masses big enough to measure means that 20 g of neon, 12 g of carbon-12, and 40 g of calcium all contain the same number of atoms. This type of relationship is at the heart of the quantitative use of chemical equations.

The mole as a unit is symbolized by mol, e.g., 2.5 mol.

Recall that one atomic mass unit (1 amu) is defined as exactly $\frac{1}{12}$ of the mass of a single atom of carbon-12 (C-12) or 1.661×10^{-24} g. Thus, 12 amu is equal to the mass of a single atom of C-12 and, by analogy, 16 amu is equal to the mass of a single O-16 atom. We could determine the mass of a single atom of any element in grams by multiplying its atomic mass in amu by the mass of 1 amu in grams. The exact mass of a single atom of any element is not particularly useful, however. For instance, who wants to remember that the mass of a single carbon atom is 1.993×10^{-23} g (obtained from 12 amu/C times 1.661×10^{-24} g/amu)? Such numbers would be incredibly difficult to remember.

We don't need all the digits for atomic weights given in the periodic table inside the front cover, so we will round them off.

To avoid the problems associated with the difficult-to-remember individual atom masses, a counting unit consisting of a certain number of atoms has been defined. You are familiar with a counting unit of one dozen, which can be used to define 12 baseballs, 12 eggs, or 12 donuts, for example. *The counting unit used for atoms is the **mole**.* One mole of anything (atoms, molecules, baseballs, eggs, pencils, etc.) is defined as 6.02×10^{23} of the object under consideration. This special number is known as **Avogadro's number** and, although it is significantly larger than 12, *it is just a counting unit.* Just as you have little trouble remembering that a dozen is 12 of something, you will soon learn to remember that *a mole is 6.02×10^{23} of something.*

Mole Unit for amount of substance; contains Avogadro's number of atoms, molecules, or other particles

Avogadro's number 6.02×10^{23}; the number of items in one mole

Molar mass Mass in grams of one mole of any substance

The mole is an important counting unit in chemistry because one mole of, for instance, carbon atoms has a mass of 12.0 g. This is the mass of Avogadro's number of carbon atoms and is known as the **molar mass**. One carbon atom has a mass of 12.0 amu and one mole of carbon atoms has a mass of 12.0 g. *Thus, the atomic weights listed below each element symbol in the periodic table may be used to refer to the mass of a single atom in amu or to the mass of Avogadro's number of atoms in grams.* Therefore, one mole of calcium atoms has a mass of 40.08 g, one mole of iodine atoms has a mass of 126.9 g, and so on.

EXAMPLE 8.2 **Molar mass**

Show that the mass of Avogadro's number of carbon atoms is 12.0 g.

SOLUTION
Recall from above that 1 amu $= 1.661 \times 10^{-24}$ g and that a single carbon atom has a mass of 12 amu. Therefore, 12 amu $\times 1.661 \times 10^{-24}$ g/amu $= 1.993 \times 10^{-23}$ g, which is the mass of 1 carbon atom in grams. Therefore, Avogadro's number of carbon atoms has a mass of $(1.993 \times 10^{-23}$ g/C atom$)(6.02 \times 10^{23}$ C atom/mol$) = 12.0$ g/mol

TRY IT 8.2A
Show that the mass of one mole (Avogadro's number) of oxygen atoms is 16.0 g.

TRY IT 8.2B
Show that the mass of one mole (Avogadro's number) of xenon atoms is 131.3 g.

This discussion and these exercises have shown that *one mole of atoms of any element is simply the atomic weight in grams listed below the element symbol in the periodic table. This is the molar mass of the element* (Figure 8.1).

We can extend this idea of the molar mass of an element to determine the molar mass of compounds. By analogy, the molar mass of a compound is simply the sum of the atomic weight of each element in grams multiplied by the subscript on the element in the formula of the compound. For instance, one mole of methane molecules (CH_4) contains one mole of carbon atoms (molar mass $= 12.0$ g) and 4 moles of hydrogen atoms (4×1.0 g/mol $= 4.0$ g). Thus, the molar mass of CH_4 is 16.0 g. To summarize, *to find the molar mass for a compound, multiply the number of atoms of each element in the formula by the atomic weight of the element and add them together.*

EXAMPLE 8.3

Determine the molar mass of rubbing alcohol (C_3H_7OH).

SOLUTION

$$3 \text{ C } (3 \text{ mol})(12.0 \text{ g/mol}) = 36.0 \text{ g}$$
$$8 \text{ H } (8 \text{ mol})(1.0 \text{ g/mol}) = 8.0 \text{ g}$$
$$1 \text{ O } (1 \text{ mol})(16.0 \text{ g/mol}) = \underline{16.0 \text{ g}}$$
$$60.0 \text{ g}$$

TRY IT 8.3A
Determine the molar mass of common table salt (NaCl).

FIGURE 8.1 One-mole quantities of some elements. The cylinders (*left to right*) hold mercury (201 g), lead (207 g), and copper (64 g). The two Erlenmeyer flasks hold sulfur (*left*, 32 g) and magnesium (*right*, 24 g). All rest on 1 mole of aluminum in the form of foil (27 g) and also on the watch glass.

Charles D. Winters

 See an animation of **One Mole of Various Substances** at http://brookscole.com/ chemistry/joesten4

TRY IT 8.3B

Determine the molar mass of the simple sugar known as glucose ($C_6H_{12}O_6$).

Because the mole is a counting unit, every balanced chemical equation can be interpreted in terms of moles—moles of atoms, moles of molecules, moles of ions, or anything else. For example,

$$H_2(g) \; + \; Cl_2(g) \longrightarrow 2\,HCl(g)$$
1 H₂ molecule   1 Cl₂ molecule   2 HCl molecules

also means

$$H_2(g) \; + \; Cl_2(g) \longrightarrow 2\,HCl(g)$$
1 mol of H₂   1 mol of Cl₂   2 mol of HCl

The molar mass of a substance provides a conversion factor that can be used to find the number of moles equal to a known mass of a substance. For example, 10 g of carbon is equal to 0.83 mol of carbon.

$$(10 \text{ g C})(1 \text{ mol}/12.01 \text{ g C}) = 0.83 \text{ mol C}$$

Also, the molar mass allows conversion of a known number of moles to the equivalent mass of a substance. For example,

$$(2 \text{ mol C})(12.01 \text{ g}/1 \text{ mol C}) = 24.02 \text{ g C}$$

Sometimes the term *molecular weight* is used instead of *molar mass*. Both terms refer to the relative mass of a substance according to the atomic weight scale. Strictly speaking, a molecular weight is for one molecule and would be in atomic mass units. We need use only molar masses.

(EXAMPLE) **8.4**   **Moles and Masses**

What is the equivalent in moles of 3.7 g of water, roughly the amount in a teaspoonful?

SOLUTION

The molar mass of water, H_2O, is 18 g. Molar mass provides the conversion factors that connect mass and numbers of moles. Therefore,

$$3.7 \text{ g } H_2O \times \frac{1 \text{ mol } H_2O}{18 \text{ g } H_2O} = 0.21 \text{ mol}$$

TRY IT 8.4

What is the mass equivalent to 50 mol of barium nitrate, $Ba(NO_3)_2$, which gives a green color to fireworks?

Masses of Reactants and Products

The mole provides us with the means to answer the "How much?" question about any chemical reaction. If all the reactants are converted to product, 2.0 g of H_2 (1 mol) should react with 71 g of Cl_2 (1 mol) to produce 73 g of HCl (2 mol).

$$H_2(g) \quad + \quad Cl_2(g) \quad \longrightarrow \quad 2\,HCl(g)$$

1 H_2 molecule	1 Cl_2 molecule	2 HCl molecules
1 mol of H_2	1 mol of Cl_2	2 mol of HCl
2.0 g of H_2	71 g of Cl_2	2 mol × 36.5 g/mol = 73 g of HCl

Have you noticed that these masses adhere to the law of conservation of matter? The 2.0 g + 71 g of reactants produce 73 g of product.

Rarely can 100% of reactants actually be converted to products. One goal of industrial preparation of chemicals is to come as close to 100% as is possible and practical.

EXAMPLE   8.5   **Information about Masses from a Chemical Equation**

Interpret the equation for a reaction that produces copper from copper ore in terms of moles, molar masses, and masses of reactants and products.

$$\underset{\text{Copper(I) sulfide}}{Cu_2S(s)} \quad + \quad \underset{\text{Copper(I) oxide}}{2\,Cu_2O(s)} \quad \overset{\text{Heat}}{\longrightarrow} \quad \underset{\text{Copper}}{6\,Cu(s)} \quad + \quad \underset{\substack{\text{Sulfur}\\\text{dioxide}}}{SO_2(g)}$$

SOLUTION

The equation in terms of moles shows by the coefficients that 1 mol of copper(I) sulfide reacts with 2 mol of copper(I) oxide to produce 6 mol of metallic copper and 1 mol of sulfur dioxide. The molar masses of the reactants and products are found from the molar masses of the elements combined in each reactant and product. Using the molar masses for copper (Cu, 64 g/mol), sulfur (S, 32 g/mol), and oxygen (O, 16 g/mol) gives for Cu_2S, 160 g/mol[$(2 \times 64$ g) + 32 g]; for SO_2, 64 g/mol [32 g + $(2 \times 16$ g)]; and for Cu_2O, 144 g/mol [$(2 \times 64$ g) + 16 g]. Thus, the equation gives the following information about reactants and products:

The number of moles (or the number of molecules) is not necessarily conserved in a chemical reaction. Mass is conserved, however, according to the law of conservation of matter.

$$Cu_2S(g) \,+ \qquad 2\,Cu_2O(g) \qquad \longrightarrow \qquad 6\,Cu(s) \quad + \; SO_2(g)$$

1 mol Cu_2S	2 mol Cu_2O	6 mol Cu	1 mol SO_2
160 g Cu_2S	2 × 144 g Cu_2O/mol = 288 g Cu_2O	6 × 64 g Cu/mol = 384 g Cu	64 g SO_2

TRY IT 8.5

Interpret the equation for making methanol in terms of moles, molar masses, and masses of reactants and products.

$$CO(g) \,+\, 2\,H_2(g) \longrightarrow CH_3OH(\ell)$$

CONCEPT CHECK   8A

1. Balancing chemical equations is an application of the _____.
2. The _____ is used by chemists the same way the _____ is used by a shopper purchasing eggs.

THE PERSONAL SIDE

Amadeo Avogadro (1776–1856)

Today, Avogadro's name is most often associated with "his" number. The mole and the modern definition of the number of particles in it, however, were not directly his invention. His fame lies in a single simple statement, made in 1811: ". . . *the number of integral molecules in gases is always the same for equal volumes.*" This concept opened the door to understanding atomic weight and the formulas for chemical compounds. (The modern version of this concept is *one mole of any gas always occupies the same volume [22.4 L] at 0°C and 1 atm of pressure.*)

Avogadro was a quiet man who was totally devoted to his studies. He was born into a prominent Italian family of lawyers and, as was surely expected of him, completed his training as a lawyer and entered government service. At age 24 he turned his back on his heritage and devoted the rest of his life to science. He became a professor of physics and mathematics and was content to pursue his studies alone, never attending scientific meetings nor seeking out colleagues with whom to exchange ideas. He published very little, but the intensity of his

© Bettmann/Corbis

AMADEO AVOGADRO

studies is shown in the 75 volumes of handwritten notes left behind when he died at age 80.

Perhaps because of Avogadro's private nature, the true value of his realization about gas volumes was not recognized until after his death, when another Italian chemist (Stanislao Cannizzaro) brought it to the attention of the scientific community. By comparing the masses of equal volumes of gases at identical temperature and pressure, the relative weights of the molecules could be compared. And by examining the volumes of gases that combined in reactions, it was possible to deduce the formulas. Before this, it was not recognized that a hydrogen molecule consisted of two hydrogen atoms rather than one, or that water molecules consist of two hydrogen atoms and one oxygen atom, rather than just one hydrogen atom and one oxygen atom.

3. Balance the following equations:
   (a) _____ $Si(s)$ + _____ $Cl_2(g)$ ⟶ $SiCl_4(g)$
   (b) _____ $Al(s)$ + _____ $O_2(g)$ ⟶ _____$Al_2O_3(s)$
   (c) _____ $(NH_4)_2CO_3(aq)$ + _____ $Cu(NO_3)_2(aq)$ ⟶ _____ $CuCO_3(s)$ + _____ $NH_4NO_3(aq)$

4. Define amu, mole, and molar mass.

5. In photosynthesis, carbon dioxide combines with water to form oxygen and the simple sugar glucose ($C_6H_{12}O_6$).
   (a) Balance the equation:

   _____ $CO_2(g)$ + _____ $H_2O(\ell)$ ⟶ _____ $C_6H_{12}O_6(aq)$ + _____ $O_2(g)$

   (b) How many molecules of CO_2 are needed to produce one molecule of glucose?
   (c) How many moles of CO_2 are needed to produce 1 mol of glucose?
   (d) What is the molar mass of glucose?
   (e) What mass in grams of CO_2 is needed to make 1 mol of glucose?

6. What is the molar mass of nitrogen dioxide (NO_2)?

7. How many grams of carbon dioxide (CO_2) are there in 186 moles of carbon dioxide? This is about the amount of carbon dioxide produced by burning 1 gallon of gasoline.

Decomposition of ammonium nitrate (NH_4NO_3); a reaction that is very fast.

8.3 RATES AND REACTION PATHWAYS: THE "HOW FAST?" QUESTION

Reaction Pathways

The chemical reactions frequently used as classroom demonstrations are usually fast. The color change, bubbles of gas, or explosion happens right away as visible proof that a reaction has occurred. Many reactions are also naturally slow, however. At everyday conditions of temperature and pressure, for example, the conversion of carbon monoxide to carbon dioxide is slow.

$$2\,CO(g) + O_2(g) \longrightarrow 2\,CO_2(g)$$

It is sometimes unfortunate that this reaction isn't fast. Breathing too high a concentration of carbon monoxide is fatal.

Out of curiosity and also for practical reasons, it is interesting to discover what makes reactions fast or slow. Ideally, to do this a chemist would like to watch the pathway of each atom from its position in the reactants to its position in the products.

What is the connection between the rate of a process and its pathway? Suppose that there are several hundred books in a storeroom on the first floor and that you have been hired to move them to the new third-floor library. Depending on the conditions, there are a variety of possible pathways. One pathway might require the following steps: (1) put 10 books (the maximum number you can lift) into a carton, (2) carry the carton up one flight of stairs to the second floor, (3) rest a bit, (4) carry the carton up another flight of stairs to the third floor, (5) empty the carton, and (6) return for another load. Another pathway might involve a different series of steps: (1) fill four cartons with books, (2) pile the cartons onto a dolly, (3) push the dolly onto the elevator, (4) ride to the third floor, (5) push the dolly off the elevator, (6) empty the four cartons, and (7) return for another load. The second pathway would probably be faster than the first. To compare them quantitatively, the rate for each pathway could be measured in books moved per hour.

Reaction of zinc (Zn) with hydrochloric acid to produce hydrogen gas; a reaction that is slower than the decomposition of ammonium nitrate.

The details of chemical reaction pathways can be extremely complex, and they are very hard to study. With the aid of computers, sophisticated electronics, and clever new techniques, however, advances in this area of observation are being announced regularly.

By using your imagination instead, what might be seen in a simple one-step chemical reaction between two different gases? The gas molecules are flying about at random. Now and then, two molecules head for a collision. As they get closer together, repulsion builds up between their negatively charged electrons. If their kinetic energies are not great enough to overcome this repulsion, the molecules just veer away from each other. No chemical reaction occurs (Figure 8.2a). But if the kinetic energy of the approaching molecules is great enough to drive them together in spite of the repulsion, then they collide. At this instant a chemical reaction might occur. If some of the original bonds break so that new bonds can form, the collision is successful and yields the product (Figure 8.2b).

In summary, the following factors are involved in any chemical reaction:

- Atoms, molecules, or ions must collide for a reaction to occur.

- The collisions between the atoms, molecules, or ions must be sufficiently energetic for reaction to occur.

- In some reactions, collisions are productive only if the reactants are properly oriented relative to each other upon collision.

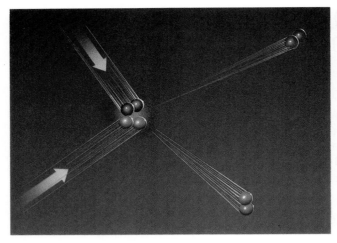

(a) Unsuccessful collision

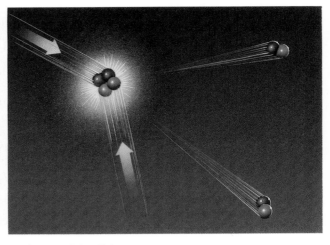

(b) Successful collision

FIGURE 8.2 **Collisions and chemical reactions.** Some collisions are successful and some are not. The difference is determined to a great extent by the kinetic energy of the particles.

- Bonds are made or broken in most chemical reactions. Some reactions may involve only electron transfer, however.
- Energy is absorbed during bond breakage and released during bond formation.

An Energy Hill to Climb

Reaction rates are usually expressed as the amount of reactant converted to product in a specific unit of time. For fast reactions the time unit might be seconds; for slow reactions it might be days. The number of successful collisions per second, minute, hour, or day determines the reaction rate.

For each reaction, there is a threshold quantity of energy needed for successful collisions, known as the **activation energy (E_{act})**. A high activation energy is like a steep mountain that must be climbed to reach the valley on the other side. In a given collection of reactant particles, only a small number have enough energy to get over a very high activation energy hill in a given time; thus, the reaction is slow. In the opposite condition, many reactant particles can get over a low activation energy hill; therefore, the reaction is fast (Figure 8.3).

Some reactions do not take place at all unless enough energy is supplied to get things started by pushing a few reactants over the activation energy barrier. After that, as in the explosion of a hydrogen–oxygen mixture or the combustion of a fuel, the reaction provides enough energy to keep itself going.

In addition to sufficient energy, successful collisions also require that the reacting parts of molecules or ions physically connect with each other with the proper orientation.

Reaction rate Amount of reactant converted to product in a specific amount of time

Activation energy (E_{act}) Quantity of energy needed for successful collision of reactants; determines reaction rate

Exergonic A chemical process in which there is a net energy release.

FIGURE 8.3 **Energy profiles of two exergonic chemical reactions.** Exergonic reactions are "downhill" processes overall in terms of energy, and the product is favored at equilibrium. Reaction (a) is faster than reaction (b) because the activation energy barrier (E_{act}) is lower for (a) than for (b).

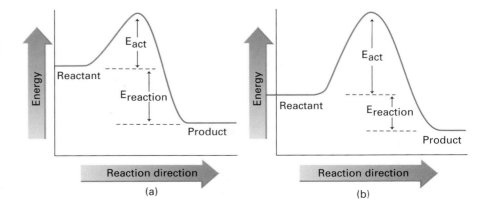

Consider the following sets of reactions in which one reaction does not occur (or occurs very slowly) at room temperature while the other occurs much more rapidly at higher temperatures.

a. $N_2(g) + O_2(g) \xrightarrow{\text{room temp.}}$ no reaction

$N_2(g) + O_2(g) \xrightarrow{\text{very high temp.}}$ 2 NO

b. $2 H_2(g) + O_2(g) \xrightarrow{\text{room temp.}}$ no reaction

$2 H_2(g) + O_2(g) \xrightarrow{\text{spark}}$ $2 H_2O(g)$

c. $4 Fe(s) + 3O_2(g) \xrightarrow{\text{room temp.}}$ $2 Fe_2O_3(s)$ (slow)

$4 Fe(s) + 3O_2(g) \xrightarrow{\text{heat}}$ $2 Fe_2O_3(s)$ (faster)

We know that nitrogen and oxygen, the major components of our atmosphere, do not react with each other at room temperature, yet these same two gases react quite readily at the high temperatures in the internal combustion engine to produce nitric oxide. Similarly, hydrogen and oxygen gas are unreactive toward each other unless a spark is added, in which case there is an explosion. Iron will react with oxygen at room temperature to slowly produce rust (Fe_2O_3) but this process is much more rapid at elevated temperatures. The rate of rusting of iron can also be increased by increasing the surface area contact between the iron and oxygen (iron filings rust more rapidly than a block of iron) or by increasing the concentration of oxygen (iron can be burned in pure oxygen).

Controlling Reaction Rates

To control the rate of a reaction requires either increasing the population of reactants with enough energy to get over the activation energy barrier or lowering the barrier. Four strategies are available: (1) adjust the temperature, (2) adjust the concentration of the reactants, (3) increase contact between the reactants, or (4) add a catalyst.

1. ***Effect of temperature on reaction rate.*** At higher temperatures, molecules (on average) move faster, so that more of them have enough energy to get over the activation energy hill and react (Figure 8.4). We make use of this principle in cooking by raising the temperature to speed up roasting a piece of meat and in preserving foods by lowering the temperature to slow down the reactions that spoil the food.

2. ***Effect of concentration on reaction rate.*** The quantity of a substance in a given quantity of a mixture is its *concentration*. For example, a solution might have a concentration of 5 g of sodium chloride in 1 L of water (5 g/L, or 5 grams per liter). A solution of 10 g of sodium chloride in 1 L of water (10 g/L) has a higher concentration.

    Increasing the concentration of reactants increases reaction rates. The more reactant atoms, molecules, or ions, the more frequent the collisions that have enough energy to be successful. There is, for example, a dramatic increase in the rate of combustion in pure oxygen compared with that in air, which is about 20% oxygen.

3. ***Effect of contact on reaction rate.*** An effect similar to a concentration increase occurs if a solid reactant is very finely divided, which essentially increases concentration by increasing the surface area at which reactions can occur. We don't think of flour as an explosive substance, but a spark can set off the explosion of flour suspended in the air. Dust explosions are a hazard in grain storage and coal mines.

See an animation of **Combustion of Hydrogen and Oxygen** at http://brookscole.com/chemistry/joesten4

FIGURE 8.4   Temperature and reaction rate. When a light stick is bent, an inner glass tube breaks, allowing reactants to mix. The result is a chemical reaction that releases energy as light. You can see by the dimmer light in the ice water (*right*) that the colder temperature slows down the reaction.

Richard Megna/Fundamental Photographs

Charles D. Winters

Reaction of chalk (calcium carbonate) with dilute hydrochloric acid. *Left:* Powdered blackboard chalk reacts faster because the greater surface area increases the amount in contact with the hydrochloric acid. *Right:* The reaction of a piece of blackboard chalk is slower than the reaction of powdered chalk.

An alligator crossing a highway. Alligators are cold-blooded animals—the rate of their metabolism depends on the temperature. This alligator crawled onto a major highway during an evening when there was a sharp temperature drop. The resulting slow-down in reaction rate put him to sleep in the middle of the road, where considerate police directed traffic around him. Finally, at midday the temperature rose and he walked off without any coaxing.

AP/Wide World Photos

Larry Cameron

FIGURE 8.5 Catalyzed decomposition of hydrogen peroxide [2 $H_2O_2(aq)$ ⟶ 2 $H_2O(\ell)$ + $O_2(g)$] on a fresh piece of liver. The liver is well supplied with an enzyme for this reaction.

Catalyst Substance that increases the rate of a chemical reaction without being changed in identity

FIGURE 8.6 Effect of a catalyst. All a catalyst does is lower the activation energy. Because more reactants have enough energy to overcome the reaction energy barrier, the reaction rate increases.

4. ***Effect of a catalyst on reaction rate.*** Sometimes a reaction speeds up dramatically when a substance other than the reactants is added to the mixture. You may have a bottle containing a dilute solution of hydrogen peroxide on your bathroom shelf—hydrogen peroxide is often used as an antiseptic. When stored in a brown or opaque bottle, it decomposes very slowly to water and oxygen (2 H_2O_2 ⟶ 2 H_2O + O_2). Have you noticed that when you put hydrogen peroxide on a cut, it bubbles vigorously? There is a substance in blood (known as *catalase*) that speeds up the decomposition of H_2O_2 (Figure 8.5).

Such substances that increase the rate of a chemical reaction without being changed themselves are known as **catalysts**. In the presence of a catalyst, an alternate pathway with a *lower* activation energy is made available (Figure 8.6). More collisions are successful because less energy is required for success. What makes catalysts so practical is that many times they can be recovered after the reaction is over and used again and again. Many industrial processes rely on rare and expensive metals as catalysts, making recovery of the catalyst an economic necessity.

Living things are even more dependent on catalysts than the chemical industry. Engineers can manipulate temperature and concentrations in an industrial process to control reaction rates. Our bodies can't. If body temperature varies far from 37°C or if the concentrations of chemicals in body fluids vary much from the normal values,

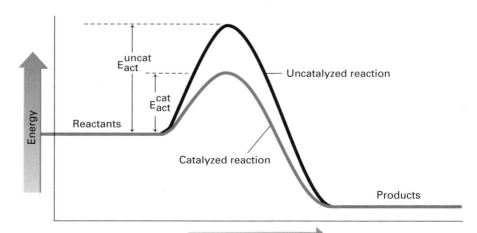

E_{act}^{uncat}

E_{act}^{cat}

Energy

Reactants

Uncatalyzed reaction

Catalyzed reaction

Products

Reaction direction

we are in serious trouble. Catalysis is the major strategy available for controlling biochemical reactions. The amazing constancy of our internal chemistry is maintained by biological catalysts known as *enzymes* (Section 15.8). There is a good reason why blood and the liver are well supplied with a fast-acting catalyst for the decomposition of hydrogen peroxide. Because it is highly reactive, hydrogen peroxide must be destroyed before it has the chance to damage essential substances in its surroundings.

See an animation of **Catalytic Decomposition of Hydrogen Peroxide** at http://brookscole.com/chemistry/joesten4

8.4 CHEMICAL EQUILIBRIUM AND THE "HOW FAR?" QUESTION

In a closed container partly filled with water the air over the liquid soon becomes mixed with water vapor. As the air starts to become saturated with water vapor, some of the water vapor starts to condense. Eventually, the evaporation and condensation of the water establish a **dynamic equilibrium**—a state of balance between exactly opposite changes occurring at the same rate. To indicate this equilibrium, a double arrow is placed between the symbols for water in the liquid and vapor states (Figure 8.7):

$$H_2O(\ell) \underset{\text{Condensation}}{\overset{\text{Evaporation}}{\rightleftarrows}} H_2O(g)$$

Chemical reactions establish the same kind of equilibrium. In **chemical equilibrium** a chemical reaction and its reverse are occurring at equal rates. Theoretically, all chemical reactions are *reversible*—able to take place in either direction and, therefore, to come to equilibrium. Studies of many, many chemical reactions have shown that reactions reach equilibrium at a characteristic and predictable point. Understanding what this is provides an answer to the "How far?" question for a reaction.

Some reactions go virtually to what we call *completion*—the conversion of such a large quantity of the reactants to products that what is unconverted is not noticeable and is unimportant. The combination of hydrogen and oxygen to form water is a reaction of this kind. Once a spark has gotten the first few molecules over the activation energy hill, the reaction continues rapidly and explosively until one or both reactants are used up. A slow reaction can also go to completion. As time passes, an iron nail exposed to the atmosphere continues to rust away gradually, until only the rust remains.

You may get the idea that all chemical reactions go to completion when you watch a hydrogen–oxygen explosion or watch a piece of wood burn in the fireplace. Chemicals do not always react to form products with the complete conversion of reactants, however. Whenever the point is reached at which the forward reaction is proceeding at the same rate as the reverse reaction, equilibrium is established and the amounts of reactants and products remain unchanged. But because equilibrium is a *dynamic* condition, the forward and reverse reactions are still happening so that each reactant or product is replaced as soon as it is consumed.

For equilibrium to be reached, it is important that none of the reactants or products can escape. An example is provided by the conversion of limestone (calcium carbonate) to lime (calcium oxide). By heating limestone in open pits, early U.S. settlers produced lime for mortar:

$$\underset{\substack{\text{Calcium carbonate}\\(\text{limestone})}}{\text{CaCO}_3(s)} \overset{\text{Heat}}{\longrightarrow} \underset{\substack{\text{Calcium oxide}\\(\text{lime})}}{\text{CaO}(s)} + \underset{\text{Carbon dioxide}}{\text{CO}_2(g)}$$

Because the CO_2 gas escapes from the open pit, the reaction keeps going until all of the calcium carbonate in the limestone is converted to lime.

FIGURE 8.7 A system at equilibrium. Water has reached equilibrium with its vapor in this ecosphere. Equilibrium also has been reached by the food and waste products of the inhabitants—a carefully balanced community of plants, shrimp, and a hundred or so kinds of microorganisms.

Ecosphere Associates, Ltd. of Tucson, Arizona

Dynamic equilibrium A state of balance between opposite changes occurring at the same rate

Chemical equilibrium Condition in which a chemical reaction and its reverse are occurring at equal rates

If instead some dry limestone is sealed in a closed container and heated, the result is different. As soon as some CO_2 accumulates in the container, the reverse reaction starts to occur. Once the concentration of CO_2 and CaO reach the appropriate values, equilibrium is established.

$$CaCO_3(s) \rightleftharpoons CaO(s) + CO_2(g)$$

If there aren't any changes in the pressure or the temperature, the forward and reverse reactions continue to take place at the same rates, and the concentration of CO_2 and the amounts of the solid $CaCO_3$ and CaO in the container remain unchanged.

Many reactions in which all reactants and products are dissolved in water occur in nature or are carried out by chemists. How far such a reaction goes before equilibrium is reached is always a matter of interest. Acids are a very important class of water-soluble chemical compounds, about which we'll have more to say in the next chapter. Acetic acid provides a good example of a reaction that reaches equilibrium with just a small amount of reactant converted to product.

$$\underset{\text{Acetic acid}}{CH_3COOH(aq)} + \underset{\text{Water}}{H_2O(\ell)} \rightleftharpoons \underset{\text{Acetate ion}}{CH_3COO^-(aq)} + \underset{\text{Hydronium ion}}{H_3O^+(aq)}$$

Since the hydronium ion (H_3O^+), which is present in all acid solutions (Section 9.1), is what makes vinegar sour, it's a good thing that this reaction doesn't go to completion. If it did, vinegar would be of little use on a salad unless it was diluted with more water to make it less acidic.

For an illustration of the importance of constant conditions to maintaining equilibrium, we can return to the decomposition of limestone that has reached equilibrium in a closed container. If the container is opened briefly to let out some of the CO_2, and then sealed again, the forward reaction will outpace the reverse reaction, and the CO_2 concentration will increase until equilibrium is again established. This type of change illustrates a very important principle that applies to all systems at equilibrium: *If a stress is applied to a system at equilibrium, the system will adjust to relieve the stress.* Known as **Le Chatelier's principle** for the French scientist, Henri Le Chatelier, who first stated it in 1884, this principle means that whatever the disruption, the reaction will shift in the direction that re-establishes equilibrium. For a chemical reaction, the stresses might be adding or taking away a reactant or product, changing the temperature, or, in some cases, changing the pressure.

As another example, consider a reaction that takes place entirely in the gaseous state, the synthesis of ammonia. Because of the need for ammonia in the production of fertilizers, this reaction is of great commercial significance (Section 19.4).

$$N_2(g) + 3\,H_2(g) \rightleftharpoons 2\,NH_3(g)$$

There are more total reactant gas molecules (4) than product gas molecules (2), which means that changing the pressure will stress the reaction (Figure 8.8). If the pressure is increased, the equilibrium will shift in the direction that decreases the pressure. To do this, the number of gas molecules must be decreased, which for ammonia synthesis means the forward reaction will be favored, and more ammonia will form.

Vinegar is 5% acetic acid by mass. Of this amount, only 0.5% is converted to acetate ion and hydronium ion at equilibrium.

 See an animation of **The Water Tank Analogy of Le Chatelier's Principle** at http://brookscole.com/chemistry/joesten4

 See an animation of **Le Chatelier's Principle** at http://brookscole.com/chemistry/joesten4

Le Chatelier's principle When a stress is applied to a system at equilibrium, the equilibrium shifts to relieve the stress

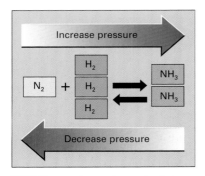

FIGURE 8.8 Ammonia synthesis. In a demonstration of Le Chatelier's principle, pressure shifts equilibrium of reaction with different amounts of gaseous reactants and products. Higher pressure shifts the reaction toward smaller amounts and therefore smaller volumes, of gas. To increase the amounts of ammonia produced, the synthesis is done under pressure.

CONCEPT CHECK   8B

1. The rate of a chemical reaction might be expressed as the amount of _____ converted to _____ per second.
2. If colliding molecules do not have enough _____, they cannot react with each other.

3. A reaction can be speeded up by increasing either the _____ of a reactant or the _____.

4. A substance that speeds up a reaction without being a reactant or product is called a _____.

5. The _____ is like a hill that must be climbed to get a reaction going.

6. Are the amounts of reactants and products always equal to each other at equilibrium?

7. Are the amounts of reactants and products present at equilibrium always the same for the same reaction under all conditions?

8. At equilibrium, are the rates of the forward and reverse reactions equal?

9. Changing the concentration of a reactant to change the equilibrium concentrations of the products is an application of _____.

10. What will happen to the decomposition of limestone ($CaCO_3$) to CaO and CO_2 in a closed container if the container is pressurized with CO_2 gas?

8.5 THE DRIVING FORCES AND THE "WHY?" QUESTION

Why does one reaction produce more product at equilibrium than another? Why do some chemicals react when they are mixed, while others have absolutely no tendency to react unless the conditions are changed? In general, *why* is one chemical reaction favorable and another not favorable?

To examine the driving forces that account for such differences requires looking at two kinds of change—changes in energy as heat and changes in the amount of order that accompany chemical reactions. We experience such changes every day. Turn on the gas stove, and energy is released as heat. Let the ice-cream freezer absorb heat from a mixture of cream and flavorings to produce ice cream. Create major disorder on your desk while rushing to finish a project. Invest the energy needed to restore order to your desk.

The potential energy change of books moved up two stories depends only on their mass and the vertical distance they were moved.

Energy Change as a Driving Force

First, it is important to note that how fast a chemical reaction proceeds and the total amount of energy associated with it are in no way connected. Our earlier analogy about moving books illustrates this principle. No matter what pathway is chosen and no matter how long the move takes, the total change in potential energy of the books is the same. In terms of a reaction, the difference between the energy stored in the reactants and products is the same, no matter how high or low the activation energy barrier.

Also, how fast or how slowly a reaction occurs has absolutely nothing to do with the position of the equilibrium for the reaction, nor does the pathway taken from reactants to products (or vice versa). The value of the activation energy or the number of reaction steps required to go from reactants to the final products also has no bearing on the position of equilibrium. The position of equilibrium is determined only by the energy difference between reactants and products, and the temperature.

A *favorable reaction* has a natural tendency to happen, like water running downhill. An *unfavorable reaction* (the reverse of a favorable reaction) can be made to occur only by the expenditure of energy.

The natural direction of chemical reactions is much like the natural direction of more familiar changes in everyday life. We all know that water going over a waterfall is favorable. Likewise, water going back up the waterfall is not favorable. To move water from the bottom of the waterfall back to the top would require energy and work—electricity could drive a mechanical pump that could move the water.

FIGURE 8.9 A favorable exergonic reaction. The favorable driving forces combine to make the reaction of sodium with water dramatically exergonic.

Endergonic A chemical process in which there is a net consumption of energy.

Water held back behind a dam has stored energy—potential energy that can be converted to kinetic energy if the gates in the dam are opened. Like the water behind a dam, chemical compounds have potential energy; it is stored in chemical bonds. In many reactions, this energy is released as heat. The reaction of sodium with water is a good example. When the reactants are mixed, the mixture bursts into flame (Figure 8.9). Any reaction that releases energy is described as *exergonic,* and most exergonic reactions, once they get going, are favorable—no input of energy is required to keep them going.

The opposite condition is found in an **endergonic** reaction, one that requires energy to take place. Most endergonic reactions, such as the decomposition of limestone in an open pit, don't happen at all without a continuous supply of energy.

So far, you've seen evidence for one answer to the "Why?" question. Like water going over a waterfall, chemical reactions are favorable when they release energy and the products thus have less potential energy than the reactants. Usually the energy is released as heat, although there are chemical reactions that generate light and, under the right conditions, electric current. A few reactions, however, are favorable but endothermic—they absorb heat from their surroundings and keep going without any outside influence. Here is evidence that there must be another driving force for change.

Entropy Change as a Driving Force

Perhaps you have seen a demonstration of the endothermic but favorable reaction of barium hydroxide and ammonium thiocyanate. This reaction absorbs so much heat from its surroundings that it can freeze water in contact with the reaction flask (Figure 8.10):

$$\text{Ba(OH)}_2 \cdot 8\,\text{H}_2\text{O}(s) + 2\,\text{NH}_4\text{SCN}(s) \longrightarrow$$
$$\text{Ba(SCN)}_2(aq) + 2\,\text{NH}_3(aq) + 10\,\text{H}_2\text{O}(\ell)$$

The reactants are both crystalline solids, and, as such, their components are held together in a repetitive, orderly arrangement. Look at the products—there is liquid water, gaseous ammonia that mostly dissolves in the water, and

(a)

(b)

FIGURE 8.10 A favorable but endothermic reaction. The reaction of barium hydroxide and ammonium thiocyanate is one of the uncommon examples of a favorable reaction that absorbs heat from its surroundings.

an ionic compound $Ba(SCN)_2$, that dissolves in the water to give separate ions (Ba^{2+} and 2 SCN^-). What a mixture! And this is the clue to the driving force. The system has moved from an ordered state to a disordered state.

The melting of ice at room temperature is another change that naturally proceeds from order to disorder. You should have no trouble identifying this tendency on a personal scale. It seems to take no effort to create a random mixture of possessions in your room, but it certainly takes energy to put them back in order.

Physical scientists have given the name **entropy** to the disorder of matter, and it can be measured. Gases have a higher entropy than liquids. Liquids have a higher entropy than crystalline solids. Large molecules often have higher entropy than small ones because their atoms can rotate around the bonds in many different ways. The more molecules or the more different kinds of molecules there are in a mixture, the higher the entropy of the mixture.

Entropy is the second factor that determines the answer to the "Why?" question. Reactions are favorable when they result in a *decrease* in energy *and* an *increase* in disorder. When one of these changes is favorable but the other is not, the greater effect controls the favorability of the reaction.

What happens when a process is unfavorable because it requires energy and creates order? Such a process is not forbidden by nature—it can be made to happen when energy is supplied from some other process. Ordering processes must be driven by favorable, disordering processes. The result is always that the net disorder of the universe is increased when an unfavorable process is driven by a favorable one.

Figure 8.11 illustrates the relationships between the energy of activation (E_{act}) and the overall energy change of a reaction ($E_{reaction}$) for an endergonic reaction in the same way that Figure 8.3 did for an exergonic reaction. In an exergonic reaction, the product is favored at equilibrium since the product has a lower energy content (i.e., is more stable) than the reactant. The energy change of a reaction ($E_{reaction}$) is a function of both **enthalpy changes** and **entropy changes**. Enthalpy changes are primarily reflected in the energy required to break bonds and the energy released upon bond formation. Entropy changes primarily reflect the amount of order or disorder created in the chemical change. In an endergonic reaction (Figure 8.11), the reactant is favored at equilibrium since it has a lower energy content (i.e., is more stable) than the product. Such reactions do occur, however, provided there is a source available to provide the additional energy required for product formation. As you will see later (Section 15.10), photosynthesis is such an example, with the Sun serving as the source of the required energy.

The First Law of Thermodynamics

The first and second laws of **thermodynamics** summarize the universal conditions for changes in energy and entropy (Table 8.1) Because they have such broad application and meaning, not just in science, they are often referred to familiarly as "the first law" and "the second law." Here is a formal statement of the **first law of thermodynamics**, sometimes known as the *law of conservation of energy:*

> *Energy can be converted from one form to another*
> *but cannot be destroyed nor created.*

When gasoline burns in an automobile engine, *all* the energy released could be accounted for if the resulting mechanical energy, the friction of moving parts,

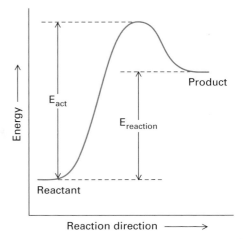

FIGURE 8.11   Energy profile of an endergonic reaction. Endergonic reactions are "uphill" processes in terms of energy, and the reactant is favored at equilibrium.

Some ionic compounds incorporate water molecules in their crystals. To indicate this, the formula is written with a dot: $Ba(OH)_2 \cdot 8 H_2O(s)$ represents a crystalline solid composed of 2 OH^- ions and 8 H_2O molecules for every Ba^{2+} ion.

Ⓦ
See an animation of **Enthalpy and Entropy Changes in Ammonium Nitrate Dissolution** at http://brookscole.com/chemistry/joesten4

Do not confuse the terms *endothermic* and *exothermic* with *endergonic* and *exergonic*. *Endothermic* and *exothermic* refer only to heat changes; *endergonic* and *exergonic* refer to energy changes that contain both heat (enthalpy) and order (entropy) components.

Entropy change A measure of the change in the disorder of a system as it undergoes chemical reaction or physical change

Enthalpy change A measure of the quantity of heat transferred into or out of a system as it undergoes a chemical or physical change

Thermodynamics The science of energy and its transformations

First law of thermodynamics Energy can be converted from one form to another but cannot be destroyed

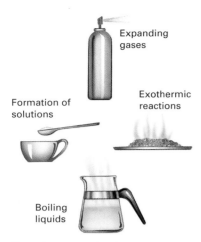

Natural processes that increase entropy.

| TABLE 8.1 | Some Statements of the First and Second Laws of Thermodynamics |

THE FIRST LAW

The energy of the universe is constant.

Energy can be converted from one form to another, but cannot be destroyed or created.

You can't get something for nothing.

There's no such thing as a free lunch.

THE SECOND LAW

The total entropy of the universe is constantly increasing.

The state of maximum entropy is the most stable state for an isolated system.

Energy is conserved in quantity but not in quality.

Every system that is left to itself will, on the average, change toward a condition of maximum probability.

You can't break even.

the energy that leaves the car in the exhaust, the energy converted to electrical and then chemical potential energy in the battery, and the increase in temperature of the engine and everything surrounding it could be measured.

Looked at another way, the first law means that *the total quantity of energy in the universe is constant.* Creation of new energy is not possible. All we can do is change it from one form to another. The first law shows up in conversation whenever someone says, "Oh well, you can't get something for nothing."

The famous insight of Albert Einstein in 1900 when he recognized that matter and energy are interconvertible created an extension of the first law:

The production of energy by nuclear fission and fusion takes advantage of Einstein's extension of the law (Section 13.9).

> ***The total amount of matter and energy in the universe is constant.***

The Second Law of Thermodynamics

Even more ways have been found to express the **second law of thermodynamics** than the first (Table 8.1). Each statement reflects the observation that although the energy of the universe is constant, once energy is converted to entropy it is never again available for useful purposes.

In chemistry textbooks, the usual statement of the second law is

 See an animation of **The Second Law of Thermodynamics** at http://brookscole.com/chemistry/joesten4

> ***The total entropy of the universe is constantly increasing.***

The universal truth of this law may be hard to accept at first. The formation of the stars and planets, the formation of continents and oceans, the formation of crystalline mineral deposits, and the growth of plants and animals are all ordering processes. Life itself is a constant struggle against entropy. The essential connection is that every ordering process in one small corner of the universe creates disorder somewhere else in the universe. As we are eating, breathing, and producing new biomolecules, we are emitting disordered waste products and contributing disorder to the atmosphere by giving off heat.

Visualizing body heat. Thermogram photos show heat in reds and yellow, illustrating that the body heats up with exercise.

Second law of thermodynamics The total entropy of the universe is constantly increasing

In fact, *every* time energy is generated and used to do work, some of the energy is converted to heat. Consider the release of energy by burning coal, petroleum, or wood. The principal products of combustion, carbon dioxide and water, will not burn and release more energy. The energy from the reactants is dispersed as heat into the random molecular motion of the surroundings, where it is not available to do more work. In the burning process, matter and energy are conserved. However, the products and the energy converted to heat are much less useful than the reactants and their stored energy. Observations of this type are the basis of one of the alternative statements of the second law:

> *Energy is conserved in quantity but not quality.*

The implications of the second law are wide ranging. Our economy is based on extracting raw materials from our surroundings, using energy to process them and, in marketplace terms, "adding value" to the raw materials. The second law reminds us that no matter how carefully a process is designed, energy is lost at each manufacturing step.

8.6 RECYCLING: NEW METAL FOR OLD

Many communities in the United States now recycle metals, paper, and plastic. There is a long way to go, however, before we can congratulate ourselves too heartily on reducing the quantity of material that is simply dumped somewhere. Between 1960 and 1990 the generation of municipal solid waste almost doubled, whereas the U.S. population increased by only one-third. In other words, each of us produced more garbage. The best remedy remains to generate less waste, rather than to expend time, energy, and money in collecting, processing, and recycling it.

How many of these items could have been repaired or manufactured to last longer?

Also, recycling alone is not the best answer. Consider what the laws of thermodynamics mean for recycled materials. Recycling counteracts the natural direction of increasing entropy and can only be done with the expenditure of energy in collecting, transporting, and remanufacturing the waste to produce newly useful materials. Remember the second law—some of the energy used at each step is lost forever to entropy.

A different and more direct approach to conserving resources is to increase the useful lifetime of our household materials and objects. Another is to diminish excess use of materials, even those that are recyclable. The three R's of waste prevention, in order of their importance, are

> **Reduce** *the amount of waste as much as possible.*
> **Reuse** *products as much as possible.*
> **Recycle** *materials as much as possible.*

To get the whole picture of the amount of waste we generate and its impact on the environment, it's important to keep in mind that municipal waste is only a small fraction of the total. Waste is generated at each stage of a product's life cycle, meaning, of course, that there are opportunities for waste reduction at each stage (Figure 8.12).

Metals are not incinerated, but are transported either to waste dumps or to recycling plants. Many factors determine the extent to which a metal is recycled.

Lead is the most recycled metal, for several reasons: It is too toxic to go to landfills; the major use is in automobile batteries, which have a predictable life span; and used batteries are collected at legally designated locations in most states. Iron and steel are second in percentage recycled, most of which is recycled within the industry rather than from consumer products. Iron and steel are used in vastly greater quantities than all other metals combined, resulting in a huge pool of scrap to be dealt with. Not recovering it would be a financial as well as an environmental burden. Also, all steel is potentially recyclable without separation of pure metals from the mixture.

Do any of you remember the time when milk and soft drinks were sold only in glass bottles that were returned, washed, and used again? Shall we make this the law?

Plastics recycling is discussed in Section 14.7.

Do you think the deposit paid on beverage cans and the resulting motivation for scavengers to collect them has contributed to their greater entry into the recycling stream? It has certainly helped to clean up public areas in some parts of the country.

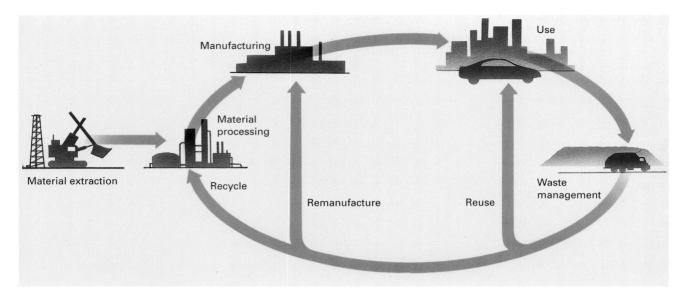

FIGURE 8.12   Stages in the production, use, and disposal of manufactured products. At each stage energy is used; there is a possibility of air or water pollution; and there is generation of waste. Therefore, there is also the possibility for conservation and recycling at each stage.

THE WORLD *of* CHEMISTRY

GREEN DESIGN

Consumer interest in environmentally friendly products has been causing change. The more desirable approach of considering environmental impact at all stages in a product's life cycle has been christened "green design." The goals of green design are summarized by the U.S. Office of Technology Assessment as shown in the diagram.

It is important to understand that decision making in green design, like that in risk management, must constantly balance positive and negative outcomes. For example, many people are happy to purchase their own telephones instead of paying a monthly rental,

as was the case before the nationwide telephone monopoly was disbanded. However, most telephones then served many users because they were designed to be remanufactured numerous times before they were retired. Today, several million phones are discarded each year after use by only one consumer. Most of these phones have broken down and were not designed to be repaired.

As applied to chemical manufacturing, green design focuses on four principles:

- Use nontoxic reactants in the manufacture of products

- Do not use solvents that are toxic and will be discarded

- Design chemical syntheses in which most of the atoms that enter the synthesis become part of the products, rather than part of wasted by-products

- During process design, consider the entire life cycle of the product, including how it can be recycled or reused.

Since 1995, a partnership between the U.S. Environmental Protection Agency (EPA) and the chemistry community has sponsored an annual competition for Presidential Green Chemistry Challenge Awards. Entries are operational green chemistry projects and are judged for health and environmental benefits, scientific innovation, and industrial applicability.

Among the 2004 Green Chemistry Challenge Awards was one honoring Optimyze, an enzyme-based technology that allows recycling of types of paper that previously could not be recycled because they contained additives that became sticky and spoiled the finished paper by creating spots and holes. Another Green Chemistry Challenge Award recognized the value of Rightfit organic pigments that replace pigments containing toxic heavy metals. Using Rightfit pigments has numerous advantages, including lower cost, better color strength, and production in water rather than in organic solvents.

CONCEPT CHECK 8C

1. In an exothermic reaction, heat is _____.
2. In an endothermic reaction, heat is _____.
3. Some favorable chemical reactions start up as soon as the reactants are mixed, but others do not start till energy is added. (**a**) True, (**b**) False.
4. The products of an exothermic reaction store less _____ than the reactants.
5. Which of the following is a system with higher entropy? (**a**) A new deck of cards as it comes from the box, (**b**) A deck of cards after the cards are scattered on the top of a table.
6. When water freezes, its entropy _____.
7. The two driving forces for favorable chemical change are a decrease in _____ and an increase in _____.
8. "You can't get something for nothing," is one way of stating the _____.
9. "You can't ever break even," is one way of stating the _____.
10. In recycling, it is hard to avoid using _____ and increasing _____.

 Assess your understanding of this chapter's topics with an online chapter quiz at **www.brookscole.com/chemistry/joesten4**

■ KEY TERMS

coefficient	activation energy	entropy
mole	catalyst	enthalpy change
Avogadro's number	dynamic equilibrium	thermodynamics
molar mass	chemical equilibrium	first law of thermodynamics
reaction rate	Le Chatelier's principle	second law of thermodynamics
exergonic	endergonic	

■ THE LANGUAGE OF CHEMISTRY

1. Number of atoms shown by the formula of $(NH_4)_3PO_4$
2. Mass of 1 mol of H_2O_2
3. Adding a catalyst
4. Cannot be destroyed
5. Mass of 1 mol of titanium
6. Balanced chemical equation
7. Lowering the temperature
8. Forward and reverse reactions proceeding at equal rates
9. Drives a reaction to completion
10. Decreases chemical potential energy
11. Number of moles of N_2 needed to produce 30 mol of ammonia, $N_2 + 3 H_2 \longrightarrow 2 NH_3$

a. Speeds up a reaction
b. 48 g
c. Same number of atoms of each element on both sides of the arrow
d. 34 g
e. Slows down a reaction
f. Dynamic equilibrium
g. Formation of gas that escapes
h. 20
i. 15
j. An exothermic reaction
k. Energy

■ APPLYING YOUR KNOWLEDGE

1. Look at the following balanced chemical equation for burning ethanol:

    $$CH_3CH_2OH(g) + 3\ O_2(g) \longrightarrow 2\ CO_2(g) + 3\ H_2O(g)$$

    (a) How many hydrogen atoms are on the product side of the equation? How many are on the reactant side of the equation?

    (b) How many oxygen atoms are on the product side of the equation? How many are on the reactant side of the equation?

    (c) What are the balancing *coefficients* in the reaction?

    (d) Explain, in your own words, how this balanced equation obeys the law of conservation of matter.

2. Equations can only be balanced by adjusting the coefficients of the reactants and products, while the subscripts within the formulas of the reactants and products cannot be changed and still keep the original sense of the equation. Explain why this is so.

3. Balance the following chemical equations:
    (a) $Al + Cl_2 \longrightarrow AlCl_3$
    (b) $Mg + N_2 \longrightarrow Mg_3N_2$
    (c) $NO + O_2 \longrightarrow NO_2$
    (d) $SO_2 + O_2 \longrightarrow SO_3$
    (e) $H_2 + N_2 \longrightarrow NH_3$

4. Balance the following chemical equations:
    (a) $CH_3OH(\ell) + O_2(g) \longrightarrow CO_2(g) + H_2O(g)$
    (b) $C_3H_8(\ell) + O_2(g) \longrightarrow CO_2(g) + H_2O(g)$
    (c) $C_6H_6(\ell) + O_2(g) \longrightarrow CO_2(g) + H_2O(g)$
    (d) $C_3H_8O(\ell) + O_2(g) \longrightarrow CO_2(g) + H_2O(g)$

5. Balance the following equations:
    (a) $Ba(s) + H_2O(\ell) \longrightarrow Ba(OH)_2(aq) + H_2(g)$
    (b) $Fe(s) + H_2O(\ell) \longrightarrow Fe_3O_4(s) + H_2(g)$
    (c) $Na(s) + H_2O(\ell) \longrightarrow NaOH(aq) + H_2(g)$
    (d) $Li(s) + H_2O(\ell) \longrightarrow LiOH(aq) + H_2(g)$

6. Balance the following equations:
    (a) $HBr(aq) + KOH(aq) \longrightarrow KBr(aq) + H_2O(\ell)$
    (b) $H_2S(g) + NaOH(aq) \longrightarrow Na_2S(aq) + H_2O(\ell)$
    (c) $HNO_3(aq) + Ca(OH)_2(aq) \longrightarrow Ca(NO_3)_2(aq) + H_2O(\ell)$
    (d) $HCl(aq) + Al(OH)_3(s) \longrightarrow AlCl_3(aq) + H_2O(\ell)$

7. Ammonia for fertilizers is made from the elements according to this equation:

    $$N_2(g) + 3\ H_2(g) \longrightarrow 2\ NH_3(g)$$

    (a) How many molecules of N_2 are needed to react with 30 molecules of H_2?

    (b) How many molecules of N_2 are needed to react with 15 H_2 molecules?

    (c) If 100 N_2 molecules and 300 H_2 molecules are allowed to react, how many NH_3 molecules will be formed?

8. Hydrogen peroxide is formed from the elements according to this equation:

    $$H_2(g) + O_2(g) \longrightarrow H_2O_2(\ell)$$

    (a) How many molecules of oxygen are needed to react with 1 molecule of $H_2(g)$?

    (b) How many hydrogen peroxide molecules are expected, if 10 oxygen molecules react with 10 hydrogen molecules?

    (c) How many hydrogen peroxide molecules are expected if 100 oxygen molecules are mixed and reacted with 10 hydrogen molecules?

9. Identify the atom used as the basis of the atomic weight scale.

10. What do 35.5 g of chlorine (Cl) and 12.011 g of carbon (C) have in common?

11. What do 103.5 g of lead (Pb) and 6.006 g of carbon (C) have in common?

12. Look up the unit called the *gross*. How many pairs of jeans are in 4 gross of jeans? In what way is the gross unit like Avogadro's number?

13. What is the numerical value of Avogadro's number?

14. What is meant by the term *molar mass*?

15. What is the definition for the mole?

16. Interpret the following equation for the complete combustion of octane in terms of moles of reactants and products, and their molar masses:

    $$2\ C_8H_{18}(\ell) + 25\ O_2(g) \longrightarrow 16\ CO_2(g) + 18\ H_2O(\ell)$$

17. Give at least one reason why some chemical reactions are fast while others are slow. Use your own analogy to explain the reason.

18. What influence does temperature usually have on the rate of a chemical reaction? Explain why this is so.

19. What effect does freezing food have on reaction rates? Why is freezing used for preservation of food, tissue samples, and biological samples?

20. Explain the term *activation energy* as it applies to chemical reactions.

21. How does the magnitude of the activation energy influence the rate of a chemical reaction?

22. What is the effect of a catalyst on the activation energy of a chemical reaction?

23. Explain why hydrogen and oxygen can remain mixed at room temperature without any noticeable reaction, yet they will combine explosively to form water if the mixture is ignited with a tiny spark.

24. Describe how each of the following changes will affect the rate of a reaction:
    (a) Increase in temperature
    (b) Increase in concentration
    (c) Introduction of a catalyst

25. Paper burns fairly rapidly in air. How would you expect paper to burn in pure oxygen? Explain.

26. What is the general name for biological catalysts?

27. What is meant by the term *reversible* when describing chemical reactions?

28. What is meant by the term *dynamic equilibrium*?

29. If you heat limestone in a closed vessel, explain which way you would expect the equilibrium to shift if
    (a) you added more CO_2
    (b) you allowed some of the CO_2 to escape from the vessel

    $$CaCO_3(s) \rightleftharpoons CaO(s) + CO_2(g)$$

30. Give a statement of Le Chatelier's principle.

31. What is meant when it is said that the reaction shifts in favor of the products of the reaction?

32. What is meant when it is said that the reaction shifts in favor of the reactants?

33. The reaction between HCl and NaOH is described as going to completion. Explain what this means in terms of how much of the reactants remain unreacted.

34. The reaction between N_2 and H_2 to produce ammonia (NH_3) is described as an equilibrium reaction

    $$N_2(g) + 3 H_2(g) \rightleftharpoons 2 NH_3(g)$$

    After the reaction reaches equilibrium, what is present in the reaction mixture?

35. Define the following terms:
    (a) Potential energy
    (b) Exothermic
    (c) Endothermic
    (d) Entropy
    (e) Favorable reaction

36. State the first law of thermodynamics. What does it mean?

37. State the second law of thermodynamics. What does it mean?

38. Does entropy increase or decrease in the following reactions? Explain your answer.
    (a) Nitrogen reacting with oxygen to form nitrogen dioxide

    $$N_2(g) + 2 O_2(g) \longrightarrow 2 NO_2(g)$$

    (b) Acetylene reacting with oxygen to form carbon dioxide and water

    $$2 HC \equiv CH(g) + 5 O_2(g) \longrightarrow$$
    $$4 CO_2(g) + 2 H_2O(g)$$

39. Calculate the molar mass for each of the following:
    (a) H_2O
    (b) I_2
    (c) KOH
    (d) NH_3

40. Calculate the molar mass for each of the following:
    (a) $C_6H_{12}O_6$
    (b) H_2SO_4
    (c) Na_2HPO_4
    (d) $Ca(NO_3)_2$

41. Which of the following is the largest mass?
    (a) 0.5 mol CO_2
    (b) 2 mol Li
    (c) 12 mol H_2

42. Ethanol in oxygenated gasoline burns as a fuel according to this equation:

    $$CH_3CH_2OH(\ell) + 3 O_2(g) \longrightarrow 2 CO_2(g) + 3 H_2O(g)$$

    (a) What is the mole ratio of ethanol to oxygen?
    (b) How many grams of oxygen are needed to burn 1 mol of ethanol?
    (c) How many grams of oxygen are needed to completely burn 500 g of ethanol?

43. Propane in gas-fired barbecues burns ideally according to this equation:

    $$CH_3CH_2CH_3(g) + 5 O_2(g) \longrightarrow 3 CO_2(g) + 4 H_2O(g)$$

    (a) How many moles of oxygen are needed to burn 100 mol of propane?
    (b) How many grams of oxygen are needed to burn 4400 g of propane?
    (c) How many pounds of oxygen are needed to burn 88 lb of propane?

44. This equation shows how water is formed from the elements:

    $$2 H_2(g) + O_2(g) \longrightarrow 2 H_2O(\ell)$$

    (a) How many grams of H_2 are needed to react with every 32 g of O_2?

(**b**) How many pounds of H_2 are needed to react with 32 lb of O_2?

(**c**) How many tons of H_2 are needed to react with 32 tons of O_2?

45. Based on the balanced equation

$$Cu(s) + Cl_2(g) \longrightarrow CuCl_2(s)$$

(**a**) How many moles of copper will be required to react with 0.5 mol of chlorine?

(**b**) How many moles of $CuCl_2$ can be formed from 1.5 mol of copper?

46. Balance the equation for the fermentation of glucose and then answer the following questions

$$C_6H_{12}O_6(aq) \longrightarrow CH_3CH_2OH(aq) + CO_2$$

(**a**) In this reaction, how many moles of ethanol (CH_3CH_2OH) can be prepared from 6 mol of glucose?

(**b**) If, during the fermentation process, 10.5 mol of carbon dioxide is found to be produced, how many moles of ethanol will have been produced?

 CHEMISTRY ON THE WEB

For up-to-date URLs, visit the text website at **www.brookscole.com/ chemistry/joesten4**

- Balancing Chemical Equations
- Chemical Equilibrium
- What Is a Mole?
- Avogadro's Number
- Recycling Steel

ACID–BASE REACTIONS

The diet soda is quite acidic, as shown by the meter reading in this laboratory test (a pH measurement, Section 9.3).

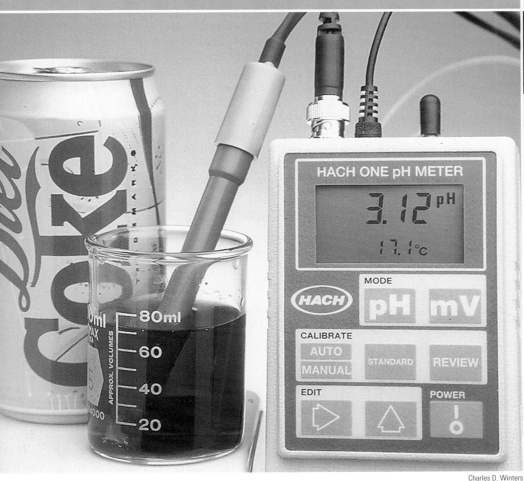

Charles D. Winters

Acid! Most people immediately think of danger when they hear this word. Base! In the United States, most people think baseball when they hear this word. To chemists and a great many other scientists, these two words call to mind chemicals that are quite common in our natural surroundings. Chemical reactions between acids and bases dramatically influence our personal health, the health of our environment and every living organism, and the health of our economy.

Some acids and bases can be quite dangerous in the hands of those inexperienced with working with such chemicals. Concentrated sulfuric acid and concentrated sodium hydroxide (a chemical base often referred to as *lye*) can cause serious chemical burns. However, batteries containing sulfuric acid and oven cleaners containing lye, both in potentially dangerous concentrations, can be found in every garage and kitchen.

Carbonated soft drinks owe their fizz to the presence of carbonic acid. We think nothing of drinking this acid. Even so, one of the authors of this book was confronted with the statement "our lawyers won't let us have acids in a shopping mall due to liability problems" when he proposed to do some chemical demonstrations involving carbonic acid as part of an observance of National Chemistry Week activities. There's that association with danger again.

Face and hand soaps contain bases, but no one who has read the warning label on a common oven cleaner would even consider getting *this* base near their eyes. If, in fact, you did get a base in your eye, you might want to be prepared to wash your eyes with boric acid. Confused?

In this chapter you will learn that the danger associated with acids and bases is a function of their concentration and identity. Most acids and bases are not inherently dangerous. They are all around us. The acid–base balance in our blood is critical to good health and many biological reactions depend on acids and bases in the body. The acid content of our rainfall (and other forms of precipitation) has a critical role in determining the health of our rivers and forests and our ability to raise the foodstuffs upon which we depend. Acids and bases are key components of many manufacturing processes. In fact, sulfuric acid is required by so many industrial processes that the amount of it sold each year is taken as a measure of a nation's economy.

By the time you finish this chapter you will be able to discuss the following questions:

- What are the chemical properties of acids and bases?
- What happens when acids and bases react with each other in a reaction known as *neutralization*?
- What is pH, and how is it used to describe acid and base solutions?
- What are some common acids and bases and their properties?
- What are buffers, and why is acid–base buffering so important to maintaining our health?

9.1 ACIDS AND BASES: CHEMICAL OPPOSITES

Everyone should know a few practical things about acids and bases, which we usually encounter as solutions in water. They can be harmful. The harm can range from the stinging sensation when you accidentally squirt lemon juice (citric acid) into your eye, to the severe and persistent burns or blindness that result if you spill battery acid (sulfuric acid) on your skin or get it in your eyes and do not flush it immediately with lots of water. The effects of Drano and some oven cleaners can be equally devastating because they contain lye (sodium hydroxide). Strongly acidic and basic solutions can also "eat" holes in clothing, very quickly if they are strong solutions. The first lesson, then, is that strongly acidic or basic substances must be handled with care.

The potential for harm from strong acids and bases, however, does not mean all acids and bases are to be avoided. The dilute solutions of acids and bases that are in everyday use would be sorely missed. Orange juice, vinegar, soda pop, household ammonia, and most of our soaps and detergents fall in this category of dilute solutions.

It turns out that all acidic water solutions, whatever their sources or applications might be, have some chemical properties in common. The same is true of all basic water solutions. Acids and bases are closely related classes of chemical compounds that are often highly reactive, both with other substances and with each other.

The word *acid* comes from the Latin *acidus,* meaning sour or tart, because in water solutions, acids have a sour or tart taste. Lemons, grapefruit, and limes taste sour because they contain citric acid and ascorbic acid (vitamin C). Vinegar is sour because it contains acetic acid. Another common property of acids is their ability to change the color of compounds known as **acid–base indicators** (Figure 9.1).

The properties that acids have in common, listed in Table 9.1, result from the ability of acids to release a hydrogen ion (H^+) in a water solution (an *aqueous solution,* symbolized by *aq*). The ionization of an acid is therefore often represented by an equation like the following for hydrochloric acid:

$$HCl(aq) \longrightarrow H^+(aq) + Cl^-(aq)$$

Hydrogen ions, because of their small size and resulting concentration of positive charge in a small area, however, do not exist as such in water solution. Instead, they bond to water molecules to give **hydronium ions** (H_3O^+). Therefore, it is more nearly correct, and as you will see more useful, to write an equation for the ionization of hydrochloric acid (and other acids) with the hydronium ion as a product.

$$HCl(aq) + H_2O(\ell) \longrightarrow H_3O^+(aq) + Cl^-(aq)$$

Hydrochloric acid, $HCl(aq)$, which is formed when hydrogen chloride gas [$HCl(g)$] dissolves in water, contains hydronium ions and chloride ions (the anions from the acid). A solution that contains H_3O^+ ions and has the properties common to acids is referred to as an **acidic solution**.

Bases also have several properties in common (Table 9.1). Water solutions of bases are slippery or soapy to the touch and change litmus from red

You should never taste anything in a laboratory.

H^+, although referred to as a hydrogen ion or proton, is simply a hydrogen atom minus its electron.

Charles D. Winters

FIGURE 9.1 Acidity of some household products. A few drops of an acid–base indicator have been added to water solutions of each of the products. The indicator color shows that the club soda is more acidic than the vinegar and the cleaning solution is slightly basic.

TABLE 9.1	Properties of Acids and Bases	

ACIDS	**BASES**
Sour taste	Bitter taste
Provide H^+ ions	Provide OH^- ions
React with active metals to give hydrogen	Slippery feeling
Change colors of indicators (e.g., litmus turns from blue to red)	Change colors of indicators (e.g., litmus turns from red to blue)
Produce CO_2 when added to limestone ($CaCO_3$)	Neutralize acids
Neutralize bases	

SOME ACIDIC SUBSTANCES	**SOME BASIC SUBSTANCES**
Vinegar	Household ammonia
Tomatoes	Baking soda
Citrus fruits	Soap
Carbonated beverages	Detergents
Black coffee	Milk of magnesia
Gastric fluid	Oven cleaners
Vitamin C	Lye
Aspirin	Drain cleaners
Ant venom	Antacids
Battery acid	

Acid–base indicators Substances that change color with changes in acidity or basicity of a solution

Hydronium ion A hydrated proton H_3O^+

$$H-\overset{+}{\underset{|}{O}}-H$$
$$|$$
$$H$$

Lewis structure

Acidic solution A solution that contains a higher concentration of H_3O^+ ions than OH^- ions

to blue, which is the reverse of the change caused by acids. The classic properties by which bases are recognized are caused by the presence in water solution of hydroxide ions (OH^-). The hydroxides of sodium and potassium (NaOH and KOH) and of calcium and magnesium [$Ca(OH)_2$ and $Mg(OH)_2$] are among the most common bases used in industry and chemical laboratories. These are ionic compounds that yield hydroxide ions when they dissolve. For example, sodium hydroxide, also known as *lye* or *caustic soda,* dissolves as represented by the equation.

$$NaOH(s) \xrightarrow{\text{Water}} Na^+(aq) + OH^-(aq)$$

A solution that contains a higher concentration of OH^- ions than hydronium ions and has the properties common to bases, is described as a **basic**, or **alkaline**, **solution**.

Neutralization Reactions

A water solution cannot be acidic and basic simultaneously. An *acidic solution* contains a higher concentration of H_3O^+ ions than OH^- ions, and a *basic solution* contains a higher concentration of OH^- ions than H_3O^+ ions. When the solution of an acid is mixed with the solution of a base, the hydronium ions from the acid react with the hydroxide ions from the base to produce water:

$$H_3O^+(aq) + OH^-(aq) \longrightarrow 2\,H_2O(\ell)$$

With bases such as potassium hydroxide and acids such as hydrochloric acid, the second product is a salt, which in this case is potassium chloride (KCl). A **salt** is a compound composed of the positive metal ion from a base and the negative ion from an acid.

$$\underset{\text{Base}}{KOH(aq)} + \underset{\text{Acid}}{HCl(aq)} \longrightarrow \underset{\text{Salt}}{KCl(aq)} + \underset{\text{Water}}{H_2O(\ell)}$$

In this reaction, KOH, HCl, and KCl are all water-soluble compounds that yield ions in solution, and water is a molecular compound. A more detailed representation of this process is

$$K^+(aq) + {}^-OH(aq) + H_3O^+(aq) + Cl^-(aq) \longrightarrow$$
$$K^+(aq) + Cl^-(aq) + 2\,H_2O(\ell)$$

When *exactly equivalent amounts* of acid and base react so that all of the H_3O^+ and OH^- ions are used up in forming water, the result is a **neutralization reaction**. The reaction of sodium hydroxide with sulfuric acid, like the reaction of potassium hydroxide with hydrochloric acid, is a neutralization.

$$\underset{\text{Base}}{2\,NaOH(aq)} + \underset{\text{Acid}}{H_2SO_4(aq)} \longrightarrow \underset{\text{Salt}}{Na_2SO_4(aq)} + \underset{\text{Water}}{2\,H_2O(\ell)}$$

In neutralization reactions, acidic and basic properties are eliminated.

TRY IT 9.1

The previous equation shows that two sodium hydroxide ions react with one molecule of sulfuric acid. Write an equation that shows that the reaction of each molecule of sulfuric acid (H_2SO_4) with water produces two hydronium ions and one sulfate anion (SO_4^{2-}). These two hydronium ions then react with two hydroxide ions to produce water, as shown in the first equation in this section on neutralization reactions.

Charles D. Winters

Cleaning products that contain bases.

Basic solution (alkaline solution) A solution that contains a higher concentration of OH^- ions than H_3O^+ ions

$$H\!-\!\ddot{\underset{..}{O}}{:}^-$$

Lewis structure

Salt Ionic compound composed of the cation from a base and the anion from an acid

Neutralization reaction Reaction of equivalent quantities of an acid and a base

Acid–Base Definitions

Our formal definitions of acids and bases take into account the role of the hydrogen ion in acid–base reactions. An **acid** is a molecule or ion able to donate a hydrogen ion to a base. A **base** is a molecule or ion able to accept a hydrogen ion from an acid. The result of these definitions is that every reaction in which a hydrogen ion is exchanged between reactants is an **acid–base reaction**. The following structural formulas show how a hydrogen ion is exchanged between water (written here as H—O—H) and hydrogen chloride gas in the formation of a hydrochloric acid solution. This is an acid–base reaction in which water is the base and HCl is the acid.

$$\text{H}-\ddot{\text{O}}-\text{H} \ + \ \text{H}-\ddot{\text{Cl}}\text{:} \longrightarrow \text{H}-\overset{+}{\underset{\underset{\text{H}}{|}}{\ddot{\text{O}}}}-\text{H} + \text{:}\ddot{\text{Cl}}\text{:}^{-}$$

Base   Acid
(H$^+$ acceptor)   (H$^+$ donor)

In the reaction between sodium hydroxide and the hydronium ion in a hydrochloric acid solution, OH$^-$ is the base and H$_3$O$^+$ is the acid.

$$\text{:}\ddot{\text{O}}-\text{H} + \text{H}-\overset{+}{\underset{\underset{\text{H}}{|}}{\ddot{\text{O}}}}-\text{H} \longrightarrow \text{H}-\ddot{\text{O}}-\text{H} + \text{H}-\ddot{\text{O}}-\text{H}$$

Base   Acid
(H$^+$ acceptor)   (H$^+$ donor)

The picture of acids and bases as hydrogen ion donors and acceptors explains not only the properties of the acids and bases we have described thus far, but also the basicity of ammonia (NH_3) and ions such as the carbonate ion (CO_3^{2-}). The neutral ammonia molecule can accept a hydrogen ion from a water molecule to produce a solution that is basic because it contains hydroxide ions. In this reaction, water donates the hydrogen ion and therefore is the acid.

$$\underset{\substack{\text{Base}\\(\text{H}^+\text{ acceptor})}}{NH_3(aq)} \ + \ \underset{\substack{\text{Acid}\\(\text{H}^+\text{ donor})}}{H_2O(\ell)} \ \rightleftharpoons \ \underset{\substack{\text{Ammonium}\\\text{ion}}}{NH_4^+(aq)} \ + \ OH^-(aq)$$

Although they contain no hydroxide ions, carbonate salts such as sodium carbonate (Na_2CO_3, washing soda) and potassium carbonate (K_2CO_3) also dissolve in water to give basic solutions. How does a carbonate salt produce hydroxide ions in water? Like ammonia, the carbonate ion accepts a hydrogen ion from a water molecule, which leaves behind a hydroxide ion. Note the omission of the Na$^+$ ions. They play no role in the acid–base reaction.

$$\underset{\substack{\text{Base}\\(\text{H}^+\text{ acceptor})}}{CO_3^{2-}(aq)} \ + \ \underset{\substack{\text{Acid}\\(\text{H}^+\text{ donor})}}{H_2O(\ell)} \ \rightleftharpoons \ \underset{\substack{\text{Bicarbonate}\\\text{ion}}}{HCO_3^-(aq)} \ + \ OH^-(aq)$$

The equations in this section illustrate a very important point. In the reaction with hydrogen chloride, water reacts as a base. In the reactions with ammonia molecules and the carbonate ion, water reacts as an acid. Because all water molecules are the same, *water must be able to react as an acid or a base depending on whether it reacts with a base or an acid.* This property of water plays an important role in our water-based world. Whenever a substance that can accept or donate a hydrogen ion dissolves in water, the result is a solution with basic or acidic properties. In removing stubborn dirt (Section 9.5), this can work to our advantage. In the production of acid rain (Section 11.2), this causes problems.

See an animation of **Proton Transfer Between an Acid and a Base** at http://brookscole.com/chemistry/joesten4

When a hydrogen atom loses an electron, the remaining charged particle is a proton. Thus, acids and bases are also sometimes described as proton donors and acceptors.

NH$_3$ is a gas at room temperature.

See an animation of **A Bronsted Acid and of A Bronsted Base Reacting with Water** at http://brookscole.com/chemistry/joesten4

The description of basic substances as *alkaline* derives from the Arabic word *al-qali*, meaning "plant ashes." Potassium carbonate, commonly known as potash, is found in ashes from wood fires. Long ago it was discovered that this compound dissolves in water to yield a solution that feels slippery, tastes bitter, and reacts with acids.

Acid Molecule or ion able to donate a hydrogen ion to a base. Acids produce H$_3$O$^+$ ions in aqueous solutions.

Base Molecule or ion able to accept a hydrogen ion from an acid. Bases release or produce $^-$OH(aq) in aqueous solutions.

Acid–base reaction Reaction in which a hydrogen ion is exchanged between an acid and a base

9.2 THE STRENGTHS OF ACIDS AND BASES

What is the difference between a strong acid such as hydrochloric acid, sold in hardware stores as *muriatic acid* and used to clean brick and concrete, and a weak acid such as the acetic acid in vinegar? *The strength of an acid or base is determined by the* **extent** *of ionization in aqueous solution.* The greater the ionization, the stronger the acid or base.

Many common acids like sulfuric acid, hydrochloric acid, and nitric acid react completely with water to give hydronium ions and anions. Therefore, like ionic compounds (Section 5.2), they are electrolytes; and because they are entirely converted to ions in solution, they are *strong electrolytes*. The solutions of these **strong acids** have high concentrations of hydronium ions and are very acidic (Table 9.2). The equation for the reation of a strong acid such as nitric acid with water is written with a single arrow, which indicates that the reaction goes to completion.

$$HNO_3(aq) + H_2O(\ell) \longrightarrow H_3O^+(aq) + NO_3^-(aq)$$

Many acids establish an equilibrium with water at a point where not all the acid molecules have been converted to ions. These are the **weak acids**— they are only slightly ionized in aqueous solution because an equilibrium is established (Section 8.4). Acetic acid is a typical weak acid:

$$\underset{\text{Acetic acid}}{CH_3COOH(aq)} + H_2O(\ell) \rightleftharpoons H_3O^+(aq) + \underset{\text{Acetate ion}}{CH_3COO^-(aq)}$$

The CH_3COOH molecules undergo ionization while H_3O^+ and CH_3COO^- ions simultaneously recombine to give CH_3COOH molecules and water. Since only a few percent of the acetic acid molecules are ionized at any given time, aqueous solutions of acetic acid contain mostly acetic acid molecules with a few hydronium ions and acetate anions, and are only weakly acidic.

As previously explained, the metal hydroxides are all ionic compounds and, therefore, **strong bases** because when dissolved in water, they are completely converted to ions (to the extent that they are soluble). Ammonia is a common **weak base**, a base that establishes an equilibrium with water to produce a solution with relatively few ammonium and hydroxide ions.

$$NH_3(g) + H_2O(\ell) \rightleftharpoons NH_4^+(aq) + OH^-(aq)$$

By "extent of ionization" we refer to the point at which ionization of an acid reaches equilibrium and the ion concentrations become constant (Section 8.4).

Although there are several H atoms in acetic acid, only the one bonded to oxygen in the —COOH group reacts with a base. This condition is typical of organic compounds that contain the —COOH group.

Strong acid Acid that is 100% ionized in aqueous solution

Weak acid Acid that is only partially ionized in aqueous solution and establishes equilibrium with nonionized acid

Strong base Base that is 100% ionized in aqueous solution

Weak base Base that is only partially ionized in aqueous solution and establishes equilibrium with nonionized base

---

TABLE 9.2 Common Acids

NAME	FORMULA	STRENGTH	USE OR OCCURRENCE
Sulfuric acid*	H_2SO_4	Strong	Cleaning steel; car batteries; making plastics, dyes, fertilizers
Hydrochloric acid	HCl	Strong	Cleaning metals and brick mortar
Nitric acid	HNO_3	Strong	Making fertilizers, explosives, plastics
Phosphoric acid	H_3PO_4	Moderate	Making fertilizers, detergents, food additives
Acetic acid	CH_3COOH	Weak	Vinegar
Propanoic acid	CH_3CH_2COOH	Weak	Swiss cheese
Citric acid	$HOC(COOH)(CH_2COOH)_2$	Weak	Fruit
Carbonic acid	H_2CO_3	Weak	Carbonated beverages
Boric acid	H_3BO_3	Weak	Eye drops, mild antiseptic

* Sulfuric acid is a polyprotic acid, meaning it has more than one acidic hydrogen. It first ionizes completely to give HSO_4^- (hydrogen sulfate ion, also known as bisulfate ion), which is a weak acid and partially ionizes to give SO_4^{2-} (sulfate ion).

TABLE 9.3	Common Bases		
NAME	**FORMULA**	**STRENGTH**	**USE**
Sodium hydroxide	NaOH	Strong	Drain cleaner; producing aluminum, rayon, soaps, detergents
Potassium hydroxide	KOH	Strong	Producing soaps, detergents, fertilizers
Calcium hydroxide	$Ca(OH)_2$	Strong	Producing bleaching powder, paper and pulp; softening water
Ammonia	NH_3	Weak	Producing fertilizer, explosives, plastics, insecticides, detergents
Sodium bicarbonate	$NaHCO_3$	Weak	Antacid
Sodium carbonate	Na_2CO_3	Weak	Detergents, glassmaking
Calcium carbonate	$CaCO_3$	Weak	Chalk

The anions of weak acids are also bases (Table 9.3). Look at the reverse of the reaction of acetic acid with water. The acetate ion (CH_3COO^-) accepts a hydrogen ion from the acidic hydronium ion and is therefore reacting as a base. You will see that this property of anions plays an important role in solutions, such as blood, in which the acid concentration must remain constant (Section 9.4). In both blood and the many household uses of sodium bicarbonate, the basic nature of the bicarbonate ion (HCO_3^-) is put to use in controlling acid concentration.

$$HCO_3^-(aq) + H_2O(\ell) \rightleftharpoons H_2CO_3(aq) + OH^-(aq)$$

Because it is not stable, the carbonic acid (H_2CO_3) formed in this reaction, breaks down to form carbon dioxide gas.

$$H_2CO_3(aq) \rightleftharpoons CO_2(g) + H_2O(\ell)$$

© 1990 M & A Doolittle/Rainbow

A fire ant from the Amazon Basin on a ginger plant. The venom of these and other ants contains the simplest organic acid, formic acid (HCOOH). In the Amazon, fire ant colonies can be so large that streams become polluted by formic acid.

CONCEPT CHECK 9A

1. An acid is a hydrogen ion _____ and a base is a hydrogen ion _____.

2. To produce a desirable sour taste, a (an) _____ is added to many carbonated beverages.

3. Soap solutions are slippery and taste bitter, which indicates that soap is a(n) (basic/acidic) substance.

4. Complete the following equation for the reaction of an acid with a base:

$$H_3O^+(aq) + \underline{\hspace{2cm}} \longrightarrow 2\,H_2O(\ell).$$

5. Identify each of the following chemical compounds as an acid, a base, or a salt: (**a**) Na_2SO_4, (**b**) $HF(aq)$, (**c**) KCl, (**d**) KOH.

6. Ammonia is a _____ base because it establishes an _____ with water in which _____ are formed.

7. Because it has a small size and a high charge, H^+ exists in water as the ion known as the _____ ion, which has the formula _____.

8. The strong acid of greatest economic importance is _____.

9. Complete the following equation:

$$\underline{\hspace{2cm}} + KOH(aq) \longrightarrow KCl(aq) + \underline{\hspace{2cm}}.$$

10. Some acids are described as _____ acids because they establish equilibrium with their ions in solution instead of being completely ionized.

Measuring the acidity of soil with a pH meter. The proper acid–base balance is essential to the health of these plants.

 See an animation of **The Autoionization of Water** at http://brookscole.com/chemistry/joesten4

Finding the number of digits to include in the answers to mathematical problems is reviewed in Appendix A.

Concentration of a solution The quantity of a solute dissolved in a specific quantity of a solvent or solution

Molarity Number of moles of solute per liter of solution

9.3 MOLARITY AND THE pH SCALE

Water, you have seen, is capable of acting as either a hydrogen ion donor or a hydrogen ion acceptor, that is, as an acid or a base, depending on the properties of the substance with which it reacts. Water can also act as an acid or a base toward itself, although the reaction occurs to only a very small extent. About 1 of every 550,000,000 water molecules is ionized at any given time.

$$H_2O + H_2O \rightleftharpoons H_3O^+ + OH^-$$

Pure water contains equal numbers of hydronium ions and hydroxide ions.

The self-ionization of water provides the basis for a convenient method for expressing numerically just how acidic or basic any water solution is. You may have seen this quantity in use for a consumer product, for example, on the label of a "pH-balanced" shampoo. (The term refers to a controlled acidity of the solution, so that it is less likely to damage hair.) To understand what pH is requires understanding how to express the **concentration of a solution**, which is the quantity of a solute dissolved in a specific quantity of solvent or solution. We might give the concentration of a solution of sodium hydroxide as 4.0 g/L, that is 4.0 g of NaOH per liter of solution. In chemistry, however, because of its relation to quantities of chemicals in reactions (Section 8.2), concentration based on the mole is preferred.

Molarity

Concentrations of solutions in chemistry are usually expressed as **molarity**, which is the number of moles of solute per liter of solution. For example, a 1 molar (1 M) solution contains 1 mol of solute per liter of solution. If the solute is sodium hydroxide, a 1.00 M solution contains 1 mol, or 40.0 g, of NaOH per liter of solution. To find the molarity of a solution requires knowing the mass of dissolved solute, the volume of the solution, and the molar mass of the solute. For example, 4.0 g of NaOH is 0.10 mol (4.0 g/40.0 g/mol). Thus, a solution with 4.0 g of NaOH dissolved in 1 L of solution is a 0.10 M NaOH solution.

EXAMPLE 9.1 Molarity

What is the molarity of a solution that contains 10.0 g of HCl (molar mass, 36.5 g/mol) per liter of solution?

SOLUTION
First, we find the number of moles of HCl equivalent to 10.0 g of HCl.

$$10.0 \text{ g HCl} \times \frac{1 \text{ mol HCl}}{36.5 \text{ g HCl}} = 0.274 \text{ mol HCl}$$

With 0.274 mol dissolved per liter of solution, the molarity is 0.274 M in HCl.

TRY IT 9.2
How many grams of NaOH have been dissolved per liter of solution to make a 4.0 M NaOH solution?

TRY IT 9.3
If 73.0 g of HCl is dissolved in 500 mL of water, what is the molarity of this solution?

TRY IT 9.4
If 20 g of NaOH is dissolved in 2 L of water, what is the molarity of this solution?

TRY IT! 9.5

Is it possible to increase the molar concentration of a solution without adding more solute? Explain why or why not.

The pH Scale

In pure water (and all neutral solutions) the concentrations of hydronium ions and hydroxide ions are the same. Expressed as molarity, their concentrations are always 1.0×10^{-7} M at 25°C.

The product of the molarity of the hydronium ions and hydroxide ions in pure water is $(1.0 \times 10^{-7})(1.0 \times 10^{-7}) = 1.0 \times 10^{-14}$. *The value 1.0×10^{-14} is important to an understanding of all aqueous solutions because it is a constant that is always the product of the molar concentrations of H_3O^+ and OH^- in the solution.* Using square brackets, as is customary to represent the molarity of a substance, this relationship is written as

$$[H_3O^+][OH^-] = 1.0 \times 10^{-14}$$

If acid is added to pure water, the concentration of H_3O^+ will be greater than 1.0×10^{-7}, and the concentration of OH^- will be less than 1.0×10^{-7}. However, the product of the two must equal 1.0×10^{-14}. This relationship is the basis for calculating the concentration of one of the two ions, hydronium or hydroxide, when the other one is known.

Consider 0.10 M NaOH. Because sodium hydroxide is a strong base that is completely ionized, the molar concentration of hydroxide ions in this solution is also 0.10 M. To compensate, the concentration of hydronium ions must be significantly less than the pure water value of 1.0×10^{-7} M. Using the preceding equation, the concentration of hydronium ions in a 0.10 M NaOH solution is found to be 1.0×10^{-13} M. In the following calculation, the value of 0.10 M is expressed in scientific notation as 1.0×10^{-1} M:

$$[OH^-][H_3O^+] = 1.0 \times 10^{-14}$$

$$(1.0 \times 10^{-1})[H_3O^+] = 1.0 \times 10^{-14}$$

$$[H_3O^+] = \frac{1.0 \times 10^{-14}}{1.0 \times 10^{-1}} = 1.0 \times 10^{-13} \text{ M}$$

An equivalent calculation for a 0.10 M HCl solution, which contains 0.10 M H_3O^+, would show that its concentration of OH^- ions is 1.0×10^{-13} M.

The difference in H_3O^+ concentration between 0.1 M HCl and 0.1 M NaOH is 1 trillion times because each change in the exponent is a power of ten, and the difference between 10^{-1} [H_3O^+] and 10^{-13} [H_3O^+] is 10^{12}, or 1 trillion. The Danish biochemist S. P. L. Sørensen proposed in 1909 that these exponents be used as a measure of acidity. He devised a scale that would be useful in his work of testing the acidity of Danish beer. Sørensen's scale came to be known as the pH scale, from the French *pouvoir hydrogene*, which means hydrogen power. **pH** is defined as *the negative logarithm of the hydronium ion concentration.*

$$pH = -\log[H_3O^+]$$

The pH of a solution is pretty easy to calculate if the hydronium ion concentration is given in a form such as 1×10^{-4} M or 1×10^{-11} M. If the hydronium ion concentration is written as 1×10 to some power, take the exponent of 10 and drop the negative sign to get the pH. Thus, we would have pH 4 and pH 11, respectively, for the solutions mentioned in the first sentence of this paragraph. This method depends on the fact that the logarithm of 1 is 0.

Scientific notation, in which large and small numbers are represented by a number between 1 and multiplied by 10 with an exponent, is reviewed in Appendix B.

The pH of 4.0 shows that the tomato juice is acidic.

With a calculator that includes base -10 logarithms, the pH is easily found from any H_3O^+ concentration by finding its log; the H_3O^+ concentration can be determined from any pH by finding the value of 10^{-pH}.

pH Negative logarithm of the hydronium ion concentration of a solution

Since pH $= -\log [H_3O^+]$, we are really asked to do the following math.

$$pH = -\log (1 \times 10^{-4}), \text{ which equates to}$$
$$pH = -\log (1) -\log(10^{-4})$$

Since the log of 1 is 0 (since $10^0 = 1$) and the log of 10^{-4} is -4 we see that

$$pH = -0 - (-4), \text{ or}$$
$$pH = 4$$

Thus, it is easy to determine pH if the concentration is written as 1 times some power of 10. This will not always be the case, however. Let's see how we could estimate the pH of a solution whose hydronium ion concentration is, for example, 6×10^{-8} M.

We need to do the following calculation.

$$pH = -\log (6 \times 10^{-8}), \text{ which equates to}$$
$$pH = -\log(6) - \log(10^{-8})$$

Now we have a minor problem because it is unlikely you've memorized the logarithm of the number 6. We could key the number 6 into a calculator and then press the log button to find out that the log of 6 is 0.778. Once we know that, we can determine that

$$pH = -0.778 -(-8), \text{ or}$$
$$pH = 7.22$$

In the absence of a calculator, we may use what we call the *bracketing method* to arrive at a rough estimate of the pH in just a few seconds. Here's how that works. The hydronium ion concentration, in proper scientific notation, will always be written as a number between 1 and 10 multiplied by some power of 10 (e.g., 6×10^{-8}). It will always be true that this number may be bracketed as shown here between 1×10^{-8} and 10×10^{-8}. Thus, if we know that the log of 1 is 0 and that the log of 10 is 1 (since $10^1 =10$) we can quickly follow the logic below.

If $[H_3O^+] = 10 \times 10^{-8}$, then the pH $= -1 -(-8)$, so pH $= 7$;

and

if $[H_3O^+] = 1 \times 10^{-8}$, then the pH $= -0 -(-8)$, so pH $= 8$.

Since the hydronium ion concentration in this example was between these two bracketing concentrations, the pH must be between 7 and 8. With a little practice, you can do this bracketing in a matter of a few seconds to arrive at an estimate of the pH. If you want, or need, a more accurate number, you will have to use a calculator.

The larger the concentration of hydronium ion in a solution, the more acidic a solution is. Therefore, as a solution gets more acidic, for example, as $[H_3O^+]$ goes from 1×10^{-3} (or 0.001) M to 1×10^{-1} (or 0.1) M, the pH becomes a smaller number—in this case it goes from pH 3 to pH 1. *The smaller the pH value, the more acidic the solution.*

In a **neutral solution**, the pH is 7 (for a hydronium ion concentration of 1×10^{-7}). As the hydronium ion concentration becomes smaller than this, a solution becomes basic. Therefore, in basic solutions the pH is larger than 7. *The larger the pH value, the more basic the solution.*

Neutral solution A solution in which the pH is 7 and $[H_3O^+] = [OH^-] = 1 \times 10^{-7}$

if pH < 7.0,	solution is acidic	$[H_3O^+] > [^-OH]$
if pH $= 7.0$,	solution is neutral	$[H_3O^+] = [^-OH]$
if pH > 7.0,	solution is basic	$[H_3O^+] < [^-OH]$

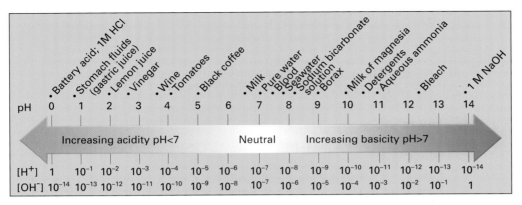

FIGURE 9.2 Relationship of pH to the concentration of hydrogen ions [H⁺ or H₃O⁺] and hydroxide ions [OH⁻] in water at 25°C. The pH values for some common substances are included in the diagram.

The relationship between pH and the hydronium ion concentration, plus the pH of some common solutions, is illustrated in Figure 9.2.

Because weak acids are only slightly ionized, the pH of a weak acid solution is not given by the concentration of the acid, but is dependent on its extent of ionization. For example, in a 1.0 M HCl solution, the hydronium ion concentration is 1.0 M, and the pH is 0, but in a 1.0 M acetic acid solution, the hydronium ion concentration is only 4.3×10^{-3} M and the pH is 2.4. Because many common substances have hydronium ion concentrations in the range of 0.1 M (1×10^{-1} M) to 10^{-14} M, the pH scale provides a more convenient way of expressing low acid concentrations.

Remember that as the hydroxide ion concentration becomes larger, the hydronium ion concentration must become smaller because of the constant relationship between their values, $[H_3O^+][OH^-] = 1 \times 10^{-14}$.

EXAMPLE 9.2 | **Finding pH**

Many soft drinks contain carbonic acid and phosphoric acid. If the hydronium ion concentration of a cola beverage is 1×10^{-3}, what is the pH?

SOLUTION
pH is defined as the negative of the logarithm of hydronium ion concentration, pH = $-\log [H_3O^+]$. For a concentration of 1×10^{-3}, the exponent -3 shows that the pH is 3.

TRY IT 9.6
Most tomatoes are acidic and have $[H_3O^+] = 1 \times 10^{-4}$ M. What is the pH of these tomatoes? Are they more or less acidic than the cola drink of Example 9.2?

TRY IT 9.7
Use the bracketing method to obtain a pH range for a sample of rainwater that has a hydronium ion concentration of 3×10^{-6} M. Once you have your estimate, determine the exact pH of this solution by using your calculator to find the exact value of the logarithm of 3.

TRY IT 9.8
Use your knowledge of the relationship between hydronium ion concentration and hydroxide ion concentration to determine the pH of a solution of seawater that has a *hydroxide* ion concentration of 2×10^{-6} M. Use your knowledge of the relationship between hydronium ion and hydroxide ion concentrations to determine the hydronium ion concentration. Then use the bracketing method to obtain a pH range, and finally, use the calculator to determine the exact pH.

Charles D. Winters

The pH of 7.46 shows that the milk is slightly alkaline.

The pH Scale Is Nonlinear

By now you may have noticed that the pH scale is not a linear scale. Logarithms do not progress linearly as we proceed from log 1 to log 2 to log 3 to log 4, etc. The logarithm of 1 is 0 and the logarithm of 10 is 1 but the logarithm of 5 is not 0.5. The logarithm of 5 is 0.699, a number that is not exactly halfway between 0 and 1. Thus, the pH scale is not linear.

TRY IT 9.9

Use your calculator to prepare a graph in which you plot the logarithm of the whole numbers 1–10 (on the y-axis) versus the numbers 1–10 (on the x-axis) to see what type of relationship emerges. This will show you the nonlinearity of this logarithmic scale and will allow you to confirm for yourself the very important fact that *a 10-fold change in hydronium ion concentration results in a pH change of 1 unit!*

TRY IT 9.10

Use your graph to see what the effect on pH would be if the hydronium ion concentration changes from 1×10^{-9} M to 4×10^{-9} M. By what factor would the hydronium ion concentration have to change to cause the pH to increase by 3 units? By what factor would the hydronium ion concentration have to change to cause the pH to decrease by 5 units?

 See an animation of **Detecting Acid–Base Reactions Using Indicators or a pH Meter** at http://brookscole.com/chemistry/joesten4

 See an animation of **pH Changes Upon Addition of HCl to Water and to a Buffer Solution** at http://brookscole.com/chemistry/joesten4

(9.4) ACID–BASE BUFFERS

An acid–base buffer is like a shock absorber—something to prevent a disturbance while retaining the original conditions or structure. The control of pH requires maintaining a steady concentration of H_3O^+ even when sudden "shocks" of acid or base are added. **Buffer solutions** contain a base that can react with an acid and an acid that can react with a base so that the pH of a solution remains close to its original value. An acid–base combination suitable for controlling pH is known as a **buffer**. You see the term on bottles of "buffered" aspirin, meaning that the tablets contain some sort of base to offset the acidity of aspirin. The principal benefit is that such tablets may dissolve faster, and thereby go to work faster. The buffering does not, however, actually buffer the highly acidic environment of the stomach, which has a pH of 1.5 to 2.

How does a buffer solution maintain its pH at a nearly constant value? Not only must the acid in the buffer react with added base and the base must react with added acid, but it is also necessary that the acid and base

Buffer solution Solution of an acid plus a base that controls pH by reacting with added base or acid

Buffer Combination of an acid and a base that can control pH

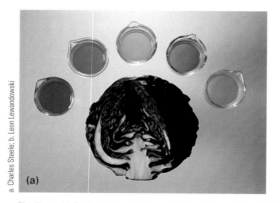
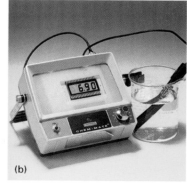

Finding pH. To find the pH of a solution, it can be tested with an acid–base indicator, for example, (a) the natural indicator in a red cabbage solution, which (from left to right) has the colors shown for solutions of pH 1, 4, 7, 10, and 13. (b) For a more accurate determination, an instrument known as a pH meter is used.

components of a buffer solution not react with each other. To meet these conditions, buffers usually are mixtures of a weak acid and its weakly basic anion (e.g., acetic acid and acetate ion, CH_3COOH and CH_3COO^-) or a weak base and its weakly acidic cation (e.g., ammonia and ammonium ion, NH_3 and NH_4^+). In a solution that contains an acetic acid–acetate ion buffer, added base will react with the acetic acid:

$$CH_3COOH(aq) + OH^-(aq) \rightleftharpoons CH_3COO^-(aq) + H_2O(\ell)$$

and added acid will react with the acetate ion:

$$CH_3COO^-(aq) + H_3O^+(aq) \rightleftharpoons CH_3COOH(aq) + H_2O(\ell)$$

Notice that in these reactions no new substances are produced. The products are always components of the buffer. Both acid and base are neutralized by different buffer components to maintain the pH at a constant value.

Buffers are very important to many industrial and natural processes. In fact, the control of the pH of your blood is essential to your health. The pH of the blood is about 7.40, and your good health depends on the ability of buffers to maintain the pH within a narrow range. If the pH falls below 7.35, a condition known as *acidosis* occurs; increasing pH above 7.45 leads to *alkalosis*. Both these conditions can be life threatening.

Carbonic acid, which forms when carbon dioxide dissolves in water, and bicarbonate ion form one of several buffer pairs that keep the pH of blood within the necessary safe range.

The blood buffering system depends on two critical equilibria. Carbon dioxide reacts with water to form carbonic acid and carbonic acid reacts with water to form hydronium ion and bicarbonate ion, as shown here.

$$CO_2(g) + H_2O(\ell) \rightleftharpoons H_2CO_3(aq)$$

$$H_2CO_3(aq) + H_2O(\ell) \rightleftharpoons H_3O^+ + HCO_3^-(aq)$$

If some condition causes the hydronium ion concentration to increase (i.e., the pH begins to decrease), the equilibrium above involving hydronium ion will shift to the left, according to LeChatelier's principle, to relieve this stress and maintain the equilibrium. This, in turn, will cause the equilibrium in the first reaction to shift to the left to expel more carbon dioxide. Thus, bicarbonate neutralizes acid in the blood buffering system.

Likewise, if some condition causes the hydronium ion concentration to decrease (i.e., the pH begins to increase), the equilibrium involving hydronium ion will shift to the right, according to LeChatelier's principle, to maintain the equilibrium and produce more hydronium ion. This is another way of saying, in this case, that an increase in hydroxide ion will be neutralized by carbonic acid.

EXAMPLE 9.3   Determining Hydronium Ion Concentration from pH

What would be the hydronium ion concentration in the blood if the pH is 7.4?

SOLUTION

Since pH = 7.4, we know, by the definition of pH, that

$$-\log [H_3O^+] = 7.4$$

$$\log [H_3O^+] = -7.4$$

We could enter 7.4 into a calculator, hit the $+/-$ button to change sign, and then hit the antilog button to find that the hydronium ion concentration that corresponds to this pH is 4.0×10^{-8} M.

Alternatively, if we don't need an exact answer, the bracketing method can be used to conclude that the hydronium ion concentration would be somewhere between 10^{-7} M and 10^{-8} M, because the pH is between 7 and 8.

TRY IT 9.11

Determine the hydronium ion concentration of a solution of household ammonia whose pH is 10.7.

⟨ 9.5 ⟩ CORROSIVE CLEANERS

> A *corrosive* substance has the ability to weaken or destroy materials by chemical reactions, often acid–base or oxidation–reduction reactions (Chapter 10).

There are some places in the home where really tough cleaning jobs exist. For these jobs, cleaners are formulated with extremes in pH, which allow the acidity or alkalinity of the cleaner to quickly attack the unwanted dirt, grease, or stain (Figure 9.3). Toilet-bowl cleaners usually contain hydrochloric acid, which can dissolve most mineral scale (mostly carbonates) and iron stains. Other acids, such as phosphoric acid and oxalic acid, are also used in these products. The pH of toilet-bowl cleaners is usually below 2, and because they contain strong acids, they can be quite harmful to skin and eyes on contact. They should be handled with extreme caution, and rubber gloves should be worn when using them.

On the other end of the pH scale are drain cleaners and oven cleaners, which have a pH of 12 or higher. These formulations almost always contain the strong base sodium hydroxide (NaOH). Drains usually clog as a result of oils, grease, and hair caught on rough edges inside the drainpipes. As the foreign matter builds up it becomes more tightly packed, until the flow of water from the drain practically stops. The only way to get rid of the material clogging the drain is to either dismantle the drain plumbing (often very difficult), use a plumber's "snake," or dissolve the material. Bases like sodium hydroxide are very good at dissolving drain clogs because they can cause rapid breaking of bonds in oils and greases of animal and vegetable origin. Once the bonds are broken, smaller, more soluble molecules are formed that can be washed down the drain. Hair and other proteins are also broken down by sodium hydroxide, so a mat of grease and hair in a clogged drain will quickly succumb to the action of the strong base. Numerous products containing sodium hydroxide are available, including solid NaOH pellets, flakes, and concentrated solutions. The solutions offer the easiest way to apply the drain cleaner, but if they become diluted, their effectiveness is diminished. Some solid drain cleaners also contain pieces of aluminum metal, which react with aqueous sodium hydroxide to form hydrogen gas (caution—it's flammable) that helps agitate the mixture and hastens the unclogging process.

$$2\,Al(s) + 2\,NaOH(aq) + 6\,H_2O(\ell) \longrightarrow$$
$$2\,Na^+(aq) + 2\,Al(OH)_4^-(aq) + 3\,H_2(g)$$

Oven cleaners also contain strong bases such as sodium hydroxide. Many oven cleaners use aerosol sprays to distribute the cleaner on the inner surface of the oven. Care must be taken not to breathe air containing these aerosols because strong bases are quite corrosive to nasal tissue, bronchial tubes, and lungs.

FIGURE 9.3 Corrosive household cleaners. Reading the labels and handling these chemicals with care are essential.

Charles D. Winters

9.6 HEARTBURN: WHY REACH FOR AN ANTACID?

The walls of a human stomach contain thousands of cells that secrete hydrochloric acid, the main purposes of which are to kill microorganisms and to aid in digestion. The stomach's inner lining is not harmed by the presence of hydrochloric acid with such a low pH (1.5–2.0), since the mucosa, the inner lining of the stomach, is replaced at the rate of about a half million cells per minute.

The uncomfortable condition known as "heartburn" occurs when the acid contents of the stomach back up into the esophagus and cause a burning sensation in the chest and throat. Among the causes are overeating; spicy, acidic, or fatty foods; and some medications, such as aspirin.

Antacids are bases used to neutralize the acid that causes heartburn. The most common antacid ingredients are magnesium and aluminum hydroxides, and bicarbonate or carbonate salts (Table 9.4). Baking soda (sodium bicarbonate) was used to relieve indigestion before many of the other commercial products became available. The bicarbonate ion, a basic anion of a weak acid, reacts with the hydronium ion from hydrochloric acid to form carbonic acid, which decomposes to give carbon dioxide and water. Note this is the same mechanism by which the blood buffering system neutralizes acid.

$$HCO_3^-(aq) + H_3O^+(aq) \longrightarrow H_2CO_3(aq) + H_2O(\ell)$$
$$H_2CO_3(aq) \longrightarrow CO_2(g) + H_2O(\ell)$$

Alka-Seltzer contains sodium bicarbonate, potassium bicarbonate, citric acid, and aspirin. The fizz of the Alka-Seltzer tablet in water is carbon dioxide gas given off by the reaction of the citric acid with the bicarbonates to give carbonic acid (the preceding reaction).

Milk of magnesia is a suspension of magnesium hydroxide (which is not very soluble) in water. Magnesium hydroxide acts as an antacid in small doses, but in large doses it is a laxative. Calcium carbonate (also known as

Until recently, it was believed that excess stomach acid caused stomach ulcers. It has now been proved that the cause is instead a bacterial infection and that ulcers can be successfully treated with antibiotics. The 2005 Nobel Prize in Physiology or Medicine was awarded to Barry J. Marshall and J. Robin Warren for their discovery of the role of *Helicobacter pylori* in gastritis and peptic ulcer disease.

Antacid Base used to neutralize excess hydrochloric acid in the stomach

---

TABLE 9.4 Some Common Antacids

Compound	Formula	Examples of Commercial Products
Magnesium hydroxide	$Mg(OH)_2$	Phillips' Milk of Magnesia
Calcium carbonate	$CaCO_3$	Tums, Titralac
Sodium bicarbonate	$NaHCO_3$	Alka-Seltzer, baking soda
Aluminum hydroxide	$Al(OH)_3$	Amphojel
Aluminum hydroxide and magnesium hydroxide		Maalox, Mylanta, Di-Gel tablets
Aluminum hydroxide, magnesium hydroxide, and magnesium carbonate	$MgCO_3$	Di-Gel liquid
Dihydroxyaluminum sodium carbonate	$NaAl(OH)_2CO_3$	Rolaids
Calcium carbonate and magnesium hydroxide		Sodium-free Rolaids

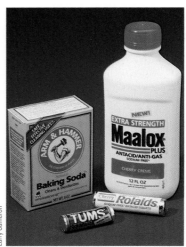

Larry Cameron

Some antacids.

chalk) is the active ingredient in Tums. The carbonate ion, a base, neutralizes hydronium ion.

$$CO_3^{2-}(aq) + 2\,H_3O^+(aq) \longrightarrow H_2CO_3(aq) + 2\,H_2O(\ell)$$

$$H_2CO_3(aq) \longrightarrow CO_2(g) + H_2O(\ell)$$

Although small amounts of calcium carbonate are safe, regular use can cause constipation. Aluminum hydroxide, the active ingredient in Amphojel, can also cause constipation in large doses. Because calcium carbonate and aluminum hydroxide can cause constipation, antacids such as Maalox and Mylanta contain aluminum hydroxide mixed with magnesium hydroxide to counteract the constipating effects of the former with the laxative action of the latter.

Another class of medications to combat heartburn has recently become more readily available by being reclassified from prescription-only to over-the-counter status. Instead of neutralizing stomach acid, these new drugs (for example, Tagamet and Pepcid) prevent its secretion.

CONCEPT CHECK **9B**

1. Which is more acidic, a pH of 6 or a pH of 2?
2. High pH means (**a**) high hydronium ion concentration, (**b**) low hydronium ion concentration.
3. Low pH means (**a**) high hydronium ion concentration, (**b**) low hydronium ion concentration.
4. In the buffer system H_2CO_3/HCO_3^-, which component neutralizes added base? Which component neutralizes added acid?
5. What is the pH of 1.0×10^{-3} M hydrochloric acid?
6. What is the pH of 1.0×10^{-3} M sodium hydroxide?
7. The pH of household ammonia is 11. What is the hydronium ion concentration? What is the hydroxide ion concentration?
8. The pH of a sodium hydroxide solution is 10. What is the hydroxide ion concentration?
9. What is the $[H_3O^+]$ for a 0.01 M HCl solution?
10. Most toilet-bowl cleaners are strongly _____ and most oven cleaners are strongly _____.
11. Which of the following compounds can function as an antacid in the treatment of heartburn? (**a**) Sodium bicarbonate, (**b**) $MgCl_2$, (**c**) $MgCO_3$, (**d**) $Al(OH)_3$, (**e**) Aluminum chloride.

 Assess your understanding of this chapter's topics with an online chapter quiz at **www.brookscole.com/chemistry/joesten4**

■ KEY TERMS

acid–base indicators	base	molarity
hydronium ion	acid–base reaction	pH
acidic solution	strong acid	neutral solution
basic solution (alkaline solution)	weak acid	buffer solution
salt	strong base	buffer
neutralization reaction	weak base	antacid
acid	concentration of a solution	

■ THE LANGUAGE OF CHEMISTRY

1. M
2. pH of pure water
3. Acidic pH
4. Strong acid
5. Strong base
6. Acid definition
7. Base definition
8. Weak acid
9. Buffer
10. Basic pH

a. 7
b. Molarity, moles of solute per liter of solution
c. Hydrogen ion donor
d. Maintains pH
e. Hydrogen ion acceptor
f. 10
g. CH_3COOH, acetic acid
h. NaOH
i. H_2SO_4
j. 3

■ APPLYING YOUR KNOWLEDGE

1. What do the following terms mean?
   (**a**) Acidic solution   (**b**) Basic solution
   (**c**) Neutral solution
2. What is a neutralization reaction?
3. What is a salt?
4. What do the following terms mean?
   (**a**) H^+ ion donor   (**b**) H^+ ion acceptor
   (**c**) Hydronium ion   (**d**) Hydroxide ion
5. What do the following terms mean?
   (**a**) Extent of ionization   (**b**) Strong acid
   (**c**) Strong base   (**d**) Weak acid
   (**e**) Weak base
6. What is the pH range for acidic solutions?
7. What is the pH range for basic solutions?
8. Label each of the following substances as acidic or basic:
   (**a**) Vinegar   (**b**) Citrus fruits
   (**c**) Aspirin   (**d**) Black coffee
9. Label each of the following substances as acidic or basic:
   (**a**) Gastric fluid   (**b**) Tomatoes
   (**c**) Oven cleaners   (**d**) Soap
   (**e**) Carbonated beverages   (**f**) Baking soda
10. Indicate which of the following is a property of an acid and which is a property of a base:
    (**a**) Sour taste   (**b**) Bitter taste
    (**c**) Slippery feeling
    (**d**) Changes color of red litmus to blue
    (**e**) Changes color of blue litmus to red
11. What do the following terms mean?
    (**a**) Molarity   (**b**) Concentration
12. What is the definition of pH?
13. What is a buffer?
14. What is an antacid?

15. Write the balanced equations for the following neutralization reactions.
    (**a**) Acetic acid, $CH_3COOH(aq)$, with potassium hydroxide, $KOH(aq)$
    (**b**) Sulfuric acid, $H_2SO_4(aq)$, with calcium hydroxide, $Ca(OH)_2(aq)$
    (**c**) Sulfuric acid, $H_2SO_4(aq)$, with sodium hydroxide, $NaOH(aq)$
16. Balance the following equations:
    (**a**) $HBr(aq) + Ca(OH)_2(aq) \longrightarrow$ _____
    (**b**) $HNO_3(aq) + Al(OH)_3(aq) \longrightarrow$ _____
17. What is the equation for the reaction between Tums, $CaCO_3(aq)$, and the $HCl(aq)$ in stomach acid?
18. Which is more basic: a solution with pH of 2 or a solution with pH of 10? Explain.
19. Which is more acidic: a solution of black coffee with a pH of 5.0 or milk with a pH of 6.5? Explain.
20. Which of the following are acidic and which are basic?
    (**a**) Cherries, pH 3.2   (**b**) Crackers, pH 8.5
    (**c**) Bananas, pH 4.6   (**d**) Drinking water, pH 8.0
21. Which is more acidic, a grapefruit, pH 3.2, or a lemon, pH 2.3?
22. Which is more acidic, soap, pH 10, or household ammonia, pH 11?
23. Which component of each of the following buffer solutions would react with added acid and which would react with added base?
    (**a**) a buffer consisting of equal amounts of ammonium ion (NH_4^+) and NH_3
    (**b**) a buffer consisting of equal amounts of hydrogen fluoride (HF) and sodium fluoride (NaF)

24. A buffer can be made using a mixture of $NaHCO_3(aq)$ and $H_2CO_3(aq)$. Describe how the pH is maintained when small amounts of acid or base are added to the combination.

25. Two solutions contain 1% acid. Solution A has a pH of 4.6 and solution B has a pH of 1.1. Which solution contains the stronger acid?

26. Moist baking soda is often put on acid burns. Why? Write an equation for the reaction, assuming the acid is hydrochloric acid (HCl).

27. How many grams of NaOH must be dissolved to make 1.0 L of 6.0 M NaOH solution?

28. How many grams of HBr must be dissolved to make 1.0 L of 2.0 M HBr solution?

29. How many moles of HCl are dissolved in 200 mL of 0.100 M HCl?

30. How many moles of KOH are dissolved in 125 mL of 0.500 M KOH?

31. What is the molarity of the solutions formed when the following are dissolved to make 1.0 L of solution?
    (a) 5.0 g HCl   (b) 40.0 g NaOH

32. How many grams of NaCl are dissolved in 1.0 L of a 1.5 M NaCl solution?

33. How many grams of H_2SO_4 are dissolved in 1.0 L of a 0.10 M H_2SO_4 solution?

34. What is the pH for each of the following solutions?
    (a) 1.0×10^{-2} M HCl   (b) 0.001 M HNO_3

35. What is the pH for each of the following solutions?
    (a) 1.0×10^{-3} M NaOH   (b) 1.0×10^{-3} M HCl

36. What is the molarity of H_3O^+ for each of the following solutions?
    (a) solution with pH 1.0   (b) Solution with pH 0.0
    (c) Solution with pH 5.0   (d) Solution with pH 3.0

37. How many milliliters of 2.0 M HCl are needed to neutralize 0.15 mol of NaOH?

38. How many milliliters of 0.50 M HBr are needed to neutralize 0.045 mol of LiOH?

39. When 25 mL of 0.10 M HCl and 35 mL of 0.10 M KOH are mixed, will the resulting 60 mL of solution be neutral? *Hint:* Compare the numbers of moles of dissolved acid and base.

40. When 31 mL of 0.50 M HI and 100 mL of 1.5 M NaOH are mixed, will the resulting 131 mL of solution be neutral? *Hint:* Compare the numbers of moles of dissolved acid and base.

Ⓦ CHEMISTRY ON THE WEB

For up-to-date URLs, visit the text website at **www.brookscole.com/chemistry/joesten4**

- Acids, Bases, Buffers and Indicators
- pH and Your Swimming Pool
- Acid and Base Chemical Cleaners
- How to Make Red Cabbage Indicator

OXIDATION–REDUCTION REACTIONS

Some oxidation–reduction reactions create flames, like the combustion of fuel in the space shuttle Discovery's engines; others create electric current and less dramatic changes.

NASA

Like the acid–base reactions discussed in Chapter 9, oxidation–reduction (or redox) reactions are common in our surroundings. Oxygen is a reactant in many of these reactions. Although oxidation and reduction reactions always occur simultaneously, most people are probably more familiar with the oxidation component. Some oxidation reactions, such as the burning of a piece of paper or the combustion of gasoline, occur quite readily while others, such as the rusting of an iron tool left out in the yard, occur more slowly. In each of these cases, the chemical or object that has been oxidized is quite obvious to a casual observer. It is not so obvious what chemical or object has been reduced, however. We rely on the oxidation of glucose for the biochemical energy we need to think, run, jump, and conduct the activities of our daily lives. Our automobile trips would not be possible without the chemical energy provided by the redox chemistry of the automobile battery. We also depend on batteries to operate flashlights, portable radios and CD-players, and cameras, to name just a few modern conveniences that we take for granted.

In this chapter the mysteries of oxidation–reduction reactions will be revealed as we explore answers to the following questions:

- What are oxidation and reduction?
- How can you recognize oxidation–reduction reactions?
- What is necessary to release the energy stored in chemical compounds as electric current?
- What is the difference between the chemical reactions in batteries and those used in electrolysis, and between their applications?
- What is corrosion and what are some ways to control it?

10.1 OXIDATION AND REDUCTION

"Oxidation" got its name from the chemical changes that occur when oxygen (O_2) combines with other elements or with compounds. Many reactive metals are mined as **oxides** (compounds of oxygen with another element) that were formed by reaction of the metals with oxygen in the air. For example, the major aluminum ore, bauxite, contains aluminum oxide (Al_2O_3); and hematite, a principal iron ore, contains the iron(III) oxide (Fe_2O_3). When materials containing carbon, nitrogen, or sulfur burn in air, the products are carbon oxides (CO_2, CO), nitrogen oxides (NO, NO_2, N_2O, and others), and sulfur oxides (SO_2, SO_3).

A substance that has combined with oxygen is described as having been *oxidized,* and the reaction is classified as *oxidation.* All **combustion** reactions are oxidations. In the combustion of the methane in natural gas, for example, the carbon is oxidized to carbon dioxide and the hydrogen is oxidized to water.

$$CH_4(g) + 2\,O_2(g) \longrightarrow CO_2(g) + 2\,H_2O(g)$$

The gradual rusting away of iron objects begins with the oxidation of iron. This gradual process is not referred to as combustion, however.

$$4\,Fe(s) + 3\,O_2(g) \longrightarrow 2\,Fe_2O_3(s)$$

In the blast furnaces that produce iron from iron ore, one of the important chemical changes is the reaction of the iron oxide in hematite with carbon monoxide.

$$Fe_2O_3(s) + 3\,CO(s) \longrightarrow 2\,Fe(\ell) + 3\,CO_2(g)$$

If the addition of oxygen to iron is oxidation, how is the removal of oxygen from iron described? It is *reduction,* a term sometimes used to mean "to bring something back." To metallurgists hundreds of years ago, reduction meant bringing a metal back from its ore. In today's chemical sense, the iron oxide has been reduced by the removal of oxygen.

Notice that in the reaction of iron oxide with carbon monoxide, oxygen has added to the carbon monoxide to produce carbon dioxide—the carbon monoxide has been oxidized. Here is a fundamental concept of chemistry— *oxidation and reduction always occur together.* If one reactant is oxidized, another must be reduced.

Over time, the definitions of oxidation and reduction have been extended to include processes other than the loss or gain of oxygen. With a more modern definition, you can better see why oxidation and reduction always occur together. An atom or ion is said to be oxidized when it loses elec-

Oxides Compounds of oxygen combined with another element

Combustion Rapid oxidation that produces heat and (usually) light

trons. Consider the reaction of sodium, a metal, with chlorine, a nonmetal (Figure 10.1):

$$2 \, Na(s) + Cl_2(\ell) \longrightarrow 2 \, NaCl(s)$$

The product is sodium chloride, a white, crystalline solid that is an ionic compound composed of equal numbers of Na^+ and Cl^- ions.

In the reaction of sodium with chlorine, sodium atoms lose their single-valence electron to produce sodium cations (Na^+). **Oxidation** is the loss of electrons, and sodium has been oxidized. Except when electrons are flowing through a wire from a negative region to a positive one, they are held to positively charged nuclei. Because we know that matter is neutral, we must ask where the electrons lost in oxidation go. The reaction of sodium with chlorine occurs because as sodium atoms lose electrons, chlorine atoms gain electrons.

Reduction is the gain of electrons, and in the conversion of chlorine to chloride ions (Cl^-) in this reaction, each chlorine atom has been reduced by the gain of a single electron. Reactions in which one reactant is oxidized and another is reduced are known as **oxidation–reduction reactions**, usually referred to as **redox reactions**. For example, every reaction in which a metallic element combines with a nonmetallic element is a redox reaction in which the metal is oxidized and the nonmetal is reduced.

Charles D. Winters

FIGURE 10.1   Oxidation of sodium by chlorine. The reaction between sodium metal and chlorine, a nonmetal, releases a large amount of energy. The product is sodium chloride, familiar to everyone as table salt.

EXAMPLE 10.1   Recognizing Oxidation and Reduction

For the reaction of copper with oxygen to give an ionic oxide,

$$2 \, Cu(s) + O_2(g) \longrightarrow 2 \, CuO(s)$$

identify which reactant is oxidized and which is reduced.

SOLUTION
The reaction of Cu with O_2 can be recognized as a redox reaction because it is the addition of oxygen to a reactant and also because it is the combination of a metal and a nonmetal. Since the product is an ionic compound, the electron gain and loss is determined by the charges on the ions. The oxide CuO must be composed of Cu^{2+} and O^{2-} ions. Therefore, the copper metal has been oxidized by the loss of two electrons from each atom, and the oxygen has been reduced by the gain of two electrons by each atom.

TRY IT 10.1
For the reaction of the very active metal lithium with oxygen,

$$4 \, Li(s) + O_2(g) \longrightarrow 2 \, Li_2O(s)$$

identify what is oxidized and what is reduced.

In oxidation the charge of an atom or ion can increase (*it's oxidized*), and in reduction the charge of an atom or ion can decrease (*it's reduced*).

See an animation of the **Redox Reaction Between Mg and HCl** at http://brookscole.com/chemistry/joesten4

See an animation of the **Redox Chemistry of MnO_4^- and Fe^{2+}** at http://brookscole.com/chemistry/joesten4

Oxidation The gain of oxygen, loss of hydrogen, or loss of electrons

Reduction The loss of oxygen, gain of hydrogen, or gain of electrons

Oxidation–reduction reactions (redox reactions) Reactions in which one reactant is oxidized and another reactant is reduced

Oxidizing agent A reactant that causes oxidation of another reactant

---

10.2   OXIDIZING AGENTS: THEY BLEACH AND THEY DISINFECT

You've seen that whenever oxygen adds to another reactant, that reactant is *oxidized*. The oxygen in such a reaction is the **oxidizing agent**, a reactant that causes an oxidation by gaining electrons. The halogens are also oxidizing agents, for example

$$2 \, Al(s) + 3 \, Cl_2(g) \longrightarrow 2 \, AlCl_3(s)$$
$$Sn(s) + F_2(g) \longrightarrow SnF_2(\ell)$$

Chlorine-containing tablets being added to a swimming pool filter.

© Yoav Levy/ Phototake NYC

TABLE 10.1		Some Oxidizing and Reducing Agents
NAME	FORMULA	USES
Oxidizing Agents		
Lead dioxide	PbO_2	Automobile batteries
Manganese dioxide	MnO_2	Batteries
Sodium hypochlorite solution	$NaOCl(aq)$	Laundry bleach, disinfection
Oxygen	O_2	Metabolism of foods, burning fuels
Ozone	O_3	Water purification, hazardous chemical destruction
Chlorine	Cl_2	Drinking water and wastewater purification, chemical synthesis
Hydrogen peroxide	H_2O_2	Bleaching, antiseptic
Reducing Agents		
Hydrogen	H_2	Fuel, chemical synthesis
Sulfur dioxide	SO_2	Chemical synthesis
Carbon	C	Iron production
Zinc	Zn	Batteries

In each of these reactions, the neutral halogen molecule has been reduced by the addition of electrons to give −1 ions. Here is another condition common to all redox reactions: *The oxidizing agent is reduced.* Some common oxidizing agents are listed in Table 10.1.

Somewhere around home, most of us have one or more oxidizing agents. The most common household oxidizing agent is sodium hypochlorite (NaOCl), a water-soluble ionic compound. In the laundry, NaOCl is a **bleaching agent**—it removes unwanted color by oxidizing colored chemical compounds to produce whiter clean clothes. If you spill an NaOCl solution (for example, Clorox bleach) on your jeans, its effectiveness as a bleach will be immediately obvious, as a white spot will appear. Substances have color because their molecular structures absorb portions of the visible spectrum of light. Many of the substances that contribute unwanted color to textiles or paper are organic compounds with long chains of alternating single and double carbon–carbon bonds (Section 5.3). The color is created by the absorption of some of the light by electrons in these bonds. A bleach disrupts the alternating pattern by breaking bonds or converting double bonds to single bonds. The result is loss of the ability to absorb visible light. A very strong bleaching solution can also break bonds in molecules that compose the fabric itself; when this occurs, the fabric develops thin spots and holes.

Sodium hypochlorite is also a *disinfectant*. When you use a liquid bleach solution to wash the bathroom floor or the kitchen counter, you are taking advantage of both the disinfecting and bleaching properties of this chemical. By oxidizing molecules in the outer surfaces of bacteria, a disinfectant is able to disrupt the structure of the cells and kill them.

Another familiar oxidizing agent is hydrogen peroxide (H_2O_2), and it too is a bleaching agent. At times, its most popular use has been in bleaching hair to produce "peroxide blonde." A dilute solution (usually 3%) of hydrogen peroxide in water is useful in the medicine cabinet because it is an *antiseptic*, a substance that can be used to cleanse a wound to prevent bacterial infection. An antiseptic acts in the same manner as a disinfectant, but, unlike a disinfectant, is mild enough for use on human tissue without doing damage.

Concentrated hydrogen peroxide solutions and pure hydrogen peroxide are extremely hazardous substances.

Bleaching agent A chemical that removes unwanted color by oxidation of the colored chemical

THE PERSONAL SIDE

Richard Feynman and the **Challenger** Explosion

Richard Feynman (1919–1988), winner of the 1965 Nobel Prize in physics, was one of the most original thinkers of this century. He wrote, "Scientific knowledge is a body of statements of varying degrees of certainty—some most unsure, some nearly sure, but *none* absolutely certain. Now, we scientists are used to this, that it is possible to live and *not* know."

Feynman gave the United States a lesson in how science works when he used a simple experiment to uncover the reason for the disastrous explosion of the space shuttle *Challenger*. On launch day, January 28, 1986, the weather was unusually cold in Florida—the temperature was 29°F. A few moments after launch, the world watched in horror as the shuttle and its rockets exploded in a gigantic fireball, killing all the astronauts aboard.

A commission was appointed to investigate the cause of the explosion. It was Feynman who reasoned that due to the cold temperature, rubber O-rings used to seal joints in the solid-fuel booster rockets had not expanded properly. This failure allowed hot flames from the booster rocket to burn through the hydrogen fuel tank. The result was

© Bettmann/Corbis

RICHARD FEYNMAN DEMONSTRATING THAT AN O-RING OF THE TYPE USED IN THE SPACE SHUTTLE *CHALLENGER* DOES NOT RETAIN ITS FLEXIBILITY WHEN COLD AND THUS WOULD ALLOW GASES TO ESCAPE FROM THE ROCKET.

the violent redox reaction that occurs when hydrogen and oxygen combine to form water. A massive amount of energy was released due to the large quantities of hydrogen and oxygen involved. To prove his point, Feynman performed a dramatic—but very simple—experiment. During a public hearing, Feynman held a sample rubber O-ring like the one in the rocket engine tightly in a clamp and immersed it in a glass of ice water. Everyone watching his experiment could see that the rubber did not spring back to its original shape. The poor low-temperature characteristics of the rubber O-ring had doomed the *Challenger* and its crew.

Chlorine itself is an oxidizing agent of vital importance in ensuring the purity of drinking water and water that is returned to natural waterways. Treatment of wastewater and drinking water is described in Sections 11.10 and 11.12.

10.3 REDUCING AGENTS: FOR METALLURGY AND GOOD HEALTH

When a metal reacts with another substance, the metal usually acts as a **reducing agent**, a reactant that causes a reduction by giving up electrons. In reactions with oxygen or the halogens, for example, the metal is the reducing agent and the oxygen or halogen is reduced.

$$2 \, Cu(s) + O_2(g) \longrightarrow 2 \, CuO(s)$$
$$2 \, Al(s) + 3 \, Br_2(g) \longrightarrow 2 \, AlBr_3(s)$$

In these, and all redox reactions, *the reducing agent is oxidized*. Some common reducing agents are included in Table 10.1.

Reducing agents are crucial in the production of metals from their ores because most metals occur in nature only in chemical compounds. Carbon, usually in the form of **coke**, is a valuable metallurgical reducing agent.

$$ZnO(s) + C(s) \longrightarrow Zn(g) + CO(g)$$

Reducing agent A reactant that causes reduction of another reactant

Coke A mostly carbon product formed by heating coal in the absence of air to drive off volatile materials

The redox reaction between aluminum and bromine releases heat and creates a cloud of vaporized bromine as some of the liquid bromine boils away.

Free radical An atom or molecule that contains an unpaired electron

FIGURE 10.2 Action of an anti-oxidant. Once a free radical (R·) has formed, it will react as soon as it can with something that will remove or pair with the unpaired electron. By intercepting free radicals, beta-carotene and other antioxidant vitamins prevent them from damaging DNA or other crucial biomolecules.

In the production of zinc, this reaction is carried out at such a high temperature that the zinc is formed as a gas that is condensed for further purification. In this process the Zn^{+2} cation in zinc oxide is reduced to zinc metal and $C(s)$ is oxidized to carbon monoxide.

Hydrogen gas, like the metals, is a reducing agent in virtually all of its reactions. Hydrogen can, for example, reduce the copper in copper(II) oxide:

$$CuO(s) + H_2(g) \longrightarrow Cu(s) + H_2O(\ell)$$

The addition of hydrogen to another compound is also classified as a reduction. In a reaction that may be of increasing importance in the production of methanol as an alternative fuel (Section 12.9), carbon monoxide is reduced to methanol by the addition of hydrogen.

$$CO(g) + 2 H_2(g) \longrightarrow CH_3OH(\ell)$$

Much has appeared in the popular media about the value to human health of consuming fruits and vegetables or vitamin supplements that contain *antioxidants*. Drugstores now offer a variety of "antioxidant vitamins," supplements that usually contain vitamin E, vitamin C, and vitamin A or beta-carotene (which converts to vitamin A in the body).

An antioxidant is, in chemical definition terms, a reducing agent. It prevents oxidation by reducing a potential oxidizing agent. The oxidizing agents of concern in the body are atoms or molecular fragments known as **free radicals**. A free radical contains an unpaired electron and does not stay around for long without grabbing another electron from a nearby molecule (Figure 10.2). If this molecule happens to have an important function, then that function will be disrupted. For example, if a free radical connects with the part of a DNA molecule that governs cell division (Section 15.11), the result might be abnormal cell division—cancer.

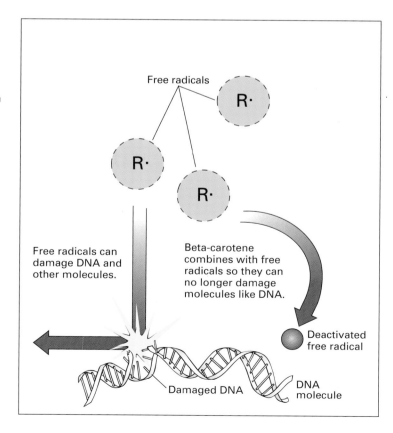

Free radicals

R·

R·

R·

Free radicals can damage DNA and other molecules.

Beta-carotene combines with free radicals so they can no longer damage molecules like DNA.

Deactivated free radical

Damaged DNA

DNA molecule

Free radicals arise in the body from normal biochemical reactions and are also produced in the presence of toxic substances from cigarette smoke, polluted air, and other sources. They are quickly deactivated by picking up an electron to form a less-reactive molecule. Thus, the antioxidants are molecules that in one way or another are able to donate an electron to a free radical before it can do any damage.

To summarize, we have illustrated three ways to recognize oxidation and reduction:

OXIDATION	REDUCTION
Addition of oxygen	Loss of oxygen
Loss of electrons	Addition of electrons
Loss of hydrogen	Addition of hydrogen

Here's another way to look at it:

OXIDIZING AGENTS ARE *REDUCED* BY:	REDUCING AGENTS ARE *OXIDIZED* BY:
Losing oxygen	Gaining oxygen
Gaining electrons	Losing electrons
Gaining hydrogen	Losing hydrogen

EXAMPLE 10.2   Oxidizing and Reducing Agents

For each of the following redox reactions, identify the oxidizing agent and the reducing agent:
(a) $2\,CO(g) + O_2(g) \longrightarrow 2\,CO_2(g)$
(b) $CuO(s) + H_2(g) \longrightarrow Cu(s) + H_2O(\ell)$

SOLUTION
(a) Oxygen is the oxidizing agent and is added to carbon monoxide. The carbon monoxide is the reducing agent.
(b) Copper oxide is the oxidizing agent as shown by its loss of oxygen. Hydrogen is the reducing agent and is oxidized by the addition of oxygen.

TRY IT 10.2
Indicate whether the reactant is being reduced or oxidized in the reactions represented by the following incomplete equations, each of which is missing a key reactant.
(a) $Cu(s) \longrightarrow CuO(s)$
(b) $CH_3C \equiv N\,(g) \longrightarrow CH_3CH_2 - NH_2(g)$
(c) $SnO(s) \longrightarrow Sn(s) + H_2O(g)$

CONCEPT CHECK   10A

1. Which is more oxidized **(a)** CO or **(b)** CO_2?
2. A chemical that causes reduction to take place is called a(n)
   _____ .
3. The conversion of Na to Na^+ is an oxidation reaction. **(a)** True,
   (b) False.
4. When coke (a form of carbon) reacts with iron ore in a blast furnace, the coke is the **(a)** oxidation product, **(b)** reducing agent, **(c)** oxidizing agent.

5. When nitrogen reacts with hydrogen to form ammonia (NH_3), the nitrogen is said to be (**a**) oxidized, (**b**) reduced.

6. A chemical that causes oxidation to occur is a(n) _____.

7. Which is the oxidized form of zinc, (**a**) Zn^{2+} ion, (**b**) Zn atom?

8. In the reaction $2\,Li(s) + S(s) \longrightarrow Li_2S(s)$, sulfur is the oxidizing agent. (**a**) True, (**b**) False.

9. In each pair, which of the two compounds is more reduced?
   (**a**) $HC \equiv N$ or CH_3NH_2
   (**b**) CH_3CH_3 or $H-C \equiv C-H$
   (**c**) CH_3CH_2OH or CH_3CO_2H

10. In chemical definition terms, an antioxidant is a(n) _____.

11. The freeing of a metal from its ore usually requires a chemical reactant that is a(n) _____.

12. A disinfectant frequently is a chemical classified as a(n) _____.

10.4 BATTERIES

A device that produces an electric current from a chemical reaction is called an **electrochemical cell**. Strictly speaking, a **battery** is a series of electrochemical cells. We will, however, stick with the everyday use of "battery" to refer to any device that converts chemical energy to electrical energy.

To function, a battery takes advantage of the relative ease with which metals lose electrons. Consider the reaction that begins to occur as soon as a piece of zinc is placed in a solution of copper ions (Cu^{2+}). The blue color of the copper ions in solution fades as metallic copper is deposited on the surface of the zinc (Figure 10.3). The copper ions are being reduced by gaining electrons and being converted to copper metal.

Since oxidation must accompany reduction, what is being oxidized? That is, what is losing electrons? Careful observation shows that the zinc strip is gradually being consumed in the reaction. The zinc is being oxidized to zinc ions in solution, which are colorless.

The reduction of copper ions by zinc can be thought of as a competition for the available electrons. From observing the reaction in Figure 10.3, it appears that the Zn atoms give up their electrons to the Cu^{2+} ions in the reaction

$$Zn(s) + Cu^{2+}(aq) \longrightarrow Zn^{2+}(aq) + Cu(s)$$

Blue   Colorless

Electrochemical cell A device in which a chemical reaction generates an electric current

Battery A series of electrochemical cells that produces an electric current; commonly refers to any electrochemical, current-producing device

FIGURE 10.3 Oxidation of zinc by copper ions. (a) A piece of zinc is immersed in a solution containing Cu^{2+} ions, which give the solution a blue color. (b) After a few minutes the blue color begins to fade and copper builds up on the remaining zinc. (c) After about an hour, the solution is almost colorless, indicating that most of the Cu^{2+} ions have been reduced to copper atoms, which have formed metallic copper on what is left of the zinc. The Zn^{2+} ions from oxidation of the zinc are colorless.

Charles D. Winters

(a)   (b)   (c)

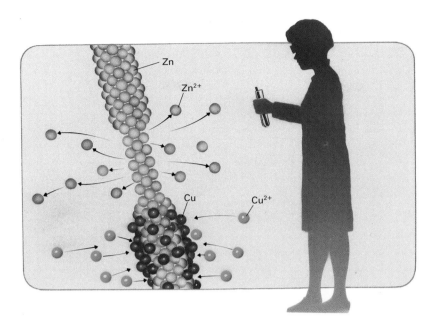

A laboratory experiment with a zinc rod immersed in a solution of copper ions, and the action at the atomic level.

If instead, Cu atoms more easily gave up their electrons to Zn^{2+} ions, the favorable reaction would be the *reverse* of the preceding one. This reaction is unfavorable, however, and does not occur on its own. Apparently, zinc is a stronger reducing agent than copper because it gives up its electrons more easily.

The electrons transferred between a metal that is a good reducing agent and the ion of another metal can provide the electron flow—the *current*—in a battery. A battery is essentially a favorable oxidation–reduction reaction occurring inside a container that has two **electrodes**. At one of these electrodes oxidation takes place as electrons flow out of the cell (it is marked with a − sign). This electrode is the **anode**. At the other electrode, reduction takes place as electrons flow into the cell (it is marked with a + sign). This electrode is the **cathode**. When all the reactants inside the battery are used up and it is impossible to convert the reaction products back to their original form, the battery is "dead" and must be discarded. On the other hand, if the reactants can be converted back to their original form, the battery can be used again.

To remember the difference between an anode and a cathode note the first letters of these words: oxidation occurs at the anode (both vowels); reduction takes place at the cathode (both consonants).

Throw-Away Batteries

The general arrangement of the parts in an electrochemical cell is diagrammed in Figure 10.4(a). The reaction between zinc and copper ions discussed at the beginning of this section provides an example of how such a cell works. The zinc is separated from the copper ion solution and the two are connected so that the reaction proceeds as the electrons are transferred through the connecting wire as shown in Figure 10.4(b). A *salt bridge* allows ions to flow from one electrode chamber to the other. Such a connection is necessary because Zn^{2+} ions are produced at the anode and negative ions must flow into that chamber to prevent positive charge from building up and stopping the reaction. In the other chamber, Cu^{2+} ions are being used up, and positive ions must move into that one for the same reason.

The electrons flow from the Zn anode through the connecting wire and then flow into the copper cathode where reduction of Cu^{2+} ions occurs. The products of the reaction in this simple battery cannot easily be converted to their original form. For this to happen, the copper deposited on the copper

Charles D. Winters

A collection of 1.5 V disposable batteries.

Electrodes Conducting materials at which electrons enter and leave an electrochemical cell

Anode Electrode at which electrons flow out of an electrochemical cell and oxidation takes place

Cathode Electrode at which electrons flow into an electrochemical cell and reduction takes place

FIGURE 10.4 **(a) Components of an electrochemical cell.** Oxidation (loss of electrons) occurs in the anode compartment and electrons flow out of the cell through the external circuit. Electrons re-enter the cell at the cathode, and reduction (gain of electrons) takes place in the cathode compartment. A porous salt bridge (or other connection) allows ions to flow between the two compartments to maintain charge balance. **(b) A simple electrochemical cell.** When the electrodes are connected by a conducting circuit, electrons flow from the zinc electrode, where zinc is oxidized, to the copper electrode, where copper is reduced. The overall reaction in this cell is $Zn(s) + Cu^{2+}(aq) \longrightarrow Zn^{2+}(aq) + Cu(s)$.

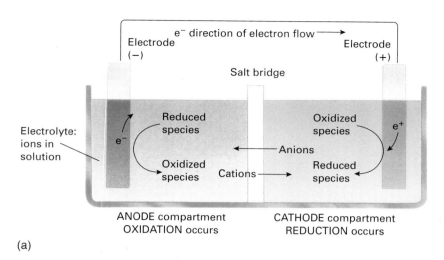

(a)

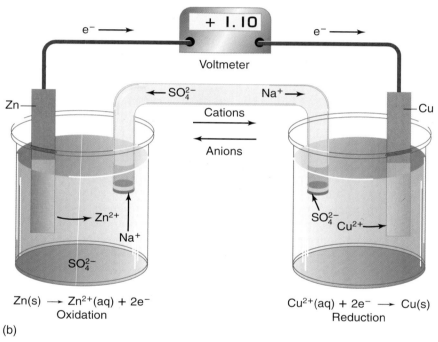

(b)

 See an animation of **Battery Operation** at http://brookscole.com/chemistry/joesten4

 See an animation of the **Voltaic Electrochemical Cell** at http://brookscole.com/chemistry/joesten4

Primary battery A battery that cannot be recharged because its reaction is not easily reversible; a throw-away battery

electrode would somehow have to be dissolved back into solution, and the Zn^{2+} ions would have to be converted back into the zinc metal strip. In other words, because it is not easily reversible, this reaction is best used in a *throwaway* battery, one that cannot be recharged. Batteries of this type are called **primary batteries**.

One of the most common primary batteries today is the *alkaline battery.* Zinc in the presence of potassium hydroxide (KOH, the alkaline substance) serves as the anode. The zinc is separated from the other chemicals (Figure 10.5) by a porous paper, which serves as the salt bridge. The cathode is made of graphite (carbon) combined with manganese dioxide (MnO_2), which is the oxidizing agent. The oxidation at the anode is conversion of Zn to Zn^{2+} (in solid ZnO) and the reduction at the cathode is conversion of Mn^{4+} (in MnO_2) to Mn^{3+} (in Mn_2O_3) so that the overall cell reaction is

$$Zn(s) + 2\,MnO_2(s) \longrightarrow ZnO(s) + Mn_2O_3(s)$$

Reusable Batteries

Some batteries allow the oxidation–reduction reactions at the electrodes to be reversed by the addition of energy, so that the battery can be recharged. Batteries of this type are called **secondary batteries**. Under favorable conditions, secondary batteries may be discharged and recharged many times over.

The lead–acid automobile battery is the most familiar secondary battery. As this battery is discharged, metallic lead is oxidized to lead sulfate at the anode, and the Pb^{4+} in lead dioxide (PbO_2) is reduced at the cathode to the Pb^{2+} in lead sulfate ($PbSO_4$). The reaction takes place in the presence of sulfuric acid (battery acid), and the equations for the reactions are

Oxidation (Anode)
$$Pb(s) + SO_4^{-2}(aq) \longrightarrow PbSO_4(s) + 2\ e^-$$

Reduction (Cathode)
$$PbO_2(s) + 4H^+(aq) + SO_4^{-2}(aq) + 2\ e^- \longrightarrow PbSO_4(s) + 2\ H_2O(\ell)$$

Net Reaction
$$Pb(s) + PbO_2(s) + 2\ H_2SO_4(aq) \longrightarrow 2\ PbSO_4(s) + 2H_2O(\ell)$$

The lead–acid battery (Figure 10.6) is reusable because the lead sulfate formed at both electrodes is insoluble and stays on the electrode surface. Then when the battery needs recharging, the lead sulfate is available for the reverse reaction.

Recharging a secondary battery requires reversing the direction of electron flow through the battery, which can be accomplished by a generator or an alternator. When recharging occurs, the reactions at the two electrodes are reversed

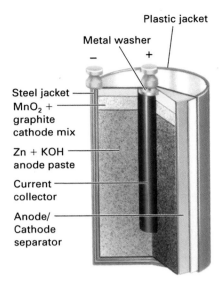

FIGURE 10.5   An alkaline battery. This type of throw-away battery provides a constant voltage of 1.54 V throughout its useful life.

© 1990 Richard Megna/Fundamental Photographs

An assortment of batteries. In front of the lead–acid automobile battery are, from left to right, three types of rechargeable nickel–cadmium batteries, three types of nonrechargeable alkaline batteries, and a zinc–graphite dry cell (also nonrechargeable).

Anode: electric current flows out from the anode upon discharge as energy is being consumed by some other source.

Cathode: electric current flows into the cathode upon charging as energy is being supplied to the battery.

Negative plates: lead grids filled with spongy lead

Positive plates: lead grids filled with PbO_2

Sulfuric acid solution

FIGURE 10.6   Lead–acid battery. The anodes are lead grids filled with spongy lead. The cathodes are lead grids filled with lead(IV) oxide. PbO_2. Each cell produces a potential of about 2 V. Six cells connected in series produce the desired overall battery voltage. Most lead batteries have a useful life of three years or less. The lead in these batteries can be a health hazard and thus there are stringent environmental requirements concerning their disposal.

Secondary battery A battery that can be recharged by reversing the flow of current, which reverses the current-producing reaction and regenerates the reactants

Discharge in a battery when starting car:

chemical energy $\longrightarrow$ electrical energy

Recharging a battery when car is running:

electrical energy $\longrightarrow$ chemical energy

Normal recharging of an automobile lead–acid battery occurs during driving. The voltage regulator senses the output from the alternator, and when the alternator voltage exceeds that of the battery, electrical energy is added back into the battery and the battery is recharged. During the recharging cycle in a lead–acid battery, some water is reduced to hydrogen at the cathode, and some water is oxidized to oxygen at the anode. The result is a potentially explosive mixture of hydrogen and oxygen at the top of the battery. Under normal driving conditions, automobile batteries do not explode; however, internal short circuits can produce explosions in older batteries. Batteries must always be protected from sparks.

All in all, the lead–acid battery is relatively inexpensive, reliable, and simple, and has an adequate life. Its high weight is its major fault. Newer secondary batteries have found use in some applications such as electronics, but none of these newer batteries can perform like the lead–acid battery does for its cost.

Fuel Cells

Normal batteries have the advantage of being relatively small, which makes them easily inserted, removed, and transported from place to place. Such batteries are limited in the amount of current they produce by the amount of the reagents inside the battery. When the oxidizable reagent in the battery is consumed, the battery is dead unless it is a rechargeable battery. One way to overcome this problem is to use fuel cells which, like batteries, have an electrode where oxidation takes place and an electrode where reduction takes place. However, fuel cells do not depend on chemicals stored inside the electrode compartments for their energy. Fuel cells produce energy from reactants that continuously flow into their compartments while the chemical reaction products flow out of them.

One type of fuel cell has been used in the space program on board the Gemini, Apollo, and space shuttle missions. On April 13, 1970, a dramatic explosion in oxygen tank 1 aboard the Apollo 13 spacecraft triggered the release of oxygen from tank 2. In addition to its use for breathing, the oxygen was necessary for the operation of the fuel cells used to produce electricity and water for the command module and the astronauts. This incident, brought to the movie screen in *Apollo 13*, led to the most dramatic astronaut rescue mission to date.

In these fuel cells, hydrogen gas is pumped onto the anode of the fuel cell, and oxygen gas is pumped onto the cathode, where the following reactions occur:

Oxidation at the anode: $2\,H_2(g) \longrightarrow 4\,H^+ + 4\,e^-$

Reduction at the cathode: $O_2(g) + 4\,H^+ + 4\,e^- \longrightarrow 2\,H_2O(g)$

Net reaction: $2\,H_2(g) + O_2(g) \longrightarrow 2\,H_2O(g) + energy$

A proton exchange membrane (PEM) separates the two halves of a fuel cell. This membrane allows protons, formed by the oxidation of hydrogen gas, to pass through and react with hydroxide ions, produced by reduction of oxygen gas at the cathode, to form water. The "waste product" of this fuel cell is simply water vapor that can be condensed for use, for instance, by astronauts.

Fuel cells deliver as much power as batteries weighing ten times as much. On a typical seven-day mission, the space shuttle fuel cells consumed 1500 pounds of hydrogen and generated 190 gallons of water.

Fuel cell technology applied to cars is currently an area of much interest. The oxygen for such a cell would come from the air, but a car which had to store and transport the hydrogen needed to operate the fuel cell would be problematic due to the explosive nature of hydrogen gas and the technological problems of storing such a light gas. Such cars have been shown to be possible, but probably not practical.

A potential solution to the hydrogen storage problem would be to generate the required hydrogen at the point of usage. Much research is currently underway to develop processes capable of converting gasoline to hydrogen and carbon monoxide by the use of appropriate catalysts. The hydrogen could then be used in a fuel cell (with oxygen supplied from the air) to produce electricity to power the car. The carbon monoxide would be converted to carbon dioxide over another catalyst so that the emissions of the car would be water vapor and carbon dioxide. This would create a so-called "pollution-free" car, although both these emission gases do contribute to global warming. Any carbon-based fuel will ultimately lead to the production of carbon dioxide, so this solution does not totally address the problem it is meant to solve, the generation of carbon dioxide and the consumption of hydrocarbon-based fuels. However, it is a start.

Gasoline–electric hybrid automobiles have been the focus of much industry effort in recent years. These *hybrid cars* rely on a combination of gasoline-generated and battery-generated electricity. Honda was the first to bring this type of car to the market. When idling or moving at low speed, this car uses its electric motor. At normal speeds, the gasoline engine and the electric motor both contribute energy. As the car slows down, the wheels drive a generator that returns energy to storage in the battery. These cars need not be plugged in for recharging and can get up to 60 miles per gallon of gasoline. As of 2005, Toyota and Honda were the major suppliers of such cars in the United States, although several other manufacturers, including Chevrolet, Ford, Lexus, Mazda, Nissan, and Saturn will introduce 2006 hybrid vehicles. These new models will include pickup trucks and some SUVs.

Fuel cells that do not depend on hydrocarbon-based fuels for the generation of hydrogen are far too costly at present to be practical, and there are still many technological problems to overcome before we see hydrogen-fueled cars lined up in the parking lots of local schools waiting to pick up children.

The 2005 Honda Civic Insight is a gasoline-electric hybrid. The energy needed to power the car comes from gasoline. The gasoline engine charges the battery that runs the electric motor. The gasoline engine switches off when the car is stopped or operating at low speeds.

10.5 ELECTROLYSIS: CHEMICAL REACTIONS CAUSED BY ELECTRON FLOW

Chemical reactions that are unfavorable can be forced to proceed by the input of energy, as in the recharging of a secondary battery. **Electrolysis reactions** are oxidation–reduction reactions driven by electrical energy from an external power supply. Where electrons flow into the cell, the electrode becomes negatively charged, positive ions in solution migrate toward that electrode, and reduction takes place. In this type of cell (Figure 10.7), the electrodes need not be in separate compartments.

The metal in some ores is so resistant to reduction that few reducing agents are strong enough to cause the reaction, and electrolysis is a good

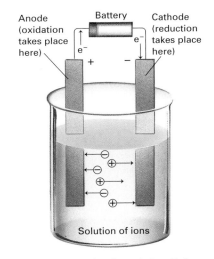

FIGURE 10.7 An electrolysis cell. For electrolysis, the electrodes need not be in separate compartments. Electrons enter the cell at the cathode, where reduction occurs. Ions flow through the electrolyte solution to maintain charge balance. At the anode, where oxidation occurs, electrons leave the cell.

Electrolysis reaction A redox reaction driven by electrical energy from an external power supply

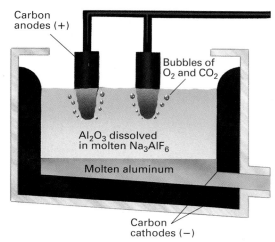

FIGURE 10.8 **Aluminum production by electrolysis.** At the cathode, aluminum ions are reduced to aluminum metal. The anode reaction is production of oxygen gas, which reacts with the carbon anodes. (The cell reaction is $2\ Al_2O_3 + 3\ C \longrightarrow 4\ Al + 3\ CO_2$.) Molten aluminum is denser than the molten salt mixture in the cell and collects at the bottom.

In electrolysis, electrical energy produces chemical change. In batteries, chemical change produces electrical energy.

The original top of the Washington Monument was aluminum, made in 1884 by the sodium-reduction method.

alternative way to provide enough energy for the reaction. Most importantly, aluminum is a metal of this type. Aluminum, in the form of Al^{3+} ions, is the third most abundant element in Earth's crust (7.4%). The quantity and commercial value of aluminum used in the United States each year is exceeded only by that of iron and steel.

From the time of its discovery in 1825 until near the turn of the century, aluminum was made by reducing $AlCl_3$ with a more active metal (potassium or sodium) but only at a very high cost. Even though there was commercial production of aluminum by 1854, aluminum was considered to be a precious metal—as gold and platinum are today—and one of its early uses was for jewelry. Napoleon III saw the possibilities of aluminum for military use, however, and commissioned studies on improving its production. The French had a ready source of aluminum-containing ore, bauxite, named for the French town of Les Baux. In 1886, a 23-year-old Frenchman, Paul Heroult, conceived the electrochemical method that is still in use today. In an interesting coincidence, an American, Charles Hall, who was 22 at the time, announced his invention of the identical process in the same year. Hence, the commercial process is now known as the Hall–Heroult process.

In the Hall–Heroult process, purified aluminum oxide from bauxite is dissolved in a molten ionic compound (cryolite, Na_3AlF_6). The mixture is electrolyzed in a cell with carbon anodes and a carbon cell lining that serves as the cathode (Figure 10.8).

Because aluminum is such an active metal, separating it from its oxide requires a lot of energy. This, combined with the energy needed to maintain the molten cryolite bath, makes aluminum production highly energy intensive. One reason for the success of aluminum recycling is the large saving in energy cost—making aluminum beverage cans from recycled aluminum requires only 5% of the energy used in making the cans from new aluminum, and the process is about 20% cheaper overall.

Another practical application of electrolysis is *electroplating*—the coating of one metal, which is made the cathode in an electrolysis cell, with another metal, which is made the anode. For example, dull iron or steel surfaces are electroplated with chromium to provide a mirror-like finish and also to protect against corrosion.

10.6 CORROSION: UNWANTED OXIDATION–REDUCTION

In the United States alone, more than $10 billion is lost each year to **corrosion**—the unwanted oxidation of metals during exposure to the environment. Much of this corrosion is the rusting of iron and steel, although other metals may corrode as well. The oxidizing agent causing all this unwanted corrosion is usually oxygen. Iron is most severely affected by corrosion because rust does not adhere strongly to the metal's surface. The continuing loss of surface iron as rust forms and then flakes off eventually causes structural weakness.

The driving forces behind metal corrosion are the activity of the metal as a reducing agent and the strength of the oxidizing agent. Whenever a strong reducing agent (the metal) and a strong oxidizing agent (such as oxygen) are in contact, a reaction between the two substances is likely. Factors governing the rates of chemical reaction, such as temperature and concentration (Section 8.3), affect the rate of corrosion as well. Consider the corrosion of an iron spike (Figure 10.9). There are tiny microcrystals of loosely bound iron atoms on the surface of the metal that can easily be oxidized.

$$Fe(s) \longrightarrow Fe^{2+}(aq) + 2\,e^-$$

Because iron is a good conductor of electricity, the electrons produced by this oxidation can migrate through the metal to a point where they can reduce something. The fact that iron is a conductor of electricity is important because corrosion would come to an abrupt halt as a result of a buildup of excessive negative charge if the electrons were not conducted away. One location on the surface of the iron where electrons can be used is any tiny drop of water containing dissolved oxygen. Here the oxygen gains electrons, forming hydroxide ions.

$$O_2(g) + 2\,H_2O(\ell) + 4\,e^- \longrightarrow 4\,OH^-(aq)$$

The Fe^{2+} ions are further oxidized to Fe^{3+} ions, which react with OH^- ions to form the hydrated iron oxide known as rust.

$$2\,Fe^{3+}(aq) + 6\,OH^-(aq) \longrightarrow Fe_2O_3 \cdot 3H_2O(s)$$
Rust

The rate of rusting is enhanced by salts, which dissolve in the water on the surface of the iron. The hydroxide ions and iron ions migrate more easily in the ion solutions produced by the presence of the dissolved salts. Automobiles rust more quickly when exposed to road salts in wintry climates. If road salts are used in your driving area, it's a good idea to wash the underside of your automobile after the snowy season ends to remove the accumulated salts.

For rust to form, three reactants are necessary. These are iron, oxygen, and water. Rusting can be prevented by protective coatings such as paint, grease, oil, enamel, or a corrosion-resistant metal such as chromium. Most of these coatings keep out moisture. Some of the metals that are more active than iron form adherent oxide coatings when they corrode. Coating iron with these metals provides corrosion protection. One of these active metals is zinc. When the zinc coating of a galvanized object is exposed to air and water, a thin film of zinc oxide forms that protects the zinc from further oxidation. If the zinc coating should get scratched so that iron is exposed to the air, zinc will quickly reduce any Fe^{2+} ions formed because zinc is more active than iron in giving up electrons.

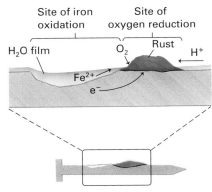

FIGURE 10.9 Corrosion of iron. The site of iron oxidation may be different from the site of oxygen reduction because iron is a conductor of electricity and electrons can move through it from one site to another.

See an animation of the **Rusting of Iron** at http://brookscole.com/chemistry/joesten4

Corrosion has made this bridge unsafe.

Corrosion The unwanted oxidation of metals during exposure to the environment

CONCEPT CHECK 10B

1. A battery contains two electrodes, named the _____ and the _____.
2. A battery is dead when _____.
3. It is necessary for _____ to flow between the electrode compartments in a battery so that _____ balance is maintained.
4. The conversion of metallic zinc to zinc ions (Zn^{2+}) is _____ and will occur at the _____ in an electrochemical cell.
5. The conversion of copper ions (Cu^{2+}) to metallic copper is _____ and will occur at the _____ in an electrochemical cell.
6. In recharging a secondary battery, _____ energy is converted to _____ energy.
7. The reduction of bauxite ore by the process of _____ produces _____ metal.
8. Why is rusting of iron an especially destructive form of corrosion?
9. Name the three reactants needed for rusting of iron.

 Assess your understanding of this chapter's topics with an online chapter quiz at www.brookscole.com/chemistry/joesten4

■ KEY TERMS

oxides

combustion

oxidation

reduction

oxidation–reduction reactions
(redox reactions)

oxidizing agent

bleaching agent

reducing agent

coke

free radical

electrochemical cell

battery

electrodes

anode

cathode

primary battery

secondary battery

electrolysis reaction

corrosion

■ THE LANGUAGE OF CHEMISTRY

1. Chemicals in automobile lead storage battery
2. Reduction of Cu^{2+} ions produces this
3. Oxidizing agent used to purify drinking water
4. Oxidation
5. Reduction
6. Gain of hydrogen
7. Oxidized form of sulfur
8. Oxidized form of nitrogen
9. Allows ions to pass from one electrode compartment in a battery to the other
10. Fuel used in some fuel cells
11. Battery used in most automobiles

a. PbO_2, Pb, and H_2SO_4
b. Hydrogen
c. Gain of oxygen
d. Reduction
e. SO_2
f. NO_2
g. Loss of oxygen
h. Lead–acid battery
i. Chlorine
j. Copper metal
k. Salt bridge

■ APPLYING YOUR KNOWLEDGE

1. Define the following terms:
   (a) Oxidation in terms of electrons
   (b) Reduction in terms of electrons
   (c) Oxidation in terms of oxygen
   (d) Reduction in terms of hydrogen
   (e) Oxidizing agent
   (f) Reducing agent

2. Give definitions for the following terms:
   (a) Anode   (b) Primary battery
   (c) Secondary battery   (d) Cathode

3. Identify the more highly oxidized substance in each of the following:
   (a) CO or CO_2 (b) NO_2 or NO
   (c) SO_3 or SO_2 (d) N_2 or NH_3

4. Identify two applications of the oxidizing properties of chlorine, Cl_2.

5. Which is the more common oxide of hydrogen, H_2O or H_2O_2?

6. Describe how combustion and oxidation differ.

7. Which oxide of carbon, CO or CO_2, is produced in incomplete combustion?

8. Which is more highly oxidized, a potassium atom or a K^+ ion?

9. Tell whether the underlined atoms in the following equations are oxidized or reduced. Justify your answers.
   (a) $2\ \underline{H}_2(g) + O_2(g) \longrightarrow 2\ H_2O(\ell)$
   (b) $2\ \underline{Cu}(s) + O_2(g) \longrightarrow 2\ CuO(s)$
   (c) $2\ \underline{Sn}O(s) \longrightarrow 2\ Sn(s) + O_2(g)$
   (d) $\underline{Fe}_2O_3(s) + 3\ C(s) \longrightarrow 2\ Fe(s) + 3\ CO(g)$
   (e) $\underline{N}_2(g) + 3\ H_2(g) \longrightarrow 2\ NH_3(g)$

10. Why can Earth's atmosphere be described as an oxidizing atmosphere?

11. Would Jupiter's atmosphere, which is rich in hydrogen (H_2) and methane (CH_4), be described as an oxidizing or a reducing atmosphere?

12. The main source of magnesium is seawater, which contains Mg^{2+} ions. Is this magnesium in an oxidized or reduced form?

13. When zinc metal reacts with an acid, Zn^{2+} ions form. Is the zinc metal oxidized or reduced?

14. Tell which substance is reduced and which is oxidized in the following equations. Label the oxidizing agent and reducing agent in each equation.
    (a) $2\ Al(s) + 3\ Cl_2(g) \longrightarrow 2\ AlCl_3(s)$
    (b) $S(s) + O_2(g) \longrightarrow SO_2(g)$
    (c) $CuO(s) + H_2(g) \longrightarrow Cu(s) + H_2O(g)$
    (d) $C_2H_4(g) + H_2(g) \longrightarrow C_2H_6(g)$

15. Why would automobile rusting be less of a problem in Arizona than in Chicago?

16. Would you expect a piece of iron to rust on the surface of the moon? Explain.

17. Oxidation always occurs at the anode of an electrochemical cell. (a) True (b) False

18. Describe a simple battery, naming three essential parts.

19. How is a fuel cell similar to a battery? How is it different?

20. Besides electricity, what do the fuel cells used on the space shuttle produce?

21. A brand-new battery has a mass of 249.6 g. After it has been fully discharged its mass is still 249.6 g. Explain this.

22. Think of ways electricity might be distributed to consumers of electric automobiles. Compare all the methods you have thought of with the methods used to distribute hydrocarbon fuels (gasoline). List as many benefits and problems as you can for each.

23. Give an advantage and disadvantage of an electric car that must be recharged by electricity from a fossil-fuel electric power plant.

24. Electric cars were popular in the early 1900s. Why were they replaced by cars that used an internal combustion engine?

25. A cell in a lead storage battery produces 2 volts, but the electrical system of a car operates at 12 volts. How can lead storage battery technology generate the 12 volts?

26. An alkaline battery produces 1.5 volts, but two batteries connected end to end produce 3.0 volts. What is happening?

27. Oxygen in the air reacting with substances in food can cause spoilage. Why are antioxidants added to food?

28. Free radicals are molecules or molecular fragments with unpaired electrons. Which of the following is a free radical — NO; CO; O_3? Justify your answer.

29. Free radicals are molecules or molecular fragments with unpaired electrons. Which of the following is a free radical — NO_2; CO_2; N_2O? Justify your answer.

30. Normally, the mass of a new battery is the same as its mass after it is fully discharged. Explain why this is true.

31. The mass of a fuel cell is constant, but the mass of reactant and product changes during operation of the cell. Explain why this is true.

32. Aqueous solutions of Cu^{2+} ion are blue. The higher the concentration, the darker the color. When a strip of zinc metal is immersed in a solution of Cu^{2+}, the blue color fades. What is happening?

33. Aqueous solutions of Cu^{2+} ion are blue—the higher the concentration, the darker the color. When a strip of copper metal is immersed in a solution of Cu^{2+}, the color stays the same. Why does this happen?

34. In addition to iron, what reactants are needed for the formation of rust?

35. Why does iron continue to corrode even after a layer of rust forms on the surface, whereas aluminum stops corroding after a corrosion layer forms?

36. Why is galvanized iron less susceptible to corrosion than bare iron?

37. The equation for the Hall–Heroult process to convert bauxite to pure Al is

$$2\ Al_2O_3 + 3\ C \longrightarrow 4\ Al + 3\ CO_2$$

What is oxidized in the reaction and what is reduced?

W CHEMISTRY ON THE WEB

For up-to-date URLs, visit the text website at **www.brookscole.com/ chemistry/joesten4**

- Fireworks
- Oxidation and Reduction
- Visualization of Some Oxidation–Reduction Reactions
- Oxidation and Reduction in the Internal Combustion Engine
- Battery Basics and Fuel Cells
- Electric Cars
- How Fuel Cells Work

WATER, WATER EVERYWHERE, BUT NOT A DROP TO DRINK?

Water, a unique liquid, is essential for life!

© Marvy!/Corbis

A single molecule of water consists of only three atoms. Yet water is so critical to life as we know it that the search for water on other planets has become a major focus of our efforts to determine if life is unique to Earth. How can it be that such a common and simple molecule, two hydrogen atoms covalently bonded to a central oxygen atom, plays such a major role in our world? Are the chemical and physical properties of water so significantly different from molecules of similar size and molecular weight as to make liquid water a key to survival of living organisms? Furthermore, why is there so much concern about the availability of pure water when more than 70% of Earth's surface is covered with water?

Too little water and we die of thirst. Too much water and we drown. Parts of Earth where water is in short supply become arid deserts, and life is difficult.

If too much water is present, normal life for land dwellers becomes impossible and aquatic life takes over. Clearly there is a healthy balance between too little and too much water, but the quality of the water, and even its physical state, are also critical. Although liquid water is abundant in the oceans, the much purer water on which many of life's activities depends is, all things considered, in relatively short supply. There is increasing evidence that the days when cattle ranchers fought over water supplies in the Old West may return to haunt us in ways people would never have imagined in those days. As Earth's population has exceeded six billion people, the demand for water for human consumption and for the activities associated with our modern lifestyles has become enormous. There is not enough pure water to satisfy the worldwide demand, and much of the pure water isn't in the areas of the world where it is needed.

Water dissolves all kinds of substances and reacts chemically with many of the atmospheric pollutants we've encountered in earlier chapters. Several nitrogen oxides and sulfur oxides react readily with water in the atmosphere to produce acidic rainfall that can have a very negative impact on our environment and on many building materials. Dissolved substances in liquid water can include natural minerals and chemicals, manmade fertilizers and pesticides, gasoline and fuel additives, toxic chemicals of all types, and even unused antibiotics that people have flushed down the toilet. To minimize environmental damage and to ensure a safe and adequate water supply, we must limit all types of pollution. We must control both the amount of pollution and the types of pollution to which we subject our water supply because what we've got is all there is. The water must be used, reused, and used again, over and over, in a never-ending cycle. If it becomes polluted, we will have to clean it up.

In this chapter we shall look at the answers to these questions:

- What are some of the unique properties of water?
- How can there be a water shortage if water is such an abundant compound?
- What is acid rainfall, and what are some of its effects?
- How does acid rainfall differ from other forms of water pollution?
- How is water used and reused ?
- What is the difference between polluted water and clean water?
- How is water pollution measured?
- What are some of the causes of water pollution?
- How can water be purified?
- How can we ensure that our water supply remains sufficient and pure enough for our needs?
- Will we ultimately be able to use the vast quantities of ocean water to meet demand?

11.1 THE UNIQUE PROPERTIES OF WATER

We explored some of the properties of water in Chapter 5 and are familiar with the shape of the molecule (Section 5.4) and the polar nature of the covalent oxygen-hydrogen bonds in the molecule (Section 5.5). We have seen that water is capable of intermolecular hydrogen bonding (Section 5.7). These features give rise to some unique physical properties of water in comparison to some other small molecules of similar size, number of atoms, and molecular weight.

TABLE 11.1 — Comparison of Water to Some Similar Sized Molecules

COMPOUND PROPERTY	WATER (H_2O)	AMMONIA (NH_3)	METHANE (CH_4)	NITROGEN (N_2)	OXYGEN (O_2)	CARBON DIOXIDE (CO_2)
Molecular Weight (g/mol)	18	17	16	28	32	44
State at Room Temperature	liquid	gas	gas	gas	gas	gas
B.p. (°C)	100	−33.4	−161.5	−195.8	−183	sublimes
M.p. (°C)	0	−77	−182.5	−210	−218.8	−56.6 @ 5.2 atm

In Table 11.1 we compare water with ammonia, methane, nitrogen, oxygen, and carbon dioxide. All these molecules contain between two and five atoms and have molecular weights between 16 g/mol and 44 g/mol. An inspection of the data in Table 11.1 shows that is where the similarities end. Water is the only one of these compounds whose physical state at room temperature is a liquid! Its boiling point is hundreds of degrees higher than the others. The melting point of water is also seen to be dramatically different than the others. Of this set of molecules, only ammonia has a boiling point and a melting point that are even remotely close to those of water. A review of the Lewis structures of these molecules reveals that water and ammonia are the only two molecules in this set that are capable of intermolecular hydrogen bonding. These physical properties also reveal that water is the only compound that exists naturally in all three physical states (solid, liquid, and gas) under normal conditions on Earth. Water appears uniquely suited for its role as the liquid essential to life on Earth.

As you read about the properties of water, try to imagine how these properties influence your daily activities and how different your lives would be were any of these properties significantly different.

Some Properties of Water

1. ***Water is a liquid at room temperature as a direct consequence of hydrogen bonding between adjacent water molecules.*** Pure water is a liquid between 0°C and 100°C.

2. ***The density of solid water (ice) is less than that of liquid water.*** Put another way, water expands when it freezes. If ice were a typical solid, it would be denser than liquid water, and lakes would freeze from the bottom up. This would have disastrous consequences for marine life, which could not survive in areas with winter seasons. The application of pressure causes ice to melt. This is a consequence of the structure of ice. The pressure causes ice to change to a form with a smaller volume, and since liquid water occupies a smaller volume than does ice, the ice converts to a liquid.

3. ***The* heat of fusion *(melting) of ice is 80 cal/g.*** This is the amount of heat required to melt 1 g of ice at 0°C into 1 g of liquid water at 0°C.

4. ***Water has a relatively high heat capacity.*** Water can absorb large quantities of heat without large changes in temperature, because the added heat can break hydrogen bonds instead of increasing the temperature. For comparison, the heat capacity of water (1.00 cal/g-°C, 4.18 J/g-°C) is about ten times that of copper (0.092 cal/g-°C, 0.385 J/g-°C) or iron (0.106 cal/g-°C, 0.444 J/g-°C). Water's heat capacity accounts for the

Charles D. Winters

Water in all three phases. Fog and mist are suspended water droplets. Water vapor cannot be seen.

The heat of fusion of water is 80 cal/g at its melting point.

The specific heat capacity of water is 1 calorie per gram per degree Celsius (1 cal/g-°C).

Specific heat capacity The amount of heat required to raise the temperature of a 1 g sample of matter of a given size by 1°C

Heat of fusion The heat required to melt a given quantity of a solid at its melting point

moderating influence of lakes and oceans on the climate. Huge bodies of water absorb heat from the Sun and release the heat at night or in cooler seasons. Earth would have extreme temperature variations if it were not for this property of water. By contrast, the temperatures on the surface of the Moon and the planet Mercury vary by hundreds of degrees through the light and dark cycles.

5. ***Water has a high heat of vaporization.*** For a liquid to vaporize, heat is required. The **heat of vaporization** of a liquid is a measure of the intermolecular attractions holding the molecules together in the liquid. Water has one of the highest heats of vaporization. A consequence of this high heat of vaporization is the cooling effect that occurs when water evaporates from moist skin. Evaporating water molecules take with them a considerable amount of energy, which was needed to overcome the attractions between the leaving molecules and those remaining behind.

6. ***Water has a high surface tension.*** Unlike gases, liquids have surface properties, and these are extremely important in the overall behavior of many liquids. Molecules beneath the surface of the liquid are completely surrounded by other molecules and experience forces in all directions due to intermolecular attractions. By contrast, molecules at the surface are attracted only by molecules below or beside them (Figure 11.1). This unevenness of attractive forces at the surface of the liquid causes the surface to contract, making it act like a skin. The energy required to overcome the "toughness" of this liquid skin is called the **surface tension**, and it is higher for liquids that have strong intermolecular attractions. Water's surface tension is high compared with those of most other liquids because of the extensive hydrogen bonding that holds water molecules to each other.

7. ***Water is an excellent solvent, often referred to as the universal solvent.*** Because it is such a good solvent, water from natural sources is not pure water but is instead a solution of substances dissolved by contact with water.

The heat of vaporization of water at 100°C is 540 cal/g.

Benzene, chloroform, ethyl alcohol, and octane—all organic compounds that are liquids at room temperature—have surface tensions about one-third as strong as that of water.

Heat of vaporization The heat required to vaporize a given quantity of liquid at its boiling point

Surface tension The amount of energy required to overcome the attraction for one another of molecules at the surface of a liquid

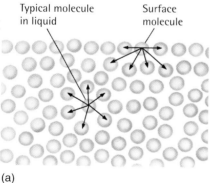

Typical molecule in liquid   Surface molecule

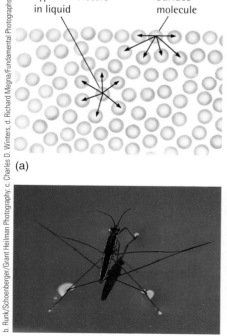

(a)

FIGURE 11.1   Surface tension. (a) The surface is strengthened by intermolecular forces attracting surface molecules. (b) The water strider, a lightweight insect that does not provide enough force per unit area to break through the surface tension. Note that the strider does not walk on the sharp ends of its "toes." (c) With care, a paper clip can be placed so that it won't sink in the water. (d) On a dirty car, this wouldn't happen. The dirt would overcome the surface tension of the water droplets.

(c)

(d)

(b)

b. Runk/Schoenberger/Grant Heilman Photography; c. Charles D. Winters; d. Richard Megna/Fundamental Photographs

> **EXAMPLE** **11.1**   **Heat Associated with Phase Changes of Water**

Calculate the total amount of heat that would be required to convert 5.00 g of ice at 0°C to 5.00 g of steam at 100°C.

SOLUTION

We need to know the following information to solve this problem:

$$\text{Heat of fusion of ice} = 80.0 \text{ cal/g}$$
$$\text{Heat capacity of water} = 1.00 \text{ cal/g-°C}$$
$$\text{Heat of vaporization of water} = 540 \text{ cal/g}$$

The heat input required to convert the solid ice at 0°C to liquid water at 0°C is found by multiplying the mass of ice (5.00 g) by the heat of fusion of ice (80 cal/g) to give 400 cal. This gives 5.00 grams of liquid water at 0°C whose temperature must be increased to 100°C before vaporization can occur. The heat required to increase the temperature is found by multiplying the mass (5.00 g) times the heat capacity of water (1.00 cal/g-°C) times the total temperature change (100°C). The temperature change requires a total of 500 cal. Lastly, the heat required to convert the 5.00 g of water at 100°C to 5.00 g of steam at 100°C is simply the mass (5.00 g) times the heat of vaporization (540 cal/g). The vaporization requires 2,700 cal. Thus, the total heat input required (an endothermic process) is 400 cal plus 500 cal plus 2,700 cal or 3,600 cal.

By analogy, the reverse of this process, conversion of 5.00 g of steam at 100°C to 5.00 g of ice at 0°C would release 3,600 cal in an exothermic process.

TRY IT 11.1

Determine the amount of heat involved in the following conversions and indicate whether the process is endothermic or exothermic.
(**a**) Conversion of 3.00 g of ice at 0°C to 3.00 g of liquid water at 70°C
(**b**) Conversion of 6.00 g of water at 50°C to 6.00 g of steam at 100°C
(**c**) Conversion of 10.0 g of steam at 100°C to 10.0 g of liquid water at 0°C

As examples of the importance of water, consider that we need liquid water for drinking, solid water for ice skating and cooling, and for water vapor helping regulate Earth's temperature in its role as a greenhouse gas. Were the density of solid water (ice) greater than that of liquid water, then lakes would freeze from the bottom up, ice skating would be more challenging, and the *Titanic* would likely have survived her maiden voyage since the iceberg that sunk her would have been at the bottom of the ocean. We'd have to more carefully monitor a full glass of iced liquid since it would overflow as the ice melted. Water's high heat capacity and high heat of vaporization are taken advantage of in many chemical and industrial processes in which water is used to regulate temperature. Humans keep cool by sweating because the evaporating water takes up heat from the body. Citrus farmers frequently spray their crops with water on nights the temperature is expected to hover very near freezing since the heat released as water freezes may be just enough to protect the fragile citrus from damage. The frozen water also serves as an insulating blanket. Water's surface tension explains why water drops are spheres and soap bubbles are round.

Water's properties as a universal solvent are both a blessing and a curse. Water is incredibly useful as a solvent for so many solutes, and we depend on this in innumerable ways. By the same token, water's ability to dissolve so many substances makes it very easily polluted. Since pure water of relatively neutral pH is critical to many human activities it will be useful to look at some of the many ways in which water is polluted and the procedures for purification of water.

11.2 ACID RAINFALL

The term **acid rain** was first used in 1872 by Robert Angus Smith, an English chemist and climatologist. He used the term to describe the acidic precipitation that fell on Manchester, England, at the start of the Industrial Revolution. Although neutral water has a pH of 7, rainwater becomes naturally acidified from dissolved carbon dioxide, a normal component of the atmosphere. The carbon dioxide, whose solubility in water at room temperature is 3.4 g/L reacts reversibly with water to form a solution of the weak acid carbonic acid.

$$CO_2\,(g) + H_2O(\ell) \rightleftharpoons H_2CO_3\,(aq)$$
$$H_2CO_3\,(aq) + H_2O(\ell) \rightleftharpoons H_3O^+ + HCO_3^-\,(aq)$$

At equilibrium, the pH of rainfall saturated with CO_2 from the air is 5.6. Any precipitation with a pH below 5.6 is considered to be acid rain.

As you have seen, NO_2, SO_2, and SO_3 can both react with water in the atmosphere to produce acids: NO_2 produces nitric acid (HNO_3) and nitrous acid (HNO_2); SO_3 and SO_2 produce sulfuric acid (H_2SO_4) and sulfurous acid (H_2SO_3). When conditions are favorable, water droplets carrying these acids precipitate as rain or snow with a low pH. Ice core samples taken in Greenland and dating back to 1900 contain sulfate (SO_4^{2-}) and nitrate (NO_3^-) ions. This indicates that at least from 1900 onward, acid rain has been commonplace.

Acid rain is a problem today due to the large amounts of acidic oxides being produced by human activities. When this precipitation falls on natural areas that cannot easily tolerate such acidity, serious environmental problems occur. The average annual pH of precipitation falling on much of the northeastern United States and northeastern Europe is between 4 and 4.5. Specific rainstorms in some areas where there are numerous sources of SO_2 and NO_x have had pH values as low as 1.5. Complicating matters further is the fact that acid rain is an international problem—rain and snow don't observe borders (Figure 11.2). Many Canadian residents are offended by the government of the United States because some of the acid rain produced in the United States falls on Canadian cities and forests (Figure 11.3).

The extent of the problems with acid rain can be seen in "dead" (fishless) ponds and lakes, dying or dead forests, and crumbling buildings. For the most part, dead lakes are still picturesque, but no fish can live in the acidified water. Lake trout and yellow perch die at pH values below 5.0, and smallmouth bass die at pH values below 6.0. Mussels die when the pH is below 6.5.

Reversibility and equilibria are discussed in Section 8.4. Weak acids are discussed in Section 9.2. See Figure 9.2 for a review of pH values.

Because dry, acid-containing particles can also settle out of the air, a more adequate term than *acid rain* is *acid deposition*.

The March 1991 eruption of Mt. Pinatubo in the Philippines injected more than 10^8 kg of SO_2 into the stratosphere. The SO_2 eventually came down as acid rain. During the period from 1991 to 1994, aerosol particles from these eruptions enhanced the beauty of sunrises and sunsets by scattering sunlight more than normal.

Effects of acid rain on a forest in one of the most polluted parts of Europe. The devastation has been caused by emission of sulfur dioxide and nitrogen oxides from factories in the former East Germany and Czechoslovakia.

Acid rain Rain that has an unusually low pH (below 5.6)

FIGURE 11.2 Air pollution can cause acid rain to fall far from where the pollutants are generated.

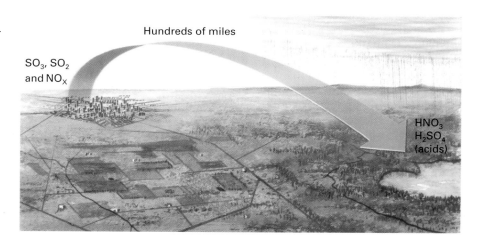

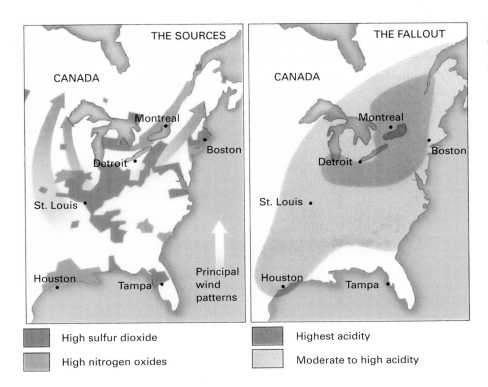

FIGURE 11.3 Distribution of sulfur dioxide, nitrogen oxides, and acid rain. Prevailing winds carry the acid droplets over the Northeast and into Canada.

Acid rain damages trees in several ways. It disturbs the stomata (openings) in tree leaves and causes increased transpiration and a water deficit in the tree. The surface structures of the bark and the leaves can also be destroyed by the acid. Acid rainfall can acidify the soil, damaging fine root hairs and thus diminishing nutrient and water uptake. In addition, acid rain dissolves minerals that are insoluble in groundwater and surface waters of normal pH, and many of these minerals contain metal ions toxic to plant life. For example, acid rain dissolves aluminum hydroxide in the soil, allowing aluminum ions (Al^{3+}) to be taken up by the roots of plants, where they have toxic effects.

$$Al(OH)_3(s) + 3\,H^+(aq) \longrightarrow Al^{3+}(aq) + 3\,H_2O(\ell)$$

The effects of acid rain and other pollution on stone and metal structures are especially devastating because of their irreversibility. By damaging stone buildings in Europe, acid rain is slowly but surely dissolving the continent's historical heritage. The bas-reliefs on the Cologne Cathedral in Germany are barely recognizable today. The Tower of London, St. Paul's Cathedral, and Lincoln Cathedral in Great Britain have suffered the same fate. Other beautifully carved statues and bas-reliefs on buildings throughout Europe and the eastern part of the United States and Canada are slowly passing into oblivion by the action of pollutants, in particular, acid rain.

What can be done about acid rain? Obviously, eliminating the emissions of the oxides of nitrogen and sulfur would be the answer. This is not easy, however. Some stopgap measures are being taken, such as spraying hydrated lime, $Ca(OH)_2$, into acidified lakes to neutralize at least some of the acid and raise the pH toward 7.

$$Ca(OH)_2(s) + 2\,H^+(aq) \longrightarrow Ca^{2+}(aq) + 2\,H_2O(\ell)$$

Some lakes in the problem areas have their own safeguard against acid rain. They have limestone-lined bottoms, which supply calcium carbonate ($CaCO_3$) for neutralizing the acid from acid rain (just as an antacid tablet relieves indigestion).

The leaching of toxic metal ions into groundwater by acid rain may also increase groundwater pollution.

Effects of acid rain on a tombstone in England that was carved in 1817.

The ultimate answers to acid rain problems lie with those industries that produce the oxides of sulfur and nitrogen and with the regulatory agencies that govern them. Methods exist for the control of SO_2 emissions, although some of these are costly. In the final analysis, the consumer will bear those costs. The control of oxides of nitrogen is more difficult because there are so many sources of combustion exhaust gases. Catalytic mufflers help control NO_x emissions from automobiles, but most home furnaces, industrial boilers, and even electrical generating plants do not have adequate controls. Fortunately for acid rain production, NO_2 is so reactive in the troposphere that it is not the major contributor that SO_2 is to acid rain.

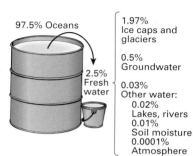

97.5% Oceans

2.5% Fresh water

1.97% Ice caps and glaciers

0.5% Groundwater

0.03% Other water:
0.02% Lakes, rivers
0.01% Soil moisture
0.0001% Atmosphere

The water supply. Of the 2.5% freshwater, less than 1% is available as groundwater or surface water for human use.

The Mississippi River discharges an estimated 400 billion gallons of water daily into the Gulf of Mexico.

A shallow well generally goes no deeper than 50 ft. A deep well is generally drilled to between 100 and 200 ft below the surface. The deepest water well in the world is in Montana—it goes 7,320 ft into the ground.

TABLE 11.2 Water Content	
	PERCENTAGE
Marine invertebrates	97
Human fetus (1 month)	93
Adult human	70
Body fluids	95
Nerve tissue	84
Muscle	77
Skin	71
Connective tissue	60
Vegetables	89
Milk	88
Fish	82
Fruit	80
Lean meat	76
Potatoes	75
Cheese	35

11.3 HOW CAN THERE BE A SHORTAGE OF SOMETHING AS ABUNDANT AS WATER?

Ensuring the quality of the water supply is a shared responsibility. The federal government enacts laws governing the quality of wastewater that can be returned to the environment and the quality of the water that can be supplied for drinking. The U.S. Environmental Protection Agency (EPA) enforces these laws. State and local governments have their own laws and shoulder the responsibility for enforcing them as well as seeing to it that local industries, sewage treatment plants, and municipal water treatment facilities meet federal standards.

The goal of these regulations is to keep water relatively pure so the next user of the water, whether that is a city wanting a water supply or a trout looking for a nice stream to live in, will find it suitable for its intended use.

Individual citizens also have a role to play in keeping the water clean. Voluntary action by an informed public is needed to halt water pollution that originates in our households. Laws alone cannot halt the disposal of hazardous waste in the municipal garbage or the pouring of potential water pollutants into our toilets, sinks, and storm drains.

You might ask why there is so much concern about polluted water when water is the most abundant substance on Earth's surface. Oceans (with an average depth of 2.5 mi) cover about 72% of Earth. They are the reservoir of 97.5% of Earth's water. Only 2.5% is freshwater. Water is also the major component of all living things (Table 11.2). For example, the water content of human adults is 70%—the same proportion as for Earth's surface.

An average of 4.35 trillion gallons of rain and snow fall on the contiguous United States each day (Figure 11.4). Of this amount, 3,100 billion gallons return to the atmosphere by evaporation and transpiration. The discharge to the sea and to underground reserves amounts to 800 billion gallons daily, leaving 450 billion gallons of surface water each day for domestic and commercial use. The United States receives enough annual precipitation to cover the entire country to a depth of 30 inches. The 48 contiguous states withdrew 40 billion gallons per day from natural sources in 1900, but that rose to over 450 billion gallons by 2004. Daily water use will increase as the world's population increases. But the total water supply does not increase. Thus, areas where water is pumped out of the ground faster than it is replenished will surely expand. This means that water must be reused on a daily basis. The water molecules you drink from a water fountain may have been in the municipal wastewater of another city just a few days earlier.

The two sources of usable water are *surface water*, such as rivers, lakes, and wetland waters, and *groundwater*, which is beneath Earth's surface. Figure 11.4 shows our water resources and the flow of groundwater. About 90 billion gallons of the total water withdrawn every day is groundwater drawn from wells

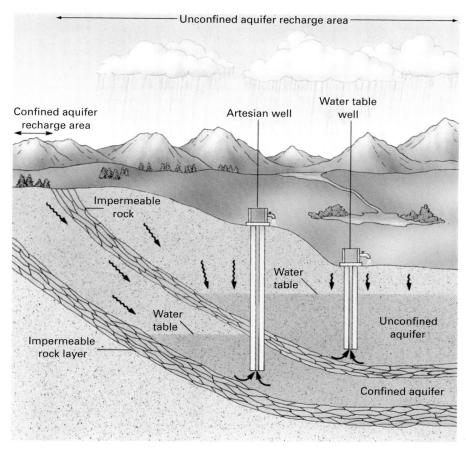

FIGURE 11.4   **Water resources.** Surface water includes water that collects in rivers, lakes, and wetlands. Groundwater may be in unconfined aquifers, which are recharged by surface water from directly above, or in confined aquifers, which lie between impermeable rock layers. Wells may tap into either kind of aquifer. Where water in a confined aquifer is under pressure, pumping of the water may not be needed.

drilled into **aquifers**, layers of water-bearing porous rock or sediment held in place by impermeable rock. These kinds of wells, which supply water to many cities in the Great Plains and along the East Coast, are called artesian wells.

In the arid West, wells used to pump water for irrigation either are going dry or require drilling so deep that irrigation is no longer economically feasible. In some places, water for agriculture has been withdrawn from the huge Ogallala aquifer, which stretches from South Dakota to Texas, 40 times faster than it is replaced by natural processes. Farms have been abandoned where taking water from shallow regions of the aquifer is no longer practical.

The depletion of a major aquifer along the Eastern seaboard has caused large sinkholes in Georgia and Florida when the limestone rock strata of the aquifer collapse as the water is withdrawn (Figure 11.5). Many coastal cities are also experiencing problems with brackish water that comes from aquifers where the freshwater has been drawn off, causing seawater to flow into the depleted aquifer. Depletion of underground sources has also caused sinkholes in Texas. The entire city of Houston has sunk several feet over the years as the result of extensive use of the underground water sources in that area.

Who's using the water? Table 11.3 shows the breakdown for water use in the United States. Thermoelectric power and irrigation account for 82% of all water used.

Industrial water usage can be directly related to production of finished products. One gallon of water, for example, is used to produce eight sheets of ordinary typing paper, while 80 gal of water are needed to produce a gallon of gasoline, and about 25 gal of water are needed to make a single box of nails. Of course, because none of these products *contain* water, the water used is either recycled into the environment or recycled within the facility.

M. Timothy O'Keufe/Tom Stack & Associates

FIGURE 11.5   **A sinkhole in Winter Park, Florida.** As the result of aquifer depletion, the top of an underground cavern has collapsed, an event that can happen in the course of just a few minutes.

Aquifer A layer of water-bearing porous rock or sediment held in place by impermeable rock

TABLE 11.3	Water Consumption in the U.S. as a Percentage of Daily Total of 408 Billion Gal
	% OF TOTAL
Households	11
Industry	5
Thermoelectric power	48
Agriculture/irrigation	34
Other	2

Source: http://water.usgs.gov/pubs/2004

In many industrial locations the largest single use of water is in plant cooling systems. Water can absorb more heat than any other readily available liquid—it has a heat capacity of 1 cal/g, or 4.18 J/g. Recirculating cooling water is an important means of water conservation. A cooling tower allows the warmed water to lose its heat to the surrounding air and helps to reduce thermal pollution of the river or lake where the used water would be discharged. In addition, the high heat capacity of water enables the industrial user to recycle heat energy captured by the cooling water.

Groundwater and surface water sources each provide half of the more than 44 billion gallons of **potable** (drinkable) water that is used each day in the United States. Table 11.4 gives the average amounts for various personal uses. Only a small fraction of municipal water really needs to be potable, that is, of drinking water quality. The largest portion of a water supply could be disinfected and made bacteriologically safe, while avoiding the more costly treatment needed to meet drinking quality standards. Water treated in this way would be suitable for irrigation of parks and golf

Water that is returned to a waterway at a higher temperature than it was withdrawn contributes to thermal pollution of the waterway. Because the solubility of oxygen in water decreases with increased temperature, thermally polluted water cannot support aquatic life as well as cooler water can.

Potable Describes water suitable for drinking

TABLE 11.4	Average Daily Per Capita Indoor Domestic Water Use[a]	
DOMESTIC INDOOR USE	GAL PER CAPITA IN AVERAGE HOMES (74 GAL)	GAL PER CAPITA IN HOMES WITH WATER-EFFICIENT FIXTURES, TOILETS, AND LEAK CONTROL (51.9 GAL)
Showers	12.6	10.0
Clothes washers	15.1	10.6
Dishwashers	1.0	1.0
Toilets	20.1	9.6
Baths	1.2	1.2
Leaks	10.0	5.0
Faucets[b]	11.1	10.8
Other domestic uses indoors[c]	1.5	1.5

[a] Source: American Water Works Association website: http://awwa.org
[b] Americans drink more than 1 billion glasses of tap water per day.
[c] On average, 50–70% of all home water is used *outdoors* for watering lawns and plants.

courses, air-conditioning, industrial cooling, and toilet flushing. Consider the data in Table 11.4, which illustrate the inefficiency of residential water systems. It is obvious that strong arguments could be made for dual water systems—one system that treats water for drinking and a different system for other uses.

Residential water conservation is a way to cut demand for freshwater supplies. Although Table 11.3 shows that residential use is a small part of the total, home water conservation can be an important step in cutting demand for freshwater supplies in large urban areas.

Almost all water molecules have been recycling since they were formed billions of years ago. Have you ever considered that your next glass of water may include some molecules that Aristotle or Abraham Lincoln drank, or some molecules that flooded into the *Titanic* when it sank? In spite of these possibilities, the idea of obtaining potable water from wastewater or even sewage, especially wastewater that was *recently* discharged, is psychologically difficult for many people to accept. Nevertheless, the technology has been developed and was used in NASA's space shuttle flights. What this means is if water (pure or not) is available, it should be considered for reuse. In the southwestern United States the rate of depletion of aquifers has led to direct recycling of water from sewage, a process called **groundwater recharge**. For example, in El Paso, Texas, 10 million gallons of pure water per day from sewage effluent is pumped into the underground aquifer that is the main source of water for El Paso.

11.4 WHAT IS THE DIFFERENCE BETWEEN CLEAN WATER AND POLLUTED WATER?

The term *pollution* is used to describe any condition that causes the natural usefulness of air, water, or soil to be diminished. Water that is judged unsuitable for drinking, washing, irrigation, or industrial uses is polluted water. The pollution (Table 11.5) may be heat, radioisotopes, toxic metal cations and anions, organic molecules, acids, alkalies, or organisms that cause disease (**pathogens**). Water suitable for some uses might be considered polluted and therefore unsuitable for other uses—you might go swimming in water you would not consider drinking. Water that is unsuitable for use has often been polluted by human activity, but natural processes can also pollute water. For example, water that contacts organic substances such as decaying leaves and

There are newer design toilets that use about 1.5 gal per flush.

Courtesy of U.S. Geological Survey

Surface level drop in San Joaquin, California. The markers on the utility pole show how the surface level has dropped over the years due to withdrawal of groundwater for irrigation.

Groundwater recharge Return to groundwater of water from treated sewage

Pathogens Disease-causing microorganisms

TABLE 11.5	U.S. Public Health Service Classes of Pollutants
POLLUTANT	**EXAMPLE**
Oxygen-demanding wastes	Plant and animal material
Infectious agents	Bacteria, viruses
Plant nutrients	Fertilizers such as nitrates and phosphates
Organic chemicals	Solvents, pesticides, detergent molecules
Other minerals and chemicals	Acids from mining operations, inorganic chemicals from metal-working operations
Sediment from land erosion	Clay silt from stream beds
Radioactive substances	Mining wastes, used radioisotopes
Heat from industry	Cooling water from electric generating plants

TABLE 11.6   Dissolved Substances Found in "Pure" Water

NAME	FORMULA	COMMENT
The following come from contact of water with the atmosphere:		
Carbon dioxide	CO_2	Makes water slightly acidic
Dust particles	—	Can be large amounts at times
Nitrogen	N_2	Along with oxygen, causes visible bubbles in hot water
Nitrogen dioxide	NO_2	Formed by lightning
Oxygen	O_2	Supports aquatic life
The following vary considerably, depending on the kinds of rock formations the water has contacted:		
Bicarbonate ions	HCO_3^-	Soils and rocks
Calcium ions	Ca^{2+}	From limestone
Chloride ions	Cl^-	Soils, clays, and rocks
Iron(II) ions	Fe^{2+}	Soils, clays, and rocks
Magnesium ions	Mg^{2+}	Soils, clays, and rocks
Potassium ions	K^+	Soils, clays, and rocks
Sodium ions	Na^+	Soils, clays, and rocks
Sulfate ions	SO_4^{2-}	Soils and rocks

Charles D. Winters

How pure is this water?

The Clean Water Act is also known as the Federal Water Pollution Control Act. It was revised in 1978, 1980, and 1988. By an act of Congress, 2002 was proclaimed the Year of Clean Water to celebrate the 30th anniversary of the Clean Water Act and to provide education and outreach on the status and importance of clean water.

animal wastes will pick up numerous organic compounds, many of which impart odors and color to the water and some of which might be pathogenic. Silt, consisting of colloidal-sized particles of dirt and sand, can also pollute water. Table 11.6 lists some of the substances that can be found in "pure" natural water. By looking at this list, it is clear that absolutely pure (100%) water is not a common commodity.

As human activities have continued to pollute water, governments have passed laws designed to keep our waters clean. The Clean Water Act of 1972 represented a major change in the thinking of Congress concerning who is responsible for keeping water clean. This act shifted the burden of producing water suitable for reuse from the user (a municipality, for example) to the wastewater discharger. Because it is easier to clean wastewater prior to dumping than to clean the water after the untreated waste has been discharged (Figure 11.6), the Clean Water Act was a major step in improving the quality of our natural waters. The Act requires the EPA to establish and monitor emission standards—the maximum amounts of water pollutants that can be discharged into natural bodies of water from factories, municipal sewage treatment plants, and other facilities. As it turns out, the wastewater effluent from an industry can now often be clean enough to be used for such purposes as irrigation or industrial cooling.

FIGURE 11.6 The EPA requires that virtually all industrial wastewaters be treated prior to discharge. This water is reasonably pure, although it is not as pure as drinking water.

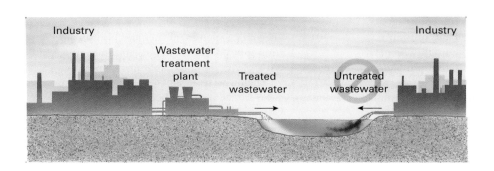

CONCEPT CHECK   11A

1. Approximately what percentage of the human body is water?
2. What is the major reservoir of water on Earth?
3. Water beneath Earth's surface is called _____.
4. Water on the surface of Earth is called _____.
5. The average person in the United States uses about _____ gallons of water per day for all purposes.
6. The actual amount of potable (drinkable) water a person needs is _____ per day.
7. Three common water pollutants are _____, _____, and _____.
8. A water-bearing stratum of porous rock, sand, or gravel is called a(n) _____.
9. What happens to most of the water that falls on the United States each day?
10. Discuss the reasons for differences in the physical properties of the compounds in Table 11.1.
11. Why do you think the heat of fusion of water (ice) is 80 cal/g but the heat of vaporization of water is 540 cal/g? Consider the disruption of the intermolecular bonding that occurs in the solid, liquid, and vapor states of water.
12. What is acid rainfall and what are some of the main sources of chemicals leading to acid rainfall?
13. What percentage of Earth's total water supply is freshwater? What percentage of Earth's freshwater is in ice caps and glaciers? What percentage of Earth's total water is readily available for human use?
14. What are the two majors uses for water in the United States?
15. What is potable water?

11.5 THE IMPACT OF HAZARDOUS INDUSTRIAL WASTES ON WATER QUALITY

Industrial processes, whether making paper, automobiles, or TV sets, produce waste materials. Table 11.7 lists some of the industrial pollutants that result from the manufacture of products important to us. For many years the disposal

USDA/Soil Conservation Service

Corroding waste barrels.

TABLE   11.7   Important Industrial Products and Pollutants Associated with Their Manufacture

PRODUCTS	POLLUTANTS
Plastics	Solvents, organic chlorine compounds
Pesticides	Organic chlorine compounds, organic phosphate compounds
Medicines	Solvents, metals such as mercury and zinc
Paints	Metals, pigments, solvents, organic residues
Petroleum products	Oils, organic solvents, acids, alkalies
Metals	Metals, fluorides, cyanide, acids, oils
Leather	Chromium, zinc
Textiles	Metals, pigments, organic chlorine compounds, solvents

The nation's first superfund site, Love Canal in Niagara Falls, N.Y., was finally declared cleaned in 2004 after two decades of work at a cost of $400 million. At the end of 2004, 1237 sites remained on the EPA's National Priorities List.

of solid wastes in *landfills* was considered good engineering practice. Many of the substances present in those wastes were partially dissolved by rainwater and became part of the groundwater, causing serious pollution of water supplies. Many of these older landfills also allowed surface runoff that contained dissolved substances from the wastes to be carried into natural bodies of water.

It was also common practice in the past to place waste and discarded chemicals into metal drums and bury them directly in the ground. After a few years the drums developed leaks due to corrosion, allowing the drum contents to leak into water that would ultimately become groundwater or surface water. In recognition of this common practice and what it was doing to water quality, the U.S. Congress in 1976 passed the Resource Conservation and Recovery Act (RCRA). In 1980, Congress established the "Superfund," a $1.6 billion program designed to clean up hazardous waste sites that were threatening to contaminate the nation's water supplies. As of 1997, it was known that an average hazardous waste site cleanup took nine years and cost $7.4 million. By 1999, 205 hazardous waste sites had been cleaned to the extent that they could be removed from the Superfund list. The need for cleanup of some remaining sites is of growing urgency. These sites were established in once-rural locations that have now been surrounded by residential development.

While the Superfund is dealing with existing hazardous waste sites, RCRA and its regulations govern the disposal of newly generated wastes that have the potential to harm the environment. The EPA defines certain industrial wastes as **hazardous wastes** and closely regulates how they are generated, stored, transported, and disposed of. RCRA was designed to give "cradle-to-grave" (origin to disposal) responsibility to *generators* of hazardous wastes. Before RCRA, an industry could hire almost anyone to haul away its waste without regard to where it was taken or how it would be disposed of. Today, a generator of a hazardous waste must know who is transporting it, where it is going, and how it will be disposed of. Each shipment of hazardous waste is accompanied by a *manifest*, a document listing the hazardous waste by name, how much is present, and how it will be disposed of. As the load of hazardous waste is transported, temporarily stored, and finally disposed of, parts of the manifests are signed by each responsible party and returned to the generator at each step. The states also receive copies of the manifests, and annual reports of hazardous waste activities are filed with EPA by everyone who handles hazardous wastes routinely.

Today, industrial hazardous wastes must be placed into *secure* landfills, incinerated, or treated in some way to render them nonhazardous. No hazardous waste is allowed to be disposed of in a way that could pollute the environment. Some hazardous wastes go to secure landfills with plastic linings (Figure 11.7) that prevent their contents from easily reaching surrounding

Hazardous waste Industrial and household wastes responsible for water pollution

FIGURE 11.7 A hazardous waste landfill. Underneath, this landfill has several feet of clay covered by three plastic liners. Barrels of hazardous waste are placed above the liners and buried in soil.

Dennis Barnes

THE WORLD OF CHEMISTRY

GREEN CHEMISTRY

Water pollutants. Air pollutants. What can be done to decrease them? Stop releasing them into the environment—that is one major approach to solving the problem.

In response to growing awareness of the need to reduce pollution, the *green chemistry* movement is gaining momentum. Industrial processes, both new and old, are being evaluated with the goal of eliminating or significantly reducing pollution. In October 2005, the Nobel Prize in Chemistry was awarded to Yves Chauvin of the French Petroleum Institute, Robert H. Grubbs of the California Institute of Technology, and Richard R. Schrock of the Massachusetts Institute of Technology for their contributions to the development of the olefin metathesis reaction for the preparation of carbon–carbon double bonds. The Swedish academy described this as "a great step forward for 'green chemistry.'" Building on the fundamental work of Chauvin and coworkers, Grubbs and Schrock developed ruthenium-based and molybdenum-based catalysts, respectively, which allow this reaction to be widely used in the preparation of many new molecules. Industrially, this reaction is now used in cleaner, less expensive, and more efficient production of polymers, pesticides, and pharmaceuticals, among other things. These attributes are cornerstones

of the twelve principles of green chemistry (http://www.epa.gov/greenchemistry/principles.html) enumerated by the Environmental Protection Agency. Green chemistry processes have been described as *benign by design*.

Consider the dry cleaning industry. Stains on clothing are dissolved in chlorine-containing solvents such as perchloroethylene (Cl_2CMCCl_2). These solvents are health hazards and persist in the environment if they enter the soil of natural waterways. A green chemistry approach to dry cleaning requires a solvent other than water, one that is nonpolar so that it will dissolve grease and dirt while not harming the fabric, and one that is neither an environmental pollutant nor a health hazard. Success in green dry cleaning has come with the use of liquid carbon dioxide as the solvent. You are familiar with gaseous carbon dioxide in soda and solid carbon dioxide as dry ice for maintaining low temperatures. At atmospheric pressure, dry ice doesn't melt; it sublimes directly to a gas. At higher pressures however, dry ice forms a liquid. With the use of pressure and the addition of surfactants (Section 15.4) designed for use in carbon dioxide, a successful new dry cleaning method has been developed.

Another facet of green chemistry is *atom economy*, which focuses

on the reactions used to prepare important organic chemicals. In many industrial syntheses, a series of reactions is needed to convert the first reactant into the desired final product. The goal of the atom economy approach is to send as few atoms as possible into unwanted by-products of these reactions. In the synthesis of propene shown here, for example, only two H atoms and one O atom are wasted.

$$CH_3CH_2CH_2OH \xrightarrow{H_2SO_4} CH_3CH{=}CH_2 + H_2O$$

The original synthesis of the pain reliever ibuprofen required six reactions that together sent 23 H atoms, 7 C atoms, 8 O atoms, and 1 Cl, 1 Na, and 1 N atoms into waste products. A new green synthesis using the same starting materials requires only three steps and creates waste products containing only 4 H atoms, 2 C atoms, and 2 O atoms.

The Presidential Green Chemistry Challenge Award program was instituted by President Clinton in 1995 and is administered by the EPA. Each year, cash awards are given to those projects deemed outstanding from among those that have applied. A complete list of past award winners and descriptions of their contributions may be found at http://www.epa.gov/greenchemistry/past.html.

water supplies. These landfills also have carefully spaced monitoring wells so any substances escaping from the landfill's contents may be detected, allowing the problem to be corrected. Other hazardous wastes may no longer be placed in secure landfills, but must be incinerated or destroyed in some other way. Although incineration seems like a logical best choice to dispose of hazardous wastes, current incinerators are operating at near capacity, and it is difficult to get proper permits for new ones due to community opposition. (Incineration of *some* hazardous wastes does produce small quantities of other, even more harmful combustion products.)

Proposition 65 states, "No person in the course of doing business shall knowingly discharge or release a chemical known to cause cancer or reproductive toxicity into water or onto or into land where such chemical passes or probably will pass into any source of drinking water."

States like California that have severe water shortage problems have taken drastic steps to protect their ground- and surface water. In California, Proposition 65, the Safe Drinking Water and Toxic Enforcement Act of 1986, requires the yearly update of a list of chemicals known to the state to cause cancer, birth defects, or other reproductive harm. Discharge of these chemicals into any water that might become a drinking water supply is prohibited, and violation of this prohibition may incur financial penalties.

11.6 HOUSEHOLD WASTES THAT AFFECT WATER QUALITY

The EPA estimates that each year 350 million gallons of waste motor oil are poured on the ground or flushed down the drain by individual citizens. That's 35 times more oil than the *Exxon Valdez* spilled in Alaska.

Often we do not think about the things we discard in our garbage, but what we throw away and how we do it can affect the quality of natural waters as much as what industry does. Household wastes that are incinerated can contribute to air pollution, but because the bulk of our household waste goes to landfills, we too can be responsible for causing pollution of groundwater as well as of rivers, streams, and lakes. Table 11.8 lists some common household products and the kinds of chemicals they contain. Because we are the consumers of industrial products, we can put the very same chemicals into the water as industry can. Although the individual amounts of harmful chemicals used in a household are less than those used in a large industry, the total amounts disposed of daily by all households can be very large, even for a medium-sized city.

Households have a greater problem disposing of hazardous wastes than industry does. Even where there is an active recycling project for glass, paper, metals, and plastics, there is often no pickup of chemicals that should be separated from the ordinary trash destined for the landfill. If these chemicals

Municipal hazardous waste day. On a designated day, citizens bring their hazardous waste to a site where trained crews sort the materials for disposal. Every community should have such days or have access to a permanent hazardous waste disposal facility.

TABLE 11.8	Some Common Household Hazardous Wastes and Recommended Disposal	
TYPE OF PRODUCT	**HARMFUL INGREDIENTS**	**DISPOSAL***
Bug sprays	Pesticides, organic solvents	Haz.
Bathroom cleaners	Bases and acids	Drain
Furniture polish	Organic solvents	Haz.
Aerosol cans (empty)	Solvents, propellants	Trash
Nail polish and remover	Organic solvents	Haz.
Antifreeze	Organic solvents, metals	Haz.
Insecticides	Pesticides, solvents	Haz.
Auto battery	Sulfuric acid, lead	Haz.
Medicine (expired)	Organic compounds	Trash
Paint (latex)	Organic polymers	Trash
Gasoline	Organic solvents	Haz.
Motor oil	Organic compounds, metals	Haz.
Drain cleaners	Bases	Drain
Paints (oil-based)	Organic solvents	Haz.
Mercury and Ni-Cd batteries	Heavy metals	Haz.

*Haz: Professional disposal as a hazardous waste.
Drain: Disposal down the kitchen or bathroom drain with plenty of water.
Trash: Treat as normal trash—will not harm groundwater. In most households, the items marked Haz. are disposed of as normal trash, which results in groundwater pollution.
Source: Water Environment Federation Fact Sheet, 2001.

are mixed with ordinary garbage, they go to the city landfill or incinerator. If they are poured out in the sink, driveway, or backyard, they will eventually reach natural waters.

How can we dispose of hazardous household wastes without danger to the groundwater supply? We can ask our city's municipal waste authorities to provide disposal sites for these wastes or to sponsor periodic household hazardous waste days when these materials can be brought to a central site. In some U.S. cities and some European countries (such as the Netherlands), special trucks routinely pick up hazardous household wastes.

Suspended particles in water are colloidal in size and might include bacteria and viruses, as well as harmless soil particles.

11.7 TOXIC ELEMENTS OFTEN FOUND IN WATER

Heavy metal ions are perhaps the most common of all water pollutants. The heavy metals include such frequently encountered elements as lead and mercury, as well as many less common ones like cadmium, chromium, nickel, and copper. These metals can, at times, be acutely toxic, causing immediate symptoms, but often they are chronically toxic in very small quantities. Chronic toxicity is characterized by nagging symptoms that lessen normal body functions. Inadequate disposal of wastes from mining or industrial activities causes these metals to find their way into water supplies. In addition, some farming activities and the disposal of household wastes can contribute to the presence of heavy metals in our water supplies.

The heavy metals are considered "heavy" because they have high atomic weights.

Mercury

Mercury is a fairly common heavy metal. Its elemental form at room temperature is that of a liquid that is volatile with the characteristic shininess of a metal. Mercury atoms are readily oxidized to Hg_2^{2+} [mercury(I) ion] and Hg^{2+} [mercury(II) ion]. Both of these ionic forms are toxic, and their effects can be cumulative (i.e., repeated exposures will increase the toxic effects because the body does not easily rid itself of the element).

Mercury is widely used in industry and can still occasionally be found in homes in mercury thermometers. The disposal of used fluorescent lamps represents a major source of mercury in water. Fluorescent lamps contain small amounts (about 60 mg per lamp) of mercury, which vaporizes inside the lamp and helps carry the electric current between the electrodes. Discarded lamps of this type are ultimately crushed when they are disposed of in landfills. This releases the mercury to become oxidized and then to dissolve in any water that comes in contact with the waste.

Once released into streams, lakes, or oceans, metallic mercury sinks to the bottom, where it is converted by microorganisms to methymercury (CH_3Hg). In this stable and persistent form, it accumulates in the flesh of large marine creatures (e.g., tuna, swordfish, marine mammals). Once there, it becomes a hazard to the health of humans who consume these creatures, and numerous health advisories have been issued about consuming mercury-contaminated seafood.

Lead

Lead is another widely encountered heavy metal. Like mercury, lead tends to accumulate in the body on repeated exposures. Lead has been used for centuries in plumbing (the Latin name for lead is *plumbum*, the word from which *plumber* is derived). Until recently, lead-based solders were routinely used in almost all plumbing fixtures, including the valves in drinking fountains as well as shower fixtures and faucets found in both the bathroom and kitchen. The current lead level for drinking water allowed by the EPA is 15 mg/liter.

In addition, lead has been widely used in various paint formulations. For the past 30 years, paints containing lead have been banned for interior use, but lead-containing paints are still available for outdoor and industrial use. When these paints or objects coated with them are discarded, lead can find its way into water supplies. The Engelhard Corporation received a 2004 Presidential Green Chemistry Challenge Award for their development of a line of paint pigments that do not contain heavy metals.

Arsenic

Arsenic occurs naturally in small amounts in many foods. Shrimp, for example, contain about 19 parts per million (ppm) arsenic, and corn may contain 0.4 ppm arsenic. The amount of naturally occurring arsenic in foods depends on the surroundings where they are grown and the metabolism of the plant or animal. While many soils contain arsenic, which causes an accumulation of the element as a plant grows, some insecticides also contain arsenic, which causes an arsenic residue when the insecticide is applied. The U.S. Food and Drug Administration (FDA) has set a limit of 76 ppm for arsenic levels in shellfish. In its ionic forms, arsenic is much more toxic than in its covalently bound compounds. The typical toxic arsenic compounds contain ions such as arsenate (AsO_4^{3-}) or arsenite (AsO_2^-).

Arsenic and the heavy metal ions are toxic primarily due to their ability to react with sulfhydryl groups (—SH) in enzymes (Section 15.8).

11.8 MEASURING WATER POLLUTION

Biochemical Oxygen Demand

Many organic compounds that find their way into water can easily be oxidized by microorganisms that are also there. This is a natural process that prevents a buildup of organic waste in natural waters. To change this organic material into simple substances (such as CO_2 and H_2O) requires oxygen. The amount of dissolved oxygen required is called the **biochemical oxygen demand (BOD)**, and it is a measure of the quantity of dissolved organic matter. The oxygen is necessary so that the bacteria and other microorganisms can metabolize the organic matter that constitutes their food. Ultimately, given near-normal conditions and enough time, the microorganisms will convert huge quantities of organic matter into the following end products:

$$\text{Organic carbon} \longrightarrow CO_2$$
$$\text{Organic hydrogen} \longrightarrow H_2O$$
$$\text{Organic oxygen} \longrightarrow H_2O$$
$$\text{Organic nitrogen} \longrightarrow NO_3^- \text{ or } N_2$$

At 68°F (20°C) the solubility of O_2 in water under normal air pressure of 1 atmosphere (atm) is only 0.0092 g O_2/L. But, a stream containing 10 ppm by weight (just 0.001%) of an organic material with the formula $C_6H_{10}O_5$ has a BOD of 0.012 g O_2/L of water. Clearly, this BOD value exceeds the equilibrium concentration of dissolved O_2 at this temperature. As the bacteria utilize the dissolved oxygen in a stream or lake with this BOD, the oxygen concentration of the water may soon drop too low to sustain any form of fish life. Whether this happens depends on the opportunities for new oxygen to become dissolved in the water. Life forms can survive in water where the

Lead has historically been used in plumbing. The Romans used lead pipes to carry water into their homes.

Biochemical oxygen demand (BOD) The amount of dissolved oxygen required by microorganisms that oxidize dissolved organic compounds to simple substances like CO_2 and H_2O

BOD exceeds the dissolved oxygen if the water is flowing vigorously in a shallow stream (this facilitates the absorption of more oxygen from the air via aeration).

BOD values can be greatly reduced by treating industrial wastes and sewage with oxygen or ozone. Numerous commercial cleanup operations now being developed and used employ this type of "burning" of the organic wastes. Another benefit of treating wastewater with oxygen is that some of the nonbiodegradable material becomes biodegradable as a result of partial oxidation.

Highly polluted water often has a high concentration of organic material, with resultant large biochemical oxygen demand. In extreme cases, more oxygen is required than is available from the environment. The result is that fish and other aquatic life can no longer survive. The aerobic bacteria (those that require oxygen for the decomposition process) die. As a result, even more lifeless organic matter results, and the BOD soars. Nature, however, has a backup system for such conditions. A whole new set of microorganisms (anaerobic bacteria) takes over; these organisms take oxygen from oxygen-containing compounds to convert organic matter to CO_2 and water.

Organic nitrogen is converted to elemental nitrogen by these bacteria. Given enough time, enough oxygen may become available, and aerobic oxidation will then return.

Characteristic BOD levels (g O_2/L):	
Untreated municipal sewage	0.1–0.4
Runoff from barnyards and feed lots	0.1–10
Food-processing wastes	0.1–10

Fish cannot live in water that has less than 0.004 g O_2/L (4 ppm).

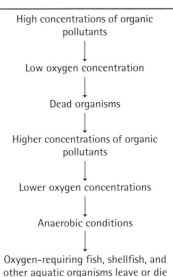

High concentrations of organic pollutants
↓
Low oxygen concentration
↓
Dead organisms
↓
Higher concentrations of organic pollutants
↓
Lower oxygen concentrations
↓
Anaerobic conditions
↓
Oxygen-requiring fish, shellfish, and other aquatic organisms leave or die

---

CONCEPT CHECK 11B

1. The federal law requiring cleanup of hazardous waste sites that can pollute water is called _____.
2. Hazardous wastes can be indiscriminately placed in landfills. (**a**) True (**b**) False
3. Name three household wastes that can contaminate groundwater with the same harmful chemicals as industrial wastes. Beside each, list the harmful chemical.

   Household Waste   **Harmful Chemical**
   _____   _____
   _____   _____
   _____   _____

4. List four household waste types that lend themselves to recycling.
5. The amount of oxygen required to oxidize a given amount of organic material is called the _____ _____ _____, which is abbreviated _____.
6. Name two industrial products whose manufacture introduces heavy metals into groundwater.
7. Name two industrial products whose manufacture introduces chlorinated organic compounds into groundwater.
8. Two heavy metals found in water are _____ and _____.

11.9 HOW WATER IS PURIFIED NATURALLY

Water is a natural resource that, within limitations, is continuously renewed. The water cycle offers a number of opportunities for nature to purify its water. The worldwide *distillation* process results in rainwater containing only traces of nonvolatile impurities, along with gases dissolved from the air. *Crystallization* of ice from ocean salt water results in relatively pure water in the form of icebergs.

Rainwater in clean air is very pure, containing only small amounts of dissolved gases such as N_2, O_2, and CO_2.

Aeration of groundwater as it trickles over rock surfaces, as in a rapidly running brook, allows volatile impurities to be released to the air and allows oxygen to be dissolved. *Sedimentation* (or *settling*) of solid particles occurs in slow-moving streams and lakes. *Filtration* of water through sand rids the water of suspended matter such as silt and algae. Of very great importance are the *oxidation* processes carried out by bacteria and other microorganisms. Practically all naturally occurring organic materials—plant and animal tissue, as well as their waste materials—can be oxidized in surface waters as long as oxygen is available and their concentration is not too high. Finally, another natural process is *dilution*. Most, if not all, pollutants found in nature are rendered harmless if reduced below certain levels of concentration by dilution.

Before the explosion of the human population and the advent of the Industrial Revolution, natural purification processes were quite adequate to provide ample water of very high purity in all but desert regions. Nature's purification processes can be thought of as massive but somewhat delicate.

Today, the activities of humans often push the natural purification processes beyond their limit, and polluted water accumulates. A simple example comes from dragging gravel from streambeds. The excavation leaves large amounts of suspended matter in the water, and for miles downstream, aquatic life is destroyed. Eventually, the solid matter settles, and normal life can be found again in the stream.

Pumping mud from the bottom to make room for deeper boats in the Mississippi River. When the bottom of a natural body of water is disturbed, the possibility exists of redistributing pollutants that had been trapped in the mud into the water.

A more complex example—one that perhaps cannot be solved by relying on natural purification processes—is pollution by organic molecules that cannot be easily oxidized by microorganisms. A **biodegradable** substance is composed of molecules that are broken down to simpler ones by microorganisms. For example, cellulose suspended in water will be converted to carbon dioxide and water. A **nonbiodegradable** substance, on the other hand, cannot be easily converted to simpler molecules by microorganisms. If the conversion process is extremely slow, or if it cannot be done at all by natural microorganisms, nonbiodegradable substances tend to accumulate in the environment.

Some organic compounds, notably some of those produced synthetically, are nonbiodegradable. When these substances are introduced into the environment, they simply stay in the natural waters or are absorbed by life forms and remain intact for a long time. Branched-chain detergent molecules, for example, cannot easily be decomposed by microorganisms.

$$CH_3-\underset{\underset{CH_3}{|}}{CH}-CH_2-\underset{\underset{CH_3}{|}}{CH}-CH_2-\underset{\underset{CH_3}{|}}{CH}-CH_2-\underset{\underset{CH_3}{|}}{CH}\text{—}SO_3^-Na^+$$

A branched-chain sodium alkylbenzenesulfonate detergent molecule

The first detergents that contained such molecules accumulated and caused noticeable foaming in rivers and streams. The branched-chain detergents were soon replaced by linear-chain alkylbenzenesulfonate detergents, which are easily decomposed by microorganisms—they are biodegradable.

$$CH_3CH_2CH_2CH_2CH_2CH_2CH_2CH_2CH_2CH_2CH_2CH_2\text{—}SO_3^-Na^+$$

A linear sodium alkylbenzenesulfonate detergent molecule

DDT is fat soluble.

Biodegradable Describes a substance that can be broken down into simple molecules by the action of microorganisms

Nonbiodegradable Describes a substance that cannot be broken down by microorganisms and therefore persists in the environment

Other examples of nonbiodegradable organic pollutants are the chlorinated and polychlorinated hydrocarbons. Many of these compounds are used as insecticides (Section 19.5). The insect-killing ability of dichlorodiphenyltrichloroethane (DDT) was first recognized in 1939. By the end of World War II, its insecticidal properties were legendary due to its ability to kill everything from malaria-causing mosquitoes to lice. This success prompted the introduction of numerous other chlorinated hydrocarbons as insecticides, and by the early 1960s their use was widespread throughout the world. These compounds are broad-spectrum insecticides, killing most insects rather effectively; however, they also biodegrade very slowly, so they tend to accumulate in the environment. This persistence is especially troublesome since their slow biodegradation allows such compounds to accumulate in the food chain. Fish-eating birds, for example, can accumulate large quantities of these insecticides that have accumulated in fish that ate smaller organisms containing these compounds. The populations of falcons, pelicans, bald eagles, ospreys, and other birds have been endangered by persistent pesticides. DDT causes reproductive failure in birds by interfering with the mechanisms that produce strong eggshells. After the use of DDT was banned in the United States in 1972, the numbers of surviving hatchlings increased rather dramatically.

A hypothetical food chain might be plant ⟶ insect ⟶ fish ⟶ hawk.

11.10 WATER PURIFICATION PROCESSES: CLASSICAL AND MODERN

The outhouses common in rural areas in years past had their counterparts in city cesspools, which were basically holes in the ground into which sewage flowed. In cesspools, organic matter is decomposed by anaerobic bacteria, producing some pretty bad-smelling chemicals such as hydrogen sulfide, which has a characteristic rotten-egg odor. The terrible job of cleaning cesspools inspired the development of cesspools that could be flush-cleaned with water, followed by a connecting series of such pools that could be flushed from time to time. City sewage systems with no holding of the wastes were the next step.

Cesspools were an early and crude form of the modern activated sludge process.

At first, city sewage systems did little but channel sewage water to rivers and streams, where natural purification processes were expected to clean the water for the next users downstream. Today, however, sewage is treated using a combination of methods that can render the treated municipal wastewater almost as clean as the natural waters into which it is being discharged. The simplest treatment method is *primary wastewater treatment*, which copies two of nature's purification methods, settling and filtration. In primary treatment, sewage goes from the primary sedimentation tank shown in Figure 11.8 to the chlorinator.

Sewage includes everything that flows from the sinks, tubs, washing machines, and toilets in our homes, factories, and public buildings. It excludes wastewater treated separately by industrial facilities.

Primary treatment removes 40 to 60% of the solids present in sewage and about 30% of the organic matter present. Calcium hydroxide and aluminum sulfate are added to produce aluminum hydroxide, which is a sticky, gelatinous precipitate that settles out slowly, carrying suspended dirt particles and bacteria with it.

Sewage is still 99.9% water.

$$3\,Ca(OH)_2(aq) + Al_2(SO_4)_3(aq) \longrightarrow 2\,Al(OH)_3(s) + 3\,CaSO_4(s)$$

For many years municipal sewage treatment plants had only primary treatment, followed by chlorination of the treated wastewater (see Section 11.12) before it was discharged into a suitable river or stream. Chlorination kills any remaining harmful pathogens. Presumably, natural processes would get rid of the remaining solids and dissolved organic matter.

Definitions of "pure water":
Chemist: "Pure H_2O—no other substance."
Parent: "Nothing harmful to my child."
Game and Fish Commission: "Nothing harmful to animals."
Sunday boater: "Pleasing to the eye and nose, no debris."
Ecologist: "Natural mixture containing necessary nutrients."

In realizing that this treatment was not sufficient to protect the public from contaminated water, the writers of the 1972 Clean Water Act required that sewage treatment plants also provide *secondary wastewater treatment*, which revives the old cesspool idea but under more controlled conditions. Modern secondary treatment operates in an oxygen-rich environment (aerobic), whereas the cesspool operates in an oxygen-poor environment (anaerobic). The results are the same: The organic molecules that will not settle are consumed by microorganisms and the resulting sludge will settle.

Carbon black is used in many processes for manufacturing extremely pure organic compounds such as pharmaceuticals or food additives. *Adsorption* is the process by which molecules are attracted and held onto a surface.

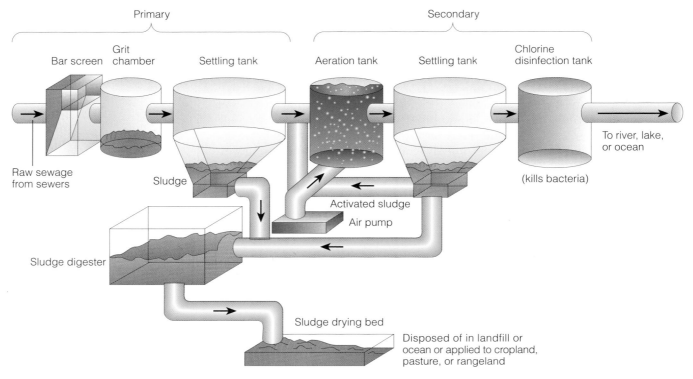

FIGURE 11.8   Primary and secondary sewage treatment.

A sewage outfall, where treated water is returned to the environment. This water should be as clean as is technologically possible.

Even a combination of primary and secondary wastewater treatment systems will not remove dissolved inorganic materials such as toxic metal ions, nutrients such as nitrate ions (NO_3^-) or ammonium ions (NH_4^+), or nonbiodegradable organic compounds such as chlorinated hydrocarbons. These materials can be removed by a variety of *tertiary wastewater treatment* methods that are selectively introduced where the nature of the wastewater requires them. One obstacle to tertiary treatment is the initial expense of modifying sewage treatment plants and the ongoing expense of additional treatment.

Filtration of the water through *carbon black* is a type of tertiary treatment effective for removing soluble organic compounds that are nonbiodegradable and thus remain in the water after secondary treatment. Carbon black consists of finely divided carbon particles with a large surface area on which solutes, including certain potentially toxic substances, can be *adsorbed*.

A different kind of tertiary treatment is needed to remove ammonia or ammonium ion. Because nitrogen is a nutrient for aquatic microorganisms, excessive nitrogen released to natural waters can cause a soaring BOD with the accompanying fish kills and other problems. The water is exposed to **denitrifying bacteria** that convert ammonium ion or ammonia to harmless nitrogen gas.

$$NH_4^+(aq) \text{ or } NH_3(aq) \xrightarrow{\overset{\text{Denitrifying}}{\text{bacteria}}} N_2(g)$$

11.11 SOFTENING HARD WATER

Denitrifying bacteria Bacteria that convert ammonia or ammonium ion to nitrogen.

The presence of Ca^{2+}, Mg^{2+}, Fe^{3+}, or Mn^{2+} ions will impart "hardness" to water. Hardness is objectionable because (1) it causes precipitates (scale) to form in boilers and hot-water systems, (2) it causes soaps to form insoluble curds (this reaction does not occur with some synthetic detergents—see Section 15.4), and (3) it can impart a disagreeable taste to the water.

Hardness due to calcium or magnesium ions, present as carbonates, is produced when slightly acidic water trickles through limestone ($CaCO_3$) or dolomite ($CaCO_3 \cdot MgCO_3$).

Such "hard water" can be softened by removing these ions. One of the methods for softening water is the lime–soda process. The lime–soda process takes advantage of the facts that calcium carbonate ($CaCO_3$) is much less soluble than calcium bicarbonate [$Ca(HCO_3)_2$] and that magnesium hydroxide is much less soluble than magnesium bicarbonate. The raw materials added to the water in this process are hydrated lime ($Ca(OH)_2$) and soda (Na_2CO_3).

The overall result of the lime–soda process is to precipitate almost all the calcium and magnesium ions in the form of calcium carbonate and magnesium hydroxide and to leave sodium ions as replacements.

Iron present as Fe^{2+} and manganese present as Mn^{2+} can be removed from water by oxidation with air (aeration) to higher oxidation states. If the pH of the water is 7 or above (either naturally or through the addition of lime), the insoluble compounds $Fe(OH)_3$ and $MnO_2(H_2O)_x$ are produced and precipitate from solution.

The desire for and achievement of soft water for domestic use has sparked a rather heated health debate. Soft water is usually acidic and contains Na^+ ions in the place of di- and trivalent metal ions. An increased intake of Na^+ is known to be related to heart disease. The acidic soft water is also more likely to attack metallic pipes, joints, and fixtures, resulting in the dissolution of toxic ions such as Pb^{2+}. One way to avoid sodium ions in drinking water and to use less soap when washing would be to drink only naturally hard water and to do your washing in soft water.

Many commercially available water-softening systems consist of an ion exchange column containing some type of insoluble, negatively charged resin material that has been loaded up with sodium cations. As water passes through the resin column, the doubly charged cations (Ca^{2+}, Mg^{2+}, and Fe^{2+}) common in hard water are attracted more tightly to the column than are the singly charged Na^+ ion. The sodium ions are released into the water and the doubly charged cations are removed. This water-softening process is illustrated in Figure 11.9. A given ion exchange column has a finite capacity to soften water since it will eventually become loaded with the hard water ions. Many modern refrigerators with drinking water dispensers and ice makers contain such water softening systems.

Sodium carbonate (Na_2CO_3), sometimes known as washing soda, can also be used to soften water. The addition of this substance to water allows for formation of the more insoluble calcium carbonate and magnesium carbonate, which will settle out as solids.

(11.12) CHLORINATION AND OZONE TREATMENT OF WATER

With the advent of chlorination of water supplies in the early 1900s, the number of deaths in the United States caused by typhoid and other waterborne diseases dropped from 35 per 100,000 population in 1900 to 3 per 100,000 population in 1930.

Chlorine is introduced into water as the gaseous element (Cl_2), and it acts as a powerful oxidizing agent for the purpose of killing bacteria in water. This process is used in treating water that will become both tap water and wastewater before it is released. Chlorination largely prevents the principal waterborne diseases spread by bacteria, which include cholera, typhoid,

Runk/Schoenberger/Grant Heilman Photography

Chlorine contact tank in a sewage treatment plant. Whatever the level of treatment, the water is chlorinated to kill disease-causing organisms before it is returned to natural waterways.

Hard water contains metal ions that react with soaps and give precipitates.

For people on low sodium diets, lime–soda treated water might represent too high a daily dose of sodium.

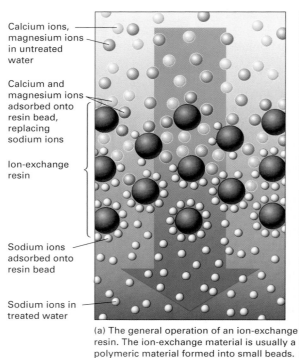

Calcium ions, magnesium ions in untreated water

Calcium and magnesium ions adsorbed onto resin bead, replacing sodium ions

Ion-exchange resin

Sodium ions adsorbed onto resin bead

Sodium ions in treated water

(a) The general operation of an ion-exchange resin. The ion-exchange material is usually a polymeric material formed into small beads.

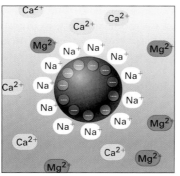

(b) An ion-exchange resin generally has negatively charged groups on the surface of the resin. Sodium ions are adsorbed onto the surface and balance the negative charge.

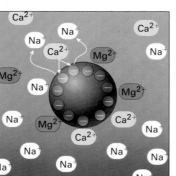

(c) Ion-exchange column after Ca^{2+} and Mg^{2+} ions have passed down the column. The more highly charged Ca^{2+} and Mg^{2+} ions are more strongly attracted to the resin beads than the sodium ions with a single positive charge. This leads to the replacement of the sodium ions on the resin by the alkaline earth ions.

FIGURE 11.9 Water softening by ion exchange in the home.

paratyphoid, and dysentery. Our municipal water supplies, however, remain vulnerable to contamination. In 1993, for example, in the Milwaukee area, a major outbreak of gastrointestinal disease was caused by *Cryptosporidium* bacteria. Lesser outbreaks have been caused by *Salmonella* bacteria. In some cases, the pathogens are found to have entered the water supply via contamination with surface water runoff.

Chlorination of industrial wastewater and city water supplies presents a potential threat because of the reaction of chlorine with residual concentrations of organic compounds to produce **disinfection by-products**.

$$\text{Water containing} \atop \text{organic compounds} \xrightarrow{\text{Chlorine}} \text{chlorinated organic} \atop \text{disinfection by-products}$$

These disinfection by-products, which may be present at levels of a few parts per million or less, include dichloromethane, chloroform, trichloroethylene, and chlorobenzene—all suspected carcinogens. The presence of chlorinated hydrocarbons can be prevented by more efficient removal of the organic matter that becomes chlorinated.

One way to eliminate chlorinated hydrocarbons as disinfection by-products is to use ozone (O_3) as the disinfectant. The ozone is produced on site by passing oxygen or air through an electric discharge. This process normally gives about a 20% ozone–oxygen mixture that is a very strong oxidizer.

Ozonation, like chlorination, is also not without potentially harmful disinfection by-products. Bromide ion (Br^-), which is found in most natural waters, is oxidized by ozone to bromate ion (BrO_3^-), which is a suspected carcinogen. Generally, this single known harmful disinfection by-product is considered less of a risk factor than the numerous chlorinated hydrocarbons produced by chlorine disinfection.

Disinfection by-products Compounds formed by the action of disinfectants like chlorine and ozone on substances found in water

THE CHEMISTRY OF

DR. STRANGELOVE, OR HOW I LEARNED TO STOP WORRYING AND LOVE THE BOMB (1964)

©2005 by Mark A. Griep, *University of Nebraska-Lincoln*

The movie *Dr. Strangelove* (released in 1964) is a comedy about cold-war paranoia. This film raises questions that are still relevant, including the debate about water fluoridation that lies at the heart of it. In the film, General Jack D. Ripper sends one wing of the Strategic Air Command to bomb their targets in the Soviet Union. Government and military authorities try to stop the bombers and, in anticipation of failure, plan for the next step.

In one clip, Group Captain Lionel Mandrake has been locked in General Ripper's office. While engaging Ripper in conversation, Mandrake learns that Ripper will only drink distilled or rain water. For Ripper, the idea that there is fluoride in the water supply is proof of a Communist plot.

Since 1956, the public water fluoridation program has been considered the greatest success of any public health program. The benefits are significant (60% fewer cavities regardless of age, gender, or economics), the risks are almost nonexistent, and the cost is very low.

The link between fluoride in drinking water and reduction of tooth decay began its 30-year journey in 1901 in Colorado Springs. A young dentist named Frederick McKay noticed that people born in the area: 1) had brown teeth and 2) were virtually free of tooth decay. McKay visited many other communities to study the brown tooth problem. He couldn't find the causative agent but he did find that it was related to the local water supply. In 1930, an industrial chemist named H. V. Churchill determined that fluoride was present at high levels in all of McKay's towns (between 2 and 14 ppm).

Further research showed that lower fluoride (about 1.0 ppm) significantly reduced tooth decay but did not cause browning. After safety trials and epidemiological studies, fluoride was added to the water supply of Grand Rapids, Michigan, in 1945. Eleven years later, the average number of cavities in 30,000 Grand Rapids children had dropped from seven to three.

$$Ca_5(PO_4)_3(OH) + F^- \rightleftharpoons Ca_5(PO_4)_3F + OH^-$$
$$\text{Hydroxylapatite} \qquad\qquad \text{Fluorapatite}$$

Hydroxylapatite is the mineral in teeth and bone that crystallizes between the collagen fibers and gives strength. When a fluoride ion displaces hydroxyl ions, it becomes fluorapatite. Fluorapatite is even stronger and prevents tooth decay by slowing demineralization and speeding remineralization.

Questions to Consider

1. What are the fluoride levels in your public water supply?

2. Since humans don't require fluoride for growth or sustenance, is fluoride an essential nutrient?

3. If this movie were remade today, what paranoid phobia might Gen. Jack D. Ripper be given?

11.13 FRESHWATER FROM THE SEA

Because seawater covers 72% of Earth, it is not surprising that it is considered a water source for areas where freshwater supplies aren't sufficient to meet the demand. The oceans contain an average 3.5% (35,000 ppm) dissolved salts by weight, a concentration too high to make ocean water useful for drinking, washing, or agricultural use (Table 11.9). The total of dissolved ions must be reduced to below 500 ppm before the water is suitable for human consumption.

Technology has been developed for the conversion of seawater to freshwater. The extent to which this technology is actually put to use depends on the availability of freshwater and the cost of the energy for the conversion.

According to the American Water Works Association, more than 12,500 desalination plants in 120 countries were in operation in 2004; 60% of these plants are in the Middle East. Two methods used to purify seawater are *reverse osmosis* and *solar distillation*.

TABLE 11.9	Ions Present in Seawater at Concentrations Greater Than 0.001 g/kg	
ION	**G/KG SEAWATER**
Cl^-	19.35
Na^+	10.76
SO_4^{2-}	2.71
Mg^{2+}	1.29
Ca^{2+}	0.41
K^+	0.40
HCO_3^-, CO_3^{2-}	0.106
Br^-	0.067
$H_2BO_3^-$	0.027
Sr^{2+}	0.008
F^-	0.001
Total	35.129

Approximately 20% of the desalination plants worldwide depend on thermal (solar) distillation. Saudi Arabia currently has the world's largest desalination plant. It produces 128 million gallons of drinkable water per day; collectively these plants produce 70% of Saudi Arabia's drinkable water. In the United States, Tampa Bay, San Diego, and El Paso already have large desalination plants that utilize reverse osmosis or have such plants nearing completion. The Tampa Bay plant, the largest plant in the United States as of this writing, has the capacity to produce 25 million gallons of drinkable water per day but has struggled to meet this capacity due to technical problems involving pretreatment of the water to prevent fouling of the reverse osmosis membranes. The San Diego plant, planned to come online in 2010, is supposed to be able to produce 50 million gallons per day of drinkable water. It is likely that more such plants will be built.

Reverse Osmosis

An extremely thin piece of material such as a sheet of synthetic polymer or animal tissue can allow some molecules to pass through it. Such a material is called a membrane and is said to be **permeable** to those molecules and ions that can pass through. Permeability is dependent on the presence of tiny passages within the membrane. A membrane permeable to water molecules but not to ions or molecules larger than water molecules is called a **semipermeable membrane**. Many membranes made from synthetic polymers have this characteristic. One such polymer is cellulose acetate. If a semipermeable membrane is placed between seawater (brine) and pure water, the pure water will pass through the membrane to dilute the seawater. This is a process called **osmosis**. The liquid level on the seawater side rises as more water molecules enter than leave, and pressure is exerted on the membrane until the rates of diffusion of water molecules in both directions are equal. **Osmotic pressure** is defined as the external pressure required to *prevent* osmosis. Figure 11.10a and 11.10b illustrates the concepts of osmosis and osmotic pressure.

Permeable Allows substances to pass through

Semipermeable membrane A membrane that allows water molecules but not ions or larger molecules to pass through

Osmosis The flow of water molecules through a semipermeable membrane from a less-concentrated to a more-concentrated solution (i.e., from the side of purer water to the side of saltier water to dilute it).

Osmotic pressure The external pressure required to prevent osmosis from taking place

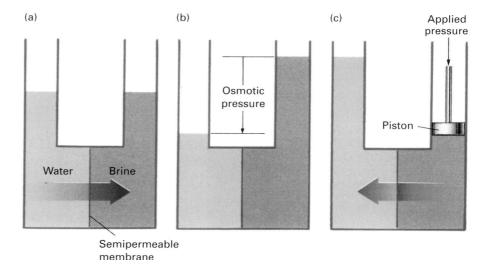

FIGURE 11.10 Osmosis (a, b) and reverse osmosis (c). (a) In normal osmosis, water molecules pass through the semipermeable membrane from the less-concentrated into the more-concentrated solution, in this case from pure water into the brine (salt water). (b) At a certain height of the brine solution in this apparatus, the pressure of the column of water is equal to the osmotic pressure and the flow stops. (c) In reverse osmosis, the application of external pressure greater than the osmotic pressure forces water molecules from the salty side to the pure water side.

Reverse osmosis is the application of pressure to cause water to pass through the membrane from the salty (aqueous solution) side to the pure-water side (Figure 11.10c). The osmotic pressure of normal seawater is 24.8 atmospheres (atm). As a result, pressures greater than 24.8 atm must be applied to cause reverse osmosis. Pressures up to 100 atm are used to provide a reasonable rate of reverse osmosis and to account for the increase in salt concentration that occurs as the process proceeds. Figure 11.11 shows a small, portable reverse osmosis device that can be used in an emergency to produce small amounts of pure water for personal use.

Courtesy of Katadyn

FIGURE 11.11 An emergency hand-operated water desalinator that works by reverse osmosis. It can produce 4.5 L of pure water per hour from seawater. Such devices can be very useful for persons adrift at sea.

11.14 PURE DRINKING WATER FOR THE HOME

In spite of all the efforts taken to purify public water supplies, many consumers are concerned about the quality of the water that comes out of the taps in their homes, schools, and places of business. Parents of small children are especially worried about chemicals such as lead and carcinogenic organic compounds that are chlorine disinfection by-products. Many have turned to bottled water or home water treatment devices.

The FDA regulates the quality of bottled water. The water passes through one or more purification steps (Figure 11.12). The three purification methods (which can also be done in home treatment systems)—distillation, carbon filtration, and reverse osmosis—have already been discussed as methods used for treating municipal water. The maximum levels of contaminants allowed by the EPA after these treatments are listed in Table 11.10. Each of these methods is expensive and results in a high cost per gallon of treated water. In the case of bottled water or home water treatment, the cost of treatment is not the major factor, because only the small amount of water needed for human consumption needs to be specially purified. Figure 11.12 shows

W See an animation of **Osmosis** at http://brookscole.com/chemistry/joesten4

In 2004, U.S. sales of bottled water reached 6.4 billion gallons at a cost of $8.3 billion. Use continues to grow at an annual rate of 10–15%.

Reverse osmosis The application of pressure on a solution to cause water molecules to flow through a semi-permeable membrane from a more-concentrated to a less-concentrated solution

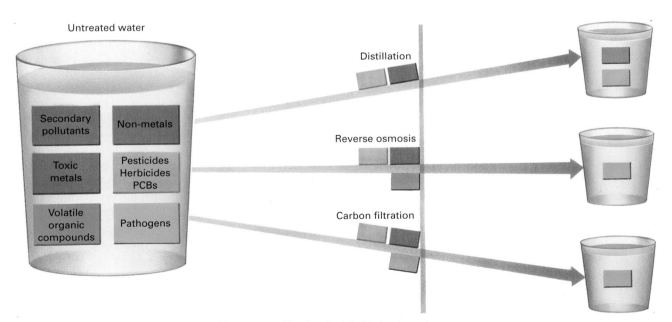

FIGURE 11.12 Final steps that can be used in water purification. Both bottled water and municipal tap water are purified in these ways. The color code shows which pollutants are removed by each method. Pathogenic bacteria can and do pass through all these methods. This is why municipal tap water must be treated with chlorine or some other disinfectant before release into the system.

TABLE 11.10 A Partial List of Maximum Contaminant Levels (MCL) for Drinking Water Allowed by the EPA*

Metals	
Beryllium	0.004
Cadmium	0.005
Chromium	0.100
Copper	1.300
Lead	0.015
Metalloids	
Arsenic	0.010
Antimony	0.006
Nonmetals	
Fluoride	4.00
Nitrate	10.00
Volatile Organic Compounds	
Benzene	0.005
Carbon tetrachloride	0.005
Vinyl chloride	0.002
Trichloroethylene	0.005
Hexachlorobenzene	0.001
Styrene	0.100
Herbicides, Pesticides, PCBs	
Chlordane	0.002
Endrin	0.0002
Heptachlor	0.0004
Lindane	0.0002
Methoxychlor	0.040
Toxaphene	0.003
PCBs	0.0005
Secondary Contaminants	
Iron	0.30
Manganese	0.05
Zinc	5.00
Chloride	250.00
Sulfate	250.00
Total dissolved solids	500.00

*Values are in milligrams per liter.
Source: www.epa.gov/safewater/mcl.html.

which trace pollutants are removed or allowed to remain in the water by each treatment method, color-coded to examples of the types of contaminants in Table 11.10. Thus, the analysis on the label of the bottled water is highly important and should be read with care before purchase.

CONCEPT CHECK 11C

1. Which water purification process is not a natural process? (**a**) Distillation, (**b**) Aeration, (**c**) Filtration, (**d**) Reverse osmosis, (**e**) Settling.
2. Which word can be used to describe organic pesticides that are not readily biodegradable? (**a**) Permanent, (**b**) Persistent, (**c**) Nonvolatile.
3. Name two water purification methods that are part of primary wastewater treatment.
4. Secondary wastewater treatment operates under (aerobic/anaerobic) conditions.
5. A cesspool operates under (aerobic/anaerobic) conditions.
6. Select the ions that may cause water to be hard: (**a**) Sodium, (**b**) Calcium, (**c**) Magnesium, (**d**) Potassium.
7. Two methods used to kill harmful microorganisms in water are _____ and _____.
8. What are the four metal ions present in seawater at concentrations of 400 ppm or higher?
9. Water flows through a semipermeable membrane from a solution of low salt concentration to a solution of higher salt concentration. This process is called _____.

The World Heath Organization (WHO) reports that drinking unclean water causes diarrhea that kills 1.6 million children under the age of five each year. An estimated 1 billion people in the world do not have access to safe, pure drinking water. By the time deaths attributable to diarrhea, cholera, schistosomiasis, and other diseases spread by contaminated water or aggravated by the lack of pure water for personal hygiene are included, the worldwide death toll is estimated at 9,300 people per day or over 3.4 million per year! These numbers are hard to fathom for most Westerners, but they are far too easy to grasp for those living in certain areas of Asia and Africa where such deaths are, unfortunately, almost routine.

Much work remains to be done to ensure that the world's limited water resources are used wisely and equitably. The problem will only become more serious as the world's population increases and as access to water becomes more and more difficult. Ultimately, the necessary water is available. It is just not pure enough for use in many cases because of anthropogenic or natural pollution or salinity and it is certainly in the wrong locations in many others. Without water from other parts of the country, cities such as Los Angeles, Phoenix, and Las Vegas would not have grown to their current size. The questions of how much longer such cities, and many others like them in the United States and around the world, can sustain this growth, and at what cost, remain to be answered. *Time* magazine reported in April of 2004 that the 2.5 billion gallons of water used to irrigate the world's golf courses each day would be enough to support 4.7 billion people at the United Nation's daily minimum of water. National Geographic reported in an excellent look at the world's water resources in September of 2002 that bone-dry Dubai had 2.6 million gallons of expensively desalinated seawater coursing through its popular Wild Wadi Water Park. It may be that money is a universal cure for this problem.

W Assess your understanding of this chapter's topics with an online chapter quiz at **www.brookscole.com/chemistry/joesten4**

■ KEY TERMS

specific heat capacity	potable	denitrifying bacteria
heat of fusion	groundwater recharge	disinfection by-products
heat of vaporization	pathogens	permeable
surface tension	hazardous waste	semipermeable membrane
acid rain	biochemical oxygen demand (BOD)	osmosis/reverse osmosis
aquifer	biodegradable/nonbiodegradable	osmotic pressure

■ THE LANGUAGE OF CHEMISTRY

1. Sedimentation
2. Biodegradable
3. Clean Water Act of 1972
4. BOD
5. Aquifer
6. Ammonium ion
7. Unused paint
8. Reverse osmosis
9. Water hardness
10. Ozone treatment
11. Recycling
12. Superfund
13. Incineration
14. Aeration
15. Carbon adsorption
16. Chlorination
17. Hydrogen bonding
18. Heat required to melt 1 gram of ice at 0°C
19. Heat required to vaporize 1 gram of water at 100°C
20. Heat capacity

a. A measure of dissolved organic material in water
b. Source of groundwater
c. Caused by metal ions such as Ca^{2+} and Mg^{2+} in solution
d. A nutrient for microorganisms living in water
e. Provides for hazardous waste site cleanup
f. Alternative to landfills but can contribute to air pollution
g. Disinfection method commonly used for drinking water and wastewater
h. Removes organic compounds from water
i. Relieves dependence on landfills
j. Common household hazardous waste
k. Primary sewage treatment process
l. Disinfectant method that can produce bromate ion from bromide ion
m. Secondary sewage treatment process
n. Naturally decomposed to simpler compounds
o. Shifted responsibility for water purity to wastewater discharger
p. Uses pressure to purify water
q. Heat of fusion
r. Heat required to raise temperature of 1 gram of water by 1°C
s. Responsible for the high boiling point of water relative to other compounds of similar molecular weight
t. Heat of vaporization

■ APPLYING YOUR KNOWLEDGE

1. Define the following terms:
   (**a**) Surface water   (**b**) Groundwater
   (**c**) Brackish water   (**d**) Pollution
   (**e**) Groundwater recharge   (**f**) Potable

2. Define the following terms:
   (**a**) Distillation   (**b**) Hard water
   (**c**) Disinfection by-products   (**d**) Semipermeable
   (**e**) Dilution   (**f**) Hazardous waste

3. Where does most of the water go that falls on the continental United States every day?

4. Explain how rainwater becomes groundwater.

5. Explain how groundwater can become contaminated with pollutants.

6. What is the origin of brackish water?

7. What activity is the largest single user of water?

8. What causes the level of an aquifer to drop? Cite some examples of the effects of aquifers dropping in level.

9. How much water do you think you use per day? List your uses and include water that might be used "for you," such as in food preparation in a restaurant.

10. What would you expect to find dissolved in "clean" water?

11. What does the term "groundwater recharge" mean? What is the source of water that is used for this purpose?

12. Name five kinds of pollution often found in water. Give a source for each.

13. Prior to the enactment of the Clean Water Act, who was responsible for ensuring that water being used was pure? After passage of this act, who is now responsible?

14. Both surface water and groundwater (natural waters) often contain dissolved ions. Name three positive and three negative ions that are often found in natural waters.

15. Name two methods of disposal for solid wastes from industry and households. Which one of these has the greater possibility to adversely impact water quality?

16. What is the "Superfund"? Explain how it is used to improve water quality.

17. Describe a way by which a landfill can be made more "secure" in terms of water quality protection.

18. Go to the hardware department in a large department store and choose five products. Then list the kinds of hazardous wastes the manufacture of these products might produce. Use Table 11.7 as a guide.

19. Name three common household wastes and the kinds of chemicals they contain that might be harmful to water quality.

20. Describe how measuring the biochemical oxygen demand (BOD) of a sample of water indicates something about its purity.

21. Describe how pure water can become contaminated with lead in the home.

22. Describe how the BOD of wastewater can be lowered.

23. Explain how distillation purifies a sample of water.

24. Explain how aeration purifies a sample of water containing dissolved organic compounds.

25. Explain how settling and filtration purify water samples.

26. What is the difference between "biodegradable" and "nonbiodegradable"? If you had a choice between using a biodegradable and a nonbiodegradable detergent to clean your clothes, which would you choose?

27. Chlorinated hydrocarbons and branched-chain hydrocarbons are nonbiodegradable. What happens to them when they are released into the environment?

28. Name the two methods of primary sewage treatment.

29. Too-high concentrations of nitrogen compounds like ammonia (NH_3) adversely affect water quality. What tertiary sewage treatment method gets rid of these compounds?

30. Name the ions commonly present in hard water. What kinds of problems do they cause?

31. How is chlorination of water similar to aeration of water? How are these different?

32. What are "disinfection by-products," and how are these potentially harmful?

33. Explain how reverse osmosis can be used to purify seawater.

34. Which method of purification of drinking water for the home would most likely get rid of dissolved organic compounds?

35. Determine the amount of heat required to convert 14.0 g of ice at 0°C to 14.0 g of liquid water at 50°C.

36. Use a search engine on your computer to discover the relationship of the following to water or water pollution.
    (a) Love Canal
    (b) *A Civil Action*, a 1995 book by Jonathan Harr
    (c) *Erin Brockovich*, the movie
    (d) Times Beach, Missouri
    (e) Minamata disease

37. It may not be immediately obvious, but a steam burn is often more serious than a burn from boiling water. Why do you think this might be?

W CHEMISTRY ON THE WEB

For up-to-date URLs, visit the text website at **www.brookscole.com/ chemistry/joesten4**

- Water Basics
- Water-Related Frauds, Quackery, and Pseudoscience
- Acid Rainfall
- General Information about Water and Water Pollution
- Drinking Water Polluted by MTBE
- Frequently Asked Questions about Bottled Water
- Water Humor for the Scientifically Illiterate

ENERGY AND HYDROCARBONS

Natural gas is flared from an offshore oil rig somewhere in the North Sea. Enormous offshore oil rigs are necessary to drill to depths of 3000 meters.

© Age Fotostock/Superstock

The average person in the United States uses far more energy (per capita energy consumption) than the citizens of any other country in the world, and the level of usage continues to rise at an alarming rate. The United States has about 5% of the world's population, but we consume around 25% of the daily supply of energy, 85% of which comes from fossil fuel combustion. We have already seen (Chapter 6) that carbon dioxide, a product of the combustion of hydrocarbons (organic compounds containing only carbon and hydrogen), may play a major role in global warming. It should come as no surprise that there is much concern over the implications of humans continuing to burn enormous quantities of fossil fuels (coal, petroleum, and natural gas) and the products that can be derived from them (e.g., gasoline, methanol, and ethanol). On one hand, there is concern that the supply of these fossil fuels, and, thus, the energy derivable from them, is limited. On the other hand, there is concern that continued exploitation of the energy derivable from the combustion of these fuels may accelerate global warming even if anthropogenic carbon dioxide emissions turn out not to be the principal cause of the phenomenon. All things considered, maybe the best use of fossil fuels is not to burn them to derive the energy contained within their chemical bonds but to exploit them for some higher purpose. We will address this issue in this chapter, consider one alternative fuel source in Chapter 13, and consider other uses of hydrocarbons in Chapter 14.

Hydrocarbons are the principal component of fossil fuels. Natural gas is primarily methane, crude petroleum is a complex mixture of thousands of hydrocarbons, and coal is an even more complex mixture of hydrocarbons. Many of the fuels we use, such as gasoline and jet fuel, are obtained from petroleum.

Fossil fuels are also the major source of hydrocarbons that are used to make thousands of consumer products. This chapter describes the chemistry of hydrocarbons and their importance to our energy needs, and Chapter 14 emphasizes the industrial uses of hydrocarbons. Alcohols and ethers are part of the energy discussion because of the need to improve emissions and reduce pollution. Questions to be considered in this chapter are:

- What are fuels?
- How do fuels produce energy?
- What are the major classes of hydrocarbons?
- What different types of isomers are possible for hydrocarbons, and why are they important?
- How is petroleum refined?
- How is high-octane gasoline produced?
- What are oxygenated gasolines, and why are they used?
- Why are methanol and ethanol receiving attention as alternate fuels?

12.1 ENERGY FROM FUELS

In our homes and our industries, we obtain most of our energy by burning fossil fuels (Figure 12.1). Why does burning fuels provide energy? Fuels are reduced forms of matter that burn readily in the presence of oxygen, and combustion is an oxidation reaction (Section 10.1) that produces heat. The burning of methane, the principal component of natural gas, can be used to illustrate how the combustion of fuels produces energy.

Fuels are burned so that we may utilize the energy released from that process for some purpose. Let's consider what determines whether energy is released (exergonic) or absorbed (endergonic) in a chemical reaction, such as the combustion of a fuel. We know that a chemical reaction involves the conversion of reactants into products. The conversion requires that some chemical bonds in the reactants be broken. Bond cleavage requires an energy input. This endothermic process is represented in mathematical terms with a plus (+) sign by chemists, as will be illustrated below. New chemical bonds must be made as product is formed. The formation of chemical bonds releases energy, and this exothermic process is represented in mathematical terms by a minus (−) sign by chemists.

Some average bond energies are shown in Table 12.1. This is the heat energy or enthalpy that is required to break one mole of the bond in question. Since bond cleavage and bond formation are the reverse of each other, this is also the heat energy that is released upon formation of one mole of the

Although the International System of Units or Système International (SI) unit of energy is the joule, the more familiar unit of heat is the calorie. A calorie is the amount of heat required to raise the temperature of 1 g of water 1°C. One calorie equals 4.18 joules (J). 1000 calories (cal) = 1 kilocalorie (kcal). For example, you use 140 kcal/h walking and 80 kcal/h even when you are asleep.

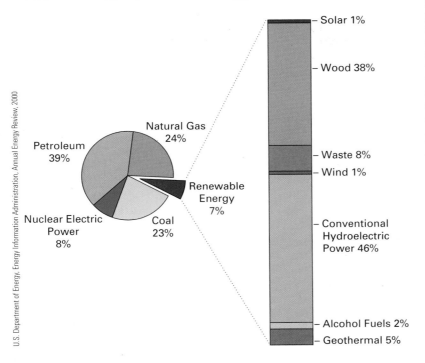

Renewable Energy as Share of Total Energy, 2000

U.S. Department of Energy, Energy Information Administration, Annual Energy Review, 2000

- Solar 1%
- Wood 38%
- Waste 8%
- Wind 1%
- Conventional Hydroelectric Power 46%
- Alcohol Fuels 2%
- Geothermal 5%

Petroleum 39%
Natural Gas 24%
Nuclear Electric Power 8%
Coal 23%
Renewable Energy 7%

FIGURE 12.1 Sources of energy used in the United States in 2000. Taken together, the fossil fuels (petroleum, natural gas, and coal) accounted for 86% of our fuel. These are not renewable sources of energy. The limit is the quantity stored within planet Earth. As fossil fuel supplies diminish, the use of renewable energy—that is, energy from sources that remain available—will become more and more important. Renewable energy comes from plants, wind, flowing water, geothermal vents, and the Sun.

bond in question. The bond energies may differ slightly from one compound to another but these average bond energies suffice for our purposes.

The energy changes associated with chemical reactions fall into three different categories:

Energy Is Consumed Overall—If the sum of the energies required to break bonds in the reactants is greater than the energy released by all new bonds formed in the products, energy will be consumed in the chemical reaction. The endothermic, bond-breaking processes (+ sign) require more energy than the exothermic, bond-forming processes (− sign) release, so the sum of these energy changes is endothermic overall (+ sign). This type of chemical reaction requires an overall energy input, and, thus, such a reaction would not be useful as a source of energy.

Energy Is Released Overall—If the sum of the energies required to break bonds in the reactants is less than the energy released by all new bonds formed in the products, energy will be released in the chemical reaction. Thus, the endothermic, bond-breaking processes (+ sign) require less energy than the exothermic, bond-forming processes (− sign) release, so the sum of these energy changes is exothermic overall (− sign). This type of chemical reaction produces energy overall, and, thus, might be a useful source of energy. In other words, the reactants might represent a usable fuel. Factors other than the amount of energy released by combustion of a fuel may also be important in determining its suitability as a fuel, however.

No Net Energy Change Occurs—If the sum of the energies required to break bonds in the reactants is exactly equal to the energy released by all new bonds formed in the products, there is no net energy change. Thus, the endothermic, bond-breaking processes (+ sign) require exactly the same amount of energy released by the exothermic, bond-forming processes (− sign) so the sum of these energy changes is zero. This is a relatively rare, although not impossible, occurrence for a chemical reaction.

TABLE 12.1 Average Bond Energies

Bond	Energy (kJ/mol)
C—H	416
O=O	498
C=O	803
O—H	467
C—C	356
H—H	436
C—O	336

One can, in principle, calculate the heat energy (enthalpy) change in a chemical reaction if the balanced chemical equation is known and the bond energies of all the bonds broken and made are known. For our purposes, we have only included a very abbreviated table (Table 12.1) of the types of bonds that might be involved in the combustion of a hydrocarbon or related fuel.

Let's begin our mathematical analysis of the combustion of some fuels by looking at methane. The complete combustion of any hydrocarbon will produce only carbon dioxide and water so it is a relatively simple matter to write such an equation for methane. By using the bond energies in Table 12.1, we see that it takes a total of 2660 kJ to break all the bonds in the reactants (1 mole of methane and 2 moles of oxygen). Since those processes are endothermic, we indicate this energy requirement as + 2660 kJ. Similarly, we see that the formation of all the bonds in the product molecules releases 3474 kJ. We designate this exothermic process as −3474 kJ. The sum of these two processes is −814 kJ (the negative sign indicates the overall process is exothermic). This is the amount of heat energy released in burning 1 mole of methane. The heat released when a fuel is burned is known as the **heat of combustion**.

$$\begin{array}{c} \text{H} \\ | \\ \text{H}-\text{C}-\text{H} \\ | \\ \text{H} \end{array} + 2\,\text{O}{=}\text{O} \xrightarrow{\text{spark}} \text{O}{=}\text{C}{=}\text{O} + 2\,\text{H}-\text{O}-\text{H}$$

Bonds Broken Endothermic (+)		Bonds Formed Exothermic (−)	
4 (C—H)	2 (O=O)	2 (C=O)	4 (H—O)
4 mol (416 kJ/mol)	2 mol (498 kJ/mol)	2 mol (−803 kJ/mol)	4 mol (−467 kJ/mol)
+1664 kJ	+996 kJ	−1606 kJ	−1868 kJ
Sum = +2660 kJ		Sum = −3474 kJ	

Net = (+2660 kJ) + (−3474 kJ)
Net = −814 kJ

By repeating this type of calculation, we can compare the energy released for several different types of fuels. However, there is something else we need to consider before comparing the amount of energy released for different fuels. Our calculations would yield the amount of energy released per mole of fuel. This is not a good way to compare fuels because one mole of methane has a mass of 16 g, but one mole of octane has a mass of 114 g. We get a better number for comparison when we divide the heat of combustion per mole by the molecular weight of the compound. This is the number we should use when comparing the heat released by different fuels. For methane, −814 kJ/mol divided by 16 g/mol for methane gives a value of −50.8 kJ/g. This number is in good agreement with the experimental number of −50.1 kJ/g in Table 12.2 that also includes the entropy contribution to the energy change. A generalization is that most hydrocarbons release about 40–50 kJ/g of energy upon complete combustion to carbon dioxide and water.

Octane and 2,2,4-trimethylpentane (both of which have the same formula, C_8H_{18}, a molecular weight of 114 g/mol) release the same amount of energy (−5,049 kJ/ mol or −44.3 kJ/g) upon combustion. Both reactions, shown below, involve the breaking and formation of exactly the same types and numbers of bonds. Note in the analysis below that the balanced

Heat of combustion The quantity of heat released when a fuel is burned; variously expressed as kcal/mol, kcal/g, kJ/mol, or kJ/g

TABLE 12.2 Experimental Heats of Combustion of Some Common Fuels

FUEL	HEAT OF COMBUSTION (KJ/MOL)	HEAT OF COMBUSTION (KJ/G)
Hydrogen	-242	-121
Methane	-802	-50.1
Ethane	-1428	-47.6
Propane	-2016	-45.8
"Oil"—a hydrocarbon mixture	—	~-45
Anthracite coal	—	-30.5
Ethanol	-1367	-29.7
Methanol	-640	-20
Lignite coal	—	-16
Glucose	-2800	-15.5 (reverse of photosynthesis)
Wood	—	-10 to -14

equation initially gives the amount of energy for combustion of 2 mole of octane so we must divide the value obtained by 2 to arrive at the heat of combustion per mole.

$$2\ CH_3(CH_2)_6CH_3 + 25\ O{=}O \xrightarrow{\text{spark}} 16\ CO_2 + 18\ H_2O$$

or

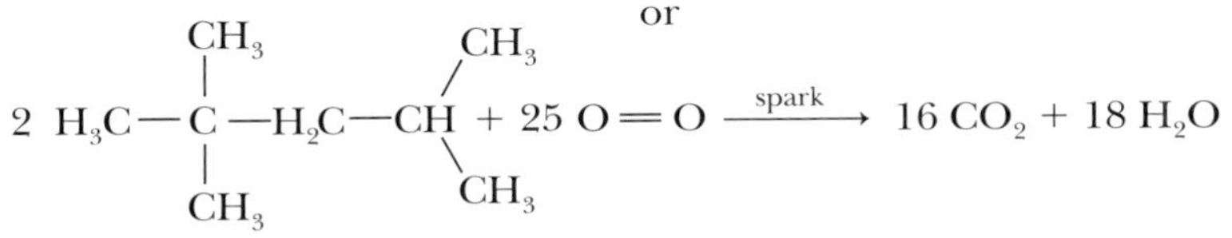

$$2\ H_3C-\underset{\underset{CH_3}{|}}{\overset{\overset{CH_3}{|}}{C}}-H_2C-CH\overset{CH_3}{\underset{CH_3}{\diagdown}} + 25\ O{=}O \xrightarrow{\text{spark}} 16\ CO_2 + 18\ H_2O$$

Bonds Broken Endothermic (+)		**Bonds Formed** Exothermic (−)	
2×7 (C—C) 14 mol (356 kJ/mol) = +4984 kJ	25 (O=O) 25 mol (498 kJ/mol) = +12,450 kJ	32 (C=O) 32 mol (-803 kJ/mol) = $-25,696$ kJ	36 (H—O) 36 mol (-467 kJ/mol) = $-16,812$ kJ

2×18 (C—H)
36 mol (416 kJ/mol)
+14,976 kJ

Sum = $-42,508$ kJ

Sum = +32,410 kJ for
2 moles C_8H_{18}

Net = (+32,410 kJ) + ($-42,508$ kJ)

Net = $-10,098$ kJ for 2 mol of C_8H_{18}

Net = $-5,049$ kJ for 1 mol of C_8H_{18}

As we shall see later (Section 12.2), octane is a very poor fuel and 2,2,4-trimethylpentane is a very good fuel even though they both release the same amount of energy upon combustion. Why is that? Clearly the quality of a fuel must be related to more than its energy content.

Let's compare four other fuels (hydrogen, ethane, methanol, and ethanol) by writing out the complete balanced equations for combustion and doing the simple mathematics using the bond energies in Table 12.1.

$$2\,H_2 \quad + \quad O_2 \quad \xrightarrow{\text{spark}} \quad 2\,H_2O$$

$$2\,CH_3CH_3 \quad + \quad 7\,O_2 \quad \xrightarrow{\text{spark}} \quad 6\,H_2O \quad + \quad 4\,CO_2$$

$$2\,CH_3OH \quad + \quad 3\,O_2 \quad \xrightarrow{\text{spark}} \quad 4\,H_2O \quad + \quad 2\,CO_2$$

$$CH_3CH_2OH \quad + \quad 3\,O_2 \quad \xrightarrow{\text{spark}} \quad 3\,H_2O \quad + \quad 2\,CO_2$$

We can determine that all of these combustion reactions are exothermic; that is, they all release heat. Table 12.2 is a summary of the heats of combustion of the fuels just considered and includes some additional numbers for comparison.

EXAMPLE 12.1   Heat of Combustion of Methanol

W See an animation of **Burning Methanol** at http://brookscole.com/chemistry/joesten4

Determine the value of the heat of combustion of methanol by using the bond energies in Table 12.1. Compare it to that of methane.

SOLUTION

Using the balanced equation above we see that the combustion of 2 moles of methanol requires 3 moles of oxygen and produces 4 moles of water and 2 moles of carbon dioxide. This requires breaking 6 moles of $C\!-\!H$ bonds (+416 kJ/mol), 2 moles of $C\!-\!O$ bonds (+336 kJ/mol), 2 moles of $O\!-\!H$ bonds (+467 kJ/mol), and 3 moles of $O\!=\!O$ bonds (+498 kJ/mol) for a total of +5596 kJ. Bond formation in the products involves 8 moles of $O\!-\!H$ bonds (−467 kJ/mol) and 4 moles of $C\!=\!O$ bonds (− 803 kJ/mol) for a total heat release of −6948 kJ. The sum of energy required (+5596 kJ) and energy released (−6948 kJ) is −1352 kJ for two moles of methanol or −676 kJ/mol. Since methanol has a molecular weight of 32 g/mol, this gives a calculated heat of combustion for methanol of −21.1 kJ/g which compares favorably with the experimental value of −20 kJ/g. Thus, the introduction of an oxygen atom into a methane molecule has reduced the heat of combustion from about −50 kJ/g to around −20 kJ/g.

TRY IT 12.1

Use the procedure in Example 12.1 to calculate the heats of combustion of ethane (C_2H_6) and ethanol (C_2H_5OH) on a per gram basis. Based upon these two examples—methane versus methanol, and ethane versus ethanol—what can you conclude about the introduction of an oxygen atom into a hydrocarbon fuel? Does this result in an increase or a decrease in the heat of combustion?

Three things are apparent from inspection of the data in Table 12.2:

- Hydrogen releases more heat per gram than any of the hydrocarbon or alcohol fuels.

- All hydrocarbons release *about* the same amount of heat per gram.

- Adding oxygen to a hydrocarbon to convert it to an alcohol reduces the amount of heat released upon combustion. This may be seen by comparing methane and methanol or ethane and ethanol.

Three other fuels (glucose, wood, and coal) are included in Table 12.2 for comparison. Note that no simple chemical equations can be written for the combustion of coal and wood because coal and wood are not pure compounds. However, we can weigh a certain mass of coal or wood, burn it, and determine the amount of heat released per gram. Most woods release about 10–14 kJ/g and coal releases about 16–30 kJ/g depending on whether we are talking about lignite (soft coal) or anthracite (hard coal). The importance of glucose will be considered later.

As has been shown, the energy content of various fuels may vary quite considerably. Factors other than the actual energy content of the fuel play a big role in determining which of these fuels we use. Availability, ease of transport and storage, pollution-producing potential, cost, and even perceived danger may affect our perception and use of the fuels. Let's look at where some of the petroleum-based fuels come from before addressing the question of "which is the best fuel?"

12.2 PETROLEUM

Crude petroleum is a complex mixture of thousands of hydrocarbon compounds, and the actual composition of petroleum varies with the location in which it is found. For example, Pennsylvania crude oils are primarily straight-chain hydrocarbons, whereas California crude oil is composed of a larger portion of aromatic hydrocarbons.

How long will petroleum be viable as a source of energy and starting materials for consumer products? This is difficult to estimate because of continued revisions of recoverable crude oil resources. For example, estimates of global petroleum reserves increased 43% between 1984 and 1994, primarily because of re-evaluation of oil reserves in the Middle East, where more than 65% of the world's oil resources are located. Counterbalancing this increase in oil reserves is the substantial increase in energy demand expected during the next 20 to 30 years from developing countries in Asia and Latin America. Current projections are for oil production to peak between 2010 and 2025. Oil production would continue for several decades after this; however, the increasing cost and lower availability will favor increased use of natural gas, coal, and alternative sources, such as wind energy and solar energy, particularly electricity obtained through the use of photovoltaic cells.

Petroleum Refining

The refining of petroleum begins with the separation of fractions according to boiling point ranges by a process called **fractional distillation**. The difference between simple distillation and fractional distillation is the degree of separation achieved for the mixture being distilled. For example, water that contains dissolved solids or other liquids can be purified by simple distillation. The impure solution is heated to boiling; the water vapor is condensed and collected in a separate container. Since petroleum contains thousands of hydrocarbons, separation of the pure compounds is not feasible or even necessary. The products obtained from distillation of petroleum are still mixtures of hundreds of hydrocarbons, so they are called **petroleum fractions**.

Figure 12.2 illustrates a fractional distillation tower used in the petroleum refining process. The crude oil is first heated to about 400°C to produce a hot vapor and liquid mixture that enters the fractionating tower. The vapor rises and condenses at various points along the tower. The lower boiling petroleum fractions (those that are more volatile) will remain in the vapor stage longer than the higher boiling fractions. These differences in boiling point

There are 42 gal of oil per barrel.

© Matthia Kulka/Corbis

Many products, including gasoline, are isolated or produced from petroleum. However, limited petroleum reserves may force us to consider how we use this resource.

Fractional distillation Separation of a mixture into fractions that differ in boiling points

Petroleum fractions Mixtures of hundreds of hydrocarbons with boiling points in a certain range that are obtained by fractional distillation of petroleum

FIGURE 12.2 Petroleum fractionation. Crude oil is first heated to 400°C in the pipe still. The vapors then enter the fractionation tower. As they rise in the tower, the vapors cool down and condense so that different fractions can be drawn off at different heights. This shows how the rising vapor is repeatedly condensed and collected at the numerous bell caps.

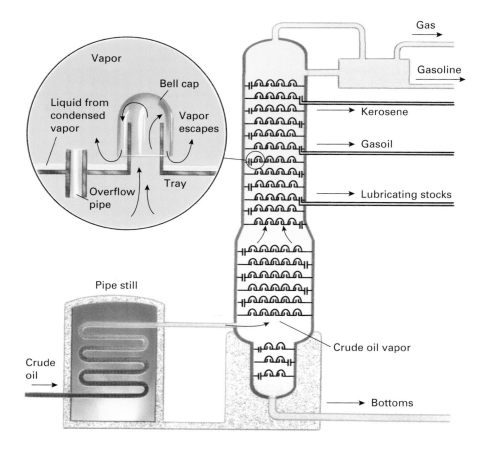

ranges allow the separation of fractions. Some of the gases do not condense and are drawn off the top of the tower, while the unvaporized residual oil is collected at the bottom of the tower. Typical products of the fractional distillation of petroleum are listed in Table 12.3.

Octane Rating

The burning properties of hydrocarbons depend on their structure. Isooctane (2,2,4-trimethylpentane) is the standard used to assign octane ratings. The octane rating is an arbitrary scale for rating the relative knocking prop-

TABLE 12.3 Hydrocarbon Fractions from Petroleum

FRACTION	SIZE RANGE OF MOLECULES	BOILING POINT RANGE (°C)	USES
Gas	C_1–C_4	0–30	Gas fuels
Straight-run gasoline	C_5–C_{12}	30–200	Motor fuel
Kerosene	C_{12}–C_{16}	180–300	Jet fuel, diesel oil
Gasoil	C_{16}–C_{18}	Over 300	Diesel fuel, cracking stock
Lubricants	C_{18}–C_{20}	Over 350	Lubricating oil, cracking stock
Paraffin wax	C_{20}–C_{40}	Low-melting solids	Candles, wax paper
Asphalt	Above C_{40}	Gummy residues	Road asphalt, roofing tar

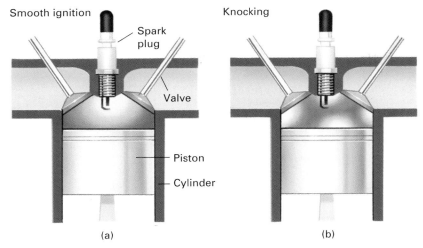

Smooth ignition

Spark plug

Valve

Piston

Cylinder

(a)

Knocking

(b)

Octane rating relates to smoothness of ignition.

erties of gasolines, and it is based on the operation of a standard test engine. Heptane knocks considerably and is assigned an octane rating of 0, whereas 2,2,4-trimethylpentane burns smoothly and is assigned an octane rating of 100. The octane rating of a gasoline is determined by first using the gasoline in a standard engine and recording its knocking properties. The test results are then compared with the behavior of mixtures of heptane and isooctane, and the percentage of isooctane in the mixture with identical knocking properties is called the *octane rating* of the gasoline. Thus, if a gasoline has the same knocking characteristics as a mixture of 13% heptane and 87% isooctane, it is assigned an octane rating of 87. This corresponds to regular unleaded gasoline. Other higher grades of gasoline available at gas stations have octane ratings of 89 (regular plus) and 92 (premium).

The "straight-run" gasoline fraction obtained from the fractional distillation of petroleum has an octane rating of only 55 and needs additional refinement because it contains primarily straight-chain hydrocarbons that burn too rapidly to be suitable for use as a fuel in internal combustion engines. Rapid burning causes uncontrolled explosion of the fuel as evidenced by a "knocking" or "pinging" sound in the engine. This reduces engine power and may damage the engine.

The octane rating of a gasoline can be increased either by increasing the percentage of branched-chain and aromatic hydrocarbon fractions or by adding octane enhancers (or a combination of both). Since the octane rating scale was established, fuels superior to isooctane have been developed, so that the scale has been extended well above 100. Table 12.4 lists octane ratings for some hydrocarbons and octane enhancers. Note that the octane rating of a fuel is not a measure of the energy content of the fuel. For instance, octane (a poor fuel) and isooctane or 2,2,4-trimethylpentane (a good fuel) release the same amount of energy but burn at different rates. One cannot merely find the fuel with the highest octane rating and decide that is the best fuel. Toluene, for instance, has a very high octane rating (118), but when other properties of toluene, including its energy content, considerable toxicity, combustion properties, and cost are considered, it is not the fuel of choice. However, most gasoline mixtures do contain about 5–7% toluene.

Charles D. Winters

Typical octane ratings for gasoline available at gas stations.

Isooctane, although technically an incorrect systematic name, is the name used in the fuel industry to refer to 2,2,4-trimethylpentane.

$$CH_3-\underset{\underset{CH_3}{|}}{\overset{\overset{CH_3}{|}}{C}}-CH_2-\underset{\underset{}{}}{\overset{\overset{CH_3}{|}}{CH}}-CH_3$$

TABLE 12.4	Octane Numbers of Some Hydrocarbons and Gasoline Additives	
NAME		**OCTANE NUMBER**
Octane		−20
Heptane		0
Pentane		62
1-Pentene		91
2,2,4-Trimethylpentane (isooctane)		100
Benzene		106
Methanol		107
Ethanol		108
Tertiary-butyl alcohol		113
Methyl *tertiary*-butyl ether (MTBE)		116
Para-xylene		116
Toluene		118

Catalytic Re-Forming

The **catalytic re-forming process** is used to increase the octane rating of straight-run gasoline by converting straight-chain hydrocarbons to branched-chain hydrocarbons and aromatics. This is accomplished by using certain catalysts, such as finely divided platinum on a support of alumina (Al_2O_3).

$$CH_3CH_2CH_2CH_2CH_3 \xrightarrow{\text{Catalyst}} \underset{\substack{\text{2-Methylbutane} \\ \text{(octane rating 94)}}}{CH_3\overset{\overset{\displaystyle CH_3}{|}}{C}HCH_2CH_3}$$

Pentane
(octane rating 62)

Review the discussion of catalysts in Section 8.3.

In this process, straight-chain hydrocarbons with low octane numbers can be re-formed into their branched-chain isomers, which have higher octane numbers. Catalytic re-forming is also used to produce aromatic hydrocarbons such as benzenes, toluene, and xylenes by using different catalysts and petroleum mixtures. For example, when the vapors of straight-run gasoline, kerosene, and light oil fractions are passed over a copper catalyst at 650°C, a high percentage of the original material is converted into a mixture of aromatic hydrocarbons, from which benzene, toluene, xylenes, and similar compounds may be separated by fractional distillation. This process can be represented by the equation for converting hexane into benzene.

$$\underset{\substack{\text{Hexane} \\ \text{(octane rating 25)}}}{CH_3CH_2CH_2CH_2CH_2CH_3} \xrightarrow{\text{Catalyst}} \underset{\substack{\text{Benzene} \\ \text{(octane rating 106)}}}{C_6H_6} + 4\,H_2$$

The catalytic re-forming process is a major source of hydrogen gas, which is also a potential fuel.

Catalytic Cracking

Catalytic re-forming process Process that increases octane rating of straight-run gasoline by converting straight-chain hydrocarbons to branched-chain hydrocarbons and aromatics

Part of the petroleum refinement process involves adjusting the percentage of each hydrocarbon fraction to match commercial demand. For example, the demand for gasoline is higher than that for kerosene. As a result, chemical reactions convert the larger kerosene-fraction molecules into molecules

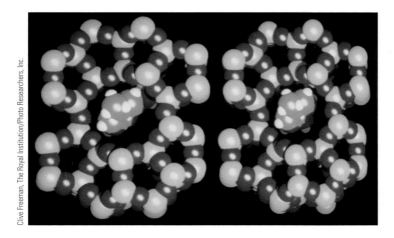

FIGURE 12.3 Computer graphics space-filling representation of the m- (left) and p- (right) isomers of xylene (page 278) in the zeolite catalyst ZSM-5. Zeolites are a group of materials, based on silicon, aluminum, phosphorus, and oxygen, that have structures based on interconnecting cavities and channels, which form an ideal environment for shape-selective catalysis. Industrial applications include the cracking of petroleum, isomerization, and hydrocarbon synthesis reactions. In this image, the proximity of methyl groups in m-xylene (left) prevent its free diffusion along ZSM-5's straight channel. In contrast p-xylene (right) is able to move freely.

in the gasoline range in a process called "cracking." The **catalytic cracking process** uses a zeolite catalyst (Figure 12.3) and involves heating saturated hydrocarbons under pressure in the absence of air. The hydrocarbons break into shorter chain hydrocarbons—both alkanes and alkenes, some of which will be in the gasoline range.

$$\underset{\text{An alkane}}{C_{16}H_{34}} \xrightarrow[\text{Heat}]{\text{Pressure}} \underset{\substack{\text{An alkane} \\ \text{in the gasoline range}}}{C_8H_{18}} + \underset{\text{An alkene}}{C_8H_{16}}$$

Since alkenes have a higher octane rating than alkanes, the catalytic cracking process also increases the octane rating of the mixture. Catalytic cracking is also important for the production of alkenes used as starting materials in the organic chemical industry.

Octane Enhancers

The octane number of a given blend of gasoline can also be increased by adding antiknock agents, or octane enhancers. Prior to 1975, the most widely used antiknock agent was tetraethyllead, $(C_2H_5)_4Pb$. The addition of 3 g of $(C_2H_5)_4Pb$ per gallon increases the octane rating by 10 to 15. Before the Environmental Protection Agency (EPA) required reductions in lead content, both regular and premium gasoline contained an average of 3 g of $(C_2H_5)_4Pb$ or tetramethyllead, $(CH_3)_4Pb$, per gallon. However, in the Clean Air Act of 1970 Congress required that 1975-model cars emit no more than 10% of the carbon monoxide and hydrocarbons emitted by 1970 models. The platinum-based catalytic converter chosen to reduce emissions of carbon monoxide and hydrocarbons required lead-free gasolines, since lead deactivates the platinum catalyst by coating its surface. For this reason, and the fact that lead released into the environment is a neurological poison, automobiles manufactured since 1975 have been required to use lead-free gasoline to protect the catalytic converter.

Since tetraethyllead can no longer be used, other octane enhancers must be added to gasoline to increase the octane rating. These have included benzene, toluene, xylenes, 2-methyl-2-propanol, methyl *tertiary*-butyl ether (MTBE), methanol, and ethanol.

Oxygenated and Reformulated Gasolines

The 1990 amendments to the Clean Air Act require cities with excessive levels of ozone and carbon monoxide pollution to use oxygenated and reformulated gasolines to reduce hydrocarbon and toxic compound emissions

Approximately one-third of the compounds in refined gasoline are aromatic compounds.

As little as two tankfuls of leaded gasoline can destroy the activity of a catalytic converter.

A side benefit of the removal of lead from gasolines has been a decrease of emissions of this toxic element into the environment.

Catalytic cracking process Process by which larger kerosene fractions are converted into hydrocarbons in the gasoline range

The difference between oxygenated gasoline and reformulated gasoline is in the refining process. Oxygenated gasoline is produced by adding oxygenated organic compounds to refined gasoline. Reformulated gasoline requires changes in the refining process to alter the percentage composition of the different types of hydrocarbons, particularly olefins and aromatics.

Sulfur in gasoline coats the catalytic converter and reduces its ability to catalyze full combustion of the fuel. This causes an increase in carbon monoxide emissions.

Oxygenated gasolines Blends of gasoline that contain oxygenated organic compounds such as alcohols or ethers to cause the gasoline to burn more cleanly

Reformulated gasoline Gasoline whose composition has been changed to reduce the percentage of olefins, aromatics, and sulfur and to add oxygenated compounds

(Section 4.9). **Oxygenated gasolines** are blends of gasoline with organic compounds that contain oxygen, such as MTBE, methanol, ethanol, and 2-methyl-2-propanol (*tertiary*-butyl alcohol). The oxygenated gasolines can be produced either by blending in additives at the refinery or by adding ethanol or methanol at the distribution terminals. MTBE has been identified as a pollutant in the water supply in many locations in the United States and its use has been severely restricted or banned in many states.

Reformulated gasoline is gasoline whose composition has been changed to reduce the percentage of unsaturated hydrocarbons, aromatics, volatile components, and sulfur, and to add oxygenated additives. This requires significant changes in the refining process, which makes reformulated gasoline more expensive to produce.

Oxygenated gasolines are required for use during the four winter months in cities that have serious carbon monoxide pollution (Section 4.10). This gasoline must contain enough oxygenated organic compounds to provide an average of 2.7% oxygen by weight. Oxygenated gasolines ignite more easily and burn more cleanly, and this reduces carbon monoxide emissions.

Nine cities with the most serious ozone pollution were required by the 1990 regulations to use reformulated gasolines, and another 87 cities that were not meeting the ozone air-quality standards were given the option of using these gasolines.

The composition of reformulated gasoline later became an issue of political and economic dissension. A ruling issued by the Environmental Protection Agency on June 30, 1994 required that 30% of the oxygenated organic compounds (called *oxygenates*) used in reformulated gasoline had to come from renewable sources. Most gasoline producers had been using MTBE, which is made from methanol, to meet the 1990 Clean Air Act regulations. Because methanol is produced from synthesis gas, it is not renewable. The new mandate would have required more use of ethanol, a renewable resource because it can be made from corn, and ethyl *tert*-butyl ether (ETBE), which is made from ethanol. Proponents of the mandate argued that the use of ethanol would reduce reliance on oil imports, cut farmers' reliance on federal farm subsidies, and increase the use of renewable resources.

In response to the EPA ruling, the American Petroleum Institute and the National Petroleum Refiners Association filed suit in the U.S. Court of Appeals for the District of Columbia. The suit accused the EPA of violating the regulatory process by ruling in favor of ethanol producers and farmers who grow corn. On April 28, 1995, the U.S. Court of Appeals approved the petition by prohibiting the EPA from requiring the use of renewable oxygenates in reformulated gasolines.

In subsequent years, a new issue that bears on the composition of reformulated gasoline has surfaced. Concentrations of MTBE, presumably from gasoline leaks and spills that wash into natural waterways, have been found in drinking water. Because MTBE is not easily decomposed by natural processes or water treatment, it tends to remain where it has accumulated. It is reported to cause off tastes and odors in water. Also, it may be a carcinogen. In response to these concerns, a blue-ribbon panel of experts reviewed the evidence. In 2000, at their recommendation, efforts were begun to eliminate use of MTBE and promote further use of renewable oxygenates like ethanol.

12.3 NATURAL GAS

Natural gas is a mixture of gases trapped with petroleum in Earth's crust and is recoverable from oil wells or gas wells where the gases have migrated through the rock. The natural gas found in North America is a mixture of C_1

to C_4 alkanes—methane (60–90%), ethane (5–9%), propane (3–18%), and butane (1–2%)—with a number of other gases, such as CO_2, N_2, H_2S, and the noble gases, present in varying amounts. In Europe and Japan, the natural gas is essentially all methane.

Natural gas is the fastest-growing energy source in the United States, and U.S. production of natural gas supplies 17% more energy than does U.S.–produced oil. About half of the homes in the United States are heated by natural gas, followed by electricity (18.5%), fuel oil (14.9%), wood (4.8%), and liquefied gas such as butane and propane (4.6%). Coal and kerosene come in at a low 0.5%, and the percentage of homes using solar heating is even lower. However, the United States has only about 5% of the known world reserves of natural gas, which at the present rate of use is enough to last until the year 2050.

Natural gas is also being used as a vehicle fuel, and worldwide there are more than 2.5 million vehicles powered by natural gas. Argentina, with more than 800,000 natural gas powered vehicles, leads the world in use of these vehicles. California and several other states are encouraging the use of natural-gas vehicles to help meet new air-quality regulations. Vehicles powered by natural gas emit minimal amounts of carbon monoxide, hydrocarbons, and particulates; and the price of natural gas is about one-third that of gasoline. The main disadvantages of natural-gas vehicles include the need for a cylindrical pressurized gas tank and the lack of service stations that sell compressed natural gas.

Although most natural gas is used as an energy source, it is also an important source of raw materials for the organic chemical industry. (Figure 14.1 shows the uses of alkanes obtained from natural gas and petroleum.)

(12.4) COAL

Coal is a complex mixture of high molecular weight hydrocarbons that are about 85% carbon by mass. It usually contains a relatively small but variable amount of sulfur. The small amount of sulfur can be a significant factor when one considers its potential as a contributor to air pollution and acid precipitation, however. By way of contrast with petroleum, coal has more fused rings of carbon atoms, and the organic structure of coal is much more complicated.

About 88% of our annual coal production is burned to produce electricity. Only 1% is used for residential and commercial heating. Although the use of coal is on the rise, its use as a heating fuel has declined because it is a relatively dirty fuel, bulky to handle, and a major cause of air pollution (because of its sulfur content). The dangers of deep coal mining and the environmental disruption caused by strip mining contributed to the decline in the use of coal.

Given our great dependence on coal for the production of electricity and our smaller but still significant dependence on coal for the production of industrial chemicals, just how much coal do we have and how long is it likely to last? World coal reserves are vast relative to supplies of the other fossil fuels. Coal represents about 91% of the world's known fossil fuel reserves with approximately equal percentages of natural gas and oil making up the remaining 9%. Only a small percentage of the world's coal has been mined to date. It is estimated that the world's proven coal reserves (Figure 12.4a) will last at least 200 years at the current rate of usage but world coal consumption (Figure 12.4b) is predicted to increase significantly as oil and natural gas become more scarce so this prediction may prove an overestimate.

Coal can be converted into a combustible gas (by coal gasification) or a liquid fuel (by coal liquefaction). In each case environmental problems can be minimized, but at additional costs per energy unit obtained from these fuels.

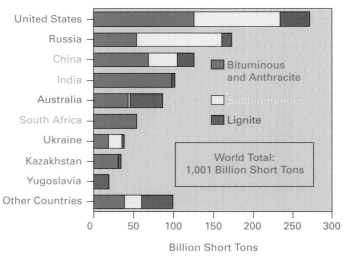

FIGURE 12.4a World Recoverable Coal Reserves.

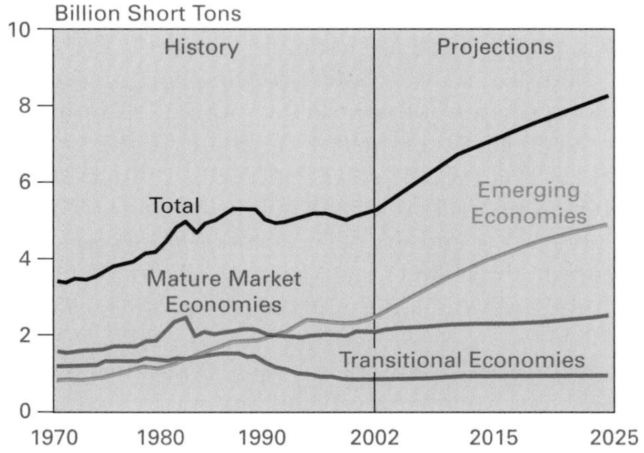

FIGURE 12.4b World Coal Consumption by Region, 1970–2025.

Coal Gasification

When coal is pulverized and treated with superheated steam, a mixture of CO and H_2 (synthesis gas) is obtained in a process known as *coal gasification.*

$$C + H_2O + 31 \text{ kcal} \longrightarrow CO + H_2$$

Synthesis gas is used both as a fuel and as a starting material for the production of organic chemicals and gasoline.

In a newer coal gasification process, methane is the end product. Crushed coal is mixed with an aqueous catalyst; the mixture is dried; and CO and H_2 are added. The resulting mixture is then heated to 700°C to produce methane and carbon dioxide. The overall reaction is

$$2\,C + 2\,H_2O + 2 \text{ kcal} \longrightarrow CH_4 + CO_2$$

Although the overall reaction is slightly endothermic, the subsequent combustion of the methane produced releases 192 kcal/mole; thus, the process is an energy-efficient way to obtain methane, an environmentally clean fuel.

> The formation of synthesis gas is an example of a reaction in which the products have higher bond energies than the reactants, as indicated by the absorption of 31 kcal in the reaction.

> The carbon atoms in coal, methane, and other hydrocarbons end up as CO_2 molecules, which contribute to global warming (see Chapter 6).

Coal Liquefaction

Liquid fuels are made from coal by reacting the coal with hydrogen gas under high pressure in the presence of catalysts (hydrogenating the coal). The process produces hydrocarbons like those in petroleum. The resulting crude oil type of material can be fractionally distilled to give fuel oil, gasoline, and certain hydrocarbons used in the manufacture of plastics, medicines, and other commodities. About 5.5 barrels of liquid are produced for each ton of coal. At the present time, the cost of a barrel of liquid from coal liquefaction is about double that of a barrel of crude oil. However, as petroleum supplies diminish and the cost of crude oil increases, coal liquefaction will become economically feasible.

12.5 METHANOL AS A FUEL

Methanol is being considered as a replacement for gasoline, especially in urban areas that have extremely high levels of air pollution caused by motor vehicles. For example, Southern California has been testing methanol-powered cars since 1981. About half of these cars use 100% methanol (M100). The other half are flexible-fueled vehicles (FFVs) that use either M85, a blend of

85% methanol and 15% gasoline, or gasoline. Although methanol fuels M85 and M100 have received more attention than corresponding ethanol fuels E85 and E100, FFVs are being built to test the use of both M85 and E85.

What are the advantages and disadvantages of switching to methanol-powered vehicles? Methanol burns more cleanly than gasoline, and levels of troublesome pollutants such as carbon monoxide, unreacted hydrocarbons, nitrogen oxides, and ozone are reduced (Chapter 4). However, there is concern about the higher exhaust emissions of carcinogenic formaldehyde from methanol-powered vehicles. Since the number of methanol-powered vehicles is limited, it is still difficult to assess the extent to which these formaldehyde emissions will contribute to the total aldehyde levels from other sources.

The technology for methanol-powered vehicles has existed for many years, particularly in racing cars that burn methanol because of its high octane rating of 100. However, methanol has only about one half the energy content of gasoline, which would require fuel tanks to be twice as large to give the same distance per tankful. This is partially compensated for by the fact that methanol costs about half as much as gasoline, so the price per mile would be competitive. Because methanol burns with a colorless flame, something needs to be added (a small amount of gasoline, for example) to methanol so that it can be seen when it burns. Another disadvantage is the tendency for methanol to corrode regular steel, so it will be necessary to use stainless steel for the fuel system or have a methanol-resistant coating. Until sufficient numbers of methanol-powered vehicles are on the road, cars equipped to run on *either* methanol or gasoline will be necessary because of the lack of service stations selling methanol. As the problems of distribution and storage are solved, better engineered methanol-fueled engines will be designed and produced, which will lead to more efficient utilization of methanol as a fuel.

Another option is to use methanol to make gasoline. Mobil Oil Company has developed a methanol-to-gasoline process although it is currently not competitive with refined gasoline prices in the United States.

Cars at the Indianapolis 500 are powered by methanol.

Unlike gasoline fires, methanol fires and ethanol fires may be extinguished using water.

$$2\ CH_3OH \xrightarrow[\text{Catalyst}]{} \underset{\text{Dimethyl ether}}{(CH_3)_2O} + H_2O$$

$$2\ (CH_3)_2O \xrightarrow[\text{Catalyst}]{} \underset{\text{Ethylene}}{2\ C_2H_4} + 2\ H_2O$$

$$\text{many } C_2H_4 \xrightarrow[\text{Catalyst}]{} \underset{\text{Gasoline}}{\text{Hydrocarbon mixture in the } C_5\text{–}Cl_2 \text{ range}}$$

CONCEPT CHECK   12A

1. Which fossil fuel furnishes the most heat energy per gram? **(a)** Coal **(b)** Petroleum **(c)** Natural gas
2. Which fuel furnishes the most heat energy per gram? **(a)** Natural gas **(b)** Hydrogen **(c)** Coal
3. Combustion of all fossil fuels gives off energy. **(a)** True **(b)** False
4. Hydrocarbons react with _____ to produce CO_2 and _____ upon combustion.
5. The fractions of petroleum are separated by _____.
6. The principal component in natural gas is _____.
7. The octane enhancer used most by gasoline producers at the present time is _____.
8. The _____ process is used to produce branched-chain and aromatic hydrocarbons from straight-chain hydrocarbons.
9. The _____ process is used in refining petroleum to convert molecules in the higher boiling fractions to molecules in the gasoline fraction.

10. Which of the following hydrocarbons would be expected to have the highest octane rating?

(a) $CH_3CH_2CH_2CH_2CH_2CH_2CH_3$

(b)
$$CH_3CH_2\overset{\displaystyle CH_3}{\underset{\displaystyle |}{C}}HCH_2CH_2CH_3$$

(c)
$$CH_3-\overset{\displaystyle CH_3}{\underset{\displaystyle CH_3}{\overset{\displaystyle |}{\underset{\displaystyle |}{C}}}}-\overset{\displaystyle CH_3}{\underset{\displaystyle H}{\overset{\displaystyle |}{\underset{\displaystyle |}{C}}}}-CH_3$$

12.6 CLASSES OF HYDROCARBONS

We have demonstrated that hydrocarbons and many related compounds are often used as fuels in our highly industrialized society but we have not looked closely at their structures and differences. At this point it is useful to take a closer look at the many types or classes of hydrocarbons and some related compounds as these have many uses beyond those as fuels. There are four classes of hydrocarbons: the *alkanes*, which contain carbon–carbon single bonds; the *alkenes*, which contain one or more carbon–carbon double bonds; the *alkynes*, which contain carbon–carbon triple bonds; and the *aromatics*, which consist of benzene, benzene derivatives, and fused benzene rings.

Alkanes: Backbone of Organic Chemistry

The simplest alkane is methane (CH_4), the principal component of natural gas. **Alkanes** are saturated hydrocarbons (Section 5.3) with the general formula C_nH_{2n+2} (Table 12.5). Notice that all hydrocarbon formulas are traditionally written with the C atom first, followed by the H atom, and that all alkanes have *-ane* as the suffix in their name. When $n = 1$ to 4, the first part of the name is something of historical origin; these are common names that we just have to remember. When $n = 5$ or more, the Greek prefixes (Table 12.5) tell how many carbon atoms are present. For example, the compound with six carbons is called hexane.

The tetrahedral structure of CH_4 has been discussed, but it is important to recognize that *every* carbon in an alkane has a tetrahedral geometry because all carbon atoms in saturated hydrocarbons have four single bonds. The tetrahedral geometry of the carbon atoms in alkanes is difficult to draw in two dimensions.

To save time and space, chains of carbon atoms are usually represented with a straight line as in Figures 12.5a, 12.6a, and 12.7a. However, keep in mind that these drawings are not an accurate representation of the tetrahedral bond angles, which are 109.5°.

Our ability to understand these tetrahedral structures is helped by the use of models. Two types of models are generally used—the "ball-and-stick" model and the "space-filling" model. In a ball-and-stick model, balls represent the atoms and short pieces of wood or plastic represent bonds. For example, the ball-and-stick model for methane (Figure 12.5b) has a blue ball representing carbon, with holes at the correct angles connected by sticks to four white balls representing hydrogen atoms. The space-filling model (Figure 12.5c) is a more realistic representation because it depicts both the relative sizes of the atoms and their spatial orientation in the molecule. This is done by scaling the pieces

Alkanes are referred to as saturated hydrocarbons because they contain the highest ratio of hydrogen to carbon possible.

Kjell B. Sandved/Visuals Unlimited

Rice fields in the Philippines. Decay of organic matter in rice fields is estimated to make up one-fifth of all methane emitted each year due to human activities.

Alkanes Hydrocarbons with carbon–carbon single bonds

TABLE 12.5 The First Eight Straight–Chain Saturated Hydrocarbons

NAME	FORMULA	BOILING POINT, °C*	STRUCTURAL FORMULA	USE
Methane	CH_4	−162	H—C—H (with H above and below)	Principal component in natural gas
Ethane	C_2H_6	−88.5	H—C—C—H	Minor component in natural gas
Propane	C_3H_8	−42	H—C—C—C—H	Bottled gas for fuel
Butane	C_4H_{10}	0	H—C—C—C—C—H	Bottled gas for fuel
Pentane	C_5H_{12}	36	H—C—C—C—C—C—H	Some of the components of gasoline
Hexane	C_6H_{14}	69	H—C—C—C—C—C—C—H	Some of the components of gasoline
Heptane	C_7H_{16}	98	H—C—C—C—C—C—C—C—H	Some of the components of gasoline
Octane	C_8H_{18}	126	H—C—C—C—C—C—C—C—C—H	Some of the components of gasoline

* Notice the gradual increase in boiling point as the molecular weight increases. Fractional distillation of petroleum is possible because of these differences (Section 12.2).

in the model according to the experimental values of atom sizes. The pieces are held together by links that are not visible when the model is assembled. The wedge-dash projection (Figure 12.5d) is frequently used by chemists to represent the geometry of molecules. Solid lines represent bonds in the plane of the paper, dashes represent bonds projecting behind the plane of the paper, and wedges represent bonds projecting out of the plane of the paper.

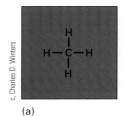

(a)

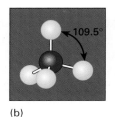

(b)

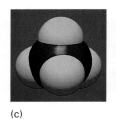

(c)

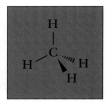

(d)

FIGURE 12.5 **The structure of methane,** as represented by (a) its structural formula, (b) a ball-and-stick model, (c) a space-filling model, and (d) a wedge-dash projection showing the geometry of the molecule.

FIGURE 12.6 The structure of ethane, as represented by (a) its structural formula, (b) a ball-and-stick model, (c) a space-filling model, and (d) a wedge-dash projection.

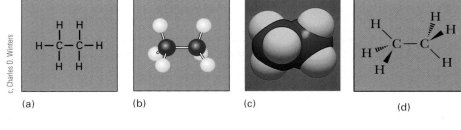

c. Charles D. Winters

(a) (b) (c) (d)

FIGURE 12.7 The structure of propane, as represented by (a) its structural formula, (b) a ball-and-stick model, (c) a space-filling model, and (d) a wedge-dash projection.

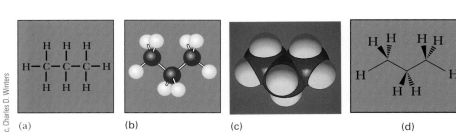

c. Charles D. Winters

(a) (b) (c) (d)

Illustrations of ball-and-stick models, such as those shown in Figures 12.5(b), 12.6(b), and 12.7(b), will be used extensively in the discussion of hydrocarbons and hydrocarbon derivatives to help you visualize the molecular geometry of molecules. For example, notice that the three carbon atoms in the ball-and-stick model of propane in Figure 12.7(b) do not lie in a straight line because of the tetrahedral geometry about each carbon atom. This illustrates why straight-chain drawings, such as the one for propane in Figure 12.7(a), are not accurate representations of the tetrahedral H—C—H and C—C—C bond angles, and why projections such as Figure 12.7(d) are used.

Straight- and Branched-Chain Isomers of Alkanes

The first three alkanes, CH_4, C_2H_6, and C_3H_8, each have only one possible structural arrangement. However, two structural arrangements are possible for C_4H_{10}—a straight-chain arrangement and a branched-chain arrangement.

Historically, straight-chain hydrocarbons were referred to as *normal* hydrocarbons, and *n-* was used as a prefix in the name of straight-chain hydrocarbons such as butane (*n*-butane). The current practice is not to use *n-*. If a name is given without indicating that the compound is a branched chain, assume the compound is a straight-chain hydrocarbon.

Structural formulas:

The condensed formulas and properties of the two compounds, butane and methylpropane, are as follows:

CONDENSED FORMULAS:	$CH_3CH_2CH_2CH_3$ BUTANE	CH_3 $\mid$ CH_3CHCH_3 METHYLPROPANE (ISOBUTANE)
Melting point	$-138.3°C$	$-160°C$
Boiling point (1 atm)	$0.5°C$	$-12°C$
Density (at 20°C)	0.579 g/mL	0.557 g/mL

These molecules have different properties even though they have the same number of atoms in the molecule. Ball-and-stick models of the two structures are shown in Figure 12.8.

Two or more compounds with the same molecular formula but different arrangements of atoms are called **isomers**. Isomers differ in one or more physical or chemical properties such as boiling point, color, solubility, reactivity, and density. Several different types of isomerism are possible for organic compounds. **Branched-chain** and **straight-chain** isomers are examples of *structural isomers* that differ in the order in which the atoms are bonded together. Structural isomerism can be compared to the results you might expect from a child building many different structures with the same collection of building blocks and using all the blocks in each structure.

The branched-chain isomer for C_4H_{10}, methylpropane, a common component of bottled gas, has a "methyl" group ($-CH_3$) attached to the central carbon atom. This is the simplest example of the fragments of alkanes known as **alkyl groups**. In this case, removal of H from methane gives a *methyl* group.

Butane

Methylpropane
(isobutane)

FIGURE 12.8 Ball-and-stick models of butane, a four-carbon straight-chain hydrocarbon, and methylpropane, a four-carbon branched-chain hydrocarbon.

$$H-\underset{\underset{H}{|}}{\overset{\overset{H}{|}}{C}}-H \xrightarrow{-H} H-\underset{\underset{H}{|}}{\overset{\overset{H}{|}}{C}}- \quad \text{or} \quad CH_3- \quad \left(\begin{array}{c}\text{also written}\\\text{as } -CH_3\end{array}\right)$$

Removal of an H from ethane gives an *ethyl* group

$$H-\underset{\underset{H}{|}}{\overset{\overset{H}{|}}{C}}-\underset{\underset{H}{|}}{\overset{\overset{H}{|}}{C}}-H \xrightarrow{-H} H-\underset{\underset{H}{|}}{\overset{\overset{H}{|}}{C}}-\underset{\underset{H}{|}}{\overset{\overset{H}{|}}{C}}- \quad \text{or} \quad CH_3CH_2- \quad \left(\begin{array}{c}\text{also written}\\\text{as } -C_2H_5\end{array}\right)$$

Notice that more than one alkyl group is possible when an H atom is removed from C_3H_8

$$H-\underset{\underset{H}{|}}{\overset{\overset{H}{|}}{C}}-\underset{\underset{H}{|}}{\overset{\overset{H}{|}}{C}}-\underset{\underset{H}{|}}{\overset{\overset{H}{|}}{C}}-H$$

−H from either end carbon

−H from middle carbon

$$H-\underset{\underset{H}{|}}{\overset{\overset{H}{|}}{C}}-\underset{\underset{H}{|}}{\overset{\overset{H}{|}}{C}}-\underset{\underset{H}{|}}{\overset{\overset{H}{|}}{C}}- \quad \text{or} \quad CH_3CH_2CH_2-$$

Propyl

$$H-\underset{\underset{H}{|}}{\overset{\overset{H}{|}}{C}}-\underset{|}{\overset{\overset{H}{|}}{C}}-\underset{\underset{H}{|}}{\overset{\overset{H}{|}}{C}}-H \quad \text{or} \quad (CH_3)_2CH$$

Isopropyl

Alkyl groups are named by dropping "-ane" from the parent alkane and adding "-yl." Theoretically, any alkane can be converted to an alkyl group. Some of the more common examples of alkyl groups are given in Table 12.6.

Isomers Two or more compounds with the same molecular formula but different arrangements of atoms

Structural isomers Isomers that differ in the order in which the atoms are bonded together

Branched-chain isomers Structural isomers of hydrocarbons that have carbon–carbon bonds in side chains

Straight-chain isomers Structural isomers of hydrocarbons with no side chains

Alkyl groups Alkanes with a hydrogen atom removed

TABLE 12.6	Some Common Alkyl Groups
NAME	**CONDENSED STRUCTURAL REPRESENTATION**
Methyl	CH_3-
Ethyl	CH_3CH_2- or C_2H_5-
Propyl	$CH_3CH_2CH_2-$ or C_3H_7-
Isopropyl	CH_3CH- or $(CH_3)_2CH-$ $\quad\quad\vert$ $\quad\quad CH_3$
Butyl	$CH_3CH_2CH_2CH_2-$ or C_4H_9-
t-Butyl*	$\quad\quad CH_3$ $\quad\quad\vert$ CH_3C- or $(CH_3)_3C-$ $\quad\quad\vert$ $\quad\quad CH_3$

*t stands for *tertiary*, sometimes abbreviated *tert*, which means that the central C atom is bonded to three other C atoms.

The number of structural isomers predicted for C_6H_{14}, C_7H_{16}, and C_8H_{18} is 5, 9, and 18, respectively. Every predicted isomer, *and no more*, has been isolated and identified for these hydrocarbons. The large number of structural isomers illustrates the complexity and variety organic chemistry can have even for simple hydrocarbons.

Naming Branched–Chain Alkanes

Many alkanes and other organic compounds have both common names and systematic names. Why are both common and systematic names used? Usually the common name came first and is widely known. Many consumer products are labeled with the common name, and when only a few isomers are possible, the common name adequately identifies the product for the consumer. For example, "isobutane," the common name for methylpropane, is sufficient because there is only one branched-chain isomer possible for C_4H_{10}. However, a system of common names quickly fails when several structural isomers are possible.

We have discussed the "octane rating" of gasoline. Octane (C_8H_{18}) has 18 possible isomers. One of these isomers, 2,2,4-trimethylpentane, known in the fuel industry as "isooctane,"

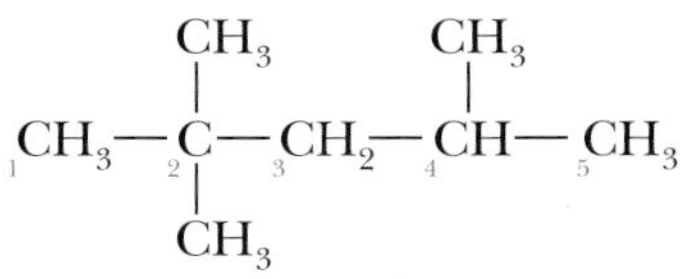

$$CH_3-\underset{\underset{CH_3}{\vert}}{\overset{\overset{CH_3}{\vert}}{C}}-CH_2-\underset{}{\overset{\overset{CH_3}{\vert}}{CH}}-CH_3$$

2,2,4-Trimethylpentane
(isooctane)

Rules for naming organic compounds are given in Appendix D.

is used as a standard in assigning octane ratings of various gasolines. In this case, a common name such as isooctane would not provide enough information about which isomer was actually being used as the standard. To a chemist the name isooctane actually refers to an entirely different structure. However, the systematic name provides complete information. The "pentane" part, which means a straight five-carbon chain, identifies the longest chain in the molecule. The numbers "2,2,4-" indicate the locations of the three groups attached to the pentane chain, and "tri" is used as a prefix for "methyl" to indicate that all three groups are methyl groups.

The numbers in an organic compound name are locators. They give the address of a group along the spine of the molecule.

EXAMPLE 12.2   Structural Isomers

Three isomers are possible for the isomeric pentanes (C_5H_{12}). Draw condensed formulas for these isomers.

SOLUTION

A good plan to follow in drawing all possible isomers—no more and no less—is to start with the straight-chain isomer and then remove one methyl group at a time, placing that methyl group on the remaining chain and checking all possibilities before removing another methyl group. In this case, start with the straight-chain five-carbon pentane as the first isomer.

$$CH_3CH_2CH_2CH_2CH_3$$
Condensed formula: pentane

You can simplify constructing possible isomers if you follow a systematic process.

Removing one methyl and placing it on the second carbon gives a second isomer, 2-methylbutane. Convince yourself that this is the only possible isomer with one methyl attached to a four-carbon chain, since putting the methyl group on the next C gives the same isomer. The third possible isomer is obtained by removing a second methyl and placing it on the second C to give 2,2-dimethylpropane.

$$\begin{array}{c} CH_3 \\ | \\ CH_3CHCH_2CH_3 \end{array} \qquad \begin{array}{c} CH_3 \\ | \\ CH_3CCH_3 \\ | \\ CH_3 \end{array}$$

Condensed formula: 2-methylbutane   Condensed formula: 2,2,-dimethylpropane

Charles D. Winters

Some products containing hydrocarbons.

TRY IT 12.2

Draw the condensed structural formulas for the following compounds:
(a) 2-methylpentane, (b) 3-methylpentane, (c) 2,2-dimethylbutane, and
(d) 2,3-dimethylbutane.

12.7 ALKENES AND ALKYNES: REACTIVE COUSINS OF ALKANES

Petroleum contains alkenes, and their presence in gasoline raises the octane rating. Alkenes for use in commercial applications are also obtained from petroleum by a cracking process. Ethene, best known by its common name, ethylene, is the first member of the **alkene** series of hydrocarbons, compounds that have one or more C=C double bonds.

The "-ene" suffix is used for hydrocarbons with one or more double bonds.

$$\begin{array}{ccc} H & & H \\ \diagdown & & \diagup \\ & C=C & \\ \diagup & & \diagdown \\ H & & H \end{array}$$

More ethylene is manufactured each year than any other organic chemical, - usually in a quantity greater than 20 million tons. Much of the ethylene is used in the production of a plastic known as polyethylene (Section 14.5).

Alkenes Hydrocarbons with one or more carbon–carbon double bonds

FIGURE 12.9 The two smallest alkenes: etheane, commonly known as ethylene, and propene, commonly known as propylene. Note the planar arrangement of atoms around the carbon–carbon double bond.

Ethene
(a)

Propene
(b)

Green tomatoes. On their way to market these tomatoes may be ripened by exposure to ethylene gas.

Ethylene is also found in plants, where it is a hormone that controls seedling growth and regulates fruit ripening. The discovery of this property led to the use of ethylene by food processors for ripening fruits and vegetables after harvest.

The general formula for alkenes with one double bond is C_nH_{2n}. The second member of the alkene series is propene (propylene). Propylene is manufactured in large quantities for use in the production of the plastic known as polypropylene (Section 14.5). Representations for ethene and propene are shown in Figure 12.9. Unlike alkanes, which undergo few chemical reactions easily (except for combustion), alkenes are quite reactive. The site of the chemical change is usually the double bond. This reactivity is essential to making many kinds of plastics, as discussed in Section 14.5. Also, addition of hydrogen to the double-bond carbon atoms converts unsaturated fats (those containing carbon–carbon double bonds) and oils, which are liquids, to the more solid consistency needed for margarine (Section 15.3).

Structural Isomers of Alkenes

In the alkene series, the possibility of locating the double bond between two different carbon atoms creates additional structural isomers. Ethene and propene have only one possible location for the double bond. However, the next alkene in the series, butene, has two possible locations for the double bond.

1-Butene

2-Butene

When groups such as methyl or ethyl are attached to carbon atoms in an alkene, the longest hydrocarbon chain is numbered from the end that will give the double bond the lowest number, and then numbers are assigned to the attached group. For example, in the following compound the longest chain has seven carbons (heptene); the double bond is between C2 and C3 (2-heptene); and the three (tri-) methyl groups are on the third, fourth, and sixth carbons (3,4,6-trimethyl-).

Hence, the name is 3,4,6-trimethyl-2-heptene.

EXAMPLE 12.3   **Drawing Alkenes**

Draw the structure of 2,3-dimethyl-2-pentene.

SOLUTION

First, draw the parent alkene and put the double bond in the correct location. In this case, 2-pentene is the parent.

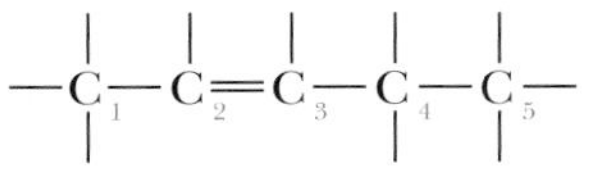

$$-\underset{1}{C}-\underset{2}{C}=\underset{3}{C}-\underset{4}{C}-\underset{5}{C}-$$

Then place the alkyl groups on the appropriate carbons. In this case, there are two methyl groups, one on C2 and one on C3. Remember that numbering the double bond takes precedence. Also check your drawing to make sure you don't have more than four bonds per carbon.

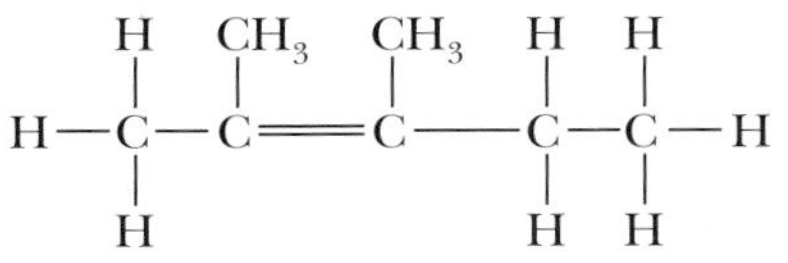

$$H-\underset{H}{\overset{H}{C}}-\underset{}{\overset{CH_3}{C}}=\underset{}{\overset{CH_3}{C}}-\underset{H}{\overset{H}{C}}-\underset{H}{\overset{H}{C}}-H$$

TRY IT 12.3

Draw the structure of 3-methyl-1-butene.

Stereoisomerism: Cis and Trans Isomers in Alkenes

Some alkenes can also have **cis** and **trans isomers**, one of two forms of **stereoisomerism**. *Here the isomers have the same molecular formulas and the same atom-to-atom bonding sequences, but the atoms differ in their arrangement in space.* The other form of stereoisomerism, optical isomerism, is discussed in Section 15.1.

An important difference between alkanes and alkenes is the degree of flexibility of the carbon–carbon bonds in the molecules. Rotation around single carbon–carbon bonds in alkanes occurs readily at room temperature, but the carbon–carbon double bond in alkenes is strong enough to prevent free rotation about the bond. Consider ethene (C_2H_4). Its six atoms lie in the same plane, with bond angles of approximately 120°.

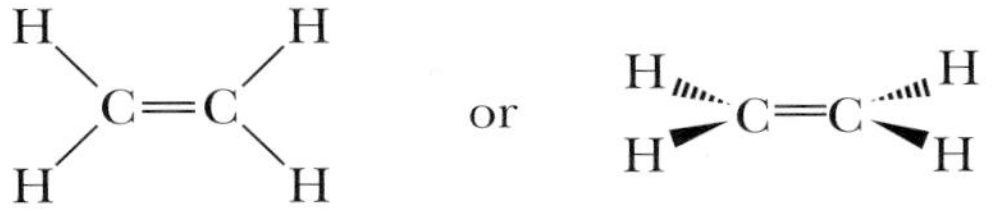

$$\underset{H}{\overset{H}{>}}C=C\underset{H}{\overset{H}{<}} \quad \text{or} \quad \underset{H}{\overset{H}{>}}C=C\underset{H}{\overset{H}{<}}$$

The health risks of *trans* fatty acids are discussed in Section 15.3.

Cis isomers Stereoisomers with groups on the same side of a carbon–carbon double bond

Trans isomers Stereoisomers with groups on opposite sides of a carbon–carbon double bond

Stereoisomerism Describes isomers with the same molecular formulas and the same atom-to-atom bonding sequence, but different arrangement of the atoms in space

If two methyl groups replace two hydrogen atoms, one on each carbon atom of ethene ($H_2C=CH_2$), the result is 2-butene ($CH_3CH=CHCH_3$). Experimental evidence confirms the existence of two compounds with the same set of bonds. The difference in the two compounds is in the location in space of the two methyl groups: the *cis* isomer has two methyl groups on the same side in the plane of the double bond and the *trans* isomer has two methyl groups on opposite sides of the double bond. The physical properties of the *cis* and *trans* isomers of 2-butene are quite different.

The third possible isomer, 1-butene (a structural isomer of the *cis* and *trans* isomers), does not have *cis* and *trans* structures. Since one carbon atom

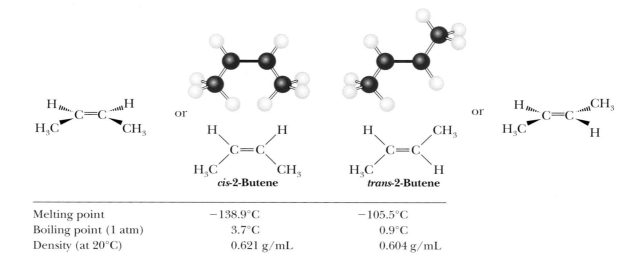

	cis-2-Butene	trans-2-Butene
Melting point	−138.9°C	−105.5°C
Boiling point (1 atm)	3.7°C	0.9°C
Density (at 20°C)	0.621 g/mL	0.604 g/mL

Many other *cis* and *trans* isomers are possible. For example, *cis*-1,2-dichloroethene and *trans*-1,2-dichloroethene are possibilities when one hydrogen atom on each carbon atom of ethene is replaced with a chlorine atom.

If free rotation occurred around a carbon–carbon double bond, these two isomers would be the same.

has two identical groups (H atoms), its properties are different from those of the 2-butene isomers.

1-Butene

Melting point	−185.3°C
Boiling point (1 atm)	−6.3°C
Density (at 20°C)	0.595 g/mL

Cis–trans isomerism in alkenes is possible only when both of the double-bond carbon atoms have two different groups.

The **alkynes** have one or more triple bonds ($-C\equiv C-$) per molecule and have the general formula C_nH_{2n-2}. The simplest one is ethyne, commonly called acetylene (C_2H_2). The naming of alkynes is similar to that of alkenes, with the lowest number possible being used for locating the triple bond.

The 180° bond angles around the triple bond make the $C-C\equiv C-C$ section of the molecule linear.

The name of the preceding compound is 4-methyl-2-pentyne. As with alkenes, changing the location of the multiple bond produces an isomer. For example, 4-methyl-1-pentyne is a different compound from 4-methyl-2-pentyne.

Alkynes Organic compounds containing a carbon–carbon triple bond

However, *cis* and *trans* isomers are not possible for alkynes because the geometry around the triple-bond carbon atoms is linear.

THE WORLD *of* CHEMISTRY

ORGANIC METALS

Organic compounds are generally good insulators, and metals conduct electricity. However researchers have been successful in making organic compounds that are conductors. Acetylene can be polymerized in the presence of a catalyst to polyacetylene, a typical plastic that does not conduct electricity.

$$2n\; H\!-\!C\!\equiv\!C\!-\!H \xrightarrow{\text{Catalyst}}$$

$$\left(\begin{array}{c} H\;\;H\;\;H\;\;H \\ |\;\;\;|\;\;\;|\;\;\;| \\ -C\!=\!C\!-\!C\!=\!C- \end{array}\right)_n$$

This polymer appears as a black powder in the usual laboratory preparation and received little attention prior to 1970. In that year, a Korean university student, having trouble understanding his Japanese instructor, Hideki Shirakawa, prepared the polymer using an excessive amount of the catalyst.

The result was a silver film that looked more like a metal than anything else. Furthermore, the polyacetylene film conducted electricity, which was a first for plastic materials.

About the same time, Alan MacDiarmid at the University of Pennsylvania was working with an inorganic polymer of sulfur nitride $(SN)_n$ that also looked like a metal and raised questions about the conductivity of polymers, MacDiarmid and Shirakawa met by chance at a seminar in Tokyo, and a collaboration was begun. Because Shirakawa knew that the optical properties of a polymer change on oxidation, they oxidized polyacetylene with iodine vapor. The product was *trans*-polyacetylene in which some electrons from the double bond had transferred to the iodine (to give I_3^-). The result was that elec-

trons from the double bonds were free to flow as electrons do in a metal, giving one million times greater conductivity.

This discovery opened the door to applications of conducting organic polymers. As in a metal, passage of a strong current through a conducting polymer causes it to glow. This property is promising for use in, for example, luminous signs and flat television displays. The ease with which polymers can be shaped is a major advantage in making such products.

In recognition of the significance of their work, Professors Shirakawa, MacDiarmid, and another collaborator, Alan J. Heeger of the University of California at Berkeley, received the Nobel Prize in chemistry in 2000.

CONCEPT CHECK 12B

1. Each carbon in a saturated hydrocarbon has _____ geometry.
2. _____ is the first member of the alkene series.
3. _____ is the first member of the alkyne series.
4. The formula for the ethyl group is_____.
5. Butane and 2-methylpropane are examples of _____ isomers.
6. The number-one organic chemical produced in the United States is often _____.
7. The rigidity of the carbon–carbon double bond allows for the possibility of _____ isomers.

12.8 THE CYCLIC HYDROCARBONS

Hydrocarbons can form rings as well as straight chains and branched chains. Two important classes of cyclic hydrocarbons found in petroleum and coal are the cycloalkanes and the aromatics.

Cycloalkanes

Cycloalkanes are saturated hydrocarbons with ring structures. The simplest cycloalkane is cyclopropane, a highly strained ring compound.

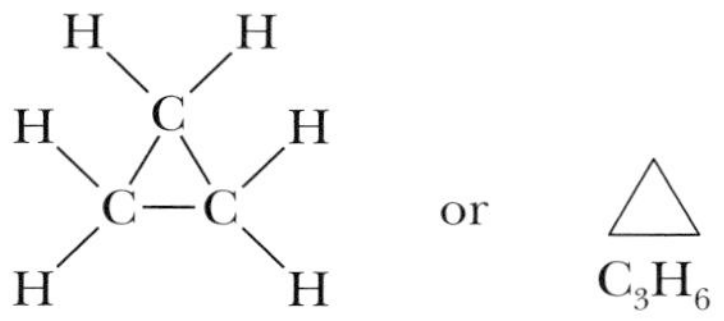

The ring is strained because of the 60° angles in the ring. Carbon that is bonded to four atoms prefers the tetrahedral angle of 109.5°. The deviation from this "ideal" angle creates the strain in some small ring systems. Cyclopropane, a volatile, flammable gas (bp -32.7°C), is a rapidly acting anesthetic. A cyclopropane–oxygen mixture is useful in surgery on babies, small children, and "bad risk" patients because of its rapid action and the rapid recovery of the patient. Helium gas is added to the cyclopropane–oxygen mixture to reduce the danger of explosion in the operating room.

The cycloalkanes are commonly represented by polygons in which each corner represents a carbon atom with two attached hydrogen atoms and the lines represent C—C bonds. The C—H bonds are not shown, but are understood. Other common cycloalkanes include cyclobutane (C_4H_8), cyclopentane (C_5H_{10}), and cyclohexane (C_6H_{12}). These cyclic compounds are represented as planar projections below even though chemists know that the rings are not planar in reality.

Cyclobutane Cyclopentane Cyclohexane

For instance, cyclohexane exists in the so-called chair conformation or shape as shown here because this allows for the preferred 109.5° bond angles.

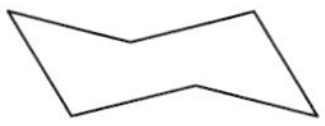

Aromatic Compounds

Hydrocarbons containing one or more benzene rings (Figure 12.10) are called **aromatic compounds**. The word *aromatic* was derived from *aroma*, which describes the rather strong and often pleasant odor of these compounds. Benzene and most other aromatic compounds, however, are toxic and often **carcinogenic**.

The main structural feature, which is responsible for the distinctive chemical properties of the benzene-like aromatic compounds, is the six-carbon benzene ring. Figure 12.10a illustrates "smearing" of some of the carbon–carbon bonding electrons above and below the plane of the ring. In other words, all the carbon–carbon bonds are equivalent and benzene is a planar molecule. The even distribution of electrons around the ring makes benzene and other aromatic compounds less reactive than alkenes. Benzene can be represented equally well by

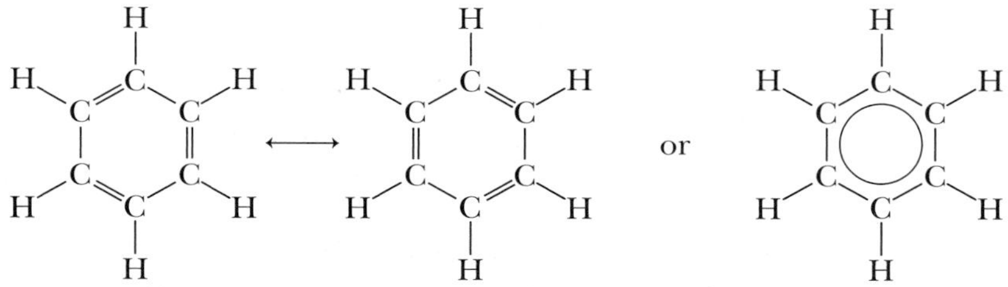

The type of cancer caused by a carcinogen may vary from one carcinogen to another. Benzene causes a form of leukemia, and benzopyrene causes skin cancer and lung cancer.

Cycloalkanes Saturated hydrocarbons with carbon atoms joined in a ring

Aromatic compounds Hydrocarbons and other compounds with one or more benzene rings or rings with benzene-like chemical properties

Carcinogenic Cancer-causing

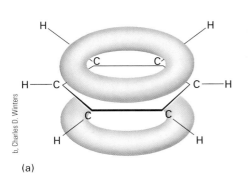
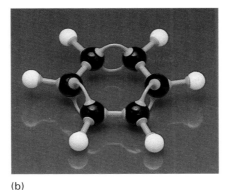

(a)

(b)

FIGURE 12.10 **Benzene, the smallest aromatic compound.** (a) The equal distribution of bonding electrons around the ring can be represented by electron clouds above and below the plane of the ring. (b) Another way to represent the bonding electrons of benzene is as alternating double and single bonds, shown here in a ball-and-stick model. Because all carbon–carbon bonds in benzene are the same, (a) is a more correct representation. In other words, there really aren't any carbon–carbon single bonds or carbon–carbon double bonds in benzene. Each carbon–carbon bond is halfway between these types.

where the circle represents the evenly distributed, smeared electrons. Chemists find the different representations useful in different situations.

When hydrogen and carbon atoms are not shown, benzene is represented by a circle in a hexagon. Each corner in the hexagon represents one carbon atom with one attached hydrogen atom. Remember that this symbol stands for C_6H_6:

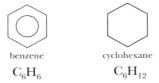

benzene

C_6H_6

cyclohexane

C_6H_{12}

and a hexagon without a circle stands for cyclohexane (C_6H_{12}).

Derivatives of Benzene

Benzene and many of its derivatives are among the most widely used chemicals (see Figure 13.2) because of their use in the manufacture of plastics, detergents, pesticides, drugs, and other organic chemicals. Several important derivatives are monosubstituted benzenes, with one atom or group replacing one of the hydrogen atoms. For example, substitution of a methyl group for one of the hydrogen atoms in benzene gives methylbenzene, usually called toluene, a common solvent. Ethylbenzene, another common chemical, contains an ethyl group substituted for one of the hydrogen atoms of benzene.

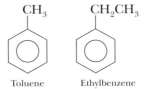

Toluene

Ethylbenzene

Structural Isomers of Aromatic Compounds

Since the benzene molecule has a planar structure, structural isomers are possible when two or more groups are substituted for hydrogen atoms on the benzene ring.

Three isomers are possible if two groups are substituted for two hydrogen atoms on the benzene ring. The prefixes *ortho-*, *meta-*, and *para-* or numbers

Benzene, toluene, and xylenes are important because they raise the octane rating of gasoline.

are used to distinguish among the isomers. When the name of the compound is written, usually only the first letter of one of these terms is given. For example, when the two groups are methyl groups, the three isomers are commonly known as *o*-xylene, *m*-xylene, and *p*-xylene, where *o*, *m*, and *p* refer to *ortho* (1,2), *meta* (1,3) and *para* (1,4) substitution.

1,2-dimethylbenzene (*ortho*-xylene) mp −25°C

1,3-dimethylbenzene (*meta*-xylene) mp −47.9°C

1,4-dimethylbenzene (*para*-xylene) mp 13.3°C

(The xylenes are used in making dyes, insecticides, and drugs)

Another type of aromatic compound has two or more benzene rings sharing ring edges. Examples include naphthalene, anthracene, benzopyrene, and phenanthrene.

Napthalene (mothballs)

Anthracene

Benzo(α)pyrene (found in charcoal smoke and cigarette smoke)

Phenanthrene

Hank Morgan/Rainbow

One source of aromatic compounds. Both the smoke and the char on the meat contain polycyclic aromatic compounds.

Many organic compounds found in nature are cyclic hydrocarbons that include both aromatic rings and cycloalkane or cycloalkene rings fused together. Steroids (Section 15.3) are good examples. Chemists who isolate organic compounds from plants and develop methods for making them in the laboratory are called *natural product chemists*. For example, Percy Julian was the first chemist to synthesize hydrocortisone, a steroid, and physostigmine, a compound useful in the treatment of glaucoma.

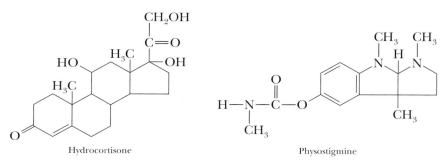

Hydrocortisone

Physostigmine

Methanol, CH_3OH

Alcohols Organic compounds containing a hydroxyl (OH) functional group

12.9 ALCOHOLS: OXYGEN COMES ON BOARD

Alcohols

Alcohols, although not strictly hydrocarbons since they also contain oxygen atoms, will be briefly considered here because we included them in our discussion of fuels. Alcohols have the general formula ROH, with R representing an alkyl group. **Alcohols** are essentially alkanes in which one of the hydrogen atoms has been replaced by the hydroxyl (—OH) functional group. A **functional group** is an atom or group of atoms in a molecule that gives the compound its characteristic chemical behavior. Thus, replacement of one of the H-atoms of methane by the —OH group gives an alcohol known as methanol (also known as methyl alcohol). The compound is named by replacing the final -*e* of methane with -*ol*. Likewise, we see that ethanol (also known as ethyl alcohol) and propanol (propyl alcohol) can be formulated from ethane and propane. We can obtain two different alcohol isomers from propane depending on which hydrogen we replace with the —OH group. Thus, we can have isomeric alcohols, propanol and 2-propanol (more commonly referred to as isopropanol or isopropyl alcohol). Isopropyl alcohol is commonly known as "rubbing alcohol" and is found in the medicine cabinet in most homes.

Methanol (CH_3OH), also called methyl alcohol, can be prepared from a mixture of carbon monoxide and hydrogen known as *synthesis gas*. High pressure, high temperature, and a catalyst are used to increase the yield.

$$\underset{\text{Coal}}{C(s)} + \underset{\text{Steam}}{H_2O(g)} \longrightarrow CO(g) + H_2(g) \quad \underset{\text{Synthesis gas}}{}$$

$$CO(g) + 2\,H_2(g) \xrightarrow[300°C]{ZnO,\ Cr_2O_3} CH_3OH(g)$$

An old method of producing methanol involved heating a hardwood such as beech, hickory, maple, or birch in the absence of air. For this reason, methanol is sometimes called *wood alcohol*. Methanol is highly toxic. Drinking as little as 30 mL can cause death, and smaller amounts (10–15 mL) cause blindness.

Ethanol (C_2H_5OH), also called ethyl alcohol or grain alcohol, can be obtained by the fermentation of carbohydrates (starch, sugars).

$$\underset{\text{Glucose}}{C_6H_{12}O_6} \xrightarrow{\text{Yeast}} 2\,\underset{\text{Ethanol}}{C_2H_5OH} + 2\,CO_2$$

The yeast contains enzymes that are catalysts for the fermentation process. A mixture of 95% ethanol and 5% water can be recovered from the fermentation products by distillation. Ethanol is the active ingredient of alcoholic beverages. Ethanol is receiving increased attention for use as an alternative fuel and as a fuel additive for oxygenated fuels. By the mid-1980s, 90% of all new auto sales in Brazil were for ethanol-only cars. Today, many people have migrated back to gasoline-powered cars due to an ethanol shortage in 1990. However, all gaso-

Additional classes of functional groups are discussed in Chapter 14.

The International Union of Pure and Applied Chemistry (IUPAC) names of alcohols include the name of the hydrocarbon to which the alcohol corresponds and indicate the number of carbon atoms; the suffix -*ol* denotes an alcohol. Common names use the name of the alkyl group (represented as R in ROH) attached to —OH. For example, methyl alcohol, ethyl alcohol, and *tertiary*-butyl alcohol are the common names for methanol, ethanol, and 2-methyl-2-propanol.

Gasohol 95, a gasoline extender made from a mixture of gasoline and ethanol.

Functional group An atom or group of atoms that is part of a larger molecule and that has a characteristic chemical reactivity

THE PERSONAL SIDE

Percy Lavon Julian (1899–1975)

© Bettmann/CORBIS

PERCY JULIAN

Percy Julian's list of achievements reads like that of others who have made it to the top in their professions: a doctorate in chemistry in Vienna in 1931 quickly followed, back home in the United States, by outstanding achievements as a researcher and university professor; 18 years as director of research in an industry where he led the way in bringing to market valuable products from soybeans; and the founding of his own research institute, the Julian Laboratories. To grasp the measure of the man, add to this brief outline dozens of scientific publications, over 100 patents granted, numerous academic honors, and positions of responsibility in many civic and humanitarian organizations.

But there were some differences from a successful career path along the way. After completing the eighth grade he had to leave his home in Montgomery, Alabama, for further studies—no more public education was available there for a black man. He enrolled as a "subfreshman" at DePauw University in Indiana. On his first day, a white student welcomed him with a handshake. Julian later related his reaction: "In the shake of a hand my life was changed, I soon learned to smile and act like I believed they all liked me, whether they wanted to or not."

Early in his career, other challenges had to be met. As a successful businessman, he and his family were the first black residents of an upscale suburb of Chicago. There, on Thanksgiving Day in 1950, his home was attacked by arsonists. The Julian family stayed on to become respected and welcome members of the community.

An organic chemist, Julian built his career around the study of chemicals of plant origin, many of them of medicinal value. The synthesis of a complicated natural molecule is a major goal in such work. Julian was first to achieve synthesis of hydrocortisone, now available in every drugstore because of its value in treating allergic skin reactions. He originated the production of soybean protein and the isolation from soybean oil of compounds from which the first synthetic sex hormone (progesterone) could be made.

Julian's talents were evident early when he and a colleague devised a series of nine chemical reactions that produced a compound identical with natural *physostigmine*. Originally isolated from the Calabar bean from Nigeria, physostigmine had already proved valuable for treating glaucoma by lowering fluid pressure in the eye.

line now sold in Brazil contains 25% ethanol. A fuel mixture of 85% ethanol and 15% gasoline, known as E85 or *gasohol*, is available in at least 22 states in the United States. These are mostly midwestern states.

© David R. Frazier Photolibrary, Inc./Alamy

Unleaded gas pump with ethanol sign.

Ethers

Ethers have the general formula R — O — R′ where R and R′ stand for alkyl groups (Table 12.6), which may be the same or different. Methyl-*tertiary*-butyl ether (MTBE) was once an important commercial ether because of its use in oxygenated and reformulated gasolines.

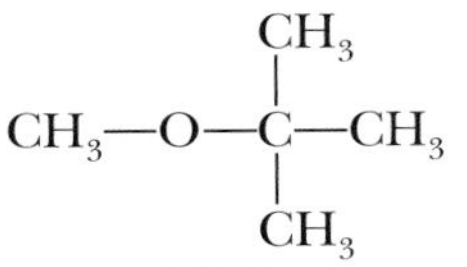

$$CH_3 - O - \underset{\underset{CH_3}{|}}{\overset{\overset{CH_3}{|}}{C}} - CH_3$$

MTBE

Diethyl ether ($C_2H_5OC_2H_5$), an organic solvent, and methyl propyl ether ($CH_3OCH_2CH_2CH_3$), an anesthetic known as neothyl, are also common and well-known ethers.

Ethers Organic compounds with the general formula R — O — R′

CONCEPT CHECK   12C

1. The difference between cyclohexane and benzene is the number of _____ atoms.

2. How many atoms does the symbol ⬡ represent?

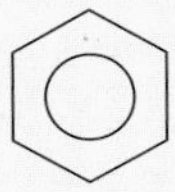

3. Synthesis gas is a mixture of_____ and_____.

4. _____ structural isomers are possible for trichlorobenzene.

5. _____ structural isomers are possible for dimethylbenzene.

6. Fermentation of carbohydrates yields (**a**) ethanol, (**b**) methanol.

7. Ethers are a class of compounds that contain the _____ linkage.

W Assess your understanding of this chapter's topics with an online chapter quiz at **www.brookscole.com/chemistry/joesten4**

■ KEY TERMS

heat of combustion	isomers	stereoisomerism
fractional distillation	structural isomers	alkynes
petroleum fractions	branched-chain isomers	cycloalkanes
catalytic re-forming process	straight-chain isomers	aromatic compounds
catalytic cracking process	alkyl groups	carcinogenic
oxygenated gasolines	alkenes	alcohols
reformulated gasoline	*cis* isomers	functional group
alkanes	*trans* isomers	ethers

■ **THE LANGUAGE OF CHEMISTRY**

1. Hydrocarbon
2. Alkane
3. Alkyl group
4. Alkene
5. Alkyne
6. Aromatic hydrocarbon
7. Methane
8. Alcohol
9. Ether
10. Synthesis gas

a. Benzene
b. R—OH
c. Major component of natural gas
d. Contains only C and H
e. R—O—R'
f. C_nH_{2n+2}
g. Mixture of CO and H_2
h. Contains C=C bond
i. Hydrocarbon that is missing a H atom
j. Contains C≡C bond

■ **APPLYING YOUR KNOWLEDGE**

1. What is the definition of a fossil fuel?
2. What are the three major fossil fuels?
3. What is a hydrocarbon?
4. What is the primary component of natural gas?
5. What is the heat of combustion?
6. Give definitions for the following terms:
   (**a**) Alkane   (**b**) Alkene
   (**c**) Alkyne   (**d**) Aromatic
7. Give definitions for the following terms:
   (**a**) Isomer
   (**b**) Straight-chain hydrocarbon
   (**c**) Branched-chain hydrocarbon
8. Write the formula for a methyl group, an ethyl group, and an alkyl group in general, $C_?H_?$.
9. Saturated hydrocarbons, the alkanes, have the general formula C_nH_{2n+2} where n is a whole number. Give the names and formulas for the first four alkane compounds in the series.
10. What is the structural formula for 1-pentene?
11. Give the structural formulas for the following:
    (**a**) 2-Methylpentane
    (**b**) 4,4-Dimethyl-5-ethyloctane
    (**c**) 2-Methyl-2-hexene
12. Draw as many different isomers as you can that have the formula C_5H_{12}.
13. Name the isomers in question 12.
14. Draw the condensed structural formulas of all possible structural isomers for C_6H_{14} and name them.
15. Draw the structures of all possible isomers that are dimethylbenzenes.
16. Draw the structure of 2,3,3-trimethyl-1-pentene.
17. Why does 2-butene have *cis* and *trans* isomers but 1-butene doesn't?
18. Draw the *cis* and *trans* isomers of 1,2-dichloroethene.
19. Explain how fractional distillation is used in the refinement of petroleum.
20. What is "straight-run" gasoline?
21. List three gasoline additives that increase the octane rating of gasoline.
22. What is gasohol? Why is gasohol controversial?
23. Explain how synthesis gas and methane can be obtained from coal, and write equations that represent these processes.
24. What do the following terms mean?
    (**a**) Catalytic re-forming   (**b**) Catalytic cracking
25. Describe what is meant by the term octane rating.
26. What types of hydrocarbons have high octane ratings?
27. What factors are likely to lead to an increased demand for methanol in the next decade?
28. How is the octane rating of a refined gasoline determined? What are oxygenated gasolines?
29. Do oxygenated gasolines cause less pollution than regular gasolines? Explain.
30. How do oxygenated and reformulated gasolines differ?
31. How is coal gasified?
32. What are the advantages and disadvantages of using methanol as an alternative fuel for vehicles?
33. What do the terms M100, E100, M85, and E85 refer to when describing alternate fuels? What is an FFV?
34. Give two reasons for and two against the use of farmland to produce grain for oxygenated fuel instead of food.

35. Label each of the following as an alkane, alkene, aromatic, or alkyne.

   (**a**) Methane (CH$_4$)

   (**b**) Benzene (C$_6$H$_6$)

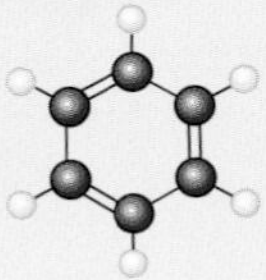

   (**c**) 1-Butene (CH$_2$CHCH$_2$CH$_3$)

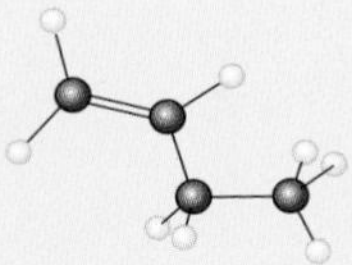

   (**d**) Acetylene (HCCH)

36. The following show the formulas and ball-and-stick models for various compounds. Label each of the following as an alkane, alkene, aromatic, alkyne, alcohol, or ether.

   (**a**) Propane (CH$_3$CH$_2$CH$_3$)

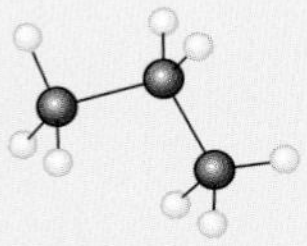

   (**b**) 1-Butyne (CHCCH$_2$CH$_3$)

   (**c**) Diethyl ether (CH$_3$CH$_2$OCH$_2$CH$_3$)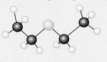

   (**d**) Ethanol (CH$_3$CH$_2$OH)

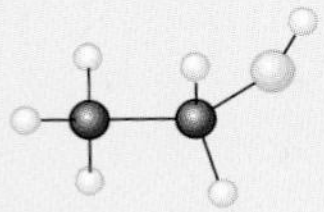

   (**e**) Ethylbenzene (CH$_3$CH$_2$C$_6$H$_5$)

W CHEMISTRY ON THE WEB

For up-to-date URLs, visit the text website at **www.brookscole.com/ chemistry/joesten4**

- Petroleum
- Crude Oil and Refinery Operation
- Gasoline
- Consumer's Guide to Gasoline
- Calculate Your Fuel Economy and See How Energy Is Used by an Automobile
- MTBE and Problems Associated with It
- MTBE and Lead in Gasoline
- Race Cars and Their Fuel
- Alkanes and Cycloalkanes
- Manipulate Common Functional Groups in 3D

NUCLEAR CHANGES AND NUCLEAR POWER

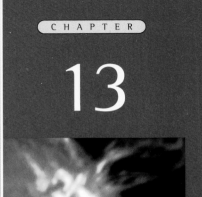

A large solar flare lasting a few hours releases energy from nuclear reactions that would provide electricity for the United States for about 100,000 years.

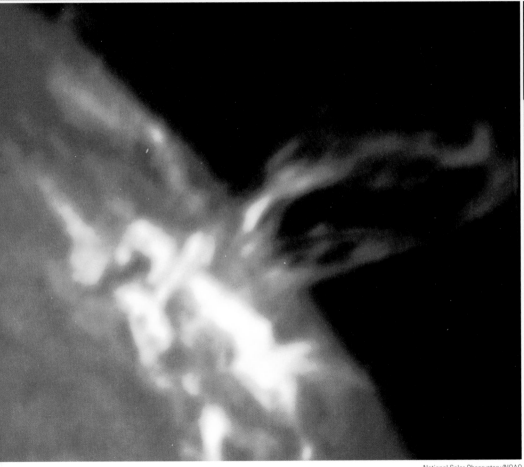

National Solar Observatory/NOAO

Radioactivity. With what do you associate that word? Hazardous waste? Nuclear power plants? Medical diagnosis? Cancer risks? Nuclear weapons? Cancer cures? All these are appropriate associations with the kind of change that happens only in atomic nuclei. Nuclear changes are very different from ordinary chemical reactions, as you will see. Nuclear changes are usually accompanied by the emission of radiation and can, in some cases, be accompanied by the release of large amounts of energy. In this chapter you will learn something about how and when nuclear changes occur.

You will also see that nuclear changes are pictured in much the same way as chemical changes: reactants going to products. Equations can be written for nuclear changes, but although there are similarities between nuclear and chemical changes, nuclear changes are different in some rather significant ways. Certainly the discovery of nuclear changes has affected our lives, some might say for the worse, however you or one of your friends may be alive today because of an application of what is known about how radioactive isotopes undergo change.

In this chapter, we will address the following questions:

- What are the characteristics of nuclear changes?
- Why do some atoms undergo spontaneous nuclear decay?
- How can an atom of one element be transformed into an atom of another element?
- Why are some radioactive isotopes more dangerous than others?
- What are some of the harmful effects of nuclear reactions and radiation?
- What are some of the useful applications of radioactive isotopes, including energy production?

The nuclear age is often considered to have started either in the late 1800s and early 1900s with the discovery of radioactive elements or in 1945 with the first explosion of an atomic bomb. It is true that early atomic theory said nothing about radioactivity, but that was because radiation cannot be detected directly by our five senses. It took the maturing of the sciences—with such diverse discoveries as how to produce a vacuum, photographic film, electricity, and magnetic fields—to lead to the knowledge that some atoms disintegrate spontaneously and, in the process, produce radiation.

(13.1) THE DISCOVERY OF RADIOACTIVITY

In February 1896, Henri Becquerel was experimenting in France with the relation between the recently discovered X rays and the phosphorescence of certain minerals. X rays had been found to penetrate substances like paper and expose photographic plates. Becquerel had already discovered that phosphorescing uranium minerals also exposed the plates. During several cloudy days some samples were left waiting in a drawer. Out of curiosity, Becquerel developed the plates. He had quite a surprise—the plates were exposed. Obviously, the emission of radiation that penetrated the paper surrounding the plates had nothing to do with phosphorescence but was a property of the mineral. In fact, all uranium compounds, and even the metal itself, exposed photographic plates. Becquerel also discovered that uranium (U) emitted radiation that was capable of causing air molecules to ionize (i.e., to lose electrons and become positively charged particles).

Before long it was recognized that the radiation from elements like uranium and radium consisted of three types, now known as alpha particles, beta particles, and gamma rays.

In 1899, Ernest Rutherford found that alpha particles could be stopped by thin pieces of paper and had a range of only about 2.5 cm to 8.5 cm in air, whereas beta particles were capable of penetrating far greater distances in air.

In 1900, Paul Villard identified a third form of natural radiation, gamma (γ) rays. These he discovered were not streams of particles, but instead had the general characteristics of light or X rays. Gamma rays, a high-energy form of electromagnetic radiation (see Figure 3.7), are extremely penetrating; they are capable of passing through more than 22 cm of steel and about 2.5 cm of lead. Figure 13.1 compares the penetrating ability of the three forms of natural radiation.

Phosphorescent minerals re-emit light after they have been exposed to light.

As you saw in Section 3.2, identification of alpha particles as helium nuclei ($^4_2$He) and beta particles as electrons played a significant role in our understanding of the structure of the atom.

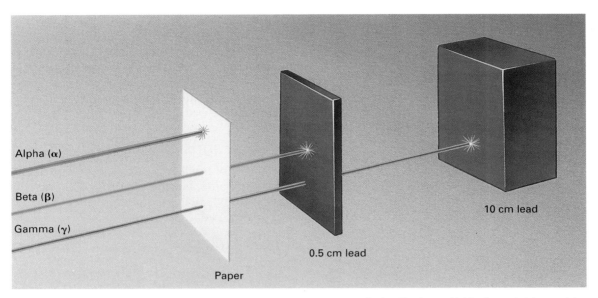

FIGURE 13.1 The relative penetrating abilities of alpha, beta, and gamma radiation. The heavy, highly charged alpha particles are stopped by a piece of paper (or the skin). The lighter, less highly charged beta particles penetrate paper, but are stopped by a 0.5-cm sheet of lead. Because gamma rays have no charge and no mass, they are the most penetrating, but can be stopped by several centimeters of lead.

13.2 NUCLEAR REACTIONS

After discovery of the natural radioactivity of uranium, thorium, and radium, many other elements were found to have radioactive isotopes. All the elements heavier than bismuth (Bi, atomic number 83) and a few lighter than bismuth have natural radioactivity. While studying radium, Rutherford found that besides emitting alpha particles, radium was also producing radioactive

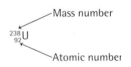

$$^{238}_{92}\text{U}$$

Mass number

Atomic number

THE PERSONAL SIDE

The Curies

Soon after Becquerel's discovery of uranium's radioactivity, Marie Sklodowska Curie (1867–1934), also working in France, studied the radioactivity of thorium (Th) and began to search systematically for new radioactive elements. She showed that the radioactivity of uranium was an atomic property—that is, its radioactivity was proportional to the amount of the element present and was not related to any particular compound. Her experiments indicated that other radioactive elements were probably also present in certain uranium samples. With painstaking technique, she and her husband Pierre Curie (1859–1906) separated the element radium (Ra) from uranium ore and found that it is more than one million times more radioactive than uranium. In 1903, Marie and Pierre Curie shared the Nobel Prize in physics with Henri Becquerel for their discovery of radioactivity. After Pierre died,

© Stock Montage

Pierre and Marie Curie with their daughter, Irene. Irene grew up to continue the study of radioactivity with her husband, Frédéric Joliot. Together, Irene and Frédéric won a Nobel Prize in 1935 for production of the first artificial radioactive isotope.

Marie continued her research and discovered polonium (Po), which she named after her native Poland. In 1911, she became the first person to win a second Nobel Prize, this one for the discoveries of radium and polonium. In 1921, Marie Curie came to the United States, where she was given 1 g of pure radium, purchased with donations from American women interested in her work.

radon gas (Rn). This led Rutherford and one of his students, Frederick Soddy, in 1902 to propose the revolutionary theory that *radioactivity is the result of a natural change of an isotope of one element into an isotope of a different element.* Such a change is a **nuclear reaction**, a process in which an unstable nucleus emits radiation and is converted into a more stable nucleus of a different element. Thus, a nuclear reaction results in a change in atomic number and often a change in mass number as well.

Equations for Nuclear Reactions

In nuclear reactions the total number of nuclear particles, called **nucleons** (protons plus neutrons), remains the same, but the identities of atoms can change. Just as with chemical equations, nuclear equations reflect the fact that matter is conserved. As a result, *the sum of the mass numbers of reacting nuclei must equal the sum of the mass numbers of the product nuclei. There must also be nuclear charge balance—the sum of the atomic numbers of the products must equal the sum of the atomic numbers of the reactants.*

Consider the equation for the nuclear reaction

$$\underset{\text{Radium-226}}{^{226}_{88}\text{Ra}} \longrightarrow \underset{\text{Alpha particle}}{^{4}_{2}\text{He}^{2+}} + \underset{\text{Radon-222}}{^{222}_{86}\text{Rn}}$$

The mass number on the left equals the sum of the mass numbers on the right. Similarly, the atomic number on the left equals the sum of the atomic numbers on the right.

$$\text{mass number: } 226 = 4 + 222$$

$$\text{atomic number: } 88 = 2 + 86$$

The isotope of uranium with atomic mass 238 is also an alpha emitter. When the $^{238}_{92}\text{U}$ nucleus gives off an alpha particle, made up of two protons and two neutrons, four units of atomic mass and two units of atomic charge are lost. The resulting nucleus has a mass of 234 and a nuclear charge of 90, showing that it is an isotope of thorium, which has 90 protons in its nucleus and an atomic number of 90.

$$\underset{\text{Uranium-238}}{^{238}_{92}\text{U}} \longrightarrow \underset{\text{Alpha particle}}{^{4}_{2}\text{He}^{2+}} + \underset{\text{Thorium-234}}{^{234}_{90}\text{Th}}$$

mass number:	238	=	4	+	234
atomic number:	92	=	2	+	90

As is seen in these two examples, loss of an alpha particle from an atom of a given element results in the formation of an atom whose atomic number is 2 less, and whose mass is 4 less, than that of the original element.

Some unstable nuclei are beta emitters. For example, uranium-235 emits a beta particle, which is an electron ($^{0}_{-1}\text{e}$).

$$\underset{\text{Uranium-235}}{^{235}_{92}\text{U}} \longrightarrow \underset{\text{Beta particle}}{^{0}_{-1}\text{e}} + \underset{\text{Neptunium-235}}{^{235}_{93}\text{Np}}$$

mass number:	235	=	0	+	235
atomic number:	92	=	−1	+	93

Knowing that some nuclei emit beta particles leads to a basic question: How can a nucleus containing protons and neutrons emit a beta particle, which is an electron? It has been established that an electron and a proton can com-

Recall that an α particle is a helium atom without its electrons and has a +2 charge. The α particle is sometimes also represented as $^{4}_{2}\alpha^{2+}$ in nuclear equations.

β particles are sometimes represented as $^{1}_{0}\beta$.

When 226 g of radium-226 has completely decayed, 222 g of radon-222 and 4 g of helium will have been formed.

Radium-226 is another way of writing $^{226}_{88}\text{Ra}$.

Nuclear reaction A process in which an isotope of one element is transformed into an isotope of another element

Nucleons Protons and neutrons

bine outside the nucleus to form a neutron. Therefore, the reverse process is proposed to occur inside the nucleus. When a beta particle is emitted from a decaying nucleus, a neutron decomposes, giving up an electron and changing itself into a proton. The ejected electron is the beta particle. The resulting proton remains in the nucleus and increases the atomic number by one. Therefore, loss of a beta particle from an atom of an element results in the formation of an atom of an element whose atomic number has increased by 1 but whose atomic mass is the same.

$$\text{Beta particle production:} \quad {}^{1}_{0}\text{n} \longrightarrow {}^{0}_{-1}\text{e} + {}^{1}_{1}\text{H}$$

Neutron   Electron   Proton

Note that this is a balanced nuclear equation.

Gamma radiation (γ) may or may not be given off simultaneously with alpha or beta particles, depending on the particular nuclear reaction involved. Gamma rays are emitted when the product nucleus must lose some additional energy to become stable. Being electromagnetic radiation, gamma rays have no charge and essentially no mass. The emission of a gamma ray, therefore, cannot alone account for production of a different element.

EXAMPLE 13.1   **Writing an Equation for an Alpha Emission**

Write an equation for alpha emission from a polonium-218 isotope.

SOLUTION
First, write the partial equation and set up a table of mass and atomic number changes under it. The atomic number of polonium is 84.

$$^{218}_{84}\text{Po} \longrightarrow {}^{4}_{2}\text{He}^{2+} + ?$$

mass number: $218 \longrightarrow$ 4 $+ ?$

atomic number: $84 \longrightarrow$ 2 $+ ?$

The mass number of the product must be 214 because its mass number plus that of the alpha particle must equal 218, the mass number of the decaying polonium-218 isotope. The atomic number of the product must be 82 because its atomic number plus that of the alpha particle must equal 84, the atomic number of the decaying isotope. Inasmuch as the element with an atomic number of 82 is lead, the product is $^{214}_{82}\text{Pb}$.

$$^{218}_{84}\text{Po} \longrightarrow {}^{4}_{2}\text{He} + {}^{214}_{82}\text{Pb}$$

TRY IT 13.1
Write an equation showing the emission of an alpha particle by an isotope of neptunium ($^{237}_{93}\text{Np}$).

EXAMPLE 13.2   **Writing an Equation for a Beta Emission**

Write an equation for beta emission from a lead-210 nucleus.

SOLUTION
First, write a partial equation that includes what is known: lead-210 is a reactant, and a beta particle is a product. Then, to aid in determining the mass and atomic number changes, set up a table like those used earlier. The atomic number of lead is 82.

$$^{210}_{82}\text{Pb} \longrightarrow {}^{0}_{-1}\text{e} + ?$$

mass number: $210 \longrightarrow$ $0 + ?$

atomic number: $82 \longrightarrow$ $-1 + ?$

THE CHEMISTRY OF

THE SAINT (1997)

©2005 by Mark A. Griep, *University of Nebraska-Lincoln*

In *The Saint* (released in 1997) electrochemist Dr. Emma Russell holds a press conference at Oxford University to announce she has achieved cold fusion. She is about to show a chalkboard full of theory when the scene changes. The chalkboard notes were supplied by the movie's science consultant Dr. Eugene Mallove, a strong proponent for cold fusion.

In the film, we learn that Dr. Russell hasn't been able to replicate her own cold fusion results. She doesn't remember the correct order of the steps because the critical formulae are on numerous sticky notes stashed in her black lace brassiere. Upon learning this, master criminal Simon Templar seduces her and steals them to meet his contract with Ivan Tretiak. By the time Simon realizes he's in love, the notes are in the hands of the evil Tretiak. Before the movie ends, Simon must prevent Tretiak's electrochemist from using the notes and, consequently, prevent the downfall of the Russian president.

On March 23, 1989, the University of Utah held a press conference that shook the energy world. Electrochemists Stanley Pons and Martin Fleischmann announced reproducible cold fusion: 10% more energy released than supplied. They passed an electric current through palladium and platinum wires in a container of heavy water and lithium sulfate. Cold fusion is nuclear fusion at ambient temperature. When the two hydrogen atoms in a water molecule are replaced with deuterium (called heavy hydrogen because it has one proton and one neutron), it is called heavy water.

$$2\,D_2O \longrightarrow 2\,He + O_2 + energy$$

The barrier to nuclear fusion is *coulombic repulsion.* The positively charged protons in each nucleus repulse the other when brought into proximity. In hot fusion, the repulsion is overcome by using either 50 million–degree temperatures or particle accelerators. In cold fusion, the negative charge of the metal is supposed to overcome it.

There are three possible deuterium fusion outcomes:

1. $D + D \longrightarrow T + H$
2. $D + D \longrightarrow {}^3He + neutron$
3. $D + D \longrightarrow {}^4He + \gamma$ (24 MeV)

Many labs have done extensive testing of the claims made by Pons and Fleischmann. They have found that no tritium, neutrons, or gamma rays are generated; that the He concentration does not change; but that there is a source of unexplained minor excess energy production. Whatever the source of that energy, it is enough to keep researchers working in this area despite the heavy negative press.

Questions to Consider

1. Prove that the overall reaction and the three outcome reactions are balanced for protons and neutrons.

2. Convert the energy for one fusion event, 24 MeV, into kilojoules using 1.602×10^{-19} J/eV.

3. Use the third fusion outcome reaction to calculate the amount of fusion energy (in kJ) in one gallon of natural freshwater (3.785 kg/gallon). The natural abundance of deuterium is 0.015%.

4. Use the following values to calculate the amount of combustion energy (in kJ) in one gallon of gasoline: 3.785 gallon/L; 43 MJ/kg; and 0.737 kg/L. Compare your answer with your answer to question 3.

The sum of the mass numbers of the products must equal 210, the mass number of the decaying lead isotope. Since the mass of the beta particle is essentially zero, the mass number of the product nucleus must be 210. The sum of the atomic numbers of the products must also equal the atomic number of lead, 82. So, the atomic number of the product must be 82 [83 + (−1) = 82], which is the atomic number of bismuth (Bi). The product nucleus is ${}^{210}_{83}Bi$.

$${}^{210}_{82}Pb \longrightarrow {}^{0}_{-1}e + {}^{210}_{83}Bi$$

TRY IT 13.2

Write an equation showing the emission of a beta particle from $^{234}_{91}\text{Pa}$.

(13.3) THE STABILITY OF ATOMIC NUCLEI

Why are some nuclei unstable and radioactive, while others are stable and not radioactive? The fact that there are strong repulsions among all those protons packed inside the nucleus of an atom has something to do with nuclear stability. In addition, the relative numbers of neutrons, which are not charged, play some role. Figure 13.2 shows a plot of the number of protons versus the number of neutrons in the known isotopes from hydrogen ($Z = 1$) to bismuth ($Z = 83$). The nonradioactive (stable) isotopes (purple and green dots) are far fewer in number than the radioactive (unstable) isotopes (red dots).

Z is the symbol for atomic number.

The stability of nuclei is apparently dependent on the relative numbers of protons and neutrons. The nucleus of the simplest atom, hydrogen, contains only a proton. Its two isotopes, deuterium ($^{2}_{1}\text{H}$) and tritium ($^{3}_{1}\text{H}$), contain one and two neutrons, respectively. By looking at Figure 13.2, you can see that the band of stable nuclei, the purple and green dots, curves upward toward the neutron axis; this shows that in stable nuclei, the number of neutrons is equal to or greater than the number of protons. From hydrogen to bismuth, except for $^{1}_{1}\text{H}$ and $^{3}_{2}\text{He}$, *the mass numbers of stable isotopes are always twice as large as the atomic number or even larger.* It appears that the larger numbers of protons in the nuclei of heavier atoms require extra neutrons to gain stability. Any unstable isotope (a red dot in Figure 13.2) will decay in such a way that its decay product falls closer to the stable band (the purple and green dots).

Perhaps because there are so many unstable (radioactive) isotopes, people sometimes think of *radioactive* when the word *isotope* is mentioned.

Beta emission occurs in isotopes that have *too many neutrons* to be stable. These isotopes appear as red dots above the stable band in Figure 13.2. When beta decay occurs, the conversion of a neutron into a proton and an electron increases the atomic number while lowering the number of neutrons, as was illustrated in Example 13.2, and the new isotope moves toward the stable region.

Beta particles result from the conversion of neutrons into protons and electrons

$$^{1}_{0}\text{n} \longrightarrow {}^{1}_{1}\text{H} + {}^{0}_{-1}\text{e}$$

Those isotopes with *too few neutrons* (red dots below the band of stability) decay as well, but in a manner that increases the number of neutrons relative to the number of protons. One way this can happen is by emission of a type of subatomic particle discovered in 1932, a **positron**—a positively charged electron, $^{0}_{+1}\text{e}$. For example, the decay of nitrogen-13, an isotope with too few neutrons, is by positron emission.

$$^{13}_{7}\text{N} \longrightarrow {}^{0}_{+1}\text{e} + {}^{13}_{6}\text{C}$$

The positron results from the decay of a proton.

$$^{1}_{1}\text{H} \longrightarrow {}^{1}_{0}\text{n} + {}^{0}_{+1}\text{e}$$
$$\text{Proton} \qquad \text{Neutron} \qquad \text{Positron}$$

The positron is sometimes called the *antielectron*. The positron is one of a group of *antimatter* particles known to exist. An electron will react with a positron to annihilate each other and produce two high-energy gamma rays.

Positron A positively charged electron

Because the positron, like the electron, has a mass number of zero, the mass number of the product nucleus is the same as that of the starting nucleus. Therefore, emission of a positron from an atom of an element results in the formation of an atom of an element whose atomic number is reduced by 1 but whose mass is unchanged.

All isotopes of the elements beyond bismuth ($Z = 83$) are unstable. Most of them decay by ejecting an alpha particle. This kind of decay, as illustrated

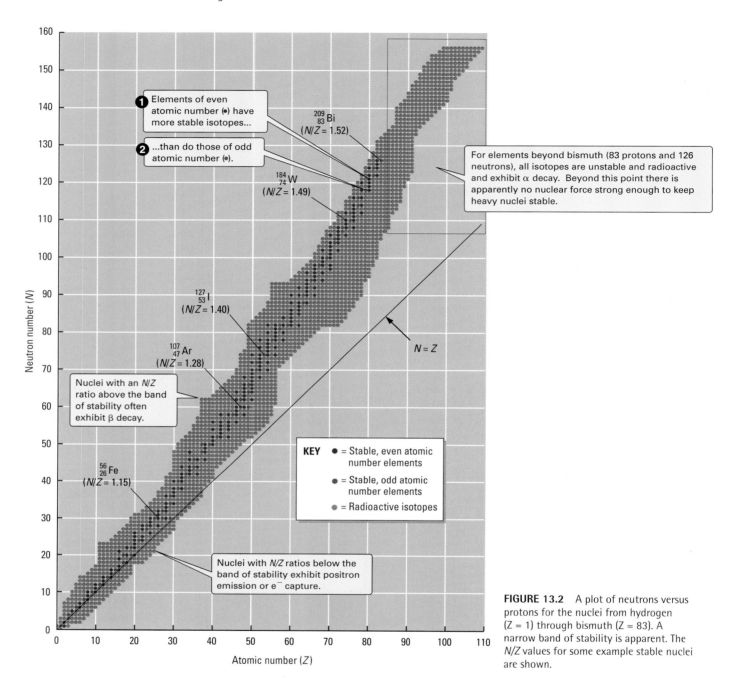

FIGURE 13.2   A plot of neutrons versus protons for the nuclei from hydrogen ($Z = 1$) through bismuth ($Z = 83$). A narrow band of stability is apparent. The N/Z values for some example stable nuclei are shown.

in Example 13.1, decreases the mass number by four and the atomic number by two. The types of radioactive decay we have discussed are summarized in Table 13.1.

13.4 ACTIVITY AND RATES OF NUCLEAR DISINTEGRATIONS

Activity The number of radioactive nuclei that disintegrate per unit of time

The number of radioactive nuclei that disintegrate in a sample per unit of time is called its **activity** (this is the "activity" in radioactivity). The activity of a sample containing radioactive isotopes depends on the number of nuclei

TABLE 13.1	Changes in Atomic Number and Mass Number Accompanying Radioactive Decay				
TYPE OF DECAY	**SYMBOL**	**CHARGE**	**MASS**	**CHANGE IN ATOMIC NUMBER**	**CHANGE IN MASS NUMBER**
Beta	$_{-1}^{0}e$	−1	0	+1	None
Positron	$_{+1}^{0}e$	+1	0	−1	None
Alpha	$_{2}^{4}He^{2+}$	+2	4	−2	−4
Gamma	$_{0}^{0}\gamma$	0	0	None	None

present and the rate at which they decay. If a sample of matter is "highly radioactive," many atoms are undergoing decay per unit of time. A small number of nuclei decaying at a rapid rate can produce the same activity as a larger number of atoms decaying at a slower rate. Radioactive disintegrations are measured in *curies* (Ci); one Ci is 37 billion disintegrations per second (dps).

To illustrate the differences in rates of decay of radioactive nuclei, consider first how cobalt-60 is used in medicine to treat cancerous tumors in the human body. When cobalt-60 decays, it produces beta particles as well as gamma rays.

$$_{27}^{60}Co \longrightarrow \ _{28}^{60}Ni + \ _{-1}^{0}e + \ _{0}^{0}\gamma$$

Although the cobalt-60 isotope is radioactive, it is stable enough so that only half of a sample will decay in 5.27 years. A cobalt-60 sample is installed in a well-shielded apparatus that emits a focused beam of gamma rays. Because the **half-life** of the cobalt-60 radioisotope is fairly long, the sample does not have to be replaced very often. By rotating the radiation source around the patient, the physician can concentrate the rays in the cancerous region being treated, while somewhat limiting the radiation to other parts of the body (Figure 13.3).

Half–Life

The rate of decay of any radioactive isotope can be represented by its characteristic half-life, the period required for one-half of the radioactive material originally present to undergo radioactive decay. Short half-lives are the results of high rates of decay, and long half-lives are the results of low rates of decay.

The half-life of an isotope is independent of the amount of radioactive material present and is essentially independent of temperature and the chemical form in which the radioactive atoms are present. Table 13.2 gives the half-lives of some radioactive isotopes. The 12.9 h half-life of $_{29}^{64}Cu$, for example, means that one-half of the original amount of copper-64 atoms will remain after 12.9 h. In another 12.9 h, half of the original half, $(\frac{1}{2})^2 = (\frac{1}{4})$, will remain. This process continues indefinitely until virtually all the copper-64 isotopes have decayed. Figure 13.4 (page 295) illustrates graphically how the concept of half-life works for a radioactive isotope. No matter what its half-life, the fraction of a radioactive isotope remaining will be one-half after one half-life, one-fourth after two half-lives, one-eighth after three half-lives, and so on.

By focusing attention on the decay products, we also are helped in understanding the concept of half-life. For example, if a sample contains one million copper-64 atoms at some beginning time, 12.9 h later only 500,000

A more suitable unit is the *microcurie* (μCi), which is 37,000 disintegrations per second (dps). Another unit used to measure radioactive disintegrations is the *becquerel* (Bq), where 1 Bq = 1 dps.

Gamma radiation is more damaging to cancer cells because they duplicate faster than normal cells.

Further examples of the use of radioisotopes in medicine are given in Section 13.8.

Mathematically, the fraction of a radioactive isotope remaining after n half-lives is $(\frac{1}{2})^n$. The fraction after two half-lives is $(\frac{1}{2})^2 = \frac{1}{4}$; after three half-lives it is $(\frac{1}{2})^3 = \frac{1}{8}$, and so on.

Half-life The period required for one-half of a sample of a radioactive substance to undergo radioactive decay

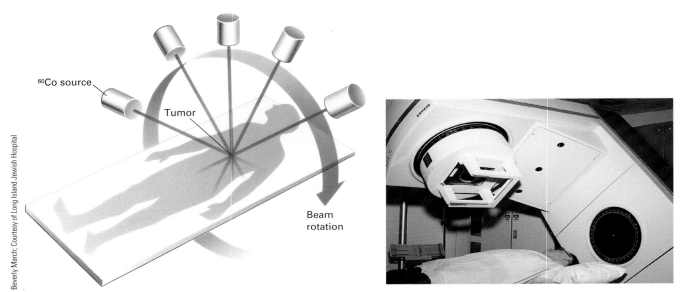

FIGURE 13.3 **Treatment for cancer with gamma radiation from cobalt-60.** By adjusting the rotation of the radiation source, the radiation is concentrated where the beams cross at the location of the diseased tissue.

copper-64 atoms would remain. However, there would be 500,000 zinc-64 atoms present that had not been there 12.9 h earlier. After 25.8 h (two half-lives) only 250,000 copper-64 atoms would remain, and there would be 750,000 zinc-64 atoms resulting from the decay of the copper atoms. After many half-lives, almost all the copper atoms will have decayed, and there will be almost one million zinc-64 atoms, which are stable and do not undergo decay.

Natural Radioactive Decay Series

As one would expect for an element with such a long half-life (Table 13.2), relatively large amounts of uranium-238 can be found in certain rocks and mineral deposits. Uranium-238 decays first to thorium-234, which then itself decays. These first two steps are part of the **uranium series**, which ends with a stable, nonradioactive isotope of lead, lead-206 (Figure 13.5, page 296). You would not expect to find much of the short-lived isotopes in a sample of rock, and indeed you have to look carefully for them, but they are there. The

Uranium series The series of steps in the naturally occurring decay of uranium-238 to lead-206

TABLE 13.2	Half-Lives of Some Radioactive Isotopes

DECAY PROCESS	HALF-LIFE
$^{238}_{92}\text{U} \longrightarrow {}^{234}_{90}\text{Th} + {}^{4}_{2}\text{He}$	4.51×10^9 years
$^{3}_{1}\text{H} \longrightarrow {}^{3}_{2}\text{He} + {}^{0}_{-1}e$	12.3 years
$^{14}_{6}\text{C} \longrightarrow {}^{14}_{7}\text{N} + {}^{0}_{-1}e$	5730 years
$^{131}_{53}\text{I} \longrightarrow {}^{131}_{54}\text{Xe} + {}^{0}_{-1}e$	8.05 days
$^{64}_{29}\text{Cu} \longrightarrow {}^{64}_{30}\text{Zn} + {}^{0}_{-1}e$	12.9 h
$^{69}_{30}\text{Zn} \longrightarrow {}^{69}_{31}\text{Ga} + {}^{0}_{-1}e$	55 min

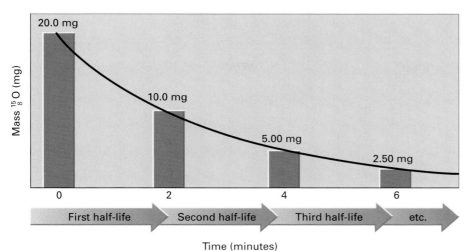

FIGURE 13.4 Radioactive decay of 20 mg of oxygen-15, which has a half-life of 2.0 min. The plotted data are given in the following table. After each half-life period, the quantity present at the beginning of the period is reduced by one-half.

longer lived isotopes such as uranium-234 and thorium-230 are readily detected. Two other similar natural decay series exist, each of which starts out with a different isotope and proceeds through a different set of radioactive decay products. Most of the naturally occurring radioactive isotopes are members of one of the three decay series.

CONCEPT CHECK   13A

1. The first evidence for radioactivity occurred when photographic films were exposed when placed near samples of uranium. (**a**) True, (**b**) False.
2. Which of these forms of radiation is the most penetrating? (**a**) Alpha particles, (**b**) Beta particles, (**c**) Gamma rays?
3. The mass number of a nucleus is unchanged after beta-particle emission. (**a**) True, (**b**) False.
4. The mass number of a uranium-238 nucleus will decrease by two units after alpha emission. (**a**) True, (**b**) False.
5. When a $^{216}_{84}$Po nucleus emits an alpha particle, the nuclear species that results is _____.
6. Nuclei with atomic numbers greater than 83 are all radioactive. (**a**) True, (**b**) False.
7. The half-life of $^{210}_{83}$Bi is 5 days. If the initial activity was 2 μCi, or 74,000 dps, the activity after 10 days is expected to be _____.

13.5 ARTIFICIAL NUCLEAR REACTIONS

In 1919, Rutherford was successful in producing the first artificial nuclear change by bombarding nitrogen (N_2) with alpha particles. All the results of the experiment could be explained if one assumed the nuclear reaction to be

$$^{14}_{7}N + {}^{4}_{2}He^{2+} \longrightarrow [{}^{18}_{9}F] \longrightarrow {}^{17}_{8}O + {}^{1}_{1}H$$

FIGURE 13.5 The uranium-238 decay series. Radium (Ra) and polonium (Po), the two elements discovered by Marie Curie, are part of this series. Radon (Rn), the radioactive gas of environmental concern, is generated as shown here wherever rocks contain uranium. Lead-206 is not radioactive. The half-lives of the isotopes in this decay series vary considerably. Uranium-238 has a half-life of 4.5 billion years, whereas the half-life of lead-210 is 22 years. Some radioactive elements, like thallium-210, have half-lives of only a few minutes.

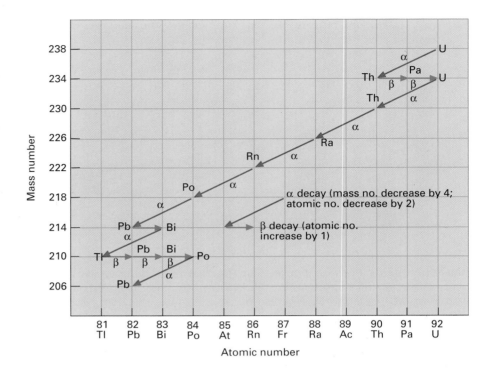

where $^{18}_{9}F$ is an unstable nucleus that quickly disintegrates to $^{17}_{8}O$ and $^{1}_{1}H$. Both product nuclei are stable. Rutherford had observed an **artificial transmutation**, the conversion of one element into another during a laboratory experiment. Following Rutherford's original experiment, there was considerable interest in discovering new nuclear reactions.

In 1934, Irene Curie Joliot, daughter of Marie and Pierre Curie, and her husband, Frédéric Joliot, bombarded aluminum (Al) with alpha particles and observed neutrons and a positron. The Joliots discovered that when the flow of alpha particles striking the Al was stopped, the neutron emissions stopped, but the positron emissions continued. They reasoned that the alpha particles reacted with aluminum nuclei to produce phosphorus-30 nuclei, which then decayed to produce positrons.

Alchemists in ancient times tried in vain to transmute one element into another—mainly lead into gold. They would have been extremely excited at the news of Rutherford's accomplishment.

$$^{27}_{13}Al + {}^{4}_{2}He^{2+} \longrightarrow {}^{30}_{15}P + {}^{1}_{0}n$$
$$^{30}_{15}P \longrightarrow {}^{30}_{14}Si + {}^{0}_{+1}e$$
$$\text{Positron}$$

The second reaction continued because the phosphorus-30 was decaying more slowly than it was being produced. Phosphorus-30 was the *first radioactive isotope to be produced artificially*. Today, more than 1000 other radioactive isotopes have been produced.

13.6 TRANSURANIUM ELEMENTS

In 1940, at the University of California, E. M. McMillan and P. H. Abelson prepared element 93, the synthetic element neptunium (Np). Neptunium was the first of the **transuranium elements**, those with atomic numbers greater than 92. The neptunium was made by directing a stream of high-energy deuterons ($^{2}_{1}H$) onto a target of uranium-238. The initial reaction was the conversion of uranium-238 to uranium-239:

Artificial transmutation Experimental conversion of one element into another

Transuranium elements Artificial elements, all with atomic numbers greater than 92

$$^{238}_{92}U + {}^{2}_{1}H \longrightarrow {}^{239}_{92}U + {}^{1}_{1}H$$

Uranium-239 has a half-life of 23.5 min and decays spontaneously to the element neptunium by the emission of beta particles:

$$^{239}_{92}\text{U} \longrightarrow {}^{239}_{93}\text{Np} + {}^{0}_{-1}\text{e}$$

Neptunium is also unstable, with a half-life of 2.33 days; it converts into a second new element, plutonium (Pu):

$$^{239}_{93}\text{Np} \longrightarrow {}^{239}_{94}\text{Pu} + {}^{0}_{-1}\text{e}$$

Plutonium-239, like neptunium-239, is radioactive, but it has a half-life of 24,100 years. Because of the relative values of the half-lives, very little neptunium could be accumulated, but the plutonium could be obtained in larger quantities. The names of neptunium and plutonium were taken from the mythological names Neptune and Pluto in the same atomic number sequence as the order of the planets Uranus (uranium), Neptune, and Pluto out from the Sun.

Although Neptune and Pluto are the last of the known planets in the solar system, their namesakes are not the last in the list of elements. The rush of experiments that followed the synthesis of plutonium produced additional man-made elements. Elements 95 (americium) through 106 (seaborgium) had been created by the end of the 1970s. Up to element 101 (mendelevium) all the elements can be made by hitting the target nuclei with small particles such as $^{4}_{2}\text{He}^{2+}$ or $^{1}_{0}\text{n}$. Beyond element 101, special techniques that use heavier bombarding particles are required. The energy of the particles is matched to the energy needed to produce the initial unstable nucleus. For example, lawrencium is made by hitting californium-252 with boron nuclei. In this reaction, five neutrons are produced.

$$^{252}_{98}\text{Cf} + {}^{10}_{5}\text{B} \longrightarrow {}^{257}_{103}\text{Lr} + 5\,{}^{1}_{0}\text{n}$$

By 1996, elements up to 112 had all been created, many with extremely short half-lives. Element 112, for example, has a half-life of 240 microseconds.

Four atoms each of element 113 and element 115 were produced in a fusion reaction between calcium-48 and americium-243 in 2004. These two elements, combined with element 114, produced earlier, extend the periodic table to 115 elements. Element 113 was produced by the alpha decay of element 115.

No one knows for certain how much plutonium has been made—probably in excess of several million kilograms. Most of this amount has been used to make nuclear warheads by the United States, the former Soviet Union, and the other nuclear nations including the United Kingdom, France, China, Israel, India, and North Korea.

Plutonium-239 is important because it is a fissionable isotope (Section 13.9) and can be used for making bombs. In addition, plutonium has the added distinction of being one of the most toxic substances known.

(13.7) RADON AND OTHER SOURCES OF BACKGROUND RADIATION

Sources of background radiation, which we receive during normal daily life, fall into two categories: man-made and natural (Figure 13.6). Of these,

Even the altitude above sea level where you live is related to your annual dose of background radiation—the higher the altitude, the greater the cosmic radiation.

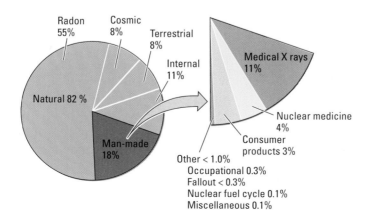

FIGURE 13.6 Sources of average background radiation exposure in the United States. The sources are expressed as percentages of the total. As seen from the figure, background radiation from natural sources far exceeds that from artificial sources.

THE PERSONAL SIDE

Glenn Theodore Seaborg (1912–1999)

Some of the most significant contributions to the modern periodic chart were made by Nobel Laureate Glenn Seaborg. Seaborg's career demonstrated the importance of maintaining the courage of one's convictions.

Thanks to his insights, it is now very well established that the transuranium elements (those with atomic numbers greater than 92), a number of which he either discovered or helped to discover, are members of the actinide series. Actinides are the elements that follow actinium. They are found in a grouping off the main periodic chart.

Until Seaborg offered his version of the periodic table, chemists were convinced that Th, Pa, and U belonged in the main body of the table, Th under Hf, Pa under Ta, and U under W. When Seaborg proposed that Th was the beginning of the actinides and that the transuranium elements belonged as a group under the rare earths, some prominent and famous inorganic chemists, many of them Seaborg's friends, tried to discourage his publication of this finding in the open literature. One very prominent inorganic chemist felt that he

Lawrence Berkeley National Laboratory

GLENN SEABORG
(1912–1999)

would ruin his scientific reputation. Nevertheless Seaborg, strongly convinced, persisted and as a result, properly placed this most important class of elements where they are today. Based on Seaborg's expansion of the periodic table, it was possible to predict accurately the properties of many of the as-yet–undiscovered transuranium elements. Subsequent preparation in atomic accelerators of these elements proved him right, and it was fitting that he was awarded the Nobel Prize in 1951 for his outstanding work. The prize was shared with E. M. McMillan (1907–1991) who started Seaborg in this area of research. Seaborg, who discovered 10 of the transuranium elements, became the only person to have an element named for him while still living. Element 106 was named seaborgium (Sg) in his honor in August 1997.

natural sources produce the bulk of our exposures. When natural sources are coupled with desirable exposures from medical applications, scarcely 3% of the exposures are truly avoidable by society as a whole. To illustrate how unavoidable most exposures to radiation are, consider the element potassium, an element essential to human life that helps regulate water balance in our bodies and is involved in nerve transmissions. A 60-kg (132-lb) person has about 200 g of potassium. It turns out that just over 0.01% of all potassium atoms are potassium-40 atoms, which are radioactive, with a half-life of 1.25 $\times$ 10^9 years. This means the 60-kg person has roughly 20 mg of radioactive potassium, which disintegrates at a rate determined by the half-life of that isotope. These disintegrations contribute to the background radiation we all receive. In addition, the beta particle from the decay of a potassium-40 disintegration in your neighbor might pass into you to contribute to your background radiation. Normal background radiation from all sources is 2 or 3 disintegrations/sec (dps).

Another element that contributes to our background radiation, and hence to our risks of radiation-caused damage, is radon. Radon-222, the most common isotope of radon, is radioactive, with a half-life of 3.82 days. It is a product of the uranium decay series (see Figure 13.5) and results from the alpha decay of radium-226.

$$\ce{^{226}_{88}Ra} \longrightarrow \ce{^{222}_{86}Rn} + \ce{^{4}_{2}He^{2+}}$$

When radon decays, it produces an alpha particle and another short-lived radioisotope, polonium-218.

Radon (Rn) is the heaviest member of the noble gas family of elements (He, Ne, Ar, Kr, Xe, and Rn).

$$^{222}_{86}\text{Rn} \longrightarrow {}^{218}_{84}\text{Po} + {}^{4}_{2}\text{He}^{2+}$$

Polonium-218 (half-life 3.05 min) also decays, producing an alpha particle and lead-214 (half-life 26.8 min).

$$^{218}_{84}\text{Po} \longrightarrow {}^{214}_{82}\text{Pb} + {}^{4}_{2}\text{He}^{2+}$$

These alpha particles from the products of radon decay make breathing air containing radon dangerous.

Radon exists as a gas, and all rocks contain *some* uranium, although in most the amount is small (one to three parts per million). Therefore, radon is constantly being formed in amounts that vary with the type of rock or soil. How much of this radon escapes into the outside air or a building on the surface depends on the porosity and moisture content of the soil and how finely divided it is (escape is easier from a small particle than from a large one). The overall geology of the site is also important.

Because radon is chemically unreactive, radon atoms in the air we breathe are inhaled and exhaled without any chemical change, although some may dissolve in lung fluids. If a radon atom happens to decay within the lungs, however, the nongaseous and radioactive *radon daughters* can remain inside the lungs, where they will continue to decay. Those up to lead-210 have short half-lives and include several alpha emitters. Because alpha particles can travel up to 0.7 mm, the approximate thickness of the epithelial cells of the lung, they can damage delicate lung tissue and create a higher than normal risk of lung cancer.

Miners in deep mines are exposed to far more than average radon levels, and as early as 1950 government agencies began monitoring radon exposures and incidences of lung cancer. Today, it is well known that radon exposure increases one's chance of developing lung cancer. If you smoke, the chances are even greater, since there seems to be a synergistic effect between smoking and radon levels in causing lung cancer.

It has been estimated that nearly 1 in 15 homes in the United States is affected by radon contamination. Levels higher than 4 picocuries per liter of air (4 pCi/L) have been designated the "action level" by the U.S. Environmental Protection Agency (EPA). Some level of radon could probably be detected in homes in almost every state.

This EPA action level is important for two reasons. First, levels of radon higher than 4 pCi/L should be reduced because levels above this are judged to lead to unacceptable risks of lung cancer. Second, it is difficult to lower the level of radon in most contaminated homes below 4 pCi/L. This last point is an acceptance of the fact that radiation exposure will always be with us.

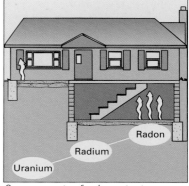

Common route of radon entry to homes—seepage through the foundation.

Daughters is a term used to describe radioactive decay products. Radon daughters come from the decay of radon.

As always, it is wise to keep in mind that all such data are based on statistical studies and risk assessment models.

If the radon radiation level is above 4 pCi/L, EPA requires that some form of remediation action be taken.

4 pCi/L is a little less than 1.5 disintegrations every 10 s.

Outdoor radon concentrations are approximately 0.1 pCi/L to 0.15 pCi/L worldwide.

CONCEPT CHECK 13B

1. When unstable uranium-239 goes through beta decay the product nucleus is _____.
2. The first transuranium element is _____.
3. Normal background radiation from all sources produces 2 dps to 3 dps. (**a**) True, (**b**) False.
4. Experiments to make elements 104 through 111 require using projectiles bigger than alpha particles or neutrons. (**a**) True, (**b**) False.
5. Radon gas comes from the decay of what naturally occurring radioactive element?
6. The EPA action level for radon gas found in homes is _____.

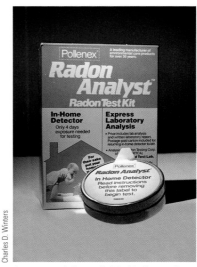

Charles D. Winters

A commercially available kit for testing for radon in the home.

USEFUL APPLICATIONS OF RADIOACTIVITY

The damaging aspects of nuclear radiation must always be kept in mind, especially when the possibilities of accidental or unintended exposures are great. However, the radiation from radioisotopes can be put to beneficial use.

Food Irradiation

In some parts of the world, stored-food spoilage may claim up to 50% of the food crop. In the United States, refrigeration, canning, and chemical additives reduce this figure considerably. Still, there are problems with food spoilage and contamination. Food protection costs amount to a sizable fraction of the final cost of food.

Foods may be irradiated to retard the growth of organisms such as bacteria, molds, and yeasts. Food irradiation for this purpose using gamma rays from sources such as cobalt-60 or cesium-137 is common in European countries, Canada, Mexico, and the United States. Such irradiation prolongs shelf life under refrigeration in much the same way that heat pasteurization protects milk. For example, chicken normally has a three-day refrigerated shelf life. After irradiation, chicken may have a three-week refrigerated shelf life.

Foods irradiated at sufficient levels will keep indefinitely when sealed in plastic or aluminum foil packages. At present, the Food and Drug Administration (FDA) has approved irradiation of many classes of foods (Table 13.3). It is important to note that in no case is there any chance of irradiated food becoming radioactive. The ongoing debate about the safety of irradiated foods revolves around the possibly harmful nature of would-be "radiolytic products"—products of chemical changes in food caused by the high energy of the radiation. For example, could irradiation of food produce a chemical capable of causing genetic damage? To date, no evidence has been found for harmful radiolytic products. Meanwhile, based on extensive review of current data, the FDA-approved radiation limits have been conservatively set at relatively low levels.

Radiocarbon Dating

Carbon has three isotopes. Two of these, carbon-12 and carbon-13, comprise 98.9% and 1.1%, respectively, of all carbon. A third isotope, carbon-

The FDA approved irradiation of beef in December 1997 in response to concerns due to food poisoning and product recalls of contaminated hamburger meat.

TABLE 13.3 Irradiated Foodstuffs Approved by U.S. FDA

FOOD	PURPOSE
Uncooked beef	Control of food-borne pathogens
Uncooked pork	Control of *Trichinella spiralis*
All fresh foods	Growth and maturation inhibition
All foods	Disinfestation of anthropod pests
Dry enzyme preparations	Microbial disinfection
Dried herbs and spices	Microbial disinfection
Uncooked poultry	Control of food-borne pathogens
Eggs	Control of *Salmonella*

Source: U.S. FDA regulations.

14, is present in very small amounts under natural conditions. Only about 1 in 10^{12} atoms of carbon is carbon-14, a radioactive isotope that emits a beta particle and has a half-life of 5730 years. Atmospheric carbon dioxide contains small amounts of this isotope. Consequently, any living plant will incorporate carbon-14 by photosynthesis since carbon dioxide is the source of carbon for these plants. Upon death, plants cease to incorporate carbon dioxide. These facts may be used to estimate the age of carbon-containing materials.

A one-gram sample of carbon taken from a living plant would give about 15 decompositions per minute. This level of decomposition will drop by half every 5730 years. Thus, with modern instrumentation, it is possible to estimate the age of dead plants or materials made from these plants. Since most animals, including humans, eat plants, this procedure may be used to calculate when a plant or animal died.

The procedure does have some limitations. It is necessary to know how the atmospheric concentration of carbon-14 has changed over thousands of years. Even with this information in hand, practical considerations dictate that the procedure may only be used to determine the age of objects in a range of about 100 years. The radiocarbon level of an object less than 100 years old will not have declined significantly enough from that of the living organism to allow an accurate determination of age. Likewise, objects older than 40,000 years (more than 7 half-lives) will have so little radioactive carbon remaining that an accurate determination will not be possible.

Radiocarbon dating has been used to estimate the age of many artifacts and its accuracy has been confirmed in tests against articles of known age. Perhaps the most famous, and controversial, example of the use of this technique involved the Shroud of Turin. This piece of cloth, which bears a remarkable human image that appears almost like a photographic negative, has long been purported by many religious groups to be the burial shroud of Jesus Christ. However, radiocarbon dating put the age of the cloth at around 650 years, which means the plant material used to make the cloth could not have been grown at the time of Jesus, about 2000 years ago. It was estimated to have been grown between 1260 and 1390 A.D. The cloth and the image on it have remained controversial in spite of this scientific analysis. In contrast to this study, the Dead Sea Scrolls, found in 1947 in a cave in the Qumran area near Jerusalem, have been authenticated as being approximately 2000 years old, an age consistent with the Hebrew writings contained within. Carbon-14 dating has also been used to determine that the remains of the Ice Man found frozen in the Italian Alps in 1991 are about 5000 years old.

Medical Imaging

Radioisotopes are used in medicine in two distinctly different ways: diagnosis and therapy. In the diagnosis of internal disorders, physicians need information on the locations of the disorders. An appropriate radioisotope is introduced into the patient's body, either alone or combined with some other chemical, and it accumulates at the site of the disorder. There the radioisotope disintegrates and emits its characteristic radiation, which can be detected. Modern medical diagnostic instruments not only determine where the radioisotope is located in the patient's body but also construct an image of the area.

Four diagnostic radioisotopes are listed in Table 13.4. Each produces gamma radiation, which in low doses is less harmful to the tissue than ionizing radiations such as beta or alpha particles. By the use of special carriers, these radioisotopes can be made to accumulate in specific areas of the body. For example, the pyrophosphate ion ($P_4O_7^{4-}$), a simple polyatomic ion, can

The half-life of $^{99m}_{43}Tc$, 6 h, makes it an ideal radioisotope for medical purposes. The more stable isotope $^{99}_{43}Tc$ has a half-life of 2.12×10^5 years. This means it has such a low activity that it will be eliminated from the patient's body before many disintegrations can occur.

TABLE 13.4	Diagnostic Radioisotopes		
RADIOISOTOPE	**NAME**	**HALF-LIFE (HOURS)**	**USES**
^{99m}Tc*	Technetium-99m	6	As TcO_4^- to the thyroid, brain, and kidneys
^{201}Tl	Thallium-201	21.5	To the heart
^{123}I	Iodine-123	13.2	To the thyroid
^{67}Ga	Gallium-67	78.3	To various tumors and abscesses

*The technetium-99m isotope is the one most commonly used for diagnostic purposes. The *m* stands for *metastable*, a term explained in the text.

bond to the technetium-99m radioisotope (the *m* denotes *metastable*, meaning the isotope decays by emitting a gamma ray to form a more stable isotope with the same mass number), and together they accumulate in the skeletal structure where abnormal bone metabolism is taking place. Such investigations often pinpoint bone tumors.

The imaging method is based on the emission of gamma rays from the target organ. As the gamma rays strike a gamma-ray camera, the signal is processed by a computer and displayed as a video image (Figure 13.7).

$$^{99m}_{43}Tc \longrightarrow {}^{99}_{43}Tc + {}^{0}_{0}\gamma$$
Gamma ray

CONCEPT CHECK 13C

1. Name two uses of radioactive isotopes.
2. The process of concentrating a radioisotope at a particular site of the body to locate and measure the extent of a disorder is called _____.
3. In the symbol for the radioisotope technetium-99m, the *m* stands for (**a**) middle, (**b**) mathematical, (**c**) metastable.

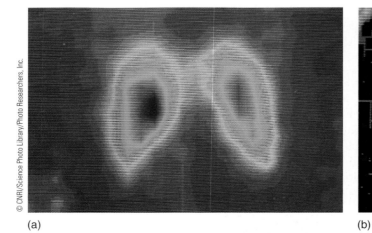
(a)

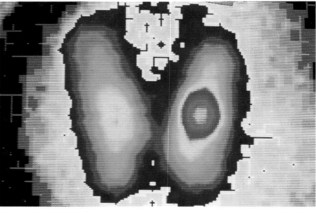
(b)

FIGURE 13.7 **Thyroid imaging.** (a) Healthy human thyroid gland. (b) Thyroid gland showing effect of hyperthyroidism. Technetium-99m concentrates in sites of high activity. Images of this gland, which is located at the base of the neck, were obtained by recording γ-ray emission after the patient was given radioactive technetium-99m. Current technology creates a computer color-enhanced scan.

4. With its half-life of approximately 6 h, how much technetium-99m would remain 18 h after injection into a patient?
   (**a**) One-eighth of the original dose
   (**b**) One-half of the original dose
   (**c**) One-sixth of the original dose
   (**d**) One-fourth of the original dose
5. If two radioisotopes available for diagnosis worked equally well and each decayed by giving off gamma rays, but one had a half-life of 13 h and the other had a half-life of 6 h, which one would you recommend?

13.9 ENERGY FROM NUCLEAR REACTIONS

A vast amount of energy is released when heavy atomic nuclei split—the nuclear **fission** process—and when small atomic nuclei combine to make heavier nuclei—the **fusion** process. In 1938, Otto Hahn, Fritz Strassman, Lise Meitner, and Otto Frisch discovered that $^{235}_{92}\text{U}$ is fissionable by neutrons (Figure 13.8). In less than a decade, this discovery led to two important applications of this energy release accompanying fission—the atomic bomb and nuclear power plants.

There is a huge difference between the amount of energy liberated in an ordinary chemical reaction, like the burning of methane in air, and the energy liberated in a nuclear fission reaction. If you compared the energy from the burning of only 16 g of methane with that from the fission of an equivalent amount of uranium-235, the fission reaction would produce almost 25 million times more energy.

Fission Reactions

Nuclear fission can occur when a neutron (^1_0n) enters a heavy nucleus. Certain heavy nuclei with an odd number of neutrons ($^{235}_{92}\text{U}$, $^{233}_{92}\text{U}$, $^{239}_{94}\text{Pu}$) will undergo fission when struck by slow-moving *thermal* neutrons (neutrons with a kinetic energy about the same as that of a gaseous molecule at ordinary temperatures). The splitting of the heavy nucleus produces two smaller nuclei, two or more neutrons (an average of 2.5 neutrons for $^{235}_{92}\text{U}$), and much energy. Uranium-235 atoms, each undergoing fission, will produce a number of different reaction products. One fission reaction that can take place follows:

$$^{235}_{92}\text{U} + ^1_0\text{n} \longrightarrow ^{236}_{92}\text{U} \longrightarrow ^{141}_{56}\text{Ba} + ^{92}_{36}\text{Kr} + 3\,^1_0\text{n} + 8.3 \times 10^{-15}\text{ kcal}$$

Unstable nucleus   Typical fission products

One kilogram of uranium fuel undergoing fission in a nuclear reactor can produce the same amount of energy as the combustion of 3000 tons of coal or 14,000 barrels of oil.

Fission Splitting of a heavy nucleus into smaller fragments, caused by bombardment with neutrons

Fusion The combination of small nuclei to make heavier nuclei

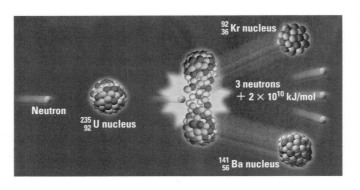

FIGURE 13.8 The fission of a $^{235}_{92}\text{U}$ nucleus from its bombardment with a neutron.

The same unstable $^{236}_{92}U$ atom may undergo fission to produce other products. For example,

$$^{235}_{92}U + {}^{1}_{0}n \longrightarrow {}^{236}_{92}U \longrightarrow$$
$$^{103}_{42}Mo + {}^{131}_{50}Sn + 2\,{}^{1}_{0}n + \text{energy}$$

The critical mass for uranium-235 is about 10 kg.

Chain reaction A process in which neutrons from one fission reaction can cause multiple fission reactions in nearby nuclei

Critical mass The mass of fissionable material able to sustain a chain reaction

Note that the same isotope may split in more than one way. The lighter nuclei produced by the fission reaction are the *fission products*. These fission products, such as $^{141}_{56}Ba$ and $^{92}_{36}Kr$, can also be unstable and emit beta particles ($_{-1}^{0}e$) and gamma rays (γ) and may have long half-lives (see Section 13.4). Eventually, the decay of these fission products leads to stable isotopes.

The neutrons emitted by the fission of one uranium-235 atom can cause the fission of other uranium-235 atoms. For example, the 3 neutrons emitted in the uranium fission (Figure 13.8) could produce fission in three more uranium atoms; the 9 neutrons emitted by those nuclei could produce nine more fissions; the 27 neutrons from these fissions could produce 81 neutrons; the 81 neutrons could produce 243 additional neutrons; those 243 neutrons could produce 729, and so on. This process is called a **chain reaction** (Figure 13.9), and it occurs at a maximum rate when the uranium sample is large enough for most of the neutrons emitted to be captured by other nuclei before passing out of the sample. A sample of fissionable material of sufficient size to self-sustain a chain reaction is termed the **critical mass**. If a critical mass of fissionable material is suddenly brought together, an explosion will occur because the energy released during each fission reaction cannot dissipate rapidly enough.

In a nuclear fission bomb the critical mass is kept separated into several smaller subcritical masses until detonation, at which time the masses are

FIGURE 13.9 A self-propagating nuclear chain reaction initiated by capture of a neutron. The fission of uranium-235 produces a variety of products. Thirty-four elements have been detected among the fission products, including those shown here. Each fission event produces two lighter nuclei plus two or three neutrons.

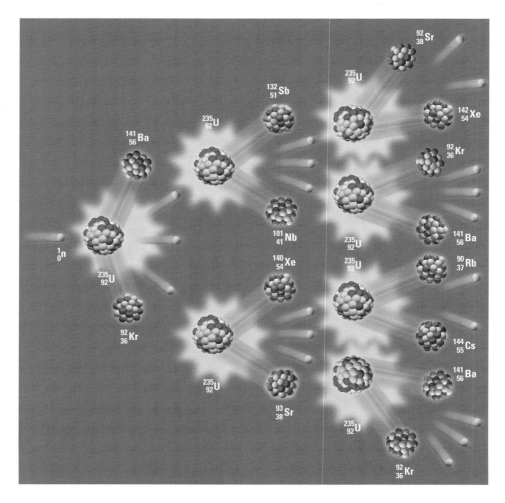

THE PERSONAL SIDE

Lise Meitner (1878–1968)

Lise Meitner was born in Vienna, Austria, one of seven children. In 1902, while studying science in Vienna, she became fascinated with the accounts of the discovery of radium by Pierre and Marie Curie and decided to pursue a career studying atomic physics. In 1906, she received her doctorate in physics from the University of Vienna. Two years later she moved to Berlin, where she became associated with Otto Hahn and Fritz Strassman, with whom she discovered atomic fission almost 20 years later. In 1918, together with Hahn, she discovered the element protactinium, which decays to form actinium. While working with Hahn and Strassman, Meitner discovered that something highly unusual was taking place when atoms of uranium-235 were struck by neutrons. In 1938, before she could solve this apparent puzzle, she fled to Sweden to escape the repression of Nazi Germany. After she established herself in Sweden, Otto Hahn announced the discovery that the uranium atom was being split into two approximately equal-sized parts by the neutron and sent Meitner his results for her analysis. Based on the data, she calculated the energy released when a uranium atom is split by the

LISE MEITNER

neutron and named the phenomenon nuclear fission. She even reported her findings in the British journal *Nature* in 1939. The uranium nucleus, she wrote, would split into a barium atom and a krypton atom under the influence of a neutron, at the same time producing more than 7.6×10^{-12} cal of energy. The enormous energy released by atomic fission was immediately recognized to have military potential, and in the United States the Manhattan Project was begun to create the first atomic bomb. In 1945, when she heard that the first atomic bomb had been dropped on Japan, Dr. Meitner said,

> I must stress that I myself have not in any way worked on the smashing of the atom with the idea of producing death-dealing weapons. You must not blame us scientists for the use to which war technicians have put our discoveries.

driven together by an explosive device (Figure 13.10). During the split second when the chain reaction occurs, the tremendous energy of billions of fission reactions is liberated, and everything in the immediate vicinity is heated to temperatures of 5 to 10 million kelvins (K) and vaporized. The sudden expansion of hot gases literally pushes aside everything nearby and scatters the radioactive fission fragments over a wide area.

Of the uranium found in nature, only 0.711% is the easily fissionable $^{235}_{92}$U. The other 99.289% is $^{238}_{92}$U, which is not fissionable by thermal neutrons. To make nuclear bombs or nuclear fuel for generation of electricity, the naturally occurring uranium must first be *enriched*, a process that increases the relative proportion of $^{235}_{92}$U atoms in a sample. It is possible to enrich uranium so that the percentage of $^{235}_{92}$U is between 2% and 5% by making use of slight differences in the volatility of uranium hexafluoride (UF_6). Of the two kinds of UF_6 ($^{238}UF_6$ and $^{235}UF_6$), the molecule containing the lighter isotope is more volatile and will pass through porous barriers faster. After multiple passes, the desired separation is achieved.

The Mass–Energy Relationship

What is the source of the tremendous energy of the fission process? It ultimately comes from the conversion of mass into energy, according to Einstein's famous equation, $E = mc^2$, where E is energy that results from the loss of an amount of mass m, and c is the speed of light (186,000 miles/s, or 3.00×10^8 m/s). If the masses of the products of the fission of a uranium-235 atom by

In 1947, the *Bulletin of Atomic Scientists* began to use this clock, set at 7 minutes before midnight, as a reminder of the potential for disaster from nuclear weapon use. The clock was reset to 17 minutes before midnight when the United States and the Soviet Union agreed to begin nuclear disarmament in 1991. In 2002, the clock was reset to 7 minutes before midnight, where it currently remains, in response to terrorist activities.

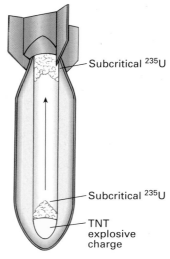

FIGURE 13.10 Implosion. By carefully shaping a conventional explosive, enough fissionable material can be brought together quickly to produce a massive chain reaction, causing many of the atoms to undergo fission almost simultaneously.

The binding energy is analogous to the concept of bond energy, in that both are a measure of the energy necessary to separate the whole (nucleus or molecule) into its parts.

It takes only about 1 kg of uranium-235 or plutonium-239 undergoing fission to produce the equivalent of about 20,000 tons (20 kilotons) of ordinary explosives such as trinitrotoluene (TNT) or dynamite.

Binding energy The force holding neutrons and protons together in a nucleus

---

FIGURE 13.11 Relative stability of nuclei with different mass numbers. Both lighter and heavier isotopes are less stable (less binding energy per nuclear particle) than those isotopes with masses between 50 atomic mass units (amu) and about 65 amu. The most stable isotope is that of iron-56. Light isotopes can be fused together to form more stable atoms (nuclear fusion), and heavier isotopes can be split into more stable, lighter atoms (nuclear fission).

a neutron are compared with the masses of the reactants, it is found that the products have less mass than the reactants. In the case of the fission reaction

$$\,^{1}_{0}\text{n} + \,^{235}_{92}\text{U} \longrightarrow \,^{93}_{37}\text{Rb} + \,^{141}_{55}\text{Cs} + 2\,^{1}_{0}\text{n} + \text{energy}$$

the difference in mass is about 0.00023 kg for every 235 g of $^{235}_{92}\text{U}$ used. This mass difference is equivalent to 5.0×10^{12} cal. The mass "lost" as a result of the fission process is the source of the tremendous energy that is released.

The small nucleus of an atom is crowded with neutrons and protons. The very fact that nuclei hold together indicates that there must be some sort of force binding the particles together. This force is called the nuclear **binding energy** and is directly related to the stability of the nucleus. The binding energy depends on the number of particles in the nucleus. For example, if separate neutrons and protons are combined to form any particular nucleus, the resulting nucleus always has less mass than the starting nucleons. This mass difference, converted into energy units according to Einstein's equation, is the binding energy and can be expressed for each atom as its *binding energy per nucleon* by dividing the total binding energy for the atom by the number of nucleons in the nucleus.

Figure 13.11 shows that light atoms and heavy atoms have lower binding energies per nucleon than atoms with mass numbers between 50 and 65. This means energy can be released when extremely light atoms are combined to make heavier atoms and when extremely heavy atoms are split to make lighter atoms. The greatest nuclear stability (greatest binding energy per nucleon) is at iron-56 ($^{56}_{26}\text{Fe}$). This is why iron is the most abundant of the heavier elements in the universe.

Because of their relative stabilities, most fission products fall into the intermediate range of mass numbers (Figure 13.11). Therefore, when fission occurs and smaller, more stable nuclei result, these nuclei will contain less mass per nuclear particle. In the process, mass must be changed into energy—the tremendous energy released in a nuclear bomb or, under controlled conditions, in a nuclear power plant.

(13.10) USEFUL NUCLEAR ENERGY

Electricity Production

Enrico Fermi (an Italian scientist who had immigrated to the United States in the late 1930s) and others believed that nuclear fission might somehow be made to proceed at a controlled rate. They reasoned that if a way could be found to control the number of thermal neutrons, their concentration could

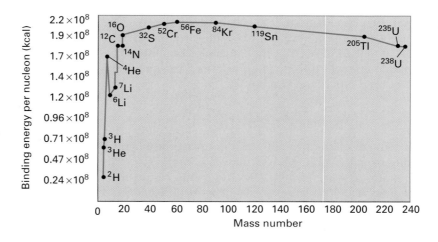

be maintained at a level sufficient to keep the fission process going but not high enough to allow an uncontrolled chain reaction. It would then be possible to drain the heat energy away on a continuing basis to do useful work. In 1942, working at the University of Chicago, Fermi was successful in building the first atomic reactor.

An atomic reactor has several essential components. The reactor fuel must contain significant concentrations of atoms such as $^{235}_{92}U$, $^{233}_{92}U$, or $^{239}_{94}Pu$ that are fissionable by slow-moving neutrons. Typically, reactor fuel will contain uranium in the form of an oxide, U_3O_8, that has been enriched to contain about 3% or 4% $^{235}_{92}U$. A *moderator* is required to slow the speed of the neutrons produced in the reactions without absorbing them. Graphite and water have been used as moderators. A *neutron-absorber*, such as cadmium or boron steel, is present to provide fine control over the neutron concentration. *Shielding* to protect the workers from dangerous radiation is an absolute necessity. Shielding tends to make reactors heavy and bulky installations. Finally, a *heat-transfer fluid* provides a large and even flow of heat energy away from the reaction center. Water is used as a heat-transfer fluid in many nuclear reactor designs. In addition to water's high heat capacity (Section 11.1), its hydrogen atoms are excellent moderators for neutrons. Conventional technology then allows the heat energy carried by the hot water from the reactor to be used to generate electricity, to power ships, or to operate any device that uses heat energy. A system for the nuclear production of electricity is illustrated in Figure 13.12.

The nuclear power plant at Indian Point, New York.

Ordinary uranium, which is mostly $^{238}_{92}U$, cannot be used as a fuel in an atomic reactor since it contains only a small concentration of the easily fissionable $^{235}_{92}U$ isotope.

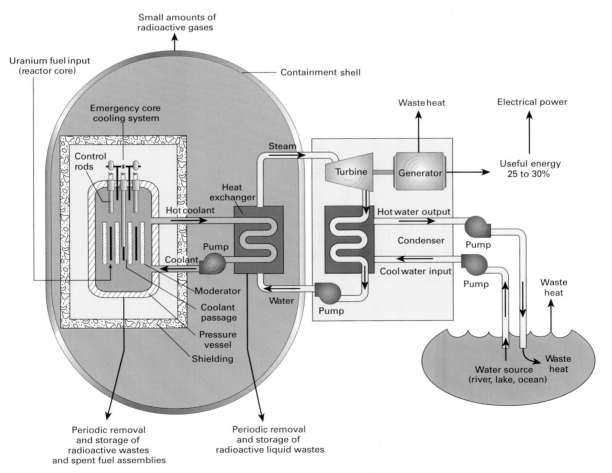

FIGURE 13.12   A nuclear power plant. In this reactor design (the most commonly used), ordinary water (called light water to differentiate it from D_2O—heavy water—used in some designs) is pressurized and allowed to carry heat energy from the reactor core to a heat exchanger, where high pressure steam is generated. This steam passes through a turbine, which generates electricity. Although simple in concept, safety considerations make the design, testing, and operation of a nuclear power plant a complex and costly operation.

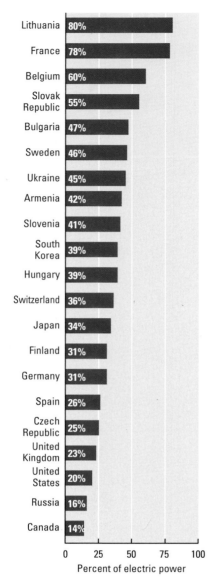

FIGURE 13.13 The approximate fraction of electricity generated by nuclear power in various countries. About 20% of the electricity in the United States is produced by nuclear power.

In a core meltdown of a nuclear reactor, the failure of cooling would allow temperatures to rise above the melting point of the metal rods containing the uranium fuel (about 1205°C). In the worst scenario, the resulting mass of highly radioactive molten metal would melt through the steel and concrete of the containment vessel beneath it. Once out of the containment vessel, the radioactivity might contaminate groundwater supplies.

Currently, electrical power is produced in 103 nuclear power plants throughout the United States. Some countries produce less energy by nuclear fission than does the United States, but in others nuclear energy provides a much larger share of the total energy production (Figure 13.13).

There are several extremely vexing problems associated with nuclear energy. One problem is the risk of a catastrophic accident at a nuclear power facility. Two such accidents, of different degrees of seriousness, are widely known and create a negative public viewpoint of all nuclear power. In late March 1979, an accident occurred at the Three Mile Island power plant near Harrisburg, Pennsylvania. A water pump failed in the reactor and caused a partial reactor core meltdown. Steam vented inside the reactor vessel, and hydrogen gas was produced by the decomposition of the steam at the very high temperatures there. This hydrogen gas, with the oxygen that was also produced, caused a risk of a chemical explosion that would have blown apart the safety containment, releasing fission products. In fact, a small amount of radioactive gases was vented into the atmosphere, but the containment building remained intact. There were no deaths directly associated with the Three Mile Island accident. Studies of individuals living within a 10-mile radius of the plant have revealed no elevation of cancer rates.

Certainly the most catastrophic nuclear accident occurred on April 26, 1986, at the Chernobyl unit 4 reactor near Kiev, Ukraine. The accident resulted in a core meltdown, explosion, and fire.

The Chernobyl reactor, built in 1983, had a design quite different from those elsewhere in the world. Graphite was used exclusively as the moderator. Although this design allows for a higher efficiency than reactors with other types of moderators, the graphite can be ignited if sufficiently high temperatures are reached. In addition, the Chernobyl reactor lacked an adequate containment vessel to withstand an explosion in its reactor core.

While engineers were running an unauthorized test on the electrical generator, the Chernobyl reactor suddenly increased its power output and before neutron absorbers could be lowered into the core, a meltdown occurred. A vast amount of steam formed, which together with burning blocks of graphite and radioactive fuel caused the entire reactor roof to blow off. The result was the release of a radioactive plume that rose almost 5000 meters into the atmosphere, scattering an estimated 100 million curies of radioisotopes, some with years-long half-lives, throughout countries west of Chernobyl and as far away as Great Britain and Norway.

The results of the released radiation are widespread and extensive. More than 170,000 people had to permanently abandon their homes. Large areas of farmland and forest are so contaminated that they may remain unusable for 100 years after the accident. More than 4300 workers who participated in cleanup at the reactor have died and cancer rates have risen among those exposed to radiation from the disaster. The Ukrainian government has spent the equivalent of billions of dollars on Chernobyl-related projects and continues to monitor the health of those in the region.

Today, the reactor is entombed in more than 300,000 tons of concrete, including an underground concrete liner to protect the groundwater. Work has been undertaken to prevent leaks in the concrete encasement. Plans are being developed for construction of a new, more stable encasement.

Radioactive Waste: A Problem to Be Solved

Nuclear waste is classified as *high-level radioactive waste* or *low-level radioactive waste* depending on the amount of ionizing radiation given off and the time period for which the waste must be safely stored. *High-level radioactive waste* is mostly spent nuclear fuel and waste associated with the weapons industry. This waste gives off large amounts of ionizing radiation and must be stored safely for thousands of years due to the long half-lives of the radioactive isotopes involved. For instance, plutonium-239 has a half-life of about 24,400 years.

As with many industrial activities that are located in a relatively localized geographic area but generate benefits that may be spread over a much broader geographic area, there are concerns about where and how to store the waste associated with the processes. Citizens are, in many cases, happy to take advantage of the benefits of nuclear power as long as they personally aren't negatively affected by the less desirable activities associated with the industry. This has led to the widespread use of NIMBY as an acronym which stands for "not in my back yard." People may be willing to utilize nuclear power generated far away from their homes and may have little concern if the waste is stored in remote areas unlikely to directly affect their health and property values. However, any proposal to build a power plant or waste storage facility (nuclear or not) just down the road from a group of politically and environmentally active citizens is often met with immediate opposition. Such concerns may multiply as we are forced to use more and more alternative energy sources in the future. The location of nuclear power plants or "farms" populated by solar panels or wind turbines is already an issue. For instance, the residents of Martha's Vineyard and Nantucket Island in Massachusetts have organized to oppose a series of large wind turbines to be built just off the coast since they feel the turbines will destroy their scenic views.

As of 2002, about 43,000 metric tons of radioactive waste from nuclear power plants was stored in the United States, an amount expected to grow by 2000 metric tons per year. The storage facilities at the power plants are considered temporary and an effort to develop a permanent storage site for high-level radioactive waste has been under way since 1982. The location and development of such a site is fraught with technical, social, and political problems.

A radioactive waste storage site must be located far from populated areas and must be free of the potential for leakage or deterioration for well over thousands of years. Any potential for environmental damage is considered unacceptable. The consensus is now that the safest storage will be in underground rock formations in isolated locations. Since 1987, as directed in a congressional act, Yucca Mountain in Nevada has been designated as the potential permanent underground storage site in the United States.

Many billions of dollars have been spent on construction and feasibility studies at Yucca Mountain, which is in the desert 145 kilometers northwest of Las Vegas. The facility would occupy 3 square miles of tunnels through dense volcanic rock 800 feet beneath the surface of an almost 5000-foot mountain. The location is almost 1000 feet above the water table in a region that gets about 6 inches of rain a year. The waste would be placed inside metal alloy containers that have titanium drip shields (titanium doesn't rust).

Nevertheless, concerns have arisen over the possibility that some rainwater may leach into the storage site through cracks in the rock. There has also been some concern because the site is near a volcano that last erupted 20,000 years ago and may erupt again, although the possibility is considered remote. Another concern is the possibility of an earthquake in the area. Scientific agreement on the stability of the site for thousands of years to come has seemed an impossible goal. There is also significant opposition to use of Yucca Mountain from the many states through which the waste would have to travel via trains or trucks to reach it.

In 2002, in a move to accelerate progress toward the date when Yucca Mountain would be receiving radioactive waste, President George W. Bush endorsed the Yucca Mountain site and the $58 billion needed for its further construction. Some predict the site can be operational by 2010; others believe it will take much longer to reach that point. Also, some argue that the uncertainties about potential hazards are too great and the site should never be used. Both the House and the Senate voted in 2002 to proceed with activation of the Yucca Mountain site. The plan is firmly opposed by the state of Nevada.

Low-level radioactive waste gives off small amounts of ionizing radiation, is usually generated in small quantities, and need only be safely stored for relatively short periods of time due to the half-lives of the radioisotopes involved. Low-level nuclear waste includes such things as contaminated laboratory clothing, cleaning equipment and supplies, medical waste that is radioactive, and discarded radioactive devices such as smoke detectors. It is only necessary to safely store this waste for periods of 100–500 years. Prior to about 1979, most waste of this type was sealed in steel drums and dumped into the ocean. Current procedures require that such waste be stored in steel drums and buried in secure sites under several feet of soil (Figure 13.14).

However, even this procedure has met with much opposition, especially from those living in the vicinity of the repositories. Even though such waste is less dangerous than high-level radioactive waste, the NIMBY philosophy comes into play when the reality of the issue of storage becomes personal.

Fusion Reactions

Nuclear fusion produces about the same amounts of energy as fission on a per-mole basis but fewer radioactive by-products; the products that are radioactive have short half-lives. Hydrogen ($_1^1H$) has two isotopes, deuterium ($_1^2H$) and tritium ($_1^3H$). The nuclei of deuterium and tritium can be fused together at high temperatures to form a helium-4 atom. The energy released is 4.1×10^8 kcal.

$$_1^2H + {}_1^3H \longrightarrow {}_2^4He^{2+} + {}_0^1n + 4.1 \times 10^8 \text{ kcal}$$

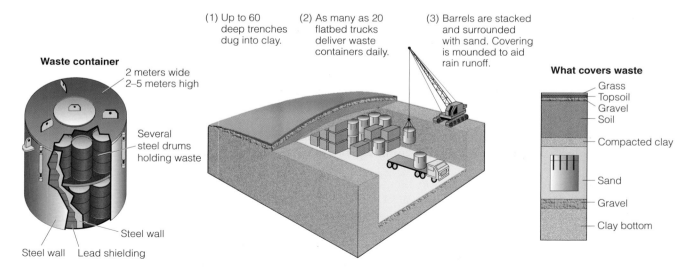

Waste container

2 meters wide
2–5 meters high

Several steel drums holding waste

Steel wall

Steel wall   Lead shielding

(1) Up to 60 deep trenches dug into clay.

(2) As many as 20 flatbed trucks deliver waste containers daily.

(3) Barrels are stacked and surrounded with sand. Covering is mounded to aid rain runoff.

What covers waste

Grass
Topsoil
Gravel
Soil

Compacted clay

Sand

Gravel

Clay bottom

FIGURE 13.14 Proposed design of a state-of-the-art low-level radioactive waste landfill. (Data from U.S. Atomic Industrial Forum)

When very light nuclei, such as those of hydrogen, helium, or lithium, are combined, or *fused*, to form an element of higher atomic number, energy equivalent to the difference between the total mass of the reacting atoms and the smaller total mass of the more stable products is released. This energy, which comes from a decrease in mass, is the source of the energy released by the sun, the stars, and hydrogen bombs. Typical examples of fusion reactions are

$$4\,_1^1\text{H} \longrightarrow \,_2^4\text{He}^{2+} + 2\,_{+1}^0\text{e} + 6.14 \times 10^8 \text{ kcal}$$
$$2\,_1^2\text{H} \longrightarrow \,_2^4\text{He}^{2+} + \,_0^1\text{n} + 7.3 \times 10^7 \text{ kcal}$$
$$2\,_1^2\text{H} \longrightarrow \,_1^3\text{H} + \,_1^1\text{H} + 9.2 \times 10^7 \text{ kcal}$$
$$_1^3\text{H} + \,_1^2\text{H} \longrightarrow \,_2^4\text{He}^{2+} + \,_0^1\text{n} + 4.04 \times 10^8 \text{ kcal}$$

If fusion were to be used as a source of energy here on Earth, a suitable source of fusible atoms (fuel) would be needed. Fortunately, the oceans are a potential source of fantastic amounts of deuterium. There are 1.03×10^{22} atoms of deuterium in a *single liter* of seawater. If all the deuterium atoms in a cubic kilometer of seawater were fused to form heavier atoms, the energy released would be equal to that released from burning 1360 billion barrels of crude oil, and this is approximately the total amount of oil originally present on this planet.

Fusion reactions occur rapidly only when the temperature is of the order of 100 million °C or more. At these high temperatures, atoms do not exist as such; instead, there is a **plasma** consisting of unbound nuclei and electrons. In this plasma nuclei can combine. The first fusion reactions that scientists were able to create artificially were produced in hydrogen bombs, or thermonuclear bombs. In a thermonuclear bomb, the high temperatures needed to initiate fusion are achieved by using the heat of a fission bomb (atomic bomb).

In one type of hydrogen bomb, lithium deuteride ($_3^6\text{Li}_1^2\text{H}$, a solid salt) is placed around an ordinary $_{92}^{235}\text{U}$ or $_{94}^{239}\text{Pu}$ fission bomb. The fission reaction is set off in the usual way. A $_3^6\text{Li}$ nucleus absorbs one of the neutrons produced and splits into tritium and helium.

$$_3^6\text{Li} + \,_0^1\text{n} \longrightarrow \,_1^3\text{H} + \,_2^4\text{He}^{2+}$$

The temperature reached by the fission of $_{92}^{235}\text{U}$ or $_{94}^{239}\text{Pu}$ is sufficiently high to bring about the fusion of tritium and deuterium.

Because there is so much available potential fuel in the oceans (as deuterium), controlled fusion seems like a natural candidate for an alternate source of energy. Another attractive feature is the rather limited production of dangerous radioisotopes. Most radioisotopes produced by fusion have short half-lives and therefore are a serious hazard for only a short time.

Three critical requirements must be met for controlled fusion to be a source of energy. First, the temperature must be high enough for fusion to occur, 100 million °C or more. Second, the plasma must be confined long enough to release a net output of energy. Third, the energy must be recoverable in some usable form.

Fusion reactions are extremely difficult to control, principally because of the difficulty of holding the hot plasma together long enough for the particles to react. A further problem is reaching the very high temperature and maintaining it long enough to sustain the reaction. Nevertheless, progress is being made. Only time will tell if fusion can be controlled to the extent that would make it practical. If it does happen, abundant and low-cost energy may truly become available to everyone.

The term *thermonuclear* refers to the extreme temperatures required to cause nuclear fusion to take place.

Plasma A high-temperature state of matter consisting of unbound nuclei and electrons

THE CHEMISTRY OF

SPIDER-MAN 2 (2004)

©2005 by Mark A. Griep, *University of Nebraska-Lincoln*

The movie *Spider-Man 2* (released in 2004) is based upon characters conceived by Stan Lee that first appeared in 1963. In the first Spider-Man movie, we learned how the shy, brilliant science major Peter Parker was transformed into Spider-Man: while on a science tour, he was bitten by a genetically modified spider. In the second movie, we learn how the brilliant scientist Dr. Otto Octavius is transformed into Spider-Man's nemesis Doc Ock. Spider-Man and Doc Ock represent the good and evil sides of scientific unintended consequences.

In the film, Octavius demonstrates his hot fusion reaction for the assembled press. His starting material is 25 lbs. of solid radioactive tritium. After he uses lasers to pump energy into the tritium, it expands, becomes molten, and looks like a miniature sun, complete with solar flares. To handle this object, he invents four intelligent mechanical arms that are impervious to heat and magnetism. The fusion experiment is a success and self-sustaining. Unfortunately, its strong magnetic field builds and attracts metal objects from around the room. Some flying objects are lethal and one causes Octavius' mechanical arms to fuse to his body.

The most successful fusion device ever created was developed by one of America's top inventors, Philo T. Farnsworth. In 1968, he and Robert Hirsch patented a functional hot fusion reactor called a Farnsworth-Hirsch fusor (patent 3,386,883). He conceived it in 1953 and built the first working model in 1959. Prior to the invention of this instrument, fuel such as deuterium or tritium was fired from small particle accelerators toward a central point where it fused upon collision. Those reactions were contained by magnetic fields, which are inherently "leaky," resulting in low temperatures and low yields.

The Farnsworth-Hirsch fusor is different from previous fusion devices in that it consists of two spherical electrodes, one inside the other. The nuclei are drawn toward the exterior negative electrode but then captured within the positively charged interior vessel. Once inside, the nuclei can't escape because they are repelled by the positively charged interior sphere. By the late 1960s, Farnsworth had occasionally observed self-sustained fusion for as long as 30 seconds after the power was turned off.

All attempts to scale up this device have failed for a variety of reasons. They never generate more energy than is applied. Today, the Farnsworth-Hirsch fusor is sold as a neutron-generating device.

Questions to Consider

1. Write balanced equations for the following fusion reactants. Each reaction forms ^{4}He and some also form neutrons.

   (a) D + D (b) D + T (c) T + T (d) ^{11}B + H

2. Which of those reactions creates the power of our sun?

3. Visit the following website to learn how to build your own desktop fusion device: **http://fusor.net/newbie/**

4. In 1927, Philo T. Farnsworth patented his most famous invention (patent 1,773,980). What was it?

CONCEPT CHECK **13D**

1. Which atom can undergo fission by slow-moving (thermal) neutrons? (**a**) uranium-238, (**b**) uranium-235, (**c**) krypton-131.

2. Complete the following nuclear equation:

$$^1_0n + ^{235}_{92}U \longrightarrow [\underline{\hspace{2cm}}] \longrightarrow ^{103}_{42}Mo + ^{131}_{50}Sn + 2\,^1_0n$$

3. Complete the following nuclear equation:

$$^1_0n + ^{239}_{94}Pu \longrightarrow ^{104}_{42}Mo + [\underline{\hspace{2cm}}] + 2\,^1_0n$$

4. If the masses of the nucleons making up an atom are summed together and compared with the mass of the atom these nucleons can make, the atom will always have a (greater/smaller) mass.

5. What is the most stable element in terms of binding energy per nucleon? (**a**) iron, (**b**) hydrogen, (**c**) uranium.
6. About what percent of the electricity generated in the United States is from nuclear energy? (**a**) 100%, (**b**) 20%, (**c**) 50%.
7. Plutonium is always produced in nuclear fission reactors. (**a**) True, (**b**) False.
8. What does NIMBY stand for?
9. Give examples of high-level nuclear waste and low-level nuclear waste.

 Assess your understanding of this chapter's topics with an online chapter quiz at **www.brookscole.com/chemistry/joesten4**

■ KEY TERMS

nuclear reaction	uranium series	chain reaction
nucleons	artificial transmutation	critical mass
positron	transuranium elements	binding energy
activity	fission	plasma
half-life	fusion	

■ THE LANGUAGE OF CHEMISTRY

1. Thermal neutrons
2. Most penetrating nuclear radiation
3. Least penetrating nuclear radiation
4. Discovered radioactivity
5. Fissionable atom
6. Co-discovered nuclear fission
7. Same as alpha particle
8. Reactor core meltdown
9. Stable isotope
10. Dangerous to the lungs
11. Source of H that may someday be used as a nuclear fuel
12. Positive electron
13. Useful in radio imaging
14. Nuclear disintegrations decrease by one half
15. Loss from the nucleus causes an increase in atomic number
16. Dangerous by-product from controlled nuclear fission

a. Half-life
b. Plutonium-239
c. He nucleus
d. Lead-206
e. Radon-222
f. The oceans
g. Positron
h. Gamma ray
i. Becquerel
j. Technetium-99m
k. Alpha particle
l. Slow moving
m. Beta particle
n. Uranium-235
o. L. Meitner
p. Chernobyl

APPLYING YOUR KNOWLEDGE

1. Which of the following three types of radiation is the most penetrating?
   (**a**) Alpha particles
   (**b**) Beta particles
   (**c**) Gamma rays

2. Give the type and approximate amount of material required to stop each of the following:
   (**a**) Alpha particles
   (**b**) Beta particles
   (**c**) Gamma rays

3. Which radioisotope sample is more hazardous, 1 g of $^{238}_{92}$U with a half-life of 4.5 billion years or 1 g of $^{222}_{86}$Rn with a half-life of 3 days? Explain your choice.

4. What is the nuclear reaction for the production of a beta particle?

5. What makes Rn-222 a health hazard? What health problem results from Rn-222 exposure?

6. Why must the sum of the mass numbers of the products of a nuclear reaction always equal the sum of the mass numbers of the reactants?

7. What is wrong with the following nuclear equation?

$$^{4}_{2}\text{He} + ^{27}_{13}\text{Al} \longrightarrow ^{49}_{15}\text{P} + ^{1}_{0}\text{n}$$

8. What kinds of projectile particles were used for the transmutations to form elements up to atomic number 101?

9. Iodine-131 has a half-life of 8 days. Approximately what fraction of iodine-131 will remain after 1 month (31 days)?

10. What are transuranium elements? How do they differ from the other elements in the periodic table?

11. What is meant by the term *uranium series*?

12. What is the final product of the uranium series?

13. What are the two major applications of radioisotopes in nuclear medicine?

14. What is the reason for irradiating food with gamma rays? What is the effect of the radiation?

15. What are the electrical charges and relative masses for the following?
    (**a**) Beta particles
    (**b**) Alpha particles
    (**c**) Gamma rays
    (**d**) Positrons
    (**e**) Neutrons

16. Tell why it is necessary to use high-energy accelerators to produce the transuranium radioisotopes from smaller nuclei.

17. Describe what effect gamma rays from cobalt-60 have on cancer cells and why this makes them useful in radiation therapy for cancer.

18. Why does airplane travel increase a person's annual dose of ionizing radiation?

19. Which source of background radiation contributes more to a person's annual dose, weapons test fallout or natural radiation in food, water, and air?

20. Fusion of two deuterium nuclei to form helium-3 and hydrogen-1 yields 1.53×10^{-16} kcal of energy. A liter of seawater is estimated to contain 1.03×10^{22} atoms of deuterium. How many kilocalories can be liberated, if all the 1.03×10^{22} deuterium atoms in a liter of seawater undergo fusion?

21. Carbon-14 dating uses the carbon-14 half-life of 5730 years to estimate the age of carbon-containing samples. How old is a wooden throne if the current carbon-14 activity is one-eighth of the original?

22. Copper-64 has a half-life of 12.9 hours. How many half-lives and hours are needed for the activity to fall to one-fourth the initial level of 1 microcurie (37,000 disintegrations per second)?

23. What was the contribution made by each of the following scientists in the study of radioactivity?
    (**a**) Becquerel
    (**b**) Marie Curie
    (**c**) Rutherford

24. Radon gas in homes is a common problem in many parts of North America because radon-222 is radioactive. This nuclide has a half-life of 3.82 days. What fraction of a radon-222 sample will remain after 15.28 days have passed?

25. What effect does the emission of a beta particle have on the mass number of a nucleus?

26. Describe nuclear fission:
    (**a**) What is the starting material?
    (**b**) What is needed to cause fission?
    (**c**) What are the products of fission?

27. Name three problems that are associated with nuclear energy from fission.

28. What happened at the Chernobyl unit 4 reactor in 1986?

29. (**a**) What does the term *reprocessing of nuclear fuels* mean?
    (**b**) What danger is associated with it?

30. What is nuclear fusion?

31. Compare nuclear fission and nuclear fusion as sources of energy. Name the fuels, benefits, problems, and current status.

32. Complete the following nuclear equations:
    (a) $^{1}_{0}n + ^{235}_{92}U \longrightarrow ^{142}_{56}Ba + [\underline{\hspace{2cm}}] + 3\,^{1}_{0}n$
    (b) $^{1}_{0}n + ^{235}_{92}U \longrightarrow [\underline{\hspace{2cm}}] + ^{129}_{50}Sn + 2\,^{1}_{0}n$
    (c) $^{1}_{0}n + ^{239}_{94}Pu \longrightarrow [\underline{\hspace{2cm}}] + ^{123}_{52}Te + 2\,^{1}_{0}n$

33. Balance the following nuclear equations, giving symbols, nuclear charges, and mass numbers:
    (a) $^{64}_{29}Cu \longrightarrow \underline{\hspace{2cm}} + ^{0}_{-1}e$
    (b) $^{69}_{30}Zn \longrightarrow \underline{\hspace{2cm}} + ^{69}_{31}Ga$
    (c) $^{131}_{53}I \longrightarrow ^{131}_{54}Xe + \underline{\hspace{2cm}}$

34. Balance the following beta decay equations, giving symbols, nuclear charges, and mass numbers:
    (a) $^{14}_{6}C \longrightarrow \underline{\hspace{2cm}} + ^{0}_{-1}e$
    (b) $^{210}_{82}Pb \longrightarrow \underline{\hspace{2cm}} + ^{0}_{-1}e$

35. (a) What is the half-life of plutonium-239?
    (b) How does this add to the dangers of this isotope?

36. In what important way are the isotopes uranium-235 and plutonium-239 similar?

37. Describe how the "magnetic bottle" is used to contain a nuclear fusion reaction.

38. How many atoms of tritium will remain after four half-lives if there were initially 300,000 atoms?

39. Radioactive decay of ^{238}U can yield ^{234}Th and an alpha particle. An initial sample contained 200,000 atoms of ^{238}U. How many alpha particles will be produced from this sample after one half-life of 4.5 billion years?

W CHEMISTRY ON THE WEB

For up-to-date URLs, visit the text website at **www.brookscole.com/chemistry/joesten4**

- The Discovery of Radioactivity: The Dawn of the Nuclear Age
- How Nuclear Radiation Works
- How Carbon-14 Dating Works
- How Nuclear Power Works
- Fear of Nuclear Power
- How Nuclear Bombs Work
- Transuranium Elements
- How Nuclear Medicine Works

ORGANIC CHEMICALS AND POLYMERS

© Claudins/zeta/CORBIS

Ethylene, a key raw material obtained from the catalytic cracking of natural gas, is processed at this plant. It is used in the production of a variety of other chemicals and consumer products.

Fossil fuels are not only our major source of energy; they are also the major source of the hydrocarbons that we use to make thousands of consumer products. About 3% of the petroleum refined today is the starting material for the synthesis of organic chemicals of commercial importance. These chemicals are essential to making plastics, synthetic rubber, synthetic fibers, drugs, dyes, pesticides, fertilizers, and thousands of other consumer products. For this reason the organic chemical industry is often referred to as the *petrochemical* industry.

A few of the major classes of organic compounds and some of their reactions, especially those used in making polymers, are described in this chapter, with the goal of introducing you to a major segment of chemistry and the chemical industry.

317

This chapter addresses the following questions:

- Why are there so many organic compounds?
- What are the characteristic functional groups?
- What are some common addition and condensation polymers?
- Will coal become a major source of chemicals?
- What types of plastics are being recycled?

Carbon compounds hold the key to life on Earth. Consider what the world would be like if all carbon compounds were removed; the result would be much like the barren surface of the Moon. If carbon compounds were removed from the human body, there would be nothing left except water and a small residue of minerals. The same would be true for all living things. Carbon compounds are also an integral part of our lifestyle. Fossil fuels, foods, and most drugs are made of carbon compounds. Since we live in an age of plastics and synthetic fibers, our clothes, appliances, and most other consumer goods contain a significant portion of carbon compounds.

Over 85% of the millions of known compounds are carbon compounds, and a separate branch of chemistry, **organic chemistry**, is devoted to the study of them. Why are there so many organic compounds? The discussion of hydrocarbons and their structural and geometric isomers in Chapter 12 indicates two reasons: (1) the ability of many carbon atoms to be linked in sequence with stable carbon–carbon single, double, and triple bonds in a single molecule and (2) the occurrence of isomers. A third reason will be discussed further in this chapter: the variety of functional groups that bond to carbon atoms.

The economic importance of the organic chemical industry can be seen by looking at the list in Table 14.1 of chemicals produced in the United States in very large quantities. Four are organic chemicals.

Catalytic cracking was described in Section 12.2.

Organic chemistry The chemistry of carbon compounds

TABLE 14.1 Major Chemicals Produced in the United States*

NAME	HOW MADE	END USES
Sulfuric acid	Burning sulfur to SO_2, oxidation of SO_2 to SO_3, reaction with water; also recovered from metal smelting	Fertilizers, petroleum refining, manufacture of metals and chemicals
Ethylene	Cracking hydrocarbons from oil and natural gas	Plastics, antifreeze production, fibers, and solvents
Ammonia	Catalytic reaction of nitrogen and hydrogen	Fertilizers, plastics, fibers, and resins
Phosphoric acid	Reaction of sulfuric acid with phosphate rock; burning of elemental phosphorus and dissolution in water	Fertilizers, detergents, and water-treating compounds
Chlorine	Electrolysis of NaCl, recovery from HCl users	Chemical production, plastics, solvents, pulp and paper
Propylene	Cracking oil and oil products	Plastics, fibers, solvents
Sodium hydroxide	Electrolysis of NaCl solution	Chemicals, pulp and paper, aluminum, textiles, oil refining
Ammonium nitrate	Reaction of ammonia and nitric acid	Explosives, fertilizers
Urea	Reaction of NH_3 and CO_2 under pressure	Fertilizers, animal feeds, adhesives, and plastics
Styrene	Dehydrogenation of ethylbenzene	Polymers, rubber, polyesters

* Organic chemicals highlighted.

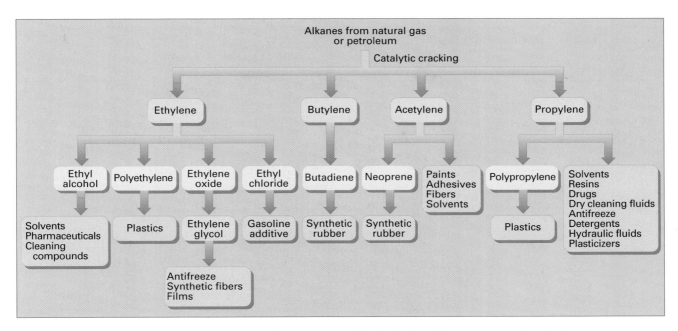

FIGURE 14.1   Hydrocarbons from petroleum and natural gas as raw materials. Catalytic cracking produces ethylene, butylene, acetylene, and propylene, which are converted into other chemical raw materials and many kinds of consumer products.

14.1   ORGANIC CHEMICALS

Many of the organic chemicals used in the chemical industry are obtained from fossil fuels. For example, ethylene, propylene, butylene, and acetylene are obtained by catalytic cracking of natural gas or petroleum; Figure 14.1 summarizes the uses of these as starting materials. Petroleum and coal tar, obtained by heating coal at high temperatures in the absence of air, are the primary sources of aromatic compounds used in the chemical industry (Figure 14.2). Distilling coal tar yields the aromatic compounds listed in Table 14.2.

Organic chemicals were once obtained only from plants, animals, and fossil fuels; and these are still direct sources of hydrocarbons (Figures 14.1

Heating coal at high temperatures in the absence of air produces a mixture of coke, coal tar, and coal gas. The process, called pyrolysis, is represented by

$$\text{Coal} \longrightarrow \text{coke} + \text{coal tar} + \text{coal gas}$$

One ton of bituminous (soft) coal yields about 1500 lb of coke, 8 gal of coal tar, and 10,000 cubic feet (ft^3) of coal gas. *Coal gas* is a mixture of H_2, CH_4, CO, C_2H_6, NH_3, CO_2, H_2S, and other gases. At one time, coal gas was used as a fuel.

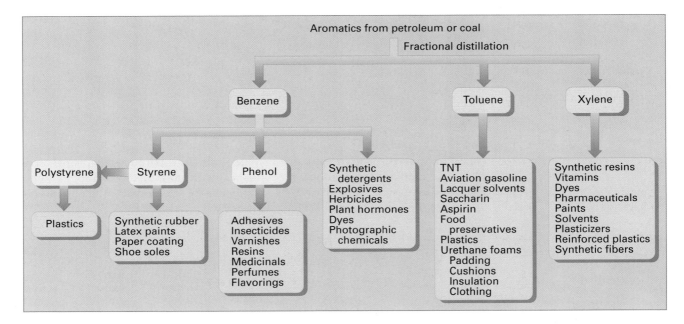

FIGURE 14.2   Aromatic compounds from petroleum and coal and their uses.

TABLE 14.2 Fractions from Distillation of Coal Tar

BOILING RANGE (°C)	NAME	TAR, MASS %	PRIMARY CONSTITUENTS
Below 200	Light oil	5	Benzene, toluene, xylenes
200–250	Middle oil (carbolic oil)	17	Naphthalene, phenol, pyridine
250–300	Heavy oil (creosote oil)	7	Naphthalenes and methylnaphthalenes, cresols
300–350	Green oil	9	Anthracene, phenanthrene
Residue		62	Pitch or tar

and 14.2) and many other important chemicals, such as sucrose from sugarcane and ethanol from fermented grain mash. However, the development of organic chemistry has led to cheaper methods for the synthesis of both naturally occurring substances and new substances.

The millions of organic compounds include classes of compounds that are obtained by replacing hydrogen atoms of hydrocarbons with atoms or groups of atoms known as *functional groups* (page 279). The functional groups for alcohols and ethers were discussed in Section 12.9. Alcohols and their oxidation products—aldehydes, ketones, and carboxylic acids—are among the most useful functional group classes of compounds.

A table of functional groups and further information on the naming of organic compounds are found in Appendix D.

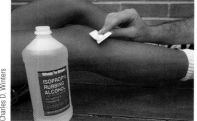
Charles D. Winters

Rubbing alcohol (isopropyl alcohol or 2-propanol) is used to clean out cuts and scrapes.

An "R" is used to represent any kind of alkyl group. The use of R, R′, and R″ indicates that the R groups may be different.

14.2 ALCOHOLS AND THEIR OXIDATION PRODUCTS

Alcohols contain one or more hydroxyl (—OH) functional groups bonded to carbon atoms and are a major class of organic compounds. The importance of methanol, ethanol, and 2-methyl-2-propanol as fuels and fuel additives was described in Sections 12.2 and 12.5. Additional uses of these and other commercially important alcohols are listed in Table 14.3. Alcohols are classified according to the number of carbon atoms bonded directly to the —C—OH carbon as primary (one other C atom), secondary (two other C atoms), or tertiary (three other C atoms). The reactivities of these classes of alcohols are different.

$$
\begin{array}{ccc}
\underset{\text{Primary}}{\overset{\displaystyle H}{\underset{\displaystyle H}{R-C-OH}}} &
\underset{\text{Secondary}}{\overset{\displaystyle R}{\underset{\displaystyle H}{R'-C-OH}}} &
\underset{\text{Tertiary}}{\overset{\displaystyle R}{\underset{\displaystyle R''}{R'-C-OH}}}
\end{array}
$$

Ethanol and 1-propanol are primary alcohols. 2-Propanol, or isopropyl alcohol, is familiar to us as rubbing alcohol, a 70% solution of 2-propanol in water, sold in drugstores and grocery stores. 2-Propanol is a secondary alcohol and is one of the two structural isomers of an alcohol with three carbon atoms.

2-Propanol   1-Propanol

TABLE 14.3 Some Important Alcohols				
CONDENSED FORMULA	**BOILING POINT (°C)**	**SYSTEMATIC NAME**	**COMMON NAME**	**USE**
CH_3OH	65.0	Methanol	Methyl alcohol	Fuel, gasoline additive, making formaldehyde
CH_3CH_2OH	78.5	Ethanol	Ethyl alcohol	Beverages, gasoline additive, solvent
$CH_3CH_2CH_2OH$	97.4	1-Propanol	Propyl alcohol	Industrial solvent
CH_3CHCH_3 | OH	82.4	2-Propanol	Isopropyl alcohol	Rubbing alcohol
$HOCH_2CH_2OH$	198.0	1,2-Ethanediol	Ethylene glycol	Antifreeze
$HOCH_2CHCH_2OH$ | OH	290.0	1,2,3-Propanetriol	Glycerol (glycerin)	Moisturizer in foods and cosmetics

For an alcohol with four carbon atoms, there are four structural isomers, including 2-methyl-2-propanol (*tertiary*-butyl alcohol), whose former use as a gasoline additive was described in Section 12.2.

$$CH_3CH_2CH_2CH_2OH$$
1-Butanol
A primary alcohol

$$CH_3\overset{\overset{\displaystyle CH_3}{\mid}}{C}HCH_2OH$$
2-Methyl-1-propanol
A primary alcohol

$$CH_3\overset{\overset{\displaystyle OH}{\mid}}{C}HCH_2CH_3$$
2-Butanol
A secondary alcohol

$$(CH_3)_3C\!-\!OH$$
2-Methyl-2-propanol
A tertiary alcohol

Alcohols can serve as the starting substances for the preparation of many other types of organic compounds. Oxidation of alcohols may yield aldehydes, ketones, or carboxylic acids

$$\overset{\overset{\displaystyle O}{\|}}{R\!-\!C\!-\!H}$$
aldehydes

$$\overset{\overset{\displaystyle O}{\|}}{R\!-\!C\!-\!R'}$$
ketones

$$\overset{\overset{\displaystyle O}{\|}}{R\!-\!C\!-\!OH}$$
carboxylic acids

depending on the alcohol used and the extent of the oxidation. If the starting compound is a primary alcohol, the first oxidation product is an aldehyde and the second oxidation product is a carboxylic acid. For example, the oxidation of ethanol, a primary alcohol, can be used to make acetaldehyde and acetic acid.

Oxidation of organic compounds is usually the addition of oxygen or the removal of hydrogen; reduction is usually the removal of oxygen or the addition of hydrogen (see Sections 10.1 and 10.2).

$$CH_3CH_2OH \xrightarrow{\text{Oxidation}} CH_3\!-\!\overset{\overset{\displaystyle O}{\|}}{C}\!-\!H \xrightarrow{\text{Oxidation}} CH_3\overset{\overset{\displaystyle O}{\|}}{C}\!-\!OH$$

Ethanol
A primary alcohol

Acetaldehyde

Acetic acid

The oxidation of ethanol to acetaldehyde is an example of how the reactivity of primary alcohols differs from the reactivity of secondary alcohols, which cannot form aldehydes.

Aldehydes Organic compounds containing a —CHO functional group as in RCHO.

Ketones Organic compounds containing a C=O (carbonyl) functional group between two carbon atoms as in RCOR′.

Carboxylic acids Organic compounds containing a —COOH functional group as in RCO_2H.

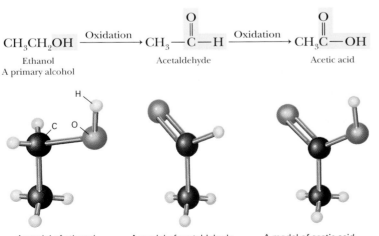

A model of ethanol   A model of acetaldehyde   A model of acetic acid

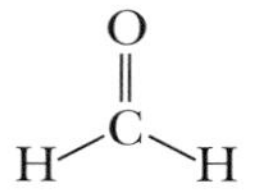

Formaldehyde

Formaldehyde (HCHO), the simplest aldehyde, is obtained by oxidation of methanol. It is a gas at room temperature but is often used as a 40% aqueous solution called formalin. Although formaldehyde is an important starting material for polymers, it presents a number of health hazards because of its toxicity and carcinogenicity. Formaldehyde is also an air pollutant, being produced in trace amounts in the incomplete combustion of fossil fuels. One of the concerns about methanol as a fuel (Section 12.5) is the potential for increased levels of formaldehyde in urban areas because of the production of formaldehyde from incomplete combustion of methanol.

Secondary alcohols are oxidized to give ketones. The oxidation of 2-propanol (isopropyl alcohol) gives acetone, a ketone widely used as an organic solvent.

Hydrogen Bonding in Alcohols

See Section 5.7 for a discussion of hydrogen bonding.

The physical properties of alcohols provide an example of the effects of hydrogen bonding between molecules in liquids. This is illustrated by the boiling points in Table 14.3. Hydrogen bonding explains why methanol (molar mass 32 g) is a liquid, whereas propane (molar mass 44 g), which has a higher molar mass but no hydrogen bonding capability, is a gas at room temperature. The boiling point of methanol is lower than that of water because methanol has only one O—H hydrogen through which it can hydrogen bond.

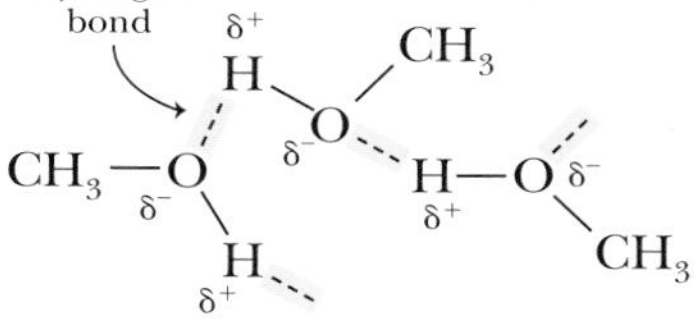

The higher boiling point of ethylene glycol can be attributed to the presence of two —OH groups per molecule. Glycerol, with three —OH groups, has an even higher boiling point.

The alcohols listed in Table 14.3 are also very soluble in water because of hydrogen bonding between water molecules and the —OH group in alcohol molecules. However, as the length of the hydrocarbon chain increases, the resulting alcohols are less soluble because the nonpolar hydrocarbon portion has greater influence on the solubility than the hydrogen bonding by the —OH group.

Ethanol

Ethanol is the "alcohol" of alcoholic beverages and is prepared for this purpose by fermentation (Section 12.9) of carbohydrates (starch, sugars) from a wide variety of plant sources (Table 14.4). The growth of yeast is inhibited at alcohol concentrations higher than about 14%, and fermentation comes to a stop. Beverages with a higher ethanol concentration are prepared either by distillation or by fortification with ethanol that has been obtained by the distillation of another fermentation product. The maximum concentration of ethanol that can be obtained by distillation of ethanol–water mixtures is

TABLE 14.4   Common Alcoholic Beverages

Name	Source of Fermented Carbohydrate	Amount of Ethyl Alcohol	Proof
Beer	Barley, wheat	5%	10
Wine	Grapes or other fruit	14% maximum, unless fortified	20–28
Brandy	Distilled wine	40–45%	80–90
Whiskey	Barley, rye, corn, etc.	45–55%	90–110
Rum	Molasses	~45%	90
Vodka	Potatoes	40–50%	80–100

95%. The "proof" of an alcoholic beverage is twice the volume percent of ethanol; 80-proof vodka, for example, contains 40% ethanol by volume; 95% ethanol is 190 proof.

Although ethanol is not as toxic as methanol (Section 12.9), 1 pint of pure ethanol, rapidly ingested, would kill most people. Ethanol is a depressant, and the effects of different blood levels of ethanol are shown in Table 14.5. Rapid consumption of two 1-ounce "shots" of 90-proof whiskey or of two 12-ounce beers can cause one's blood alcohol level to reach 0.05%. Ethanol is quickly absorbed into the bloodstream and metabolized by enzymes produced in the liver. The rate of detoxification is about 1 ounce of pure ethanol per hour. Ethanol is oxidized to acetaldehyde, which is further oxidized to acetic acid; eventually, CO_2 and H_2O are produced and eliminated through the lungs and kidneys.

Proof = 2 × volume percent.

Many consumer products like Listerine™ and Nyquil™ contain ethanol.

$$
\underset{\text{Ethanol}}{
\begin{array}{c}
\text{H}\quad\text{H} \\
\text{H}-\text{C}-\text{C}-\text{OH} \\
\text{H}\quad\text{H}
\end{array}}
\xrightarrow[\text{enzymes}]{\text{Liver}}
\underset{\text{Acetaldehyde}}{
\begin{array}{c}
\text{H}\quad\text{O} \\
\text{H}-\text{C}-\text{C}-\text{H} \\
\text{H}
\end{array}}
$$

The federal tax on alcoholic beverages is about $20 per gallon. Since the cost of producing ethanol is only about $1 per gallon, ethanol intended for industrial use must be *denatured* to avoid the beverage tax. **Denatured alcohol** contains small amounts of a toxic substance, such as methanol or gasoline, that cannot be removed easily by chemical or physical means.

Charles D. Winters

A small structure difference makes a big difference in properties. Antifreeze contains ethylene glycol, which is composed of two carbon atoms with an —OH group on each one, and is extremely poisonous. Glycerol, which is composed of three carbon atoms with an —OH group on each one, is nonpoisonous, sweet, syrupy, and holds moisture to the skin. It is used in soap and also as a food additive (Section 16.9).

Denatured alcohol Ethanol with small added amounts of a toxic substance, such as methanol, that cannot be removed easily by chemical and physical means

TABLE 14.5   Effects of Blood Alcohol Level*

Number of Drinks‡	% Alcohol by Volume	Effect
1 to 4	0.05–0.15	Lack of coordination, altered judgment, exaggerated emotions
5 to 10	0.15–0.20	Intoxication (slurred speech, altered perception and equilibrium)
10 to 15	0.30–0.40	Unconsciousness, coma
Over 15	0.50	Possible death

* In some states a person with a blood alcohol level of 0.08% or higher is legally intoxicated.
‡ 1 drink = 12 oz of 4.5% beer
　　　　 4 oz of 14% wine
　　　　 1 to 1½ oz of 45% alcohol liquor

Apart from being used in the alcoholic beverage industry, ethanol is used widely in solvents, in the preparation of many other organic compounds, and as a gasoline additive (Section 12.9). For many years industrial ethanol was also made by fermentation. However, in the last several decades, it has become cheaper to make the ethanol from petroleum by-products, specifically by the catalyzed addition of water to ethylene. More than 1 billion pounds of ethanol are produced each year by this process.

Ethylene Glycol and Glycerol

More than one alcohol group can be present in a single molecule. Ethylene glycol, a di-alcohol (Table 14.3) is used in antifreeze and in the synthesis of polymers.

Ethylene glycol works very well as a major ingredient in antifreeze but its use is not without problems. Just as ethanol may be oxidized to the carboxylic acid known as acetic acid, so may ethylene glycol's two hydroxyl groups be oxidized by liver enzymes known as *alcohol dehydrogenases* to give oxalic acid. At physiological pH, oxalic acid is converted to its calcium salt, calcium oxalate.

$$\text{HOCH}_2\text{CH}_2\text{OH} \xrightarrow{\text{liver enzymes}} \underset{\text{oxalic acid}}{\text{HO} \overset{\text{O} \quad \text{O}}{\diagup\diagdown} \text{OH}} \xrightarrow[\text{conditions}]{\text{under physiological}} \text{Ca}^{2+} \left(\underset{\text{calcium oxalate}}{{}_-\text{O} \overset{\text{O} \quad \text{O}}{\diagup\diagdown} \text{O}_-} \right)$$

ethylene glycol

Calcium oxalate crystallizes in the kidneys and in the brain, leading to serious illness or even death. Most such poisonings occur when radiator fluid is changed. Carelessly stored antifreeze or antifreeze spilled on a driveway is frequently ingested by pets and even small children. Many antifreezes are sold as brightly colored, often green, yellow, or red, mixtures that attract small children. Children and pets also find the sweet taste of ethylene glycol attractive, and these combined effects can lead to deadly consequences. As little as 5 mL can kill a cat, a small child can die from drinking as little as 30 mL, and 60 mL can kill an adult or a large dog. Coma and death often result if treatment is not begun shortly after ingestion.

One treatment for ethylene glycol poisoning involves flooding the system with ethanol. Alcohol dehydrogenases in the liver have a natural preference for metabolizing ethanol over ethylene glycol. If these enzymes can be kept busy metabolizing ethanol, the unmetabolized ethylene glycol is excreted from the kidneys with a half-life of about 14 hours. After 5 half-lives (70 hours) only about 3% of the initial amount of ethylene glycol remains. Thus, routine treatment for ethylene glycol poisoning involves administering ethanol intravenously under carefully controlled conditions if the victim cannot be induced to regurgitate the ingested antifreeze shortly after ingestion. Careful monitoring by a physician or veterinarian is required because the acetic acid produced by ethanol metabolism can lead to excessive buildup of acid in the bloodstream resulting in a condition known as metabolic acidosis, which can be fatal. You may wish to review the blood buffering system in Section 9.4 to see how this would work. Doctors often administer sodium bicarbonate intravenously in a liquid form to combat this buildup of acidity.

Thousands of pets, and more than a few children and adults, die each year from ethylene glycol poisoning. Most of these deaths are accidental, but there are, unfortunately, many recorded cases in which ethylene glycol has been used to poison an unsuspecting spouse or child.

Ethylene glycol's use as a de-icer for airplanes preparing to fly in icing conditions has been problematic because hundreds of thousands of gallons of the solvent were used each year at large airports around the world. Much of this solvent often found its way into the local ecosystem around the airports, causing serious environmental problems. In recent years, much more attention has been given to restricting the amount of ethylene glycol, and other chemicals used for de-icing, released into the environment. Recovery and recycling systems have been put in place and there now is much more stringent monitoring of stormwater runoff from airports.

Propylene glycol is gaining popularity as an antifreeze because its chemical nature renders it

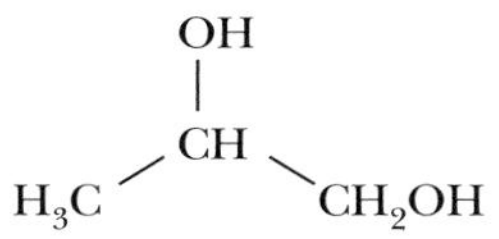

$$
\begin{array}{c}
OH \\
| \\
H_3C \diagup CH \diagdown CH_2OH
\end{array}
$$

propylene glycol

incapable of oxidation to oxalic acid. More environmentally friendly antifreezes, such as Sierra® brand, use this nontoxic alcohol. These antifreezes are very popular at zoos and other places where large numbers of animals may be exposed to leaking antifreeze. One of the drawbacks of propylene glycol, however, is that it costs more than ethylene glycol.

Glycerol, a tri-alcohol (Table 14.3), is a by-product of the manufacture of soaps. Because of its moisture-holding properties, glycerol has many uses in foods and tobacco as a digestible and nontoxic humectant (a substance that gathers and holds moisture), and in the manufacture of drugs and cosmetics. It is also used in the manufacture of nitroglycerin and numerous other chemicals. Perhaps the most important compounds of glycerol are its natural esters, which are the fats and oils found in plants and animals (Section 15.3).

CONCEPT CHECK 14A

1. What is the R group of 1-propanol, $CH_3CH_2CH_2OH$?
2. Coal tar is the principal source of what major class of hydrocarbons?
3. Ethanol is quickly absorbed into the bloodstream and oxidized to _____ in the liver.
4. 2-Propanol is commonly known as _____.
5. The primary component of antifreeze is _____.
6. Gin that is 84 proof contains what percentage of ethanol?
7. Ethanol intended for industrial use is _____ by the addition of small amounts of a toxic substance.
8. Glycerol has _____ alcohol groups.
9. Fats and oils are esters of the alcohol _____.
10. _____ is the formula of the alcohol that may be oxidized to formaldehyde.
11. What alcohol is oxidized to oxalic acid by liver enzymes? This alcohol is a component of many antifreezes.
12. What is the antidote for ethylene glycol poisoning?

Three ways of representing the carboxylic acid group are —COOH, —CO₂H, and

$$\underset{\underset{\displaystyle}{\parallel}}{\overset{\displaystyle O}{-C}}-OH.$$

14.3 CARBOXYLIC ACIDS AND ESTERS

Carboxylic Acids

Carboxylic acids contain the —COOH functional group and are prepared by the oxidation of alcohols or aldehydes. These reactions occur quite easily, as evidenced by the souring of wine to form vinegar, which results from the oxidation of ethanol to acetic acid in the presence of oxygen from the air.

Carboxylic acids are polar and readily form hydrogen bonds. This hydrogen bonding results in relatively high boiling points for the acids, even higher than those of alcohols of comparable molar mass. For example, formic acid (46 g/mol) has a boiling point of 101°C, whereas ethanol (46 g/mol) has a boiling point of only 78°C.

All carboxylic acids are weak acids (see Section 9.2) and react with bases to form salts, for example,

$$\underset{\text{Acetic acid}}{CH_3\overset{O}{\overset{\parallel}{C}}-OH(aq)} + NaOH(aq) \longrightarrow \underset{\text{Sodium acetate}}{CH_3\overset{O}{\overset{\parallel}{C}}O^-Na^+(aq)} + H_2O(\ell)$$

A number of carboxylic acids are found in nature and have been known for many years. As a result, some of the familiar carboxylic acids are almost always referred to by their common names (Table 14.6).

Acetic acid, the acid found in about 5% concentration in vinegar, is produced in large quantities for use in making cellulose acetate, a polymer used in the manufacture of photographic film base, synthetic fibers, plastics, and other products (Section 14.5).

Three other acids produced in large quantity have two carboxylic acid groups and are known as dicarboxylic acids.

Charles D. Winters

Spinach and rhubarb contain oxalic acid (Table 14.7). Individuals prone to kidney stones composed of highly insoluble calcium oxalate must limit their intake of foods containing oxalic acid.

$$\underset{\text{Adipic acid}}{HO\overset{O}{\overset{\parallel}{C}}(CH_2)_4\overset{O}{\overset{\parallel}{C}}OH} \qquad \underset{\text{Terephthalic acid}}{HO-\overset{O}{\overset{\parallel}{C}}-\overset{O}{\overset{\parallel}{C}}-OH} \qquad \underset{\text{Phthalic acid}}{\overset{O}{\overset{\parallel}{\underset{\overset{\parallel}{O}}{C-OH}}}}$$

TABLE 14.6	Some Simple Carboxylic Acids			
STRUCTURE	**ODOR**	**COMMON NAME**	**SYSTEMATIC NAME**	**BOILING POINT (°C)**
$HC\overset{O}{\overset{\parallel}{O}}H$	Sharp	Formic acid	Methanoic acid	101
$CH_3C\overset{O}{\overset{\parallel}{O}}H$	Vinegar	Acetic acid	Ethanoic acid	118
$CH_3CH_2C\overset{O}{\overset{\parallel}{O}}H$	Swiss cheese	Propionic acid	Propanoic acid	141
$CH_3(CH_2)_2C\overset{O}{\overset{\parallel}{O}}H$	Rancid butter	Butyric acid	Butanoic acid	163
$CH_3(CH_2)_3C\overset{O}{\overset{\parallel}{O}}H$	Manure	Valeric acid	Pentanoic acid	187

TABLE 14.7	Some Naturally Occurring Carboxylic Acids	

NAME	STRUCTURE	NATURAL SOURCE
Glycolic acid	$HO-CH_2-COOH$	Sugarcane
Citric acid	$HOOC-CH_2-\underset{\underset{COOH}{\overset{OH}{\mid}}}{C}-CH_2-COOH$	Citrus fruits
Lactic acid	$CH_3-\underset{\underset{OH}{\mid}}{CH}-COOH$	Sour milk
Oleic acid	$CH_3(CH_2)_7-CH=CH-(CH_2)_7-COOH$	Vegetable oils
Oxalic acid	$HOOC-COOH$	Rhubarb, spinach, cabbage, tomatoes
Stearic acid	$CH_3(CH_2)_{16}-COOH$	Animal fats
Tartaric acid	$HOOC-\underset{\underset{OH}{\mid}}{CH}-\underset{\underset{OH}{\mid}}{CH}-COOH$	Grape juice, wine

These three acids are used to manufacture polymers (Section 14.5). There are other acids, however, whose sources are familiar to you (Table 14.7), since they occur in nature.

Esters

Carboxylic acids react with alcohols in the presence of strong acids to produce **esters**, which contain the —COOR functional group. In an ester the —OH of the carboxylic acid is replaced by the OR group from the alcohol. For example, when ethanol is mixed with acetic acid in the presence of sulfuric acid, ethyl acetate is formed. This reaction is a dehydration (loss of water) in which sulfuric acid acts as a catalyst and dehydrator.

$$\underset{\text{Acetic acid}}{CH_3\overset{\overset{O}{\|}}{C}-OH} + \underset{\text{Ethanol}}{H-OCH_2CH_3} \xrightarrow{H_2SO_4} \underset{\text{Ethyl acetate}}{CH_3\overset{\overset{O}{\|}}{C}-OCH_2CH_3} + H_2O$$

Names of esters are derived from the names of the alcohol and the acid used to prepare the ester. The alkyl group from the alcohol is named first followed by the name of the acid changed to end in -*ate*. For example, ethyl acetate is the name of the ester prepared from the reaction of ethanol and acetic acid. Ethyl acetate is a common solvent for lacquers and plastics and is often used as fingernail polish remover.

Unlike the acids from which they are derived, esters often have pleasant odors (Table 14.8). The characteristic odors and flavors of many flowers and fruits are due to the presence of natural esters. For example, the odor and flavor of bananas is primarily due to the ester 3-methylbutyl acetate (also known as isoamyl acetate).

Food and beverage manufacturers often use mixtures of esters as food additives. The ingredient label of a brand of imitation banana extract reads

Three ways of representing the ester group are

$$-\overset{\overset{O}{\|}}{C}-OR, \quad -CO_2R, \quad \text{and} \quad -COOR.$$

Charles D. Winters

Some household products containing ethyl acetate, which is an excellent solvent with a pleasant odor.

Esters Organic compounds containing a —COOR functional group

TABLE 14.8	Some Acids, Alcohols, and Their Esters		

ACID	ALCOHOL	ESTER	ODOR OF ESTER
CH_3COOH Acetic acid	$\overset{\displaystyle CH_3}{\overset{\displaystyle |}{CH_3CHCH_2CH_2OH}}$ 3-Methyl-1-butanol	$\overset{\displaystyle O}{\overset{\displaystyle \|}{CH_3COCH_2CH_2}}\overset{\displaystyle CH_3}{\overset{\displaystyle |}{CHCH_3}}$ 3-Methylbutyl acetate	Banana
$CH_3CH_2CH_2CH_2COOH$ Pentanoic acid	$\overset{\displaystyle CH_3}{\overset{\displaystyle |}{CH_3CHCH_2CH_2OH}}$ 3-Methyl-1-butanol	$CH_3CH_2CH_2CH_2\overset{\displaystyle O}{\overset{\displaystyle \|}{C}}OCH_2CH_2\overset{\displaystyle CH_3}{\overset{\displaystyle |}{CHCH_3}}$ 3-Methylbutyl pentanoate	Apple
$CH_3CH_2CH_2COOH$ Butanoic acid	$CH_3CH_2CH_2CH_2OH$ 1-Butanol	$CH_3CH_2CH_2\overset{\displaystyle O}{\overset{\displaystyle \|}{C}}OCH_2CH_2CH_2CH_3$ Butyl butanoate	Pineapple
$CH_3CH_2CH_2COOH$ Butanoic acid	 ⬡—CH_2OH Benzyl alcohol	$CH_3CH_2CH_2\overset{\displaystyle O}{\overset{\displaystyle \|}{C}}OCH_2$—⬡  Benzyl butanoate	Rose

"water, alcohol (40%), isoamyl acetate and other esters, orange oil and other essential oils, and FD&C Yellow #5." Except for the water, these are all organic compounds.

CONCEPT CHECK 14B

1. The simplest aldehyde is _____.
2. The organic acid found in vinegar is _____.
3. Ethyl acetate can be prepared by reacting _____ with _____.
4. The name of the ester obtained by reacting methanol with acetic acid is _____.
5. Citric acid found in citrus fruits has _____ carboxylic acid groups.
6. The simplest ketone is _____.
7. The characteristic odors and flavors of many fruits are caused by (**a**) carboxylic acids, (**b**) esters, (**c**) alcohols.
8. What is the major source of most organic chemicals?

Both natural and synthetic polymers are known. Examples of natural polymers include proteins, nucleic acids, starch, cellulose, and rubber. Natural polymers are discussed in Chapter 15.

Polymers Very large molecules composed of repeating units derived from small molecules

14.4 SYNTHETIC ORGANIC POLYMERS

It is impossible for us to get through a day without using a dozen or more synthetic organic **polymers**. The word *polymer* means "many parts" (Greek *poly*, meaning "many," and *meros*, meaning "parts"). Polymers are giant molecules with molar masses ranging from thousands to millions. Our clothes are polymers; our food is packaged in polymers; our appliances and cars contain a number of polymer components (Figure 14.3).

Approximately 80% of the organic chemical industry is devoted to the production of synthetic polymers. The prominence of synthetic polymers in consumer products is indicated by the fact that about half of the top 50 chemicals produced in the United States are used in the production of plastics, fibers, and rubbers. The average production of synthetic polymers in the United States exceeds 200 lb per person annually. Many synthetic organic polymers are *plastics* of one sort or another. *All plastics are polymers, but not all polymers are plastics.* Examples of items often made of plastics include dishes and cups, containers, telephones, plastic bags for packaging and wastes, plastic pipes and fittings, automobile steering wheels and seat covers, and cabinets for appliances, radios, and television sets. In fact, such plastic items, along with textile fibers and synthetic rubbers, are so widely used that they are commonly taken for granted.

Some of our most useful polymer chemistry has resulted from copying giant molecules in nature. Rayon is remanufactured cellulose; synthetic rubber is copied from natural latex rubber. As useful as these polymers may be, however, polymer chemistry is not restricted to nature's models. Polystyrene, nylon, and Dacron are a few examples of synthetic molecules that do not have exact duplicates in nature.

Polymers are made by chemically joining together many small molecules into one giant molecule, or macromolecule. The small molecules used to synthesize polymers are called **monomers**. Synthetic polymers can be classified as **addition polymers**, made by monomer units directly joined together, or **condensation polymers**, made by monomer units combining so that a small molecule, usually water, is split out between them.

As you proceed to study polymers in this chapter, you may find it particularly fascinating to couple your reading of this text with an inspection of a website called The Macrogalleria (**http://www.pslc.ws/macrog.htm**) devoted solely to polymers. This site, developed and maintained by the polymer department at the University of Southern Mississippi, uses the stores of a modern shopping mall to illustrate the variety of polymers and their practical uses before allowing you, should you so choose, to explore the chemistry of these polymers, including structure, properties, and synthesis, in as much detail as you wish.

FIGURE 14.3 Just a few everyday items made from synthetic polymers.

> A plastic is a substance that will flow under heat and pressure and hence is capable of being molded into various shapes. *All plastics are polymers, but not all polymers are plastics.*

> A *macromolecule* is a molecule with a very high molecular weight.

Addition Polymers

Polyethylene

The monomer for addition polymers normally contains one or more double or triple bonds. The simplest monomer of this group is ethylene (CH_2=CH_2). When ethylene is heated to between 100°C and 250°C at a pressure of 1000 atmospheres (atm) to 3000 atm in the presence of a catalyst, polymers are formed with molar masses of up to several million. A reaction of ethylene begins with breaking of one of the bonds in the carbon–carbon double bond, so an unpaired electron, a reactive site, remains at each end of the molecule. This step, the *initiation* of the polymerization, can be accomplished with initiator chemicals such as organic peroxides that are unstable and easily break apart into free radicals, which have unpaired electrons (Section 10.3). The free radicals readily add to molecules containing carbon–carbon double bonds to produce new free radicals. This destroys the double bond.

> An organic peroxide, RO—OR, produces free radicals, RO·, each with an unpaired electron.

> **Monomers** The small molecules that combine to form polymers.

> **Addition polymers** Polymers made from monomers (usually containing carbon–carbon double bonds) joined directly with one another

> **Condensation polymers** Polymers made from monomers (each with two functional groups) that combine so that small molecules are split out between their functional groups

See an animation of **Addition Polymerization of Ethylene to Form Polyethylene** at http://brookscole.com/chemistry/joesten4

The growth of the polyethylene chain then begins as the unpaired electron combines with a double bond electron in an unreacted ethylene molecule. This leaves another unpaired electron to react with yet another ethylene molecule. For example,

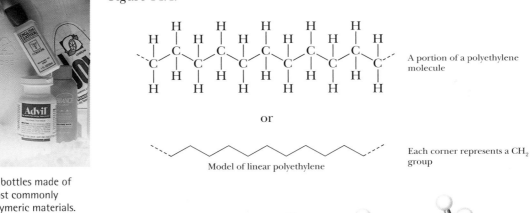

Polyethylene
n ranges from 1000 to 50,000

In the process, the unsaturated hydrocarbon monomer, ethylene, is changed to a saturated hydrocarbon polymer, *polyethylene.*

Polyethylene is the world's most widely used polymer, with many millions of tons produced annually in the United States alone. What are some of the reasons for this popularity? The wide range of properties of polyethylene leads to many uses.

Polyethylenes formed under various pressures and catalytic conditions have different molecular structures and hence different physical properties. For example, chromium oxide as a catalyst yields almost exclusively the linear polyethylene shown at the end of this paragraph—a polymer with no branches on the carbon chain. The zigzag structure represents the shape of the chain more closely because of the tetrahedral arrangement of bonds around each carbon in the saturated polyethylene chain. Long, linear chains of polyethylene can pack closely together to give a material with high density (0.97 g/mL) and high molar mass, referred to as high-density polyethylene (HDPE). This material is hard, tough, and rigid. A plastic milk bottle is a good example of an application of HDPE. Other HDPE containers are shown in Figure 14.4.

Representation of the relationship between the polymer chains of high-density polyethylene (HDPE).

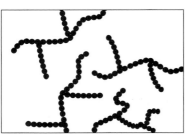

FIGURE 14.4 Some bottles made of HDPE, one of the most commonly recycled kinds of polymeric materials.

A portion of a polyethylene molecule

or

Model of linear polyethylene

Each corner represents a CH_2 group

If ethylene is heated to 230°C at a pressure of 200 atm, free radicals attack the polyethylene chain at random positions, causing irregular branching (Figure 14.5a). Branched chains of polyethylene cannot pack closely together, so the resulting material has a lower density (0.92 g/mL) and is called low-density polyethylene (LDPE). This material is soft and flexible (Figure 14.6). Sandwich bags are made from LDPE. Other conditions can lead to cross-

Representation of the relationship between the branched polymer chains of low-density polyethylene (LDPE).

(a)

(b)

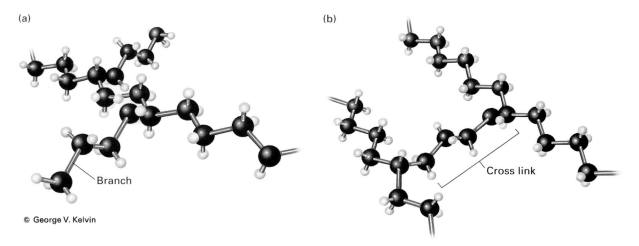

Branch

Cross link

© George V. Kelvin

FIGURE 14.5 Models of (a) branched and (b) cross–linked polyethylene.

linked polyethylene, in which branches connect long chains to each other (Figure 14.5b). If the linear chains of polyethylene are treated in a way that causes cross-links between chains to form cross-linked polyethylene (CLPE), a very tough form of polyethylene is produced. The plastic caps on soft-drink bottles are made from CLPE.

Polymers of Ethylene Derivatives

Many different kinds of addition polymers are made from monomers in which one or more of the hydrogen atoms in ethylene have been replaced

(a)

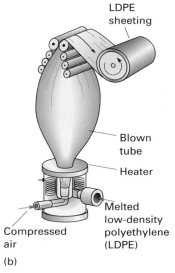

Rolled thin
LDPE
sheeting

Blown
tube

Heater

Melted
low-density
polyethylene
(LDPE)

Compressed
air

(b)

FIGURE 14.6 Making LDPE film. (a) The blown tube of polyethylene film. (b) Diagram showing how the tube is blown and the film collected in a roll. The process is similar to blowing a bubble gum bubble.

with either halogen atoms or a variety of organic groups. If the formation of polyethylene is represented as

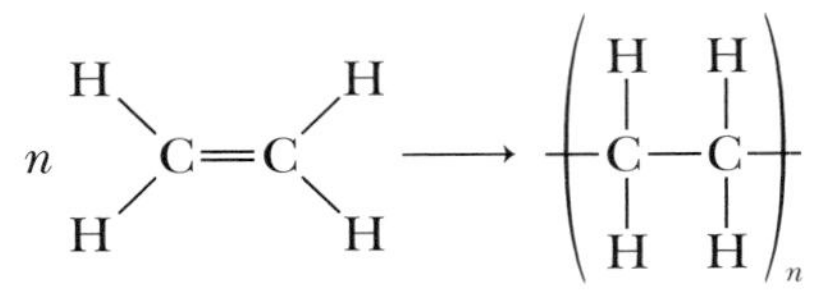

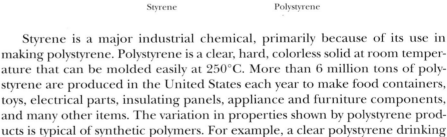

then the general reaction

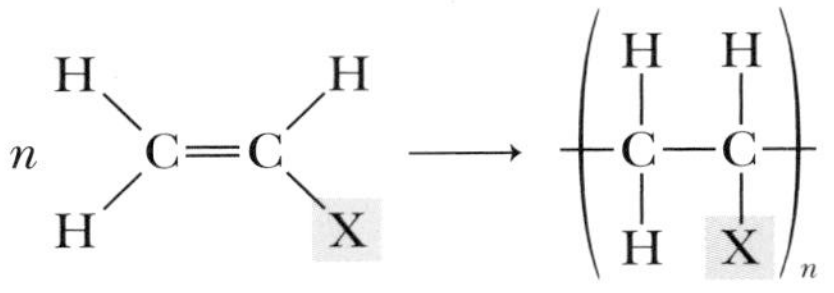

can be used to represent a number of other important addition polymers, where X is Cl, F, or an organic group (Table 14.9).

For example, the monomer for making **polystyrene** is styrene, and n is about 5700.

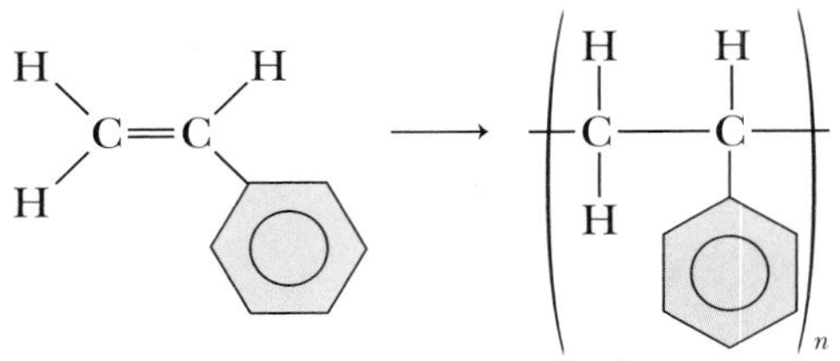

Styrene → Polystyrene

Polystyrene. Expanded polystyrene coffee cup (left) is soft. Clear polystyrene cup (right) is brittle.

Styrene is a major industrial chemical, primarily because of its use in making polystyrene. Polystyrene is a clear, hard, colorless solid at room temperature that can be molded easily at 250°C. More than 6 million tons of polystyrene are produced in the United States each year to make food containers, toys, electrical parts, insulating panels, appliance and furniture components, and many other items. The variation in properties shown by polystyrene products is typical of synthetic polymers. For example, a clear polystyrene drinking glass that is brittle and breaks into sharp pieces somewhat like glass is much different from a polystyrene coffee cup that is soft and pliable.

A major use of polystyrene is in the production of Styrofoam by "expansion molding." In this process, polystyrene beads are placed in a mold and heated with steam or hot air. The beads, 0.25 mm to 1.5 mm in diameter, contain 4 to 7% by weight of a low-boiling liquid such as pentane. CFCs used to be used for this purpose before their negative impact on the ozone layer was discovered. The steam causes the low-boiling liquid to vaporize; this expands the beads, and as the foamed particles expand, they are molded in the shape of the mold cavity. Styrofoam is used for egg cartons, meat trays, coffee cups, and packing material (Figure 14.7).

Polypropylene, used in indoor–outdoor carpeting, bottles, fabrics, and battery cases, is made from propylene.

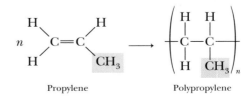

Propylene → Polypropylene

TABLE 14.9 Ethylene Derivatives that Undergo Addition Polymerization

FORMULA	MONOMER COMMON NAME	POLYMER NAME (TRADE NAMES)	USES
$H_2C{=}CH_2$	Ethylene	Polyethylene (Polythene)	Squeeze bottles, bags, films, toys and molded objects, electrical insulation
$H_2C{=}CH{-}CH_3$	Propylene	Polypropylene (Vectra, Herculon)	Bottles, films, indoor–outdoor carpets
$H_2C{=}CH{-}Cl$	Vinyl chloride	Poly(vinyl chloride) (PVC)	Floor tile, raincoats, pipe
$H_2C{=}CH{-}CN$	Acrylonitrile	Polyacrylonitrile (Orlon, Acrilan)	Rugs, fabrics
$H_2C{=}CH{-}C_6H_5$	Styrene	Polystyrene (Styrofoam, Styron)	Food and drink coolers, building material insulation
$H_2C{=}CH{-}O{-}C({=}O){-}CH_3$	Vinyl acetate	Poly(vinyl acetate) (PVA)	Latex paint, adhesives, textile coatings
$H_2C{=}C(CH_3){-}C({=}O){-}O{-}CH_3$	Methyl methacrylate	Poly(methyl methacrylate) (Plexiglas, Lucite)	High-quality transparent objects, latex paints, contact lenses
$F_2C{=}CF_2$	Tetrafluoroethylene	Polytetrafluoroethylene (Teflon)	Gaskets, insulation, bearings, pan coatings

Poly(vinyl chloride) (PVC), used in floor tiles, garden hoses, plumbing pipes, and trash bags, has a chlorine atom substituted for one of the hydrogen atoms in ethylene.

$$n\ \overset{H}{\underset{H}{C}}{=}\overset{H}{\underset{Cl}{C}} \longrightarrow \left(\overset{H}{\underset{H}{C}}{-}\overset{H}{\underset{Cl}{C}} \right)_n$$

Vinyl chloride   Poly(vinyl chloride)

Although the representation

$$\left(\overset{H}{\underset{H}{C}}{-}\overset{H}{\underset{H}{C}} \right)_n$$

FIGURE 14.7 Some familiar applications of polystyrene foam, which is an excellent thermal insulator.

Charles D. Winters

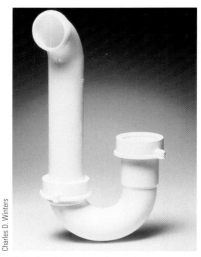

PVC plumbing pipes. Such pipes are common in repairs and new construction.

Charles D. Winters

saves space, keep in mind how large the polymer molecules are. Generally n is 500 to 50,000, and this gives molecules with molecular weights ranging from 10,000 to several million. The molecules that make up a given polymer sample are of different lengths and thus are not all of the same molecular weight. As a result, only the average molecular weight can be determined.

In summary, the numerous variations in substituents, length, branching, and cross-linking make it possible to produce a variety of properties for each type of addition polymer. Chemists and chemical engineers can fine-tune the properties of a polymer to match desired properties. Appropriate selection of monomer and reaction conditions accounts for the widespread use of these giant molecules.

EXAMPLE 14.1 Addition Polymers

Draw the structural formula of the repeating unit for the following addition polymers: (**a**) polypropylene, (**b**) poly(vinyl acetate), and (**c**) poly(vinyl alcohol).

SOLUTION
The names show that the monomers for these polymers are propylene (CH_2=$CHCH_3$), vinyl acetate (CH_2=$CHOOCCH_3$), and vinyl alcohol (CH_2=$CHOH$). The repeating units in the polymers therefore have the same structures, but without the double bonds.

(**a**)
$$\left(\begin{array}{cc} H & H \\ | & | \\ -C & -C- \\ | & | \\ H & CH_3 \end{array}\right)_n$$
(**b**)
$$\left(\begin{array}{cc} H & H \\ | & | \\ -C & -C- \\ | & | \\ H & O \\ & | \\ & O=C \\ & | \\ & CH_3 \end{array}\right)_n$$
(**c**)
$$\left(\begin{array}{cc} H & H \\ | & | \\ -C & -C- \\ | & | \\ H & OH \end{array}\right)_n$$

TRY IT 14.1
Draw the structural formula of the monomers used to prepare the following polymers: (**a**) polyethylene, (**b**) poly(vinyl chloride), and (**c**) polystyrene.

Although there are many uses of addition polymers, perhaps one of the most interesting applications is the use of sodium polyacrylate in superabsorbent diapers and some feminine hygiene products. The monomer for this polymer is sodium acrylate. This

$$n\ H_2C = CH \overset{\displaystyle CO_2^- \ Na^+}{\diagup} \xrightarrow{\text{polymerize}} \left(CH_2 - CH\right) \overset{\displaystyle CO_2^- \ Na^+}{|}$$

sodium acrylate   sodium polyacrylate

polymer, available in powder form, is found in most modern disposable diapers in a layer consisting of paper fluff and the polymer. The polymer has been marketed under the trade name Waterlock because of its ability to absorb up to 800 times its own weight in water and convert the liquid water to a gooey gel. Since there is actually no chemical reaction of the water with the polymer, the water will, over relatively long periods of time, be released from the polymer. For this reason, a crystalline version of the polymer is sold in garden stores for use in the soil surrounding plants to help cut down on watering frequency. These crystals can also be placed around the base of live Christmas trees to prevent premature drying. The technology has also been used in the toy industry, where small figures such as animals and fish can be expanded to many times their original size simply by placing them in water for a period of time. Distilled water works best for this since salts dissolved in the water inhibit the swelling. In fact, the gelatinous goo created by placing powdered Waterlock polymer in distilled water can be quickly converted back to liquid just by adding common table salt (NaCl).

Rubber: Natural and Synthetic

Natural rubber, a product of the *Hevea brasiliensis* tree, is a hydrocarbon with the empirical formula C_5H_8. When rubber is decomposed in the absence of oxygen, the monomer isoprene is obtained

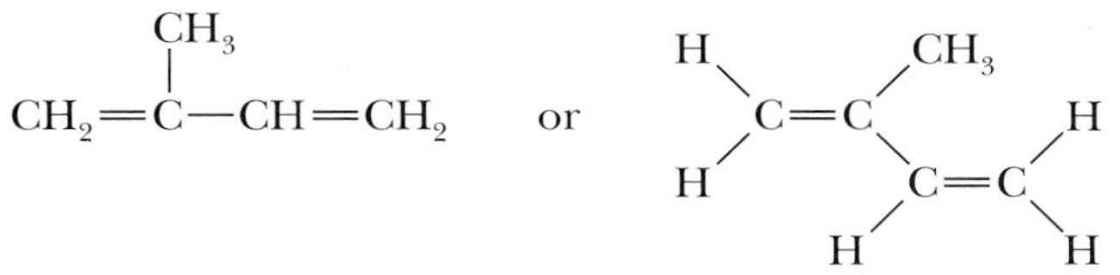

Isoprene (2-methyl-1,3-butadiene)

Natural rubber occurs as *latex* (an emulsion of rubber particles in water), which oozes from rubber trees when they are cut. Precipitation of the rubber particles yields a gummy mass that is not only elastic and water-repellent but also very sticky, especially when warm. In 1839, after five years of work on this material, Charles Goodyear (1800–1860) discovered that heating gum rubber with sulfur produces a material that is no longer sticky but is still elastic, water-repellent, and resilient.

Vulcanized rubber, as the type of rubber Goodyear discovered is now known, contains short chains of sulfur atoms that bond together the polymer chains of the natural rubber and reduce its unsaturation. The sulfur chains help to link the polymer chains, so the material does not undergo a permanent change when stretched but springs back to its original shape and size when the stress is removed (Figure 14.8). Substances that behave this way are called *elastomers*.

In later years chemists searched for ways to make a synthetic rubber so we would not be completely dependent on imported natural rubber during emergencies, such as during the first years of World War II. In the mid-1920s, German chemists polymerized butadiene (obtained from petroleum and structurally similar to isoprene, but without the methyl-group side chain).

Natural rubber Poly-*cis*-isoprene from the *Hevea brasiliensis* tree

Vulcanized rubber Rubber with short chains of sulfur atoms that bond together the polymer of natural rubber

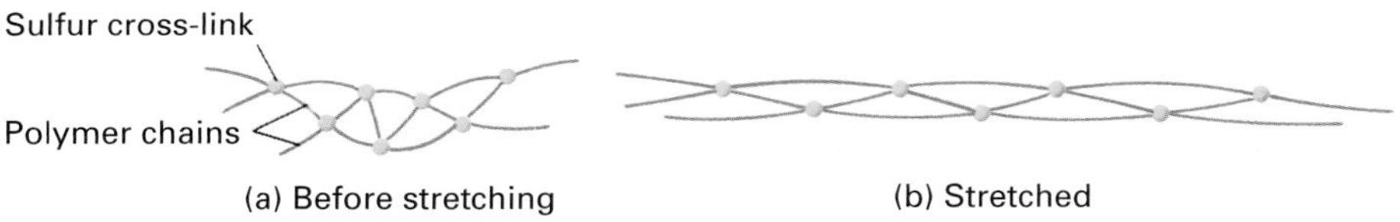

Sulfur cross-link
Polymer chains

(a) Before stretching   (b) Stretched

FIGURE 14.8 Stretching vulcanized rubber. After stretching, the cross-links pull it back into the original arrangement.

THE CHEMISTRY OF

ROMY & MICHELLE'S HIGH SCHOOL REUNION (1995)

©2005 by Mark A. Griep, University of Nebraska-Lincoln

"Wow! We have a whole class of inventors," says one girl in *Romy & Michelle's High School Reunion* (released in 1995). She has just learned that Heather Mooney invented the quick-burning paper in Lady Fair cigarettes: "Twice the Taste in Half the Time for the Gal on the Go." While in high school, Heather realized there was a market for a cigarette that "you can smoke all the way through between classes."

Another inventor classmate was Sandy Frink. He "invented some special kind of rubber that's used in every tennis shoe in North America." Both Sandy and Heather had been loners at Tucson's Sage Brush High School and both were good at science. One flashback scene shows them at their science fair poster.

In the "dream sequence" Michelle convincingly describes the special glue that she invented for sticky notes. Overall, the subtext of the film is that any loner can become an inventor-scientist. It also implies that the more ordinary and ubiquitous the invention, the greater the wealth that the inventor will accrue.

There was no market for sticky notes (Post-its is the 3M trade name) when they were released nationwide in 1980. By 1990, they were hailed as the top consumer product of the decade. In 2004, they were among the 122 everyday items in the "Humble Masterpieces" exhibit at the Museum of Modern Art.

The sticky notes story began in 1968 when 3M chemist Spencer Silver was preparing a series of acrylate copolymers in his search for a new adhesive. One of his preparations formed paper-thin microspheres that could be sprayed onto surfaces. Since these spheres didn't coat the paper surface uniformly, the paper was removable and restickable. He spent the next six years trying to develop a commercial product. 3M chemical engineer Art Fry had learned about the glue at one of Silver's presentations and, in 1974, realized that he could make a sticky but removable bookmark for his hymnal. Since 3M allows developers to spend 15% of their time on the project of their choosing, Silver and Fry started working on the project together. They made sure that the sticky bookmarks didn't cost too much, were recyclable, etc. While doing so, Fry realized that you could also write on them and they became the sticky notes that we know today.

Many movies have been made about real and fictional inventor-scientists. Examples include *You're Telling Me* (1934) with W. C. Fields, *Edison The Man* (1940) with Spencer Tracy, *Monkey Business* (1952) with Cary Grant, and the recent documentary *Me & Isaac Newton* (1999) featuring Dr. Gertrude Elion among others.

Questions to Consider

1. Discover the molecules, chemists, and stories behind other office products: adhesive tape, ballpoint pens, bubble wrap, electric light bulbs, fountain pens, liquid paper (or wite-out), rubber bands, the telephone, and xerography. What products did each of these replace?

1,3-Butadiene → Addition polymerization → Poly-*cis*-butadiene (the —CH₂—CH₂— groups are *cis*)

Polybutadiene is used in the production of tires, hoses, and belts.

The behavior of natural rubber (polyisoprene), it was learned later, is due to the specific molecular geometry within the polymer chain. We can write the formula for polyisoprene with the CH_2 groups on opposite sides of the double bond (the *trans* arrangement)

Poly-*trans*-isoprene (the —CH₂—CH₂— groups are *trans*)

or with the CH_2 groups on the same side of the double bond (the *cis* arrangement).

Poly-*cis*-isoprene (the —CH_2—CH_2— groups are *cis*)

Golf balls with gutta-percha covers.

Natural rubber is poly-*cis*-isoprene. However, the *trans* material also occurs in nature in the leaves and bark of the sapotacea tree and is known as *gutta-percha*. It is brittle and hard and is used for golf ball covers, electrical insulation, and other applications not requiring the stretching properties of rubber. Without an appropriate catalyst, polymerization of isoprene yields a solid that is like neither rubber nor gutta-percha because it is a random mixture of the *cis* and *trans* geometries. Neither the *trans* polymer nor the randomly arranged material is as good as natural rubber (*cis*) for making automobile tires.

In 1955, chemists at the Goodyear and Firestone companies almost simultaneously discovered how to use *stereoregulation* catalysts to prepare synthetic poly-*cis*-isoprene. This material is structurally identical to natural rubber. More than 10 million tons of synthetic rubber are produced worldwide every year.

Stereoregulation catalysts catalyze reactions that favor the formation of one geometric isomer.

CONCEPT CHECK) 14C

1. The individual molecules from which polymers are made are called _____.

2. The initiation step of an addition polymerization uses _____ _____ such as _____ _____.

3. Monomers must have a(n) _____ bond in their structure if they are to participate in an addition reaction.

4. The double bond in ethylene is converted to a(n) _____ bond during an addition reaction.

5. The monomer in Teflon is CF_2CF_2, $\begin{smallmatrix} F \\ F \end{smallmatrix}C=C\begin{smallmatrix} F \\ F \end{smallmatrix}$. (a) True, (b) False.

6. PVC is poly(vinyl chloride). It is built from the monomer _____.

7. Natural rubber is poly- _____ -isoprene.

8. Three different types of polyethylene are LDPE, HDPE, and CLPE.
   (a) LDPE stands for _____.
   (b) HDPE stands for _____.
   (c) CLPE stands for _____.

Condensation Polymers

A **condensation reaction** is one in which two molecules, each containing at least one reactive functional group, react by forming a larger molecule and eliminating a small molecule, frequently water. The reaction of alcohols with carboxylic acids to give esters and water (Section 14.3) is an example of condensation. This important type of chemical reaction does not depend on the presence of a carbon–carbon double bond in the reacting molecules. Instead, it requires the presence of two different kinds of functional groups on two different molecules. If each reacting molecule has two functional groups, both of which can react, it is possible for condensation reactions to produce long-chain polymers.

Condensation reaction Reaction between two molecules in which a larger molecule is formed and a small molecule is eliminated

Ⓦ See an animation of **Condensation Polymerization of Ethylene Glycol and Terephthalic Acid to Form a Polyester** at http://brookscole .com/chemistry/joesten4

Polyesters

A molecule with two carboxylic acid groups, such as terephthalic acid, and another molecule with two alcohol groups, such as ethylene glycol, can react with each other at both ends to form ester linkages (Section 14.3).

$$2 \ HO-\overset{\overset{O}{\|}}{C}-\bigcirc-\overset{\overset{O}{\|}}{C}-OH \ + \ 2 \ HO-CH_2-CH_2-OH \longrightarrow$$

Terephthalic acid   Ethylene glycol

$$HO-\overset{\overset{O}{\|}}{C}-\bigcirc-\overset{\overset{O}{\|}}{C}-O-CH_2-CH_2-O-\overset{\overset{O}{\|}}{C}-\bigcirc-\overset{\overset{O}{\|}}{C}-O-CH_2-CH_2-OH \ + \ 2 \ H_2O$$

Ester links   Ester link

A collection of polyester threads.

FIGURE 14.9 Medical uses of Dacron. A Dacron patch is used to close an atrial septal defect in a heart patient.

If n molecules of acid and alcohol can react in this manner, the process will continue until a large polymer molecule, known as a *polyester*, is produced.

$$\left(\overset{\overset{O}{\|}}{C}-\bigcirc-\overset{\overset{O}{\|}}{C}-O-CH_2-CH_2-O\right)_n$$

Poly(ethylene terephthalate)

More than 2 million tons of poly(ethylene terephthalate), commonly referred to as PET or PETE, are produced in the United States each year for use in beverage bottles, apparel, tire cord, film for photography, food packaging, coatings for microwave and conventional ovens, and home furnishings. Polyester textile fibers are marketed under such names as Dacron and Terylene. Films of the same polyester, when magnetically coated, are used to make audiotapes and videotapes. This film, Mylar, has unusual strength and can be rolled into sheets 1/30 the thickness of a human hair.

The inert, nontoxic, nonallergenic, noninflammatory, and non-blood-clotting properties of Dacron polymers make Dacron tubing an excellent substitute for human blood vessels (Figure 14.9) in heart bypass operations. Dacron sheets are used as a skin substitute for burn victims.

Polyamides (Nylons)

Another useful and important type of condensation reaction is that between a carboxylic acid and a **primary amine**, which is an organic compound containing an —NH₂ functional group. Amines can be considered derivatives of ammonia (NH₃), and most of them are weak bases, similar in strength to ammonia. An amine reacts with a carboxylic acid to split out a water molecule and form an **amide**, for example,

$$R-\overset{\overset{O}{\|}}{C}-OH \ + \ H-\underset{\underset{H}{|}}{N}-R \xrightarrow{\text{Heat}} R-\overset{\overset{O}{\|}}{C}-\underset{\underset{H}{|}}{N}-R \ + \ H_2O$$

Carboxylic acid   Amine   Amide

Polymers are produced when diamines react with dicarboxylic acids. Reactions of this type yield a group of polymers that perhaps have had a greater impact on society than any other type. These are the *polyamides*, or nylons.

In 1928, the DuPont Company embarked on a program of basic research headed by Dr. Wallace Carothers (1896–1937), who came to DuPont from the

Amine (primary) An organic compound containing an —NH₂ functional group

Amide An organic compound containing a —CONH₂ (or —CONHR) functional group

Harvard University faculty. His research interests were high-molecular-weight compounds, such as rubber, proteins, and resins, and the reaction mechanisms that produced these compounds. In February 1935, his research yielded a product known as nylon-6,6 (Figure 14.10), prepared from adipic acid (a diacid) and hexamethylenediamine (a diamine).

$$n\ HO-\overset{\overset{\displaystyle O}{\|}}{C}-(CH_2)_4-\overset{\overset{\displaystyle O}{\|}}{C}-OH + n\ H_2N-(CH_2)_6-NH_2 \longrightarrow$$

Adipic acid   Hexamethylenediamine

$$\left(\!-N-(CH_2)_6-\overset{}{\underset{H}{N}}-\overset{\overset{\displaystyle O}{\|}}{C}-(CH_2)_4-\overset{\overset{\displaystyle O}{\|}}{C}\!-\right)_{\!n} + n\ H_2O$$

$$\overset{}{\underset{H}{|}}$$

Nylon-6,6

> The name of nylon-6,6 is based on the number of carbon atoms in the diamine and diacid, respectively, that are used to make the polymer. Since both hexamethylenediamine and adipic acid have six carbon atoms, nylon-6,6 is the product.

This material could easily be extruded into fibers that were stronger than natural fibers and chemically more inert. The discovery of nylon jolted the American textile industry at almost precisely the right time. Natural fibers were not meeting the needs of 20th-century Americans. Silk was not durable and was very expensive; wool was scratchy; linen crushed easily; and cotton did not have a high-fashion image. All four had to be pressed after cleaning. As women's hemlines rose in the mid-1930s, silk stockings were in great demand, but they were very expensive and short-lived. Nylon changed all that almost overnight. Nylon could be knitted into the sheer hosiery women wanted, and it was much more durable than silk. The first public sale of nylon hose took place in Wilmington, Delaware (the location of DuPont's main office), on October 24, 1939. World War II caused all commercial use of nylon to be abandoned until 1945, as the industry turned to making parachutes and other war materials. Not until 1952 was the nylon industry able to meet the demands of the hosiery industry and to release nylon for other uses.

> In extrusion, a pliable substance is given a shape by being pushed through an opening. Toothpaste is extruded from a toothpaste tube.

Figure 14.11 illustrates another facet of the structure of hydrogen bonding in nylon which explains why nylons make such good fibers. To have good tensile strength, the chains of atoms in a polymer should be able to attract one another, but not so strongly that the plastic cannot be initially extended to form the fibers. Ordinary covalent chemical bonds linking the chains

> The amide linkage in nylon is the same linkage found in proteins, where it is called the peptide linkage (Section 15.7).

> Hair, wool, and silk are examples of nature's version of nylon. However, these natural polymers have only one carbon
>
> between each pair of $-\overset{\overset{\displaystyle O}{\|}}{C}-\overset{}{\underset{\underset{H}{|}}{N}}-$ units
>
> instead of the half dozen or so found in synthetic nylons.

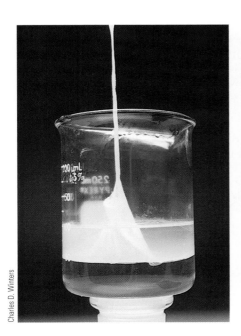

Charles D. Winters

FIGURE 14.10 Making nylon. The diacid is dissolved in hexane and the diamine is dissolved in water. They do not mix with each other. The layers form because hexane and water are immiscible. Water, which is more dense than hexane, forms the bottom layer. The nylon polymer forms at the interface between the two reactants and continues to form there as the nylon "rope" is pulled away.

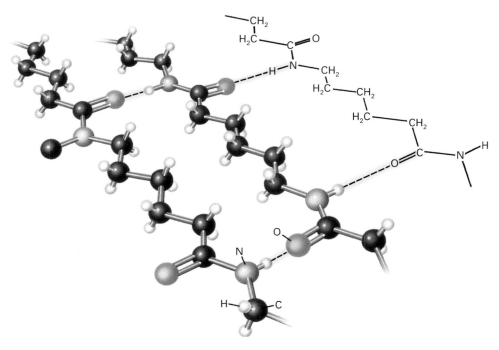

FIGURE 14.11 Hydrogen bonding in nylon-6,6. Hydrogen bonds are highlighted in blue shading.
(© George V. Kelvin)

FIGURE 14.12 Vests made of Kevlar have saved many police officers' lives. (Kevlar® is a registered trademark of DuPont for its aramid fiber.)

together would be too strong. Hydrogen bonds, with a strength about one-tenth that of an ordinary covalent bond, link the nylon chains in the desired manner. Kevlar, another polyamide, is used to make bulletproof vests, (Figure 14.12), vests worn by bull riders, military helmets, and fireproof garments. Kevlar is made from *p*-phenylenediamine and terephthalic acid.

$$\left(\!\!\begin{array}{c}\overset{O}{\underset{}{\underset{\parallel}{C}}}\end{array}\!\!-\!\!\bigcirc\!\!-\!\!\overset{O}{\underset{\parallel}{C}}\!\!-\!\!\overset{}{\underset{H}{N}}\!\!-\!\!\bigcirc\!\!-\!\!\overset{}{\underset{H}{N}}\!\!\right)_{n}$$

Kevlar

(**EXAMPLE** **14.2**)   **Condensation Polymers**

Write the repeating unit of the condensation polymer obtained by combining $HOOCCH_2CH_2COOH$ and $H_2NCH_2CH_2NH_2$. Identify the amide linkage.

SOLUTION

The dicarboxylic acid reacts with the diamine to split out water molecules and form a polyamide. The repeating unit is

$$\left(\!\!-\!\!\overset{O}{\underset{\parallel}{C}}CH_2CH_2\overset{O}{\underset{\parallel}{C}}\overset{\text{amide linkage}}{\underset{\underset{H}{\mid}}{N}}CH_2CH_2\underset{\underset{H}{\mid}}{N}-\!\!\right)_{n}$$

TRY IT 14.2

Draw the structure of the repeating unit of the condensation polymer obtained from reacting terephthalic acid with ethylene glycol.

THE PERSONAL SIDE

Stephanie Louise Kwolek (1923–)

Stephanie Kwolek received a Bachelor of Science degree from Carnegie–Mellon University in 1946. Although she wanted to study medicine, she couldn't afford it and decided to take a temporary job at DuPont. She liked her work so well that she stayed for 40 years, retiring in 1986. During her career at DuPont, she received 17 U.S. patents and 86 foreign patents for her work on a variety of polymeric fibers. However, she is best known for her work on the development of Kevlar fiber, which is five times stronger than steel on a weight basis. The patent for Kevlar fiber was issued in 1965, but 15 years passed before it was fully commercialized. Although Kevlar is best known for its use in bulletproof vests, it has a number of other important uses that include brake linings, underwater cables, and high-performance composite materials. In 1994, DuPont featured Stephanie Kwolek in a television commercial about the use of Kevlar in bulletproof vests, which resulted in name recog-

Courtesy of DuPont

STEPHANIE KWOLEK

nition for both Kevlar and Kwolek by the American public. However, what isn't widely known are the accomplishments of a woman during a time when women often didn't receive appropriate recognition for their work. She has received many awards, including an honorary Doctor of Science degree from Worcester Polytechnic Institute in 1981 for her contributions to polymer and fiber chemistry; the American Chemical Society Award for Creative Invention in 1980; election to the Engineering and Science Hall of Fame in Dayton, Ohio; and the 1997 Perkin Medal from the Society of Chemical Industry for her outstanding achievements in applied chemistry.

14.5 NEW POLYMER MATERIALS

Few plastics produced today find end uses without some kind of modification. For example, body panels for the GM Saturn and Corvette automobiles are made of **reinforced plastics**, which contain fibers embedded in a matrix of a polymer. These are often referred to as **composites**. The strongest geometry for a solid is a wire or a fiber, and the use of a polymer matrix prevents the fiber from bending or buckling. As a result, reinforced plastics are stronger than steel on a weight basis. In addition, the composites have a low density—from 1.5 g/cm^3 to 2.25 g/cm^3 compared with 2.7 g/cm^3 for aluminum, 7.9 g/cm^3 for steel, and 2.5 g/cm^3 for concrete. The only structural material with a lower density is wood, which has an average value of 0.5 g/cm^3. In addition, polymers do not corrode. The low density, high strength, and high chemical resistance of composites are the basis for their use in the automobile, airplane, construction, and sporting goods industries.

Glass fibers currently account for more than 90% of the fibrous material used in reinforced plastics because glass is inexpensive and glass fibers possess high strength, low density, good chemical resistance, and good insulating properties. In principle, any polymer can be used for the matrix material. Polyesters are the number-one polymer matrix at the present time. Glass-reinforced polyester composites have been used in structural applications such as boat hulls, airplanes, missile casings, and automobile body panels.

Other fibers and polymers have also been used, and the trend is toward increased utilization of composites in automobiles and aircraft. For example, a composite of graphite fibers in a polymer matrix is used in the construction of the F-117 Stealth fighter and other military aircraft. Graphite–polymer composites are used in a number of sporting goods such as golf club shafts, tennis racquets, fishing rods, and skis. The F-16 military aircraft was the first

Reinforced plastics (composites) Plastics with fibers of a substance, such as graphite, imbedded in a polymer matrix

a. Tom Pantages b. © Ann Kotowicz/Rainbow
c. © Tom Hollyman/Photo Researchers, Inc.

New polymer materials. (a) The flexible fender panel in a Saturn automobile. (b) An array of skis. (c) An F-16 fighter plane.

to contain graphite–polymer composite material, and the technology has advanced to the point where many aircraft, such as the F-18, use graphite composites.

Although few automobiles contain exterior body panels made of plastics, a number of components are plastic. Examples include bumpers, trim, light lenses, grilles, dashboards, seat covers, and steering wheels—enough plastics to account for an average of 250 lb per car. The increased emphasis on

THE WORLD OF CHEMISTRY

CHEMISTRY IN OUTER SPACE

To some chemists, the reactions of greatest interest are happening very far away, in the interstellar dust clouds that lie between the stars. The region is dark and cold, with an average temperature of 10 kelvins; that's almost 500 degrees below zero on the Fahrenheit scale.

The raw materials of organic compounds—hydrogen, carbon, nitrogen, and oxygen atoms—are present in these clouds, as are molecules and molecular fragments containing these atoms. Radioastronomy, which can be done from Earth's surface, and data from a satellite (the Infrared Space Observatory) have provided the evidence.

One hundred and twenty different kinds of molecules and molecular fragments have been identified. Some of the organic molecules that

are recognizable based on the chemistry we have described include the following:

H_2CO	CH_3COOH
HCOOH	CH_3CH_2CN
CH_3OH	CH_3CH_2OH
CH_3CN	NH_2CH_2COOH

There may also be larger aromatic hydrocarbons and fullerenes.

Like all atoms and molecules in a gas, the atoms and molecules in outer space are in constant motion and so can collide with each other. Compared to the billions of collisions per second in the gas phase at atmospheric pressure on Earth, however, the interstellar particles can collide only occasionally, perhaps only once a day. They exist in a better vacuum than we can create

on Earth, even with our best vacuum pumps.

What our Earth-bound chemists would like to know is, How do the chemical reactions that create organic molecules and fragments in space occur? There are smoke-like dust particles in space, and reactions can occur between fragments briefly trapped on the particle surfaces. The surfaces can act like catalysts, making the reactions possible. Water ice may play a role. Every relevant tool available to our Earth-bound chemists is brought to bear on creating the conditions and experiments to study the reactions. As one chemist has described it, "It's a messy problem."

Source: *Chemical and Engineering News*, July 15, 2002.

improving fuel efficiency will likely lead to the use of greater amounts of plastics in the construction of automobiles, both in interior components and exterior body panels.

14.6 RECYCLING PLASTICS

Recycling metals such as aluminum, iron, and lead has been occurring for years (Section 8.6). However, programs for recycling plastics developed much more slowly because of the costs associated with separating different types of plastics and producing usable recycled products from the used plastics.

Disposal of plastics has been the subject of considerable debate as municipalities face increasing problems in locating sufficient landfill space. The number-one waste is paper products, which make up about 40% of the volume in landfills. (Newspaper alone accounts for 16% of the volume.) Next is plastics, which make up about 20% of the volume in landfills.

Four phases are needed for a successful recycling of any waste material: *collection*, *sorting*, *reclamation*, and *end use*. Public enthusiasm for recycling and state laws requiring recycling have resulted in a dramatic increase in the collection of recycled items. This led to an annual average growth in recycling plastics of 32% per year for the 1988–1993 period, but it has slowed since then (Figure 14.13), primarily because of the continuing high cost of collecting and sorting recyclable plastics.

Codes are stamped on plastic containers to help consumers identify and sort their recyclable plastics (Figure 14.14). Polyethylene terephthalate (PET) and high density polyethylene (HDPE) are both widely used for making soft-drink bottles and, for that reason, are the two most commonly recycled plastics. In 2003, just under 20% of all PET bottles and just under 25% of all HDPE bottles were recycled, according to the American Plastics Council. Bottles made from PVC, LDPE, and polypropylene were recycled at far lower rates, with a 3.4% rate of recycling for polypropylene bottles being by far the largest for these three types of bottles. Overall, about 21% of plastic bottles were recycled in 2003. Figure 14.13 also shows the recent trend in recycling of plastic bottles and Figure 14.15 shows that recycled PET bottles are used largely for fibers for carpets, clothing, and furniture. Their use for new food and beverage containers and industrial strapping is also quite common. HDPE is recycled for similar uses, as well as trash containers, drainage pipe, garbage bags, and plastic lumber.

Patagonia® has recycled over 86 million plastic soda bottles in making post-consumer recycled (PCR®) clothing such as this women's Synchilla® jacket and notes that no synchillas are harmed in producing their clothing line.

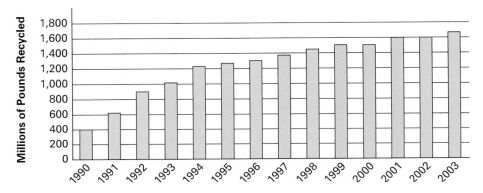

FIGURE 14.13   Growth in Post-Consumer Plastic Bottle Recycling

Code	Material	Percent of total bottles recycled
1 PETE	Poly(ethylene terephthalate) (PET or PETE)*	20–30
2 HDPE	High-density polyethylene	50–60
3 V	Poly(vinyl chloride) (PVC)*	5–10
4 LDPE	Low-density polyethylene	5–10
5 PP	Polypropylene	5–10
6 PS	Polystyrene	5–10
7 OTHER	All other resins and layered multi-material	5–10

*Bottle codes are different from standard industrial identification to avoid confusion with registered trademarks.

FIGURE 14.14 Plastic container codes, used to identify types of plastic so that containers can be sorted for recycling.

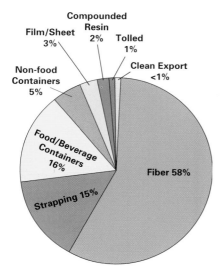

FIGURE 14.15 Domestic Recycled PET Bottle End Use

Compounded Resin 2%
Film/Sheet 3%
Tolled 1%
Non-food Containers 5%
Clean Export <1%
Food/Beverage Containers 16%
Strapping 15%
Fiber 58%

CONCEPT CHECK 14D

1. An example of the formation of a condensation polymer is the reaction of _____ with _____ to give _____ with the elimination of _____.
2. Nylon is an example of a (an) _____ polymer.
3. Polyamides are formed when _____ is split out from the reaction of many organic groups and many amine groups.
4. Polyesters are formed by (**a**) addition reactions, (**b**) condensation reactions.
5. Successful recycling involves the four phases of collection, sorting, reclamation, and end use. (**a**) True, (**b**) False.
6. _____ fibers currently account for over 90% of the fibrous material used in reinforced plastics.
7. Graphite fiber in a _____ matrix is a composite used to make tennis racquets, fishing rods, and aircraft components.
8. What two polymers, used to make soft-drink bottles, are most often recycled?

THE CHEMISTRY OF

THE MAN IN THE WHITE SUIT (1951)

©2005 by Mark A. Griep, University of Nebraska–Lincoln

The movie *The Man in the White Suit* (released in 1951) is fictional but based on the real story of nylon. It is about the fear of change brought on by research and it is about the drive that scientists have to carry out their research despite obstacles. In this story, the mill owners unite with the mill workers to oppose the use of a new synthetic fiber developed by Sydney Stratton (Alec Guinness). The new fiber doesn't stain, dye, or tear. The owners and workers worry that no one will need to buy more than one set of clothes, thereby putting them out of business.

In the opening scenes, we learn that Sydney Stratton lost his Cambridge University scholarship for mysterious reasons. We soon suspect that it was due to the explosions that accompany his polymerization process. After losing his eighth mill job due to lack of credentials and overspending, he is accidentally asked to train researchers at Birnley Mill Works.

In the clip, the mill owner's daughter, Daphne Birnley, discovers that Sydney is working at her father's mill under cover. Sydney tries to stop her from exposing him to her father by explaining his new fibers. They are very long copolymers of amino acids and carbohydrates (see Chapter 15) with ionic crosslinks for extra strength plus the ability to repel dirt. She becomes his ally.

Sydney's polymerization reaction requires "heavy hydrogen" and "radioactive thorium" and his reactions go so fast that they explode. The pun of this movie is that it deliberately confuses polymerization chain reactions with nuclear chain reactions. When polymers are formed from monomers, they do so by a chain reaction, which is a sequential series of events.

By the end of the movie, Sydney's fiber breaks down, he loses his job, and he leaves Daphne behind. Despite those failures, he shrugs it off like his copolymer repels dirt. We hear the sound of his machine and we know he has a new idea to test.

Questions to Consider

1. Cellulose is a repeating polymer of the carbohydrate named glucose. Find the special hydrogen bond that gives cellulose its amazing strength.

2. The stretchiness of nylon occurs because each polymer makes special hydrogen bonds with the adjacent strand. Find the special hydrogen bonds. Why do the hydrogen bonds of cellulose add strength but the hydrogen bonds of nylon make it stretchy?

3. Which of the following fibers are synthetic polymers based on cellulose? a) acetate; b) Kevlar; c) Mylar; d) polyester; e) rayon.

4. Sydney's copolymer fiber combines the incredible strength of cotton with the beauty and versatility of silk. If you actually made a random copolymer of glucose and an amino acid, why wouldn't it be as strong as Sydney's copolymer?

5. Why could the moviemakers of 1951 assume that their audience had heard of "chain reactions" so that they could make a joke out of it?

Assess your understanding of this chapter's topics with an online chapter quiz at www.brookscole.com/chemistry/joesten4

■ KEY TERMS

organic chemistry

aldehydes

ketones

carboxylic acids

denatured alcohol

esters

polymers

monomers

addition polymers

condensation polymers

natural rubber

vulcanized rubber

condensation reaction

amine

amide

reinforced plastics (composites)

■ THE LANGUAGE OF CHEMISTRY

1. Aldehyde
2. Ketone
3. Carboxylic acid
4. Ester
5. Monomer
6. Present primary source of organic chemicals used as raw materials
7. Poly-*cis*-isoprene
8. Cross-linking via reaction with sulfur
9. Polyester
10. Nylon
11. Alcohol

a. Contains ROH
b. Polyamide
c. Vulcanization
d. Contains $R-\overset{\overset{\displaystyle O}{\|}}{C}-R'$
e. Contains $R-\overset{\overset{\displaystyle O}{\|}}{C}-H$
f. Contains $R-\overset{\overset{\displaystyle O}{\|}}{C}-OH$
g. Contains $R-\overset{\overset{\displaystyle O}{\|}}{C}-O-R'$
h. Petroleum
i. Formed from a dialcohol and a diacid
j. Natural rubber
k. Building block for polymer

■ APPLYING YOUR KNOWLEDGE

1. Why are there more than 11 million organic compounds?
2. Draw the structures of four alcohols that have the formula $C_4H_{10}O$.
3. Identify the class of each of the following compounds:
   (a) $CH_3CH_2CH_2CH_2COOH$
   (b) $CH_3CH=CHCH_2CH_3$
   (c) $C_2H_5OC_2H_5$
   (d) $CH_3CH_2CHOHCH_2CH_3$
   (e) $(CH_3)_3COH$
   (f) $CH_3\overset{\overset{\displaystyle O}{\|}}{C}H$
4. Name the compounds in question 3.

5. Draw the structures, classify, and name the following
   (a) CH_3COOH (b) CH_3OCH_3
   (c) CH_3CH_2OH (d) CH_3COCH_3
   (e) HCHO   (f) CH_3COOCH_3
6. What is meant by the following terms?
   (a) Proof rating of an alcohol
   (b) Denatured alcohol
7. What is the volume percent ethanol in 90 proof gin?
8. What is the proof of pure ethanol?
9. How do the structures for ethanol, ethylene glycol, and glycerol differ? Give one use for each.

10. Many naturally occurring carboxylic acids have more than one acid group (Table 14.7). What other functional group is often present?

11. Identify each of the following molecules as an alcohol, an ether, an aldehyde, a ketone, a carboxylic acid, or an ester:

    (a) $\overset{\displaystyle O}{\overset{\|}{H}}CCH_2CH_3$

    (b) CH_3CH_2OH

    (c) $CH_3\overset{\displaystyle O}{\overset{\|}{C}}CH_2CH_3$

    (d) $CH_3CH_2CH_2\overset{\displaystyle O}{\overset{\|}{C}}-OH$

    (e) $CH_3\overset{\displaystyle O}{\overset{\|}{C}}-OCH_2CH_3$

    (f) CH_3OCH_3

    (g) $CH_3\overset{\displaystyle O}{\overset{\|}{C}}H$

12. Write the equation for the formation of an ester from acetic acid (CH_3COOH) and ethanol (C_2H_5OH), and give its name.

13. How do the structures of primary, secondary, and tertiary alcohols differ?

14. Give examples of
    (a) two naturally occurring esters and where they are found.
    (b) two naturally occurring carboxylic acids and where they are found.

15. Describe how the liver detoxifies ethanol.

16. Sketch the structure for the monomer of polystyrene.

17. In what ways is a railroad train like polystyrene?

18. What is the origin of the word *polymer*?

19. What is meant by the term *macromolecule*?

20. What do the following terms mean?
    (a) Monomer   (b) Polymer

21. What structural features must a molecule have to undergo addition polymerization?

22. What do the following mean?
    (a) Polyethylene   (b) HDPE
    (c) LDPE   (d) CLPE

23. What is the repeating unit of natural rubber?

24. What is the difference in the structures of poly-*trans*-isoprene and poly-*cis*-isoprene? Sketch the arrangements for the two polymers. How do the two differ in their physical properties?

25. What is the role of sulfur in the vulcanization process?

26. What is the effect of vulcanization on the physical properties of natural rubber?

27. What property does a polymer have when it is extensively cross-linked?

28. By using polyethylene as an example, draw a portion of
    (a) a linear polymer.
    (b) a branched polymer.
    (c) a cross-linked polymer.

29. Indicate which of the following can undergo an addition reaction and which cannot. Explain your choices.

    (a) Styrene ($C_6H_5CH=CH_2$)

    (b) Propene ($CH_2=CHCH_3$)

    (c) Ethane (CH_3CH_3)

30. Explain how polymers could be prepared from each of the following compounds (other substances may be used):
    (a) $CH_3CH=CHCH_3$
    (b) $HOOCCH_2CH_2COOH$
    (c) $H_2NCH_2CH_2CH_2CH_2NH_2$

31. Draw the structural formula of the monomer used to prepare the following polymers:
    (a) Poly(vinyl chloride)   (b) Polystyrene
    (c) Polypropylene

32. Orlon has the polymeric chain structure shown here:

    $$-CH_2-\underset{\underset{CN}{|}}{CH}-CH_2-\underset{\underset{CN}{|}}{CH}-CH_2-\underset{\underset{CN}{|}}{CH}-$$

    What is the monomer from which this structure can be made?

33. Give definitions for the following terms:
    (a) Polyester   (b) Polyamide
    (c) Nylon-6,6   (d) Diacid
    (e) Diamine   (f) Peptide linkage

34. What polymer is identified as PET?

35. What feature do all condensation polymerization reactions have in common?

36. What are the starting materials for nylon-6,6?

37. Which do you think is the source of most polymers used today, green plants or petroleum? Do you think this will ever change? Explain.

38. Do you think that plastic production will increase in the future? What advantages, if any, do you see when plastics are used in place of other materials?

39. What are the four phases involved in a successful recycling program? What happens if the public is cooperative in turning in recyclable materials but there isn't an adequate infrastructure to process the collected materials?

40. Composite materials are typically reinforced plastics. What properties do the embedded glass fibers or graphite fibers give to composites? Do you foresee any recycling problems with these materials?

41. Many organic compounds have more than one type of functional group. Identify the functional groups in the following naturally occurring molecules:

(a)

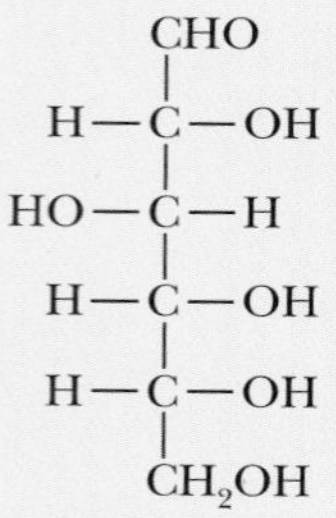

$$
\begin{array}{c}
CHO \\
| \\
H-C-OH \\
| \\
HO-C-H \\
| \\
H-C-OH \\
| \\
H-C-OH \\
| \\
CH_2OH
\end{array}
$$

D-Glucose

(b)

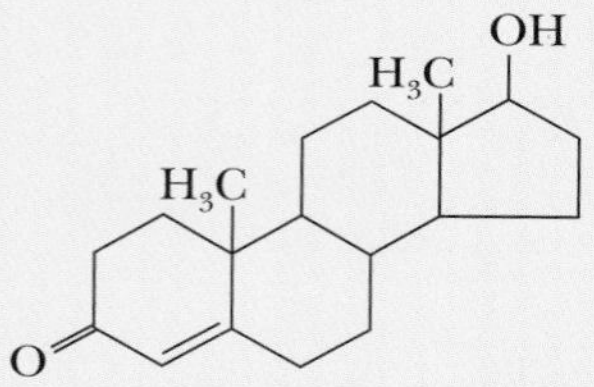

Testosterone

(c)

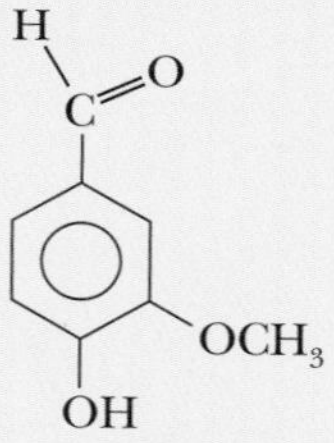

Vanillin
(vanilla bean)

(d)

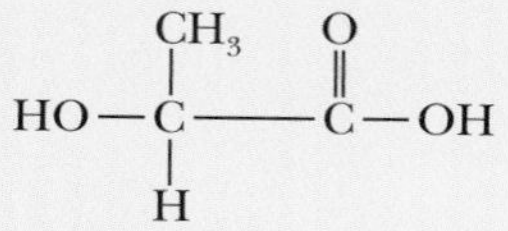

$$
\begin{array}{c}
CH_3 \quad O \\
| \quad\quad || \\
HO-C-\!\!-\!\!-C-OH \\
| \\
H
\end{array}
$$

Lactic acid

42. Discuss what plastics are currently being recycled, and give examples of some products being made from these recycled plastics.

43. What are polymer composite materials? Give two examples that illustrate the importance of polymer composites in the manufacture of consumer products.

44. What is the polymer formed by the reaction of the following monomers?

(a) Ethylene glycol (HOCH$_2$CH$_2$OH)

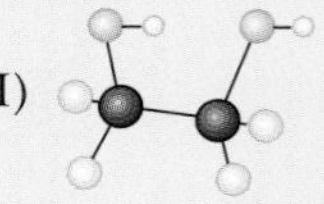

and terephthalic acid, HOOC 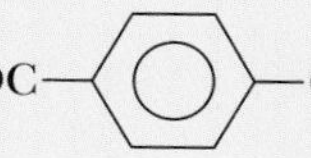 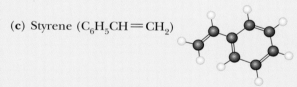 COOH

(b) Ethylene (CH$_2$=CH$_2$)

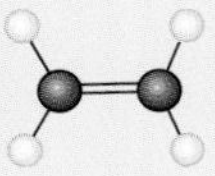

(c) Styrene (C$_6$H$_5$CH=CH$_2$)

45. How many ethylene monomer units, CH$_2$=CH$_2$, are linked together in a polyethylene molecule with a molecular weight of 280,000?

46. How many propylene monomer units of propene,

(CH$_2$=CHCH$_3$) 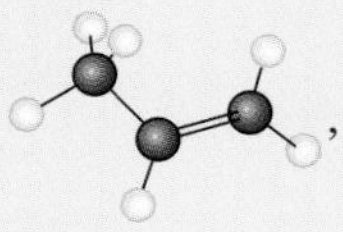 , are linked to-

gether in a polymer molecule with molecular weight of 84,000?

47. What is the predicted molecular weight for a polymer molecule made up of 400,000 styrene units (C$_6$H$_5$CH=CH$_2$)?

48. What is the predicted molecular weight for a polymer molecule made up of 200,000 1,3-butadiene units (CH$_2$=CHCH=CH$_2$)?

CHEMISTRY ON THE WEB

For up-to-date URLs, visit the text website at **www.brookscole.com/ chemistry/joesten4**

- Polymers (Description, Preparation, and Uses)
- Organic Functional Groups
- A Look at Functional Groups in NutraSweet®
- Recycling Plastics and History of Plastics

THE CHEMISTRY OF LIFE

A model of a segment of DNA, the molecule that controls the chemistry of life.

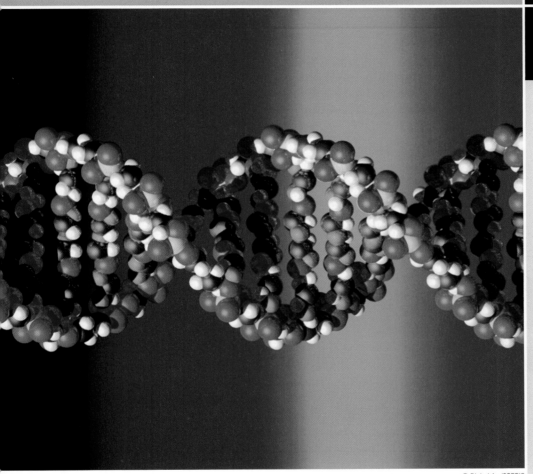

© Digital Art/CORBIS

Cooking, nutrition, personal health, drugs, medicine and dentistry, agriculture, our natural environment—biochemistry is fundamental to them all. Sometimes biochemistry is applied in a practical fashion, for example, when a cook scrambles an egg or someone grabs an aspirin tablet to quiet a headache. Sometimes it is applied by a practicing professional or research scientist, for example, when an agriculture expert recommends the appropriate pesticide to a farmer or a pharmaceutical chemist designs a molecule to combat disease. In this chapter you will be introduced to the major classes of biochemicals and their functions in the body. Some of the applied aspects of biochemistry are also included. Questions that will be addressed in this chapter include:

- What are optical isomers, and why are they important in biochemical reactions?

- What are the different types of sugars?
- What is the difference between starch and cellulose?
- What are triglycerides?
- What are the differences among saturated, monounsaturated, and polyunsaturated fatty acids?
- What are steroids?
- What are soaps and synthetic detergents, and in what kinds of products do we find them?
- What are amino acids and proteins?
- What is the structure of hair?
- What do enzymes do?
- What are DNA and RNA, and how do they relate to the genetic code?

The chemistry of life is referred to as **biochemistry**, and the organic chemicals found in living things are *biochemicals.*

A petrochemical manufacturing plant and the cells in your body are doing the same thing—taking in raw materials, using energy and carefully controlled conditions to perform chemical reactions, and putting out valuable products and waste materials. Many biochemicals are polymers. Starches are condensation polymers of simple sugars; proteins are condensation polymers of amino acids; and nucleic acids are condensation polymers of simple sugars, nitrogenous bases, and phosphoric acid groups. All these very large biomolecules and other essential, smaller ones must be assembled from the raw materials in food. At the same time, burning food must provide energy. To carry out these functions simultaneously, living things have evolved an exquisite system for breaking down food molecules and putting them back together again. The overview of biochemistry in this chapter focuses on the molecular structure of the most important kinds of biomolecules. Many of them exhibit "handedness," another type of isomerism, so that's where we begin.

Charles D. Winters

FIGURE 15.1   Mirror images. The mirror image of your left hand, as shown at the top, looks like your right hand, shown at the bottom. But if you place one hand over the other with palms up, they are not identical. This shows that they are nonsuperimposable mirror images.

Biochemistry The chemistry of living things

Chiral Cannot be superimposed on its mirror image

Achiral Can be superimposed on its mirror image

Enantiomers (optical isomers) A chiral molecule and its nonsuperimposable, mirror-image molecule. Enantiomers are one type of stereoisomer

Stereoisomers Molecules with the same molecular formula but different three-dimensional structures

⬭15.1⬭ HANDEDNESS AND OPTICAL ISOMERISM

Are you right-handed or left-handed? Regardless of our preference, we learn at a very early age that a right-handed glove doesn't fit the left hand, and vice versa. Our hands are not identical; they are mirror images of one another and are not superimposable (Figure 15.1). *An object that cannot be superimposed on its mirror image is called* **chiral**. Objects that are superimposable on their mirror images are **achiral**. Stop and think about the extent to which chirality is a part of our everyday life. We've already discussed the chirality of your hands (and feet). Helical seashells are chiral, and most spiral to the right like a right-handed screw. Golf clubs are chiral but baseball bats are achiral.

What is not as well known is that a large number of the molecules in plants and animals are chiral, and usually only one form of the chiral molecule (left-handed or right-handed) is found in nature. For example, all but 1 of the 20 naturally occurring amino acids in our proteins are chiral, and they have the same "handedness." Most of the natural sugars exhibit the opposite "handedness" when compared to amino acids.

A chiral molecule and its non-super-imposable mirror image are called **enantiomers**. Enantiomers are two different molecules, just as your left hand and right hand are different. To have enantiomers a molecular structure must be asymmetric (without symmetry). The simplest case is a tetrahedral carbon atom bonded to four *different* atoms or groups of atoms.

Such a carbon atom is asymmetric and is said to be chiral. A molecule that contains a chiral atom is likely to be a chiral molecule, although this is not always so.

Some compounds are found in nature in both enantiomeric forms. For example, both forms of lactic acid are found in nature. During the contraction of muscles the body produces only one enantiomer of lactic acid; the other enantiomer is produced when milk sours. Let's look at the structure of lactic acid to see why two different isomers or chiral molecules might be possible. The central carbon atom of lactic acid has four different groups bonded to it: $-CH_3$, $-OH$, $-H$, and $-COOH$.

A chiral seashell with a right-hand spiral.

$$\underset{\text{Lactic acid}}{\overset{\displaystyle H}{\underset{\displaystyle CH_3}{HO\diagup\overset{\displaystyle |}{\underset{\displaystyle |}{C}}\diagdown COOH}}}$$

As a result of the tetrahedral arrangement around the central carbon atom, it is possible to have two different arrangements of the four groups. If a lactic acid molecule is placed so the C—H bond is vertical, as illustrated in Figure 15.2, one possible arrangement of the remaining groups would be that $-OH$, $-CH_3$, and $-COOH$ are attached in a clockwise manner (isomer I). Alternatively, these groups can be attached in a counterclockwise fashion (isomer II). To see further that the arrangements are different, we place isomer I in front of a mirror (Figure 15.2b). Now you see that isomer II is the mirror image of isomer I. What is important, however, is that these mirror-image molecules *cannot be superimposed* on one another, no matter how the molecules are rotated. *These two nonsuperimposable, mirror-image chiral molecules are enantiomers.*

Enantiomers of a chiral compound have the same melting point, the same boiling point, the same density, and many other identical physical and chemical properties. They differ with respect to only one physical property: They rotate a beam of *plane-polarized light* in opposite directions, as shown in Figure 15.3. For this reason, chiral molecules are sometimes referred to as **optical isomers** and are said to be *optically active*.

One must be able to specify the "handedness" of chiral centers in order to distinguish between them. Enantiomers are mirror-image molecules that have

> Enantiomers have the same set of atoms connected by the same set of chemical bonds, but the atoms have a different three-dimensional arrangement in space—just like your left and right hands do.

> Superimposable objects, including molecules, are identical to each other.

> *Plane-polarized light* consists of electromagnetic waves with their electric and magnetic fields oscillating in the same direction.

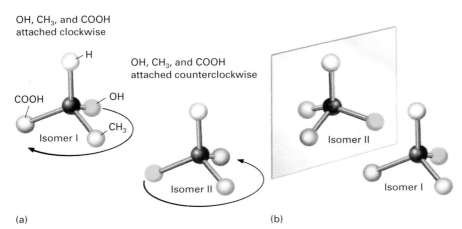
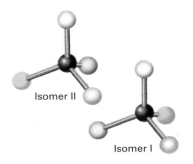

FIGURE 15.2 The enantiomers of lactic acid. (a) In isomer I, the $-OH$, $-CH_3$, and $-COOH$ groups are attached in clockwise order. In isomer II, the same groups are attached in counterclockwise order. (b) Isomers I and II are mirror images, a result of their having four different groups on the same carbon atom. (c) The two isomers cannot be turned in any way to make them superimposable.

FIGURE 15.3 Rotation of plane-polarized light by an optical isomer. (*Top*) Monochromatic light (light of only one wavelength) is produced by a sodium lamp. After it passes through a polarizing filter, the light vibrates in only one direction—it is polarized. Polarized light will pass through a second polarizing filter if this filter is positioned parallel to the first filter, but not if the second filter is perpendicular. (*Bottom*) A solution of an optical isomer placed between the first and second polarizing filters causes rotation of the plane of polarized light. The angle of rotation can be determined by rotating the second filter until maximum light transmission occurs. The magnitude and direction of rotation are unique physical properties of the optical isomer being tested.

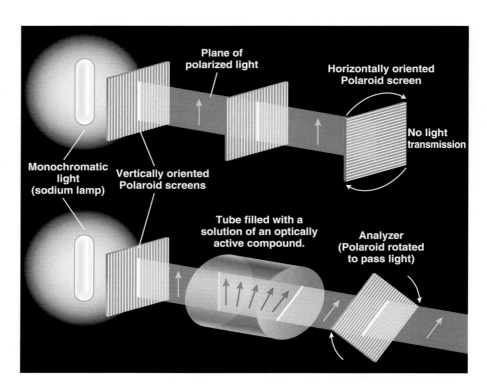

all the same groups attached but arranged differently in space. It has been necessary to develop an artificial system to describe the arrangement of these atoms so that scientists may be certain they are talking about the same molecule. Historically, enantiomers were referred to as D-enantiomers if they had a spatial arrangement of groups similar to that of the enantiomer of glyceraldehyde (D-glyceraldehyde), which rotated the plane of plane-polarized light in a clockwise fashion, referred to as dextrorotatory (after the Latin *dexter,* meaning "right"). Enantiomers were designated as L-enantiomers if they had a spatial arrangement of groups similar to that of the enantiomer of glyceraldehyde (L-glyceraldehyde), which rotated plane-polarized light in a counterclockwise fashion, referred to as levorotatory (after the Latin *laves* meaning "left"). It was soon apparent that this simple system would be difficult to apply to more complex molecules with multiple chiral centers or to molecules bearing little similarity to glyceraldehyde. This system is still used in referring to naturally-occurring amino acids and naturally-occurring sugars, however. Natural amino acids are L-amino acids and natural sugars are D-sugars because of their structural relationship to the glyceraldehyde enantiomers, but it is not true that all D-sugars are dextrorotatory, nor are all L-amino acids levorotatory. This confusion can be illustrated with lactic acid. D-lactic acid is found in souring milk and it is levorotatory. L-Lactic acid, which accumulates in muscle tissue during exercise and may cause cramps, is dextrorotatory.

D-glyceraldehyde

L-glyceraldehyde

Enantiomers may also differ with respect to their biological properties. They react at different rates in a *chiral environment*, and many of the molecules in plants and animals are chiral. To understand why this difference in activity might exist, think of the hand-in-glove analogy. Although you can put a right-handed glove on your left hand, it takes longer to put it on and it doesn't fit very well. Since nature has a preference for L-isomers of amino acids and since enzymes, the catalysts for biochemical reactions, are proteins made from L-amino acids, enzymes are chiral molecules. The catalytic activity of enzymes is dependent on their three-dimensional structure, which in turn depends on their L-amino acid sequence. As a result, enzymes have a binding preference for one enantiomer of a reactant.

Even though nature usually has a preference for one enantiomer, laboratory synthesis of a chiral compound normally gives a mixture of equal amounts of the enantiomers, which is called a **racemic mixture**. The separation and purification of enantiomers are difficult because of the identity of their physical properties. The first separation of enantiomers by Louis Pasteur in 1848 was based on his observation of mirror-image crystals in a racemic mixture. This method is rarely an option, since racemic modifications seldom form mixtures of crystals recognizable as mirror images. The usual method of separating enantiomers is to react them with optically active reagents that have a greater affinity for one enantiomer than the other. Pasteur's discovery of a mold that would selectively destroy one enantiomer of tartaric acid was the first example of this approach.

The chirality of sugars, proteins, and DNA makes the human body highly sensitive to enantiomers. For example, one enantiomer of a drug may be more active (or more toxic) than the other. A tragic example that called attention to the need for testing both enantiomers occurred in 1963, when horrible birth defects were induced by thalidomide use by pregnant women. After this was discovered, it was determined that one enantiomer was useful for treating morning sickness and the other enantiomer was a **teratogen**. The teratogenic enantiomer caused serious birth defects in the children of women who took thalidomide during the first trimester of their pregnancy.

As a result of these incidents, thalidomide became the most notorious drug ever marketed and the so-called "thalidomide children" were, in most cases, forced to face life with stunted flippers for arms or legs. They and their families, as well as scientists around the world, vowed that this type of incident would never again occur. Remarkably, the tragedy was limited primarily to Great Britain and Europe, thanks, in large part, to some stubborn skepticism about the safety of the drug voiced by Dr. Frances Kelsey at the U.S. Food and Drug Administration. Assigned to review the thalidomide drug approval application, she fought a dogged battle against the drug company that sought to market thalidomide in the United States until the disastrous news from Europe was indisputable.

It is important to consider how the European thalidomide tragedy occurred. Scientists knew before this particular series of incidents that enantiomers could, in principle, exhibit quite different biological properties, but nothing of this magnitude had ever occurred before. The safety testing done on thalidomide before it was marketed did not reveal its teratogenic effect, although more extensive testing would have done so. As a result of this event, pharmaceutical companies around the world initiated much more stringent testing of both enantiomers of drugs and such testing continues today.

It may surprise you to learn that there are currently many drugs marketed as racemic mixtures in the United States. In these cases, scientists have been able to show that one enantiomer elicits a desired positive therapeutic response while the other enantiomer shows no biological activity or only a

Since enantiomers rotate the plane of polarized light an equal amount but in opposite directions, a solution of a racemic mixture will not rotate the plane of polarized light.

Thalidomide

Racemic mixture Mixture of equal amounts of enantiomers; does not rotate polarized light

Teratogen A chemical or factor that causes malformation of an embryo

THE WORLD of CHEMISTRY

CHIRAL DRUGS

One enantiomer of a chiral drug is usually more active than the other in fighting a disease. In 1992, the U.S. Food and Drug Administration (FDA) released long-awaited guidelines for the marketing of chiral drugs. The decision about whether to sell a chiral drug in the racemic mixture or the enantiomerically pure form has been left to the drug's manufacturer, although the decision is subject to FDA approval. With the regulations in place, drug companies have been faced with major scientific, technical, and economic choices. They must decide whether to market a drug as an easier-to-synthesize racemic mixture or to tackle the more costly and difficult synthesis of the pure active enantiomer. Which will give them the competitive edge?

In some cases, a choice is made based on the biochemical activity of the enantiomers. For example, ibuprofen, the pain reliever contained in Advil, Motrin, and Nuprin, is sold as a racemic mixture. The D-isomer of ibuprofen is the active pain reliever; the L-isomer is inactive. Since L-ibuprofen is converted to D-ibuprofen in the body, there is no therapeutic advantage to the patient in a switch from the racemic mixture to what would be the more costly D-ibuprofen.

(a)

Charles D. Winters

Ibuprofen (Advil) is sold as a racemic mixture and naproxen (Aleve) is sold as a single enantiomer.

By contrast naproxen, like ibuprofen a medication to combat pain, fever, and inflammation, is an example of a chiral drug sold as an enantiomer instead of the racemic mixture. In this case, one enantiomer is a pain reliever and the other causes liver damage. Originally available only by prescription, naproxen is now available over the counter (as Aleve). The sale of chiral drugs is growing rapidly.

(b)

Naproxen

*Chiral carbon

Meanwhile, drug companies are facing up to the challenge of bringing chiral drugs to market. One motivation is replacement of a drug sold as a racemic mixture with the active enantiomer when the original patent expires. Use of biochemical catalysts (enzymes) is under study and chiral intermediates for drug synthesis are under development. The time needed to develop a new drug is increasing, a change attributed by some to the increasing complexity of the diseases under attack.

modest undesired biological activity. In such cases, the drugs may be marketed as racemic mixtures. In other cases, where one enantiomer elicits an intolerable biological response, it is necessary for scientists to use available technology to either separate the good enantiomer from the bad or to develop methods which allow the synthesis of only the good enantiomer. Pharmaceutical companies continue to invest hundreds of millions of dollars in developing these technologies to market enantiomerically pure drugs. It is interesting to note that many drugs currently in use are isolated from natural sources where they are almost always found as single enantiomers. This is because nature generally produces single enantiomers to the exclusion of other possible stereoisomers.

One might think the story of thalidomide is over. However, like many drugs, thalidomide has recently been shown to have other medical uses. It is currently marketed in the United States under the trade name Thalomid for use in treating people with HIV, certain neuroblastoma cancers, and leprosy, among other things. However, the teratogenic effect of thalidomide still exists and the situation became even more complicated when it was shown that the "good" enantiomer of thalidomide is actually converted into the teratogenic enantiomer in the body. Thus, even if the thalidomide is administered only as the pure enantiomer, the teratogenic effect will still be manifested. Fortunately, this interconversion of enantiomers in a biological system is not universal with all drugs.

For this reason, thalidomide is said to be the most regulated drug ever marketed, and women who are given the drug must agree in writing to use two separate forms of birth control so they will avoid the consequences of a thalidomide-influenced pregnancy. In spite of the potential for good that can come from thalidomide treatment in some cases, there are still people and organizations who take the approach that because it is known that there is great potential for harm from this drug, it should forever be banned from use. The benefits, according to these people, who include many thalidomide victims and their families, do not outweigh the risks under any circumstances.

How do you feel about this issue and its broader applicability to other drugs and technologies?

Often, the biological difference between enantiomers is one of activity or effectiveness, with no difference in toxicity. Aspartame (NutraSweet), widely used as an artificial sweetener, has two enantiomers. However, one enantiomer has a sweet taste, while the other enantiomer is bitter. This indicates that the receptor sites on our taste buds must be chiral, since they respond differently to the "handedness" of aspartame enantiomers. This becomes clearer when looking at the properties of the simple sugars. D-Glucose is sweet and nutritious, whereas L-glucose is tasteless and cannot be metabolized by the body.

Metabolism is a general term for the sum of all the chemical and physical processes in a living organism. When something is "metabolized," it is changed by these processes

---

EXAMPLE 15.1   Chiral Molecules

For each of the following molecules, decide whether the underlined carbon atom is or is not a chiral center: (a) $\underline{C}H_2Cl_2$, (b) $H_2N-\underline{C}H(CH_3)-COOH$, and (c) $Cl-\underline{C}H(OH)-CH_2Cl$.

SOLUTION

To be a chiral center, an atom must be bonded to four different groups. The underlined carbon atoms in molecules (**b**) and (**c**) meet this condition and are chiral centers. The underlined carbon in (**a**) is bonded to a pair of H atoms and a pair of Cl atoms and is therefore not chiral.

TRY IT 15.1

Which of the following molecules is chiral? Draw the enantiomers for any chiral molecule.

(a)   (b)   (c)   (d)

Ibuprofen

15.2   CARBOHYDRATES

The word "carbohydrate" literally means "hydrate of carbon." Carbohydrates have the general formula $C_x(H_2O)_y$, in which x and y are whole numbers. However, even though the reaction of the carbohydrate sucrose with sulfuric acid produces carbon (Figure 15.4), this does not mean that sucrose is a

FIGURE 15.4 Reaction of sulfuric acid with sugar. When sulfuric acid is poured onto the sugar, carbon can be seen forming as water is lost from the sugar.

Hydrolysis is the term used for chemical reactions in which chemicals are decomposed through their reaction with water.

Simple sugars such as glucose are optical isomers, as indicated by the "D" prefix in D-glucose.

Carbohydrates Biomolecules composed of simple sugars (monosaccharides, $C_6H_{12}O_6$) and including disaccharides, starch, and cellulose

simple combination of carbon and water. The carbon, hydrogen, and oxygen in carbohydrates are arranged primarily into three organic functional groups

$$—O—H \qquad \overset{\overset{\displaystyle O}{\parallel}}{—C—H} \qquad \overset{\overset{\displaystyle O}{\parallel}}{—C—}$$

Alcohol hydroxy group   Aldehyde group   Ketone group

Carbohydrates are divided into groups depending on how many monomers are combined by condensation polymerization: *monosaccharides* (Latin *saccharum*, "sugar"), *disaccharides, trisaccharides* (etc.), and *polysaccharides*. Monosaccharides are simple sugars that cannot be broken down into smaller carbohydrate units by acid hydrolysis. In contrast, hydrolysis of a disaccharide or trisaccharide yields two or three monosaccharides (either the same or different), while complete hydrolysis of a polysaccharide produces many monosaccharides (sometimes thousands of them).

Carbohydrates, which make up about half the average human diet, form an essential part of the energy cycle for living things (Section 15.10). Besides energy storage in plants and energy production in animals, carbohydrates serve many other biological purposes. Cellulose is the main structural component of plants. The nucleic acids (Section 15.11) incorporate carbohydrate units in their repeating structure.

Monosaccharides

The most common simple sugar is D-glucose (also known as dextrose, grape sugar, and blood sugar), which is found in fruit, blood, and living cells. A solution of glucose is often given intravenously when a source of quick energy is needed to sustain life. D-Glucose, along with D-galactose and D-fructose (fruit sugar), are the three common monosaccharides found in the body. They have different structures but the same molecular formula, $C_6H_{12}O_6$. D-Galactose and D-glucose are stereoisomers of each other and differ at only a single chiral center.

D-Glucose   D-Galactose   D-Fructose

Planar representations of the monosaccharides, as shown above, are frequently used for convenience, but it should be understood that the tetrahedral bonding of the carbon atoms in these molecules results in the three-dimensional structures shown below. This should be kept in mind when viewing planar representations of monosaccharides and similar or more complex molecules.

D-Glucose   D-Galactose

Disaccharides

Disaccharides result when two monosaccharides are joined together with the elimination of water to give compounds with the general formula, $C_{12}H_{22}O_{11}$. The three most common disaccharides are

Sucrose (from sugarcane or sugar beets), composed of a D-glucose monomer and a D-fructose monomer—table sugar;

Maltose (from starch), composed of two D-glucose monomers—used as a sweetener in prepared foods; and

Lactose (from milk), composed of a D-glucose monomer and a D-galactose monomer—used in formulating drugs and infant foods, in baking, and in making yeast.

The formula for these disaccharides ($C_{12}H_{22}O_{11}$) is not simply the sum of two monosaccharides, $C_6H_{12}O_6 + C_6H_{12}O_6$. A water molecule is eliminated as two monosaccharides are united to form the disaccharide. In the body, enzymes catalyze the breakdown (hydrolysis) of disaccharides to their monosaccharides.

Sucrose is produced in a high state of purity on an enormous scale—more than 80 million tons per year. About 40% of the world's sucrose production comes from sugar beets and 60% comes from sugarcane. A comparison of the sweetness of common sugars and artificial sweeteners relative to sucrose is given in Table 15.1. Honey, which is a mixture of the monosaccharides glucose and fructose, has been used for centuries as a natural sweetener for foods and is sweeter than sucrose, or cane sugar (Table 15.1). To convert cane

TABLE 15.1	Sweetness of Common Sugars and Artificial Sweeteners Relative to Sucrose
SUBSTANCE	**SWEETNESS RELATIVE TO SUCROSE AS 1.0**
Lactose	0.16
Galactose	0.32
Maltose	0.33
Glucose	0.74
Sucrose	1.00
Fructose	1.17
Aspartame*	180
Saccharin*	300
Splenda*	600

* Artificial sweeteners.

Sugar beets.

sugar into glucose and fructose requires treatment with acid or with a natural enzyme called "invertase." The product, known as *invert sugar*, is often used as a sweetener in commercial food products.

Artificial Sweeteners

Saccharin

Saccharin was the first common artificial sweetener. Saccharin passes through the body undigested and consequently has no caloric value. It has a somewhat bitter aftertaste that is offset in commercial products by the addition of small amounts of naturally occurring sweeteners. Such products do have a small caloric value because of the natural sweeteners added.

Aspartame (NutraSweet), which has replaced saccharin as the principal artificial sweetener, is used in more than 3000 products and accounts for 75% of the one-billion-dollar worldwide artificial sweetener market. It is a dipeptide derivative made from aspartic acid and the methyl ester of phenylalanine. Aspartame can be digested, and its caloric value is approximately equal to that of proteins. However, since much smaller amounts of aspartame than of table sugar are needed for sweetness, many fewer calories are consumed in the sweetened food. Aspartame is not stable at cooking temperatures, which limits its use as a sugar substitute to cold foods and soft drinks.

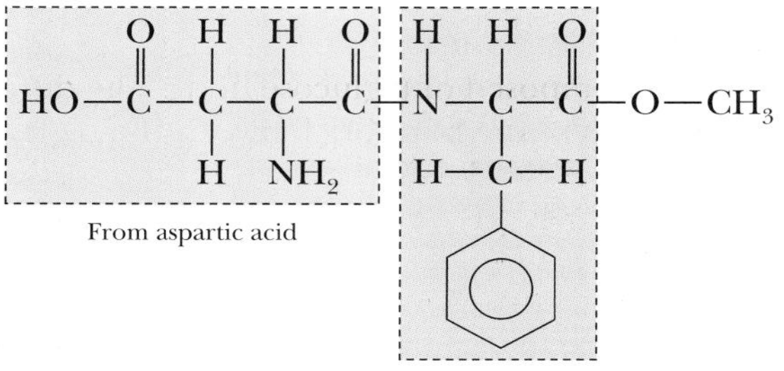

Aspartame

Sucralose (Splenda) is a relatively new artificial sweetener that was approved for use in the United States in 1998 after several years of use in other countries. It is purported to be as much as 600 times sweeter than sucrose. A comparison of its structure to that of sucrose reveals that sucralose is prepared by the replacement of three OH groups in sucrose by three Cl atoms. Like that of other artificial sweeteners, the safety of sucralose has been publicly questioned by some consumer advocate groups.

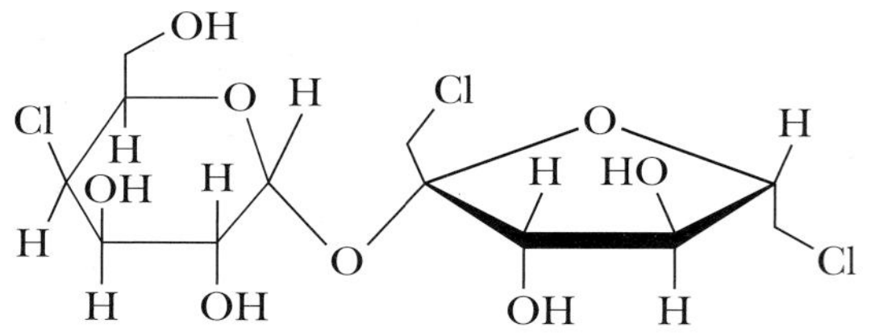

Sucralose (Splenda)

Polysaccharides

Nature's most abundant polysaccharides are the starches, glycogen, and cellulose. Some polysaccharides have more than 5000 monosaccharide monomers and molar masses over 1 million. The monosaccharide most commonly used to build polysaccharides is D-glucose.

Some foods that are high in starch.

Starches and Glycogen

Plant starch is found in protein-covered granules. If these granules are ruptured by heat, they yield a starch, *amylose*, that is soluble in hot water. Amylose constitutes about 25% of most natural starches.

Structurally, amylose is a straight-chain condensation polymer with an average of about 200 glucose monomers per molecule. Each monomer is bonded to the next with the loss of a water molecule, just as the two units are bonded in maltose. A representative portion of the structure of amylose is shown in Figure 15.5a.

Glycogen is an energy reservoir in animals, just as starch is in plants. Glycogen is stored in the liver and muscle tissues and is used for "instant" energy until the process of fat metabolism can take over and serve as the energy source.

Cellulose

Cellulose is the most abundant organic compound on Earth, and its purest natural form is cotton. The woody part of trees, the supporting material in plants and leaves, and the paper we make from them are also mainly cellulose. Like amylose, it is composed of D-glucose units. The difference between cellulose and amylose lies in the bonding between the D-glucose units (Figure 15.5b). This subtle structural difference in the connectivity between the glucose units in amylose and cellulose is the reason we cannot digest cellulose. Human beings do not have the necessary enzymes to hydrolyze cellulose but they do have enzymes that can recognize and hydrolyze the linkage between glucose units in amylose. On the other hand, termites, a few species of cockroaches, and ruminant mammals (such as cows, sheep, goats, and camels) are able to digest cellulose because bacteria living in their guts provide the necessary enzyme. D-Glucose can be obtained from cellulose by heating a suspension of the polysaccharide in the presence of a strong acid.

Small groups of cellulose are held together by hydrogen bonding.

(a) Amylose

(b) Cellulose

FIGURE 15.5 **(a) Amylose structure.** From 60 to 300 glucose units bond together in the manner shown. **(b) Cellulose structure.** From 900 to 6000 glucose units bond together in the pattern shown here.

15.3 LIPIDS

A **lipid** is an organic substance found in living systems that has limited solubility in water but is soluble in organic solvents. Because their classification is based on insolubility in water rather than on a structural feature such as a functional group, lipids vary widely in their structure and, unlike proteins and polysaccharides, are not polymers. Lipids include fats, oils, steroids, and waxes. The predominant lipids are fats and oils, which make up 95% of the lipids in our diet. The other 5% are steroids and several other lipids that are important to cell function.

Charles D. Winters

An assortment of vegetable oils.

Fats and Oils

Fats and oils are **triglycerides**—triesters of glycerol (glycerin) and fatty acids. The general equation for the formation of a triester of glycerol is

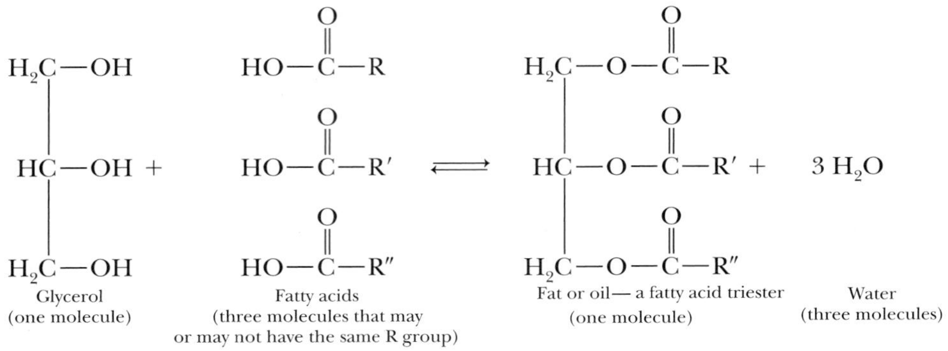

$$
\begin{array}{llll}
\text{H}_2\text{C—OH} & \text{HO—C(=O)—R} & \text{H}_2\text{C—O—C(=O)—R} \\
\text{HC—OH} + & \text{HO—C(=O)—R}' & \rightleftharpoons & \text{HC—O—C(=O)—R}' + & 3\,\text{H}_2\text{O} \\
\text{H}_2\text{C—OH} & \text{HO—C(=O)—R}'' & \text{H}_2\text{C—O—C(=O)—R}''
\end{array}
$$

Glycerol (one molecule) Fatty acids (three molecules that may or may not have the same R group) Fat or oil— a fatty acid triester (one molecule) Water (three molecules)

Lipids Class of biomolecules not soluble in water but soluble in organic solvents; includes fats, oils, steroids, and waxes

Triglycerides Triesters of glycerol and fatty acids; fats and oils

The three R groups can be the same or different groups within the same fat or oil, and they can be saturated or unsaturated. The most common fatty acids in fats and oils are listed in Table 15.2.

Fatty acids such as oleic acid, which contain only one double bond, are referred to as *monounsaturated acids*. One of the unsaturated acids, linoleic acid, is an *essential fatty acid*. The human body cannot produce this acid, but

TABLE 15.2 Common Fatty Acids in Fats and Oils

ACIDS		MP °C	SOURCE
Saturated (All Solids at Room Temperature)			
Lauric	$CH_3(CH_2)_{10}COOH$	44	Coconut oil
Palmitic	$CH_3(CH_2)_{14}COOH$	63	Animal and vegetable fats
Stearic	$CH_3(CH_2)_{16}COOH$	69	Animal and vegetable fats
Unsaturated (All Liquids at Room Temperature)			
Oleic	$CH_3(CH_2)_7CH{=}CH(CH_2)_7COOH$	4	Animal and vegetable fats
Linoleic*	$CH_3(CH_2)_4CH{=}CHCH_2CH{=}CH(CH_2)_7COOH$	−5	Linseed oil, cottonseed oil
Linolenic	$CH_3CH_2CH{=}CHCH_2CH{=}CHCH_2CH{=}CH(CH_2)_7COOH$	−11	Linseed oil

* An essential fatty acid that must be part of the human diet.

it is required for the synthesis of the *prostaglandins*, an important group of more than a dozen related compounds. The prostaglandins have potent effects on physiological activities such as blood pressure, relaxation and contraction of smooth muscle, gastric acid secretion, body temperature, food intake, and blood platelet aggregation. Many prostaglandins cause inflammation and fever. The fever-reducing effect of aspirin results from the inhibition of cyclooxygenase, the enzyme that catalyzes the synthesis of prostaglandins.

The term *fat* is usually reserved for *solid* triglycerides (such as butter and lard), and *oil* is the term for *liquid* triglycerides (olive, soybean, and corn oils, for example). The R groups in the fatty acid portions of fats are generally saturated, with only C—C single bonds. The R groups in oils are usually either monounsaturated (one C=C double bond) or polyunsaturated (two or more C=C double bonds). Since the C=C bonds interrupt the zigzag pattern of tetrahedral angles with 120° angles, the molecules are irregular in shape and do not pack together efficiently enough to form a solid. Table 15.3 illustrates the percentages of saturated and unsaturated fat found in common dietary oils and fats.

Hydrogen can be added catalytically to the double bonds of an oil to convert it into a semisolid fat. For example, liquid soybean and other vegetable oils are *hydrogenated* to produce cooking fats and margarine.

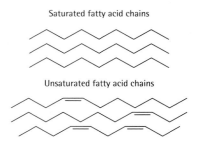

Saturated fatty acid chains

Unsaturated fatty acid chains

The hydrogenation process also forms unnatural *trans* fatty acids from natural *cis* fatty acids (see The World of Chemistry, "*Cis* and *Trans* Fatty Acids and Your Health").

$$H_2C—O—\underset{\overset{\|}{O}}{C}—(CH_2)_7CH=CH(CH_2)_7CH_3$$

$$HC—O—\underset{\overset{\|}{O}}{C}—(CH_2)_7CH=CH(CH_2)_7CH_3 \xrightarrow[200°C]{H_2, Ni}$$

$$H_2C—O—\underset{\overset{\|}{O}}{C}—(CH_2)_7CH=CH(CH_2)_7CH_3$$

Triolein (a liquid oil)

$$H_2C—O—\underset{\overset{\|}{O}}{C}—(CH_2)_7CH_2CH_2(CH_2)_7CH_3$$

$$HC—O—\underset{\overset{\|}{O}}{C}—(CH_2)_7CH_2CH_2(CH_2)_7CH_3$$

$$H_2C—O—\underset{\overset{\|}{O}}{C}—(CH_2)_7CH_2CH_2(CH_2)_7CH_3$$

Tristearin (a solid fat)

TABLE 15.3 Amounts of Saturated and Unsaturated Fatty Acids in Fats and Oils

DIETARY OIL/FAT	SATURATED FAT	POLYUNSATURATED FAT	MONOUNSATURATED FAT
Canola oil	6%	36%	58%
Safflower oil	9%	78%	13%
Sunflower oil	11%	69%	20%
Corn oil	13%	62%	25%
Olive oil	14%	9%	77%
Soybean oil	15%	61%	24%
Peanut oil	18%	34%	48%
Lard	41%	12%	47%
Palm oil	51%	10%	39%
Butterfat	66%	4%	30%
Coconut oil	92%	2%	6%

THE WORLD *of* CHEMISTRY

CIS AND TRANS FATTY ACIDS AND YOUR HEALTH

The direct link between diets high in saturated fats and heart disease is well known. Saturated fatty acids are usually found in solid or semisolid fats, whereas unsaturated fatty acids are usually found in oils. As a result, nutritionists recommend the use of liquid vegetable oils for cooking, and people are generally aware of this. However, what is less well known is that partially hydrogenated oils are a health hazard because of the formation of *trans* fatty acids during the process by which margarine and semisolid cooking fats are made. Hydrogen is added to the C=C double bonds in unsaturated fats (oils) to convert them to a solid or semisolid fat that has better consistency and less chance for spoilage. This hydrogenation process decreases the number of double bonds but also forms unnatural *trans* fatty acids from natural *cis* fatty acids. The *trans* fatty acids are not easily metabolized in the human system. Since the *trans* fatty acids are "straight" molecular structures, they pack together like the saturated fatty acids. By contrast, the *cis* fatty acids are bent and do not pack well [see (b) and (c) that follow].

You can reduce the health risks of *trans* fatty acids by not eating processed vegetable fats. How do you know which products were made with processed vegetable fats? Read the label. For example, a box of cookies or crackers may say on the label "made with 100% pure vegetable shortening . . . (partially hydrogenated soybean oil with hydrogenated cottonseed oil)." Although soybean oil and cottonseed oil are low in saturated fat, the hydrogenation process converts cottonseed oil to a saturated fat, and the partially hydrogenated soybean oil contains *trans* fatty acid.

Some of the many products that contain partially hydrogenated vegetable oils.

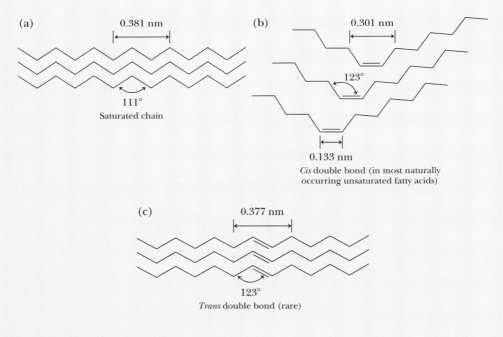

(a)

0.381 nm

111°
Saturated chain

(b)

0.301 nm

123°

0.133 nm

Cis double bond (in most naturally occurring unsaturated fatty acids)

(c)

0.377 nm

123°

Trans double bond (rare)

Saturated fatty acids (a) and *trans* fatty acids (c) pack more tightly than *cis* fatty acids (b).

If it is better to consume unsaturated fats instead of saturated ones, why do food companies hydrogenate oils to reduce their unsaturation? There are several answers to this question. First, the double bonds in the fatty acid are reactive, and oxygen can attack the fat at these bonds. When the oil is oxidized, unpleasant odors and flavors develop. Hydrogenating an oil reduces the likelihood that the food will oxidize and become rancid. Second, hydrogenating an oil makes it less liquid. There are many times when a food processor needs a solid fat in a food to improve its texture and consistency. For example, if liquid vegetable oil were used in a cake icing, the icing would slide off the cake. Instead of using animal fat, which also contains cholesterol, the manufacturer turns to a hydrogenated or partially hydrogenated oil.

Steroids

Steroids are found in all plants and animals and all contain the following four-ring skeleton.

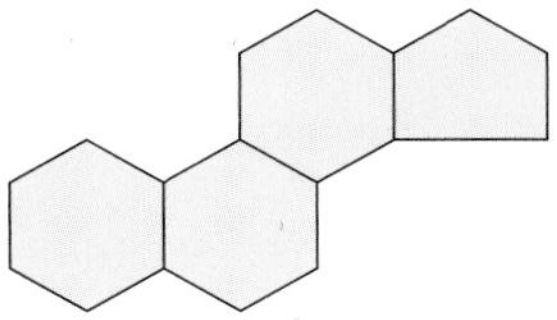

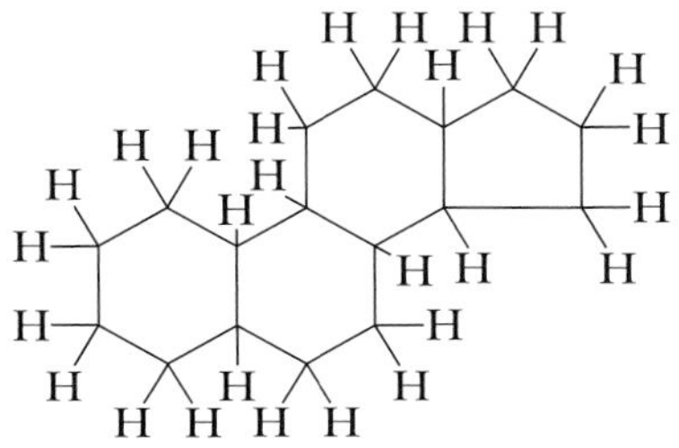

The skeletal four-ring structural drawing on the left is chemical shorthand similar to that described for cyclic hydrocarbons in Section 12.8. There is a carbon at each corner, and the lines represent C—C bonds. Since every carbon atom forms four bonds, additional bonds between carbon atoms and hydrogen atoms are understood to be present whenever the skeletal structure shows fewer than four bonds. The structure on the right shows the hydrogen atoms understood to be present in the four-ring structure shown on the left. Although all the rings in the skeletal drawing are shown as saturated rings, steroids often have one ring that is unsaturated or aromatic. For example, cholesterol has one double bond in the second ring. Note that its structure also includes alkyl groups and an alcohol group. These groups replace hydrogen atoms in the skeletal representation.

Aromatic means a benzene-like structure.

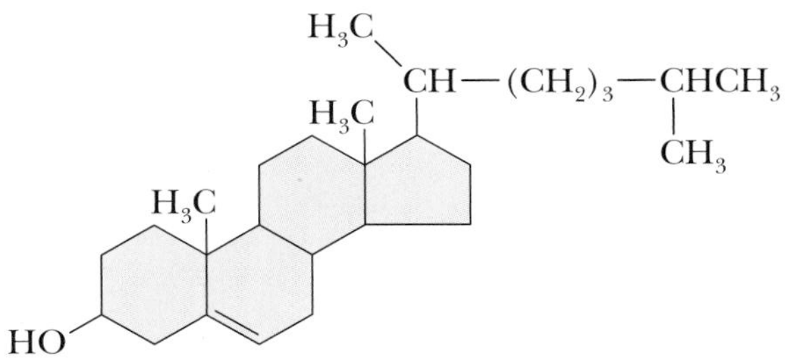

Heinrich Otto Wieland (1877–1957) and Adolf Wintaus (1876–1959) received Nobel Prizes in chemistry in 1927 and 1928, respectively, for their work leading to the determination of the structure of cholesterol.

Cholesterol

Cholesterol is the most abundant animal steroid. The human body synthesizes cholesterol and readily absorbs dietary cholesterol through the intestinal wall. Therefore, a diet with absolutely no cholesterol does not preclude the presence of cholesterol in the body. An adult human contains about 250 g of cholesterol. Cholesterol receives a lot of attention because high blood

Steroids Lipids with a four-ring structure; they include cholesterol and male and female sex hormones

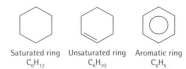

Saturated ring   Unsaturated ring   Aromatic ring
C_6H_{12} C_6H_{10} C_6H_6

cholesterol levels are associated with heart disease (Section 17.11). Proper amounts of cholesterol are essential to our health, because cholesterol undergoes biochemical modification to give milligram amounts of many important **hormones**, such as vitamin D, cortisone, and sex hormones.

Sex Hormones

Cholesterol is the starting material (the precursor molecule) for the synthesis of steroid sex hormones. One female sex hormone, *progesterone*, differs only slightly in structure from the male hormone, *testosterone*. In fact, relatively minor structural differences between many female and male sex hormones account for significantly different biological activities.

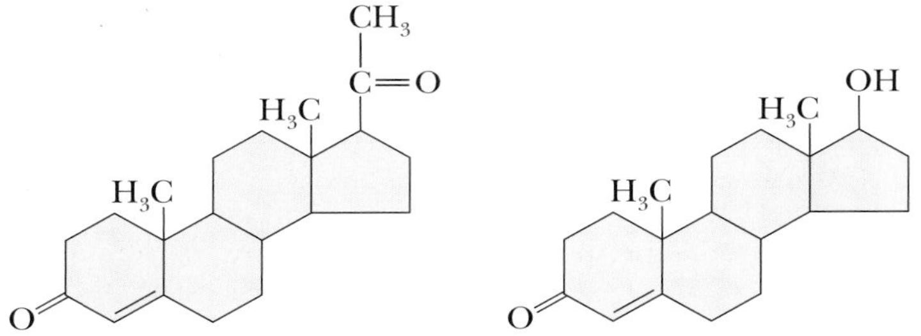

Progesterone

Testosterone

Other female hormones are estradiol and estrone, together called *estrogens*. The estrogens contain an aromatic ring, which differentiates them from the steroids progesterone and testosterone.

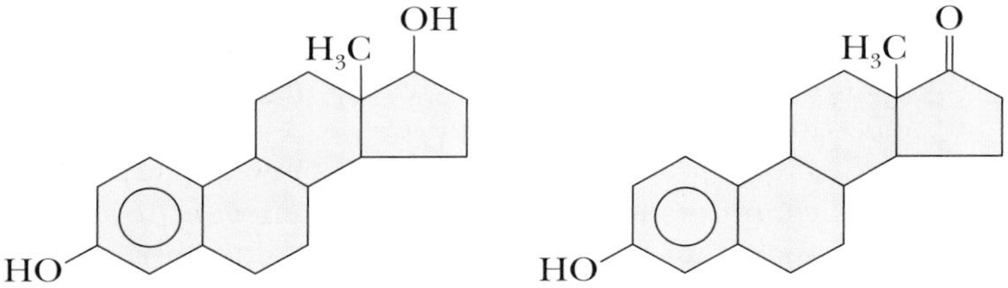

> The term *estrogens* is used to collectively refer to female sex hormones.

Estradiol

Estrone

The estrogens and progesterone are produced by the ovaries. Estrogens are important to the development of the egg in the ovary, whereas progesterone causes changes in the wall of the uterus and, after fertilization, prevents release of a new egg from the ovary (ovulation). Birth control drugs use derivatives of estrogens and progesterone to simulate the hormonal process resulting from pregnancy and thereby prevent ovulation (Section 17.4).

Testosterone and *androsterone* are the two most important male sex hormones, known as *androgens*. They are synthesized in the testes from cholesterol and are responsible for the development of male secondary sex characteristics and for promoting muscle and tissue growth.

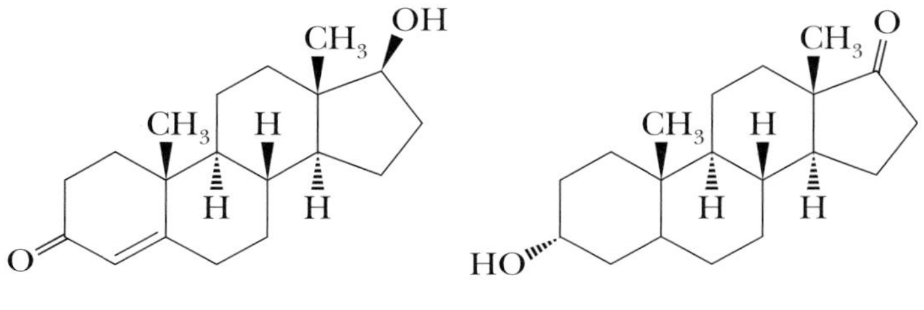

> The term *androgens* is used to collectively refer to male sex hormones.

Hormones Chemical messengers secreted in the bloodstream by endocrine glands

Testosterone

Androsterone

Androstenedione, another male sex hormone, also has muscle-building properties that have been exploited, both legally and illegally. In fact, these *anabolic* (muscle-building) properties have led to the development of a wide variety of synthetic (not naturally occurring) steroids to mimic the tissue-building effects of the natural male sex hormones.

Androstenedione   Methandrostenolone
(Dianabol)

Synthetic steroids such as methandrostenolone (Dianabol) have been abused by professional athletes and, increasingly, amateur athletes as young as junior high school age to gain a competitive advantage. The variety of such steroids readily available on the open market and the Internet, without a prescription in many cases, is frightening when one considers the possible consequences of abuse of these powerful chemical agents. Among the side effects of steroid abuse in men are shrinking testes, enlarged breasts and feminization, increased balding, high blood pressure, unpredictable and rapid mood changes (sometimes known as "roid rage"), and even death. Women who abuse these steroids tend to become very masculine, often to the point of being visually almost unrecognizable as women. The competitive advantages to be gained often seem to outweigh the risks of steroid abuse in the minds of many athletes. *Sports Illustrated* magazine did a survey of world-class athletes regarding the use of banned steroids in competition. Over 90% of these athletes said they would use the banned drugs if they knew they would win all of their competitions and were guaranteed to go uncaught. A follow-up question revealed that 50% would still do so if they knew they would not be caught and would win all of their competitions for a five-year period, even if told they would die from the side effects of using the banned drugs after only five years.

Waxes

Waxes are esters formed from long-chain (16 or more carbon atoms) fatty acids and long-chain alcohols. The general formula of a wax is the same as that of a simple ester, RCOOR′, with the qualification that R and R′ are limited to alkyl groups with a large number of carbon atoms. Natural waxes are usually mixtures of several esters. Wax coatings on leaves help to protect the leaves from disease and also help the plant to conserve water. The feathers of birds are also coated with wax. Our ears are protected by wax. Several natural waxes have been used in consumer products. These include carnauba wax (from a Brazilian palm tree), which is used in floor waxes, automobile waxes, and shoe polishes, and lanolin (from lamb's wool), which is used in cosmetics and ointments. Lanolin also contains cholesterol.

CONCEPT CHECK **15A**

1. To have optical isomers in carbon compounds, a carbon atom must have _____ different groups attached.
2. The complete hydrolysis of a polysaccharide yields _____.

3. When a molecule of sucrose is hydrolyzed, the products are one molecule each of the monosaccharides _____ and _____.

4. Starch and cellulose are condensation polymers built of _____ monomers.

5. _____ bonding holds polysaccharide chains together, side by side, in cellulose.

6. The sugar referred to as blood sugar, grape sugar, or dextrose is actually the compound _____.

7. Fats and oils are esters of _____ and _____.

8. The structural difference between a saturated fat and an unsaturated fat is _____.

9. Cholesterol is a (**a**) steroid, (**b**) protein, (**c**) sex hormone, (**d**) carbohydrate.

10. Cholesterol is essential to our health. (**a**) True, (**b**) False.

11. Waxes are esters of _____ and _____.

12. Androgens are _____ and estrogens are _____.

North Wind Picture Archives

Making soap the old-fashioned way.

15.4 SOAPS, DETERGENTS, AND SHAMPOOS

In strongly basic solutions, fats and oils undergo hydrolysis to produce glycerol and salts of fatty acids. Such reactions are called *saponification* reactions, and the sodium or potassium salts of the fatty acids formed are *soaps*.

$$
\begin{array}{l}
CH_3-(CH_2)_{16}-COO-CH_2 \\
CH_3-(CH_2)_{16}-COO-CH \quad + \; 3\,NaOH \longrightarrow 3\,CH_3-(CH_2)_{16}-COO^-Na^+ \; + \\
CH_3-(CH_2)_{16}-COO-CH_2
\end{array}
\qquad
\begin{array}{l}
HO-CH_2 \\
HO-CH \\
HO-CH_2
\end{array}
$$

| Tristearin (an animal fat) | Sodium hydroxide | Sodium stearate (a soap) | Glycerol (glycerin) |

Principal fats and oils for soap making are tallow from beef and mutton, coconut oil, palm oil, olive oil, bone grease, and cottonseed oil.

Lye is a term used to refer to sodium hydroxide and potassium hydroxide solutions which are highly alkaline.

—COO⁻ or —CO₂⁻ is known as carboxylate.

Pioneers prepared their soap by boiling animal fat with an alkaline solution obtained from the ashes of hardwood. The resulting "lye" soap could be "salted out" by adding sodium chloride, because soap is less soluble in a salt solution than in water. The crude soap made this way contained considerable caustic material (sodium hydroxide or potassium hydroxide) in addition to the soap molecules, but it did its job of cleaning quite well.

Substances that are water soluble can readily be removed from the skin or a surface by simply washing with an excess of water. To remove a sticky sugar syrup from your hands, you can dissolve the sugar in water and rinse it away. Many times the material to be removed is oily, and water will merely run over the surface of the oil. Since the skin has natural oils, even substances such as ordinary dirt that are not oily themselves can adhere quite strongly to the skin and to clothing containing these oils. The hydrogen bonding holding water molecules together is too large to allow the oil and water to intermingle (Figure 15.6), so something like soap is needed to loosen the dirt and wash it away.

The cleaning action of soap is explained by its molecular structure. When present in an oil–water system, the fatty acid anion in the soap moves to the interface between the oil and the water.

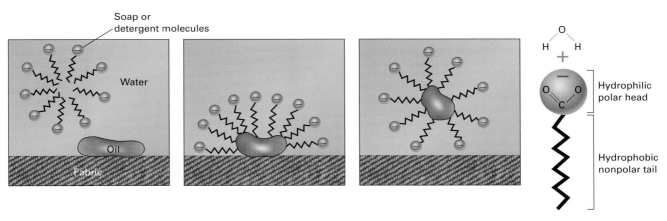

FIGURE 15.6   **How soaps and detergents work.** The hydrophilic, ionic ends of the molecules interact strongly with water. The hydrophobic, hydrocarbon ends of the molecules avoid water and are drawn to the oily portion of the dirt. As greasy dirt is broken up by agitation, the particles are surrounded and isolated from one another. This prevents them from coming together again and allows them to be carried away in the wash water.

$$\text{CH}_3\text{CH}_2\text{CH}_2\text{CH}_2\text{CH}_2\text{CH}_2\text{CH}_2\text{CH}_2\text{CH}_2\text{CH}_2\text{CH}_2\text{CH}_2\text{CH}_2\text{CH}_2\text{CH}_2\text{CH}_2\text{CH}_2\overset{\displaystyle O}{\overset{\displaystyle \|}{\text{C}}}\text{O}^-\text{Na}^+$$
Sodium stearate molecule

The molecular structure of a soap or synthetic detergent molecule consists of a long oil-soluble (**hydrophobic**, meaning not liking, or fearing water) group and a water-soluble (**hydrophilic**, meaning liking or loving water) group. The hydrophobic, nonpolar hydrocarbon chain or tail mixes readily with the non-polar oil or grease molecules, whereas the highly polar hydrophilic carboxylate head group is attracted to water molecules (Figure 15.6) where it can be solubilized by hydrogen bonding with the water molecules.

One undesirable property of soaps is their tendency to form precipitates with Ca^{2+}, Mg^{2+}, and Fe^{2+} ions found in "hard" water. The resulting fatty-acid salts of these doubly positive ions are not as soluble in water as the Na^+ ion salts. These less-soluble molecules appear as a scum that sticks to laundry and bathtubs, often containing trapped dirt, which makes it appear even worse.

Synthetic detergents are derived from organic molecules designed to have even better cleaning action than soaps but less reaction with the doubly positive ions found in hard water. As a consequence, synthetic detergents are often more economical to use and are more effective in hard water than soap. There are many different synthetic detergents on the market. An inventory of cleaning materials in a typical household might include half a dozen or more formulated products designed to be the most suitable for a specific job—cleaning skin, hair, clothes, floors, or the family car.

Synthetic detergents generally have a benzene ring at one end. This allows greater variability in the types of head groups.

$$\text{CH}_3\text{CH}_2\text{CH}_2\text{CH}_2\text{CH}_2\text{CH}_2\text{CH}_2\text{CH}_2\text{CH}_2\text{CH}_2\text{CH}_2\text{CH}_2\text{CH}_2\text{CH}_2\!-\!\!\bigcirc\!\!-\!\text{SO}_3^-\text{Na}^+$$

Oil-soluble tail   Water-soluble head
(hydrophobic; non-polar)   (hydrophilic; polar)
A typical synthetic detergent molecule

Typical hydrophilic head groups include negatively charged sulfate ($-\text{OSO}_3^-$), sulfonate ($-\text{SO}_3^-$), and phosphate ($-\text{OPO}_3^{2-}$) groups. Compounds with these groups are *anionic surfactants*.

Stearate soaps include sodium cations—hard soap; potassium cations—soft soap; and ammonium cations—liquid soap.

Floating soaps float because of trapped air.

Hydrophobic Water fearing; does not dissolve in water

Hydrophilic Water loving; dissolves in water

Most of us have a shelf like this. We rely on a wide variety of chemical surfactants to keep things clean.

Charles D. Winters

Cationic (positively charged) *surfactants* are almost all quaternary ammonium halides (four groups attached to the central nitrogen atom) with the general formula

$$R_1-\overset{\overset{\textstyle R_2}{|}}{\underset{\underset{\textstyle R_3}{|}}{N}}{}^+\!-R_4\;X^-$$

where one of the R groups is a long hydrocarbon chain and another frequently includes an —OH group. The X^- in the formula represents a halide ion such as chloride (Cl^-) or bromide (Br^-) ion.

Shampoos are generally more complex formulations than simple soap solutions, with a number of ingredients to satisfy different requirements for maintaining clean and healthy-looking hair. If you use soap to wash your hair, incomplete rinsing will result in a "soap film" on the hair, which makes it appear dull. If soap is used to wash hair in hard water, a very noticeable film can usually be seen and is very difficult to remove. In addition, soaps tend to produce solutions with basic pH values. These harsh conditions are damaging to the hair.

Most shampoos contain anionic detergents. These give the product good foaming characteristics because anionic detergents generally foam more than cationic and nonionic detergents. Nonionic surfactants, such as the products obtained by reacting diethanolamine and lauric acid, are often also present in shampoos. While not as good surfactants as anionic surfactants, the nonionics are useful as thickeners and foam stabilizers. They make a shampoo pour from the bottle more slowly and cause the lather to remain thick for a longer period.

A little knowledge of chemistry goes a long way when it comes to shampooing your hair, doing the dishes, or doing the laundry. Anyone who has ever tried to wash oily, dirty hair using only water quickly reaffirms the old adage

Lather has nothing to do with cleaning efficiency, but consumers have been taught to expect a good lather whenever they use shampoo.

$$\underset{\text{Diethanolamine}}{HN(CH_2CH_2OH)_2} + \underset{\text{Lauric acid}}{CH_3(CH_2)_{10}COOH} \longrightarrow$$

$$\underset{\substack{\text{Lauric diethanolamide}\\\text{(an amide detergent)}}}{CH_3(CH_2)_{10}-\overset{\overset{\textstyle O}{\|}}{C}-N(CH_2CH_2OH)_2} + H_2O$$

that oil and water don't mix because the hair remains quite oily upon drying. However, the addition of a little shampoo solves the problem. Anionic detergents and soaps form roughly spherical clusters when placed in water. These spherical clusters (Figure 15.6), known as *micelles*, may contain hundreds of individual detergent or soap molecules that cluster together with the hydrophobic tails pointing toward the interior of the cluster and the hydrophilic head groups on the surface of the cluster where they are solubilized by hydrogen bonding to the many water molecules with which they are in contact. Manual agitation of the shampoo through the hair allows the hydrophobic oils and dirt on the hair to become solubilized in the hydrophobic, oily interior of the micelles. These oils and dirt are then flushed away when the hair is rinsed in running water.

Similar processes take place when detergents are used to remove grease and oils from dishes or clothing. Some clothing stains do not lend themselves to this type of chemical removal and other methods of cleaning must be tried.

Hair is more manageable, has a better sheen, and has less tendency to attract static charges (causing "fly-away" hair) if all the shampoo is removed after washing. An anionic detergent can be removed from the hair by using a *rinse*, or conditioner, containing a dilute solution of a cationic detergent, usually a quaternary ammonium compound, which electrically attracts the anions and facilitates their removal. The positive charged end of the quaternary ammonium surfactant also neutralizes negative charges on damaged hair ($—COO^-$ groups from disrupted protein chains), while the alkyl chain attaches to the hair and gives it a smooth feel.

> Caution should be taken with a cationic rinse because it can irritate the eyes.

An after-shampoo conditioner also attempts to put some of the oils that were removed by the detergent back into the hair. A typical conditioner has a water–alcohol dispersing medium as well as skin softeners, oils, waxes, resins, and even short amino acid polymers that can adhere to the hair to produce a more pliable and elastic fiber that is not as likely to become dry or be affected by atmospheric conditions. Holding the correct amount of moisture is the key to hair control, because too much water causes the hair to be limp and too little causes the individual hairs to attract static charge. Although many hair preparations make direct or indirect claims that various proteins and other beneficial ingredients can penetrate the hair and "repair" and even strengthen it, there is no scientific evidence for these claims. Protein molecules, for example, are simply too large to pass through the surface of the hair. Only much smaller molecules can do this. In fact, *if* hair preparations did function in this way, they would have to be classified as drugs by the Food and Drug Administration (FDA).

Lanolin and mineral oil (or their substitutes) are often added to shampoos to replace the natural oils in the scalp, thus preventing it from drying out and scaling. The presence of oil additives and stabilizers sometimes gives the shampoo a pearlescent appearance. These ingredients also make the shampoo less foamy, a quality that is popular in European countries.

15.5   CREAMS AND LOTIONS

If dry skin is treated with an oily substance after washing, the skin will be protected until enough natural oils have been regenerated. The oily substance used can be derived from animal oils, vegetable oils, or even oils from petroleum (the mineral oils). It is not uncommon to see some kind of rare animal oil, such as mink oil, used in a skin preparation; yet the oil from a mink is not any more effective at holding the skin's moisture than is a less expensive vegetable or mineral oil. When choosing a skin product containing an oil, perhaps a more important factor than the kind of oil is whether it will be soothing or irritating to the skin.

Any substance that holds moisture in the skin can be called a *moisturizer*. Since these substances are all oily in nature, they can lubricate the skin (restore the feeling of normal oiliness) and have the noticeable effect of softening and soothing the skin. These substances are also called *emollients*. All the oils listed in the previous paragraph that help to hold moisture in the skin would be called emollients.

Getting an emollient distributed evenly on your skin is not as easy as it may seem. Instead of just pouring oil over your body after a shower (this would leave your skin too oily), it would be better to use a mixture containing the emollient. Two kinds of mixtures are commonly used: *creams* and *lotions* are made by mixing an oily component with water and other ingredients in the right proportions to form a stable mixture that can be more like a solid (a cream) (Figure 15.7) or more like a liquid (a lotion). Mixtures like these are called **emulsions**. Emulsions consist of two substances that would

Charles D. Winters

FIGURE 15.7 Cold cream, which has a thicker consistency than a lotion.

Emulsions Stable mixtures of water and an oily component; may have a creamy appearance

normally not mix, such as an oil and water, but that are made to mix by an **emulsifying agent**, a compound whose molecules have a part that is soluble in water and a part that is soluble in oil. With its dual solubility, the emulsifying agent stabilizes the mixture. Thus, a cream can contain a rather high percentage of an oil and not feel oily or greasy because of the presence of the water. When the cream is applied to the skin, some of the water is absorbed by the skin, and the oil (emollient) remains on the surface to hold in the moisture.

Emulsions are examples of **colloids**, which are quite common. Fog, foams, foods such as milk, and aerosol sprays are all colloids (Table 15.4). Colloidal mixtures differ from solutions. The colloid particles (called the *dispersed phase*) distributed in the solvent-like medium (called the *continuous phase*) are much larger than the molecules or ions that are the solutes in true solutions. Two kinds of emulsions can be formed between oil and water: oil droplets of colloid size dispersed in water and water droplets of colloid size dispersed in oil. As you might expect, oil-in-water emulsions have more of the properties of an aqueous solution, while the water-in-oil emulsions have more oil-like properties; for example, they tend to feel more greasy. An oil-in-water emulsion has tiny droplets of an oily or waxy substance dispersed throughout a water medium; homogenized milk is an example. A water-in-oil emulsion has tiny droplets of a water solution dispersed throughout an oil; examples are natural petroleum and butter. An oil-in-water emulsion can be washed off the skin surface with tap water, whereas a water-in-oil emulsion gives skin a greasy, water-repellent surface that resists being washed off by running water. With careful formulation of the emulsion, a *barrier cream* can be made that will effectively resist aqueous solutions that might contain harmful ingredients.

The ingredient listed first on a cream or lotion label is an indication of the kind of emulsion present. If water is listed first, the emulsion is probably an oil-in-water type. If an oil is listed first, the emulsion is probably the water-in-oil type (Figure 15.8). A lotion might contain the same emollient as a cream but contain a different emulsifier and a different ratio of water to oil. Whether an emollient is distributed on the skin in a cream or a lotion

---

The name *colloid* comes from the Greek word meaning "glue." A close look at a typical glue will show that there are large particles dispersed in water. These particles are large, colloidal-sized aggregates of similar molecules.

The molecules in the continuous phase of colloidal mixtures can have molar masses as high as several hundred thousand.

Emulsifying agent A compound that has a water-soluble part as well as an oil-soluble part and that stabilizes an emulsion

Colloids Particles larger than most molecules or ions and dispersed in a solvent-like medium

---

TABLE 15.4		Types of Colloids	
CONTINUOUS PHASE	**DISPERSED PHASE**	**TYPE**	**EXAMPLES**
Gas	Liquid	Aerosol	Fog, clouds, aerosol sprays
Gas	Solid	Aerosol	Smoke, airborne viruses, automotive exhaust
Liquid	Gas	Foam	Shaving cream, whipped cream
Liquid	Liquid	Emulsion	Mayonnaise, milk, face creams
Liquid	Solid	Solution	Milk of magnesia, mud
Solid	Liquid	Gel	Jelly, cheese, butter
Solid	Solid	Solid solution	Milk glass, some gemstones, many alloys such as steel

Note: Most colloids will separate unless stabilized. Food and cosmetic emulsions, foams, and aerosols are usually stabilized with emulsifying agents to give them a long shelf life and consistent properties, such as color and texture, throughout their life.

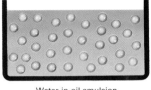

Water-in-oil emulsion

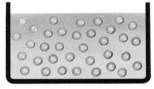

Oil-in-water emulsion

FIGURE 15.8 Two moisturizing lotions, one with a water-in-oil emulsion (Nivea) and the other an oil-in-water emulsion (Keri).

depends more on how the product is perceived by the user than on which kind of product is more effective. All creams and lotions have "shelf lives." Creams and lotions, like all colloids, can tend to "settle out" over a period of time, a property not observed in solutions.

CONCEPT CHECK 15B

1. To make soap, a fat is treated with _____.
2. The hydrocarbon end of a soap or detergent molecule is (hydrophilic, hydrophobic) while the polar end is (hydrophilic, hydrophobic).
3. Which is more likely to precipitate the hard-water ions (Ca^{2+}, Mg^{2+}, Fe^{3+}) as a sticky precipitate? (**a**) Traditional soaps (**b**) Synthetic detergents.
4. Another name for a skin softener is a(n) _____.
5. A skin cream is either an oil-in-water or water-in-oil _____.
6. An example of a foam colloid is _____.
7. An example of a solid aerosol colloid is _____.
8. An example of a liquid aerosol colloid is _____.
9. Milk is an example of a(n) _____ colloid.

15.6 AMINO ACIDS

All proteins are condensation polymers of **amino acids**. A large number of proteins exist in nature. For example, the human body is estimated to have 100,000 different proteins. It is amazing that all these proteins are derived from only 20 different amino acids (Table 15.5). Even more amazing is

Amino acids Biomolecules containing an alpha–amino group and a carboxyl group; building blocks of proteins

TABLE 15.5 Common L-Amino Acids Found in Proteins (R Groups Highlighted)

AMINO ACID	ABBRE-VIATION	STRUCTURE
		Nonpolar R Groups
Glycine	Gly	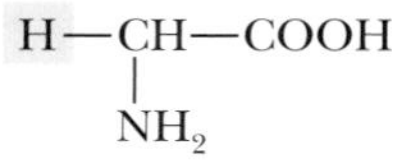 H—CH—COOH / NH$_2$
Alanine	Ala	CH$_3$—CH—COOH / NH$_2$
Valine*	Val	CH$_3$—CH—CH—COOH / CH$_3$ NH$_2$
Leucine*	Leu	CH$_3$—CH—CH$_2$—CH—COOH / CH$_3$ NH$_2$
Isoleucine*	Ile	CH$_3$—CH$_2$—CH—CH—COOH / CH$_3$ NH$_2$
Proline	Pro	H$_2$C—CH$_2$ / H$_2$C CHCOOH / N / H
Phenylalanine*	Phe	⬡—CH$_2$—CH—COOH / NH$_2$
Methionine*	Met	CH$_3$—S—CH$_2$CH$_2$—CH—COOH / NH$_2$
Tryptophan*	Trp	(indole)CH$_2$—CH—COOH / NH$_2$ / N / H
		Polar but Neutral R Groups
Serine	Ser	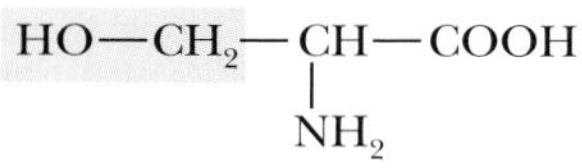 HO—CH$_2$—CH—COOH / NH$_2$
Threonine*	Thr	CH$_3$—CH—CH—COOH / OH NH$_2$
Cysteine	Cys	HS—CH$_2$—CH—COOH / NH$_2$
Asparagine	Asn	H$_2$N—C—CH$_2$—CH—COOH / ‖ / O NH$_2$

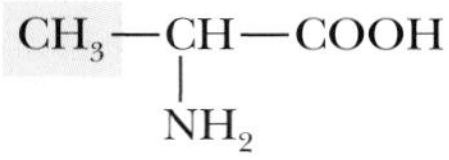

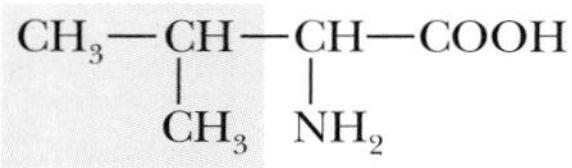

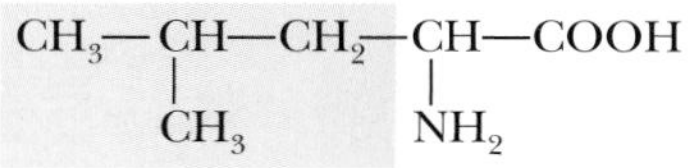

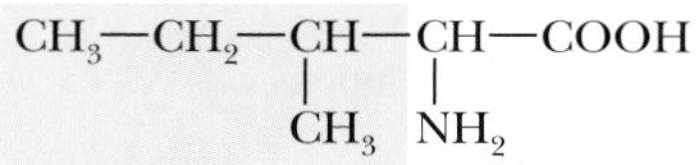

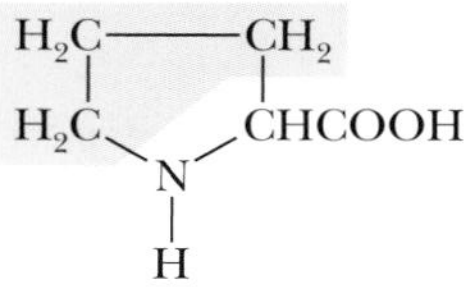

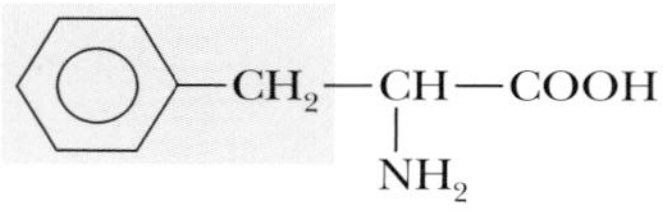

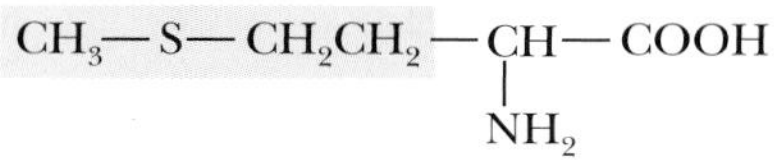

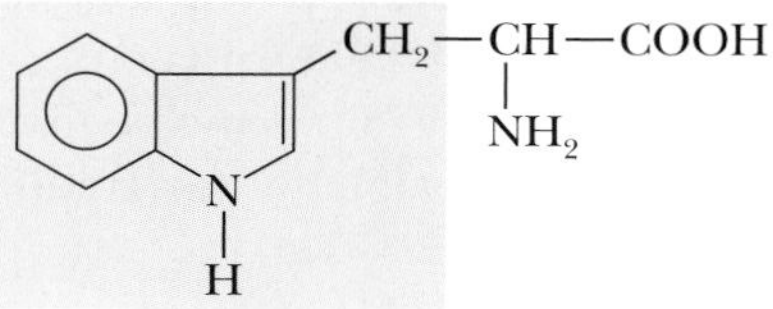

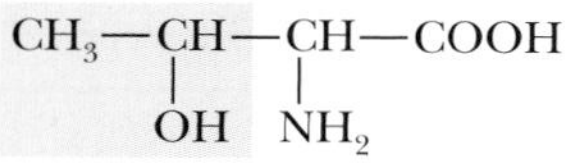

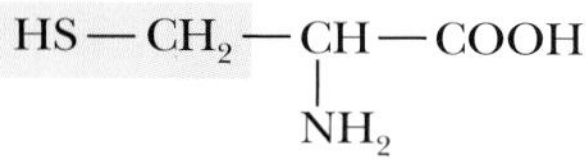

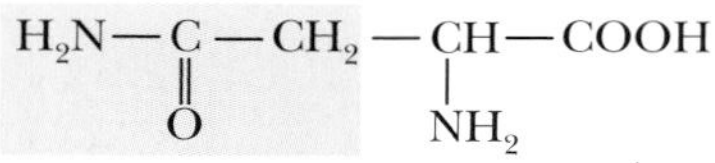

TABLE 15.5	Continued	

AMINO ACID	ABBRE-VIATION	STRUCTURE
		Polar but Neutral R Groups (cont.)
Glutamine	Gln	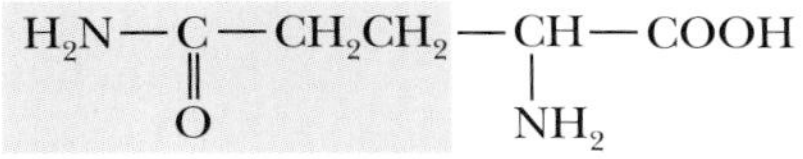 H₂N—C—CH₂CH₂—CH—COOH (C double bonded to O; CH bonded to NH₂)
Tyrosine	Tyr	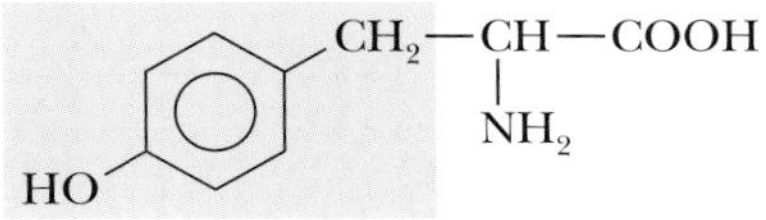 (benzene ring with HO) CH₂—CH—COOH (CH bonded to NH₂)
		Acidic R Groups
Glutamic acid	Glu	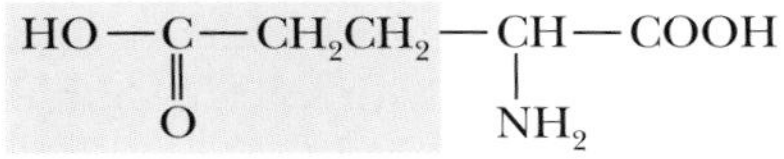 HO—C—CH₂CH₂—CH—COOH (C double bonded to O; CH bonded to NH₂)
Aspartic acid	Asp	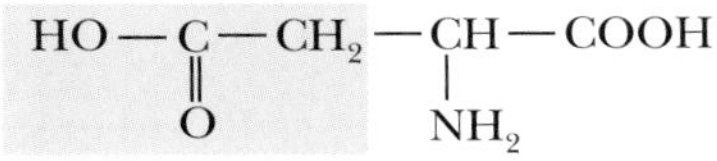 HO—C—CH₂—CH—COOH (C double bonded to O; CH bonded to NH₂)
		Basic R Groups
Lysine*	Lys	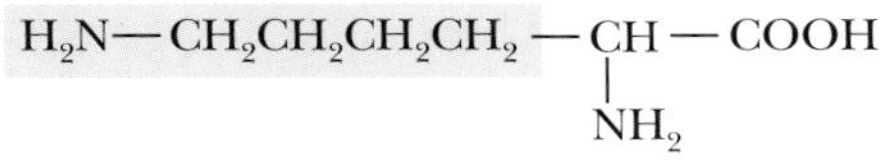 H₂N—CH₂CH₂CH₂CH₂—CH—COOH (CH bonded to NH₂)
Arginine†	Arg	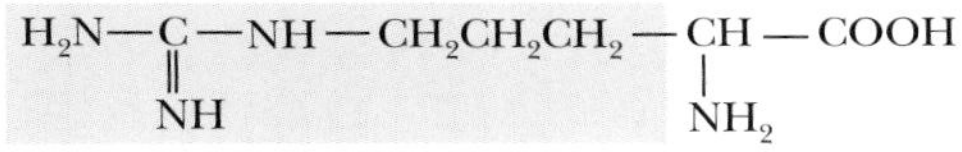 H₂N—C—NH—CH₂CH₂CH₂—CH—COOH (C double bonded to NH; CH bonded to NH₂)
Histidine	His	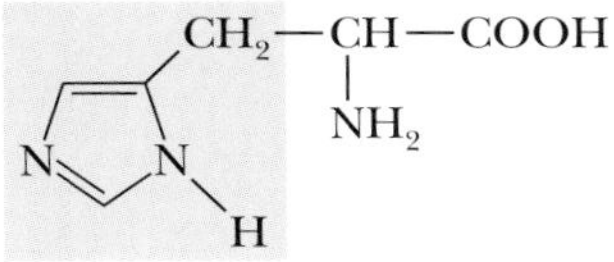 (imidazole ring with N, N—H) CH₂—CH—COOH (CH bonded to NH₂)

* Essential amino acids that must be part of the human diet. The other amino acids can be synthesized by the body.
† Growing children also require arginine in their diet.

nature's preference for only the L-enantiomer of these amino acids. All but one of the 20 amino acids found in nature have the general formula

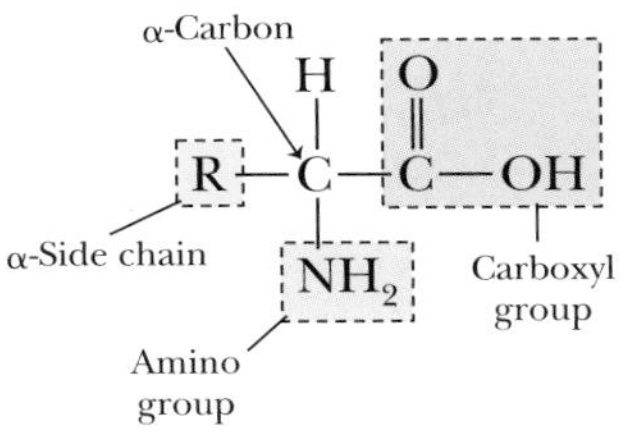

An amino acid

with an amino ($-NH_2$) group, a carboxylic acid group ($-COOH$), and an R group attached to the *alpha*-carbon (the first carbon next to the $-COOH$). R is a characteristic group for each amino acid (see Table 15.5), and the α-carbon is a chiral carbon atom, with one exception—glycine. R is a hydrogen atom in glycine, the simplest amino acid. Therefore, glycine is achiral (not chiral) and is the only naturally occurring amino acid that does not have enantiomers. The polarity of the R groups in amino acids affects the structure and function of proteins. The amino acids are grouped in Table 15.5 according to whether the R group is nonpolar, polar, acidic, or basic.

The **essential amino acids** must be ingested from food; they are indicated by asterisks in Table 15.5. A diet that includes meat, milk, eggs, or cheese provides all the essential amino acids. The other amino acids can be synthesized by the human body.

For good nutrition we require all the essential amino acids in our daily diet, but the amount required does not exceed 1.5 g per day for any of them.

15.7 PEPTIDES AND PROTEINS

How are amino acids polymerized to give proteins? The formation of an amide from the reaction of an amine and a carboxylic acid is described in Section 14.5 in the discussion of polyamides such as nylons.

$$R-\overset{\overset{\displaystyle O}{\|}}{C}-OH + H_2NR' \longrightarrow R-\overset{\overset{\displaystyle O}{\|}}{C}-\overset{\overset{\displaystyle H}{|}}{N}-R' + H_2O$$

Because amino acids have both an amine group and a carboxylic acid group, the —COOH of one amino acid can combine with the —NH_2 of a second amino acid.

A peptide bond

Names of peptides are written from left to right starting with the amino- or N-terminal end. The -ine ending of all amino acid residues (except the carboxyl or C-terminal end) is changed to -yl. For example, Gly-Ala-Ser is the tripeptide glycylalanylserine.

In the condensation reaction, one molecule of water is eliminated between the carboxylic acid of one amino acid and the amine group of another. The result is a **peptide bond** (called an *amide group* in simpler molecules), and the molecule is a *dipeptide*. When two different amino acids are linked by amide bonds, two different combinations are possible, depending on which amine reacts with which acid group. For example, when glycine and alanine react, both glycylalanine and alanylglycine can be formed. Either end of the dipeptide can then react with another amino acid.

Glycylalanine (Gly-Ala)   or   Alanylglycine (Ala-Gly)

Essential amino acids Amino acids that are not synthesized by the body and, therefore, must be obtained from the diet

Peptide bond Amide group that connects two amino acids; found in proteins

Polypeptide Condensation polymer of amino acids

Since each dipeptide has a —COOH and an —NH_2, group, a tripeptide can be formed from each dipeptide by reaction at either end, and the polymerization process can continue until a large **polypeptide** chain is formed (Figure 15.9). Note that, as above, peptides are always written with the N-terminal end at the left.

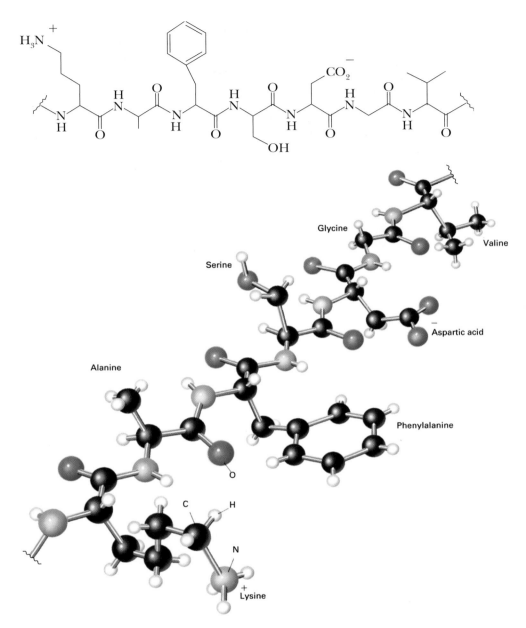

FIGURE 15.9 **Two different representations of the same polypeptide chain.** The color coding of atoms in the model is carbon (black), nitrogen (blue), oxygen (red), and hydrogen (white). Identify each peptide bond, the R groups of each amino acid, and the N-terminus and C-terminus in this fragment.

Proteins are polypeptides containing from 50 to thousands of amino acid residues, and they vary greatly in structure and composition. They can be divided into two classes: simple and conjugated. *Simple proteins* consist only of amino acids. Two examples of simple proteins are insulin and chymotrypsin. Insulin, a hormone that is essential to controlling the concentration of glucose in the blood, has 51 amino acid residues in two linked chains (Figure 15.10). Chymotrypsin, an enzyme that aids in the digestion of proteins in our diet, contains 245 amino acid residues.

Conjugated proteins are much more common than simple proteins and contain non-protein parts called *prosthetic groups*. Prosthetic groups are small non-protein molecules covalently bonded to the protein. Both myoglobin (Figure 15.11a) and hemoglobin (Figure 15.11b) contain a prosthetic group known as

Proteins Biomolecules that are polymers of amino acids; may also include nonamino acid parts

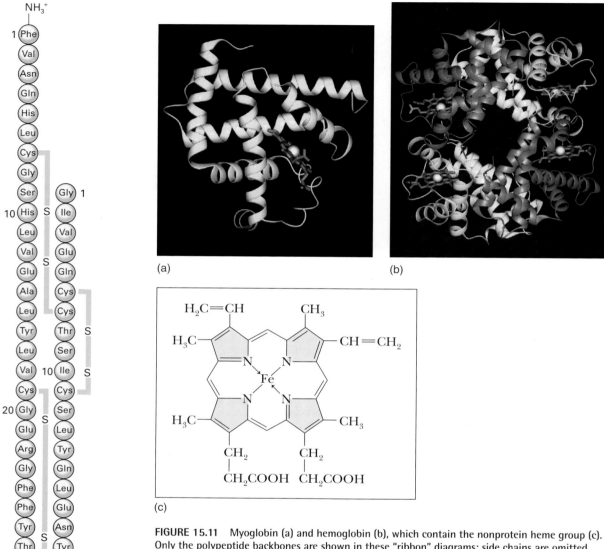

FIGURE 15.10 The amino acid sequence of insulin, which contains 51 amino acids in two chains linked by disulfide bonds (Section 15.9) between the side chains of cysteine residues.

FIGURE 15.11 Myoglobin (a) and hemoglobin (b), which contain the nonprotein heme group (c). Only the polypeptide backbones are shown in these "ribbon" diagrams; side chains are omitted. Myoglobin has a single protein chain (yellow) and one heme group (Fe ion shown in white). Hemoglobin has four protein chains of two different kinds (yellow and purple), with one heme group in each protein chain. In this type of representation of proteins, the alpha-helix parts of the protein chain are shown as ribbon-like spirals.

a *heme group* (Figure 15.11c). Myoglobin, which binds oxygen in muscles, consists of a single protein chain (Figure 15.11a) with one heme group (Figure 15.11c). Hemoglobin binds oxygen in blood. Human hemoglobin contains four protein chains (Figure 15.11b), each of which contains a heme group (Figure 15.11c). Two of these four chains are identical with 141 amino acid residues and the other two are identical with 146 amino acid residues. In both myoglobin and hemoglobin, the oxygen binds reversibly with an iron ion (Fe^{2+}) in the center of the heme group. Carbon monoxide binds to the iron in the heme of hemoglobin much more strongly than does oxygen. Hemoglobin complexed with carbon monoxide cannot carry the oxygen necessary for metabolism and death may result in severe cases of carbon monoxide poisoning.

Proteins known as enzymes (Section 11.8) often have small non-protein parts called *cofactors* that are necessary for biological activity. These cofactors are frequently inorganic cations but may be small organic molecules that are referred to as *coenzymes*. The distinction between a prosthetic group and a coenzyme is basically associated with the type of protein involved; conjugated proteins contain prosthetic groups and enzymes contain coenzymes.

To summarize, then, *proteins are polypeptides that are condensation polymers of amino acids.* They vary greatly in size, and some proteins include non–amino acid groups.

15.8 PROTEIN STRUCTURE AND FUNCTION

The order of the amino acid residues in a peptide or protein molecule is essential to its function. As the length of the chain increases, the number of variations in the sequence of amino acids quickly increases. Six tripeptides are possible if three amino acids (for example, glycine, Gly; alanine, Ala; and serine, Ser) are linked in all possible combinations. They are:

- Gly-Ala-Ser
- Gly-Ser-Ala
- Ser-Gly-Ala
- Ser-Ala-Gly
- Ala-Gly-Ser
- Ala-Ser-Gly

If n amino acids are all different, the number of arrangements is $n!$. For four different amino acids, the number of different arrangements is $4!$, or $4 \times 3 \times 2 \times 1 = 24$. For five different amino acids, the number of different arrangements is $5!$, or 120. If all 20 different naturally occurring amino acids are bonded, the sequences alone make 2.43×10^{18} (2.43 quintillion) uniquely different 20-monomer molecules. *Since proteins can also include more than one molecule of a given amino acid, the possible combinations are essentially infinite.* However, of the many different possible proteins that could be made from a set of amino acids, a living cell will make only the relatively small, select number it needs.

Many short-chain peptides are important biochemicals. For example, enkephalins and endorphins (Figure 15.12) are referred to as "natural opiates" because they moderate pain in the same manner as opium derivatives. Our bodies synthesize enkephalins and endorphins to moderate pain, and our pain threshold is related to levels of these neuropeptides in our central nervous system. Individuals with a high tolerance for pain produce more of these neuropeptides and consequently tie up more receptor sites than normal; hence, they feel less pain. A dose of heroin temporarily bonds to a high percentage of the sites, resulting in little or no pain. Continued use of heroin causes the body to reduce or cease its production of enkephalins and endorphins. If use of the narcotic is stopped, the receptor sites become empty, and withdrawal symptoms occur.

> The notation $n!$, called n *factorial*, is mathematical shorthand for the result of multiplying n by all the numbers between it and zero.

> Neuropeptides are *neurotransmitters*, which are biomolecules that transmit chemical messages along nerve pathways. They act by connecting with other molecules (or parts of molecules) called *receptors*. The neurotransmitter–receptor interaction plays an important role in the effects on the body of both poisons and medicines.

Peptide	Amino acid sequence
Met-enkephalin	Tyr—Gly—Gly—Phe—Met
Leu-enkephalin	Tyr—Gly—Gly—Phe—Leu
β-endorphin	Lys—Arg—Tyr—Gly—Gly—Phe—Met—Thr—Ser—Glu—Lys—Ser—Glu—Thr—Pro—Leu—Val—Thr—Leu—Phe—Lys—Asn—Ala—Ile—Ile—Lys—Asp—Ala—Tyr—Lys—Lys—Gly—Glu

FIGURE 15.12 Enkephalins and endorphins, polypeptides that are natural opiates. Note that all three polypeptides have one identical amino acid sequence.

Disc-shaped

Sickle-shaped

Red blood cells (erythrocytes). Normally they are disc-shaped, but with a single amino acid defect in two of the protein chains, they adopt the pointed sickle shape. Sickle cell anemia is an inherited (genetic) disease caused by a flaw in DNA.

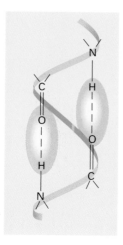

FIGURE 15.13   Helical protein structure. An illustration of how hydrogen bonding connects a peptide bond nitrogen atom to an oxygen atom in the third amino acid unit down the chain, resulting in a coiled structure.

A coiled spring is helical in structure.

Primary protein structure Amino acid sequence in a protein

Secondary protein structure Shape of the backbone structure of a protein

---

EXAMPLE 15.2   **Peptides**

Use the structures of amino acids in Table 15.5 to draw the structure of the tripeptide represented by Ala-Ser-Gly, and give its name.

SOLUTION

The amino acid sequence in the abbreviated name shows that alanine should be written at the left with a free $H_2N—$ group, glycine should be written at the right with a free $—COOH$ group, and both should be connected to serine by peptide bonds:

$$H_2N—\overset{\overset{\displaystyle H}{|}}{\underset{\underset{\displaystyle CH_3}{|}}{C}}—\overset{\overset{\displaystyle O}{\|}}{C}—\overset{\overset{\displaystyle H}{|}}{\underset{\underset{\displaystyle H}{|}}{N}}—\overset{\overset{\displaystyle H}{|}}{\underset{\underset{\displaystyle CH_2OH}{|}}{C}}—\overset{\overset{\displaystyle O}{\|}}{C}—\overset{\overset{\displaystyle H}{|}}{\underset{\underset{\displaystyle H}{|}}{N}}—\overset{\overset{\displaystyle H}{|}}{\underset{\underset{\displaystyle H}{|}}{C}}—\overset{\overset{\displaystyle O}{\|}}{C}—OH$$

The name is alanylserylglycine.

TRY IT 15.2

Use the structures of amino acids in Table 15.5 to draw the structure of the tetrapeptide Cys-Phe-Ser-Ala.

Proteins are important in a wider variety of ways than other kinds of biomolecules. As *enzymes* they serve as catalysts in biological synthesis and degradation reactions. As *hormones* they serve a regulatory role, and as *antibodies* they protect us against disease. They make up the muscle fibers that contract so that we can move. And proteins are the major constituents of cellular and intracellular membranes, skin, hair, muscle, and tendons. Each individual protein has its own group of amino acids arranged in a definite molecular structure that is specific to the function of that protein.

The sequence of amino acids bonded to one another in a protein by peptide bonds is the protein's **primary structure**. Changing the sequence alters the properties of a protein, and just one change may produce a new protein unable to function like the original one. For example, *sickle cell anemia* manifests itself when a mutant variation of normal hemoglobin known as sickle cell hemoglobin is present in the body. Sickle cell hemoglobin results when a single acidic amino acid (glutamic acid) is replaced by a single nonpolar amino acid (valine) in two of the four protein chains of normal hemoglobin. This change of only two amino acids, out of a total of 574 amino acids in the four protein chains, changes the shape of hemoglobin from disc-shaped to sickle-shaped. The red blood cells with sickle-shaped hemoglobin are unusually fragile. The inability of these cells to flow easily through capillaries causes pain and inflammation and can lead to severe anemia, organ damage, and even an early death.

In some entire proteins and in parts of others, the shape of the backbone of the molecule (the chain containing peptide bonds) has a regular, repetitive pattern that is referred to as its **secondary structure**. The two most common secondary structures are the α-helix and the β-pleated sheet. The α-helix is held together by *intramolecular* (within the molecule) hydrogen bonding between backbone peptide bonds. An N—H group of one amino acid forms a hydrogen bond with the oxygen atom in the third amino acid down the chain (Figure 15.13).

The α-helix is the basic structural unit of the α-keratins in wool, hair, skin, beaks, nails, and claws.

Silk has the β-sheet structure (Figure 15.14) in which several chains of amino acids are joined side-to-side by *intermolecular* (between protein

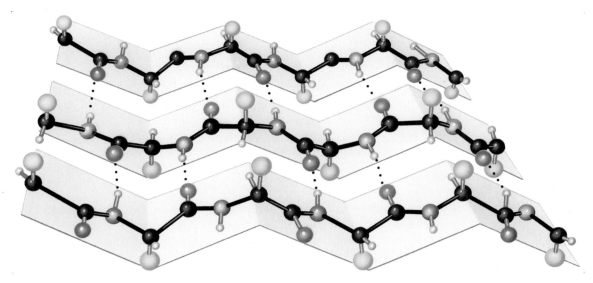

FIGURE 15.14 *β*-Pleated sheet protein structure. Hydrogen bonds are shown as dotted lines.

chains) hydrogen bonds. The resulting structure is not elastic, because stretching the fibers would break either covalent bonds or the many hydrogen bonds holding the individual protein strands in the sheet. However, just as you can bend the stack of pages in this book, so too can the stack of protein sheets be bent.

Tertiary structure refers to how a protein molecule is folded. One kind of tertiary structure is found in collagen, a fibrous protein: three amino acid chains twist into left-handed helices, which in turn are twisted into a right-handed superhelix to form an extremely strong fibril (Figure 15.15). Bundles of fibrils make up the tough collagen.

A second kind of tertiary structure is found in globular proteins, which contain regions where the polypeptide chain forms *α*-helices, other regions where parallel parts of the chain are organized into *β*-pleated sheets, and some parts that are just coiled randomly.

Quaternary structure is the shape assumed by the entire group of chains in a protein composed of two or more chains. Hemoglobin (Figure 15.11) is a prime example of quaternary structure—its four protein chains, together with the nonprotein heme molecule carried by each, can function only when combined in precisely the right shape.

All the structural features—primary, secondary, tertiary, and quaternary—are critical to the proper functioning of a protein and give a protein its "native" or "natural" structure (Figure 15.16). Any physical or chemical process that changes the native protein structure and makes it incapable of performing its normal function is called a **denaturation** process. For example, heating an aqueous solution of a protein breaks hydrogen bonds in the secondary and tertiary structure and causes the protein molecule to unfold. Denaturing chemicals include reducing agents, which break disulfide linkages, and acids and bases, which affect the hydrogen bonds and ionic interactions between polypeptide chains. Whether denaturation is reversible depends on the protein and the extent of denaturation.

Enzymes

Enzymes function as catalysts for chemical reactions in living systems. Many enzymes are globular proteins. One outcome of the highly organized,

FIGURE 15.15 Collagen, a fibrous protein. Collagen is composed of three polypeptide chains twisted together into a rope-like fiber.

Tertiary protein structure Shape of an entire folded protein chain

Quaternary protein structure Structure of a protein with more than one polypeptide

Denaturation Destruction of proper functioning of a protein by changing its structure

Enzymes Biomolecules that function as catalysts for biochemical reactions

One way to denature a protein.

The general equation for hydrolysis of an ester is

$$RCOOR + H_2O \longrightarrow RCOOH + ROH$$

Active site The region of an enzyme where the catalytic chemistry occurs. This region is usually composed of only a few critical amino acid residues

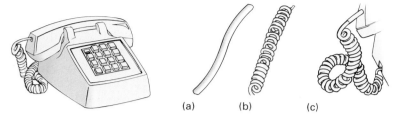

(a)   (b)   (c)

FIGURE 15.16   Structure of a telephone cord. (a) The straight cord is like the primary structure of a protein. (b) The curling of the cord is like the secondary structure of a protein. (c) When the curled cord is twisted up it is like the tertiary structure of a protein (which may include in its twisted form regions of α-helix and β-sheet).

though seemingly random, structure of globular proteins is creation of the region (the **active site**) that allows an enzyme to function as a catalyst. Like all catalysts, enzymes increase the rate of a reaction by lowering the energy of activation (Figure 8.3). This lowering occurs because the reaction path or process is changed. Enzymes are very effective catalysts and typically increase reaction rates by anywhere from 10^6 to 10^{16} times.

Many biomolecules are broken down during digestion by hydrolysis reactions, which are essentially the reverse of condensation reactions. In hydrolysis, a larger molecule is split into smaller molecules with the addition of H— and —OH of water where a bond was broken. The enzyme maltase catalyzes the hydrolysis of the sugar maltose into two molecules of D-glucose. This is the only function of maltase, and no other enzyme can substitute for it. Sucrase, another enzyme, hydrolyzes only sucrose. Lysozyme catalyzes hydrolysis of sugar polymers (polysaccharides) found in bacterial cell walls (Figure 15.17). Some enzymes are less specific. The digestive enzyme trypsin, for example, primarily hydrolyzes peptide bonds in proteins. However, the structure and polarity of trypsin are such that it can also catalyze the hydrolysis of some esters.

Since the structure of the active site of an enzyme is important, the same factors that cause denaturation will also destroy the activity of the enzyme. For example, most enzymes are effective only over a narrow temperature range and a narrow pH range. Enzymes are denatured irreversibly at high temperatures or at pH values outside their effective range.

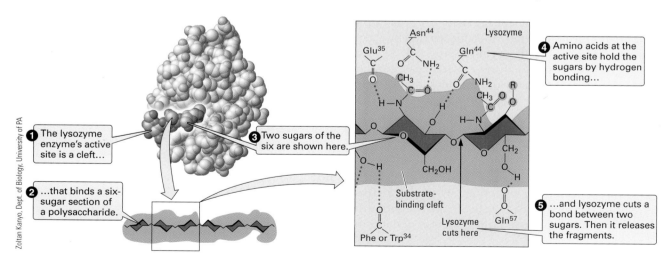

❶ The lysozyme enzyme's active site is a cleft…

❷ …that binds a six-sugar section of a polysaccharide.

❸ Two sugars of the six are shown here.

Asn⁴⁴   Lysozyme

Glu³⁵   Gln⁴⁴

❹ Amino acids at the active site hold the sugars by hydrogen bonding…

Substrate-binding cleft

Phe or Trp³⁴

Lysozyme cuts here   Gln⁵⁷

❺ …and lysozyme cuts a bond between two sugars. Then it releases the fragments.

FIGURE 15.17   Lysozyme with substrate in the active site. The structure of lysozyme is shown at (1) as a space-filling model. The active site is a cleft in the surface of the lysozyme that stretches horizontally across the middle of the enzyme. The active site is occupied by a portion of a polysaccharide molecule, the substrate (*green atoms*). (The part of the polysaccharide not bound to the active site has been omitted so that you can see the enzyme better.) The diagram at (4) shows noncovalent interactions (*red dotted lines*) that hold the substrate to the enzyme. The bond that will be broken when the substrate is hydrolyzed is marked by an arrow (5).

Many inherited diseases or defects affect how enzymes function. An example is found in *lactose intolerance*, which is an inability to digest lactose, the sugar present in milk from all mammals. Well over half the world's population has this problem to some degree. The pattern of inheritance is noticed largely in those with Asian or African ancestry, Native Americans and Latinos, and to a lesser extent in those of Northern European ancestry. When they are infants, people with lactose intolerance manufacture the enzyme lactase, which is necessary to digest lactose. As they grow older, their bodies stop producing this enzyme, and the ingestion of milk products containing lactose can lead to considerable discomfort in the form of stomach aches and diarrhea. The problem is avoided by eliminating milk products from the diet or by taking enzyme-containing tablets before eating any milk product.

Over-the-counter medications containing the lactase enzyme are available without prescription.

15.9 HAIR PROTEIN AND PERMANENT WAVES

Hair is composed principally of keratin. An important difference between hair keratin and other proteins is its high content of the amino acid cysteine, which contains the —SH (sulfhydryl) group in its side chain.

Fingernails and toenails are composed of "hard" keratin, a very dense type of this protein. The epidermal cells of nails grow from epithelial cells lying under the white crescent at the growing end of the nail. Like hair, the nail tissue beyond the growing cells is dead.

Cysteine plays an important role in the structure of hair by forming disulfide bonds (—S—S—) between protein chains. These protein chains are twisted together into spirals that group together to constitute individual strands of hair (Figure 15.8).

In addition to the disulfide bonds between hair protein structures, ionic bonds and hydrogen bonds between protein side chains affect the behavior of hair. In fact, these bonds explain "bad hair days." Consider, for example, the interaction between a lysine —NH_2 group and a —COOH of glutamic acid on a neighboring protein chain. The acidic —COOH groups lose their protons, forming negatively charged —COO^- groups, while the basic NH_2 groups gain protons to form positively charged —NH_3^+ groups. When the —NH_3^+ and —COO^- groups on adjacent chains approach each other, an ionic bond is formed that helps hold the two protein chains together.

Charles D. Winters

A freshly created permanent wave.

Ionic bond between two protein chains

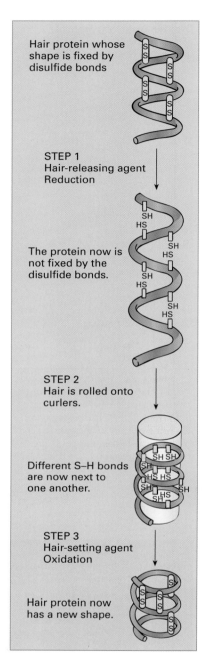

Hair protein whose shape is fixed by disulfide bonds

STEP 1
Hair-releasing agent
Reduction

The protein now is not fixed by the disulfide bonds.

STEP 2
Hair is rolled onto curlers.

Different S–H bonds are now next to one another.

STEP 3
Hair-setting agent
Oxidation

Hair protein now has a new shape.

FIGURE 15.18 Permanent waving, a chemical oxidation–reduction process. The disulfide bonds in hair protein are broken by a reducing agent (Step 1). The protein strands are twisted into a more curly shape (Step 2). Then the new shape is fixed by an oxidizing agent that re-forms the disulfide bonds (Step 3).

Or consider how a hydrogen bond might form between side groups.

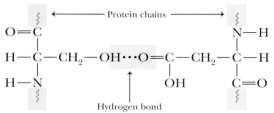

Hydrogen bond between two protein chains

Moisture affects hydrogen bonds between protein chains, and when their number and arrangement changes, hair changes. When hair is wet, it can be stretched to one and one-half times its dry length. On a very moist day, enough bonds will break simultaneously for hair to lose its curl. On a very dry day, enough static electrical charge can build up for individual hairs to repel each other. Different people have "bad hair" on different days because the bonding patterns in their hair vary.

For more permanent curling, a more permanent chemical change in the hair structure is needed. The disulfide bonds between protein chains hold hair in its natural shape. In "permanent waving," these cross-links are broken by a reducing agent (Step 1, Figure 15.18), which relaxes the tension. An oxidizing agent generates new cross-links, and the hair retains the shape of the roller, so it appears curly.

The most commonly used hair-waving reducing agent is the ammonium salt of thioglycolic acid ($HSCH_2COO^-NH_4^+$). A typical waving solution contains 5.7% thioglycolic acid, 2.0% ammonia, and 92.3% water. The usual oxidizing agent is hydrogen peroxide (H_2O_2) or a perborate compound that releases hydrogen peroxide in solution ($NaBO_3 \cdot 4H_2O$).

CONCEPT CHECK 15C

1. All amino acids except _____ have optical isomers.
2. Only _____ isomers of amino acids are found in your body.
3. Amino acids that the body cannot synthesize from other molecules are called _____ amino acids.
4. The peptide linkage that bonds amino acids together in protein chains has the structure _____.
5. (a) If we have three different amino acids and can use each one three times in any given tripeptide, we can make a total of _____ different tripeptides.
   (b) If we can use each amino acid only once, there are still _____ possible different tripeptides.
6. (a) The primary structure of a protein refers to its _____.
   (b) The secondary structure refers to its _____.
   (c) The tertiary structure refers to _____.
   (d) The quaternary structure refers to _____.
7. The helical structure of proteins is caused by _____ bonding.
8. That portion of the enzyme at which the reaction is catalyzed is called the _____.
9. The best term to describe the general function of enzymes is (a) catalyst, (b) intermediate, (c) oxidant.
10. The activation energy of many biological reactions is decreased if a(n) _____ is present.

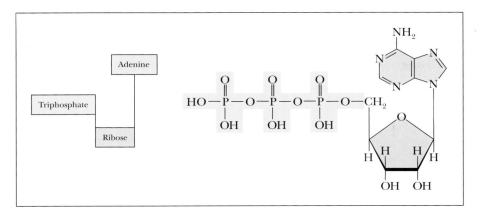

FIGURE 15.19   Stylized structure (left) and molecular structure (right) of adenosine triphosphate (ATP).

11. Hair protein side chains are held together by _____ and _____ bonds.
12. Water easily breaks _____ bonds found in hair structure, making wet hair more pliable.

15.10 ENERGY AND BIOCHEMICAL SYSTEMS

The energy to sustain all but a few forms of life comes from the sun. During photosynthesis, green plants absorb energy from the sun to make glucose and oxygen from carbon dioxide and water. Glucose is a major energy source for all living organisms. During metabolism, much of the protein, fat, and carbohydrate in our diets is converted to glucose (Section 16.2). The energy stored in glucose is eventually transferred to the bonds in molecules such as adenosine triphosphate (ATP, Figure 15.19). When living organisms need energy, phosphate bonds in the ATP molecules are hydrolyzed to give adenosine diphosphate (ADP, Figure 15.20) and energy for other biochemical reactions.

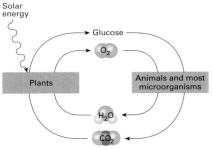

The flow of carbon atoms, oxygen atoms, and energy between plants and animals.

FIGURE 15.20   Hydrolysis of ATP to produce ADP (adenosine diphosphate).

THE CHEMISTRY OF

MEN IN BLACK II (2002)

© 2005 by Mark A. Griep, *University of Nebraska-Lincoln*

The second movie in the *Men in Black* series (released in 2002) concerns an evil Kylothian named Serleena. She is searching the Earth for the Light of Zartha, an item with the power to destroy the third planet. Early in her search, Serleena angrily uses one of her lethal tentacles to split another alien in half. He was a humanoid and had owned Ben's Pizza of SoHo for twenty years.

The Men in Black are from a secret government agency responsible for monitoring and policing extraterrestrial activity on Earth. They are investigating this alienicide when Agents Jay and Tee show up. First, they learn that there is "phosphorus residue on the walls and floor." Then, Agent Jay explains to waitress Laura Vasquez that the dead alien didn't have skin but a "protoplasm polymer" similar to the chemical makeup of the chewing gum packaged with baseball cards. Elsewhere in the movie, we see quite a variety of aliens and learn that some mask their normal shapes with more humanoid ones.

The Earth's oxidizing atmosphere ensures that most phosphorus is in the form of phosphate. In Earthlings, it is found in important molecules such as DNA, RNA, and ATP. It is also an important part of vertebrate bone, along with calcium and some special proteins.

Is there a connection between chewing gum and skin? Like all chewing gum, the kind packaged with trading cards gets its chewiness from *latex*. Latex can be used for many things, including the manufacture of rubber tires. In 1869, American inventor Thomas Adams developed the modern chewing gum base from a South American rubber plant called chicle. He found that chicle latex was unsuitable for making tires. In the 1950s, synthetic gum base alternatives began to appear on the market, including two types of co-polymers: *butadiene-styrene* and *isobutylene–isoprene*. These co-polymers have longer and smoother chewing properties than natural latexes.

Questions to Consider

(These questions are best answered as group projects.)

1. Was the dead alien's outer layer (**a**) a natural protective "skin," or (**b**) a humanoid body mask?

2. Use the Drake equation found at **http://www.pbs.org/lifebeyondearth/listening/drake.html** to predict the likelihood that intelligent extraterrestrials exist and are trying to contact us from within the Milky Way galaxy.

3. Why is it likely that all of the aliens in the movie are made of the same elements as Earthlings?

4. Do the aliens in the Men in Black series resemble life forms on Earth because (**a**) we share a common ancestor or (**b**) similar environments produce similar adaptations? Explain your reasoning.

5. All Earthlings use the same set of chiral biomolecules. Why would it be a good thing for us if protein-hungry aliens used biomolecules with the opposite chirality?

Elemental Number Abundance

VERTEBRATES	SOLAR SYSTEM
H (62%)	H (96%)
O (24%)	He (3%)
C (12%)	O (0.4%)
N (1.1%)	C (0.4%)
P (0.2%)	N
Ca (0.2%)	Ne
S	Mg
Na	Si
K	Fe
Cl	S

In the complex process of photosynthesis, carbon dioxide is reduced to make sugar and water is oxidized to oxygen

$$6\,CO_2 + 6\,H_2O + 688\ \text{kcal} \longrightarrow C_6H_{12}O_6 + 6\,O_2$$

Carbon dioxide   Water   Energy (sunlight)   Glucose   Oxygen

The oxygen produced in photosynthesis is the continuing source of all the oxygen in our atmosphere. We are dependent on the plant life of our planet, and we must live in balance with the oxygen output of that plant life, as well as with the food output of the same plant life. Photosynthesis is absolutely vital to life on Earth, and this endothermic reaction requires energy from the Sun.

15.11 NUCLEIC ACIDS

The genetic information that makes each organism's offspring look and behave like its parents is encoded in molecules called **nucleic acids**. Together with a set of specialized enzymes that catalyze their synthesis and decomposition, the nucleic acids constitute a remarkable system that accurately copies millions of pieces of data with very few mistakes.

Like polysaccharides and polypeptides, nucleic acids are condensation polymers. Each monomer in these polymers includes one of two simple sugars, one phosphoric acid group, and one of a group of heterocyclic nitrogen compounds that behave chemically as bases. A particular nucleic acid is a **deoxyribonucleic acid (DNA)** if it contains the sugar 2-deoxy-D-ribose, and it is a **ribonucleic acid (RNA)** if it contains the sugar D-ribose.

The five organic bases that play a key role in the mechanism for information storage are *adenine* (A), *guanine* (G), *thymine* (T), *cytosine* (C), and *uracil* (U). These bases are mentioned so often in any discussion of nucleic acid chemistry that to save space they are usually referred to only by the first letter of each name.

D-Ribose   2-Deoxy-D-ribose

Notice the presence of the OH group on the 2 carbon in D-ribose. The 2-deoxy-D-ribose does not have this OH group.

Adenine (A)   Guanine (G)   Thymine (T)   Cytosine (C)   Uracil (U)

Nucleic acids are found in all living cells, with the exception of the red blood cells of mammals. DNA occurs primarily in the nucleus of the cell, and RNA is found mainly in the cytoplasm, outside the nucleus. There are three major types of RNA, each with its own characteristic size, base composition, and function in protein synthesis (as described later in this section): messenger RNA (mRNA), transfer RNA (tRNA), and ribosomal RNA (rRNA).

The monomers that polymerize to make both DNA and RNA are known as **nucleotides**. They have structures like the one shown in Figure 15.21a. Each nucleotide contains a phosphoric acid unit, a ribose sugar unit, and one of the five bases. The nucleotides of DNA and RNA have two structural differences: (1) they contain different sugars (deoxyribose and ribose); and (2) the base uracil occurs only in RNA, whereas the base thymine occurs only in DNA. The other bases—adenine, guanine, and cytosine—are found in both DNA and RNA.

Polynucleotides have molar masses ranging from about 25,000 for tRNA molecules to billions for human DNA. The sequence of nucleotides in the polymer chain (as shown by the base sequence) is its primary structure. Polynucleotides are formed by the polymerization of nucleotides to make esters. As an example, Figure 15.21b shows three monomers condensed to a trinucleotide.

In 1953, James D. Watson and Francis H. C. Crick proposed a secondary structure for DNA that revolutionized our understanding of heredity and

Nucleic acids Biomolecules that control heredity (DNA, RNA); polymers of nucleotides that contain deoxyribose or ribose, nitrogen bases, and phosphate groups

Deoxyribonucleic acids (DNAs) Nucleic acids that function as genetic information storage molecules; contain deoxyribose

Ribonucleic acids (RNAs) Nucleic acids that transmit genetic information and direct protein synthesis; contain ribose

Nucleotide Biomolecule with a five-carbon sugar bonded to a nucleic acid base and a phosphate group; monomer in DNA

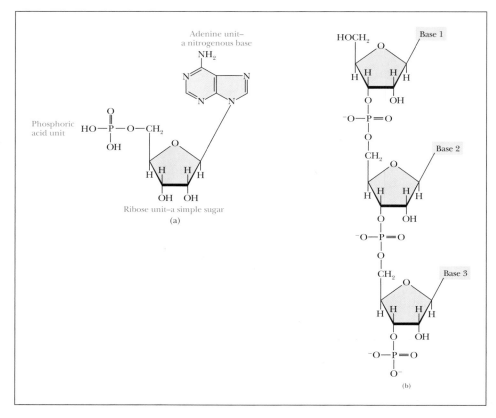

FIGURE 15.21 (a) **A nucleotide.** Because its sugar unit is ribose, this is a ribonucleotide. (b) **Bonding in a trinucleotide.** The sugars could also be deoxyribose units and the bases can be any of the five bases shown in the text. The primary structures of both DNA and RNA are extensions of this structure.

Hydrogen bonding between two of the nucleic acid bases, thymine and adenine.

RNA is generally a single strand of helical polynucleotide.

The inherited traits of an organism are controlled by DNA molecules.

Genes Segments of DNA that direct synthesis of proteins

Genome Total sequence of base pairs in the DNA of a plant or animal

genetic diseases. They proposed that pairs of polynucleotides are arranged in a double helix stabilized by hydrogen bonding between the base groups lying opposite each other in the two chains. The critical point of the Watson–Crick model is that hydrogen bonding can best occur between specific bases. Adenine–thymine (A—T) and guanine–cytosine (G—C) pairs occur exclusively because they are very tightly hydrogen bonded. The complementary hydrogen bonding between these specific bases is illustrated in Figure 15.22.

The function of polynucleotides is to transcribe hereditary information so that *like begets like*. The almost infinite variety of primary structures of polynucleotides likewise allows an almost infinite variety of information to be recorded in the molecular structures of the strands of nucleic acids. The different arrangements of just a few bases produce a wide diversity of structures. In a somewhat similar fashion, multiple arrangements of just a few language symbols convey the many ideas in this book. The coded information in the polynucleotide controls the inherited characteristics of the next generation as well as most of the continuous life processes of the organism.

Double-stranded DNA forms the 46 human chromosomes, which have specialty heredity areas called **genes**. Genes are segments of DNA that have as few as 1000 or as many as 100,000 base pairs such as those shown in Figure 15.22. Each gene holds the information needed for the synthesis of a single protein. Human DNA (the human **genome**) is estimated to have up to 40,000 genes and about 3 billion pairs of bases. However, genes are estimated to make up only 3% of DNA, with each gene sandwiched between noncoding

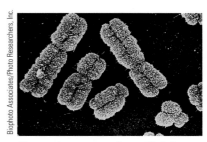

Human chromosomes (magnified about 8,000 times). DNA assembles into chromosomes when a cell is about to divide.

FIGURE 15.22   Complementary hydrogen bonding in the DNA double helix. Hydrogen bonds in the thymine–adenine (T—A) and cytosine-guanine (C—G) pairs stabilize the double helix.

DNA sequences. There are also short segments that act as switches to signal where the coding sequence begins.

Replication of DNA: Heredity

The transfer of coded information begins with the replication of DNA and continues with natural protein synthesis. Almost all nuclei in an organism's cells contain the same chromosomal composition. This composition remains constant regardless of whether the cell is starving or has an ample supply of food materials. Each organism begins life as a single cell with this same chromosomal composition; in sexual reproduction, half of a chromosome comes from each parent. The DNA structure is faithfully copied during normal cell division (mitosis—both strands); only half is copied in cell division that produces reproductive cells (meiosis—one strand).

In **replication** the double helix of the DNA structure unwinds, and each half of the structure serves as a template, or pattern, from which the other complementary half can be reproduced from the molecules in the cell environment (Figure 15.23). Replication of DNA occurs in the nucleus of the cell before the cell divides.

Natural Protein Synthesis

The proteins of the body are continually being replaced and resynthesized from the amino acids available to the body. The use of isotopically labeled amino acids has made possible studies of the average lifetimes of amino acids as constituents in proteins—that is, the time it takes the body to replace a protein in a tissue. Considering that this process must be extremely complex, replacement is very rapid. Only minutes after radioactive amino acids are injected into animals, radioactive protein can be found. Although

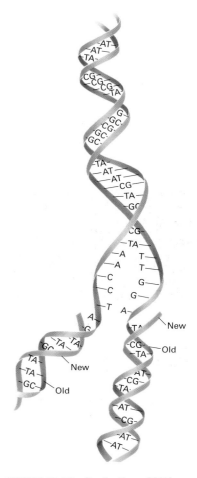

FIGURE 15.23   Replication of DNA. When the double helix of DNA (blue) unwinds, each half serves as a template on which to assemble nucleotides to form new DNA strands.

Replication The process by which DNA is copied when a cell divides all the proteins in the body are continually being replaced, the rates of replacement vary. Half the proteins in the liver and plasma are replaced in 6 days; the time needed for replacement of muscle proteins is about 180 days; and replacement of protein in other tissues, such as bone collagen, takes even longer.

Recall that each organism has its own kinds of proteins. The number of possible unique arrangements of 20 amino acid units is 2.43×10^{18}, yet proteins characteristic of a given organism can be synthesized by the organism in a matter of a few minutes.

The DNA in the cell nucleus holds the code for protein synthesis, which is carried out in a series of steps summarized in Figure 15.24. First, messenger RNA, like all forms of RNA, is synthesized in the cell nucleus. The sequence of bases in one strand of the chromosomal DNA serves as the template from

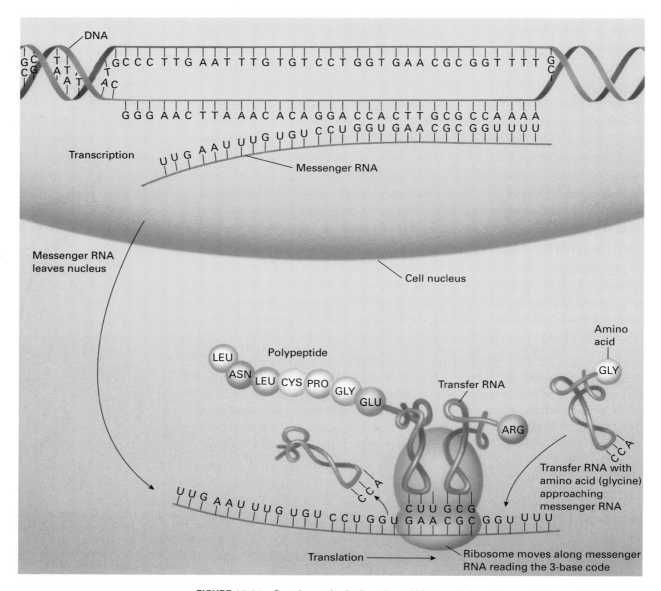

FIGURE 15.24 **Protein synthesis.** A section of DNA unwinds and transcription results in production of messenger RNA. Outside the cell nucleus, at a ribosome, the code carried by messenger RNA is translated into the amino acid sequence of a newly synthesized protein. Transfer RNA brings the amino acids into position one at a time.

which a single strand of a messenger ribonucleotide (mRNA) is made in a process known as **transcription**. The bases of the mRNA strand complement those of the DNA strand. Two bases are complementary when each one fits the other and forms one or more hydrogen bonds. Messenger RNA contains only the four bases: adenine (A), guanine (G), cytosine (C), and uracil (U). DNA contains the four bases adenine (A), guanine (G), cytosine (C), and thymine (T). The bases in DNA are transcribed into bases in mRNA as follows:

DNA	MRNA
A	U
G	C
C	G
T	A

This means that, provided the necessary enzymes and energy are present, wherever a DNA has an adenine base (A), the mRNA will transcribe a uracil base (U), and so on.

After transcription, mRNA passes from the nucleus of the cell to a ribosome (which contains the rRNA), where mRNA serves as the template for the sequential ordering of amino acids during protein synthesis by the process known as **translation**. As its name implies, messenger RNA contains the sequence message, in the form of a three-base code (called a **codon**), for ordering amino acids into proteins (Table 15.6). Each of the thousands of different proteins synthesized by cells is coded by a specific mRNA or segment of an mRNA molecule.

Transcription The process by which the information in DNA is read and used to synthesize RNA

Translation Process for sequential ordering of amino acids, directed by RNA during protein synthesis

Codon A three-base sequence on mRNA that codes a specific amino acid in protein synthesis

TABLE 15.6 Messenger RNA Codes for Amino Acids*

FIRST LETTER OF CODE	SECOND LETTER OF CODE				THIRD LETTER OF CODE
	U	**C**	**A**	**G**	
U	Phenylalanine	Serine	Tyrosine	Cysteine	U
	Phenylalanine	Serine	Tyrosine	Cysteine	C
	Leucine	Serine	STOP	STOP	A
	Leucine	Serine	STOP	Tryptophan	G
C	Leucine	Proline	Histidine	Arginine	U
	Leucine	Proline	Histidine	Arginine	C
	Leucine	Proline	Glutamine	Arginine	A
	Leucine	Proline	Glutamine	Arginine	G
A	Isoleucine	Threonine	Asparagine	Serine	U
	Isoleucine	Threonine	Asparagine	Serine	C
	Isoleucine	Threonine	Lysine	Arginine	A
	START or methionine	Threonine	Lysine	Arginine	G
G	Valine	Alanine	Aspartic acid	Glycine	U
	Valine	Alanine	Aspartic acid	Glycine	C
	Valine	Alanine	Glutamic acid	Glycine	A
	Valine	Alanine	Glutamic acid	Glycine	G

* In groups of three (called *codons*), bases of mRNA code the order of amino acids in a polypeptide chain. A, C, G, and U represent adenine, cytosine, guanine, and uracil, respectively. Some amino acids have more than one codon, and hence more than one tRNA can bring the amino acid to mRNA.

Transfer RNAs carry the amino acids to the mRNA one by one. Each of the 20 amino acids found in proteins has at least one corresponding tRNA, and some have multiple tRNAs (Table 15.6). Table 15.6 lists the RNA codes and shows in the first line that, for example, UUU codes for phenylalanine and UCU codes for serine.

In the schematic illustration in Figure 15.24, which summarizes DNA ⟶ RNA ⟶ protein, pick a three-base sequence on the bottom DNA strand of the two DNA strands at the top of the figure and follow it through to the bottom, where a tRNA attaches to a three-base codon on mRNA. Assume the DNA strand you have selected serves as the template for the synthesis of the single strand of mRNA shown in the figure. Do you agree with the changes that occur in the letters representing the three-base sequence?

Protein synthesis is initiated by START codons and terminated by STOP codons (Table 15.6).

EXAMPLE 15.3 Nucleic Acids

If the base sequence in a DNA segment is ...GCTGTA..., what is the base sequence in the complementary mRNA? What is the order in tRNA?

SOLUTION
The base pairs between DNA and mRNA are G...C, A...U, and T...A. Therefore, the complementary mRNA segment for a DNA segment of GCTGTA is determined by using these allowed base pairs. Start with GCTGTA and place the correct base below it. The resulting mRNA segment is CGACAU. The order in tRNA is the complement of the mRNA segment. The allowed base pairs are G...C and A...U, so the order in tRNA is GCUGUA.

TRY IT 15.3
If the sequence of bases along a mRNA strand is ...UCCGAU..., what was the sequence along the DNA template?

CONCEPT CHECK 15D

1. The reactants in the photosynthesis process are _____ and _____; _____ must also be supplied.
2. Most of the energy obtained by the oxidation of food is used immediately to synthesize the molecule _____.
3. The hydrolysis of ATP produces _____ and phosphoric acid, _____ is also released.
4. The basic code for the synthesis of protein is contained in the _____ molecule.
5. The sugar in RNA is _____, whereas the one in DNA is _____.
6. A nucleotide contains _____, _____, and _____.
7. The secondary structure of DNA is in the shape of a (an) _____.
8. When DNA replicates itself, each base in the chain is matched to another one via _____ bonds.
9. Base pairs in DNA are formed between A and _____, G and _____, T and _____, and C and _____.
10. Human DNA is estimated to have _____ base pairs.

W Assess your understanding of this chapter's topics with an online chapter quiz at **www.brookscole.com/chemistry/joesten4**

■ KEY TERMS

biochemistry	hydrophilic	enzymes
chiral	emulsions	active site
achiral	emulsifying agent	nucleic acids
enantiomers	colloids	deoxyribonucleic acids (DNAs)
optical isomers	amino acids	ribonucleic acids (RNAs)
stereoisomers	essential amino acids	nucleotide
racemic mixture	peptide bond	genes
teratogen	polypeptide	genome
carbohydrates	proteins	replication
lipids	primary protein structure	transcription
triglycerides	secondary protein structure	translation
hormones	tertiary protein structure	codon
steroids	quaternary protein structure	
hydrophobic	denaturation	

■ THE LANGUAGE OF CHEMISTRY

1. Energy "cash" in the living cell
2. Natural protein
3. Product of ATP hydrolysis
4. Saponification
5. D-Glucose
6. Enzymes
7. Carbohydrate stored in animals
8. Keratin
9. Polypeptides
10. DNA
11. Fibrous protein
12. Cysteine

a. ADP + energy
b. Amino acid common in hair
c. Structure determined by DNA and RNA
d. ATP
e. Hair protein
f. Proteins
g. Main sugar present in the blood
h. Polynucleotide
i. Biochemical catalysts
j. Glycogen
k. Collagen
l. Hydrolysis reaction in which a fat or oil produces glycerol and one or more fatty acids

■ APPLYING YOUR KNOWLEDGE

1. What is meant by the word *chiral*?
2. What do the following terms mean?
   (**a**) L-Isomers   (**b**) D-Isomers
   (**c**) Optically active (**d**) Polarimeter
3. What is a racemic mixture?
4. Give definitions for the following:
   (**a**) Carbohydrate   (**b**) Monosaccharide
   (**c**) Disaccharide   (**d**) Polysaccharide

5. What is an artificial sweetener?
6. What are the following substances?
   (**a**) Starches   (**b**) Glycogen
   (**c**) Cellulose
7. What is a lipid?
8. How do the structures of unsaturated fatty acids and saturated fatty acids differ?
9. What is a partially hydrogenated fatty acid?

10. What is a *trans* fatty acid?

11. Describe how a surfactant such as a soap or detergent molecule works.

12. It is possible to mix oil and water and get a stable emulsion using a soap. It is also possible to make a stable oil-in-water emulsion using the yolk of an egg (mayonnaise). Explain how these two emulsifiers work.

13. Explain why synthetic detergents are more effective than soaps in hard water.

14. A certain cream imparts a somewhat greasy feel to the skin after its application. What kind of emulsion do you suspect this cream to be based on?
    (**a**) Water-in-oil   (**b**) Oil-in-water

15. A skin cream has its ingredients listed on the back of the jar. Water is listed *after* several other ingredients, some of which have the word *oil* in their names. On what kind of emulsion do you think this cream is based?

16. What is an amino acid? Draw and label a generalized structure for an amino acid.

17. What is a peptide bond?

18. What is an essential amino acid?

19. What functional groups are always present in each molecule of an amino acid?

20. What is meant by the following terms?
    (**a**) Dipeptide   (**b**) Polypeptide
    (**c**) Amino acid residue   (**d**) Protein

21. What is meant by the term "amino acid sequence"?

22. Define "neuropeptides" and give two examples.

23. What are the meanings of the terms "primary," "secondary," "tertiary," and "quaternary" in reference to the structures of proteins?

24. Which of the following biochemicals are polymers?
    (**a**) Starch   (**b**) Cellulose
    (**c**) Glucose   (**d**) Fats
    (**e**) Glycylalanylcysteine   (**f**) Proteins
    (**g**) DNA   (**h**) RNA

25. What is meant by the following terms?
    (**a**) α-Helix   (**b**) β-Pleated sheet

26. What does it mean to "denature" a protein?

27. What is the active site of an enzyme?

28. Glutathione is an important tripeptide found in all living tissues. It is also named glutamylcystylglycine. Draw the structure of glutathione. Which enantiomeric forms of the amino acids would you predict are used to synthesize this peptide?

29. (**a**) How many tetrapeptides are possible if four amino acids are linked in different combinations that contain all four amino acids?
    (**b**) Write these combinations for tetrapeptides made from glycine, alanine, serine, and cys-

tine. Use three-letter abbreviations for the amino acids. For example, one combination is Gly-Ala-Ser-Cys.

30. Use three-letter abbreviations to write the possible tripeptides that can be formed from phenylalanine, serine, and valine if each tripeptide contains all three amino acids.

31. Name the following tripeptide:

32. Draw the structure of alanylglycylphenylalanine.

33. Explain why the human body can metabolize D-glucose but not L-glucose.

34. What is lactose intolerance?

35. Describe the two kinds of bonding that hold protein strands together in hair.

36. What is the effect of water on the natural protein bonding in hair?

37. What is the disulfide linkage, and what role does it play in protein structure?

38. What three molecular units are found in nucleotides?

39. What are the differences in structure between DNA and RNA?

40. What is the purpose of ATP?

41. What are the fundamental components of nucleic acids?

42. Explain the word *codon*.

43. What is meant by the term "complementary bases"?

44. How many hydrogen bonds are possible between the following complementary base pairs?
    (**a**) Cytosine and   (**b**) Thymine and
        guanine   adenine

45. A segment of a DNA strand has the base sequence ...GCTGTAACCGAT...
    (**a**) What is the base sequence in the complementary mRNA?
    (**b**) What is the order in tRNA?
    (**c**) Consult Table 15.6 and give the amino acid sequence in the portion of the peptide being synthesized.

46. A segment of a DNA strand has the base sequence ...TGTCAGTGGGCCGCT...
    (**a**) What is the base sequence in the complementary mRNA?

(**b**) What is the order in tRNA?

(**c**) Consult Table 15.6 and give the amino acid sequence in the portion of the peptide being synthesized.

47. Write the overall equation for photosynthesis.

48. What stabilizing forces hold the double helix together in the secondary structure of DNA proposed by Watson and Crick?

49. (**a**) What are the three major types of RNA?
    (**b**) What are their functions?

50. A base sequence in a DNA segment is ...GTAGC... What is the base sequence in the complementary mRNA?

51. (**a**) What are the three bases present in both messenger RNA and DNA?

(**b**) What is the fourth base found in mRNA?

(**c**) What is the fourth base found in DNA but not in mRNA?

52. Give the mRNA complementary base sequence that matches the following DNA segment base sequence:

DNA SEQUENCE	MRNA SEQUENCE
G...	
A...	
C...	
A...	

 # CHEMISTRY ON THE WEB

For up-to-date URLs, visit the text website at **www.brookscole.com/ chemistry/joesten4**

- Carbohydrates
- Artificial Sweeteners
- Lipids
- How Cholesterol Works
- Steroids and Sex Hormones
- Steroid Abuse
- Side-by-Side Images of Amino Acids
- Primary through Quaternary Structure of Proteins
- How Does DNA Work?
- DNA from the Beginning
- Understanding Genetics
- DNA and Forensics

NUTRITION: THE BASIS OF HEALTHY LIVING

How much do you know about the nutritional value of this breakfast?

© Stockbyte Platinum/Getty Images

We are living in a health-conscious society. Newspapers, magazines, and television programs constantly offer us advice on what to eat. The advice ranges from responsible journalism to scare tactics and sensationalism. Fads about what to eat and what not to eat are rapidly translated into marketable products. And advertisements bombard us from every direction with information about food. The advertisements and food stories in the media are bound to attract our attention. After all, many times every day, we make choices about food.

How can we sort out the information so that these choices are good ones? The task is impossible without knowing something about the basic principles of nutrition, which requires starting with the fundamentals of biochemistry presented in Chapter 15. The questions addressed here are the following:

- How is our biochemical energy generated?
- What roles do carbohydrates, fats, and proteins play in our diets and health?
- What are current recommendations for a healthy diet?
- What is the relationship between diet and heart disease?
- What are the categories of vitamins and minerals?
- What are the major functions of vitamins and minerals in our diets?
- What are the types and functions of food additives?

Nutrition is the science that deals with diet and health. The old saying "we are what we eat" is true. The skin that covers us now is not the same skin that covered us seven years ago. The fat beneath our skin is not the same fat that was there just a year ago. Our oldest red blood cells are 120 days old. The entire lining of our digestive tract is renewed every three days. Many chemical reactions are required to replace these tissues, and the energy and raw materials for these reactions are supplied by what we eat.

Nutrition, then, is concerned with the chemical requirements of the body—the nutrients and the chemical energy we get from them. **Nutrients** are chemical substances in foods that provide the energy and raw materials required by biochemical reactions. *The five classes of nutrients are carbohydrates, fats, proteins, vitamins, and minerals.* In addition, an average adult needs about 2.5 L of water each day.

⟨16.1⟩ DIGESTION: IT'S JUST CHEMICAL DECOMPOSITION

Even before you swallow your lunch, chemical reactions begin the process of **digestion**. Like all biochemical processes, digestion is under the control of enzymes (biochemical catalysts, Section 15.8). Amylase, an enzyme in saliva, begins the digestion of the carbohydrates in starch. Carbohydrates are polymers of simple sugars (Section 15.2), and their digestion is the reverse of their formation—bonds are broken to set free the simple sugars. The same is true for digestion of proteins (polymers of amino acids) and the triglycerides in fats and oils (composed of glycerol and fatty acids).

$$\text{Carbohydrates} \xrightarrow{\text{Digestion}} \text{simple sugars (monosaccharides)}$$

$$\text{Proteins} \xrightarrow{\text{Digestion}} \text{amino acids}$$

$$\text{Fats (triglycerides)} \xrightarrow{\text{Digestion}} \text{glycerol + fatty acids}$$

Digestion begins in the mouth, continues in the stomach, and is completed in the small intestine. By the time a meal has been digested, all the nutrient molecules too large to cross cell membranes and enter the bloodstream have been converted into smaller molecules and absorbed.

Suppose you have a peanut butter and jelly sandwich for lunch. Digestion of starch from the bread begins as you chew and continues in your stomach. It is completed in the small intestine, where sugars from the jelly are also converted to monosaccharides. The indigestible carbohydrates (dietary fiber) from the bread and peanut butter pass unchanged through the small *and* large intestines.

Nutrition The science that deals with diet and health

Nutrient Substance in food that provides energy or biochemical raw material (fats, proteins, carbohydrates, vitamins, minerals)

Digestion The breakdown of food into substances small enough to be absorbed from the digestive tract

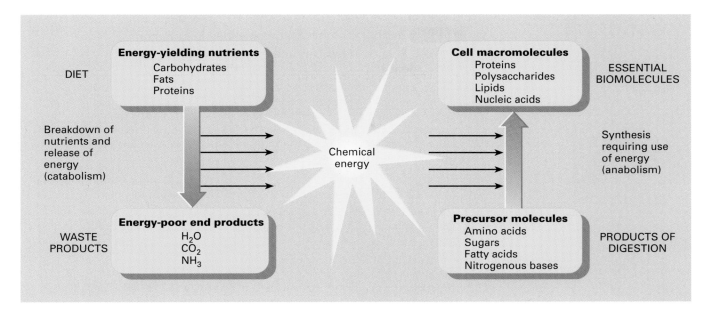

FIGURE 16.1 The release and use of chemical energy in the breakdown and synthesis of biomolecules.

Digestion of protein from the bread and peanut butter starts in your stomach, where large polypeptides are broken down into smaller ones. The job is finished in the small intestine, where the individual amino acids enter the bloodstream. Lipids, mainly fats and oils from the peanut butter, are digested mainly in the small intestine.

Once digestion is finished, what happens to the small molecules that have been produced? They may undergo (1) complete breakdown to produce energy, carbon dioxide, and water; (2) recycling into new biomolecules; or (3) placement into storage as triglycerides (fat) for future use (Figure 16.1). The energy yield from foods is described in the next section. The following sections examine the roles of the nutrients in our diets.

A peanut butter and jelly sandwich on firm, toasted white bread contains approximately 20 g of protein, 86 g of carbohydrate, and 34 g of fat.

16.2) ENERGY: USE IT OR STORE IT

How We Use Energy

A certain amount of energy is needed just to stay alive. While you are sitting completely still, your heart beats, your chest rises and falls as you breathe, your body temperature is maintained, chemical reactions proceed in cells, and messages that control these activities flow through your nervous system. The energy needed for these activities is the **basal metabolic rate (BMR)**. It is measured when a person is at rest at a comfortable temperature but not asleep, has not eaten for 12 hours, and has not engaged in vigorous activity for several hours.

The BMR is affected by many factors including body weight and activity level. An increased BMR can come from anxiety, stress, lack of sleep, low food intake, congestive heart failure, fever, increased heart activity, and the ingestion of drugs (including caffeine, amphetamine, and epinephrine). A decreased BMR can result from malnutrition, inactive tissue due to obesity, and low-functioning adrenal glands. Infants and children have higher BMRs than adults, and after early adulthood the BMR decreases about 2 to 3% per decade.

Basal metabolic rate (BMR) The minimum energy required by an awake but resting body

TABLE 16.1 Calories/Hour Expended in Common Physical Activities

Some examples of physical activities commonly engaged in and the average amount of food Calories a 154-pound individual will expend by engaging in each activity for 1 hour. The expenditure value encompasses both resting metabolic rate food Calories and activity expenditure. Some of the activities can constitute either moderate- or vigorous-intensity physical activity depending on the rate at which they are carried out (for walking and bicycling).

MODERATE PHYSICAL ACTIVITY	FOOD CALORIES/HOUR EXPENDED IN COMMON PHYSICAL ACTIVITY[a]
Hiking	370
Light gardening/yard work	330
Dancing	330
Golf (walking and carrying clubs)	330
Bicycling (<10 mph)	290
Walking (3.5 mph)	280
Weight lifting (general light workout)	220
Stretching	180
VIGOROUS PHYSICAL ACTIVITY	
Running/jogging (5 mph)	590
Bicycling (>10 mph)	590
Swimming (slow freestyle laps)	510
Aerobics	480
Walking (4.5 mph)	460
Heavy yard work (chopping wood)	440
Weight lifting (vigorous effort)	440
Basketball (vigorous)	440

[a] Food Calories burned per hour will be higher for persons who weigh more than 154 lbs (70 kg) and lower for persons who weigh less.
Source: Adapted from the 2005 Dietary Guidelines for Americans: **www.healthierus.gov/dietaryguideline**

As soon as voluntary activity begins, the metabolic rate speeds up. Some examples are given in Table 16.1.

Energy from Foods

Fats in the diet provide more than twice as many calories per gram as do carbohydrates and proteins.

Fats—9 kcal/g; 9 Cal/g   **Carbohydrates—4 kcal/g; 4 Cal/g**   **Proteins—4 kcal/g; 4 Cal/g**

For example, if a steak is 49% water, 15% protein, 0% carbohydrate, 36% fat, and 0.7% minerals, then 3.5 ounces of steak (about 100 g) would provide about 384 kcal, or 384 food Cal. Note that a food Calorie (Cal) is indicated with a capital C and is actually equivalent to 1000 calories (cal) or 1 kilocalorie (kcal). The caloric values of most foods are found as illustrated in the margin, and these are the values that are listed in diet books and on food labels. (Some representative values are given in Table 16.2.)

The human body, like everything else, is subject to the law of conservation of energy. In our case it translates into the following equation:

Energy taken in (food) = energy used + energy stored (fat)

---

1 food Calorie (Cal) = 1 kilocalorie (kcal) = 1000 calories (cal). Nutritional labels list food values in terms of Calories with a capital C.

Fifteen percent (15%) of 100 g is 0.15×100 g.

Energy from protein
100 g of steak $\times$ 0.15 = 15 g protein

$$15 \text{ g protein} \times \frac{4 \text{ kcal}}{\text{g protein}} = 60 \text{ kcal}$$

Energy from fat
100 g of steak $\times$ 0.36 = 36 g fat

$$36 \text{ g fat} \times \frac{9 \text{ kcal}}{\text{g fat}} = 324 \text{ kcal}$$

Total energy $\qquad$ 384 kcal

TABLE 16.2 The Approximate Percentages of Carbohydrates, Fats, Proteins, and Water in Some Whole Foods as Normally Eaten

FOOD	WATER	PROTEIN	FAT	CARBO-HYDRATES	kcal/100 g	FOOD	WATER	PROTEIN	FAT	CARBO-HYDRATES	kcal/100 g
Vegetables						*Meats and Fish (cont'd)*					
Spinach, raw	90.7	3.2	0.3	4.3	26	Salmon, broiled	63.4	27.0	7.4	0	182
Collard greens, cooked	89.6	3.6	0.7	5.1	33	Freshwater perch, raw	79.2	19.5	0.9	0	91
Lettuce, Boston, raw	91.1	2.4	0.3	4.6	25	Oysters, raw	84.6	8.4	1.8	3.4	66
Cabbage, cooked	93.9	1.1	0.2	4.3	20	*Grains and Grain Products*					
Potatoes, cooked	75.1	2.6	0.1	21.1	93	Wheat grain, hard	13.0	14.0	2.2	69.1	330
Turnips, cooked	93.6	0.8	0.2	4.9	23	Brown rice, dry	12.0	7.5	1.9	77.4	360
Carrots, raw	88.2	1.1	0.2	9.7	42	Brown rice, cooked	70.3	2.5	0.6	25.5	119
Squash, summer, raw	94.0	1.1	0.1	4.2	19	Whole wheat bread	36.4	10.5	3.0	47.7	243
Tomatoes, raw	93.5	1.1	0.2	4.7	22	White bread	35.8	8.7	3.2	50.4	269
Corn kernels, cooked on cob	74.1	3.3	1.0	21.0	91	Whole wheat flour	12.0	14.1	2.5	78.0	361
Snap beans, cooked	92.4	1.6	0.2	5.4	25	White cake flour	12.0	7.5	0.8	79.4	364
Green peas, cooked	81.5	5.4	0.4	12.1	71	*Dairy Products and Eggs*					
Lima beans, cooked	70.1	7.6	0.5	21.1	111	Milk, whole	87.4	3.5	3.5	4.9	65
Red kidney beans, cooked	69.0	7.8	0.5	21.4	118	Yogurt, whole milk	89.0	3.4	1.7	5.2	50
Soybeans, cooked	73.8	9.8	5.1	10.1	118	Ice cream	62.1	4.0	12.5	20.6	207
Meats and Fish						Cottage cheese	79.0	17.0	0.3	2.7	86
Lean beef, broiled	61.6	31.7	5.3	0	183	Cheddar cheese	37.0	25.0	32.2	2.1	398
Beef fat, raw	14.4	5.5	79.9	0	744	Eggs	73.7	12.9	11.5	0.9	163
Lean lamb chops, broiled	61.3	28.0	8.6	0	197	*Fruits, Berries, and Nuts*					
Lean pork chops, broiled	69.3	17.8	10.5	0	171	Apples, raw	84.4	0.2	0.6	14.5	58
Lard, rendered	0	0	100.0	0	902	Pears, raw	83.2	0.7	0.4	15.3	61
Calf's liver, cooked	51.4	29.5	13.2	4.0	261	Oranges, raw	86.0	1.0	0.2	12.2	49
Beef heart, cooked	61.3	31.3	5.7	0.7	188	Cherries, sweet	80.4	1.3	0.3	17.4	70
Brains	78.9	10.4	8.6	0.8	125	Bananas, raw	75.7	1.1	0.2	22.2	85
Chicken, whole, broiled	71.0	23.8	3.8	0	136	Blueberries, raw	83.2	0.7	0.5	15.3	62
Cod, raw	81.2	17.6	0.3	0	78	Red raspberries, raw	84.2	1.2	0.5	13.6	57
						Strawberries, raw	89.9	0.7	0.5	8.4	37
						Almonds	4.7	18.6	54.2	19.5	598
						Pecans	3.4	9.2	71.2	14.6	689
						Walnuts	3.5	14.8	64.0	15.8	651

For most people, the secret to dieting is little more than applying this equation. When energy taken in exceeds energy used, the excess enters storage, mostly as triglyceride molecules in fat cells. When energy taken in remains the same, but more energy is used, less is stored as fat. When energy used is more than energy taken in, some must be removed from storage. To be genuinely successful at providing weight loss, a program must integrate management of both diet and exercise.

Glucose is our major fuel molecule. It is the end product of carbohydrate digestion, and if more is needed there are biochemical pathways for making it from stored fat and even protein. These pathways are essential because glucose is the only fuel that the brain and red blood cells can use. To provide energy, glucose molecules are "burned" throughout our bodies by a tightly integrated series of biochemical reactions whose final products are carbon dioxide and chemical energy stored in adenosine triphosphate (ATP), whose structure was given in Figure 15.19. ATP is often called the body's energy currency. It carries energy to be spent wherever in the body energy is needed.

To estimate your basal metabolic rate, multiply your weight in pounds by 10 kcal/lb. Then, to get a rough estimate of your daily caloric needs, choose your general level of activity according to Table 16.3 and multiply your basal metabolic rate by the appropriate factor given in the table.

EXAMPLE 16.1 Daily Energy Needs

Julie weighs 110 lb and is a college freshman—she plays tennis or skates several times a week and keeps in shape by running every day. Using Table 16.3, estimate her daily caloric needs.

SOLUTION

Based on her weight, Julie's estimated basal metabolism rate (BMR) is

$$110 \text{ lb} \times 10 \text{ kcal/lb} = 1100 \text{ kcal} = 1100 \text{ food Calories}$$

From what we know about Julie, her activity level is moderate. Therefore, using the factor of 1.6 from Table 16.3, her estimated daily caloric needs are

$$1100 \text{ kcal} \times 1.6 = 1760 \text{ kcal} = 1760 \text{ food Calories}$$

TRY IT 16.1

Jack weighs 160 lb. As a graduate student he spends most of his time reading in the library or working at the computer. Estimate his daily caloric needs.

These two individuals have different levels of activity and different caloric needs.

Section 16.6 explains how to interpret food labels. Some additional examples of food and energy arithmetic are given in Section 16.10.

TABLE 16.3 Daily Energy Needs According to Physical Activity

Activity Level	Factor	
Very light		To estimate your daily energy needs, multiply
Men	1.3	your estimated BMR (10 kcal/lb × your
Women	1.3	weight in pounds) by the factor listed in
Light		the table that best represents your general
Men	1.6	level of daily activity:
Women	1.5	**Very light:** mostly sitting and standing activities
Moderate		**Light:** mostly walking activities
Men	1.7	**Moderate:** cycling, tennis, dancing
Women	1.6	**Heavy:** heavy manual digging, climbing,
Heavy		basketball, soccer
Men	2.1	
Women	1.9	

1. Digestion converts carbohydrates, proteins, and triglycerides into _____, _____, and _____ plus _____, respectively.
2. The quantity of heat required to operate the body at rest is the _____, which in kilocalories is about 10 times your _____.
3. A food Calorie has the same value as _____ kilocalorie(s).
4. The kilocalories per gram are 4 kcal, 9 kcal, and 4 kcal for _____, _____, and _____, respectively.
5. Our daily caloric needs are determined mainly by our weight and level of physical activity. (**a**) True (**b**) False
6. Which of the following is the body's major fuel molecule?
   (**a**) Triglyceride (**b**) Glucose (**c**) Maltose
7. Energy is carried to where it is needed by the molecule abbreviated as _____.

Foods high in dietary fiber.

16.3 SUGAR AND POLYSACCHARIDES: DIGESTIBLE AND INDIGESTIBLE

The major kinds of digestible carbohydrates in foods are the simple sugars (glucose and fructose), disaccharides (sucrose, maltose, and lactose), and polysaccharides (amylose and amylopectin in starch from plants, and glycogen from meat). The indigestible carbohydrates include cellulose and its derivatives, pectin (the substance that makes jam and jelly gel), and plant gums.

Carbohydrate structures are given in Section 15.2.

The indigestible polysaccharides are collectively referred to as **dietary fiber**. All dietary fiber comes from plants. There is insoluble fiber, mainly from the structural cellulose parts of plants, and soluble fiber—the gums and pectins. Barley, legumes, apples, and citrus fruits are foods with a high content of gums and pectins.

It has long been known that dietary fiber prevents constipation. Evidence has been accumulating that dietary fiber has significant other benefits. Adding soluble fiber to the diet decreases blood cholesterol levels, thereby decreasing the risk of heart disease. There is also evidence of a connection between insufficient dietary fiber and colorectal cancer. In our increasingly health-conscious society, such information provides a marketing advantage. We have all seen evidence of this in the bran muffins now found on every breakfast counter.

16.4 LIPIDS, MOSTLY FATS AND OILS

Most of the lipids in the diet, which we usually refer to as "fats," are triglycerides. We get them from meats and fish, vegetables and vegetable oils, and dairy products. They may be solid fats or oils, and they may incorporate saturated or unsaturated fatty acids (Section 12.3). The fat content of some common foods is shown in Figure 16.2.

Fatty tissue is composed mainly of specialized cells, each featuring a large globule of triglycerides. When their energy is needed, the triglycerides in fat cells are hydrolyzed to give glycerol and free fatty acids, which leave the cells and are transported to the liver. There, the fatty acids are broken down to two-carbon molecular fragments that can enter the main energy-producing pathway.

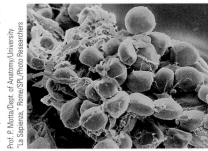

Where the fat goes. The yellow cells are adipocytes, the fat-storing cells of our bodies. Each adipocyte is almost entirely composed of a single droplet of triglycerides.

Dietary fiber Food polysaccharides that are not broken down by digestion

THE CHEMISTRY OF

LORENZO'S OIL (1992)

©2005 by Mark A. Griep, *University of Nebraska-Lincoln*

The movie *Lorenzo's Oil* (released in 1992) is a true story about the parents of a young boy named Lorenzo Odone. After Lorenzo is diagnosed in 1984 with adrenoleukodystrophy (ALD) and given less than two years to live, Lorenzo's parents learn everything about the disease, bring researchers and clinicians together, and devise a nutritional cure called Lorenzo's Oil. The movie ends with real boys saying that they were helped by Lorenzo's Oil.

The earliest symptom of ALD is attention deficit disorder, followed by rapid impairment of cognition, vision, hearing, and motor functions. Any one of 200 different mutations to the ABCD1 gene can cause ALD. The gene codes for ALD protein and is located on one end of the X chromosome.

The ALD protein is found primarily in liver peroxisomes, organelles that degrade fatty acids two carbons at a time to generate energy. When normal ALD protein dimerizes in the membrane, it transports very long chain fatty acids (VLCFA) into the peroxisome. The disease forms of ALD protein can't do this and blood levels of saturated VLCFA increase, namely C24:0 (lignoceric acid) and C26:0 (cerotic acid). Somehow, this causes demyelination of nerves and destruction of the adrenocortex. The myelin sheath allows nerves to conduct action potentials. The adrenocortex produces some hormones.

VLCFAs enter the body from two sources: diet and biosynthesis. There are only a few foods that are high in VLCFA (fatty meat, nuts, and the waxy coating of fruits and vegetables) so it is relatively easy to restrict dietary intake. The more difficult problem is turning off the natural biosynthetic pathway.

Lorenzo's Oil is a 4:1 mixture of the triglyerides of C22:1 (erucic acid with *cis*-C13) and C18:1 (oleic acid with *cis*-C9). It was prepared by cosmetic chemist Don Suddaby of Croda International (who played himself). His most difficult task was to isolate large quantities of erucic acid from rapeseed oil. Lorenzo's Oil inhibits the biosynthesis of the detrimental saturated VLCFA but not the helpful unsaturated VLCFA. It did not reverse demyelination.

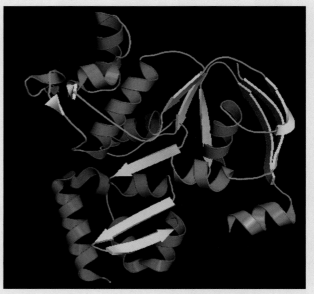

Protein with a structure that is similar to ALD protein.

Specifics to Consider

1. Draw erucic triglyceride and oleic triglyceride.

2. In the clip, Lorenzo's father used models such as paper clips to help him understand how the enzyme worked. Explain how Lorenzo's Oil alters the function of the fatty acid synthesizing enzyme.

3. When rats are fed a diet of erucic acid triglyceride, they soon die of heart disease. Lorenzo's mother and maternal aunt volunteered to establish its human safety. It was safe. Rats have a different fatty acid degradation pathway. What made these women good test subjects? Poor test subjects?

4. How do chemists synthesize triglycerides of fatty acids? Why would a cosmetic chemist be an expert at synthesizing esters?

5. Visit The Myelin Project website (**www.myelin.org**) to read about the latest ALD research.

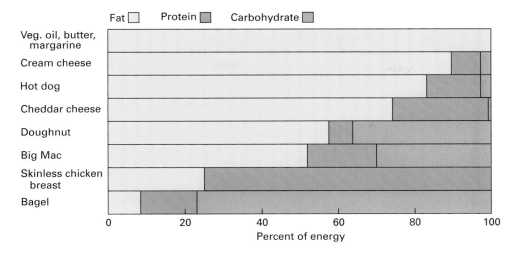

FIGURE 16.2 Percentage breakdown of energy provided by fats, proteins, and carbohydrates in some common foods.

A fat content of 20–35% is strongly recommended in the 2005 dietary guidelines for Americans released by the U.S. Department of Health and Human Services. The fat in today's diet is about 40% saturated, 40% monounsaturated, and 20% polyunsaturated. Lowering the saturated and monounsaturated fat and raising the polyunsaturated fat content of the diet is also strongly recommended. What is the basis for these recommendations? Heart disease is the primary cause of death in the United States (Section 17.1), and *atherosclerosis*, the buildup of fatty deposits called *plaque* on the inner walls of arteries, reduces the flow of blood to the heart. If a coronary artery is blocked by plaque, a heart attack occurs as a result of the reduced blood flow carrying oxygen to the heart. About 98% of all heart attack victims have atherosclerosis, and the major components of atherosclerotic plaque are saturated fatty acids and cholesterol.

The relation between blood levels of cholesterol and heart disease is well established. The more saturated fats and cholesterol in the diet, the higher the blood cholesterol is likely to be. Cholesterol, a lipid, has a waxy consistency, so to be transported in the bloodstream it must bond to a more water-soluble substance. Cholesterol combines with proteins to form **lipoproteins**, which are water soluble because of their many $-NH_3^+$ and $-COO^-$ ions. About 65% of the cholesterol in the blood is carried by **low-density lipoproteins (LDLs)**, whereas about 25% is carried by **high-density lipoproteins (HDLs)**. (The density difference is caused by the ratios of lipid to protein.)

LDLs are the "bad" cholesterol and HDLs are the "good" cholesterol referred to in discussions of heart disease. LDLs transport cholesterol away from the liver and throughout the body; they are therefore "bad" because they distribute cholesterol to arteries, where it can form the deposits of atherosclerosis. HDLs are "good" because they transport excess cholesterol from body tissues to the liver, where it is converted to bile acids that are needed in digestion.

Cholesterol occurs naturally *only* in dairy products, meat, and fish. There is no cholesterol in fresh vegetables, fruits, or vegetable oils.

The heart disease risk–blood cholesterol connection. The values given here are blood levels for persons more than 20 years of age. The values for children and adolescents are about 15% lower.

Risk Level	Total Cholesterol (mg/ 100 mL)	LDL Cholesterol (mg/ 100 mL)
Low	200 or lower	130 or lower
Moderate	200–239	130–159
High	240 or higher	160 or higher

Lipoproteins Assemblages of lipids, cholesterol, and water-soluble proteins

Low-density lipoproteins (LDLs) Lipoproteins that transport cholesterol from the liver to tissues throughout the body (the "bad" lipoproteins)

High-density lipoproteins (HDLs) Lipoproteins that transport cholesterol from body tissues to liver (the "good" lipoproteins)

16.5 PROTEINS IN THE DIET

Meat, fish, eggs, cheese, and beans are high-protein foods. The major function of dietary protein is to provide amino acids for new protein synthesis. Proteins are also the major dietary source of nitrogen for the synthesis of other kinds of nitrogen-containing biomolecules. Although the air we breathe is 78% nitrogen gas, this nitrogen cannot be used directly for biosynthesis.

THE WORLD *of* CHEMISTRY

THE DIETARY GUIDELINES FOR AMERICANS 2005

The Dietary Guidelines for Americans 2005 were released by the U.S. Department of Health and Human Services in January 2005. These guidelines focused more attention than ever on efforts that individuals should take to reduce calorie consumption and to increase physical activity. The report identified 41 key recommendations, of which 23 are for the general public and 18 are for special populations. They are grouped into nine general topics whose key recommendations are summarized here.

Adequate Nutrients Within Calorie Needs

- Consume a variety of nutrient-dense foods and beverages within and among the basic food groups while choosing foods that limit the intake of saturated and trans fats, cholesterol, added sugars, salt, and alcohol.

- Meet recommended intakes within energy needs by adopting a balanced eating pattern, such as the U.S. Department of Agriculture (USDA) Food Guide or the Dietary Approaches to Stop Hypertension (DASH) Eating Plan.

Weight Management

- To maintain body weight in a healthy range, balance calories consumed from foods and beverages against calories expended.

- To prevent gradual weight gain over time, make small decreases in food and beverage calories and increase physical activity.

Physical Activity

- Engage in regular physical activity and reduce sedentary behavior to promote health, psychological well-being, and a healthy body weight.

- To reduce the risk of chronic disease in adulthood, engage in at least 30 minutes of moderate-intensity physical activity, above usual activity at work or home, on most days of the week.

- For most people, greater health benefits can be obtained by engaging in physical activity of more vigorous intensity or longer duration.

- To help manage body weight and prevent gradual, unhealthy body weight gain in adulthood, engage in approximately 60 minutes of moderate- to vigorous-intensity activity on most days of the week and stay within caloric intake requirements.

- To sustain weight loss in adulthood, participate in at least 60 to 90 minutes of daily moderate-intensity physical activity and stay within caloric intake requirements. Some people may need to consult with a health-care provider before participating in this level of activity.

- Achieve physical fitness by including cardiovascular conditioning, stretching exercises for flexibility, and resistance exercises or calisthenics for muscle strength and endurance.

Food Groups to Encourage

- Consume a sufficient amount of fruits and vegetables while staying within energy needs. Two cups of fruit and 2½ cups of vegetables per day are recommended for a reference 2,000-Calorie intake; higher or lower amounts may be appropriate depending on recommended Calorie intake.

- Choose a variety of fruits and vegetables each day. In particular, select from all five vegetable subgroups (dark green, orange, legumes, starchy vegetables, and other vegetables) several times a week.

- Consume 3 or more ounce-equivalents of whole-grain products per day, with the rest of the recommended grains coming from enriched or whole-grain products. In general, at least half the grains should come from whole grains.

- Consume 3 cups per day of fat-free or low-fat milk or equivalent milk products.

Fats

- Consume less than 10% of Calories from saturated fatty acids and less than 300 mg/day of cholesterol, and keep trans fatty acid consumption as low as possible.

- Keep total fat intake between 20 to 35% of Calories, with most fats coming from sources of polyunsaturated and monounsaturated fatty acids, such as fish, nuts, and vegetable oils.

- When selecting and preparing meat, poultry, dry beans, and milk or milk products, make choices that are lean, low-fat, or fat-free.

- Limit intake of fats and oils high in saturated or trans fatty acids, and choose products low in such fats and oils.

Carbohydrates

- Choose fiber-rich fruits, vegetables, and whole grains often.

- Choose and prepare foods and beverages with few added sugars or caloric sweeteners; follow amounts suggested by the USDA Food Guide and the DASH Eating Plan.

- Reduce the incidence of dental caries by practicing good oral hygiene and consuming sugar- and starch-containing foods and beverages less frequently.

Sodium and Potassium

- Consume less than 2300 mg (approximately 1 tsp of salt) of sodium per day.
- Choose and prepare foods with little salt. At the same time, consume potassium-rich foods, such as fruits and vegetables.

Alcoholic Beverages

- Those who choose to drink alcoholic beverages should do so sensibly and in moderation—defined as the consumption of up to one drink per day for women and up to two drinks per day for men.

- Alcoholic beverages should not be consumed by some individuals, including those who cannot restrict their alcohol intake, women of childbearing age who may become pregnant, pregnant and lactating women, children and adolescents, individuals taking medications that can interact with alcohol, and those with specific medical conditions.
- Alcoholic beverages should be avoided by individuals engaging in activities that require attention, skill, or coordination, such as driving or operating machinery.

Food Safety

- To avoid microbial food-borne illness:

  Clean hands, food contact surfaces, and fruits and vegetables.

Meat and poultry should not be washed or rinsed.

Separate raw, cooked, and ready-to-eat foods when shopping for, preparing, or storing foods.

Cook foods to a safe temperature to kill microorganisms.

Chill (refrigerate) perishable food promptly and defrost foods properly.

Avoid raw (unpasteurized) milk or any products made from unpasteurized milk, raw or partially cooked eggs, foods containing raw eggs, raw or undercooked meat and poultry, unpasteurized juices, and raw sprouts.

Source: U.S. Food and Drug Administration Center for Food Safety and Applied Nutrition: **http://www.cfsan.fda.gov/ ~dms/lab–cons.html**

Excess amino acids from dietary proteins are not stored in the body. The nitrogen is converted to ammonia and then excreted as urea, while the carbon atoms are cycled to glucose and energy generation or stored as fat. Therefore, it is necessary to eat some protein every day. The average amount of protein in the American diet is, however, well beyond what is necessary, and a protein-deficient diet is rare in the United States. *Protein–energy malnutrition*, a group of disorders due to various combinations of deficient protein and energy intake, is most common in children in underdeveloped countries.

16.6 OUR DAILY DIET

Following the implementation of a comprehensive food-labeling law in 1994, the most readily available information about the nutritive value of food is on food packages. An example of the Nutrition Facts label is given in Figure 16.3. The total food Calories and food Calories from fat are listed at the top. Next comes a section that lists the weight in grams or milligrams per serving, plus the % Daily Value for the *macronutrients* judged most important to health. It is mandatory that this section include total fat, saturated fat, cholesterol, sodium, total carbohydrate, dietary fiber, sugars, and protein, as illustrated. Except for dietary fiber and protein, these are nutrients that should be limited.

The *reference daily values* are established as listed at the bottom of the food label and are further explained in Table 16.4. The percentage of daily values on a label (the *DVs*) are calculated from these values, as represented in a 2000-Cal daily diet. This many calories is about right for moderately active women, teenage girls, and sedentary men. Teenage boys, active men, and very active or pregnant women have higher caloric requirements.

The condition of extreme protein deficiency is called *kwashiorkor*. The name comes from an African dialect and translates as "the evil spirit that infects the first child when the second one is born." Kwashiorkor begins when the earlier child is no longer breast-fed and is switched to a carbohydrate-based diet. On this carbohydrate-based diet the older child no longer receives the necessary amino acids. If nutrition is improved before the condition has progressed too far, health can be restored.

Macronutrients are those nutrients needed by the body in large amounts.

Micronutrients are those needed by the body in small amounts—mostly vitamins and minerals.

One reason the FDA settled on listings in terms of 2000 Cal per day is because it is a nice round number. Others must adjust their interpretation of the labels according to their own needs. Labels on large packages, where there is room, include the data for a 2500 Cal per day diet.

The appendices to Dietary Guidelines for Americans 2005 contain recommendations of food from each food group for different food Calorie levels ranging from 1000–3200 food Calories/day.

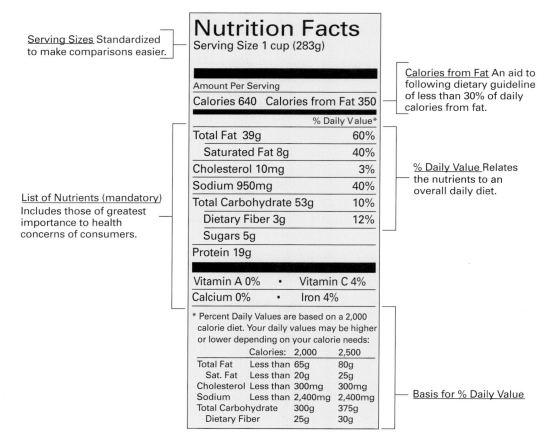

Serving Sizes Standardized to make comparisons easier.

List of Nutrients (mandatory) Includes those of greatest importance to health concerns of consumers.

Calories from Fat An aid to following dietary guideline of less than 30% of daily calories from fat.

% Daily Value Relates the nutrients to an overall daily diet.

Basis for % Daily Value

Nutrition Facts
Serving Size 1 cup (283g)

Amount Per Serving

Calories 640   Calories from Fat 350

% Daily Value*

Total Fat 39g	60%
Saturated Fat 8g	40%
Cholesterol 10mg	3%
Sodium 950mg	40%
Total Carbohydrate 53g	10%
Dietary Fiber 3g	12%
Sugars 5g	

Protein 19g

| Vitamin A 0% | • | Vitamin C 4% |
| Calcium 0% | • | Iron 4% |

* Percent Daily Values are based on a 2,000 calorie diet. Your daily values may be higher or lower depending on your calorie needs:

		Calories: 2,000	2,500
Total Fat	Less than	65g	80g
Sat. Fat	Less than	20g	25g
Cholesterol	Less than	300mg	300mg
Sodium	Less than	2,400mg	2,400mg
Total Carbohydrate		300g	375g
Dietary Fiber		25g	30g

Nutrients that may be listed at the manufacturer's option:

Calories from saturated fat
Stearic acid (meat and poultry products only)
Polyunsaturated fat
Monounsaturated fat
Potassium

Soluble fiber
Insoluble fiber
Sugar alcohol (e.g., the sugar substitute sorbitol)
Other carbohydrate
% Vitamin A as beta-carotene
Other essential vitamins and minerals

FIGURE 16.3 The Nutrition Facts label. An example of the label required by the law since 1994. Variations are allowed for small packages or foods that do not contain significant amounts of certain nutrients (e.g., vitamins and minerals need not be listed on canned soda).

TABLE 16.4	Basis for the Reference Daily Values on Nutrition Labels
Total Fat	30% of daily food Calories—an upper limit (65 g/2000 Calories daily diet)
Saturated Fat	10% of daily food Calories—an upper limit (20 g/2000 Calories)
Cholesterol	300 mg—a daily maximum for all diets
Sodium	2400 mg—a daily maximum for all diets
Total Carbohydrate	60% of daily food Calories (300 g/2000 Calories)
Dietary Fiber	23 g (based on 11.5 g/1000 Calories)
Sugars	No specified daily value
Protein	Listing the percentage daily value is not mandatory, but daily value is 10% of dietary food Calories from protein (50 g/2000 Cal) for adults (excluding pregnant women and nursing mothers) and children over 4 years old

Beneath the heavy line on every nutrition facts label are mandatory listings for vitamin A, vitamin C, calcium, and iron. Here also, these required listings are for those vitamins and minerals deemed to be of greatest public health concern. The major deficiency diseases of today are anemia, associated with iron deficiency, and osteoporosis, associated with calcium deficiency. The two vitamins in the mandatory list are those that decrease cancer risk.

CONCEPT CHECK 16B

1. Indigestible polysaccharides are known as _____ _____.
2. Most of the lipids in our diet are _____.
3. The _____ are "good" cholesterol because they _____.
4. The _____ are "bad" cholesterol because they _____.
5. Proteins are the major source of _____ in the diet.
6. Which nutrient is recommended to be decreased in a healthy diet?
7. Which kinds of foods are recommended to be increased in a healthy diet?
8. A major goal in a healthy diet is to have no more than _____ % of food Calories from fat.

16.7 VITAMINS IN THE DIET

A **vitamin** is an organic compound essential to health that must be supplied in small amounts by the diet. Vitamins provide no energy and are unchanged by digestion. The body does not synthesize vitamins, but needs them to function as **coenzymes**. There are two kinds of vitamins, fat soluble and water soluble.

The fat-soluble vitamins—A, D, E, and K—can be stored in the fatty tissues of the body (especially the liver). They have nonpolar hydrocarbon chains and rings that are compatible with nonpolar oils and fats. It is important to store enough fat-soluble vitamins, but not too much because excesses are not excreted and can be toxic.

The water-soluble vitamins are the eight B vitamins and vitamin C. Because excesses are excreted rather than stored, it had previously been assumed that overdosing on water-soluble vitamins has no toxic effects. With the rise in popularity of megadoses of vitamins, some toxic reactions have been observed. It does, however, take much greater quantities of water-soluble vitamins to create harmful effects than for the fat-soluble vitamins.

Fat-Soluble Vitamins

Vitamin A   Vitamin A, also known as retinol, is essential to vision because it is a component of vision receptors in the eye. Carrots, long associated with good vision, provide beta-carotene, not vitamin A itself. As it passes through the intestinal wall, beta-carotene is converted into vitamin A. Vitamin A also aids in the prevention of infection by barring bacteria from entering and passing through cell membranes. A single dose of vitamin A greater than 200 mg (660,000 international units [IU], a unit for vitamin doses used by nutritionists) can cause acute *hypervitaminosis A* in an adult, with symptoms that include vomiting, fatigue, and headache.

We have our choice of a wide variety of vitamin supplements.

Foods high in beta-carotene, which, once consumed, is converted to vitamin A.

Vitamin Organic compound essential to health that must be supplied in small amounts by the diet

Coenzyme A nonprotein molecule needed by an enzyme to make the enzyme function possible

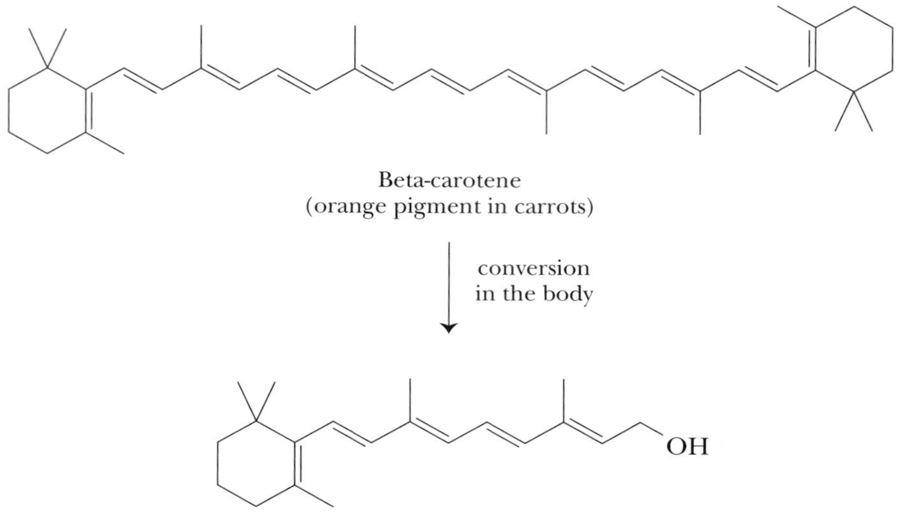

Beta-carotene
(orange pigment in carrots)

conversion
in the body

Vitamin A
(retinol)

Vitamin D Vitamin D is produced when ultraviolet light (UV) shines on the skin and triggers the conversion of a steroid known as *ergosterol* to vitamin D. Its major role is to help the body use calcium, and a deficiency causes rickets in children, the same condition caused by calcium deficiency. Vitamin D supplements are rarely needed, except by those who are almost never exposed to the sun. Both vitamin A and vitamin D are essential to normal growth and development. Overdosage of vitamin D can have serious consequences, however. Calcium deposits can form in the kidney, lungs, or tympanic membrane of the ear (leading to deafness). Infants and small children are especially susceptible to vitamin D toxicity.

Vitamin E Vitamin E is now well established as an *antioxidant* (Section 10.3). It is particularly effective in preventing the oxidation of polyunsaturated fatty acids to form peroxides (which contain —O—O— groups). Perhaps this is why vitamin E is always found distributed among fats in nature. The fatty acid peroxides are particularly damaging because they can cause runaway oxidation in the cells. Vitamin E protects the integrity of cell membranes, which are composed mainly of triglycerides. No toxicity has been specifically associated with large doses of vitamin E, although caution in taking megadoses is advised.

Vitamin E is the only vitamin destroyed by the freezing of food. However, it survives cooking in water.

Vitamin K We need a steady supply of vitamin K because the body uses it up rapidly. Its major and essential role is in regulation of blood clotting, which goes on to repair not only cut skin but also small tears in blood vessels (on a daily basis). Because overdoses cause excessive blood clotting and danger of brain damage, vitamin K is the only vitamin for which a prescription is needed for supplements containing the vitamin alone.

Water–Soluble Vitamins

All the water-soluble vitamins are coenzymes that have polar —OH, —NH$_2$, and —COOH groups, as illustrated on the following page by the structures of vitamin C and several of the B vitamins.

Charles D. Winters

Broccoli, a good source of vitamin K.

THE WORLD *of* CHEMISTRY

OLESTRA, A "FAT" FOR CALORIE COUNTERS

Take a look back at the structure of sucrose (a disaccharide of glucose and fructose, p. 359) and picture a long hydrocarbon chain attached to every —OH group in the molecule:

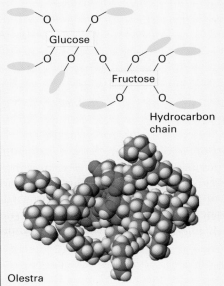

Olestra

The result is the compound olestra, developed by Procter and Gamble at a cost of $200 million dollars. The long hydrocarbon chains give it the properties of a fat or oil that are desirable in foods, but it is sufficiently different from natural triglycerides that digestive enzymes cannot break it down. Olestra (trade name, Olean) was the first synthetic nutrient to receive FDA approval for use in food products since aspartame (NutraSweet), which was allowed into the marketplace in 1983. Unlike some artificial fats that have been studied, a major advantage of olestra is that it does not break down at cooking temperatures.

Not only is the olestra molecule immune to enzyme attack, but it is also so large that it is not absorbed into the bloodstream through the walls of the intestine. It passes through the body unchanged, and therefore adds zero calories to a food. The first approval for olestra was limited to use in salty snack foods such as potato chips, crackers, and tortilla chips. A 1-oz serving of potato chips made with olestra adds no digestible fat and 75 Cal to the diet, as compared with 10 g of digestible fat and 150 Cal for regular potato chips. On further FDA review, olestra was approved for other fat-laden foods such as pastries and doughnuts.

During its development, long-term studies with laboratory animals revealed no tendency for olestra to cause cancer or other major health problems. In addition, tests showed that despite humans' inability to digest it, microorganisms common in soil and sewage can fully degrade it to carbon dioxide.

The dietary calorie reduction in olestra-containing foods does not come without a couple of potential difficulties, however, and its use has been vigorously criticized by some watchdog groups for this reason. One possible negative result of consuming large quantities of olestra is inhibition of absorption by the body of the fat-soluble vitamins, A, D, E, and K. To counter this effect, the FDA requires that these vitamins be added to olestra-containing foods. Another possible problem for some people, and especially if they consume large quantities of olestra, is abdominal cramping and diarrhea, difficulties that result from the nondigestible but oily nature of the compound. (Mineral oil, which contains no "minerals," is an old, traditional laxative that is an indigestible petroleum product.)

Olestra has now been on the market for almost ten years. Packages for foods containing it must carry the following information: "This product contains olestra. Olestra may cause abdominal cramping and loose stools. Olestra inhibits the absorption of some vitamins and other nutrients. Vitamins A, D, E, and K have been added."

Reference E. Doyle, *Journal of Chemical Education*, Vol. 74, pp. 370–372, April, 1997.

B Vitamins Vitamin B_6, considered the master vitamin, is known to be involved in 60 enzymatic reactions, many in the metabolism and synthesis of proteins. It is also needed for the synthesis of hemoglobin and the white blood cells of the immune system. Consuming more than 60 times the recommended daily quantity of vitamin B_6 causes nerve damage, and in huge doses (2–6 g per day) it can cause paralysis.

Large doses of niacin (50 mg per day or more) can also be toxic, with reactions ranging from skin rash and nausea to abnormal liver function and changes in heart rhythm. Niacin deficiency is common in countries with a corn-based diet.

When the hulls of wheat, rice, and other grains are removed and the kernels ground to produce flour, a large proportion of riboflavin and certain

Once thought to be a single substance, the eight B vitamins are often found together and interact so that a deficiency of one can cause deficiencies of others. The B vitamins are B_6, B_{12}, folic acid, niacin, thiamin, riboflavin, biotin, and pantothenic acid.

other vitamins and minerals is lost, as is the soluble fiber. To counteract this loss, since the 1940s flour sold in the United States has been *enriched* by adding back some of the lost vitamins and minerals. Customarily, enriched wheat flour contains added riboflavin, thiamin, niacin, and iron. When enriched flour came into use in the United States, the incidence of pellagra, the niacin deficiency disease, almost disappeared.

Thiamine

Riboflavin

In 1996, based on growing scientific evidence, the FDA ruled that folic acid should also be added to enriched flour. Folic acid can reduce risks of certain birth defects, anemia, and heart disease.

Vitamin C Vitamin C, *ascorbic acid*, helps destroy invading bacteria; aids the synthesis and activity of interferon, which prevents the entry of viruses into cells; and combats the ill effects of toxic substances, including drugs and pollutants. It also aids in the synthesis of collagen, which is present in cartilage, bone, tendons, and connective tissue, and is therefore important in the healing of wounds and for infants and pregnant women. In addition, vitamin C is an antioxidant.

Vitamin C is a strong enough acid to damage tooth enamel if vitamin capsules are chewed.

English sailors have been called "limeys" because the British admiralty in 1835 ordered a daily ration of lime juice to prevent scurvy, the vitamin C deficiency disease. The message did not hit home in America, however, where scurvy was common among troops during the Civil War.

Vitamin C

The role of vitamin C in preventing the common cold has long been debated, and numerous studies have found it to be either effective or ineffective. Currently, there is more evidence in favor of its ability to decrease the severity of cold symptoms than for any ability to prevent colds. It is reported, however, that one-third of the U.S. population takes vitamin C supplements to ward off colds. Daily doses of 1 g or more can cause diarrhea, nausea, and abdominal cramps in some people.

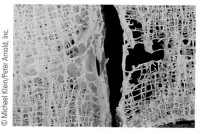

Osteoporosis. The effect of calcium loss on bone is shown by the comparison between a normal bone (*left*) and a bone afflicted with osteoporosis (*right*).

Mineral (as dietary nutrient) An element other than carbon, hydrogen, nitrogen, and oxygen that is needed for good health

⏺16.8 MINERALS IN THE DIET

As nutrients, **minerals** are elements, other than carbon, hydrogen, nitrogen, and oxygen, that are needed for good health. Although referred to on labels as "potassium" or "iodine" or so on, most minerals are present in foods, food supplements, and our bodies in ionic form and are not present in their elemental state. Not only does the human body need minerals, but these minerals must be maintained in balanced amounts, with no deficiencies and no excesses. Many of the body's minerals are excreted daily and must therefore be replenished each day.

The *seven macronutrient minerals* make up about 4% of body weight. They are *calcium, phosphorus, magnesium, sodium, potassium, chlorine,* and *sulfur;* their

TABLE 16.5   The Macrominerals*

NAME	SOURCES	MAJOR FUNCTIONS	DEFICIENCY SYMPTOMS	GROUPS AT RISK OF DEFICIENCY
Sodium (Na^+)	Table salt, processed foods	Major extracellular ion, nerve transmission, regulates fluid balance	Muscle cramps	Those consuming a severely sodium-restricted diet
Potassium (K^+)	Fruits, vegetables, grains	Major intracellular ion, nerve transmission	Irregular heartbeat, fatigue, muscle cramps	Those consuming diets high in processed foods, those taking high blood pressure medication
Chloride (Cl^-)	Table salt	Major extracellular ion	Unlikely	No one
Calcium (Ca^{2+})	Milk, cheese, bony fish, leafy green vegetables	Bone and tooth structure, nerve transmission, muscle contraction, blood clotting	Increased risk of osteoporosis	Postmenopausal women, teenage girls, those with kidney disease
Phosphorus†	Meat, dairy, cereals, and baked goods	Bone and tooth structure, buffers, membranes, ATP, DNA	Bone loss, weakness, lack of appetite	Premature infants, elderly
Magnesium (Mg^{2+})	Nuts, greens, whole grains	Reactions involving ATP, nerve and muscle function	Nausea, vomiting, weakness	Alcoholics, those with kidney and gastrointestinal disease
Sulfur‡	Protein foods, preservatives	Part of amino acids and vitamins, glutathione, acid–base balance	None when protein needs are met	No one

* Adapted from L. A. Smolin and M. B. Grosvenor: *Nutrition: Science & Applications.* Philadelphia: Saunders College Publishing, 1997.
† Phosphorus is covalently bonded in organic compounds and is also present in phosphate ions (PO_4^{3-}).
‡ Sulfur is covalently bonded in organic compounds and is also present in sulfate ions (SO_4^{2-}).

sources and functions are listed in Table 16.5. Except for the role of long-term calcium deficiency in osteoporosis, deficiencies of these minerals are rare because of their abundance in a variety of foods.

Sodium, potassium, and *chloride,* as ions (Na^+, K^+, and Cl^-), are essential to *electrolyte balance* in body fluids. Electrolyte balance, in turn, is essential for fluid balance, acid–base balance, and transmission of nerve impulses. Table salt is the principal source of sodium and chloride ions, and dietary deficiencies are unlikely. When there is extreme fluid loss through vomiting, diarrhea, or traumatic injury, electrolytes must be supplied to restore their concentration in body fluids.

About 99% of the *calcium ions* in the body are in bones and teeth. Together with sodium and potassium ions, calcium ions also participate in transmission of nerve impulses and regulation of the heartbeat. To be absorbed, calcium must be present in solution as Ca^{2+}. Absorption is enhanced in the presence of lactose (milk sugar) and also by a fatty meal, which passes through the intestine more slowly, allowing more time for absorption. A deficiency of vitamin D decreases calcium ion absorption and contributes to the bone deformities that accompany rickets.

A gradual loss of bone mass and density during adulthood is a normal process, and one of every three people older than 65 has some degree of

Electrolytes are substances that dissolve in water to produce ions that conduct electricity.

Potassium chloride (KCl) is sold as a salt (NaCl) substitute.

More than 10 years ago, an undergraduate chemistry student discovered that many calcium and vitamin pills hardly dissolved at all. Subsequent media reports claimed that undissolved vitamin pills could be observed in large-intestine X rays and septic tanks. The manufacturers of nutrition supplements have since adopted voluntary standards for disintegration and solubility of their products.

By definition, the micronutrient minerals, or trace elements, are those needed in quantities of 100 mg or less per day, or present in the body at less than 0.01% of body weight.

Charles D. Winters

A breakfast-cereal flake and iron, which the cereal contains. A close inspection of the labels of some breakfast cereals will reveal that some contain reduced (or elemental) iron, while others contain iron in its Fe^{2+} state as $Fe_3(PO_4)_2$.

Iron supplement pills should be stored out of reach of children, who may be poisoned by eating them.

A thorough medical evaluation of an individual with anemia is necessary because the condition can result from a variety of diseases as well as from nutritional deficiencies. Vitamin B_{12} and folic acid deficiencies can also cause anemia.

osteoporosis. The role of a calcium-deficient diet over a lifetime in hastening osteoporosis is sufficiently well established that a food label health claim on this basis is allowed. Postmenopausal white or Asian women of slight body build with inactive lifestyles are at greatest risk for osteoporosis. After menopause, the body produces less estrogen, which has the ability to slow bone dissolution. A preventive measure against later osteoporosis is consumption of adequate calcium during the adolescent years (ages 12 to 18), when bone is forming. Estrogen replacement for postmenopausal women is effective in slowing bone loss (Section 17.4).

Several minerals are essential in small amounts (micronutrient minerals). Of these, four—iron, copper, zinc, and iodine—are allowed to be listed on Nutrition Facts labels, iron being mandatory and the others optional. Other minerals known to be essential are selenium, manganese, fluorine, chromium, and molybdenum. Still other minerals (10 or more) are known to be essential in other mammals and are present and probably essential in humans. The list of trace minerals and our knowledge of their functions are constantly evolving.

Iodide ion has long been known to be essential to thyroid gland function. Hormones produced by the thyroid gland contain iodine and are responsible for growth, development, and maintenance of all body tissues. If there is a deficiency of iodine, the thyroid glands can become extremely enlarged, a condition called *goiter*. The routine use of iodized salt (0.1% potassium iodide) is the best way to ensure adequate iodine in the diet. More than half the table salt sold in the United States is iodized, and the use of such salt is mandatory in some countries.

Iron deficiency is associated with *anemia*—a decrease in the oxygen-carrying capacity of the blood, as indicated by low red blood cell or heme concentrations. Iron-deficiency anemia is the most common nutritional deficiency disease in both developed and underdeveloped nations. Early symptoms include muscle fatigue and lethargy. The ability to fight off invading bacteria is also diminished. Continued anemia creates defects in the structure and function of skin, fingernails, mouth, and stomach.

Zinc deficiency is known to cause poor growth and development, decreased immune function, and poor wound healing. Because meat is a better source of zinc than plant-based foods, vegetarians are among those persons at greatest risk of zinc deficiencies (others are the elderly and children with poor diets). Zinc supplements are often touted as able to improve immune function, enhance sexual performance, and increase fertility. It's a good idea to bring a degree of skepticism to such claims for zinc and other minerals. In the case of zinc, adding it to the diet of a person with a mild deficiency might improve wound healing, immune function, and appetite. Currently, however, there is little evidence for such changes in a healthy person with no existing zinc deficiency.

CONCEPT CHECK 16C

1. Vitamins are synthesized by cells of the body. (**a**) True, (**b**) False.
2. Vitamins A, D, E, and K are _____ soluble, while the B-complex vitamins and vitamin C are _____ soluble.
3. In the body, beta-carotene is converted into vitamin _____.
4. How many carbon atoms are there in beta-carotene and vitamin A, respectively?
5. Vitamins A, C, and E, and beta-carotene are all _____.
6. The B-complex vitamins function in the body as parts of _____.

Osteoporosis Reduction of bone mass associated with calcium loss

7. When vitamins are added to flour or other foods, the food is described as _____.

8. The chemical form of minerals in the diet is mostly as _____.

9. Sodium and potassium are essential to _____ and _____ balance and the transmission of _____ signals.

10. _____ is caused by iron deficiency, and _____ is caused by calcium loss.

16.9 FOOD ADDITIVES

Many chemicals with little or no nutritive value are added to commercially processed food (Figure 16.4). Some of these food additives serve to protect the food from being spoiled by oxidation, bacterial attack, or aging. Others add and enhance flavor or color. Still others control pH; prevent caking; or stabilize, thicken, emulsify, sweeten, leaven, or tenderize the food.

The GRAS List   The Food and Drug Administration lists hundreds of substances "generally recognized as safe" (GRAS) for their intended use by the FDA. The GRAS list was published in several installments in 1959 and 1960. It was compiled from the results of a questionnaire asking experts in nutrition, toxicology, and related fields to give their opinions about the safety of various seasonings, artificial flavorings, and other substances customarily added to foods. Since its publication, some substances have been removed, mostly due to suspicion that they are cancer-causing agents. *New* food additives—substances not on the original GRAS list—must receive FDA approval for use in foods based on test results submitted by the manufacturers.

Food Preservation   Oxidation and microorganisms (bacteria, fungi, and others) are the major enemies in the decomposition of food. Any process that prevents the growth of microorganisms or retards oxidation is generally an effective preservation process. Drying grains, fruits, and meat is one of the oldest

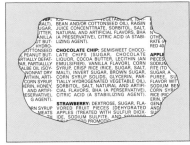

Note the variety of food additives described on the label of a package of cookies.

Some substances on the GRAS list.

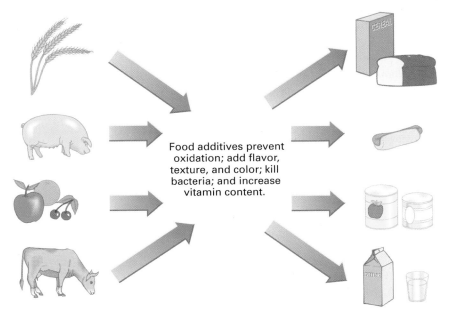

Food additives prevent oxidation; add flavor, texture, and color; kill bacteria; and increase vitamin content.

FIGURE 16.4 Food additives are needed to preserve and enhance many processed foods.

CH₃CH₂CO⁻ Na⁺

Sodium propionate

Sodium benzoate

A preservative must interfere with microbes but be harmless to the human system—a delicate balance. Two common preservatives are shown here.

Charles D. Winters

Because many individuals are allergic to sulfites, foods that contain more than 10 parts per million (ppm) of sulfites must have them listed on the label. The sulfites include chemicals such as sulfur dioxide (SO_2), sodium sulfide (Na_2S), and sodium bisulfite ($NaHSO_3$).

preservation techniques. Drying is effective because water is necessary for both the growth of microorganisms and the chemical reactions of oxidation.

There are also chemical additives that can preserve food. Salted meat and fruit in a concentrated sugar solution are protected from microorganisms. The dissolved sodium chloride or sucrose creates a **hypertonic solution** in which water flows by **osmosis** (page 245) from the microorganism to its environment. Thus, salt and sucrose have the same effect on microorganisms as dryness; both dehydrate them.

A preservative is effective if it prevents multiplication of microorganisms during the shelf life of the product. In general, food is spoiled by toxic substances secreted by the microorganisms. Sterilization by heat or radiation, or inactivation by freezing, is often undesirable, since they may affect the quality of the food. Chemical agents seldom achieve sterile conditions but can preserve foods for considerable lengths of time. Two common chemical preservatives in packaged foods are sodium benzoate (which is permitted in nonalcoholic beverages and in some fruit juices, fountain syrups, margarines, pickles, relishes, olives, salads, pie fillings, jams, jellies, and preserves) and sodium propionate (which can be used in bread, chocolate products, cheese, pie crust, and fillings).

Antioxidant additives in foods are serving the same purpose as antioxidant vitamins in the body (Section 10.3).

Antioxidants   The direct action of oxygen in the air is the chief cause of the destruction of the fats in food. Carbon–carbon double bonds in polyunsaturated fatty acids are particularly susceptible. Oxidation produces a complex mixture of volatile aldehydes, ketones, and acids that causes a rancid odor and taste. Foods kept wrapped, cold, and dry are relatively protected from air oxidation. The most common antioxidant food additives are butylated hydroxyanisole (BHA) and butylated hydroxytoluene (BHT), which act by releasing a hydrogen atom from their —OH groups as a free radical (H·).

Hypertonic solution A solution with a concentration of solute higher than that found in its surroundings

BHA (two isomers)   BHT

EDTA

FIGURE 16.5 Ethylenediaminetetraacetic acid (EDTA). The EDTA molecule has the remarkable ability to "sequester," or isolate, a metal ion (shown as M^{3+}) by forming six bonds to it—two from nitrogen atoms in amino groups and four from oxygen atoms in ionized carboxyl groups ($—COO^-$).

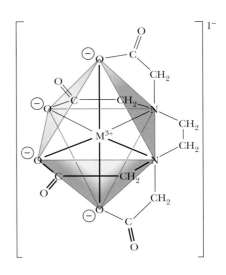

To sequester means "to withdraw from use." The sequestering ability of EDTA accounts for its use in treating heavy-metal poisoning.

Sequestrants Trace amounts of metals get into food from the soil and from machinery during harvesting and processing. Copper, iron, and nickel, as well as their ions, catalyze the oxidation of fats. The class of food additives known as **sequestrants** react with trace metals in foods, tying them up in complexes so the metals will not catalyze decomposition of the food. With the competitor metal ions tied up (sequestered), antioxidants such as BHA and BHT can accomplish their task much more effectively.

The sodium and calcium salts of ethylenediaminetetraacetic acid (EDTA) are common sequestrants in many kinds of foods and beverages. The structural formulas of free EDTA and EDTA bonded to a metal ion are shown in Figure 16.5.

Some 1700 natural and synthetic substances are used to flavor foods, making flavors the largest category of food additives.

Food Flavors Much of the sensation of taste in food is from our sense of smell. For example, the flavor of coffee is determined largely by its aroma, which in turn is due to a mixture of more than 500 compounds, mostly volatile oils. Most flavor additives, like many perfume ingredients, were originally derived from plants. The plants were crushed, and the compounds were extracted with various solvents such as ethanol or carbon tetrachloride. Sometimes a single compound was extracted; more often, the residue contained a mixture of several compounds. By repeated efforts, relatively pure oils were obtained. Oils of wintergreen, peppermint, orange, lemon, and ginger, among others, are still obtained in this way. These oils, alone or in combination, are then added to foods to produce the desired flavor. Today synthetic preparations of flavors are also common food additives.

In some people, MSG causes the so-called Chinese restaurant syndrome, an unpleasant reaction characterized by headaches and sweating that usually occurs after an MSG-rich Chinese meal. MSG is a natural constituent of many foods, such as tomatoes, strawberries, and mushrooms. These foods affect some individuals in the same way.

Flavor Enhancers Flavor enhancers have little or no taste of their own but amplify the flavors of other substances. They exert synergistic and potentiation effects. *Synergism* is the cooperative action of two different substances (or people, or anything) such that the total effect is greater than the sum of the effects of each used alone. *Potentiators* do not have an effect themselves but exaggerate the effects of other chemicals. Some nucleotides (Section 15.11), for example, have no taste but enhance the flavor of meat and the effectiveness of salt. Flavor enhancers were first used in meat and fish but now are also used to intensify flavors or mask unwanted flavors in vegetables, bread, cakes, fruits, and beverages. Three common flavor enhancers are monosodium glutamate (MSG), 5'-nucleotides, and maltol.

Monosodium glutamate
(commonly MSG)

Food Colors Some 30 chemicals are used as food colors. About half of them are laboratory synthesized and half are extracted from natural materials. Most food colors are large organic molecules with several double bonds and aromatic rings. Such structures have electrons that can absorb certain wavelengths of light and pass the rest; the wavelengths passed give the substances their characteristic colors. Beta-carotene (page 410), the orange-red

Sequestrant Chemical that ties up metal atoms or ions by forming bonds with them; used as food additives

substance in carrots and a variety of plants (and also an antioxidant), is an example of a natural food color that may be noted on the ingredient list as "artificial color." Synthetic food colors must be "certified food colors," tested by the FDA to ensure safety, quality, consistency, and strength of color; these must be listed by name.

The acid–base chemistry of buffers was discussed in Section 9.4. Buffers react as a base with an acid or as an acid with a base. The result is to maintain constant pH. (High pH is basic; low pH is acid.)

pH Control in Foods　Weak organic acids are added to such foods as cheese, beverages, and salad dressings to give a mildly acidic taste. They often mask undesirable aftertastes, but in some cases, such as in fruit-flavored sodas, the acidic taste is expected. Weak acids and acid salts react with bicarbonate to form CO_2 in the baking process. Buffers are also added to adjust and maintain a desired pH. Potassium acid tartrate, for example, is a buffer because it is a salt of an organic acid that can act as either an acid or a base.

Acidulants are food additives used to impart a sharp flavor. Citric acid is the most common acidulant.

The versatile *acidulants* also function as preservatives to prevent the growth of microorganisms, as antioxidants to prevent rancidity and browning, as viscosity modifiers in dough, and as melting point modifiers in such food products as cheese spreads and hard candy.

Hygroscopic substances absorb moisture from the air. You've probably had trouble getting salt out of the shaker during humid weather.

Anticaking Agents　Anticaking agents are added to *hygroscopic* foods (in amounts of 1% or less) to prevent caking in humid weather. Table salt is particularly prone to caking unless an anticaking agent is present. The additive (magnesium silicate, $MgSiO_3$, for example) incorporates water into its structure as water of hydration and does not appear wet like sodium chloride does when it absorbs water physically on the surface of its crystals. As a result, the anticaking agent keeps the surface of sodium chloride crystals dry and prevents crystal surfaces from co-dissolving and joining together.

One use of FDA-approved food dyes.

Stabilizers and Thickeners　Stabilizers and thickeners improve the texture and blends of foods. Like carrageenan, most stabilizers and thickeners are polysaccharides, which have numerous hydroxyl groups as a part of their structure. The hydroxyl groups form hydrogen bonds with water and help to provide a more even blend of the water and oils throughout the food. Stabilizers and thickeners are particularly effective in icings, frozen desserts, salad dressings, whipped cream, confections, and cheeses. In reduced-fat foods, they replace some of the "fatty" texture that would otherwise be missing. The plant gum thickeners (e.g., gum arabic, guar gum, locust bean gum, gum tragacanth) are also good sources of soluble dietary fiber (Section 16.3).

Foods containing modified food starch, pectin, and carbohydrate gums as stabilizers and thickeners.

16.10　SOME DAILY DIET ARITHMETIC

Sometimes doing a little arithmetic can help you interpret food labels or identify your own food needs. A few examples of very practical kinds of diet and food calculations follow.

EXAMPLE　16.2　**Daily Grams of Fat**

A good way to keep track of food Calories from fat is to monitor the total grams of fat in your diet every day. A moderately active 100-lb woman would get 30% of her daily 1750 food Calories by consuming 53 g of fat per day. She is deciding how many Yummie Bite cookies she can eat. Their nutrition label says that there are two cookies per serving and 45 Cal from fat per serving. How many cookies can she eat and leave 33 g of fat for other foods to be eaten that day? What percentage of her 53 g of fat will the cookies provide?

SOLUTION
To leave 33 g of fat for other foods means that she can eat enough cookies to contain

$$53 \text{ g fat} - 33 \text{ g fat} = 20 \text{ g fat}$$

With 9 Cal per gram of fat, the grams of fat per "serving" of two cookies is

$$\frac{45 \text{ Cal}}{\text{serving of 2 cookies}} \times \frac{1 \text{ g fat}}{9 \text{ Cal}} = 5 \text{ g fat per serving of 2 cookies}$$

With two cookies and 5 g of fat per serving, she can enjoy eight Yummie Bite cookies, which contain a total of 20 g of fat. Other foods eaten that day will have to be pretty low in fat content. The 20 g of fat in the eight cookies provides 38% of her daily fat allowance.

$$\frac{20 \text{ g fat (cookies)}}{53 \text{ g fat (daily fat)}} \times 100 = 38\%$$

TRY IT 16.2
Comparison of the labels gives the following information per serving of two kinds of soup

Healthy Chicken:	Calories	90	Calories from fat	15
Cream of Mushroom:	Calories	140	Calories from fat	80

How many grams of fat are there in one serving of each soup? For persons who wish to limit their daily fat consumption to 65 g, what percentage of this 65 g is in one serving of each?

EXAMPLE 16.3 **Caloric Value of Food and Exercise**

A slice of pepperoni pizza contains 10 g of protein, 20 g of carbohydrate, and 7 g of fat. (**a**) How many Calories does the pizza slice provide? (**b**) What percentage of the total food Calories is from fat? (**c**) How long would a person have to run at 5 mi/h to burn off the food Calories in the pizza slice (consult Table 16.1)?

SOLUTION
(**a**) The total caloric value of the pizza is calculated from the quantities of protein, carbohydrate, and fat

$$10 \text{ g protein} \times \frac{4 \text{ Cal}}{\text{g protein}} = 40 \text{ Cal}$$

$$20 \text{ g carbohydrate} \times \frac{4 \text{ Cal}}{\text{g carbohydrate}} = 80 \text{ Cal}$$

$$7 \text{ g fat} \times \frac{9 \text{ Cal}}{\text{g fat}} = 63 \text{ Cal}$$

$$\text{Total} \quad \overline{183 \text{ Cal}}$$

(**b**) Percentage of food Calories from fat

$$\frac{63 \text{ Cal (fat)}}{183 \text{ Cal (total)}} \times 100 = 34\%$$

Charles D. Winters

How much of these popular foods should you eat?

There is growing concern about the increasing population of obese children and adults in the United States. The rise in obesity (a body weight 20% or more above the defined healthy standard) is attributed to an excess of high-calorie foods and a shortage of exercise.

(**c**) Running at 5 mi/h burns ~10 Cal (total) per minute (590 Cal/hr). Therefore, to burn off this slice of pizza would require running for 18.3 minutes.

$$183 \text{ Cal} \times \frac{1 \text{ min}}{10 \text{ Cal burned}} = 18.3 \text{ min}$$

TRY IT 16.3

A slice of devil's food cake with chocolate frosting contains 3 g of protein, 8 g of fat, and 40 g of carbohydrate. (**a**) How many Calories does the slice of cake provide? (**b**) What percentage of the Calories is from fat? (**c**) How long could a person listen to a lecture fueled by the energy from this piece of cake?

CONCEPT CHECK **16D**

1. Two of the oldest means of preserving foods are _____ and adding _____.
2. GRAS is an acronym for _____.
3. Antimicrobial preservatives make food sterile. (**a**) True, (**b**) False.
4. BHA and BHT are very common food additives because they function as _____.
5. An important sequestrant goes by the initials _____.
6. The flavor of a food can usually be traced to a single compound. (**a**) True, (**b**) False.
7. Most stabilizers and thickeners (e.g., carageenan) are from the class of nutrients known as _____.
8. If a food yields 230 Cal per serving, of which 100 Cal are from fat, the percentage from fat is found from (_____ divided by _____) × 100%.
9. If you require 2000 Cal per day, then to find how many Calories from fat per day will give the maximum of 30%, you would do the following calculation: _____ × 2000 Cal.

Ⓦ Assess your understanding of this chapter's topics with an online chapter quiz at **www.brookscole.com/chemistry/joesten4**

■ KEY TERMS

nutrition	lipoproteins	mineral
nutrient	low-density lipoproteins (LDLs)	osteoporosis
digestion	high-density lipoproteins (HDLs)	hypertonic solution
basal metabolic rate (BMR)	vitamin	sequestrant
dietary fiber	coenzyme	acidulant

THE LANGUAGE OF CHEMISTRY

1. Chemical decomposition
2. Glucose
3. Calorie value per gram of fat
4. Calorie value per gram of protein
5. Causes atherosclerosis
6. Result of deficient dietary iodine
7. LDL
8. HDL
9. Butylated hydroxyanisole
10. Nutrient and food coloring
11. Sodium benzoate
12. Flavor enhancer
13. Loss of calcium from bone
14. Result of deficient dietary iron
15. Physical activity level
16. Sequesters metals

a. Dietary saturated fat
b. Goiter
c. Anemia
d. Digestion
e. Sodium EDTA
f. Monosodium glutamate
g. 4 Cal
h. Beta-carotene
i. Major fuel molecule
j. Affects daily caloric need
k. Good cholesterol
l. Osteoporosis
m. Food preservative
n. 9 Cal
o. Bad cholesterol
p. Antioxidant food additive

APPLYING YOUR KNOWLEDGE

1. What is meant by the word *digestion*?
2. What is amylase? To what class of biochemical molecules does it belong?
3. What are the relationships between mass and energy for the following substances?
   (**a**) Fats (**b**) Proteins
   (**c**) Carbohydrates
4. What are the products formed by digestion of the following nutrients?
   (**a**) Carbohydrates (**b**) Fats
   (**c**) Proteins
5. What is meant by *BMR* (or basal metabolic rate)?
6. What is meant by the following words and terms?
   (**a**) Simple sugars
   (**b**) Digestible carbohydrates
   (**c**) Indigestible carbohydrates
   (**d**) Disaccharides
   (**e**) Polysaccharides
   (**f**) Dietary fiber
7. What are triglycerides?
8. Give definitions for the following words and terms as applied to fats and fatty acids:
   (**a**) Saturated (**b**) Monounsaturated
   (**c**) Polyunsaturated (**d**) Partially hydrogenated

9. What do the following terms mean?
   (**a**) Atherosclerosis (**b**) Cholesterol
   (**c**) Plaque (**d**) Lipoproteins
   (**e**) Low-density lipoproteins
   (**f**) High-density lipoproteins
10. What are the following?
    (**a**) FDA (**b**) reference daily values
11. Give definitions for the following words:
    (**a**) Macronutrients (**b**) Micronutrients
12. What are the major categories of information required on a Nutrition Facts label?
13. Where do the following substances occur in nature?
    (**a**) Starches
    (**b**) Glycogen
    (**c**) Cellulose
14. What are micronutrient minerals? Which one must be listed on a Nutrition Facts label?
15. What is a vitamin? What is the function of a vitamin?
16. Overdoses of which of the following vitamins should definitely be avoided? Why?
    (**a**) Riboflavin (**b**) Vitamin C
    (**c**) Vitamin A (**d**) Vitamin B
    (**e**) Vitamin D

17. What is an electrolyte? What is meant by the *electrolyte balance* of the body?

18. What is the GRAS list?

19. What are the seven most abundant macronutrient minerals in the body?

20. What is the function of each of the following food additives?
    (**a**) Sodium benzoate
    (**b**) Sodium propionate
    (**c**) BHA (butylated hydroxyanisole)
    (**d**) BHT (butylated hydroxytoluene)
    (**e**) Sodium EDTA

21. Why are sequestrants needed as food additives?

22. What is the function of each of the following food additives?
    (**a**) Monosodium glutamate (MSG)
    (**b**) Yellow no. 5
    (**c**) Acetic acid
    (**d**) Calcium silicate
    (**e**) Carrageenan
    (**f**) Lecithin

23. What percentage of fat is currently recommended for today's diet by most public health authorities?

24. Lactose is sometimes referred to as "milk sugar" because it is the principal carbohydrate in milk. The lactose structure follows. Is this a monosaccharide, disaccharide, or polysaccharide? Explain.

25. Lactose intolerance occurs in people who are unable to hydrolyze milk sugar (lactose). This happens because they do not produce enough of the enzyme lactase. Normally infants and children have adequate amounts of lactase, but many adolescents and adults produce less. Should milk be promoted as a healthy food for the general population? Give one argument for and one against.

26. The body stores excess energy in small globules of fat in specialized cells of adipose tissue. The energy available from a gram of fat is 9 Cal. What would happen to the volume of these cells if the body stored energy by collecting carbohydrates instead of fats? Assume that the density of carbohydrate and fat are the same. (Recall that 1 g carbohydrate = 4 Cal.)

27. Cholesterol has the following structure. Why is cholesterol insoluble in water? Why is it recommended that people monitor their blood serum cholesterol levels? How many carbon atoms does a cholesterol molecule contain?

28. Folic acid is found in citrus fruits and leafy green vegetables. The FDA is now requiring that folic acid be added to cereals, breads, and pastas. The *American Journal of Public Health* says that such grain fortification would prevent 300 to 700 birth defects per year. At the level recommended for addition, approximately 3.25 million adults over 50 would receive too much folic acid. What question would you want answered before you would support such a nationwide program?

29. Why is osteoporosis a more significant problem for women after menopause? What preventive measures can be taken to reduce the chances of experiencing osteoporosis?

30. Should the government embark on a national program to add more calcium to our diets to reduce our chances of calcium deficiency and osteoporosis? Give one reason for such a program and one against.

31. What happens to the human thyroid gland if a person has an iodine deficiency? What is this condition called? What is the link between this problem and iodized salt?

32. When fats and fatty portions of food become rancid, what kind of reactions have probably occurred? What food additives are used to minimize these reactions?

33. Choose a label from a food item and try to identify the purpose of each food additive.

34. What foods have you eaten during the past week that did not contain any food additives?

35. What is the basal metabolic rate (BMR) for a 110-lb woman who has a moderate activity level?

36. What is the BMR for a 145-lb man who is sedentary, with very light activity? What would this person's BMR be if he started to exercise regularly so his activity increased to the moderate level?

37. How much time bicycling at 5 mi/h is needed to consume the 100 Cal available from a serving of tomato soup? (*Note:* Energy expended bicycling at 5 mi/h is 5 Cal/min.)

38. A 100-g serving of ice cream contains 207 Cal. How long would you have to run at 5 mi/h to consume the energy available from the ice cream?

39. Walking on a road at 3.5 mi/h consumes 3.5 Cal/min. How long would you have to walk to consume the energy available from 100 g of white bread? (*Note:* 100 g white bread = 269 Cal.)

40. A blood serum cholesterol level of 200 mg/100 mL is supposed to give a low heart disease risk level. How much total cholesterol would be carried by the blood serum of an adult with a blood volume of 12 pints? (1 pint = 473 mL.)

41. A blood serum cholesterol level of 240 mg/100 mL is supposed to give a high heart disease risk level. How much total cholesterol would be carried by the blood serum of an adult with a blood volume of 13 pints? (1 pint = 473 mL.)

42. The quantity of fat in a food is important to people who need to restrict fat intake. Which of the following would be the better low-fat choice, a piece of pecan pie weighing 113 g providing 459 Cal with 180 Cal from fat, or a piece of apple pie weighing 125 g providing 305 Cal with 108 Cal from fat?

43. How many Calories are in a fast-food chicken sandwich that contains 27 g of protein, 46 g of carbohydrate, and 34 g of fat?

44. How many kcal are in a fast-food bacon cheeseburger supreme if it contains 34 g of protein, 44 g of carbohydrate, and 46 g of fat?

45. An exceptionally active female volleyball player who weighs 145 lb wants to maintain a diet of 30% calories from fat, 60% from carbohydrates, and 10% from protein. How many Cal does the player need per day? How many grams of fat, carbohydrate, and protein does she need?

46. By using the following Nutrition Facts label, determine what percentage of the Calories in the product come from fat. What percentage of your daily Calorie need (1800 Cal) is provided by a serving? Would this product be a good food item for a person on a low-sodium diet? Justify your answers.

Nutrition Facts

Serving Size 1/2 cup (120mL) condensed soup
Servings Per Container About 6

Amount Per Serving

Calories 100   Calories from Fat 20

	% Daily Value*
Total Fat 2g	**3%**
Saturated Fat 0g	**0%**
Cholesterol 0mg	**0%**
Sodium 730mg	**30%**
Total Carbohydrate 18g	**6%**
Dietary Fiber 2g	**8%**
Sugars 10g	
Protein 2g	

Vitamin A 10%	•	Vitamin C 30%
Calcium 2%	•	Iron 4%

CHEMISTRY ON THE WEB

For up-to-date URLs, visit the text website at **www.brookscole.com/chemistry/joesten4**

- Vitamin and Mineral Fact Sheets
- Olestra
- Dietary Guidelines for America
- Beta-Carotene
- Vitamin B$_{12}$
- How Vitamin C Works

CHEMISTRY AND MEDICINE

A pharmacist dispenses a prescription to a customer. Almost 93 million prescriptions were written for the top prescription drug in 2004 in the United States alone.

© Royalty-free/CORBIS

Everyone is interested in developments that influence public health and, most particularly, their own personal health. Half of all Americans take at least one prescription medication and the demand for non-prescription, over-the-counter medications is enormous. For even a general understanding of the major public health issues the amount to be learned is large. Knowledge, however, is our personal first line of defense. With chemistry as the framework, in this chapter we take a look at the most important classes of diseases and the medications available to treat them. The major questions addressed are the following:

• What are some of the major health issues in the United States?

• How do antibiotics function?

- What is the cause of acquired immunodeficiency syndrome (AIDS)?
- What roles do hormones and neurotransmitters play in body chemistry?
- How do drugs enhance or counteract hormones and neurotransmitter actions?
- What are some important pain-killing and mood-altering drugs?
- What conditions can cold medications attack?
- What are the major chemical weapons against heart disease and cancer, and how do they work?
- What is homeopathic medicine, and is there a scientific basis for it?

The average life expectancy for men in the United States rose from 53.6 years in 1920 to 74.8 years in 2003. During this same period, the life expectancy for women rose from 54.6 years to 80.1 years.

What roles do chemistry and chemical technology play in this ongoing increase in life expectancy and the accompanying improvements in the quality of life?

Most importantly, remarkable progress has been made in recent years in understanding the chemical reactions that regulate biological processes. As this understanding grows, the old trial-and-error method of screening chemicals for use as drugs is being replaced. Instead, drug molecules are being designed to have exactly the molecular shape and chemical reactivity needed to counteract disease. Another positive outcome is a growing understanding of what makes for a healthy lifestyle.

17.1 MEDICINES, PRESCRIPTION DRUGS, AND DISEASES: THE TOP TENS

Americans spent more than $162 billion on prescription drugs in 2002, four times more than the amount spent in 1990. More than $15 billion of this is for **over-the-counter drugs**—those that can be bought in any supermarket or drugstore. The rest is for **prescription drugs**, which cannot be purchased without instructions from a physician. A drug is classified as one or the other by the U.S. Food and Drug Administration (FDA). In general, a substance is available only by prescription if it has potentially dangerous side effects, if it should be used only by people with specific medical problems, or if it treats a condition so serious that a person with that condition should be under a doctor's care.

The top ten prescription drugs, based on the number of prescriptions written during 2004 (Table 17.1), show an interesting cross-section of medicinal uses. In first place is a mixture of pain and cough-suppressant drugs, each available in lower dosages without a prescription. Five of the top 10 drugs prescribed (Atenolol, Furosemide, Hydrochlorothiazide, Norvasc, and Lisinopril) treat heart disease and a related cardiovascular condition of high blood pressure (**hypertension**). The sixth and eighth most-prescribed drugs (amoxicillin and Zithromax) are antibiotics used to treat common bacterial infections.

As illustrated in Table 17.1, all drugs have a **trade name** and a **generic name**. The trade name (brand name) is the name used by the drug manufacturer. For example, the antibiotic amoxicillin is sold under trade names such as Amoxil, Amoxidall, Amoxibiotic, Infectomycin, Moxaline, Trimox, Utimox, and Wymox. *Amoxicillin* is a generic name, which is a drug's generally accepted common chemical name. Once the patent protection on a drug has expired, it can be manufactured and marketed competitively by many companies and prescribed by its generic name. Often, prescriptions written by the generic name are cheaper to fill.

A systematic chemical name for amoxicillin, from which a chemist should be able to write its complete structure, is 6-(p-hydroxy-α-aminophenylacetamido)penicillinic acid.

Over-the-counter drugs Drugs that can be purchased without a prescription

Prescription drugs Drugs that can be purchased only at the direction of a physician

Hypertension Elevated blood pressure

Trade name The manufacturer's name for a drug

Generic name The generally accepted chemical name of a drug

TABLE 17.1 Top Ten Drugs Prescribed in 2004

TRADE NAME	GENERIC NAME	DRUG CLASS
Hydrocodone w/APAP	Hydrocodone w/APAP	Analgesic, narcotic
Lipitor	Atorvastatin	Cholesterol-lowering agent
Lisinopril	Lisinopril	Antihypertensive
Atenolol	Atenolol	Antihypertensive
Synthroid	Levothyroxine	Thyroid hormone
Amoxicillin	Amoxicillin	Antibiotic
Hydrochlorothiazide	Hydrochlorothiazide	Diuretic/Antihypertensive
Zithromax	Azithromycin	Antibiotic
Furosemide	Furosemide	Diuretic/Antihypertensive
Norvasc	Amlodipine	Antihypertensive

SOURCE: **http://www.rxlist.com/top200.htm**

In 1900, 5 of the 10 leading causes of death for individuals of all ages in the United States were infectious diseases (pneumonia and influenza, tuberculosis, gastrointestinal infections, kidney infections, diphtheria). Currently, pneumonia and influenza (tabulated together) are the only infectious diseases from this group remaining in this top 10, causing 2.7% of all deaths of individuals of all ages (Table 17.2). This dramatic decrease in deaths from infectious diseases can be attributed in large measure to the development of antibiotics.

TABLE 17.2 Leading Causes of Death and Total Deaths in 2003*

TOTAL DEATHS	ALL AGES 2,443,930		AGES 15–24 33,022		AGES 25–34 128,924	
	%	Rank	%	Rank	%	Rank
Heart disease	28.0	1	3.3	5	12.6	3
Cancer	22.7	2	4.9	4	14.8	2
Stroke	6.5	3	0.6	8	2.3	8
Chronic lower respiratory disease	5.2	4	0.5	9	2.6	7
Accidents	4.3	5	45.1	1	21.6	1
Diabetes mellitus	3.0	6	—	—	2.1	9
Influenza and pneumonia	2.7	7	0.7	7	1.0	10
Alzheimer's disease	2.6	8	—	—	—	—
Kidney disease	1.7	9	—	—	—	—
Blood poisoning	1.4	10	—	—	—	—
Suicide	—	—	11.9	3	8.7	4
Homicide	—	—	15.6	2	5.7	5
HIV	—	—	0.5	10	5.3	6
Congenital and genetic abnormalities	—	—	1.3	6	—	—

* Data in each group are percentages of all deaths. Deaths from other causes are not included, so values do not add up to 100%.
SOURCE: National Vital Statistics Report, vol. 53, no. 15, 28 Feb. 2005.
www.cdc.gov/nchs/data/nvsr53/nvsr53_15.pdf.

Meanwhile, AIDS (acquired immunodeficiency syndrome), an infectious disease that is caused by the human immunodeficiency virus (HIV), has come onto the scene. In 1993, HIV infection hit the news by moving ahead of all other causes of death for those in the 25- to 44-year-old age group. Only three years earlier, in 1990, HIV infection placed third as the cause of death in this age group, with accidents and cancer coming first and second. Although AIDS continues to be a very serious health issue, AIDS deaths in the United States have dropped dramatically thanks to intensive educational programs regarding AIDS prevention and the development of drugs that, although unable to cure AIDS, can prolong life expectancy dramatically. AIDS deaths now rank as the sixth most-common cause of death in the 24–44 age group in the United States. Unfortunately, the cost of these drugs has, at least in part, prevented their use on a worldwide basis. AIDS continues to be a devastating disease, especially in sub-Saharan Africa.

A sad reflection of the social condition in the United States is the fact that accidents, homicide, and suicide rank as the first, second, and third most common causes of death, respectively, in the 15–24 age group. In the 24–44 age group, accidents still rank as the primary cause of death, with suicide now occupying the fourth position, followed by homicide in fifth place.

17.2 DRUGS FOR INFECTIOUS DISEASES

Modern chemotherapy—the treatment of disease with chemical agents—began with the work of Paul Ehrlich (see *The Personal Side*). In 1904, he concluded that **infectious diseases** could be conquered if chemicals could be found that attack disease-causing microorganisms without harming the host. After observing that dyes used to stain bacteria also killed the bacteria, he developed arsenic compounds similar to the dyes. One of these compounds (*arsphenamine*) was the first effective drug for an infectious disease. It revolutionized the treatment of syphilis at the time it was introduced. Syphilis is now treated with penicillin, the oldest of the antibiotics that are now our principal weapons against infectious diseases.

Strictly speaking, an **antibiotic** is a substance produced by a microorganism that inhibits the growth of other microorganisms. Any compound, whether natural or synthetic, that acts in this manner is now referred to as an "antibiotic," however. A person falls victim to an infectious disease when invading microorganisms multiply faster than the body's immune system can destroy them. Antibiotics help the immune system by either destroying invaders or preventing their multiplication.

Viruses, bacteria, fungi, and other parasites cause infectious diseases. Ehrlich referred to his chemotherapeutic drugs for infectious diseases as "magic bullets"—they killed the microorganism but not the host.

Infectious diseases Diseases caused by microorganisms

Antibiotic A substance that inhibits the growth of microorganisms

The enemies: bacteria and viruses. (a) The blue rods are *Haemophilus influenza* bacteria lying on human nasal tissue. This bacterium causes bronchitis, pneumonia, and a wide range of diseases in children, including meningitis and blood poisoning. (b) The dots are *Rubella* viruses emerging from the surface of an infected cell, where the viruses are being reproduced. *Rubella* is the virus that causes German measles.

a. © Dr. Tony Brain/SPL/Photo Researchers, Inc.
b. © NIBSC/Photo Researchers, Inc.

(a)

(b)

THE PERSONAL SIDE

Paul Ehrlich (1854–1915)

Paul Ehrlich, a German bacteriologist, was a pioneer in the application of chemistry to medicine. While he was a medical student, he became fascinated with chemistry and pursued learning about it on his own. He is an excellent example of a scientist driven by curiosity to do experiments that had incredibly successful practical applications. While investigating the biological properties of the many synthetic dyes available from the German chemical industry, he originated the microscopic study of blood cells. His work with the serum of immunized animals was fundamental to the then-young science of immunology (study of the body's response to and defense against disease-causing invaders). As noted in the text, he created the concept of chemotherapy.

© SPL/Photo Researchers, Inc.

PAUL EHRLICH

Ehrlich was the first to explain the action of toxic substances and drugs. In what he called his "side-chain" theory, he explained that these substances affect only cells that have matching molecular fragments extending from their surfaces. Today we call these side chains *receptors*—the parts of cell surface molecules that extend outside the cell and interact with external messengers and drugs. They are still the object of intensive research.

Penicillins   The penicillins were discovered by Sir Alexander Fleming, a bacteriologist at the University of London. He was working in 1928 with cultures of *Staphylococcus aureus*, a bacterium that causes boils and some other infections. One day he noticed that one culture was contaminated by a blue-green mold. For some distance around the mold growth, the bacterial colonies had been destroyed. On further investigation, Fleming found that the broth in which this mold had grown had a similar lethal effect on many **pathogenic** (disease-causing) bacteria. The mold was later identified as *Penicillium notatum* (the spores sprout and branch out in pencil shapes, hence the name). Although Fleming showed that the mold contained an antibacterial agent, which he called *penicillin*, he was not able to purify the active substance.

In 1940, the active ingredient, penicillin G, was identified, and by 1943 it was available for clinical use. As World War II drew to a close, penicillin G was saving many lives threatened by pneumonia, bone infections, gonorrhea, gangrene, and other infectious conditions. Since then, numerous penicillins have been developed. Amoxicillin, the most-prescribed antibiotic, is a penicillin.

Howard Florey and Ernest Chain, who isolated penicillin, and Sir Alexander Fleming, who discovered it, shared the Nobel Prize for medicine and physiology in 1945.

Another *Penicillium* strain that proved to be an excellent source of a new antibiotic was discovered on a moldy cantaloupe in a Peoria, Illinois market.

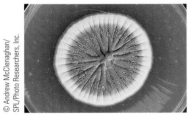

© Andrew McClenaghan/ SPL/Photo Researchers, Inc.

A *Penicillium notatum* culture.

Penicillin G, the first penicillin

$$R = -NH-\overset{\overset{\displaystyle O}{\|}}{C}-CH_2-\bigcirc$$

Amoxicillin, a broad-spectrum antibiotic— active against a variety of bacteria

$$R = -NH-\overset{\overset{\displaystyle O}{\|}}{C}-\overset{\overset{\displaystyle H}{|}}{\underset{\underset{\displaystyle NH_2}{|}}{C}}-\bigcirc-OH$$

General penicillin structure

Pathogenic Capable of causing disease

All penicillins kill growing bacteria by preventing normal development of their cell walls. Unlike our cells and those of other mammals, bacteria rely on a rigid cell wall. The rigidity is maintained by cross-linking bonds between peptide chains. Penicillins inhibit the enzyme that forms these cross-links and prevent the reaction from occurring, causing the cell to burst.

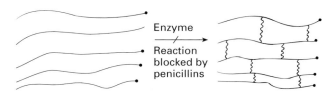

Other Classes of Antibiotics The *cephalosporins* are similar in structure to the penicillins and also act by disrupting cell wall synthesis. They are widely used in hospitals because of their low toxicity and broad range of antibacterial activity. All cephalosporins have the same general structure in which R_1 is often a hydrogen and R_2 and R_3 are substituents with several functional groups.

General cephalosporin structure

R groups in cephalexin, used to treat streptococcal infections

$R_1 =$ —H

$R_2 =$ —CH_3

> Some other antibiotics that act by inhibiting cell wall synthesis are bacitracin (used only on the skin, not internally) and vancomycin (important for treating bacteria resistant to other antibiotics).

Other kinds of antibiotics attack bacteria in different ways from the penicillins and cephalosporins. Many interfere with the synthesis or functioning of bacterial DNA. The *tetracyclines* (e.g., Achromycin and Terramycin) and *erythromycin*, for example, inhibit bacterial protein synthesis, and *rifampin* (important in treating tuberculosis) inhibits RNA synthesis from DNA.

The defeat of infectious diseases by antibiotics was once thought to be complete. That was a mistake. After more than a half-century of use, many antibiotics are not nearly as effective as they once were. Strains of malaria, typhoid fever, gonorrhea, and tuberculosis have emerged that are resistant to the antibiotics that once defeated them. Researchers have begun to hunt for new antibiotics, in some cases resuming a search that had been abandoned because it was believed unnecessary. The urgency of the search was heightened with the discovery in 1997 of a strain of vancomycin-resistant staph (*Staphylococcus aureus*), the bacteria responsible for deadly infections among the seriously ill in hospitals. Vancomycin has been the only antibiotic able to cure such staph infections. In fact, vancomycin has been designated as the antibiotic of last resort, only to be used when all others fail.

According to the Centers for Disease Control and Prevention (CDC), almost two million patients in the United States pick up an infection *in the hospital* each year. Of those patients, about 90,000 die each year as a result of their infection. This number was about 13,300 in 1992. More than 70 percent of the bacteria that cause hospital-acquired infections are resistant to at least one of the drugs commonly used to treat them.

Strains of *S. aureus* resistant to methicillin are common in hospitals and these bacteria are increasingly being found in non-hospital settings such as locker rooms. Minor scrapes incurred by high school football players and wrestlers have become infected and have led to amputation of limbs and even death in very short periods of time. The problem has become so serious that it was the focus of an article in *Sports Illustrated* in the spring of 2005.

THE CHEMISTRY OF

ME & ISAAC NEWTON (1999)

© 2005 by Mark A. Griep, University of Nebraska-Lincoln

M*e & Isaac Newton* (released in 1999) is an inspiring documentary in which scientists from seven diverse fields explain why they do what they do. Gertrude Elion represents pharmaceutical chemists and their industry. The other six scientists and engineers are Ashok Gadgil, water purification; Michio Kaku, string theory; Maja Mataric, robotics; Steven Pinker, language disorders; Karol Sikora, gene therapy; and Patricia Wright, lemurs. Each of the scientists describes how they became interested in science, what type of work they do, their first "eureka!" moment, taking risks in science, and their thoughts about the future of science.

In the film, Elion describes the first major drug she and her senior colleague George Hitchings developed in the early 1950s. She says that it has cured many cases of childhood leukemia but she does not name the drug or show its structure. In 1988, she and Hitchings shared the Nobel Prize in Medicine for their discovery and their nucleoside metabolism methods.

Leukemia is a cancer of the blood-forming cells (usually white blood cells) located in bone marrow that spreads to the blood. Leukemia accounts for one-third of all cancers in children. Every year about 2800 American children develop leukemia. The exact cause is unknown, although it sometimes occurs in conjunc-

Gertrude Elion in 1988.

tion with certain rare genetic disorders. Today, the 5-year survival rate is 85% because of chemotherapies such as the one discovered by Hitchings and Elion.

Questions to Consider

1. Find the full chemical name, the drug name, and chemical structure of the anti-leukemic developed by Hitchings and Elion.

2. Learn about the symptoms, diagnoses, and treatments for childhood leukemia, which often strikes before age three, from the American Cancer Society at **www.cancer.org**.

While the race for new antibiotics is on, the medical community endeavors to make everyone aware of some basic facts:

- Doctors should take care not to overprescribe antibiotics.
- People should not demand antibiotic treatment in cases where it is not necessary (e.g., antibiotics have no effect on viral infections).
- Antibiotics must always be taken for the full period of the prescription.
- Everyone should wash their hands frequently and thoroughly.

17.3 AIDS, A VIRAL DISEASE

Since its first appearance in the 1980s, HIV (human immunodeficiency virus) infection has spread rapidly and worldwide. In the United States, the first 100,000 cases of HIV infection were reported over the course of the 8 years from 1981 to 1989. The second 100,000 cases were detected in the following two years. An estimated 850,000 to 950,000 persons in the United States were living with HIV in 2003. About 25–35% of these people did not know they are infected. Worldwide, as estimated by the World Health Organization, about 40 million individuals were HIV infected as of 2004.

Fewer than a dozen drugs are available for attacking viral diseases (as opposed to treating their symptoms).

From a chemical perspective, it is important to understand that very few ways are known to combat viral diseases (other than vaccination). The reason for this is the method of attack used by viruses—they are essentially chemical parasites. They take over the DNA of human cells and put it to work for their own reproduction. It is difficult to attack the cells reproducing the virus without also attacking the host's own cells.

HIV is a *retrovirus*. It consists of an outer double lipid layer surrounding a matrix containing proteins, an enzyme called *reverse transcriptase*, and RNA. The word *retrovirus* is used because the virus enzyme carries out RNA-directed synthesis of DNA rather than the usual DNA-directed synthesis of RNA (see Figure 15.23).

The AIDS retrovirus penetrates the T cells of the immune system (1 in Figure 17.1). Once inside, the virus releases its contents, and the *reverse transcriptase* of the AIDS virus translates the RNA code of the virus into double-stranded DNA (2 in Figure 17.1). The virus DNA enters the T cell nucleus and is incorporated into the cell's own DNA (3 in Figure 17.1) Then the T cell makes RNA from viral DNA, the proteins needed for a new virus are made from this RNA (4 in Figure 17.1), and the new virus is released (5 in Figure 17.1). Eventually the T cell swells and dies, releasing more AIDS viruses to attack other T cells. As their T cells are destroyed, individuals are attacked by diseases that are normally defeated by the body's immune system and thus are rare in healthy individuals.

The search for weapons against HIV reached a turning point in 1996. Earlier efforts had yielded drugs that somewhat slow the progress of AIDS, but in no way approach a cure of the disease. In early 1996, three members of a new class of drugs directed at HIV, the *protease inhibitors*, were approved in rapid succession. Convincing clinical trials had shown that a protease inhibitor taken in *combination* with older AIDS drugs dramatically lowers the concentration of the virus in patients' blood, often to below detectable levels. The drug combination diminishes the number of serious complications and decreases the death rate among seriously ill individuals. Apparently, a combination of drugs overcomes the major problem with single-drug therapy—if even a few virus particles are left alive by a drug treatment, the virus can very quickly adapt by converting to a drug-resistant form.

The United Nations has warned that the number of people in China with AIDS could rise to 10 million by 2010.

FIGURE 17.1 Attack of HIV on a T cell (a lymphocyte). The virus enters the lymphocyte (1) and produces its own DNA (2). The DNA then enters the lymphocyte nucleus (3), where it combines with the lymphocyte's DNA. Then, through the cell's normal processing, DNA is transcribed to messenger RNA and, back outside the nucleus, the viral RNA provides the template for the production of proteins (4) that assemble into a new virus particle (5).

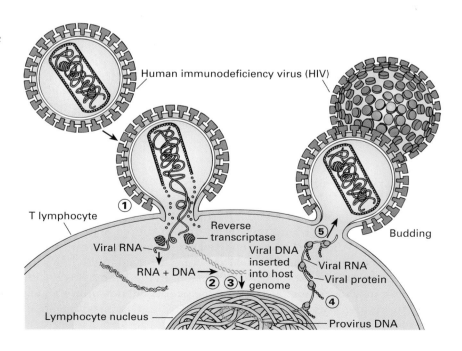

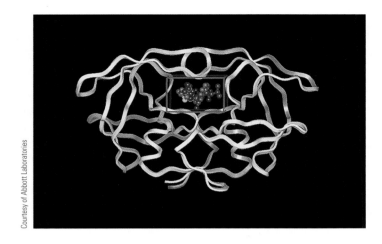

Courtesy of Abbott Laboratories

As their name implies, protease inhibitors act by inhibiting a protease enzyme. By fitting into the active site of the enzyme (Figure 17.2), the inhibitors prevent the enzyme from functioning. Because this enzyme cuts a long protein chain into smaller proteins essential to survival of the HIV virus, halting its action kills the virus.

The most successful older anti-AIDS drug, *azidothymidine* (AZT, also called *zidovudine*), is one of the drugs used in combination with a protease inhibitor. AZT is a derivative of deoxythymidine, a nucleoside. (Nucleosides are nucleotides without the phosphate group; see Section 15.11.) Because AZT resembles a natural nucleoside, it is accepted by the HIV reverse transcriptase enzyme and placed into the viral DNA. Once there, its structure prevents additional nucleosides being added to the chain, and cell division is halted. AZT has been found to reduce the mortality of babies born to HIV-positive mothers to close to zero.

Despite the new air of optimism, it is important to keep the situation in perspective relative to several facts. The new drug combination is not yet a cure. It is extremely expensive, it is unavailable in many third world countries, and it can have unpleasant side effects. An ominous further problem is that one must take up to 20 pills a day on a precisely timed schedule; wavering from the schedule may provide opportunities for new drug-resistant variations of the virus to develop. As of 2004, three new AIDS drugs (Enfuvirtide, Atazanavir, and Fosamprenavir) became available for use where more traditional treatments fail. However, they are inconvenient to administer and are very expensive. Therefore, they are usually not used until three or four other treatments have proven unsuccessful.

AZT differs from the natural nucleoside by having an —N_3 group instead of an —OH group.

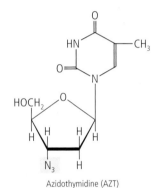

Azidothymidine (AZT)

CONCEPT CHECK 17A

1. The _____ name applies to the same drug no matter who makes it, but every manufacturer has its own _____ name for the drug.
2. The infectious diseases now in the top 10 leading causes of death for individuals of all ages in the United States are _____, _____, and _____.
3. Infectious diseases are caused by _____.
4. The three leading causes of death for 15–24-year-olds in the United States are _____, _____, and _____.
5. The general structures for penicillin and cephalosporin antibiotics have in common atoms of _____ and _____ in addition to

the expected C, H, and O atoms, and also a _____-membered ring.

6. Penicillins act by blocking the enzyme that causes _____ of peptides in bacterial cell walls.

7. Penicillin G, the first antibiotic, and amoxicillin, currently the most prescribed antibiotic, differ in structure only by a (an) _____ group and a (an) _____ group.

8. Viruses are difficult to combat because they attack from _____ host cells rather than from _____ the cells.

9. Because the human immunodeficiency virus carries out RNA-directed synthesis of DNA rather than DNA-directed synthesis of RNA, it is called a (an) _____.

10. AZT, which is used to treat _____, fools the virus by becoming part of its _____.

11. The treatment of disease with chemical agents is referred to as _____.

12. The newest and most successful drugs to combat AIDS are known as _____ and the target of their action is a(n) _____ molecule.

17.4 STEROID HORMONES

In considering hormones, we turn our attention from drugs that fight invading organisms to drugs that make up deficiencies in natural biochemicals or mimic their action.

Hormones are produced by glands and secreted directly into the blood. They serve as chemical messengers, regulating biological processes by interacting with **receptors** that are sometimes distant from where they are secreted. For example, the adrenal gland just above the kidney releases a group of hormones that act throughout the body to regulate the availability of glucose in the blood. Hormones are chemically diverse but are mostly proteins or steroids.

One of the most revolutionary medical developments of the 1950s was the worldwide introduction and use of "the pill." The development of synthetic analogs of the female sex hormones made reliable birth control available to women for the first time. The ongoing search for an equivalent birth control pill for use by men has not yet been equally successful.

Most oral contraceptive pills used by women today contain a combination of two synthetic hormones. One, most commonly ethynyl estradiol, has estrogen-like activity in regulating the menstrual cycle. The other, most commonly norethindrone, has progesterone-like activity and establishes a state of false pregnancy that prevents ovulation. In a theme that is common in drug development, successful contraceptives are very similar in molecular structure to those natural molecules with similar activity. (Compare the following structures with those in Section 15.3.)

A receptor is often on the surface of a cell and interacts with another molecule to cause some change in biochemical activity. Some drugs block receptors and prevent an undesirable change in activity. Other drugs activate a receptor to cause a desirable change in activity.

Cortisol, also called *hydrocortisone*, is a steroid hormone. The ring structure common to all steroids is highlighted here. Synthetic cortisol is used here medically as an anti-inflammatory agent applied to the skin or injected into joints.

Cortisol

Statistics show that the health risks of using birth control pills are significantly less than the risks associated with pregnancy.

Receptor A molecule or portion of a molecule that interacts with another molecule to cause a change in biochemical activity

Ethynyl estradiol
(synthetic estrogen)

Norethindrone
(synthetic progesterone)

On a monthly cycle in nonpregnant women, release of a hormone by the pituitary gland triggers ovulation, which results in the release of an egg with the potential for fertilization. Once a woman is pregnant, a hormone known as *progesterone* (the "pregnancy hormone") is released. This hormone carries a number of chemical messages that help prepare the woman's uterus for implantation of the fetus and block further ovulation during pregnancy. Synthetic steroid-based contraceptives act by occupying the progesterone binding site. This occupation effectively fools the woman's body into "thinking" she is pregnant. Thus, ovulation does not occur and the woman does not become pregnant.

Even though there is some controversy surrounding the use of contraceptives that prohibit ovulation, another use of this chemical knowledge has resulted in even more controversy: the development of post-fertilization birth control chemicals. A drug commonly known as RU-486 (mifepristone), developed by the French pharmaceutical company Roussel-Uclaf and used in Europe since 1988 to induce abortion, became available in the United States in 2000 under the brand name Mifeprex after Rousell-Uclaf donated the patent rights to the nonprofit Population Council to avoid boycotts and lawsuits. Mifeprex is actually a mixture of mifepristone and misoprostol. Mifepristone is a progesterone antagonist that occupies the progesterone-binding site but is able to send the appropriate chemical message to implant the fetus and misoprotol is an abortifacient, so the fetus is spontaneously aborted. Patients must agree that they will have a surgical abortion if this procedure fails.

An alternative to drug-induced abortion involves the use of methotrexate in combination with misoprostol. Methotrexate inhibits normal cell growth and division to inhibit the development of the embryo and placenta. It is worth noting these drugs have uses other than those discussed here, but they remain controversial because of their potential use as abortion agents.

So-called "morning-after pills" or emergency contraceptive pills actually consist of large doses of ordinary oral contraceptives to be taken shortly after sexual intercourse. These were initially used for rape victims but are now approved for emergency contraception when taken within 72 hours of unprotected intercourse. A physician's prescription is required for these drugs.

Synthetic steroids with estrogen-like activity are also used medically to replace natural hormones where there is a disease-related deficiency, after a hysterectomy, or after menopause. A major benefit of these drugs is a slowing of the normal bone loss (osteoporosis) that occurs with age.

(17.5) NEUROTRANSMITTERS

Hormones and neurotransmitters have in common their roles as chemical messengers within the body. Neurotransmitters, as illustrated in Figure 17.3, carry nerve impulses from one nerve to the next or to the location where a response to the message will occur.

The body has a variety of neurotransmitters, each with its own distinctive molecular receptors and functions. New information about neurotransmitters is being reported frequently and is of great interest because of its usefulness to medicine. Most drugs that affect the brain or the nervous system interact with neurotransmitters or their receptors.

Norepinephrine, Serotonin, and Antidepressive Drugs

Norepinephrine and serotonin are neurotransmitters with receptors throughout the brain. Norepinephrine helps to control the fine coordination of body movement and balance, alertness, and emotion; it also affects

FIGURE 17.3 Transmission of a nerve impulse. The neurotransmitter is stored in the vesicles until it is needed. When a nerve impulse arrives, the vesicles move to the cell membrane and join with it so that the neurotransmitter is released. By crossing the synapse and binding to receptors on the surface of the adjacent nerve cell, the neurotransmitter transmits the impulse. The receptors in turn initiate chemical changes that allow the impulse to proceed.

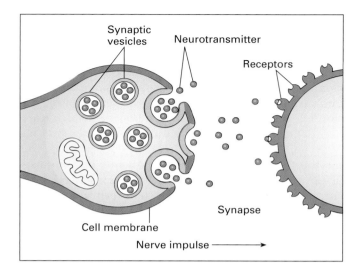

mood, dreaming, and the sense of satisfaction. Serotonin is involved in temperature and blood pressure regulation, pain perception, and mood. Serotonin and norepinephrine appear to work together to control the sleeping and waking cycle.

Norepinephrine   Serotonin

The normal cycle of neurotransmitter action at nerve *synapses*, the gaps between nerve endings (Figure 17.3) is as follows:

1. The neurotransmitter is released from the neuron.

2. The neurotransmitter crosses the synapse to interact with the receptor.

3. The neurotransmitter is inactivated, either by re-uptake by the neuron it came from or by conversion to an inactive form by an enzyme.

The biochemistry of mental depression is not fully understood, but a deficiency of norepinephrine and serotonin (possibly also dopamine; see next section) almost certainly plays a role. Evidence is provided by the manner in which three classes of drugs, illustrated by the following three compounds, influence the action of these neurotransmitters.

Amitriptyline, a tricyclic antidepressant
(Elavil)

Phenelzine, an MAO inhibitor
(Nardil)

Fluoxetine, an SSRI
(Prozac)

The tricyclic antidepressants such as Elavil (the trade name) prevent inactivation of neurotransmitters by preventing their re-uptake by neurons, which increases the concentration of the neurotransmitter in the synapse. The monoamine oxidase (MAO) inhibitors such as Nardil diminish the action of

MAO, which is the enzyme that inactivates norepinephrine and serotonin. The third type of drug action is represented by Prozac, which prevents the recapture of serotonin by neurons that release it. Drugs of this type, (Zoloft is another), known as *selective serotonin re-uptake inhibitors (SSRIs)*, have become the drugs of choice for treating serious clinical depression. For each of these three classes of drugs, the major mechanism of action is to increase the concentration of neurotransmitters at synapses.

Dopamine

Dopamine is produced in several areas of the brain, where it helps to integrate fine muscular movement as well as to control memory and emotion. An understanding of the brain chemistry of dopamine led to development of an effective treatment for Parkinson's disease. Patients with this disease experience trembling and muscular rigidity, among other symptoms, because of a deficiency of dopamine. Dopamine does not cross the blood–brain barrier (Figure 17.4) and thus cannot be administered as a drug. L-Dopa, it was found, can cross the barrier and then be converted to dopamine in the brain. While it does not cure Parkinson's disease, L-dopa can completely alleviate its symptoms for several years.

Dopamine has also been identified as the neurotransmitter that produces the feelings of well-being and reward associated with drug addiction (see *The World of Chemistry*, Section 17.8). Drugs that block dopamine receptors have been used to treat schizophrenia, which is, however, a complex condition not attributable solely to dopamine activity.

Epinephrine and the Fight-or-Flight Response

Epinephrine, also know as adrenaline, is both a neurotransmitter in the brain and a hormone released from the adrenal gland. Its sudden discharge when we are frightened produces the *fight-or-flight response*, which includes increased blood pressure, dilation of blood vessels, widening of the pupils, and erection of the hair. Because of its widespread and rapid effects, epinephrine has a number of medical uses, notably in crisis situations. It is administered to counteract cardiac arrest (by stimulating heart rate), to elevate dangerously low blood pressure (by constricting blood vessels), to halt acute asthma attacks (by dilating bronchial tubes), and to treat the extreme allergic reaction known as *anaphylactic shock*.

Sertraline hydrochloride, an SSRI (Zoloft)

Dopamine

L-Dopa

Epinephrine (Adrenaline)

Chemicals known as beta blockers are sometimes used to counteract epinephrine effects in patients suffering from heart disease as discussed in Section 15.11.

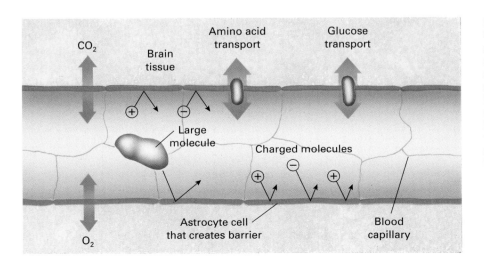

FIGURE 17.4 The blood–brain barrier. Openings in brain capillary membranes are small enough to keep out large molecules, and because the membrane contains mostly lipid molecules, ions cannot cross the membrane either. Only lipid (fat)-soluble molecules can cross unaided. These limitations provide the barrier that protects the brain from toxic compounds, but sometimes present a problem by also keeping out beneficial drugs.

CONCEPT CHECK 17B

1. Two main classes of chemical messengers within the body are _____ and _____.
2. Chemical messengers act only on cells that contain the appropriate _____.
3. Chemically, most hormones are _____ or _____.
4. Which hormone is known as the pregnancy hormone?
5. Most oral contraceptives contain synthetic derivatives of which two of the following? **(a)** Progesterone, **(b)** An androgen, **(c)** An antihistamine, **(d)** An estrogen.
6. Patients with Parkinson's disease have a deficiency of the neurotransmitter _____.
7. Our bodies react to a sudden fright with a rapid production of _____.
8. Hormones are produced by _____ and secreted directly into the blood.
9. Several antidepressant drugs act by (increasing/decreasing) the concentration of neurotransmitters at synapses.

17.6 THE DOSE MAKES THE POISON

The amount of a chemical substance that enters the body is known as a *dose*. The substance might be a life-saving medication or it might be a poison. What accounts for the difference? *The dose makes the poison.* A German physician and chemist who referred to himself as Paracelsus recognized this in the 16th century and it is still true. Whether a dose of a given substance is poisonous or not depends on the size of the dose. Most substances can be poisonous in a sufficiently large dose. For a given individual, age, gender, weight, and general state of health also play a role in the effect of a given dose.

Doses of medications and poisons are customarily expressed as milligrams per kilogram of body weight (mg/kg). Aspirin, for example, is used to treat rheumatoid arthritis at a dosage of about 110 mg/kg per day. (In contrast, a typical aspirin tablet contains just 325 mg; for an average 70-kg adult, two aspirin tablets amount to a dose of 9 mg/kg.)

A quantitative measure of toxicity is obtained by administering various doses of substances to be tested to laboratory animals (such as rats). The dose found to be lethal in 50% of a large number of test animals under controlled conditions is called the LD_{50} (lethal dose—50%) and is usually reported in milligrams of the substance per kilogram of body weight. Thus, if a statistical analysis of data on a large population of rats showed that a dose of 1 mg/kg was lethal to 50% of the population tested, the LD_{50} for this poison would be 1 mg/kg. Obviously, species differences can produce different LD_{50} values for a given poison. For this reason, defining risk to human beings based on animal data is difficult. It is, however, generally safe to assume that a chemical with a low LD_{50} value for several species will also be quite toxic to human beings. In Table 17.3, compare the LD_{50} values for substances you have probably ingested at one time or another with those for substances you would want to avoid.

Every chemical substance has an LD_{50} associated with it. It's just that we normally think of LD_{50} values when we are talking about medicines or toxic substances. However, one should always be aware of the potential toxicity associated with substances we normally would not consider dangerous. Relatively small doses of these substances may elicit the same negative responses

TABLE 17.3	LD$_{50}$ Values for Several Chemicals
CHEMICAL	**LD$_{50}$ (mg/kg ADMINISTERED ORALLY TO RAT)**
Sodium chloride	3750
Aspirin	1750
Ethanol	1000
Morphine	500
Caffeine	200
Heroin	150
Lead	20
Cocaine	17.5
Parathion	13
Aflatoxin	10
Sodium cyanide	10
Nicotine	2
Strychnine	0.8
Sarin	0.4
Batrachotoxin	0.002*
Tetanus toxin	5×10^{-6}
Botulinum toxin	3×10^{-8}

* From a poisonous frog. LD$_{50}$ in mice.

© Dr. Jeremy Burgess/SPL/Photo Researchers, Inc.

Seed capsule of an opium poppy. Crude opium is harvested from the sticky liquid that oozes through slits in the capsule.

as smaller doses of more toxic substances. Some chemicals commonly used as pesticides, insecticides, and nerve agents have incredibly small LD$_{50}$ values, as shown in Table 17.3. These substances should always be used with caution.

17.7 PAINKILLERS OF ALL KINDS

Drugs that relieve pain are known as **analgesics**. They range from cocaine to morphine to aspirin. Some are illegal drugs, some are prescription drugs, and some are over-the-counter drugs.

Opium and Its Relatives

Opium, obtained from the unripened seed pods of opium poppies, contains at least 20 different compounds. Chemically, they are *alkaloids*—organic compounds that contain nitrogen, are bases, and are produced by plants. About 10% of crude opium is *morphine*, which is primarily responsible for the effects of opium. Morphine is medically valuable as a strong painkiller able to produce sedation and loss of consciousness. The term *opioid* is now applied to all compounds with morphine-like activity.

Heroin, the diacetate ester of morphine, does not occur in nature but can be synthesized from morphine. As shown in Figure 17.5, their structures differ in only one kind of functional group. Heroin is much more addictive than morphine and for that reason has no legal use in the United States. *Codeine*, a methyl ether of morphine, is one of the alkaloids in opium and is used in cough syrup and for relief of moderate pain. Codeine is less addictive than morphine, but its analgesic activity is only about one-fifth that of morphine. One of the most effective substitutes for morphine is *meperidine*, whose structure was identified in 1931, is now sold as Demerol. It is less addictive than morphine.

In an earlier time, when few medications were available, Sir William Osler, a famous physician, called morphine "God's own medicine." It was named for the Greek god of dreams, Morpheus, by the German pharmacist who first isolated it from opium in 1803.

Oxycontin and Vicodin, both codeine derivatives, along with Demerol, are among the most commonly abused prescription drugs.

Meperidine
(Demerol)

Analgesics Drugs that relieve pain

FIGURE 17.5 Opioids. Morphine and codeine are natural alkaloids in the opium poppy. Heroin is a synthetic derivative with similar activity as a drug, but it is dangerously addictive.

Opioid structure

	R₁	R₂
Morphine	—O—H	—O—H
Codeine	—O—CH₃	—O—H
Heroin	—O—C(=O)—CH₃	—O—C(=O)—CH₃

Mild Analgesics

When milder general analgesics are required, few compounds work as well for many people as *aspirin*. Each year about 40 million pounds of aspirin are manufactured in the United States. Aspirin is also an **antipyretic** (fever reducer) and an **anti-inflammatory** agent. Aspirin inhibits cyclooxygenase, the enzyme that catalyzes the reaction of oxygen with polyunsaturated fatty acids to produce prostaglandins. Excessive prostaglandin production causes fever, pain, and inflammation—just the symptoms aspirin relieves.

Aspirin is not without its problems, however. It is known to reduce the ability of blood to clot and frequent use of aspirin is a known factor in the incidence of peptic ulcers. These ulcers are crater-like sores that develop when the digestive juices produced by the stomach eat away the lining of the digestive tract. The combination of these two effects suggests caution should be used by people with a history of gastrointestinal bleeding or clotting disorders, among other things, since aspirin may exacerbate bleeding problems. Health professionals often recommend that patients cease using aspirin at least five days before surgery or dental work that could lead to bleeding. Aspirin has also been implicated as a factor in the development of a condition known as Reye's syndrome. This is an unusual reaction to a viral infection that results in brain swelling and fatty disease of the liver and kidneys and can result in death. It tends to occur, primarily in children, during the recovery from flu or chicken pox. The exact cause of Reye's syndrome is unknown but studies have shown that using aspirin or chemically-similar salicylate medications to treat symptoms of viral illnesses increases the risk of developing the syndrome. Despite these concerns, aspirin remains a highly recommended pain reliever.

A bottle of aspirin tablets that has developed the vinegar-like odor of acetic acid should be discarded. Acetic acid forms as aspirin ages and breaks down by reacting with moisture in the air.

Willow trees, source of natural salicylates. Extracts from willow bark were known as painkillers in ancient times. Aspirin, first synthesized in the later 1800s, is a derivative of the natural salicylates, but with milder side effects.

Acetylsalicylic acid (aspirin) →(H₂O) Salicylic acid + Acetic acid

Antipyretic Reduces fever

Anti-inflammatory Reduces inflammation

Several over-the-counter aspirin alternatives are now available for pain sufferers. The three principal ones are ***acetaminophen (Tylenol), ibuprofen (Advil, Nuprin, Motrin),*** and ***naproxen (Aleve).*** All, except acetaminophen, contain a carboxylic acid group as does aspirin.

Acetaminophen
(Tylenol)

Ibuprofen
(Advil, Motrin)

Naproxen
(Aleve)

Acetaminophen is the only one that does not promote some bleeding in the stomach. Acetaminophen is an effective analgesic and antipyretic, but it is not an anti-inflammatory agent. Aspirin, ibuprofen, and naproxen are *nonsteroidal anti-inflammatory drugs (NSAIDs)*, as distinguished from anti-inflammatory drugs that are steroids, such as cortisone. Ibuprofen, originally available only by prescription (Motrin), is similar to aspirin in its effectiveness but causes less bleeding. Naproxen, also originally a prescription drug (Anaprox or Naprosyn), has as its principal advantage a long period of activity, making twice-a-day administration possible for round-the-clock pain relief.

A group of nonsteroidal anti-inflammatory drugs known as COX–2 inhibitors recently came under considerable scrutiny because of potential side effects. These drugs block an enzyme known as *cyclooxygenase-2* and this action inhibits the production of chemical messengers known as prostaglandins that cause the pain and swelling associated with arthritis inflammation. COX–2 inhibitors included celocoxib (Celebrex), rofecoxib (Vioxx), and valdecoxib (Bextra). In late 2004, these drugs were shown to significantly increase the risk of major fatal and nonfatal heart attacks in clinical trial participants taking these drugs. While the absolute risk of a COX–2 induced heart attack remained small, the 50% increase in risk caused considerable concern. The initial response by the FDA was to recommend their removal from the market, but significant demands by arthritis sufferers have resulted in a reconsideration of this decision. As this text is being written, doctors may still prescribe some COX–2 inhibitors, but Merck, the manufacturer of Vioxx, was dealt a major blow when a Texas jury awarded more than $250 million to a Texas woman whose husband died of a heart attack. He had been taking the arthritis painkiller. Although Merck has vowed to appeal this award, it still faces thousands of lawsuits filed by some of the more than 20 million people who took the drug. Ultimately, the increased risk of heart attack will be balanced against the benefits in pain relief available to users of these drugs.

The reason that acetaminophen is the most used mild analgesic in hospitals is simple—it doesn't cause bleeding, which for hospital patients could be hazardous.

In some individuals, ibuprofen causes gastrointestinal problems, with cramping and diarrhea.

17.8 MOOD-ALTERING DRUGS, LEGAL AND ILLEGAL

Everyone who drinks coffee or alcoholic beverages has personal experience with the mild effects of drugs classified as stimulants (the caffeine in coffee) or depressants (ethyl alcohol). Stimulants and depressants with stronger activity are among the drugs that are often *abused*—used in ways that are socially unacceptable or harmful. Although moderate alcohol consumption is an accepted social custom in the United States, alcohol is an abused drug for many individuals. Other kinds of drugs that are subject to abuse are opioids and hallucinogens.

Although the list of drugs of abuse is wide ranging, a central theme is their relationship to brain chemistry and the effect of the drug on neurotransmitters or their receptor sites. Before discussing the mood-altering drugs, it's

TABLE 17.4	Classification of Drugs	
DESIGNATION	**DESCRIPTION**	**EXAMPLES**
Over-the-counter (OTC)	Available to anyone	Antacids, aspirin, most cough medicines
Prescription drugs	Available only by prescription	Antibiotics
Unregulated nonmedical drugs	Available in beverages, foods, or tobacco	Ethanol, caffeine, nicotine
Controlled substances*		
Schedule I	Abused drugs with a lack of accepted safety and there is no accepted medical use.	Heroin, LSD, marijuana, methaqualone
Schedule II	Abuse may lead to severe physical and psychological dependence but there is a currently accepted medical use with severe restrictions.	Morphine, PCP, cocaine, methadone, methamphetamine
Schedule III	Abuse may lead to moderate or low physical dependence or high pyschological dependence but there is a currently accepted medical use.	Anabolic steroids, codeine, hydrocodone with aspirin or Tylenol and some barbiturates
Schedule IV	Abuse may lead to limited physical or psychological dependence relative to Schedule III drugs but there is a currently accepted medical use.	Darvon, Talwin, Equanil, Valium, and Xanax
Schedule V	Abuse may lead to limited physical or psychological dependence relative to Schedule IV drugs but there is a currently accepted medical use.	Cough medicines with codeine

* The sale, distribution, and possession of these drugs or substances classified as controlled substances are controlled by the Drug Enforcement Administration of the U.S. Department of Justice. Detailed information about scheduling and abused drugs may be found at the following websites: **http://www.usdoj.gov/dea/pubs/csa/812.htm#c** and **http://www.usdoj.gov/dea/concern/concern.htm**

important to understand the legal control of abused drugs. Table 17.4 lists the various classifications of the U.S. Drug Enforcement Administration. Schedule I drugs are not legally available in any manner. Schedule II drugs can be dispensed only once from a written prescription. A refill requires a new written prescription.

Depressants

The effect of central nervous system depressants depends more on the dose than on the particular drug. The sequence proceeds from **sedation**, or relaxation, to sleep, to general anesthesia, to coma and death. Medically, depressants are used to treat anxiety and insomnia.

The best-known depressants are the *barbiturates*. Variations in chemical structure produce a range of barbiturates from very short acting to very long acting. Barbiturates are potential drugs of abuse, and overdoses cause many deaths. They are especially dangerous when ingested with ethyl alcohol, another depressant, because the two together give a *synergistic* effect.

The second major class of depressants is the *benzodiazepines*. The familiar *tranquilizers* Librium (chlordiazepoxide) and Valium (diazepam) are members of this group. Both the barbiturates and the benzodiazepines

Synergism is the working together of two things to produce an effect greater than the sum of the individual effects.

Sedation Induction of a relaxed state, usually by a medication

act at receptors for the neurotransmitter gamma-aminobutyric acid (GABA, $HOOCCH_2CH_2CH_2NH_2$), which normally *inhibits* rather than excites transmission of nerve impulses. Because of differences in their molecular mechanism of action, benzodiazepines are safer and less subject to abuse than barbiturates.

Ethyl alcohol (CH_3CH_2OH), like barbiturates, enhances the action of the neurotransmitter GABA. Unlike barbiturates, however, the effects are dose dependent. At low doses, higher brain functions are affected, producing decreased inhibitions, altered judgment, and impaired control of motion. As dosage increases, reflexes diminish, and consciousness diminishes to the point of coma and death.

General barbiturate structure

$R_1 = \text{—}CH_2CH_3$

$R_2 =$

Phenobarbital (Luminal), a long-acting barbiturate used to treat insomnia and seizures

Stimulants

Stimulants are drugs that excite the central nervous system. They stimulate production of the neurotransmitters norepinephrine, serotonin, and dopamine in the brain, heart (which is stimulated to beat faster), and veins (which become constricted). Note the similarity in the structures of amphetamine and methamphetamine to epinephrine and norepinephrine (Section 17.5), which helps to explain their action.

Diazepam (Valium), a benzodiazepam tranquilizer

Amphetamine

Methamphetamine

Amphetamines were once available in over-the-counter preparations used to stay awake. Now, the only approved ingredient for "stay-awake" pills is caffeine. Amphetamines have become controlled substances because of their great potential for abuse. The only generally accepted medical use of amphetamines is to treat *narcolepsy*, a condition of uncontrollable attacks of sleep.

Abuse of *methamphetamine* has become a major problem in the United States in recent years. Hardly a week goes by without a story about methamphetamine use making the front page of the newspaper or the evening news. Often these stories focus on relatively rural areas where the manufacture of methamphetamine and its distribution have become a cottage industry of the very worst kind. The ready availability of a chemical that can easily be made into methamphetamine using information found on the Internet has created a nightmare drug problem of epidemic proportions. That chemical is *pseudoephedrine*, the major ingredient in many over-the-counter cold medications such as Sudafed and Claritin-D.

A similar chemical known as *phenylpropanolamine* (simply replace the CH_3 group on the nitrogen atom of pseudoephedrine with a hydrogen atom) was, until recently, the active ingredient in Dexatrim, a common over-the-counter diuretic.

Pseudoephedrine is a decongestant that shrinks blood vessels in the nose, lungs, and other mucous membranes. Drug dealers have learned how to isolate this ingredient from cold pills. Then, using common chemicals such as anhydrous ammonia, which is often stolen from farm storage tanks, they easily convert it into methamphetamine in houses, garages, and even in traveling meth labs in cars, vans, and pickup trucks. The problem has become so serious that in 2004 Oklahoma became the first state in the nation to classify these common cold remedies as Schedule V narcotics. Oklahoma

Pseudoephedrine

restricts their sale to pharmacies, requires the pills to be placed behind glass enclosures, limits the amount sold per customer, and requires purchasers to show a photo ID and sign a computer register whose data is available statewide. Other states, including the border state of Texas, where drug dealers from Oklahoma simply have to cross the Red River to purchase these medications, are considering similar legislation. In fact, as of this writing, Texas is considering requiring a prescription for such medications and almost every state is currently trying to address this problem. Some law enforcement officers have suggested that any state in the Midwest that doesn't pass such legislation quickly has the opportunity to become the meth capital of the region. One pharmaceutical company has begun to use a different active ingredient in its abused decongestant to help address this problem. This ingredient cannot easily be converted into methamphetamine. Despite these efforts at remediation, in all probability, the problem will continue for many years to come.

Cocaine, derived from the leaves of the coca plant of South America, is a stimulant and a Schedule II drug (used medically as a local anesthetic). By preventing the removal of norepinephrine from nerve endings, it causes uncontrolled firing of the nerves. *Crack* is a form of cocaine obtained by heating a mixture of cocaine and sodium bicarbonate. The reaction is an acid–base reaction since the base, sodium bicarbonate, neutralizes cocaine hydrochloride, the usual form in which cocaine is isolated. The term "crack" refers to the crackling sound made by the heated mixture during the release of carbon dioxide as bicarbonate reacts with acid. The appearance of crack cocaine on the illegal drug market has caused an increase in the number of cocaine addicts because crack is much more addictive than cocaine. The "high" lasts less than 10 minutes, creating the need to use crack repeatedly over a short period. Many users become addicted after only one try, and there is a high risk of taking a lethal dose.

Cocaine

Cocaine hydrochloride

Hallucinogens

Hallucinogens are chemicals that cause vivid illusions, fantasies, and hallucinations. Many have been found in plants, including *mescaline*, which comes from the fruit of the peyote cactus, and *lysergic acid diethylamide (LSD)*, which is made from lysergic acid derived from either the morning glory or ergot, a fungus that grows on grasses. These drugs are not addictive in the same manner as cocaine, but sometimes cause destructive behavior and lingering psychological problems. There are no therapeutic uses of the hallucinogens, which are believed to activate serotonin receptors.

Marijuana is a mild hallucinogen and sedative made from the hemp plant, *Cannabis sativa*. Although the millions of marijuana users regard it as a "safe" drug, this is a controversial conclusion. At high doses, marijuana is moderately addictive and can create paranoia and intense anxiety. Long-term use at moderate doses causes a general state of disinterest in personal achievements. In addition, tetrahydrocannabinol (THC), the active ingredient in marijuana smoke, can cause damage to the lungs, impede brain function, and hamper the immune system. Animal studies also suggest it may produce birth defects in offspring.

Phencyclidine (PCP, known as "angel dust") is an especially dangerous drug with a unique pattern of effects. At low doses, its effects resemble those of alcohol. With higher doses, hallucinations set in and behavior can become hostile and self-destructive, promoting psychoses that can last for weeks. Physical effects include seizures, coma, and death from cardiac arrest.

© R. Kong/Jacana/Photo Researchers, Inc.

Peyote cactus, source of mescaline.

© Carolina Biological Supply Company, Phototake NYC

Morning glory, whose seeds contain lysergic acid.

Hallucinogens Drugs that cause perceptions with no basis in the real world

THE WORLD OF CHEMISTRY

ADDICTIVE DRUGS

What role does brain chemistry play in the action of habit-forming drugs? Answering this question may open the door to combating the social, financial, and emotional costs of drug addiction.

Evidence is accumulating that all addictive drugs have an end result in common. In one way or another they increase the concentration of dopamine at nerve synapses in the brain. For example, cocaine blocks re-uptake of dopamine from the nerve synapse and amphetamines accelerate release of dopamine. In numerous studies, it has been demonstrated that dopamine is the neurotransmitter responsible for feelings of reward and satisfaction. The higher the concentration of dopamine in the brain, the greater the reward—the "high" of drug addiction.

Not surprisingly, the first studies of the brain chemistry of addictive drugs were performed with laboratory animals. Thanks to modern noninvasive techniques for imaging the brain, the connection between dopamine and cocaine addiction in humans was verified early in 1997. Brain scans were performed as co-

A marijuana plant.
© Scott Camazine/Photo Researchers, Inc.

caine addicts received infusions of cocaine. The addicts receiving cocaine reported the expected pleasant feelings after at least half of their dopamine receptors were occupied by dopamine, as revealed in the scans.

There is an ongoing debate about whether marijuana is an addictive drug. One argument that it is not addictive is based on the fact that traumatic withdrawal symptoms do not result from stopping its use. A probable explanation of this absence of withdrawal lies in the relatively long time that THC, the active ingredient of marijuana, persists in the bloodstream. Because its concentration decreases

gradually rather than very suddenly, withdrawal is not experienced in the same manner as with drugs that are rapidly decomposed in the body. Research reported in 1997 demonstrated that in at least one respect—a surge in concentrations of brain dopamine—marijuana is no different from cocaine or heroin. The levels of dopamine in the brains of rats were observed to double when they were injected with THC, the active ingredient in marijuana.

In the hope of combating addiction of all kinds, the search is on for a new generation of drugs that will block the effects of increased dopamine in the brain.

Tetrahydrocannabinol (THC), the active ingredient in marijuana

Reference N. D. Volkow et al.: *Nature*, Vol. 386, pp. 827, 830, April 24, 1997; G. Tanda et al.: *Science*, Vol. 276, pp. 2048–2050, July 4, 1997.

17.9 COLDS, ALLERGIES, AND OTHER "OVER-THE-COUNTER" CONDITIONS

What is the first thing we do when we feel sick? According to estimates, 75% of all illnesses are treated with products from the drugstore shelves. There are more than 300,000 over-the-counter (OTC) products on the market. But note that there are about only 800 active ingredients packaged in all these different combinations. The top three categories of nonprescription medications in terms of sales dollars are used to treat allergies and colds, pain, and gastrointestinal problems.

About one person in 10 suffers from some form of allergy. Allergic symptoms occur when special cells in the nose and breathing passages release *histamine*. Histamine is a neurotransmitter that accounts for most of the symptoms of hay fever, bronchial asthma, and other allergies. In a familiar theme in drug design, *antihistamines* are medications that block histamine receptors. Chlorpheniramine and compounds with very similar chemical structures (e.g., brompheniramine and diphenhydramine) are ingredients in many OTC preparations for treating hay fever and the "stuffy" noses associated with allergies. A frequent side effect of many antihistamines is drowsiness. In fact, many OTC sleep aids contain antihistamines.

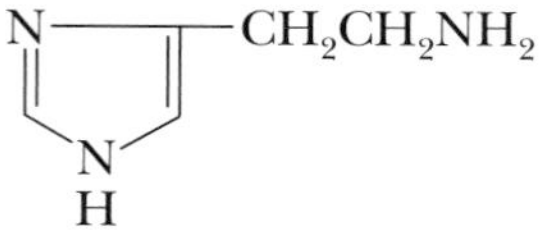

Histamine
(neurotransmitter that
causes allergic reaction)

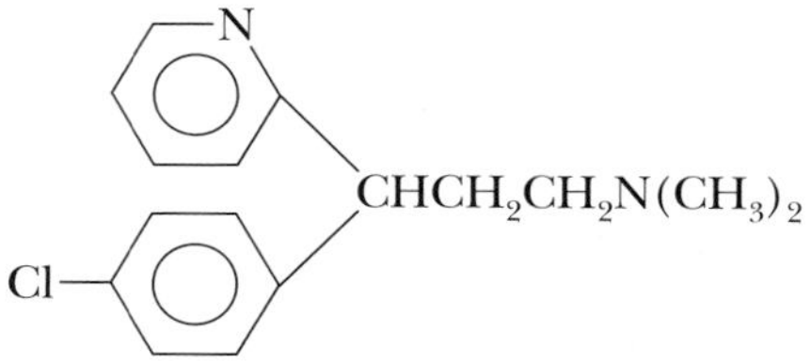

Chlorpheniramine
(antihistamine, e.g., in
Chlor-Trimeton)

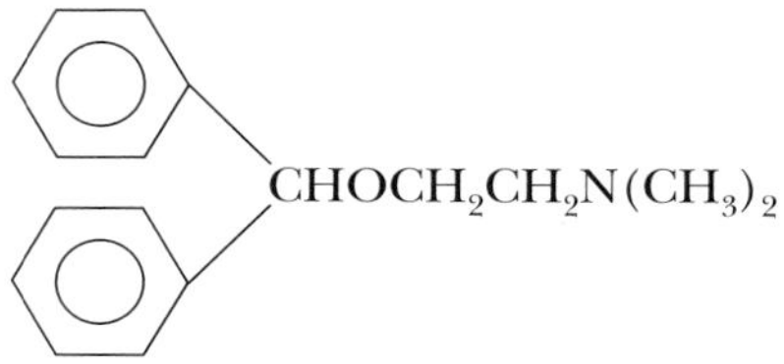

Diphenhydramine
(antihistamine used in sleep aids,
e.g., Sleep-Eze)

The common cold is caused by a virus, and like other, more serious viral diseases, it cannot truly be cured by any known method. The best we can do is treat the symptoms. Most OTC preparations for colds contain a variety of ingredients, usually two or more of those listed in Table 17.5. We have already

TABLE 17.5 Typical Ingredients in a Combination OTC Cold Medication

TYPE OF INGREDIENT (PURPOSE)	COMMON EXAMPLE	
	Name	*Structure*
Decongestant (shrink nasal tissues)	Pseudoephedrine	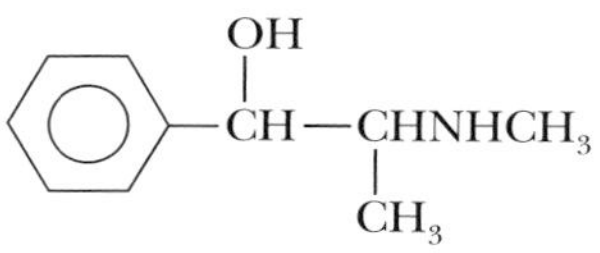
Antitussive (prevent cough)	Dextromethorphan	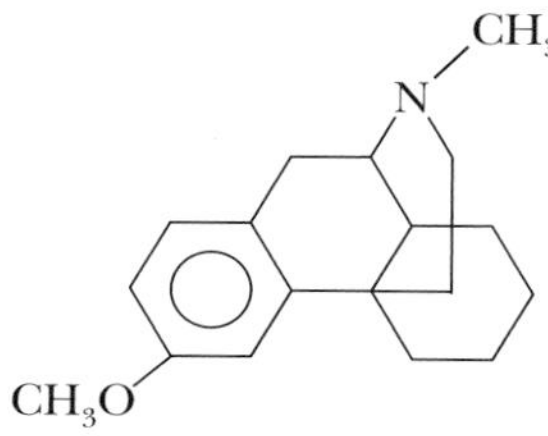
Expectorant (loosen fluids in cough)	Guaifenesin	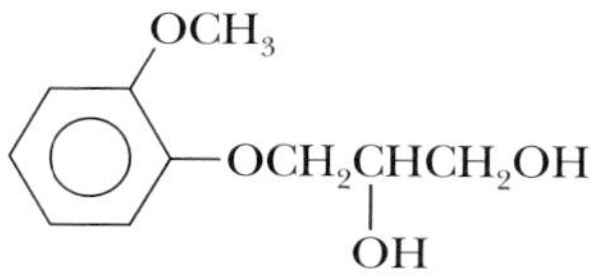
Analgesic (diminish pain, fever)	Acetaminophen	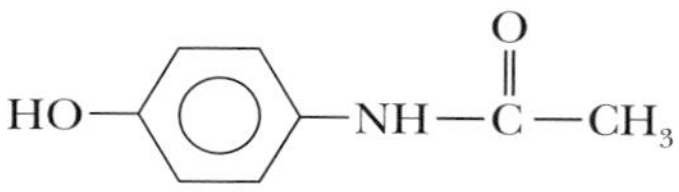
Antihistamine (counteract allergic reaction)	Chlorpheniramine	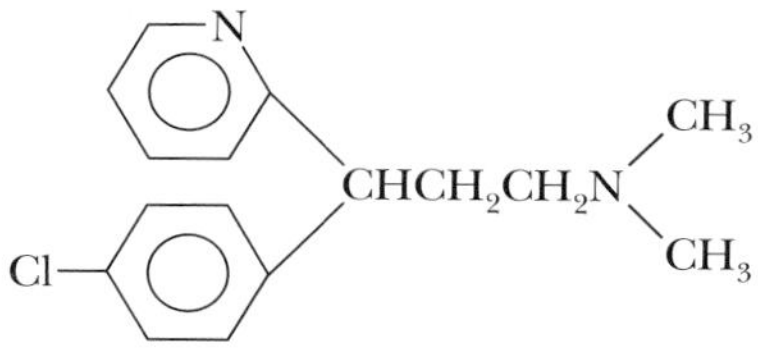

Two plants that send many people to the drugstore: (a) goldenrod, a major cause of hay fever; (b) poison ivy ("leaflets three, let it be").

(a)　　　　　　　　　　　　　　　(b)

discussed antihistamines and analgesics. There is some doubt about how useful an antihistamine is in treating a cold.

Decongestants shrink nasal passages to relieve the stuffiness that accompanies colds and allergies. The decongestants activate receptors for epinephrine (an ingredient in some cold medications) and similar neurotransmitters. *Antitussives* suppress coughing. Opioids all have antitussive activity, and codeine is often used for this purpose. Dextromethorphan (see Table 17.5) is an opioid-related compound that has antitussive activity without the analgesic and other effects of opioids. Dextromethorphan is a nonaddictive opiate. Its name derives from the fact that it is the optically active stereoisomer of methorphan that rotates plane-polarized light in a clockwise (dextrorotatory) direction (Section 15.1). Its mirror-image stereoisomer, levomethorphan, is an addictive opiate that rotates plane-polarized light in a counterclockwise direction. *Expectorants* are meant to stimulate secretions in the respiratory tract so that mucus is dislodged in coughing. Most likely, guaifenesin is the only effective ingredient of this type.

Over-the-counter drugs—we have a great many to choose from.

A few guidelines for selecting and using OTC products are recommended by numerous groups concerned with public health:

- Choose single-ingredient products specific to the condition you have.
- Cut down on unnecessary expense by choosing generic products. (The chemical ingredients are the same.)
- *Read* labels and *follow instructions* for dosage.
- *Pay attention* to cautions with respect to drowsiness, or interactions with alcohol or other medications.

17.10 PREVENTIVE MAINTENANCE: SUNSCREENS AND TOOTHPASTE

Many drugstore products are directed toward preventing bodily harm rather than curing an existing condition. Among the most important are sunscreens, which protect against skin cancer, and toothpaste, which prevents not only cavity formation but also deterioration of the gums and eventual tooth loss.

Our world is bathed in ultraviolet (UV) radiation that is sufficiently energetic to harm living things exposed to it. The natural protective mechanism against UV radiation is an increase in the skin of the pigment known as melanin, producing what we call a "tan." The melanin molecules absorb some of the UV energy and convert it to heat, thus diminishing damage to the molecular structure of the skin. However, even with melanin's protection, exposure to sunlight causes trouble. Especially in fair-skinned people whose skin contains smaller amounts of melanin, the result is a visible reddening—*erythema* or, in everyday language, sunburn. Although it is less visibly noticeable, dark-skinned people can also experience sunburn. Evidence has been mounting that the risk of developing skin cancer rises with the amount of lifetime exposure to the sun's rays (Section 7.5).

In 2004, the ultraviolet (UV) index developed by the National Weather Service and the U.S. Environmental Protection Agency (EPA) in 1994, was replaced by the Global Solar UV Index. The new scale rates the level of UV exposure expected on a given day as measured at noon, although the actual UV rises and falls as the day progresses. The color-coded scale is shown in Table 17.6.

If you are going to be exposed to direct sunlight, it is a good idea to protect yourself from as much UV exposure as you can. Besides physical barriers such as long-sleeve shirts and wide-brimmed hats, there is a variety of chemicals that will selectively absorb UV light. These *sunscreens* can be applied as oils, creams, or lotions. The sunscreen must function in a manner similar to melanin by absorbing UV light and converting it into heat energy. Many sunscreens contain *p*-aminobenzoic acid (PABA) and other chemicals with similar structures as the active ingredients. These compounds absorb an appreciable amount of the most harmful UV radiation.

Sunscreens have sun protection factors listed prominently on their labels. The *sun protection factor (SPF)* is defined as the ratio of time required to produce a perceptible erythema on a site protected by a specified dose of the sunscreen to the time required for minimal erythema development on unprotected skin. An SPF of 4, for example, would provide four times the skin's natural sunburn protection. The time required for sunburn depends on an individual's skin

p-Aminobenzoic acid
(PABA)

TABLE 17.6 The Global Solar UV Index Established in 2004*

EXPOSURE CATEGORY	INDEX NUMBER	SUN PROTECTION MESSAGES
LOW	1–2	• Wear sunglasses on bright days. In winter, reflection off snow can nearly double UV strength. • If you burn easily, cover up and use sunscreen.
MODERATE	3–5	• Take precautions, such as covering up and using sunscreen, if you will be outside. • Stay in shade near midday when the sun is strongest.
HIGH	6–7	• Protection against sunburn is needed. • Reduce time in the sun between 11 a.m. and 4 p.m. • Cover up, wear a hat and sunglasses, and use sunscreen.
VERY HIGH	8–10	• Take extra precautions. Unprotected skin will be damaged and can burn quickly. • Try to avoid the sun between 11 a.m. and 4 p.m. Otherwise, seek shade, cover up, wear a hat and sunglasses, and use sunscreen.
EXTREME	11+	• Take all precautions. Unprotected skin can burn in minutes. Beachgoers should know that white sand and other bright surfaces reflect UV and will increase UV exposure. • Avoid the sun between 11 a.m. and 4 p.m. • Seek shade, cover up, wear a hat and sunglasses, and use sunscreen.

* UV rays are only about half as intense 3 h before and after the peak. Physical surroundings such as snow, sand, and water reflect more UV and intensify exposure. Latitude and altitude also play a role; exposure increases with proximity to the equator and with altitude.
SOURCE: **http://www.epa.gov/sunwise/doc/uviguide.pdf**

type and on the intensity of the UV radiation. The EPA recommends against any communication of the "time it takes to sunburn" since that implies a safe period in which no protection is required. *There is no safe period since any change in your skin's natural color is a sign of damage to the skin.* Tanning does not protect you from skin cancer. Always use a sunscreen with an SPF of at least 15.

Keeping teeth clean requires *toothpaste*, a mixture of detergents and *abrasives*, which are hard substances that help remove unwanted materials on the tooth surface. The structure of tooth enamel is essentially that of a stone composed of calcium carbonate ($CaCO_3$) and calcium hydroxy phosphate [apatite—$Ca_{10}(PO_4)_6(OH)_2$]. Despite being the hardest substance in the human body, tooth enamel is readily attacked by acids. Because the decay of some food particles produces acids, it is important to keep teeth clean.

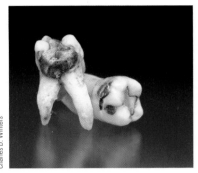

Teeth with large fillings, showing the effects of prolonged decay.

Within moments after you clean your teeth, a transparent film composed of proteins from saliva begins to coat the teeth and gums. This coating offers a place for food debris to collect and for oral bacteria to multiply. These bacteria convert dextrins, from the breakdown of sugars, into acids. At the same time, a tenacious film composed of these bacteria, food particles, and their breakdown products begins to form. As the film hardens, it becomes dental *plaque*. If this plaque is not removed regularly and completely from the surface of the teeth and beneath the gum line by brushing and flossing, the generation of acids and other harmful substances continues, destroying the tooth or the gum and eventually the bone that holds it in place.

Plaque that is not removed from the teeth becomes calcified from minerals in the saliva. The calcified plaque is known as *tartar*. It is possible to control tartar buildup by using toothpastes containing sodium pyrophosphate ($Na_4P_2O_7$), which interferes with the mineral crystallization that causes tartar buildup. Beneath the gum line, tartar is a special problem because its presence makes it easier for plaque to grow, which irritates gum tissue and allows the gum to become diseased. Only a dentist or oral hygienist can remove tartar from beneath the gum line. By keeping teeth free from plaque and from prolonged contact with the acids produced by plaque bacteria, we can preserve the hard, stonelike enamel of the tooth.

You can't see it, but it's there. Bacterial plaque on the surface of a tooth.

The abrasive material in toothpaste serves to cut into the surface deposits, and the detergent assists in suspending the particles in the rinse water. Abrasives commonly used in toothpastes include hydrated silica (a form of sand, $SiO_2 \cdot nH_2O$), hydrated alumina ($Al_2O_3 \cdot n\,H_2O$), and calcium carbonate ($CaCO_3$). It is difficult to select an abrasive that is hard enough to cut the surface contamination yet not so hard as to cut the tooth enamel. The choice of detergent is easier; any good detergent will do quite well. Because the necessary ingredients in toothpaste are not very palatable, it is not surprising to see that various flavorings, sweeteners, thickeners, and colors are included to appeal to our senses.

Tooth decay occurs when bacteria eat food particles that remain in tiny fissures and crevices in the teeth and produce acids that attack the tooth enamel. Fortunately, it is possible to modify the crystalline structure of tooth enamel and make it more resistant to decay by the addition of fluoride ion (F^-) to toothpaste. When regularly applied, some of the fluoride ions actually replace the hydroxide ions in the hydroxyapatite structure to form fluoroapatite [$Ca_{10}(PO_4)_6F_2$]. The fluoride ion forms a stronger ionic bond in the crystalline structure than the hydroxide ion because of its high concentration of negative charge; as a result, the fluoroapatite is harder and less subject to acid attack than the hydroxyapatite. Hence, there is less tooth decay. Fluoride ion is introduced into essentially all the public water supplies in the United States for this purpose. Concentrations of fluoride ion of one part per million (ppm) have proved safe and efficient for reducing tooth decay. About 80% of all toothpaste sold in this country contains fluoride ions in some form. Compounds such as

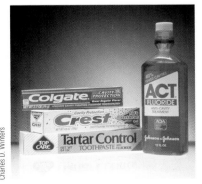

Some products containing fluoride.

stannous fluoride (SnF_2) and sodium monofluorophosphate (Na_5FPO_3) provide a low fluoride ion concentration in toothpastes.

In the United States more teeth are now lost as a result of gum disease than from decay. Gum disease results from the lack of proper massage, from irritating deposits below the gum line, from bacterial infection, and from poor nutrition. More attention is being given to toothpastes containing disinfectants such as peroxides in addition to the other ingredients.

CONCEPT CHECK 17C

1. _____ is a derivative of morphine that is not found in nature.
2. Codeine is a derivative of morphine that is found in nature and that (does/does not) have a medical use.
3. Which of these three terms refers to each of the following products? (i) a pain killer, (ii) a medication that combats fever, (iii) a medication that combats the condition that causes muscle pain, swelling, and other symptoms
   (**a**) Antipyretic   (**b**) Analgesic   (**c**) Anti-inflammatory
4. The chemical name for aspirin is _____.
5. _____ is a major side effect associated with the use of aspirin.
6. Acetaminophen differs from ibuprofen and aspirin in two important ways: It does not contain a(an) _____ functional group, which means it does not cause intestinal bleeding, and it does not act as an anti-_____.
7. A Schedule I controlled substance is a drug that is _____ and has no _____.
8. Barbiturates and benzodiazapines are two major classes of _____.
9. What active ingredient in many common cold medications is easily converted to methamphetamine?
10. Crack is a purified form of _____ and is dangerously even more _____.
11. Histamine is a neurotransmitter that causes which of the following? (**a**) Pain, (**b**) Upset stomach, (**c**) Allergic reaction.
12. A cold remedy can cure a cold. (**a**) True, (**b**) False.
13. Name the five classes of ingredients found in cold medications.
14. Sunscreens function by absorbing _____ _____.
15. The SPF, which stands for _____ _____ _____, is a guide to choosing the correct sunscreen for yourself.
16. A toothpaste may contain (**a**) A detergent, (**b**) An abrasive, (**c**) A fluoride, (**d**) Sodium pyrophosphate. Match each of these ingredients with the condition it helps to prevent. (i) Tooth decay (cavities), (ii) Tartar buildup, (iii) Accumulation of food debris and bacteria.
17. Whether a chemical is a drug or a poison is determined by the _____.

17.11 HEART DISEASE

Many drugs and surgical techniques now in use are able to decrease the death rate and improve the quality of life for persons suffering from heart disease. However, heart disease remains the number-one killer of Americans. Known medically as **cardiovascular disease**, heart disease results from any condition that decreases the flow of blood, and consequently oxygen, to the heart or diminishes the ability of the heart to beat regularly and function in a normal manner.

Cardiovascular disease Disease of the heart or blood vessels

The most common cause of heart disease is the plaque buildup on artery walls known as *atherosclerosis*, which was discussed in Section 16.4 in conjunction with the role of diet in plaque formation. If changes in lifestyle do not successfully combat this condition, the next step is cholesterol-lowering drugs, which include *lovastatin* and *cholestyramine*. Lovastatin acts by interfering with cholesterol synthesis in the liver. Cholestyramine acts by binding to bile acids in the intestines and accelerating their excretion. This causes the liver to convert more cholesterol into bile acids, leaving less to enter the circulatory system.

A result of plaque buildup in blood vessels can be the chest pain during exertion known as *angina*, which occurs because of insufficient oxygen delivery to the heart muscle. The attacks are brought on when the heart must work harder and thus increase its oxygen demand. To treat angina, *vasodilators*, drugs that *dilate* veins (make them open wider), are used. By dilating the veins, the blood pressure against which the heart must work is reduced. The classic vasodilators are organic nitrogen compounds such as nitroglycerin or amyl nitrite.

High blood pressure also contributes to heart disease. For this condition the next step after lifestyle changes is use of a **diuretic**, most commonly a *thiazide* (e.g., Diazide), which stimulates the production of urine and excretion of Na^+. With increased urine output, blood volume and consequently blood pressure are decreased.

The development of beta-blockers, drugs used to treat angina and other aspects of heart disease, illustrates how understanding the biochemistry of disease can lead to design of drugs.

In the 1960s two types of receptors that are part of the natural regulatory system for heart rate were discovered and named *beta-receptors*. The beta-1 receptors are located primarily in the heart—stimulation of these sites speeds up the rate at which the heart beats. The beta-2 receptors are located in the peripheral blood vessels and the bronchial tubes. Stimulation of the beta-2 receptors relaxes muscle fibers, opening up the blood vessels and bronchial tubes so that blood flows more easily, making it easier to breathe deeply and quickly. These receptors are stimulated by the natural hormones epinephrine and norepinephrine during the fight-or-flight response (Section 17.5).

Armed with this information, chemists began to search for chemicals that would compete with epinephrine and norepinephrine at the beta-receptor sites. If these sites could be blocked, stimulation of the heart muscle could be prevented. For a heart already overworked from the buildup of plaque in the arteries, this might produce enough relaxation to avoid an impending heart attack. In addition, these drugs might be able to relieve high blood pressure.

The first successful drug of this type, a *beta-blocker*, was *propranolol (Inderal)*, now used to treat cardiac arrhythmias, angina, and hypertension. Look back at Section 17.5 to see its similarity in structure to the compounds whose action it blocks. Propranolol and the other beta-blockers have become widely prescribed drugs.

OH
|
OCH₂CHCH₂NHCH⟨CH₃ / CH₃⟩

Propranolol (Inderal)

A heart attack (a *myocardial infarction*) results from a reduction in blood flow to the heart muscle. About 98% of all heart attack victims have atherosclerosis, and the reduction in blood flow can usually be traced to a clot in the plaque of a coronary artery. If prolonged, the blockage causes part

Bile acids are secreted into the small intestine to aid in the digestion of food. They are steroids (Section 15.3) and, like all steroids, are synthesized from cholesterol.

Isn't it interesting that nitroglycerin is also a powerful explosive?

CH₂—ONO₂
|
CH—ONO₂
|
CH₂—ONO₂   C₅H₁₁NO₂
Nitroglycerin   Amyl nitrite

Diuretic Causes excretion of fluid from the body

It is important to realize that although death from heart attacks is most common among older people, the condition of atherosclerosis begins many years earlier.

of the heart muscle to die from lack of oxygen; if the damage is sufficient, the heart attack can be fatal.

New, clot-dissolving drugs given to heart attack victims in the emergency room or the ambulance show promise in reducing the death rate. These drugs are enzymes that act on *plasminogen*, a natural factor in the blood, by converting it to *plasmin*. Once this happens, plasmin proceeds to dissolve blood clots by the body's own natural mechanism. Three enzymes have been developed as drugs that catalyze the plasminogen ⟶ plasmin conversion: (1) *urokinase*, a natural enzyme isolated from human urine; (2) *streptokinase*, isolated from a *Streptococcus* bacterium, and (3) *tissue plasminogen activator (TPA)*, one of the first drugs produced by recombinant DNA technology to receive government approval. TPA is made in genetically altered hamster ovary cells and is identical to the natural human enzyme that activates plasminogen.

(17.12) CANCER, CARCINOGENS, AND ANTICANCER DRUGS

Cancer is not one but perhaps 100 different diseases. A cancer begins when a cell in the body starts to multiply without restraint and produces descendants that invade other tissues. It seems reasonable then that drugs might be able either to stop this undesirable spreading of cancer cells or to prevent cancer from happening at all. A major obstacle to successful drug treatment for cancer is that its biochemistry is not well understood. There is, however, general agreement that cancer is initiated by damage to DNA, which may be done by physical (e.g., ionizing radiation), biological (e.g., viruses), or chemical agents (e.g., compounds in cigarette smoke).

Every cancer comes from a single cell—one that is a modification of a normal cell. A normal cell functions according to directions stored in its genetic data bank, the DNA, and when a cell divides, each new cell gets its own exact copy of the parent DNA. If anything disrupts this DNA replication process, the genetic code in one of the descendant cells may cause that cell to grow and function differently from a normal cell. **Carcinogens** are chemicals that cause cancer, which manifests itself in at least three ways: (1) The *rate of cell growth* (that is, the rate of cellular multiplication) in cancerous tissue differs from the rate in normal tissue. Cancerous cells may divide more rapidly or more slowly than normal cells. (2) Cancerous cells *spread to other tissues*; they know no bounds. Normal liver cells divide and remain a part of the liver. Cancerous liver cells may leave the liver and be found, for example, in the lung. (3) Most cancer cells show *partial or complete loss of specialized functions*. Although located in the liver, cancerous cells no longer perform the functions of the liver.

Attempts to determine the chemical causes of cancer have evolved from early studies in which the disease was linked to a person's occupation. We now know that a person's lifestyle plays a role as well. In 1775, Dr. Percivall Pott, an English physician, first noticed that people employed as chimney sweeps had a higher rate of skin cancer than the general population. It was not until 1933 that benzo(α)pyrene ($C_{20}H_{12}$, an aromatic hydrocarbon containing five fused carbon rings) was isolated from coal dust and shown to be metabolized in the body to produce one or more carcinogens.

Carcinogenesis is often a two-stage process. In the first stage, *initiation*, a chemical, physical, or viral agent alters the cell's DNA. Sometimes a single exposure to some carcinogen causes a rapid onset of a tumor that is composed of rapidly growing, uncontrolled cells, but usually the abnormal cells continue to reproduce in about the same way as normal cells around them. Then a *promotion* occurs. This is the second stage and may occur days, months, or years after the initiation. This promotion may be a physical irritation or

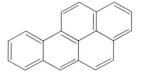

Benzo (α) pyrene

When a cancer spreads from one site to another, the process is called *metastasis*.

Carcinogen An agent that causes cancer

exposure to some toxic chemical that is itself not a carcinogen. In either case, the promotion results in the killing of a large number of cells. The destruction of cells is almost always compensated for by a sudden growth of new cells, and the abnormal cells begin to grow in ways the original DNA coding never intended. The cancer has started.

To illustrate the initiation and promotion aspects of carcinogenesis, consider some experiments performed in 1947 at Oxford University in England. First, very small doses of dimethylbenzanthracene (DMBA), a known carcinogenic component of coal tar, were applied to the skin of a group of mice. These mice were then separated into two groups. In one group the exposed skin was daubed with croton oil, a strongly irritating natural oil. The other group of mice had their skin daubed with croton oil four months later. Almost every mouse in both groups developed a tumor where the DMBA had been applied. In other groups of mice tested, neither croton oil nor DMBA alone produced any tumors, and if croton oil was applied first to the skin, followed by the DMBA, tumors failed to appear. Apparently DMBA had an initiation effect, whereas croton oil had a promotion effect.

Cancers are treated by (1) surgical removal of cancerous growths and surrounding tissue; (2) irradiation to kill the cancer cells; and (3) chemicals that kill the cancer cells, a process referred to as cancer *chemotherapy*. Cancer patients are considered cured if, after their treatment, they die at about the same rate as the general population. Another definition of success in cancer therapy is given by the number of patients who survive for five years after the treatment. In the 1930s fewer than 20 cancer patients in 100 were alive 5 years after treatment; in the 1940s, it was 25 in 100; in the 1960s, it was 33 in 100; and today it is close to 60 in 100.

Men and women show some significant variations in the types of cancers to which they are susceptible. Figure 17.6 shows the estimated number of cancers by type predicted for men and for women in the United States in 2005 along with the estimated numbers of deaths expected from each of these cancer types. Prostate cancer accounts for almost one-third of the cancer cases in men, but lung and bronchial cancers cause the most deaths, with prostate cancer deaths accounting for only 10% of total male cancer deaths. In women, breast cancer accounts for almost one-third of the cancer cases, but only about 15% of the deaths. Once again, lung and bronchial cancers account for the largest percentage (27%) of cancer deaths in women. Figures 17.7 and 17.8 show the trend in the number of cancer deaths by type for men and women from 1930–2001. It is encouraging that, in the past few years, deaths from all types of cancers have declined for both men and women except in one instance. The death rate due to lung and bronchial cancers in women had been increasing for many years but now appears to be, at best, leveling off.

During World War I the toxic effects of the mustard gases were found to include damage to bone marrow and changes in DNA (mutations) that created abnormal offspring. In addition, the so-called *nitrogen mustards* caused cancers in some animals.

Mustard gas   Nitrogen mustard   A nitrogen mustard
                           (general formula)

When the wartime-imposed secrecy surrounding these chemicals ended, it occurred to some researchers that cancers might be treated with similar compounds. The result might be to alter the DNA in cancer cells to the extent that these cells could be destroyed selectively.

Smoking is thought to play both an initiation and a promotion role in cancer causation.

The mouse has come to be the classic animal for studies of carcinogenicity. Strains of inbred mice and rats have been developed that are genetically uniform and show a standard response to various chemicals.

The five-year cancer survival rate data are useful for tracking progress in detection and treatment of cancer. The American Cancer Society notes that the five-year rate is "less informative when used to predict individual progress."

The term *mustard gas* comes from its mustard-like odor; mustard "gas," however, is not a gas but a high-boiling liquid that was dispersed as a mist of tiny droplets.

FIGURE 17.6 Ten leading cancer types for the estimated new cancer cases and deaths by gender, U.S. 2005.

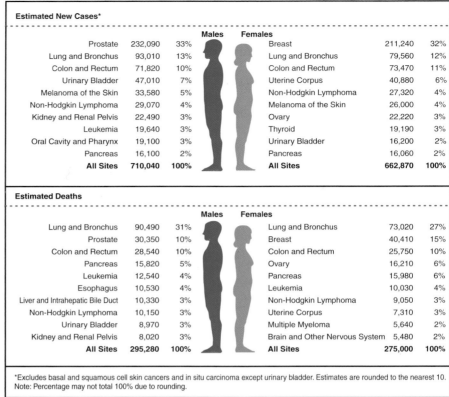

Estimated New Cases*

					Males	Females			
Prostate	232,090	33%					Breast	211,240	32%
Lung and Bronchus	93,010	13%					Lung and Bronchus	79,560	12%
Colon and Rectum	71,820	10%					Colon and Rectum	73,470	11%
Urinary Bladder	47,010	7%					Uterine Corpus	40,880	6%
Melanoma of the Skin	33,580	5%					Non-Hodgkin Lymphoma	27,320	4%
Non-Hodgkin Lymphoma	29,070	4%					Melanoma of the Skin	26,000	4%
Kidney and Renal Pelvis	22,490	3%					Ovary	22,220	3%
Leukemia	19,640	3%					Thyroid	19,190	3%
Oral Cavity and Pharynx	19,100	3%					Urinary Bladder	16,200	2%
Pancreas	16,100	2%					Pancreas	16,060	2%
All Sites	**710,040**	**100%**					**All Sites**	**662,870**	**100%**

Estimated Deaths

					Males	Females			
Lung and Bronchus	90,490	31%					Lung and Bronchus	73,020	27%
Prostate	30,350	10%					Breast	40,410	15%
Colon and Rectum	28,540	10%					Colon and Rectum	25,750	10%
Pancreas	15,820	5%					Ovary	16,210	6%
Leukemia	12,540	4%					Pancreas	15,980	6%
Esophagus	10,530	4%					Leukemia	10,030	4%
Liver and Intrahepatic Bile Duct	10,330	3%					Non-Hodgkin Lymphoma	9,050	3%
Non-Hodgkin Lymphoma	10,150	3%					Uterine Corpus	7,310	3%
Urinary Bladder	8,970	3%					Multiple Myeloma	5,640	2%
Kidney and Renal Pelvis	8,020	3%					Brain and Other Nervous System	5,480	2%
All Sites	**295,280**	**100%**					**All Sites**	**275,000**	**100%**

*Excludes basal and squamous cell skin cancers and in situ carcinoma except urinary bladder. Estimates are rounded to the nearest 10.
Note: Percentage may not total 100% due to rounding.

Source: Cancer Statistics 2005: A Presentation from the American Cancer Society

FIGURE 17.7 Annual age-adjusted cancer death rates* among males for selected cancer types, U.S. 1930 to 2001.

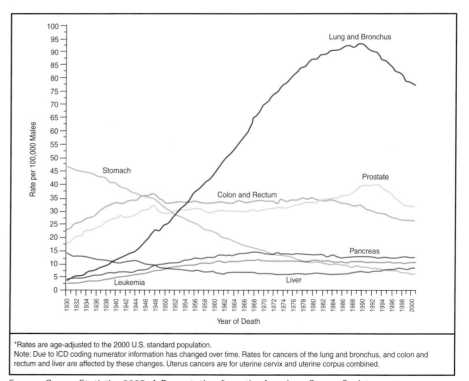

*Rates are age-adjusted to the 2000 U.S. standard population.
Note: Due to ICD coding numerator information has changed over time. Rates for cancers of the lung and bronchus, and colon and rectum and liver are affected by these changes. Uterus cancers are for uterine cervix and uterine corpus combined.

Source: Cancer Statistics 2005: A Presentation from the American Cancer Society

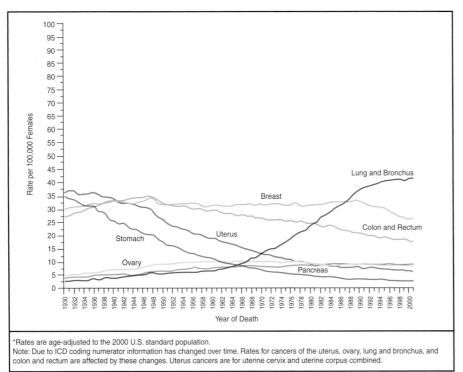

FIGURE 17.8 Annual age-adjusted cancer death rates* among females for selected cancer types, U.S. 1930 to 2001.

*Rates are age-adjusted to the 2000 U.S. standard population.
Note: Due to ICD coding numerator information has changed over time. Rates for cancers of the uterus, ovary, lung and bronchus, and colon and rectum are affected by these changes. Uterus cancers are for uterine cervix and uterine corpus combined.

Source: Cancer Statistics 2005: A Presentation from the American Cancer Society

One of the most widely used anticancer drugs is now cyclophosphamide, a compound that contains the nitrogen mustard group.

Cyclophosphamide, and other anticancer drugs that act in the same manner, are *alkylating agents*—reactive organic compounds that transfer alkyl groups in chemical reactions. Their anticancer activity results from the transfer of alkyl groups (e.g., CH_3CH_2—) to the nitrogen bases in DNA, often to guanine. The alkyl group physically gets in the way of base pairing and prevents DNA replication, which stops cell division. Although alkylating agents attack both normal cells and cancer cells, the effect is greater for rapidly dividing cancer cells, a criterion that applies to all cancer chemotherapy agents.

An alkylated guanine in DNA. You can see how the group alters the size of the guanine residue.

Another class of chemotherapy drug, the *antimetabolites*, interferes with DNA synthesis because these drugs are similar in molecular structure to compounds required for DNA synthesis (*metabolites*). Methotrexate, for example, is an antimetabolite similar in structure to folic acid (a B vitamin), used in the synthesis of nucleic acids. Methotrexate prevents reduction of folic acid in the first step of nucleic acid synthesis by strongly binding to the enzyme

Because cells in hair follicles divide rapidly, they are often killed during cancer chemotherapy, resulting in hair loss. (When new cells are generated, hair returns.)

Doxorubicin

for that reaction. (Methotrexate and folic acid are large molecules. Methotrexate differs from folic acid by the addition of one —CH_3 group and replacement of a $C=O$ by a $C—NH_2$.)

Yet another way to attack DNA is by physically disrupting its shape. Drugs accomplish this by fitting a planar ring portion of their molecular structure in between the base pairs that form the ladder structure of DNA (see Figure 15.22). Doxorubicin (structure in margin) is a most valuable chemotherapeutic agent of this type because it is active against an unusually large number of kinds of cancer.

All cancer chemotherapy is tedious and has risks and unpleasant side effects. The problem is that no agent has yet been found that kills only cancer cells, so there is always a balance to be struck between killing healthy cells and killing cancer cells. Because chemotherapy drugs kill actively dividing cells, the goal is to kill many more cancer cells than normal cells (ideally, of course, 100% of the cancer cells). In addition to being highly toxic, most of the cancer chemotherapy drugs are themselves carcinogenic and very high doses are usually necessary. As a result, single-agent chemotherapy has largely given way to combination chemotherapy because of positive additive, or even synergistic, effects when two or more drugs are used together. Because of their synergistic action, lower doses of each compound can be used when they are given together; this reduces the harmful side effects. Chemotherapy alone is most successful in treating cancers such as leukemia or lymphoma, in which the cancer cells are widely dispersed in the body.

17.13 HOMEOPATHIC MEDICINE

In recent years, Americans and others have tended to rely less and less on modern, scientific medical practices and remedies for treating many medical conditions. Popular magazines, television shows, and newspapers routinely include stories about alternative medical treatments, aromatherapy, herbal and natural remedies, magnet therapy, touch therapy, psychic healing, acupuncture, and homeopathy. The debate over the efficacy of such treatments continues. It is safe to say, however, that most, if not all, of these treatment methods have not been subjected to the same rigorous scientific standards as most modern medical practices and treatments.

A discussion of all types of alternative medical treatments is beyond the scope of this book. We will, instead, focus on one area where it can be easily demonstrated that the science does not support the claims. That area is **homeopathy**, a system of medical treatment based on the use of minute quantities of remedies that in larger doses produces effects similar to those of the disease or condition being treated.

Homeopathy is often confused by consumers with "natural remedies" or "herbal remedies," but homeopathy is a very particular approach to treating disease that has little or no basis in any kind of traditional, alternative, or natural medicine. Modern medicine depends on the *dose-response relationship*, in which there is a direct relationship between the dosage of medicine taken and the effect it elicits. Most scientists believe that a chemical agent must be physically present in the biological system for it to cause an observable effect. The science of **pharmacology** relies on this accepted scientific principle, as does the science of chemistry. Even seemingly mundane activities such as adding sweetener to soft drinks or sugar to iced tea reveal that there is a relationship between concentration and effect.

Dr. Steven Barrett describes homeopathy in the following terms.

Proponents call homeopathy's defining principle the "Law of Similars" or "like cures like." This holds that substances that cause healthy people to get symptoms of a certain medical condition or disease can cure conditions or

Homeopathy The treatment of a disease system by using minute doses of natural substances that, in a healthy person, would produce symptoms of the same disease

Pharmacology The branch of medicine concerned with the uses, effects, and modes of action of drugs

diseases that have these symptoms. This idea is a form of "sympathetic magic" similar to the primitive idea that eating the heart of a lion will make a person brave.

Homeopathy's founder, Samuel Hahnemann, M.D. (1775–1843), is said to have based his theory on an experience in which he ingested cinchona bark, the source of quinine used to treat malaria. After taking it, he experienced thirst, throbbing in the head, and fever—symptoms common to malaria. He decided that the drug's power to cure the disease arose from its ability to produce symptoms similar to the disease itself. He and his early followers then conducted "provings" in which they administered herbs, minerals, and other substances to healthy people, including themselves, and kept detailed records of what they observed.

Hahnemann was troubled by the toxic side effects of many of the substances he used and he began experimenting with dilution of the medication. He claims to have discovered that although dilution eliminated the side effects, it did not eliminate the positive effects of the substances. This "fact" became known as "the law of infinitesimals." He also developed ritualistic shaking (*succession*) of the medications after each successive serial dilution and claimed the method of shaking could affect the outcome of the treatment. Hahnemann lived at roughly the same time as Avogadro and was, thus, unaware of Avogadro's ideas concerning the number of molecules (6.02×10^{23}) in one mole of a substance. This may account for the fact that the excessive dilution factors that should worry modern scientists and advocates of homeopathy did not concern Hahnemann. Simple calculations show that many true homeopathic remedies do not contain a single molecule of the purported active ingredient. For example, a highly advertised homeopathic cold and flu cure known as Occillococcinum, made from an extract of duck liver, is sold at a dilution that would require the consumption of 10^{377} liters of the medicine to ensure the ingestion of only six molecules of the active ingredient!

True homeopathy has no scientific basis, in contrast to modern medicine. Many scientists and consumer protection groups have called for increased control and regulation of homeopathic and herbal cures. *Caveat emptor* should be stamped on the label of every homeopathic medicine. Consumers spend billions of dollars yearly on natural supplements and something well over $200 million on true homeopathic medicines. As the distinction between natural supplements and homeopathic remedies becomes increasingly blurred, one must urge caution in this area. Many natural herbal supplements and treatments are not homeopathic in nature, even though many have added the word *homeopathic* to their advertising to cash in on the cachet of the word. Even though it is well known that many proven medicines have their origin in folklore and indigenous natural remedies, few of the natural supplements currently on the market have been subjected to the rigorous standards required by the FDA. Such testing often reveals that the supplements contain significant impurities and highly variable quantities of the purported active ingredient. In many cases, the natural supplements contain none of the purported active ingredient. These issues should be of concern to anyone taking their health seriously. Finally, patients are often reluctant to tell physicians they are seeing that they are taking such supplements. This is a serious issue because supplements may interact badly with physician-prescribed medications.

CONCEPT CHECK 17D

1. Heart disease is known medically as _____.
2. During a heart attack, heart muscle can die from lack of _____.
3. When beta-receptors in the heart are stimulated by epinephrine, the heart beats **(a)** faster, **(b)** more slowly.

4. Clot-dissolving drugs used in emergency treatment of heart attacks are classified biochemically as _____.
5. Which of the following can initiate cancer? **(a)** Smoking, **(b)** Viruses, **(c)** Ionizing radiation.
6. Most anticancer drugs can also cause cancer. **(a)** True, **(b)** False.
7. The nitrogen mustards act on cancer cells by blocking _____ replication.
8. An ideal cancer chemotherapy agent would be able to _____ cancer cells without doing the same to noncancerous cells.
9. _____ and _____ cancers account for the largest percentage of cancer deaths among men in the United States.
10. Modern medicine depends on the _____ relationship of drugs.
11. Many of the solutions or pills used in _____ treatments can be shown to contain none of the active ingredient as a result of prescribed serial dilution procedures used.

 Assess your understanding of this chapter's topics with an online chapter quiz at www.brookscole.com/chemistry/joesten4

■ KEY TERMS

over-the-counter drugs	pathogenic	cardiovascular disease
prescription drugs	receptor	diuretic
hypertension	analgesics	carcinogen
trade name	antipyretic	homeopathy
generic name	anti-inflammatory	pharmacology
infectious diseases	sedation	
antibiotic	hallucinogens	

■ THE LANGUAGE OF CHEMISTRY

1. Can be bought without a prescription
2. Interfere with DNA synthesis
3. Beta-blockers
4. Prevents serotonin re-uptake
5. Receptor
6. Amoxicillin
7. Action of a penicillin
8. Neurotransmitter that causes allergic reaction
9. Dopamine
10. Common chemical name of a drug
11. Morphine, codeine, and heroin
12. Infectious disease
13. Acetaminophen
14. A very poisonous substance
15. Protease inhibitor
16. Acetylsalicylic acid

a. Treated with antibiotics
b. Frequently the site of action of a drug
c. Class of drugs for treating angina
d. Generic name
e. Antimetabolite for cancer chemotherapy
f. Over-the-counter drug
g. Opioids with very similar chemical structures
h. Most-prescribed antibiotic in 2004
i. New type of drug for AIDS treatment
j. Has a very small LD_{50}
k. Histamine
l. Neurotransmitter associated with drug addiction and Parkinson's disease
m. Prozac
n. Aspirin
o. Inhibit formation of cross-linking bonds in bacterial cell wall
p. An analgesic

■ APPLYING YOUR KNOWLEDGE

1. Give examples of a bacterial disease and a viral disease.

2. Give definitions of the following terms:
   (a) Over-the-counter drug
   (b) Prescription drug
   (c) Generic name for a drug
   (d) Trade name for a drug

3. Name the agency responsible for classifying drugs in the United States.

4. What are antibiotics?

5. Name three major classes of antibiotics.

6. What do the following three acronyms represent?
   (a) AIDS   (b) HIV
   (c) AZT

7. What is chemotherapy?

8. Describe how penicillin kills bacteria.

9. What is a retrovirus?

10. For what disease or condition is each of the following classes of drugs used?
    (a) Analgesics   (b) Antipyretics
    (c) Antibiotics   (d) Antihistamines

11. For what disease or condition is each of the following classes of drugs used?
    (a) Vasodilators   (b) Alkylating agents
    (c) Beta-blockers   (d) Antimetabolites

12. Describe the role of a receptor in biochemistry.

13. What two classes of natural biomolecules require receptors for their action?

14. To what classes of drugs do the following compounds belong?
    (a) Barbiturates   (b) Amphetamines

15. To what classes of drugs do the following compounds belong?
    (a) Methotrexate   (b) Chlorphenirimine

16. Describe the following and give an example of each:
    (a) Schedule I drugs   (b) Schedule II drugs

17. Which of the following terms apply to codeine?
    (a) Analgesic   (b) Antibiotic
    (c) Opioid
    (d) A scheduled drug with a potential for abuse
    (e) An antitussive

18. How do antihistamines work in the body? What, if any, side effects do antihistamines have?

19. What are the symptoms experienced in angina?

20. Name a drug that might be used to treat angina.

21. What is a barbiturate? What are the physiological effects of barbiturates?

22. Nitrogen mustards are alkylating agents, drugs that interfere with DNA replication. Explain what this means.

23. What happens when beta receptor sites in heart muscle are stimulated?

24. Name the class of biomolecule that includes dopamine, norepinephrine, and serotonin.

25. The estrogen estradiol has the following structure:

   What functional groups are different in ethynyl estradiol?

26. What is the function of the hormone progesterone?

27. Give the functions for the following:
    (a) Dopamine   (b) Epinephrine

28. What are the applications for the following drugs?
    (a) Tylenol   (b) Ibuprofen

29. What are the four classifications of drugs in terms of Drug Enforcement Administration regulations?

30. How are cocaine and crack related?

31. Classify each of the following substances as either an hallucinogen, an antidepressant, or a depressant:
    (a) Mescaline
    (b) Lysergic acid diethylamide (LSD)
    (c) *Cannabis sativa*   (d) Phencyclidine (PCP)
    (e) Barbiturates   (f) Amphetamines

32. What are the functions of the following over-the-counter drugs?
    (a) Antihistamines   (b) Analgesics
    (c) Decongestants   (d) Antitussives

33. What disease is treated with each of the following drugs? Tell the function of each drug.
    (a) Vasodilators   (b) Diuretics

34. What are nitrogen mustards? What was their original purpose? What is their current medical use?

35. What are the three modes of treatment for cancers?

36. Penicillins have the general formula shown as follows. What is the R group in penicillin G? Why is it necessary to have a number of different penicillins?

37. What is the effect of a hallucinogen? Name two examples of hallucinogens.

38. Describe the normal steps in the action of a neurotransmitter at a nerve synapse.

39. Of the three compounds heroin, morphine, and codeine,
    (**a**) which is the most effective pain killer?
    (**b**) which is not a natural alkaloid?
    (**c**) which is so addictive that its sale and use are illegal in the United States?

40. Identify by chemical name the oxygen-containing functional groups in morphine, codeine, and heroin (see Figure 17.5).

41. What is the physiological effect of nitroglycerine? What disease is it used to treat?

42. The FDA requires extensive testing of a prospective drug. The process can take almost 12 years. Other countries have much shorter approval processes. Imagine that you have a close relative who is suffering from a serious illness and an effective drug is available only outside the United States. What are the possible courses of action? What would you do?

43. Explain the role of the blood–brain barrier. What does it mean for the design of a drug that must be active in the brain?

W CHEMISTRY ON THE WEB

For up-to-date URLs, visit the text website at **www.brookscole.com/ chemistry/joesten4**

- Prescription Drugs
- Antibiotic Resistance
- HIV and AIDS
- Painkiller Risks
- General Medical Information
- Discussions of Pros/Cons of Homeopathy
- Steroids and Steroid Abuse
- Check Out the Structures and Properties of Some Steroids and Other Molecules
- Dopamine—A Sample Neurotransmitter
- The Pill
- Methamphetamine Dangers
- Cancer Statistics in the United States

THE CHEMISTRY OF USEFUL MATERIALS

Courtesy of Fraunhofer-Institut

An organic light-emitting diode (OLED). The organic layer is emissive so it does not require a backlight and can be used in television screens, computer displays, and portable electronic devices.

The long view from space has dramatized what we already knew—the crust of Earth is a very unusual environment, uniquely suited, at least in this solar system, for the production and support of life. Our environment is also quite heterogeneous. Mixtures abound; everywhere we look, the elements and compounds are almost lost in the complicated array of mixtures produced by natural forces acting over very long periods.

Throughout most of history, we had not developed the power to alter our environment significantly. Early everyday objects, such as stone hammers or wooden plows, were only physically changed from the natural material. Then came the chemical reduction of copper from natural minerals, followed by iron, and now the flood of new materials produced each year. We have developed, beyond question, the power to change Earth's natural chemical mixtures in almost any way we choose.

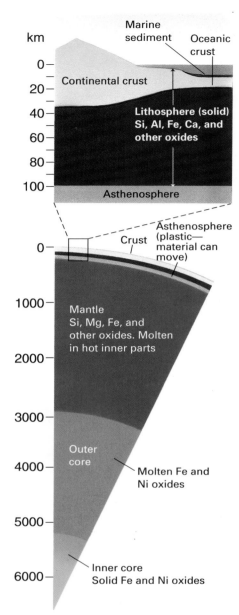

FIGURE 18.1 A cross-section of Earth. Geologists customarily list the composition of Earth in terms of oxides, as shown here.

Hydrosphere All freshwater and saltwater that is part of planet Earth

Minerals Naturally occurring, solid inorganic compounds

In this chapter, our focus is on inorganic substances—the elements other than carbon, and their compounds. We look at the origins of the raw materials for our pots and pans, homes and office buildings, automobiles and airplanes, and a multitude of other manufactured items. We examine these types of questions:

- How are chlorine and sodium hydroxide made from salt?
- How are metals extracted from their minerals and ores?
- What causes metals to be conductors of electricity?
- What are semiconductors and superconductors, and what are some of their uses?
- What are the general compositions, properties, and methods of production of glass, ceramics, and cement?

18.1 THE WHOLE EARTH

This chapter illustrates how elements and compounds are separated from the **hydrosphere** and crust of Earth, and put to use. The hydrosphere, which includes saltwater and freshwater above and below Earth's surface, must supply the water necessary to sustain life. The soluble salts in the oceans are a commercial source of magnesium, bromine, and sodium chloride, which is not only table salt but also an essential chemical raw material.

The portion of the solid Earth available to us is a very small part of the whole, less than 1% by mass. The deepest mine extends only 3.8 km beneath the surface. Geologists define Earth's crust as a region between the surface and a depth of about 5 km to 35 km that lies over regions of greater density (Figure 18.1).

Three major types of rocks are found in Earth's crust: *igneous rocks*, formed by solidification of molten rock (e.g., basalt); *sedimentary rocks* (e.g., sandstone, which is cemented sand), formed by deposition of dissolved or suspended substances from oceans and rivers; and *metamorphic rocks* (e.g., marble), formed by the action of heat and pressure on existing rocks. Figure 18.2 gives the average composition of Earth's crust. The most abundant substances in rocks are silicates, which are composed of silicon, oxygen, and positive metal ions (Section 18.5). The more than 2000 kinds of known **minerals** fall into a few major classes (Table 18.1).

TABLE 18.1 Major Mineral Groups in Earth's Crust

MINERAL GROUP	EXAMPLE	FORMULA	USES
Silicates	Quartz	SiO_2	Glass, ceramics, alloys
	Feldspar	$KAlSi_3O_8$	Ceramics
Oxides	Hematite	Fe_2O_3	Iron ore, paint pigment
Carbonates	Calcite	$CaCO_3$	Optical instruments (pure crystals), industrial chemicals
Sulfides	Galena	PbS	Lead ore, semiconductors
Sulfates	Gypsum	$CaSO_4 \cdot 2\,H_2O$	Cement, plaster of paris, wallboard, paper sizing
Halides	Fluorite	CaF_2	Lasers and electronics (pure crystals), source of fluorine (F_2)

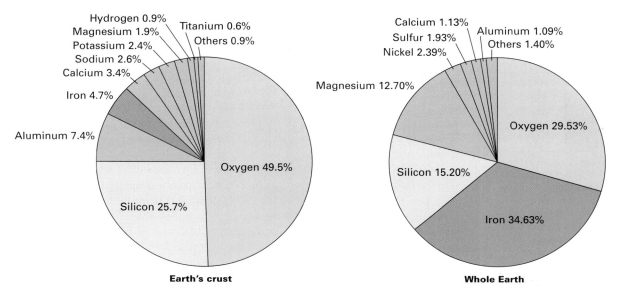

FIGURE 18.2 Relative abundance (by mass) of elements in Earth's crust compared to abundance in the whole Earth.

Fortunately for the mining industry, the composition of the crust is not uniform. Natural forces have concentrated different minerals in different places. For example, as molten rock gradually cools, the minerals that solidify first (those with the higher melting points) can sink in the remaining liquid and become concentrated. Or minerals can be redistributed according to variations in their solubility in natural waters.

The western United States was once covered by a large, landlocked sea. The water evaporated, leaving huge deposits of sodium carbonate, a soluble salt that is a valuable chemical raw material. While most nations must manufacture sodium carbonate from other chemicals, the United States meets a large proportion of its needs by mining. Other minerals are concentrated elsewhere, of course. Most of the known deposits of nickel are in New Caledonia and Zimbabwe. Most of the chromium is in Botswana and Turkey. Each country must rely on imports of one kind or another.

18.2 CHEMICALS FROM THE HYDROSPHERE

A single mouthful of seawater is enough to convince anyone that it is salty. Indeed, sodium chloride is the major mineral in seawater. But consider that other dissolved minerals are constantly being deposited into the oceans from rivers, undersea volcanoes, and thermal vents. In addition to sodium and chlorine, the major elements present in ions in solution are magnesium, sulfur, calcium, potassium, bromine, carbon, nitrogen, and strontium. Lower concentrations of other elements in seawater could also provide huge quantities of economically important metals such as uranium, copper, manganese, and gold.

The lower the concentration of a metal ion in seawater, of course, the higher the cost of isolating the metal is likely to be. Nevertheless, as high-quality mineral deposits on the land are depleted, the economics of mining the sea might become more attractive. Marine organisms may help to solve the problem. For example, one family of such organisms (the *tunicates*) accumulates vanadium to more than 280,000 times its concentration in seawater. Perhaps aquatic farming of such creatures could be put to work extracting metals.

Minerals have definite internal structure and chemical composition. Most of Earth's crust consists of *rocks*, which may be single minerals or mixtures of minerals. A natural material that contains a sufficient concentration of an element or compound to economically justify mining it is an *ore*.

Some other uses for sodium chloride include food processing; highway snow melting; animal feed; domestic table salt; and rubber, oil, paper, and textile manufacture.

Separation of salt from seawater by evaporation in San Francisco Bay.

Salt and the Chloralkali Industry

The majority of our salt is obtained from natural waters. In coastal regions salt is separated from seawater by evaporation of the water in large lagoons open to the sun. The natural *brines*, or salty waters, found in wells and lakes such as the Great Salt Lake in Utah are other sources of salt. After isolation, the salt is purified to the degree required for its use. Much of it is destined for the *chloralkali industry*, which produces chlorine gas and sodium hydroxide.

Chlorine gas is used to disinfect drinking water and sewage, and in the production of organic chemicals such as pesticides and vinyl chloride, the building block of plastics called polyvinyl chlorides (PVCs, Section 14.5). Chlorine gas is commonly among the top chemicals produced each year in the United States. Almost all chlorine gas is made by electrolysis of aqueous sodium chloride. The other product of sodium chloride electrolysis, sodium hydroxide, is equally valuable because it is the most commonly used base in industrial processes. The reaction in electrolysis of aqueous NaCl is

$$2\ NaCl(aq) + 2\ H_2O(\ell) \xrightarrow{\text{Electrical energy}} 2\ NaOH(aq) + H_2(g) + Cl_2(g)$$

A complicating factor in designing electrochemical cells for this reaction is that the chlorine and sodium hydroxide, if they remain in contact, react with each other.

A modern electrochemical cell for producing chlorine and sodium hydroxide is illustrated in Figure 18.3. The two electrode compartments are separated by a synthetic membrane that allows only sodium ions that enter as part of the brine to pass through it. The reactions are

$$2\ Cl^-(aq) \longrightarrow Cl_2(g) + 2\ e^- \qquad \text{Oxidation}$$

$$2\ H_2O(\ell) + 2e^- \longrightarrow H_2(g) + 2\ OH^-(aq) \qquad \text{Reduction}$$

$$\overline{2\ Cl^-(aq) + 2\ H_2O(\ell) \longrightarrow H_2(g) + Cl_2(g) + 2\ OH^-(aq)} \qquad \text{Overall reaction}$$

Brine is introduced into the anode compartment, and chloride ion oxidation occurs there. To maintain charge balance within the cell, as Cl^- ions are oxidized, Na^+ ions must pass from the anode to the cathode compartment. Since OH^- ions are produced in the cathode compartment, the product there is aqueous NaOH, with a concentration of 20 to 35% by weight. Most sodium hydroxide is used in industrial processes, although it is also present in some oven, drain, and sewer-pipe cleaners.

Salt has been important throughout history. The Romans gave a *salarium* to those who were "worth their salt"—the origin of the word *salary*.

For many years, cells with mercury cathodes were used in the chloralkali industry. Mercury is not very reactive or soluble and was thought to be harmless in the environment. Then some individuals who ate fish from mercury-contaminated waters became seriously ill. This event brought to light the fact that aquatic microorganisms convert metallic mercury to a toxic, water-soluble compound that enters the food chain (Section 15.6).

Electrochemical cells and electrolysis are described in Section 10.5.

FIGURE 18.3 A chloralkali cell. Large banks of these cells produce gaseous chlorine and aqueous sodium hydroxide solution, both important industrial chemicals. Because of the need for cheap electricity, chloralkali plants are located near hydroelectric plants at, for example, Niagara Falls.

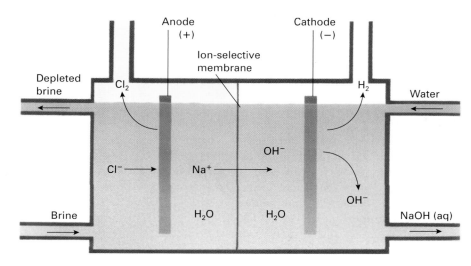

Magnesium from the Sea

Magnesium, with a density of 1.74 g/cm^3, is the lightest structural metal in common use. It is 36% lighter than aluminum and 78% lighter than iron. Many **alloys** designed for light weight and great strength contain magnesium. Most manufactured aluminum objects, for example, contain about 5% magnesium, which is added to improve the mechanical properties and corrosion resistance of the aluminum under alkaline conditions. There are also alloys that have the reverse formulation, that is, more magnesium than aluminum. These alloys are used where a high strength-to-weight ratio is needed and where corrosion resistance is especially important. The amount of magnesium used in American cars is expected to rise from about 4 kg per vehicle in 2002 to as much as 150 kg in some models by 2020. This usage is expected to increase as manufacturers produce lighter-weight cars to meet new federal fuel–economy standards.

Because there are 6 million tons of magnesium present as Mg^{2+} in every cubic mile of seawater, which is about 0.14% magnesium, the sea can furnish an almost limitless amount of this element. The recovery of magnesium from seawater begins with the **precipitation** of insoluble magnesium hydroxide $[Mg(OH)_2]$. The only thing needed for this step is a ready supply of an inexpensive base (a source of OH^- ions), a need fulfilled nicely by seashells, which contain calcium carbonate $(CaCO_3)$. Heating the calcium carbonate converts it to lime, which then reacts with water to give calcium hydroxide, the base used in the precipitation.

Some alloys are homogeneous mixtures and others are heterogeneous.

$$CaCO_3(s) \xrightarrow{\text{Heat}} CaO(s) + CO_2(g)$$
$$\underset{\text{Seashells}}{} \qquad \underset{\text{Lime}}{}$$

$$CaO(s) + H_2O(\ell) \longrightarrow Ca(OH)_2(aq)$$

$$Mg^{2+}(aq) + Ca(OH)_2(aq) \longrightarrow Mg(OH)_2(s) + Ca^{2+}(aq)$$

The solid magnesium hydroxide is isolated by filtration and then neutralized by another inexpensive chemical, hydrochloric acid.

$$Mg(OH)_2(s) + 2\,HCl(aq) \longrightarrow MgCl_2(aq) + 2\,H_2O(\ell)$$

When the water is evaporated, solid hydrated magnesium chloride is left. After drying, it is melted (at $708°C$) and then electrolyzed in a huge steel pot that serves as the cathode. Graphite bars serve as the anodes. The electrode reactions are

Notice how many kinds of chemical reactions are organized into the production of magnesium—decomposition by heat, neutralization, precipitation, and oxidation–reduction by using electrical current (electrolysis).

$$2\,Cl^- \longrightarrow Cl_2(g) + 2\,e^- \qquad \text{Oxidation}$$
$$\underline{Mg^{2+} + 2\,e^- \longrightarrow Mg(\ell) \qquad \text{Reduction}}$$
$$Mg^{2+} + 2\,Cl^- \longrightarrow Mg(\ell) + Cl_2(g) \qquad \text{Overall reaction}$$

As the molten magnesium forms, it is removed. The chlorine produced in the electrolysis is recycled into the process by reacting it with hydrogen to produce hydrochloric acid.

The electrolysis of aluminum ions is discussed in Section 10.5.

CONCEPT CHECK ⟩ **18A**

1. The most abundant element in Earth's crust is _____.
2. The most abundant metal in Earth's crust is _____.
3. Identify the source of each of the following substances as (i) the hydrosphere, or (ii) Earth's crust:
   - **(a)** Chlorine
   - **(b)** Sodium chloride
   - **(c)** Magnesium
   - **(d)** Sand
   - **(e)** Marble

Alloy A mixture of two or more metals

Precipitation Formation of a solid product during a chemical reaction between reactants in solution

4. Which metal is sufficiently concentrated in the oceans so that it is currently extracted commercially: aluminum, copper, magnesium, or iron?

5. Magnesium ions are (reduced/oxidized) to produce magnesium metal.

18.3 METALS AND THEIR ORES

According to the U.S. Geological Survey, the average American will require the quantities of minerals listed in Table 18.2 on an annual basis. In this section, we focus on the metals. Some of the chemistry of familiar silicon-based materials—glass, ceramics, cement—is discussed in Section 18.5.

Less reactive metals such as copper, silver, and gold can be found as free elements. The somewhat more reactive metals are present as sulfides formed early in Earth's existence (e.g., CuS, PbS, and ZnS). Because of their extremely low water solubility, sulfides resist oxidation and reactions with water and other ions. The still more reactive metals have been converted over millennia into oxides (e.g., MnO_2, Al_2O_3, and TiO_2) and are mined in that form.

The most reactive metals, such as sodium and potassium, are present in nature as soluble salts in the ocean and mineral springs; in solid deposits of these salts; or in insoluble, stable aluminosilicates, such as albite ($NaAlSi_3O_8$) and orthoclase ($KAlSi_3O_8$). Such silicates are found in all parts of the world, but because of their great stability they are not currently used as sources of the metals they contain. These minerals, like the ocean, represent a resource that may have to be tapped when richer and more easily processed ores are depleted.

A selection of beautiful minerals that are also ores is pictured in Figure 18.4. (A mineral is an *ore* if separating the metal from it is possible and economically practical.) The preparation of metals from minerals requires chemical reduction—the conversion of positive metal ions to free metals. To reduce

TABLE 18.2	Approximate Per Capita Yearly Requirement of New Raw Materials from Earth's Crust for an American	
ELEMENT OR MINERAL	**QUANTITY (POUNDS)**	**MAJOR USES**
Stone, sand, gravel	19,000	Roads and buildings
Coal	7,650	Generating electricity; iron, steel, and chemicals manufacture
Iron ore	600	Automobiles and ships, structural support
Clays	280	Bricks, paper, paint, glass, pottery
Salt	420	De-icing, detergents, cooking, chemical manufacturing
Aluminum	75	Food and beverage cans, household items, vehicles
Copper	25	Electrical motors, wiring
Zinc	15	Brass, galvanized iron, steel
Lead	15	Auto batteries, solder, electronics parts
Manganese	6	Iron and steel production; glass decolorizer; brick and ceramic colorant

SOURCE: **http://minerals.usgs.gov/west/**

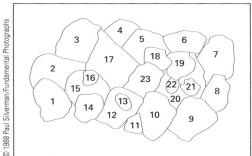

© 1988 Paul Silverman/Fundamental Photographs

FIGURE 18.4 A collection of native metals and minerals. The minerals serve as ores for the metals indicated in the following key.

1. Bornite (iridescent)—COPPER
2. Dolomite (pink)—MAGNESIUM
3. Molybdenite (gray)—MOLYBDENUM
4. Skutterudite (gray)—COBALT, NICKEL
5. Zincite (mottled red)—ZINC
6. Chromite (gray)—CHROMIUM
7. Stibnite (*top right,* gray)—ANTIMONY
8. Gummite (yellow)—URANIUM
9. Cassiterite (rust, *bottom right*)—TIN
10. Vanadinite crystal on goethite (red crystal)—VANADIUM
11. Cinnabar (red)—MERCURY
12. Galena (gray)—LEAD
13. Monazite (white)—RARE EARTHS: cesium, lanthium, neodymium, thorium
14. Bauxite (gold)—ALUMINUM
15. Strontianite (white, spiny)—STRONTIUM
16. Cobaltite (gray cube)—COBALT
17. Pyrite (gold)—IRON
18. Columbinite (tan, gray stripe)—NIOBIUM, TANTALUM
19. Bismuth (shiny)
20. Rhodochrosite (pink)—MAGNESIUM
21. Rutile (shiny twin crystal)—TITANIUM
22. NATIVE SILVER filigree on quartz
23. Pyrolusite (black, powdery)—MANGANESE

a metal ion requires a source of electrons, which can be either an electrical current (as in the case of magnesium production mentioned earlier) or a chemical reducing agent.

Iron

Iron is the fourth most abundant element in Earth's crust and the second most abundant metal. Our economy depends on iron and its alloys, particularly steel. Most of the world's iron is located in large deposits of iron oxides in Minnesota, Sweden, France, Venezuela, the former Soviet Union, Australia, and the United Kingdom.

Iron ore is reduced in a blast furnace (Figure 18.5a). The solid material fed into the top of the furnace is a mixture of iron oxide (Fe_2O_3), coke (C), and limestone ($CaCO_3$). A blast of heated air is forced into the furnace near the bottom. The major reactions that occur within the blast furnace result in reduction of iron oxide

$$2\,C(s) + O_2(g) \longrightarrow 2\,CO(g) + heat$$

$$Fe_2O_3(s) + 3\,CO(g) \longrightarrow 2\,Fe(s) + 3\,CO_2(g) + heat$$

and conversion of silica present in the ore to molten calcium silicate ($CaSiO_3$).

$$CaCO_3(s) \xrightarrow{\text{Heat}} CaO(s) + CO_2(g)$$

$$CaO(s) + SiO_2(s) \xrightarrow{\text{Heat}} CaSiO_3(\ell)$$
$$\text{Slag}$$

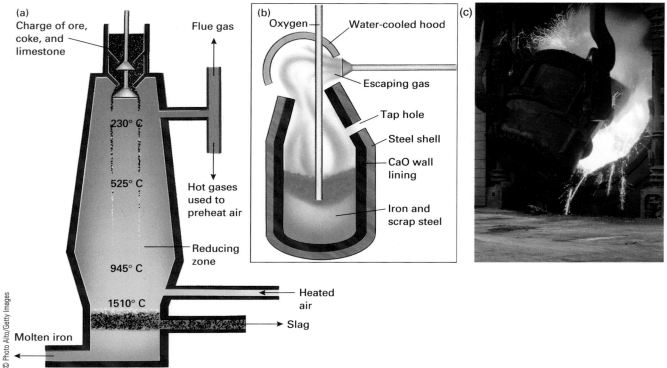

(a)
Charge of ore, coke, and limestone

230° C

525° C

945° C

1510° C

Molten iron

Flue gas

Hot gases used to preheat air

Reducing zone

Heated air

Slag

© Photo Alto/Getty Images

(b)
Oxygen

Water-cooled hood

Escaping gas

Tap hole

Steel shell

CaO wall lining

Iron and scrap steel

(c)

FIGURE 18.5   Iron and steel production. In the blast furnace (a) the descending materials in the charge are hit by a blast of hot air from coke burning in the heated air. The basic oxygen furnace for steel production (b) is charged with a mixture of molten pig iron from the blast furnace, steel scrap to be recycled, and other metals according to the type of steel being made. After oxygen is blown in for about 20 min, the finished steel is poured off through the tap hole, and the furnace is ready for another charge. The photo (c) shows molten iron being poured into a basic oxygen furnace.

Consequently, as the blast furnace operates, two molten layers collect in the bottom. The lower, denser layer is mostly liquid iron. The upper, less dense layer is the **slag**. From time to time the furnace is tapped at the bottom, and the molten iron is drawn off. Another outlet somewhat higher in the blast furnace can be opened to remove the liquid slag.

The iron that comes from the blast furnace, known as *pig iron*, contains many impurities (up to 4.5% carbon, 1.7% manganese, 0.3% phosphorus, 0.04% sulfur, and 1% silicon). Iron reacts with the carbon impurity at the temperatures of the blast furnace to form cementite, an iron carbide (Fe_3C), which causes pig iron to be brittle.

$$3 \, Fe(s) + C(s) \longrightarrow Fe_3C(s)$$

About 75% of the weight of an automobile is iron and steel.

When molten pig iron is poured into molds of a desired shape (engine blocks, brake drums, transmission housings), it is called **cast iron**. However, pig iron and cast iron contain too much carbon and other impurities for most uses. The structurally stronger material known as **steel** is obtained by removing the phosphorus, sulfur, and silicon impurities and decreasing the carbon content.

Slag Mixture of nonmetallic waste products separated from a metal during its refining

Cast iron Molded pig iron or other carbon–iron alloy

Steel Malleable iron-based alloy with a relatively low percentage of carbon

Steels

Many kinds of iron alloys are collectively known as *steels*. The most common is *carbon steel*, an alloy of iron with about 1.3% carbon. To convert pig iron into carbon steel, the excess carbon is burned out with oxygen.

In the *basic oxygen process* for steel production (Figure 18.5b), pure oxygen is blown into molten iron through a refractory tube, which is pushed below the surface of the iron. A *refractory* is a material that withstands high temperatures without melting. At elevated temperatures, the dissolved carbon reacts very rapidly with the oxygen to give gaseous carbon monoxide and carbon dioxide, which escape.

During steelmaking, silicon or transition metals such as chromium, manganese, and nickel can be added to give alloys with specific physical, chemical, and mechanical properties.

The properties of steel can also be adjusted by the temperature and rate of cooling used in its production. If the steel is cooled rapidly by quenching in water or oil, the carbon in the steel remains in the form of cementite (Fe_3C) and the steel will be hard, brittle, and light colored. Slow cooling favors the formation of crystals of carbon (graphite) instead of cementite. The resulting steel is more *ductile* (easily drawn into shape).

All the processes in steelmaking, from the blast furnace to the final heat treatment, use tremendous quantities of energy, mostly in the form of heat. In the production of one ton of steel, approximately one ton of coal or its energy equivalent is consumed.

Steel structures in Chicago. Alexander Calder's steel sculpture, *Flamingo*, stands in a plaza surrounded by steel-frame office buildings.

In carbon steel, the carbon atoms fit in the small spaces (interstices) between the larger iron atoms of the metal. The small structural change of even one part per thousand carbon makes slipping of the iron atoms more difficult. That's why steel is stronger than iron.

Copper

Although copper metal occurs in the free (elemental) state in some parts of the world, the supply available from such sources is quite insufficient for the world's needs. The majority of the copper used today is obtained from various copper sulfide ores, such as chalcopyrite ($CuFeS_2$), chalcocite (Cu_2S), and covellite (CuS). Because the copper content of these ores is about 1–2%, the powdered ore is first concentrated by the flotation process.

In the flotation process, the powdered ore is mixed with water and a frothing agent such as pine oil. A stream of air is blown through the mixture to produce froth. The **gangue** in the ore, which is composed of sand, rock, and clay, is easily wetted by the water and sinks to the bottom of the container. In contrast, a copper sulfide particle is hydrophobic—it is not wetted by the water. The copper sulfide particle becomes coated with oil and is carried to the top of the container in the froth. The froth is removed continuously, and the floating copper sulfide minerals are recovered from it.

The preparation of copper metal from copper sulfide ore involves **roasting** the ore in air to convert some of the copper sulfide and any iron sulfide present to the oxides.

$$2\,Cu_2S(s) + 3\,O_2(g) \longrightarrow 2\,Cu_2O(s) + 2\,SO_2(g)$$
$$2\,FeS(s) + 3\,O_2(g) \longrightarrow 2\,FeO(s) + 2\,SO_2(g)$$

Subsequently the mixture is heated to a higher temperature, and some copper is produced by the reaction

$$Cu_2S(s) + 2\,Cu_2O(s) \xrightarrow{\text{Heat}} 6\,Cu(s) + SO_2(g)$$

The product of this operation is a mixture of copper metal, sulfides of copper, iron and other ore constituents, and slag. The molten mixture is heated in a converter with silica materials. When air is blown through the molten material in the converter, two reaction sequences occur. In one, the iron is converted to slag.

$$2\,FeS(s) + 3\,O_2(g) \longrightarrow 2\,FeO(s) + 2\,SO_2(g)$$
$$FeO(s) + SiO_2 \longrightarrow FeSiO_3(\ell)$$
Molten slag

In the other, the remaining copper sulfide is converted to copper metal by the preceding two reactions shown for Cu_2S. The copper produced in this manner

Roasting is a common step in recovery of metals from their ores. It consists of heating the ore, often a metal sulfide, in air to convert the sulfide to the oxide. If the metal oxide product from roasting is less stable than SO_2, the free metal is obtained by sulfide roasting. This is the case for Cu_2S, where further heating of the mixture of Cu_2S and Cu_2O produces Cu and SO_2. Unfortunately, if the SO_2 is vented to the atmosphere, it contributes to air pollution and is one source of "acid rain" (Section 11.2).

Gangue Unwanted substances mixed with a mineral during production of metal from ore

Roasting (of ore) Heating an ore in air, usually to produce an oxide

THE WORLD *of* CHEMISTRY

SWORDS OF DAMASCUS LEAVE LEGACY OF SUPERPLASTIC STEEL

About 3000 years ago, a type of steel called wootz steel was produced in India by heating a mixture of pure iron ore and wood in a sealed pot or crucible. Some of the carbon reduced the iron ore to metallic iron, which then absorbed some of the remaining carbon to form an excellent steel. The wootz steel later became famous as Damascus steel, used for making swords that retained their sharpness and strength after countless battles. The blacksmiths who forged this steel into swords used a process that they carefully kept secret—one that produced a steel that was much more pliable at high temperatures than normal steel. The knowledge of how to make this special steel was lost sometime in the 19th century.

Professor Oleg Sherby of Stanford University and specialists at the Lawrence Livermore National Laboratory have developed types of steels with properties similar to those of Damascus steel. The new steels have a higher percentage of carbon (about 1.8%) than normal carbon steels and are referred to as ultrahigh carbon steel alloys or superplastic steels.

Unlike most steels, which fail after being stretched to 2 times their original size, superplastic steel at elevated temperatures can be stretched to 11 times its size without cracking or pulling apart. As a result, heated superplastic steel can be formed into complex shapes. It can even flow like molasses and be poured into a mold. This property eliminates the need for machining, which typically results in about 50% scrap. Superplastic steel is also similar to stainless steel in its resistance to corrosion, but is made with less scarce and expensive materials than the nickel and chromium in stainless steel.

The superior properties of superplastic steel are attributed to a much finer grain structure than that in ordinary carbon steels, which are quite brittle. Once its manufacture was understood, a group of industrial firms banded together to develop its use. As

Stanford University Professor Oleg D. Sherby holds a 300-year-old sword of Damascus steel. (For further information, see *Chemecology*, p. 6 March 1992; *R&D Innovator*, Vol. 3, No. 10, 1994.)

noted by Oleg Sherby, their success "will mark a remarkable reincarnation of a material that once helped conquer the known world."

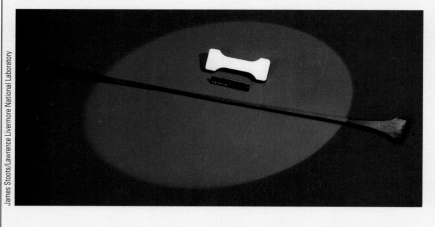

Superplastic, ultrahigh carbon steel can be stretched to 11 times its size at 900°C. The original size of the bottom steel piece was 1 inch, but it was pulled to a length of 14 inches. Conventional steels collapse when they are extended to two times their size.

is crude or "blister" copper, (96–99.5% Cu), the blistered surface resulting from the escaping gas. The blister copper is later purified electrolytically.

In the electrolytic purification of copper, the anodes are blister copper bars and the cathodes are made of pure copper. As electrolysis proceeds, copper is oxidized at the anode, moves through the solution as Cu^{2+} ions, and is deposited on the cathode. The voltage of the cell is regulated so that more active impurities (such as iron) are left in the solution, and less active ones are not oxidized at all. The less active impurities include gold and silver, which collect as "anode slime," an insoluble residue beneath the anode. The anode slime is subsequently treated to recover the valuable metals.

The copper produced by the electrolytic cell is 99.95% pure and is suitable for use as an electrical conductor. Copper for this purpose must be pure because very small amounts of impurities, such as arsenic, considerably reduce the electrical conductivity of copper.

Open-pit copper mining near Bagdad, Arizona.

A piece of native copper.

CONCEPT CHECK 18B

1. The metal consumed in largest quantity during a person's lifetime is _____.

2. The principal impurity in pig iron is _____.

3. The reducing agent in the production of iron from iron ore is _____.

4. Pig iron is brittle. (**a**) True (**b**) False

5. Copper usually occurs in its ores combined with the element _____.

6. Copper can be purified using electrolysis. (**a**) True (**b**) False

18.4 CONDUCTORS, SEMICONDUCTORS, AND SUPERCONDUCTORS

Metals have some properties totally unlike those of other substances. Except for mercury, which is a liquid at room temperature, and gallium, which is a liquid at slightly above room temperature, all metals are solids. Some remain solids even at very high temperatures. Tungsten has a melting point of 3410°C. There are several properties common to the metals:

Section 8.6 discusses the recycling of metals.

- *High electrical conductivity*. Metal wires easily carry electrical currents. The electrons in metals are quite mobile.

- *High thermal conductivity*. Some metals are much better conductors of heat than others. Try a sterling silver spoon in a cup of hot coffee or tea. Compare its thermal, or heat, conductivity with that of a stainless steel spoon.

- *Ductility and malleability*. Most metals can be drawn into wire (ductility) or hammered into thin sheets (malleability); gold is the most malleable metal. Extremely thin sheets of gold are used for decoration.

- *Luster*. Polished metal surfaces reflect light; most metals have a silvery white color because they reflect all wavelengths of light equally well.

Electroplating, Section 10.5, is used to deposit protective and decorative coating of metals on various objects.

- *Insolubility in water and other common solvents*. No metal dissolves in water, but a few, such as the Group 1A and 2A metals, react with water to form hydrogen gas and solutions of metal hydroxides.

Any theory of the bonding of metal atoms must be consistent with these properties. Structural investigations of metals have led to the conclusion that solid metals are composed of regular arrays, or *lattices*, of metal ions in which

Putting the properties of metals to use. (a) Installation of electrically conducting wires. (b) A flask coated, like a mirror, with silver. (c) Very thin and stable gold leaf being applied to a surface.

(a)   (b)   (c)

the bonding electrons are loosely held. Figure 18.6 illustrates one model for metallic bonding in which the regular array, or lattice, of positively charged metal ions is embedded in a "sea" of mobile electrons. These mobile valence electrons are delocalized over the entire metal crystal, and the freedom of these electrons to move throughout the solid is responsible for the properties associated with metals. In contrast to those in metals, the valence electrons in nonmetals are fixed in bonds between like atoms. This means that nonmetals are nonconductors of electricity.

Semiconductors: The Basis of Our Modern World

You should not be surprised that somewhere between the excellent electrical conductivity of most metals and the nonconductivity of nonmetals, there are some elements that are **semiconductors**. This means they will conduct electricity under certain conditions. Silicon, when it is in a highly purified state, is a semiconductor. Silicon acts like a nonmetal and fails to conduct a current until a certain voltage is applied; then it begins to conduct moderately well. This behavior interested electrical engineers, who recognized that silicon might act like a "gate" for electron flow in electrical circuits. This electron

The elements shown in blue in the periodic table inside the front cover of this book are referred to as the *metalloids* (Section 3.6).

Semiconductors Materials with electrical conductivity intermediate between that of metals and insulators

FIGURE 18.6 "Electron sea" model of bonding in metals. The positively charged metal atom nuclei are surrounded by a "sea" of negatively charged electrons.

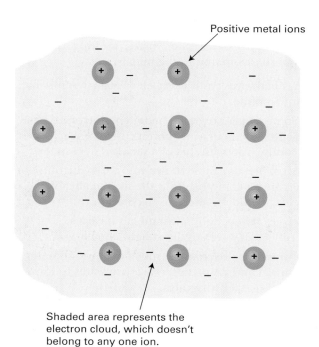

Positive metal ions

Shaded area represents the electron cloud, which doesn't belong to any one ion.

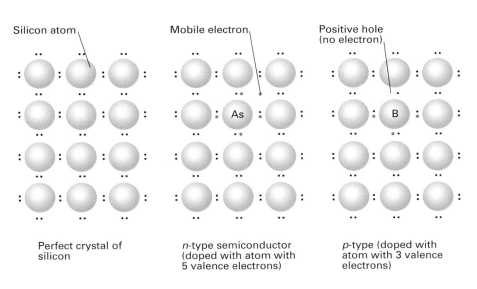

Silicon atom   Mobile electron   Positive hole (no electron)

Perfect crystal of silicon

n-type semiconductor (doped with atom with 5 valence electrons)

p-type (doped with atom with 3 valence electrons)

FIGURE 18.7 **Doping of silicon in semiconducting devices.** Adding atoms with five valence electrons (e.g., arsenic) introduces extra electrons that can move through the crystal. Adding atoms with three valence electrons introduces holes that can also move through the crystal. Note that each silicon atom has four valence electrons, so a perfect crystal has each silicon atom surrounded by an octet of electrons.

gatelike activity was not realized until a process known as **doping** was discovered. One common dopant is boron, a Group 3A element just to the upper left of silicon in the periodic table. When boron is added to pure silicon, the boron atoms, with one fewer valence electron than silicon, introduce *positive holes* in the lattice arrangement of silicon atoms. The presence of these positive holes (which can move about in the solid just like electrons, but in the opposite direction) makes the doped silicon somewhat more conductive—in effect, it becomes a better electron gate. Silicon doped with an element that creates positive holes is called a ***p*-type** semiconductor. Another dopant is arsenic, from Group 5A, which contains one *more* valence electron than silicon. Silicon doped with arsenic is called an ***n*-type** semiconductor because there are extra negative electrons present in the solid. These, of course, enhance the conductivity of the doped silicon. Through careful control of the amount and type of dopant, the conductivity of the silicon can be adjusted to a fine degree (Figure 18.7).

In 1947, a device consisting of a layer of *p*-type silicon sandwiched between two *n*-type layers was constructed by John Bardeen, Walter Brattain, and William Shockley at the Bell Laboratories. This device, called the **transistor**, has revolutionized our world (Figure 18.8). Because the transistor can control electron flow in circuits with such accuracy, yet is so small and requires so little power to operate, it is now possible to design electronic

Bardeen, Brattain, and Shockley shared the Nobel Prize in physics in 1956 for the discovery of the transistor. The importance of the transistor was recognized as soon as it was discovered. Although it was first demonstrated at the Bell Laboratories in December 1947, it wasn't announced until July 1, 1948, after patent applications had been filed. John Bardeen was awarded a second Nobel Prize in physics in 1972, along with J. R. Schrieffer and Leon N. Cooper, for work on the theory of superconductivity.

Doping (of semiconductor) The addition of atoms of an element with extra or fewer valence electrons than the principal element of the semiconductor material.

Transistor Semiconductor device that controls electron flow in circuits

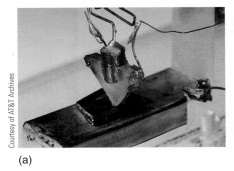

(a)   (b)

FIGURE 18.8 **The transistor and its inventors.** (a) The first transistor, constructed in 1947 at Bell Laboratories. Electrical contact is made at a single point and the signal is amplified as it passes through a solid semiconductor; modern junction transistors amplify in a similar manner. (b) Envelope and stamp commemorating 25 years of the transistor, with portraits of its inventors, Walter Brattain, William Shockley, and John Bardeen.

FIGURE 18.9 Modern computer chip on a finger tip.

The land area of the United States is about 9×10^9 km^2.

circuits to fit into extremely small volumes. Such objects as TV cameras as small as a pea, radios small enough to strap to the back of an ant, and other amazing devices can be made using transistors based on doped silicon. Of course, more mundane items such as automatic cameras, microwave ovens, fax machines, and cellular phones also owe their existence to the transistor. The central processing unit (CPU) of computers consists of millions of transistors and other circuit elements fabricated on wafers of pure silicon (Figure 18.9). These devices are called integrated circuits.

Solar Energy to Electricity

The *solar cell* or photovoltaic cell converts energy from the sun into electron flow. Solar cells are now used in calculators, watches, spaceflight applications, communication satellites, and signals for automobiles and trains, and as the source of electricity in undeveloped regions throughout the world, where electrical power grids are virtually nonexistent. It has been estimated that all the electricity used in the United States could be made by solar cells having only 10% efficiency and covering about 13,000 km^2, which is 0.13% of the land area in the United States.

One type of solar cell consists of two layers of almost pure silicon. The lower, thicker layer contains a trace of boron (B); and the upper, thinner layer has a trace of arsenic (As). As pointed out earlier, the As-enriched layer is an *n*-type semiconductor, with mobile electrons; and the B-enriched layer is a *p*-type semiconductor, with the electron deficiencies known as holes (see Figure 18.7). There is a strong tendency for the mobile electrons in the *n*-type layer to pair with the unpaired electrons in the holes in the *p*-type layer. If the two layers are connected by an external circuit (Figure 18.10) and light of sufficient energy strikes the surface and is absorbed, excited electrons can leave the *n*-type layer and flow through the external circuit to the *p*-type layer. As this layer becomes more negative because of added electrons, electrons are repelled internally back into the *n*-type layer, which is now positive and attracts the electrons. The process can continue indefinitely as long as the cell is exposed to sunlight.

Can Anything Conduct Better Than a Metal?

When metals are heated, their electrical conductivity decreases. Lower conductivity at higher temperatures can be explained if the movement of the valence electrons is considered to be limited by rapidly vibrating atoms in the metal lattice. The kinetic molecular theory says that higher temperatures

FIGURE 18.10 A solar cell. Beneath the outer glass is a metal grid that allows as much light as possible to strike the *n*-type semiconductor layer, while serving as the electrode at which electrons leave the cell. The *n*-type semiconductor layer is almost transparent. Beneath it is the *p*-type semiconductor layer and the electrode at which electrons re-enter the cell.

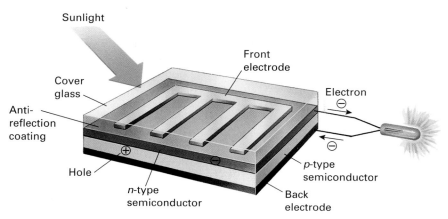

mean more motion, and this applies even when the atoms are in fixed positions, as they are in a metal. When the metal atoms are relatively stationary, as they are when the metal is cool, the electrons can move through the lattice much like a person moving through a room filled with a large number of other people quietly chatting with one another. When the temperature is elevated, the metal atoms begin to vibrate wildly, and the electrons have more trouble getting through the lattice, in much the same way that a person would have trouble moving through the room if all the other occupants suddenly became agitated.

From this picture of electrical conductivity of metals, you might assume that if a sufficiently low temperature were reached, conductivity might be quite high (almost zero resistance, in other words). In fact, the conductivity of a pure metal crystal does approach infinity (zero resistance) as absolute zero is approached. In many metals, however, a more interesting thing happens before absolute zero is reached. At a certain low temperature, the conductivity suddenly increases, as though absolute zero had already been reached. At this temperature, called the *superconducting transition temperature*, the metal becomes a **superconductor** of electricity. The superconductor offers no resistance whatever to electrical flow. This phenomenon means that electric motors made of superconducting wires would be 100% efficient, and electrical transmission lines could be made 100% efficient. Resistance to electron flow causes energy loss in motors, transmission lines, and other electrical devices. In superconductors the hindrance of electron flow by vibrating atoms in the metal lattice has been replaced by some kind of cooperative action that allows electron movement. Table 18.3 lists some of the metals that have superconducting transition temperatures. Not all metals display superconducting properties.

The relatively low transition temperatures of the metals shown in Table 18.3 mean that it would be impractical to make superconducting motors or transmission lines from them. Shortly after the superconductivity of metals was discovered, certain alloys (mixtures of metals) were prepared that had much higher transition temperatures than the metals themselves. Niobium alloys showed the most promise, but they still had to be cooled to below 23 K (−250°C) to exhibit superconductivity. To maintain such a low temperature would require liquid helium, which costs about $6 per liter—an expensive proposition for all but the most exotic applications.

In January 1986, K. Alex Müller and J. Georg Bednorz, scientists at an IBM laboratory in Switzerland, discovered that a barium–lanthanum–copper oxide became superconducting at 35 K, a temperature that must be maintained with liquid helium. This discovery provoked a flurry of activity that quickly resulted in a substance that became superconducting at 90 K. $LaBa_2Cu_3O_x$ was then

A refrigerator powered by the sun. Photovoltaic cells at the top of the refrigerator generate the electricity needed to keep vaccines cold while they are being delivered to remote locations.

In the Kelvin or Absolute Temperature Scale $T(K) = T(°C) + 273.15$

The boiling point of helium is 4.35 K (−268.8°C).

Müller and Bednorz shared the 1987 Nobel Prize in physics for their discovery of superconducting $LaBa_2Cu_3O_x$.

Superconductor A substance that offers no resistance to electrical flow

| TABLE 18.3 | Superconducting Transition Temperatures of Some Metals |

METAL	TRANSITION TEMPERATURE (K)
Aluminum	1.183
Gallium	1.087
Lanthanum	4.8
Lead	7.23
Niobium	9.17

found to superconduct at temperatures above the boiling point of nitrogen (77 K). At less than 50 cents per liter, liquid nitrogen is a much cheaper refrigerant than liquid helium. As of 2005, the record was held by $HgBa_2Ca_2Cu_3O_9$, which is a superconductor at 135 K.

Why the great excitement over the potentialities of superconductivity? Superconducting materials are being used to build more powerful electromagnets, such as those used in nuclear particle accelerators (Section 13.6) and in magnetic resonance imaging (MRI) machines, which are used in medical diagnosis. One of the main factors affecting the efficiency of MRI machines is the heating of the electromagnet due to electrical resistance. Many scientists are saying that the discovery of high-temperature superconductors may prove to be more important than the discovery of the transistor because of its potential effect on electrical and electronic technology. For example, the use of superconducting materials for transmission of electric power could save as much as 30% of the energy now lost because of the resistance of the wire.

CONCEPT CHECK **18C**

1. Many solar cells contain two types of semiconductors sandwiched together. These types are called _____ and _____ types.
2. A substance whose resistance falls to almost zero at a certain low temperature is called a(an) _____.
3. As the temperature of a metal increases, its resistance to electrical flow (increases/decreases).
4. Arsenic, a Group 5A element, can be used to dope silicon, a Group 4A element. This doping will produce (*n*-type/*p*-type) doped silicon.
5. Boron, a Group 3A element, will produce (*n*-type/*p*-type) doped silicon.
6. Doped silicon is used to make (**a**) magnets, (**b**) transistors, (**c**) electrical transmission wires.
7. Which would offer more promise, a compound whose resistance dropped to zero at or just above the boiling point of nitrogen, or one whose resistance dropped to zero at or just above the boiling point of helium?

18.5 FROM ROCKS TO GLASS, CERAMICS, AND CEMENT

Earth's crust is largely held together by chemical bonds between silicon and oxygen, in either pure **silica** (SiO_2) or **silicate** minerals in which silicon-oxygen anions are combined with metal cations.

The most common of the many crystalline forms of pure silica is quartz. It is a major component of granite and sandstone and also occurs as pure crystals. The basic structural unit of quartz and most silicates is the tetrahedron. In quartz, every silicon atom is bonded to four oxygen atoms, and every oxygen atom is bonded to two silicon atoms. The result is an infinite array of tetrahedra sharing corners.

Most silicates consist of networks of silicon–oxygen tetrahedra linked together in ways that range from chains, rings, and sheets to three-dimensional networks.

The simplest network silicates are the *pyroxenes*, which contain extended chains of linked SiO_4 tetrahedra (Figure 18.11). If two such chains are laid

Silica Naturally occuring silicon dioxide

Silicates Minerals composed of metal cations and silicon–oxygen anions

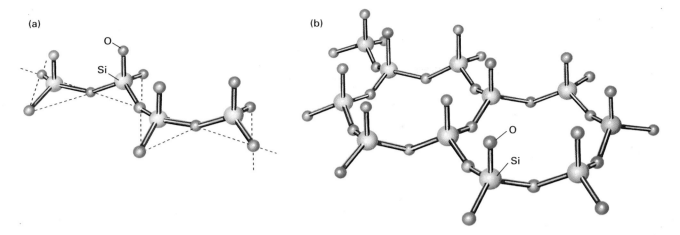

FIGURE 18.11 Silicate structures with oxygen atoms shown in red and silicon atoms shown in yellow. (a) Pyroxene, Tetrahedral SiO_4 units are joined in chains by silicon–oxygen–silicon bonds. (b) Amphibole, which is asbestos. Chains of SiO_4 units are joined side-by-side by silicon–oxygen–silicon bonds.

The serpentine form of asbestos is composed of two-dimensional sheets curled over into fibrous tubes.

side by side, they may link up by sharing oxygen atoms in adjoining chains. The result is an *amphibole*. Because of their double-stranded chain structure, the amphiboles are fibrous materials.

If the linking of silicate chains continues in two dimensions, sheets of SiO_4 tetrahedral units result (Table 18.4). Various clays and mica have this sheetlike structure. Clays, which are essential components of soils, are *aluminosilicates*—some Si^{4+} ions are replaced by Al^{3+} ions plus other cations that take up the additional positive charge. Feldspar, a component of many rocks and a network silicate, is weathered in the following reaction to form clay.

$$2\ KAlSi_3O_8(s) + CO_2(g) + 2\ H_2O(\ell) \longrightarrow$$

Feldspar

$$Al_2(OH)_4Si_2O_5(s) + 4\ SiO_2(s) + K_2CO_3(aq)$$

Kaolinite (a clay)

Some medications sold in the United States (e.g., Kaopectate) contain highly purified clay that absorbs excess stomach acid and possibly harmful bacteria that cause stomach upset.

Naturally occurring silicates: sheets of mica in the background, sand in the front, and, clockwise from the left, light green talc, clear quartz, and sandstone.

Feldspars make up 60% of Earth's crust.

Glass

When silica is melted, some of the bonds are broken and the tetrahedral SiO_4 units move with respect to each other. On cooling, reorganization into the same orderly arrangement in crystalline silica is hard to achieve because of the difficulty the groups experience in moving past one another. Instead, cooling produces a **glass**—a hard, noncrystalline transparent substance with an internal structure like that of a liquid. The random liquid-like molecular arrangement accounts for one of the typical properties of a glass: It breaks irregularly instead of splitting along a plane like a crystal.

Common window glass is made by melting a mixture of silica with sodium and calcium carbonates. Bubbles of carbon dioxide are evolved, and the cooled mixture is a glass composed of sodium and calcium silicates.

$$n\ Na_2CO_3(\ell) + n\ SiO_2(s) \xrightarrow{\text{Heat}} Na_{2n}(SiO_3)_n + n\ CO_2(g)$$

$$n\ CaCO_3(\ell) + n\ SiO_2(s) \xrightarrow{\text{Heat}} Ca_n(SiO_3)_n + n\ CO_2(g)$$

Glass and other solid substances (e.g., tar) that lack the internal order and properties of crystals are known as *amorphous substances.*

The n in these formulas represents a large number and is necessary to show that glass contains SiO_3 groups linked together rather than individual SiO_3^{2-} ions.

Glass A hard, noncrystalline, usually transparent substance made by melting silicates with other substances

---

TABLE 18.4	Silicates	
SI—O TETRAHEDRA PRESENT AS	**CLASS NAME**	**STRUCTURE**
Individual anions	Orthosilicates	
Chains	Pyroxenes (linear chains)	
	Amphiboles (double chains)	
Sheets	Mica, talc, clays	
Three-dimensional networks	Silica	—
	Feldspars and zeolites	

White sand is the source of the silica for glassmaking. Even the best grade of sand contains a small proportion of iron(III) compounds that gives it a brown or yellow color. When this sand is made into glass, the iron is converted to a mixture of light green iron(II) silicates, explaining the green tint of some old bottles. Adding a manganese compound to the melt produces pink manganese silicates, which offset the green of the iron silicates, making the glass appear colorless.

Glass. (a) Natural ingredients for making glass: sand (SiO_2), seashells for lime ($CaCO_3$), and seaweed for soda ash (Na_2CO_3). (b) Molten glass ready to be shaped. (c) A hand-blown goblet being turned in the annealing oven.

a. © James L. Amos/Peter Arnold, Inc.; b and c, Tom Pantages

(a)

(b)

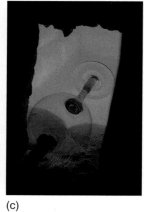

(c)

THE WORLD OF CHEMISTRY

ASBESTOS

The word *asbestos* is generally applied to two forms of fibrous natural silicates: the amphiboles and serpentine. The asbestos minerals do not burn, do not rot, and have low thermal and electrical conductivity. They can be woven into fabrics, compressed into mats, and mingled with such binders as rubber and asphalt to produce strong and dimensionally stable composites. These properties have led to widespread use of asbestos in fireproofing materials, brake linings, floor tiles, pipes, and roofing materials.

The serpentine form of asbestos, known as *chrysotile*, is mined chiefly in Canada and the former Soviet Union; more than 90% of the asbestos used in the United States is in this form. The amphibole *crocidolite* is mined in small quantities, mainly in South Africa. The two minerals differ greatly in composition, color, shape, solubility, and persistence in human tissue. Crocidolite is blue, relatively insoluble, and persists in tissue. Its fibers are long, thin, and straight; and they penetrate narrow lung passages. In contrast, chrysotile is white, and it tends to be soluble and disappear in tissue. Its fibers are curly; they ball up like yarn and are more easily rejected by the body. Scientific studies of many types and by groups in many countries have shown that chrysotile asbestos is significantly less of a health hazard than other types. It is important to note that almost all manufactured materials in the United States contain only this form of asbestos.

Long-term occupational exposure to airborne asbestos fibers can, however, lead to a greater-than-average risk of lung cancer and certain other health problems, risks greatly enhanced by cigarette smoking. An understanding of this risk has fostered tight controls during the mining of asbestos and fabrication of asbestos-containing products.

The perceived risk also fostered an assortment of drives in the United States to ban all uses of asbestos and to remove all asbestos-containing products in existing public buildings and homes. One outcome of this activity came in 1989 when the U.S. Court of Appeals struck down the proposed U.S. Environmental Protection Agency (EPA) ban on asbestos, citing failure to provide sufficient evidence and failure to adequately consider alternative measures. The EPA bypassed the opportunity to appeal this ruling. Another outcome was a congressionally mandated study of asbestos in buildings, which concluded in 1991 (after seven years of study) that asbestos-containing material in good repair inside buildings creates no higher asbestos fiber concentration in the air inside the buildings than in the air outside the buildings. Subsequently, an EPA advisory, while stressing assessment of asbestos risk in public buildings, stated:

Removal is often not a school district's or other building owner's best course of action to reduce asbestos exposure. In fact, an improper removal can create a dangerous situation where none previously existed.

"An Advisory to the Public on Asbestos in Buildings," U.S. Environmental Protection Agency, 1991.

Public fears about asbestos exposure were still in the news in 1993, when a school opening in New York City was delayed because of delays in removal. At that time, a group of 17 scientists and physicians from throughout the United States and the United Kingdom, all of them actively involved in asbestos study, stated:

Except under unusual conditions, such as demolition or extensive renovation, we strongly advise against asbestos removal in schools. It is scientifically unsound, economically wasteful, and medically imprudent . . . since it carries potentially unnecessary future risks.

Letter to the Editor, *New York Times*, December 23, 1993.

Armed with information of this kind, what would you do if your town proposed to spend $10 million of taxpayers' money on asbestos removal in your local high school? What questions should be asked of the decision makers?

Charles D. Winters

Chrysotile asbestos.

© Fulvio Roiter/CORBIS

An assortment of differently colored art glass perfume bottles. Different metal compounds produce the different colors.

TABLE 18.5	Substances Used to Color Glass
SUBSTANCE	**COLOR**
Copper(I) oxide	Red, green, blue
Tin(IV) oxide	Opaque
Calcium fluoride	Milky white
Manganese(IV) oxide	Violet
Cobalt(II) oxide	Blue
Finely divided gold	Red, purple, blue
Uranium compounds	Yellow, green
Iron(II) compounds	Green
Iron(III) compounds	Yellow

Countless variations in glass composition and properties are possible. If part of the silica is replaced by boron oxide, the glass has less tendency to crack with changes in temperature. Pyrex, the trademarked glass common in kitchens and laboratories, is a borosilicate glass. Many beautiful colors can be produced by adding the substances listed in Table 18.5. The composition and properties of some other types of glasses are listed in Table 18.6.

In the manufacture of glass, proper **annealing** is important. Annealing is the cooling schedule that a glass is put through on its way from a viscous, liquid state to a solid at room temperature. If a glass is cooled too quickly, bonding forces become uneven in local regions as small areas of crystallinity develop. This results in strain that will cause the glass to crack or shatter when subjected to mechanical shocks or sudden temperature changes. High-quality glass, such as that used in optics, must be annealed very carefully. The huge Mt. Palomar, California, observatory mirror was annealed from 500°C to 300°C over a period of nine months.

Ceramics

Annealing Heating and cooling a metal, glass, or alloy in a manner that makes it less brittle

What do you know about "ceramics"? Perhaps you associate the term with pottery vases that you see at a crafts show, bathroom tile, or components of

TABLE 18.6	Some Special Glasses
SPECIAL ADDITION OR COMPOSITION	**DESIRED PROPERTY**
Large amounts of PbO with SiO_2 and Na_2CO_3	Brilliance, clarity, suitable for optical structures; crystal or flint glass
SiO_2, B_2O_3, and small amounts of Al_2O_3	Small coefficient of thermal expansion; borosilicate glass: Pyrex, Kimax, and others
One part SiO_2 and four parts PbO	Ability to stop (absorb) large amounts of X-rays and gamma rays: lead glass
Large concentrations of CdO	Ability to absorb neutrons
Large concentrations of As_2O_3	Transparency to infrared radiation

electrical equipment. **Ceramics** are a large and diverse class of materials with the properties of nonmetals. What they have in common is that all are made by baking or firing minerals or other substances, often including silicates and metal oxides.

Ceramic materials have been made since well before the dawn of recorded history. They are generally fashioned from clay or other natural earths at room temperature and then permanently hardened by heat. Silicate ceramics include objects made from clays, such as pottery, bricks, and table china. The three major ingredients of common pottery are clay (from weathering of feldspar as described previously), sand (silica), and feldspar (aluminosilicates). Clays mixed with water form a moldable paste because they consist of many tiny silicate sheets that can easily slide past one another. When the clay–water mixture is heated, the water is driven off, and new Si—O—Si bonds are formed so that the mass of platelets becomes permanently rigid.

A new type of glasslike ceramics with unusual properties has become widely available for home and industrial use. Ordinary glass breaks because once a crack starts, there is nothing to stop the crack from spreading. It was discovered that if glass objects produced in the usual manner are heated until many tiny crystals develop, the resulting material, when cooled, is much more resistant to breaking than normal glass. In molecular terms, the randomness of the glass structure has been partially replaced by the order of a crystalline silicate. The materials produced in this way (an example is Pyroceram) are generally opaque and are used for kitchenware and in other applications in which the material is subjected to stress or high temperatures.

Ceramic materials are attractive for several reasons. The starting materials for making them are readily available and cheap. Ceramics are lightweight in comparison with metals and retain their strength at temperatures above 1000°C, where metal parts tend to fail. They also have electrical, optical, and magnetic properties of value in the computer and electronic industries.

The one severely limiting problem in utilizing ceramics is their brittle nature. Ceramics deform very little before they fail catastrophically, the failure resulting from a weak point in the bonding within the ceramic matrix. However, such weak points are not consistent from object to object, so the predictability of failure is poor. Since the stress failure of ceramic materials is due to molecular abnormalities resulting from impurities or disorder in the basic atomic arrangements, much attention is now being given to purer starting materials and the control of the processing steps. In addition, *ceramic composite materials*, mixtures of ceramic materials or of ceramic fibers with plastics, can overcome the tendency of plain ceramics to crack.

A potter at his wheel, with already-fired pieces in the background.

Cement and Concrete

A **cement** is a material that can bond mineral fragments into a solid mass. The most common cement, known as portland cement, is made by roasting a powdered mixture of calcium carbonate (limestone or chalk), silica (sand), aluminosilicate mineral (kaolin, clay, or shale), and iron oxide at a high temperature in a rotating kiln. As the materials pass through the kiln, they lose water and carbon dioxide and ultimately form "clinker," in which the materials are partially melted together. The clinker is ground to a very fine powder after the addition of a small amount of calcium sulfate (gypsum). A typical composition of a portland cement, expressed in terms of oxides, is 60 to 67% CaO; 17 to 25% SiO_2; 3 to 8% Al_2O_3; up to 6% Fe_2O_3; and small amounts of magnesium oxide, magnesium sulfate, and potassium and sodium oxides. As

Portland cement, patented in England in 1824, is named for its similarity to a rock native to the Isle of Portland in England. It is also called *hydraulic cement* because it sets even under water.

Ceramics Materials and products made by firing mixtures of nonmetallic minerals

Cement A material that bonds mineral fragments into a solid mass; any material that bonds other materials together

in glass, the oxides are not isolated into molecules or ionic crystals, and the submicroscopic structure is quite complex.

Many different reactions occur during the setting of cement. Initially the calcium silicates react with water to give a sticky gel. The gel has very large surface area and is responsible for the strength of concrete. Reactions with carbon dioxide in the air also occur at the surface. After the initial solidification, small, densely interlocked crystals begin to form, a process that continues for a long time and increases the compressive strength of the cement.

More than 800 million tons of cement are manufactured each year, most of which is used to make concrete. Concrete, like many other materials containing Si—O bonds, is virtually noncompressible but lacks tensile strength. If concrete is to be used where it will be subject to tension, it must be reinforced with steel.

Mortar is a mixture of cement, sand, water, and lime. *Concrete* is a mixture of cement, sand, and aggregate (crushed stone or pebbles).

Tensile strength refers to the resistance of a material to being stretched.

CONCEPT CHECK 18D

1. The two principal nonmetals in glass are _____ and _____.
2. Which of the following apply to (i) silica and (ii) silicates?
   (a) SiO_2
   (b) Quartz
   (c) Contain metal ions
   (d) Clay
   (e) Built up from tetrahedral units
3. How does glass differ from silica?
4. What three silicate minerals are used to make common pottery?
5. A distinctive property of ceramics is their ability to withstand high _____. Another is their tendency to fail in response to _____.
6. The three principal ingredients of concrete are _____, _____, and _____.

 Assess your understanding of this chapter's topics with an online chapter quiz at www.brookscole.com/chemistry/joesten4

■ KEY TERMS

hydrosphere	gangue	silicates
minerals	roasting	glass
alloy	semiconductors	annealing
precipitation	doping	ceramics
slag	transistor	cement
cast iron	superconductor	
steel	silica	

THE LANGUAGE OF CHEMISTRY

1. Glass
2. SiO_4
3. Steel
4. Sodium hydroxide
5. Product of the blast furnace
6. $Fe^{3+} \longrightarrow Fe$
7. Valuable mixture of minerals
8. Ceramics and glass
9. Superconductor
10. Seashells
11. Produced from seawater
12. Metal from the ocean
13. p-type semiconductor
14. Portland cement
15. $Fe \longrightarrow Fe^{3+}$

a. Reduction
b. Chlorine
c. Magnesium
d. Noncrystalline
e. Ore
f. In mortar and concrete
g. Pig iron
h. Product of chloralkali industry
i. No resistance to electrical flow
j. Building block of Earth's crust
k. B-enriched Si
l. Oxidation
m. Source of calcium carbonate
n. Alloy
o. Made of silicates

APPLYING YOUR KNOWLEDGE

1. Define the following terms:
   (a) Sedimentary rock   (b) Igneous rock

2. Define the following terms:
   (a) Slag   (b) Ductile
   (c) Brine   (d) Alloy

3. Define the following terms:
   (a) Annealing   (b) Amorphous
   (c) Ceramic   (d) Cement

4. What are the two products of the "chloralkali" process? Name two common uses for each of the products.

5. Name a common metal taken from seawater. Give several uses for this metal.

6. Complete and balance (if needed) the following equations:
   (a) $2 H_2O(\ell) + 2 e^- \longrightarrow \underline{\hspace{1cm}} + OH^-(aq)$
   (b) $Mg^{2+}(aq) + 2 OH^-(aq) \longrightarrow \underline{\hspace{1cm}}$
   (c) $Cl^-(aq) \longrightarrow Cl_2(g) + \underline{\hspace{1cm}}$
   (d) $CaCO_3(s) \longrightarrow \underline{\hspace{1cm}} + CO_2(g)$

7. Name two elements that occur in the free metallic state in nature.

8. What are the three ingredients used in a blast furnace to make iron? Which one of these is the reducing agent?

9. Explain why the molten mixture in a blast furnace separates naturally into a layer of slag and a layer of molten iron.

10. What are the differences between pig iron, cast iron, and steel? Mention composition as well as properties and uses.

11. What is the principal impurity in pig iron?

12. What element is used to convert pig iron into steel?

13. Complete and balance (if needed) the following equations:
    (a) $Cu_2S(s) + O_2(g) \longrightarrow \underline{\hspace{1cm}}$
    (b) $FeS(s) + O_2(g) \longrightarrow \underline{\hspace{1cm}}$

14. How is pure copper produced?

15. Name four properties of metals and give an example of the use of a metal taking advantage of each property.

16. Describe the electron structure of metals in terms of their conductivity of electricity.

17. Describe a p-type semiconductor made from silicon. Name two elements that might be used as dopants.

18. Name an important application of p- and n-type semiconductors in addition to their application in the manufacture of transistors and integrated circuits.

19. Draw the structural unit SiO_4, found in many silicate minerals. What is the name for the shape of this structural unit?

20. Draw a structure consisting of SiO_4 units arranged in a chain with each unit sharing two of its oxygen atoms with neighboring SiO_4 units. What is this silicate structure named?

21. What is a superconductor and why is the discovery of a "room-temperature" superconductor an important goal?

22. The element silicon occurs in all clays. Name two other elements common in all clays.

23. What element causes glass to have a brown to yellow color? (*Hint:* This is the same element that causes some clays to be red in color.)

24. What is the purpose of annealing glass?

25. Compare ceramics with metals by filling in the following table with a *yes* or a *no* depending on whether the material has that property.

	Metal	Ceramic
Hardness	_____	_____
Strength	_____	_____
Ductility	_____	_____
Electrical conductivity	_____	_____
Brittleness	_____	_____

26. Write the formula for limestone.

27. Cements contain calcium oxide (lime), SiO_2 (sand), and various other metal oxides. Which of the oxides present reacts with CO_2 in the air, and which helps bond everything together?

28. How is the structure of a common glass different from the structure of a solid like sodium chloride?

29. When a molten mixture of different materials cools, will all the substances crystallize at the same time? Silicates and quartz (SiO_2) have the highest melting points. Which will crystallize first, the metallic elements or the silicates?

30. Would it make any difference in our ability to extract pure elements if they were uniformly distributed on Earth instead of being concentrated as many of them are?

 CHEMISTRY ON THE WEB

For up-to-date URLs, visit the text website at **www.brookscole.com/chemistry/joesten4**

- Metals and Metal Links
- The Chemistry of Steelmaking
- Copper Mining
- How Semiconductors Work
- The History of Glassmaking

FEEDING THE WORLD

To successfully bring crops like this corn to market requires conditions provided by nature, by the farmer, and by the economy.

© EyeWire/Getty Images

When hunting and gathering from nature's bounty were the primary means of obtaining food, only catastrophic events could void the fundamental relationship between the effort and a satisfied stomach. As time passed, population increases and the concentration of people in larger and larger towns and cities forced the development of basic agriculture, which has mostly succeeded but has sometimes dramatically failed. To help grow the enormous amount of food needed to feed the world population, currently 6.5 billion and growing at the rate of 1.2% per year, the chemical industry supplies modern scientific agriculture with a large assortment of chemicals—the *agrichemicals*. These include fertilizers, medicine and food supplements for livestock, and chemicals to destroy unwanted pests and plant diseases. However, even with the use of new agrichemicals, sound agricultural practices, and biotechnology advances, the world's food supply cannot keep pace with the rate of growth of the world's human population. To have sustained food security for the world's population, global cooperation is essential in maintaining the recent reduction in the rate of population growth and reducing both environmental problems that threaten food productivity and problems of food distribution.

In this chapter we will address questions like:

- What are current projections for feeding the growing world population?
- What factors affect the productivity of soil?
- What agricultural practices are essential to sustaining soil productivity?
- How dependent is modern agriculture on insecticides and herbicides?
- What biotechnology advances in agriculture are helping to increase food production?
- What is sustainable agriculture?

Through the 8000 to 10,000 years of recorded human history, food production techniques have developed enormously. It is generally estimated that 90% of the U.S. population worked to provide food and fiber during most of the 19th century. Although we now have more efficient agricultural methods, the world's farmers are falling behind in their ability to feed the growing world population. More than 1 billion people in the world depend on agricultural lands that are not productive enough to support them adequately.

A report released by the United Nations Environment Program (UNEP) in 1992 stated that 4.84 billion acres of soil—an area the size of China and India combined—had been degraded to the point at which it will be difficult or impossible to reclaim them. By 2002 UNEP was describing soil as a "threatened natural resource" and reporting that 17% of the land surface worldwide is strongly degraded.

The main causes of soil degradation are erosion, overgrazing, and deforestation. Urbanization and industrialization also contribute to cropland loss. In addition, as populations grow, prolonged cultivation prevents cropland from maintaining its productivity.

One of the main reasons that the world grain harvest tripled between 1950 and 1990 was a 2.5-fold increase in irrigated land. However, farmers of irrigated land are now facing the prospect of water scarcity as competition with urban areas for shrinking water supplies increases.

19.1 WORLD POPULATION GROWTH

Global pollution, extinction of wildlife, degradation and loss of natural resources, and depletion of energy reserves are all related to the increasing human population. How close are we to reaching Earth's capacity to support its human population? What is the maximum sustainable population level? What factors determine Earth's capacity? These questions have been studied by organizations such as the Worldwatch Institute and at world conferences such as the 2002 World Summit on Sustainable Development held in Johannesburg, South Africa. All such studies recognize that controlling the growth rate of the world's population is essential to maintaining adequate food supplies to feed the world.

The world's population rose to more than 6.5 billion in 2005 (Figure 19.1). This is more than 3.5 times the size of Earth's population at the beginning of the 20th century and double the size in 1960. However, according to the *Global Population Profile: 2002* issued by the U.S. Census Bureau in 2004, the population increased by only 74 million people in 2002–2003 compared

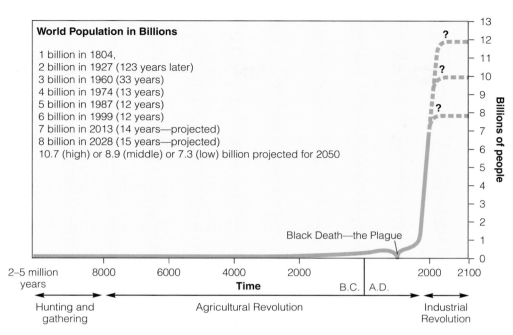

World Population in Billions

1 billion in 1804,
2 billion in 1927 (123 years later)
3 billion in 1960 (33 years)
4 billion in 1974 (13 years)
5 billion in 1987 (12 years)
6 billion in 1999 (12 years)
7 billion in 2013 (14 years—projected)
8 billion in 2028 (15 years—projected)
10.7 (high) or 8.9 (middle) or 7.3 (low) billion projected for 2050

FIGURE 19.1 The *J*-shaped curve of past exponential world population growth, with projections to 2100. Notice that exponential growth starts off slowly, but as time passes the curve becomes increasingly steep. The current world population of 6.5 billion people is projected to reach 8–12 billion people some time in this century. (This figure is not to scale.) (Data from World Bank and United Nations)

with the high annual increment of 87 million people in 1989–90. Furthermore, the annual growth rate in 2002 was only 1.19% compared with the high of 2.2 percent in 1963–64. Census experts believe that this slowdown in population growth will continue into the foreseeable future. This slowdown is largely attributable to a decline in the fertility of women around the world. In 1990, women around the world, on average, gave birth to 3.3 children during their lifetime. This number had dropped to an average of 2.6 by 2002. This number varies dramatically from country to country, however, with 2004 data showing a high of 8 children per woman in Niger and a low of 1.2 children per woman in Poland, with the United States coming in at exactly 2 children per woman. These encouraging numbers have led to Census Bureau projections that the level of fertility will drop below the replacement level of slightly more than 2.1 children per woman by 2050. Anything higher than this will result in an exponential population growth. Even so, the world's population is expected to exceed 9 billion people by 2050. The assumptions in such projections include birth rates and death rates, which, of course, change as time passes. What is not considered in the projections is the quality of life for those alive at any period.

In 2002 noted biologist Edward O. Wilson provided the following statement in his book *The Future of Life.*

> The ecological footprint—the average amount of productive land and shallow sea appropriated by each person in bits and pieces around the world for food, water, housing, energy, transportation, commerce and waste absorption—is about 2.5 acres in developing nations but about 24 acres in the U.S. The footprint for the total human population is 5.2 acres. For every person in the world to reach present U.S. levels of consumption with existing technology would require four more planet Earths.

The ultimate question becomes, What is the *carrying capacity* of Earth—that is, what is the maximum population our planet can support? Here is

A handful of humus, which is partially decomposed organic material from plants and animals. Where soil contains ample humus, it is spongy, holds water well, and is a healthy environment for plants and organisms that live in the soil.

where the quality of life and human values become important. If everyone were to live as we do in highly developed countries, the high level of resource consumption would quickly reduce the carrying capacity of Earth. Currently, the consumption of resources is vastly different in developed and developing countries, with many in developing countries trapped in poverty and malnourishment. It will be difficult to raise their standard of living without significant resource depletion.

A factor that affects Earth's carrying capacity is food-use efficiency. Up to half of the world's grain is fed to animals each year. A kilogram of feedlot-produced beef requires 7 kg of grain compared to 4 kg for 1 kg of pork, and 2 kg for 1 kg of poultry or 1 kg of fish. Just reducing the world's consumption of beef would free up grain either for direct consumption or as feed for pork, poultry, and fish, which are more efficient meat producers than beef. Consider an average American who uses 800 kg of grain per year, the great bulk of it consumed indirectly in the form of meat, eggs, milk, cheese, yogurt, and ice cream. If Americans cut their annual grain intake in half, 105 million tons of grain would be saved, which is enough to feed two-thirds of the population of India for one year. The good news is that the greater awareness among Americans for lower-fat diets (Section 16.6) has resulted in a decrease in beef consumption and a growing preference for pork, poultry, and fish. But this development is offset by the fact that as developing countries strive to advance, their diets move toward increased meat consumption and decreased grain consumption.

19.2 WHAT IS SOIL?

Soil is a mixture of four components—mineral particles, organic matter, water, and air (Figure 19.2). Weathering processes in nature over thousands of years break rock into small mineral particles found in soil. Organic matter

FIGURE 19.2 Soil formation. Weather, plants and their litter, earthworms and other organisms, and topography interact to produce soil.

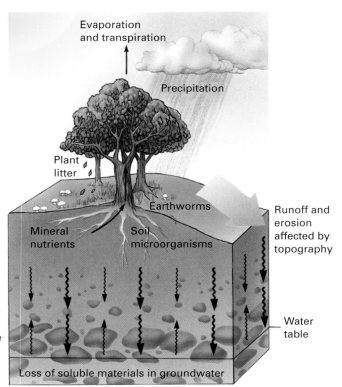

Evaporation and transpiration

Precipitation

Plant litter

Earthworms

Runoff and erosion affected by topography

Mineral nutrients

Soil microorganisms

Organic additions (underground) from soil organisms

Fine particles and solubles are washed downward

Capillarity and evaporation cause some materials to rise

Water table

Loss of soluble materials in groundwater

in soil is a mixture that includes leaves, twigs, plant and animal parts in various stages of decomposition, and microorganisms. **Humus**, the dark-colored decomposed organic material, is important to a good soil structure. As a source of nutrients for plants, humus is almost like a time-release capsule, slowly releasing its contents.

Maintaining humus in the soil is of major concern to the agriculturist. Humus such as peat moss or organic fertilizer can be added. However, there is no real substitute for natural plant growth that is returned to the ground for humus formation. Clover is often grown for this purpose and plowed under at the point of its maximum growth. The compost pile of the gardener is another effort to maintain humus for a productive soil.

In addition to being a source of plant nutrients, humus is important in maintaining good soil structure, often keeping it *friable*. Soil rich in humus may contain as much as 5% organic matter. Soils in the grasslands of North America have humus to a considerable depth, in contrast to rainforest regions, where there is only a thin film of humus on the ground surface.

Friable material crumbles easily under slight pressure.

Soil Profile

Layers within the soil are called **horizons** (Figure 19.3). The **topsoil** contains most of the living matter and the humus from dead organisms. Topsoil is usually several inches thick, and in some locations more than three feet of topsoil can be found. The **subsoil**, up to several feet in thickness, contains the inorganic materials from the parent rocks as well as organic matter, salts, and clay particles washed out of the topsoil.

Because healthy topsoil has abundant life forms, it must also contain an abundant supply of oxygen. Soil that supports vegetative growth and serves as a host for insects, worms, and microbes is typically full of pores; such soil is likely to have as much as 25% of its volume occupied by air. The ability of soil to hold air depends on soil particle size and how well the particles pack and cling together to form a solid mass. The particle size groups in soils vary from clays (the finest) through silt and sand to gravel (the coarsest). The particle size of a clay is 0.005 mm or less. The small particles in a clay deposit pack closely together to eliminate essentially all air and thus support little or no life. A typical soil horizon is composed of several particle size groups. A **loam**, for example, is a soil consisting of a friable mixture of varying proportions of clay, sand, and organic matter; a loam has a high air content.

Humus Decomposed organic matter in soil

Horizons Layers within the soil

Topsoil The layer of soil exposed to the air

Subsoil The layer of soil just beneath the topsoil

Loam Soil consisting of a friable mixture of varying proportions of clay, sand, and organic matter

FIGURE 19.3   Structure of a sandy soil. The layers, known as *horizons*, are built up through weathering of rock and interaction with water, air, plants, and animals as shown in Figure 19.2.

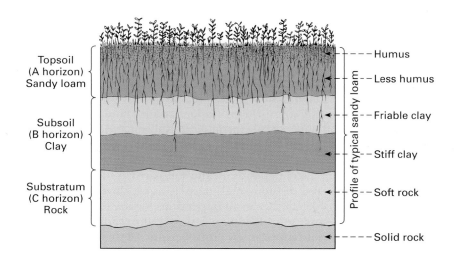

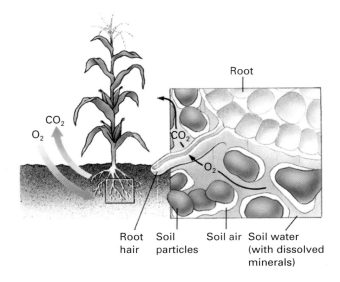

Exchanges between water and air in soil pores. Plant roots absorb oxygen and water, and release carbon dioxide.

Root

Root hair   Soil particles   Soil air   Soil water (with dissolved minerals)

Air in soil has a different composition from the air we breathe. Normal dry air at sea level contains about 21% oxygen (O_2) and 0.04% carbon dioxide (CO_2). In soil the percentage of oxygen may drop to as low as 15%, and the percentage of carbon dioxide may rise above 5%. This results from the partial oxidation of organic matter in the closed space. The carbon in the organic material combines with oxygen to form carbon dioxide. This increased concentration of carbon dioxide tends to cause groundwater to become acidic; acidic soils are described as *sour* soils because of the presence of aqueous acids.

$$CO_2(g) + 2H_2O(\ell) \longleftarrow H_3O^+(aq) + HCO_3^-(aq)$$

Crushed limestone ($CaCO_3$) applied to soil combines with hydrogen ions to form bicarbonate ions, thus raising the pH.

$$H_3O^+(aq) + CO_3^{2-}(aq) \longleftarrow HCO_3^-(aq) + H_2O$$

A slightly basic soil is a "sweet" soil.

If enough limestone is added to neutralize the acid in the soil and leave an excess of limestone, the pH of the soil becomes alkaline (basic).

Water in the Soil: Too Much, Too Little, or Just Right

Water can be held in soil in three ways: It can be *absorbed* into the structure of the particulate material, it can be *adsorbed* onto the surface of the soil particles, and it can occupy the pores ordinarily filled with air.

It takes several hundred pounds of water for the typical food crop to make 1 lb of food.

Water is removed from soil in four ways: Plants transpire water while carrying on life processes, soil surfaces evaporate water, water is carried away in plant products, and water moves through the subsoil and rock formations below in a process called **percolation**. Soils with good percolation naturally drain water from all but the small pores.

The percolation of a soil depends on the soil particle size and its chemical composition. Because of their small particle sizes, clays, and to a lesser degree silts, tend to pack together in an impervious mass with little or no percolation. Of course, sand, gravel, and rock pass water readily. Waterlogged soils that do not percolate support few crops because of their lack of air and oxygen. Rice, which requires water-logged soil, is an important exception. A

Percolation Movement of water through subsoil and rock formations

negative aspect of the massive flow of water through soil is the **leaching effect**. Water, known as the "universal solvent" because of its ability to dissolve so many different materials, dissolves away, or leaches, many of the chemicals needed to make a soil productive. If the leached material is not replaced, the soil becomes increasingly unproductive.

Soils become acidic, or sour, not only because of the oxidation of organic matter but also because of *selective leaching* by the passing groundwater. Salts of Group 1A and 2A metals are more soluble than salts of the Group 3A and transition metals. For example, a soil containing calcium, magnesium, iron, and aluminum ions is likely to be slightly alkaline, or sweet, before leaching with water. After the selective removal of calcium and magnesium salts, the soil becomes acidic because the iron and aluminum ions each tie up hydroxide ions from water and release hydrogen ions.

$$Fe^{3+} + H_2O \longleftarrow FeOH^{2+} + H^+$$

$$Al^{3+} + H_2O \longleftarrow AlOH^{2+} + H^+$$

A natural method of nitrogen fixation. The energy in a bolt of lightning is sufficient to disrupt the very stable triple bond in a nitrogen molecule (N_2). The result is the reaction with oxygen in the air to produce NO. This is the same reaction that takes place in an internal combustion engine.

$$N_2 + O_2 \overset{\text{high temp}}{\longleftarrow} 2\,NO$$

19.3 NUTRIENTS

At least 18 known elemental nutrients are required for normal green plant growth. Three of these, the **nonmineral nutrients**—carbon, hydrogen, and oxygen—are obtained from air and water (Table 19.1). The **mineral nutrients** must be absorbed through the plant root system as solutes in water. The 15 known mineral nutrients fall into 3 groups: **primary nutrients, secondary nutrients**, and **micronutrients**, depending on the amounts necessary for healthy plant growth.

Primary Nutrients

The primary nutrients are nitrogen, phosphorus, and potassium. Although bathed in an atmosphere of nitrogen, most plants are unable to use the air as a supply of this vital element. They require nitrogen fixation, the process of changing atmospheric nitrogen into water-soluble compounds that can be absorbed through the plant's roots and assimilated by the plant.

Nature fixes nitrogen in two ways. In the first method, bolts of lightning provide the high energy needed to oxidize nitrogen to nitric oxide (NO). The NO is further oxidized to NO_2 (Section 4.6), which reacts with water to

Leaching effect The dissolving of soil material as water moves through the soil

Nonmineral nutrients Carbon, hydrogen, and oxygen in soil

Mineral nutrients Plant nutrients absorbed through roots as solutes in water

Primary nutrients Major elements required by plants (nitrogen, phosphorus, potassium)

Secondary nutrients Elements required by plants in moderate amounts (calcium, magnesium, sulfur)

Micronutrients Elements required by plants in relatively small amounts (less than 100 mg/day)

TABLE 19.1	**Essential Plant Nutrients**		
NONMINERAL	**PRIMARY**	**SECONDARY**	**MICRONUTRIENTS**
Carbon	Nitrogen	Calcium	Boron
Hydrogen	Phosphorus	Magnesium	Chlorine
Oxygen	Potassium	Sulfur	Copper
			Iron
			Manganese
			Molybdenum
			Sodium
			Vanadium
			Zinc

A balanced ecosystem. There is a smooth cycling of nutrients from the soil to plants and animals, and then back to the soil.

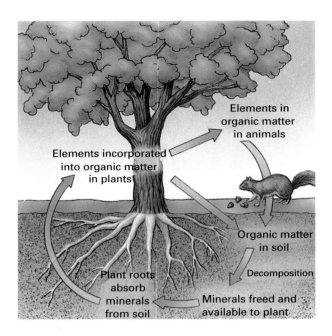

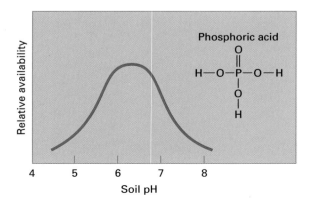

Nitrogen fixation requires breaking the strong triple bond connecting the two atoms in the nitrogen molecule N_2 ($:N\equiv N:$), as described in Section 5.3.

Another major source of nitrogen replenishment in soil is dead organisms and animal wastes. Even in the absence of legumes, this can be an adequate source of nitrogen.

Potassium is absorbed as the free ion, $K^+(aq)$.

Nitrogenase Bacterial enzyme that catalyzes nitrogen fixation

form nitric acid (HNO_3). This method is estimated to provide less than 10% of the nitrogen fixed by nature. Nitric acid is readily soluble in rain, clouds, or ground moisture and thus increases nitrate concentration in soil.

In the most important method, nitrogen-fixing bacteria that live in the roots of legumes, such as soybeans and alfalfa, use an enzyme, **nitrogenase**, to catalyze a complex series of reactions that convert atmospheric nitrogen into ammonia under normal atmospheric conditions. Legume nitrogen fixation can add more than 100 lb of nitrogen per acre of soil in one growing season.

Like nitrogen, phosphorus must be in a soluble mineral or inorganic form before it can be used by plants. Unlike nitrogen, phosphorus comes totally from the mineral content of the soil. Salts of the dihydrogen phosphate ion ($H_2PO_4^-$) and monohydrogen phosphate ion (HPO_4^{2-}) are the dominant phosphate ions in soils of normal pH (Figure 19.4). Because of the great concentration of electric charge associated with the trivalent phosphate ion (PO_4^{3-}), phosphates are more tightly held to positive ions such as Ca^{2+} and Fe^{3+} and are not as easily leached by groundwater as are nitrate salts, which are generally soluble in water.

Potassium in the form of K^+ ion is a key element in the enzymatic control of the interchange of sugars, starches, and cellulose. Although potassium is the seventh most abundant element in Earth's crust, soil used heavily in crop

FIGURE 19.4 Availability of phosphate in the soil as a function of pH. The ions present vary with the pH. At pH 5 to pH 8, $H_2PO_4^-$ and HPO_4^{2-} predominate. At very low pH values, phosphorus is in the form of the nonionized acid H_3PO_4. At a very high pH, all three protons are removed and the phosphorus is in the form of the phosphate ion (PO_4^{3-}). Low soil temperatures in temperate regions significantly reduce phosphorus uptake by plants.

production can be depleted of this important metabolic element, especially if the soil is regularly fertilized with nitrate, with no regard to potassium content.

Secondary Nutrients

Calcium and magnesium are available in small amounts as Ca^{2+} and Mg^{2+} ions as well as in complex ions and crystalline formations. These abundant elements are bound tightly enough by soil so they are not readily leached yet are held loosely enough to be available to plants. When held in the soil as sulfate salts (SO_4^{2-}), sulfur is readily available to plants.

Micronutrients

Only very small amounts of micronutrients are required by plants; therefore, unless extensive cropping or other factors deplete the soil of these nutrients, sufficient quantities are usually available.

Iron is also an essential component of the enzyme involved in the formation of chlorophyll. When the soil is iron deficient or when too much lime is present in the soil, iron availability decreases. Often a gardener or lawn worker will apply phosphate and lime to adjust soil acidity, only to see green plants turn yellow because of **chlorosis**. What happens in such cases is that both phosphate and the hydroxide from the lime tie up the iron and make iron unavailable to the plants.

$$Fe^{3+}(aq) + 2\,PO_4^{3-}(aq) \longleftarrow Fe(PO_4)_2^{3-}(aq)$$
$$\text{Phosphate} \qquad\qquad \text{Tightly bound complex}$$

$$Fe^{3+}(aq) + 3\,OH^-(aq) \longleftarrow Fe(OH)_3(s)$$
$$\qquad\qquad\qquad \text{Insoluble hydroxide}$$

Charles D. Winters

A leaf on a blackberry plant suffering from chlorosis. Chlorophyll, the green plant pigment, requires nitrogen, magnesium, and iron from the soil. Deficiencies of any of these nutrients cause chlorosis, a condition of low chlorophyll content, indicated by yellowing of the leaves.

Chlorosis Yellowing of plant leaves due to nutrient deficiency

THE WORLD *of* CHEMISTRY

TRYING TO MIMIC NATURE

The process for making ammonia from the reaction of nitrogen and hydrogen at high temperatures and pressures in the presence of a catalyst was invented by Fritz Haber in 1908 and is still the major industrial process for the synthesis of ammonia (see Figure 19.6). However, the Haber process is far from doing the job of fixing nitrogen as well as nature does. Fixation of nitrogen by bacteria on root nodules of legume crops occurs at atmospheric temperature and pressure. The function of nitrogenase, the biological catalyst for this process, is still not fully understood, but iron and molybdenum are known to be important to the process by which the high-energy triple bond in N_2 is broken during the reduction of N_2 to NH_3

in biological nitrogen fixation. Agriculturists know that a deficiency of molybdenum creates chlorosis in crops (see photo above) and that addition of molybdenum to the soil can be a cure for this condition.

For years, chemists have been studying reactions of transition metal compounds with nitrogen in an attempt to understand the function of the metal ions in nitrogenase. Understanding the structure of the nitrogenase molecule is also a major research focus. One result has been the discovery of a molybdenum compound that breaks the triple bond of N_2 at room temperature and atmospheric pressure to give another molybdenum compound with a single nitrogen atom bonded directly to molybdenum.

Here is a major breakthrough in synthetic chemistry. The discovery could lead to both lower energy requirements for industrial nitrogen fixation and a better understanding of the role of molybdenum in the function of nitrogenase in biological nitrogen fixation. Ongoing studies pursue the goal of making nitrogen available to crops in a more direct method than via the Haber process and fertilizer synthesis from ammonia.

References C. E. Laplaza and C. C. Cummins, Dinitrogen cleavage by a three-coordinate molybdenum (III) complex, *Science*, Vol. 268, pp. 861–863, 1995. C. E. Laplaza et al., Dinitrogen cleavage by three-coordinate molybdenum(III) complexes: Mechanistic and structural data, *J. Am. Chem. Soc.*, Vol. 118, p. 8623, 1996.

CONCEPT CHECK) **19A**

1. The world population in 2005 had grown to _____ and is expected to exceed _____ by 2050.
2. Rank the following types of soils from those with the smallest soil particles to those with the largest: silts, sandy soils, loams, clays.
3. Carbon dioxide causes soils to be (**a**) acidic, (**b**) basic. Limestone ($CaCO_3$) causes soils to be (**a**) acidic, (**b**) basic.
4. The two factors that determine the percolation of a soil are _____ and _____.
5. Which is more acidic, a monovalent ion such as Na^+ or a trivalent ion such as Fe^{3+}?
6. A well-decomposed, dark-colored plant residue that is relatively resistant to further decomposition is known as _____.
7. The primary elemental plant nutrients necessary in the soil for healthy plant growth are _____, _____, and _____.
8. The secondary elemental plant nutrients are _____, _____, and _____.
9. Nitrogen fixation by lightning discharges involves breaking a nitrogen–nitrogen triple bond and combining nitrogen with _____.

19.4) FERTILIZERS SUPPLEMENT NATURAL SOILS

Primitive peoples raised crops on a cultivated plot until the land lost its fertility; then they moved to a virgin piece of ground where they cut down ("slash") natural vegetation and burned off the stubble to clear the land. In many cases, the slash–burn–cultivate cycle was no more than a year in length, and few found a piece of ground anywhere that could support successful cropping for more than five years without fertilization. In farming villages, developed in ancient times and prevalent throughout the Middle Ages, innovation in fertilization was demanded, because the same land had to be used for many years. With the use of legumes in crop rotations, manures, dead fish, or almost any organic matter available, the land was kept in production.

The acreage used worldwide in the cultivation of food crops would probably be sufficient if modern synthetic chemical fertilization were employed on all of it. For example, chemical fertilization accounted for crop yield explosions including (1) U.S. corn—25 bushels per acre in 1800, 110 bushels per acre in the 1980s, 130 bushels per acre in the 1990s; (2) English wheat—less than 10 bushels per acre from A.D. 800 to 1600, more than 75 bushels per acre in the 1980s. However, the cost of synthetic chemical fertilization is beyond what developing countries can bear, and the environmental impact of such a large dispersion of fertilizer chemicals would probably be massive. For example, aquifer contamination with nitrates due to corn crop fertilization renders well water from these natural underground basins unfit for drinking in large areas of the U.S. corn belt.

Fertilizers that contain only one nutrient are called **straight fertilizers**. Potassium chloride for potassium is an example of a straight fertilizer. Those containing a mixture of the three primary nutrients are called **complete**, or **mixed**, **fertilizers**. The primary nutrients are absorbed by plant roots as simple inorganic ions: nitrogen in the form of nitrates (NO_3^-), phosphorus as phosphates ($H_2PO_4^-$ or HPO_4^{2-}), and potassium as the K^+ ion. Organic

Straight fertilizers Contain only one nutrient

Complete (mixed) fertilizers Contain the three primary nutrients (nitrogen, phosphorus, potassium)

fertilizers, which have very little mineral content, can supply these ions, but only when used in large quantities over a long time. For example, a manure might be a 0.5–0.24–0.5 fertilizer, in contrast to a typical chemical fertilizer, which might carry the numbers 6–12–12. These numbers indicate the analysis in order of the percentage of nitrogen as N, phosphorus as P_2O_5, and potassium as K_2O (commonly known as potash) in the fertilizer (Figure 19.5). In addition to containing the desired ions, chemical fertilizers place the ions in the soil in a form that can be absorbed directly by plants. The problem is that these inorganic ions are relatively easily leached from the soil and may pose pollution problems if not contained. The much slower organic fertilizer tends to stay put. **Quick-release fertilizers** are water soluble, as opposed to **slow-release fertilizers**, which require days or weeks for the material to dissolve completely. Table 19.2 lists the necessary plant nutrients and suitable chemical sources of each.

Nitrogen Fertilizers

The cheapest source of nitrogen is the air, but it must be combined with relatively expensive hydrogen, obtained from petroleum, to form ammonia in the Haber process.

The complex interactions among social, economic, and political necessities are well illustrated by the development of the industrial synthesis of ammonia. The need for an industrial process for nitrogen fixation was recognized as early as 1890. Scientists in England noted that the world's future food supply would be determined by the amount of nitrogen compounds

FIGURE 19.5   How to read a fertilizer label. The three numbers, in order, refer to the percentage by weight of N (nitrogen), P_2O_5 (phosphate), and K_2O (potash). Following the lead of J. von Liebig, his German mentor, who was the first scientist to suggest adding nutrients to soils, Samuel William Johnson, an American, burned plants and analyzed their ashes. He expressed the nutrient concentrations as the oxides present in the ashes, a practice that has continued to this day.

Quick-release fertilizers Water soluble

Slow-release fertilizers Combinations of plant nutrients that are slow to dissolve in water

TABLE 19.2	Some Chemical Sources of Plant Nutrients
ELEMENT	**SOURCE COMPOUND(S)**
Nonmineral Nutrients	
C	CO_2 (carbon dioxide)
H	H_2O (water)
O	H_2O (water)
Primary Nutrients	
N	NH_3 (ammonia), NH_4NO_3 (ammonium nitrate), H_2NCONH_2 (urea)
P	$Ca(H_2PO_4)_2$ (calcium dihydrogen phosphate)
K	KCl (potassium chloride)
Secondary Nutrients	
Ca	$Ca(OH)_2$ (calcium hydroxide, slaked lime), $CaCO_3$ (calcium carbonate, limestone), $CaSO_4$ (calcium sulfate, gypsum)
Mg	$MgCO_3$ (magnesium carbonate), $MgSO_4$ (magnesium sulfate, epsom salts)
S	Elemental sulfur, metallic sulfates
Micronutrients	
B	$Na_2B_4O_7 \cdot 10H_2O$ (borax)
Cl	KCl (potassium chloride)
Cu	$CuSO_4 \cdot 5H_2O$ (copper sulfate pentahydrate)
Fe	$FeSO_4$ (iron(II) sulfate, iron chelates)
Mn	$MnSO_4$ (manganese(II) sulfate, manganese chelates)
Mo	$(NH_4)_2MoO_4$ (ammonium molybdate)
Na	$NaCl$ (sodium chloride)
V	V_2O_5, VO_2 (vanadium oxides)
Zn	$ZnSO_4$ (zinc sulfate, zinc chelates)

© Scott T. Smith/CORBIS

Anhydrous ammonia, a fertilizer. At normal temperatures, anhydrous ammonia (ammonia gas containing no water) can be held as a liquid under tank pressure. On release from pressure, the ammonia returns to the gaseous state and is injected into the soil by an "ammonia knife." In slightly acid soil, the ammonia is immediately converted to the water-soluble ammonium ion and enters the natural nitrogen pathways of the soil.

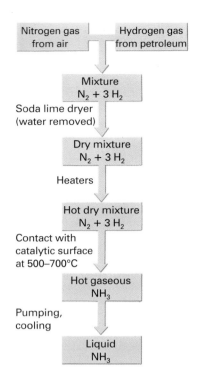

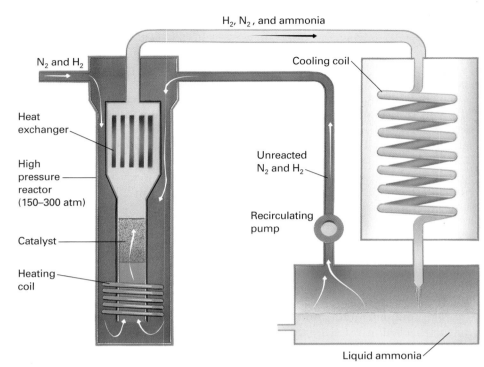

FIGURE 19.6 **The Haber process for ammonia production.** A mixture of N_2 and H_2 in the proper proportions for reaction is heated and passed under pressure over the catalyst. The ammonia is collected as a liquid and unreacted gases are recycled.

The Haber process is based on a delicate balancing act among the reaction energy, the reaction rate, and the reaction equilibrium. The combination of nitrogen with hydrogen is exothermic, but slow. The rate is increased by a higher temperature, but the yield of ammonia decreases with increasing temperature (because the reaction is exothermic). Therefore, a catalyst that acts at a moderate temperature is necessary. Also, because 4 mol of reactant give 2 mol of product, the extent of the forward reaction can be increased by increasing the pressure (Le Chatelier's principle; Section 8.4).

Ammonia is a gas at normal temperatures and pressures. It is very soluble in water.

The structural formula of urea is

$$ H_2N\overset{\overset{\displaystyle O}{\displaystyle \|}}{-}C-NH_2 $$

Haber process Industrial process for the reaction of nitrogen and hydrogen gases to form ammonia

available for fertilizers. At the time, the sources of such compounds were limited to sodium nitrate from Chile and rapidly depleting supplies of guano (bird droppings) from Peru.

Shortly after 1900, many researchers in England and Germany began investigating methods of preparing ammonia, and in 1908 the German chemist Fritz Haber developed a feasible process for the direct synthesis of ammonia from its elements.

$$ N_2(g) + 3 H_2(g) \rightleftharpoons 2 NH_3(g) $$

The first industrial plant based on the **Haber process** (see Figure 19.6) for ammonia synthesis began operation in Germany in 1911 and in the United States industrial production was begun in 1921.

Subsequently, the Haber process has provided ammonia for fertilizers on a huge scale and is now largely responsible for our ability to feed a world population of more than 6 billion. The Haber process has been so well developed that ammonia is very inexpensive (about $400 per ton) and is usually among the top 10 industrial chemicals in terms of quantity produced each year.

Ammonia can be applied directly to the soil or converted into numerous agrichemicals (Figure 19.7). Two common straight fertilizers for nitrogen are ammonium nitrate (NH_4NO_3) and urea. A slurry of water, urea, and ammonium nitrate is often applied to crops under the name "liquid nitrogen." Such a solution can contain up to 30% nitrogen and is easy to store and apply.

When applied to the surface of the ground around plants, urea is subject to considerable nitrogen loss unless it is washed into the soil by rain or irrigation. When urea decomposes, ammonia is formed, some of which is lost to

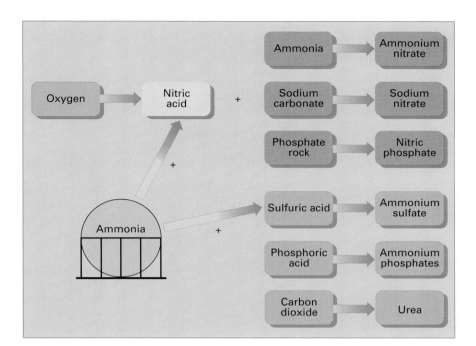

FIGURE 19.7 Nitrogen fertilizers produced from anhydrous ammonia.

the air and some is absorbed by moist soil particles. As much as half of the nitrogen applied to the soil can be lost in this way.

Phosphate Fertilizers

Phosphorus is readily available in the form of phosphate rock, which can be transformed into the needed fertilizers (Figure 19.8). World deposits of phosphate rock are limited, and the costs of supplying phosphorus fertilizers will increase as deposits are depleted. The phosphate rock, $Ca_3(PO_4)_2$, is not very useful because of its exceedingly low solubility. When treated with sulfuric acid, however, phosphate rock becomes more soluble; the resulting mixture of phosphate and sulfate salts of calcium is called "superphosphate."

A phosphate mine in Florida. In the background is a part of this mine that was restored after mining was completed and is now used as a pasture.

$$Ca_3(PO_4)_2 + 2\,H_2SO_4 \longleftarrow Ca(H_2PO_4)_2 + 2\,CaSO_4$$
Phosphate rock   Superphosphate

THE PERSONAL SIDE

Fritz Haber (1868–1934)

Although Fritz Haber was a fine scientist and the success of his ammonia synthesis made him a rich man, he ultimately had a tragic life. At the start of World War I he joined the German Chemical Warfare Service, where he supervised the use of chlorine as a chemical weapon during the battle of Ypres in France. This first use of a chemical weapon led to further tragic developments in chemical warfare and also to personal tragedy for Haber. His wife pleaded with him to stop his work in this area; and when he refused, she committed suicide. In 1918, he was awarded the Nobel Prize for the ammonia synthesis, but the Nobel Committee's choice was criticized because of Haber's role in developing chemical warfare. After World War I, Haber continued his academic research in Germany until, shortly before his death in 1934, he left the country to avoid danger because of his Jewish background.

FIGURE 19.8 Fertilizers produced from phosphate rock.

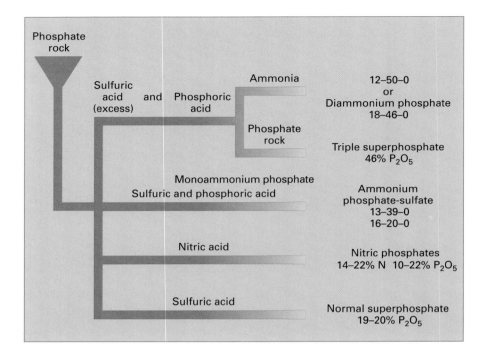

⬭ **EXAMPLE** **19.1** ⬭ **Fertilizer Labels**

The label on a fertilizer bag lists the numbers 22–6–8. What do these numbers mean? Calculate the pounds of each nutrient present in a 50.0-lb bag of fertilizer.

SOLUTION

The numbers mean that the fertilizer contains 22.0% nitrogen (N) in the form of some nitrogen-containing compound, 6.00% of a phosphorus-containing compound (calculated as a percentage of P_2O_5), and 8.00% of a potassium-containing compound (calculated as a percentage of K_2O). The pounds of each nutrient present are calculated as follows:

Nitrogen: The pounds of nitrogen present are calculated by multiplying the percentage of nitrogen times 50.0 lb. Since the unit in this problem is pounds, 22.0% of N can be represented as 22.0 lb of N per 100 lb of fertilizer.

$$\frac{22.0 \text{ lb N}}{100 \text{ lb fertilizer}} \times 50.0 \text{ lb fertilizer} = 11.0 \text{ lb N}$$

Phosphorus: Since the amount of phosphorus present is represented as 6.00% P_2O_5, the actual percentage present as elemental phosphorus is smaller. To calculate the fraction of phosphorus in P_2O_5 requires using the atomic masses of P and O in a conversion factor

$$\frac{2 \times \text{atomic mass P}}{(2 \times \text{atomic mass P}) + (5 \times \text{atomic mass O})}$$

$$2 \text{ P} = 2 \times 31.0 \text{ amu} = 62.0 \text{ amu}$$

$$5 \text{ O} = 5 \times 16.0 \text{ amu} = 80.0 \text{ amu}$$

Total for P_2O_3 = 142.0 amu, so factor is

$$\frac{62.0 \text{ amu P}}{142 \text{ amu } P_2O_5}$$

The relative masses are the same no matter what units are used. Since the unit in this problem is pounds, the conversion factor can be written as

$$\frac{62.0 \text{ lb P}}{142 \text{ lb P}_2\text{O}_5}$$

$$\frac{6.00 \text{ lb P}_2\text{O}_5}{100 \text{ lb fertilizer}} \times \frac{62.0 \text{ lb P}}{142 \text{ lb P}_2\text{O}_5} \times 50.0 \text{ lb fertilizer} = 1.31 \text{ lb P}$$

Potassium: Since the amount of potassium present is represented as 8.00% K_2O, calculating the pounds of potassium present requires using the atomic masses of K and O in a conversion factor

$$\frac{2 \times \text{atomic mass K}}{(2 \times \text{atomic mass K}) + (\text{atomic mass O})}$$

2 K = 2 × 39.0 amu = 78.0 amu; 1 O = 16.0 amu; total of 2 K and 1 O = 94.0 amu so the conversion factor showing the ratios in pounds is

$$\frac{78.0 \text{ lb K}}{94.0 \text{ lb K}_2\text{O}}$$

$$\frac{8.00 \text{ lb K}_2\text{O}}{100 \text{ lb fertilizer}} \times \frac{78.0 \text{ lb K}}{94.0 \text{ lb K}_2\text{O}} \times 50.00 \text{ lb fertilizer} = 3.32 \text{ lb K}$$

TRY IT 19.1

Ammonium nitrate (NH_4NO_3) is widely used as a fertilizer. A bag of ammonium nitrate sold as fertilizer lists the numbers 35–0–0. Use the formula weight for NH_4NO_3 to verify that this is the correct number.

19.5 PROTECTING FOOD CROPS

The natural enemies of crops include more than 80,000 diseases brought on by viruses, bacteria, fungi, algae, and similar organisms; 30,000 species of weeds; 3000 species of nematodes; and about 10,000 species of plant-eating insects. About one-third of the food crops in the world are lost to pests each year, with the loss rising above 40% in some developing countries. A "pest" is any organism that in some way reduces crop yields or human health. **Pesticides**, chemicals used to control pests, are classified according to the pests they control: **insecticides** kill insects, **herbicides** kill weeds, and **fungicides** kill fungi.

Some pesticides available for a home garden.

Charles D. Winters

Pesticides

Pesticides are the chemical answer to pest control. Eighteen common classes of pesticides are fortified with more than 2600 active ingredients to fight the battle with pests. More than 5 billion pounds of pesticides are produced worldwide each year. In 1993, pesticide sales in the United States totaled $6.8 billion. Although the dollar cost was up to $11 billion in 2001, according to the EPA, the actual poundage use began a decline in 1987. Worldwide expenditures were around $32 billion in 2001. There are three reasons for a decrease in the demand for pesticides in the United States: (1) Cropland planted was less than 330 million acres, down from a high of 383 million acres in 1982. (2) Farming is becoming more cost-effective as farmers learn to use a minimum of pesticide for the desired effect. (3) Farmers are becoming more concerned with environmental and health issues related to the use of pesticides.

Pesticides Chemicals used to kill pests

Insecticides Pesticides that kill insects

Herbicides Pesticides that kill weeds

Fungicides Pesticides that kill fungi

Paul Müller was awarded the Nobel Prize in medicine and physiology in 1948 for his discovery of the insecticidal properties of DDT.

DDT

LD$_{50}$ is defined in Section 17.6.

Insecticides

Before World War II, the list of insecticides included only a few compounds of arsenic, petroleum oils, nicotine, pyrethrum (obtained from dried chrysanthemum flowers), rotenone (obtained from roots of derris vines), sulfur, hydrogen cyanide gas, and cryolite (the mineral Na_3AlF_6). Dichlorodiphenyltrichloroethane (DDT), the first of the chlorinated organic insecticides, was originally prepared in 1873, but it was not until 1939 that Paul Müller of Geigy Pharmaceutical in Switzerland discovered the effectiveness of DDT as an insecticide.

The use of DDT increased enormously on a worldwide basis after World War II, primarily because of its effectiveness against mosquitoes that spread malaria and lice that carry typhus. The World Health Organization estimates that approximately 25 million lives have been saved through the use of DDT. DDT seemed to be the ideal insecticide—it is cheap and has relatively low toxicity to mammals (oral LD$_{50}$ is 300–500 mg/kg). However, problems related to extensive use of DDT began to appear in the late 1940s. Many species of insects developed resistance to DDT, and DDT was also discovered to have a high toxicity toward fish. The chemical stability of DDT and its fat solubility compounded the problem.

DDT is not metabolized very rapidly by animals; instead, it is deposited and stored in the fatty tissues. The biological half-life of DDT is about eight years; only half the amount of DDT an animal assimilates today will be metabolized in eight years. If ingestion continues at a steady rate, DDT builds up within the animal over time.

The use of DDT was banned in the United States in 1973, although it is still in use in some other parts of the world. The buildup of DDT in natural waters is a reversible process; the EPA reported a 90% reduction of DDT in Lake Michigan fish by 1978 as a result of the ban on the use of the insecticide in the United States.

The most important insecticides can be grouped in three structure classes—chlorinated hydrocarbons, organophosphorus compounds, and carbamates. Chlorinated hydrocarbons such as DDT are referred to as *persistent pesticides* because they persist in the environment for years after their use.

Malathion is a widely used member of a class of pesticides known as *organophospate derivatives*. It is biodegradable and less toxic (LD$_{50}$ for female rats is 1000 mg/kg) than DDT. Several other organophosphorus compounds are modifications of this basic structure. They include dimethoate (LD$_{50}$ = 387 mg/kg), chlorpyrifos (LD$_{50}$ = 95–270 mg/kg), and methyl parathion (LD$_{50}$ = 18–50 mg/kg). These organophosphates often are used in the treatment of many pests that are found on vegetables, or, in the case of methyl parathion, on cotton plants.

Malathion

Dimethoate

Methyl parathion

Chlorpyrifos

The carbamate insecticides are derivatives of carbamic acid.

$$
HO-\overset{\overset{\displaystyle O}{\|}}{C}-\overset{\overset{\displaystyle H}{|}}{N}-H
$$

Carbamic acid

$$
O-\overset{\overset{\displaystyle O}{\|}}{C}-\overset{\overset{\displaystyle H}{|}}{N}-CH_3
$$

Carbaryl

Pirimicarb

They are also nonpersistent insecticides. The most widely used carbamate is *carbaryl*, a general-purpose insecticide with a relatively low mammalian toxicity (oral LD_{50} for rats is 250 mg/kg). A serious drawback to the use of carbaryl for spraying crops is its high toxicity toward honeybees. Another carbamate of interest is *pirimicarb* (oral dose for female rats is 147 mg/kg), which is a selective insecticide for aphids and has the advantage of controlling strains that have developed resistance to organic phosphates. Pirimicarb has a low toxicity to predators such as bees. It is rapidly metabolized and leaves no lasting residues in plant materials. Pirimicarb is expensive, however; its preparation requires five or six steps and special handling procedures.

The choice of solutions to the problems of insecticide use is not an easy one. Their use introduces trace amounts of them into our environment and our water supplies. If we fail to use them, we must tolerate malaria, plague, sleeping sickness, and the consumption of a large part of our food supply by insects. Continuing research in the development of more effective and safer insecticides is intense, and new products are introduced each year.

Gilbert S. Grant/Photo Researchers, Inc.

A corn earworm at work.

The goal of the insecticide quest is a selectively toxic chemical that is quickly biodegradable.

Herbicides

Herbicides kill plants. They may be **selective** and kill only a particular group of plants, such as the broad-leaved plants or the grasses; or they may be **nonselective**, making the ground barren of all plant life.

Selective herbicides act like hormones, very selective biochemicals that control a particular chemical change in a particular type of organism at a particular stage in its development. Most selective herbicides in use today are growth hormones; they cause cells to swell, so that leaves become too thick for chemicals to be transported through them and roots become too thick to absorb needed water and nutrients. Nonselective herbicides usually interfere with photosynthesis and thereby starve the plant to death. On application, the plant quickly loses its green color, withers, and dies.

The most widely used herbicide is 2,4-dichlorophenoxyacetic acid (2,4-D). The corresponding trichloro- compound (common name: 2,4,5-T) has also been shown to be highly effective, but it was banned by the EPA because of a number of health problems associated with its use.

$$
Cl-\text{(ring)}-O-\overset{\overset{\displaystyle H}{|}}{\underset{\underset{\displaystyle H}{|}}{C}}-\overset{\overset{\displaystyle O}{\|}}{C}-OH
$$

2,4-D
(2,4-Dichlorophenoxyacetic acid)

$$
Cl-\text{(ring)}-O-\overset{\overset{\displaystyle H}{|}}{\underset{\underset{\displaystyle H}{|}}{C}}-\overset{\overset{\displaystyle O}{\|}}{C}-OH
$$

2,4,5-T
(2,4,5-Trichlorophenoxyacetic acid)

The only structural difference between 2,4,5-T and 2,4-D is the additional chlorine atom on the benzene ring in the fifth position. Agent Orange, widely used as a defoliant during the Vietnam War, is a mixture of these two

Selective herbicides Herbicides that kill only a particular group of plants

Nonselective herbicides Herbicides that kill all plant life

compounds. Many veterans of the Vietnam War have attributed their chronic illnesses to exposure to Agent Orange.

Both 2,4-D and 2,4,5-T result in an abnormally high level of RNA (Section 15.11) in the cells of the affected plants, causing the plants to grow themselves to death.

Triazines have been found to be effective as herbicides; the most important one is atrazine.

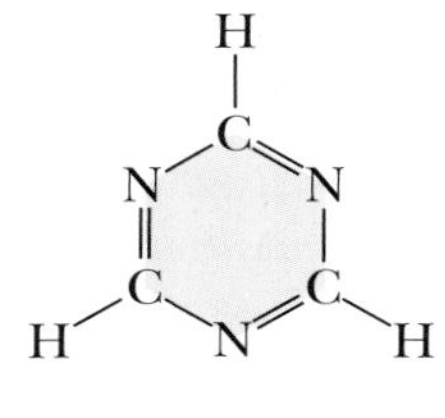

1,3,5-Triazine

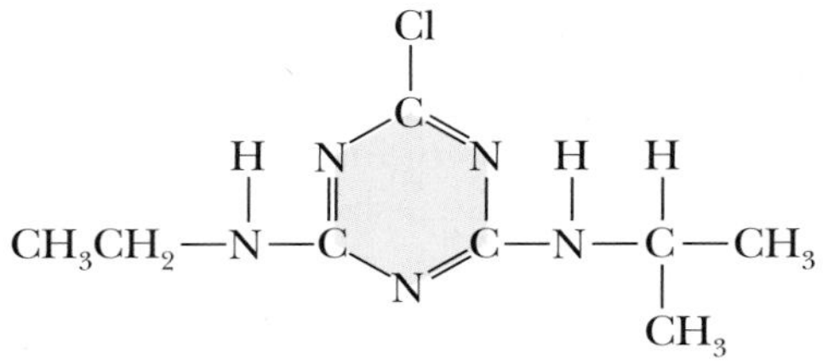

Atrazine
(2-chloro-4-ethylamino-6-isopropylamino-triazine)

Tillage weed control requires tilling or plowing soil between furrows to uproot weeds.

Atrazine is widely used in no-till corn production and for weed control in minimum tillage. Atrazine is a poison to any green plant if it is not quickly metabolized. Corn and certain other crops have the ability to render atrazine harmless, which weeds cannot do. Hence, the weeds die, and the corn shows no ill effect.

Several herbicides work by inhibiting plant enzymes that catalyze the synthesis of amino acids in plants that are essential amino acids in animals (Section 15.6). Unlike animals, plants can synthesize all 20 amino acids. Since animals do not synthesize essential amino acids, developing inhibitors for amino acids that are synthesized by plants but not by animals would produce safer herbicides. Glyphosate and sulfonylureas are herbicides with this mechanism of action. Glyphosate, the active ingredient in RoundUp, is a phosphate derivative of glycine that inhibits the synthesis of the essential amino acids tyrosine and phenylalanine, and is used to control perennial grasses.

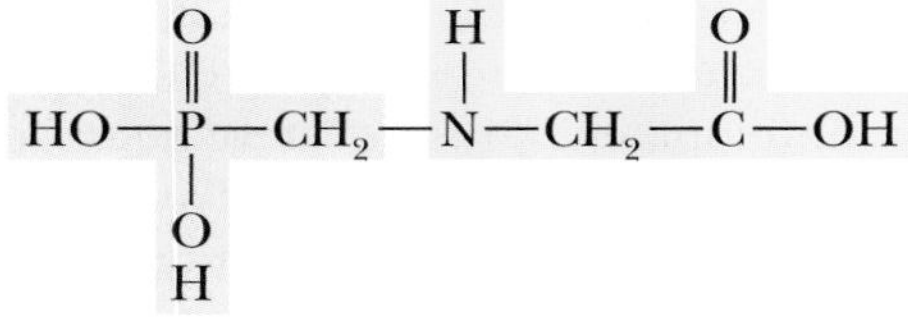

Glyphosate

Sulfometuron methyl, the active ingredient of Oust, is a sulfonylurea that inhibits the synthesis of the essential amino acids valine, leucine, and isoleucine.

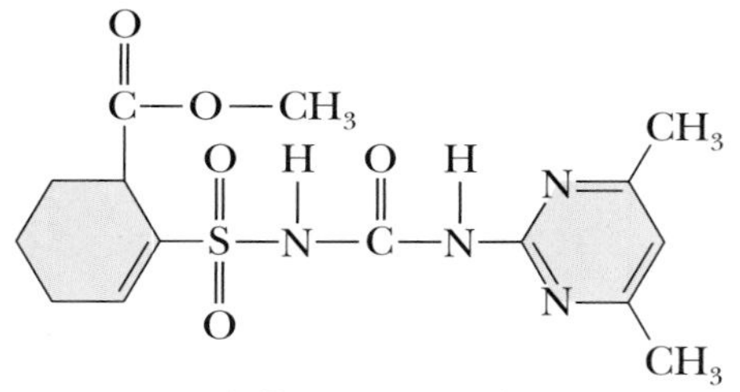

Sulfometuron methyl

Paraquat is a herbicide and can be used to kill weeds before the crop sprouts. Such herbicides are called "pre-emergent" herbicides. When applied directly to susceptible plants, paraquat quickly causes a frostbitten appearance and death. Paraquat has a nitrogen atom in each aromatic ring of the two-ring system.

$$H_3C—{}^+N \bigcirc —\bigcirc N^+—CH_3$$
$$Cl^- \qquad\qquad Cl^-$$

Paraquat
(1,1'-dimethyl-4,4'-bipyridinium dichloride)

Paraquat has received considerable attention because it was used to spray illegal poppy and marijuana fields in Mexico and elsewhere, which caused drug users to suffer lung damage from residual paraquat.

The amount of energy saved by herbicides used in no-till farming is enormous. The saving of topsoil is also considerable, because the cover from the previous crop holds the soil against wind and water runoff. However, agriculturists who use herbicides are highly dependent on agricultural research institutions for the selection of herbicides that will do the desired job without harmful side effects. Such selections depend on considerable research, much of which is carried out on a trial-and-error basis on test plots. A procedure that is recommended today may be outdated by the next growing season.

Fungicides

About 200 of the 100,000 classified fungal species are known to cause serious plant disease. Agricultural fungicide application accounts for about 20% of all pesticide use. Most fungicides are applied to the seed or foliage of the growing plant or to harvested produce to prevent storage losses. The earliest fungicides were inorganic substances such as elemental sulfur and compounds of copper and mercury. In the 19th century, sulfur was used to control mildew on fruit and grapes.

Methyl bromide (CH_3Br) was widely used as a soil fumigant in strawberry and tomato production and in some fruit tree nurseries until it was shown that it was contributing to the destruction of the ozone layer (Section 7.4). Methyl bromide production and use began to be severely regulated and restricted after this discovery with planned total phaseout by 2005. However, in 2005 the parties to the Montreal Protocol granted the United States critical use exemptions of 37.5% of the historic baseline for methyl bromide use. An additional critical use exemption was granted for new production of methyl bromide in 2006 for 27% of the historic baseline production of this compound. It is unclear how long these special critical use exemptions will be granted. The California Department of Pesticide Regulation began an investigation of the risk assessment of methyl iodide (CH_3I) after it received an application to register products containing this compound as a fumigant in 2004.

In recent times, different types of organic compounds have been used. An important class of fungicides is the dithiocarbamates and their derivatives, which are widely used on many crops such as fruits and field vegetables.

$$\begin{array}{ccccc} & S & & S & \\ & \| & & \| & \\ (CH_3)_2N—C—S—S—C—N(CH_3)_2 \end{array}$$
Thiram, a dithiocarbamate derivative

Sulfur is a solid at room temperature. Powdered sulfur is used to control fungi.

(19.6) SUSTAINABLE AGRICULTURE

Extensive use of fertilizers and pesticides, and development of higher yielding varieties of crops, have made it possible to increase crop yields dramatically. However, poor farming practices have resulted in soil erosion and loss

T. McCabe/Visuals Unlimited

Soil erosion caused by water runoff, which carries soil, fertilizer, and pesticides with it.

of soil fertility. The annual amount of topsoil lost to wind and water erosion is from 2 to 4 tons per acre in Africa, Europe, and Australia; 4 to 8 tons per acre in North, Central, and South America; and nearly 12 tons in Asia. These amounts are about 20 times the rate of replenishment of the topsoil. Compaction of soil, which decreases soil fertility, is caused by (1) growing shallow-rooted crops year after year; (2) not incorporating enough organic matter into the soil; (3) practices that alter soil microbial and earthworm populations; and (4) using heavy machinery, particularly on wet soils. Compacted soil, when dry, looks like brick. It restricts root growth and has soil oxygen levels below those necessary for optimum uptake of nutrients.

Another major problem is the increasing resistance of pests to pesticides. About 500 different insect and mite species, 80 fungus species, and 80 weed species are now resistant to commonly used pesticides.

These problems have led to considerable variations of traditional farming methods sometimes referred to as **organic farming**, **alternative farming**, and **sustainable agriculture**. Even the U.S. Department of Agriculture concedes that it is difficult to define these terms, so it is little wonder that there is confusion over their meaning. However, most agree that one goal of agriculture would be to sustain a viable system of food production with minimal negative impact on the environment. This will, in all probability, require a combination of efforts which might fall under one definition or another.

Organic farming is farming without chemical fertilizers and pesticides. Organic farming uses only about 40% of the energy required for modern farming with synthetic chemicals and produces about 90% of the yield. The costs of energy saved in organic farming are offset by the costs of human labor required by the use of natural fertilizers. Many claims are made that organic farming produces a better product for human consumption. However, there is no real evidence that these claims are generally true. Organic farming does have one clear advantage, however; it is definitely less of a threat to the environment than regular farming if agrichemicals are not very carefully controlled.

Sustainable agriculture, as defined by the U.S. Department of Agriculture, means an integrated system of plant and animal production practices having a site-specific application that will, over the long term:

- satisfy human food and fiber needs
- enhance environmental quality and the natural resource base upon which the agricultural economy depends
- make the most efficient use of nonrenewable resources and on-farm resources and integrate, where appropriate, natural biological cycles and controls
- sustain the economic viability of farm operations
- enhance the quality of life for farmers and society as a whole.

One goal of sustainable agriculture would be to limit the use of agrichemicals to that which is absolutely necessary by increasing the use of environmentally friendly procedures, where possible, to fight pests and to produce food and fiber. Other examples of sustainable practices include (1) expanding crop rotation because the same pests do not attack every crop, (2) using multiple crops in alternate plantings with a given planting field (Figure 19.9), (3) using as much natural fertilizer as possible before resorting to agrichemicals, (4) increasing the use of biological pest controls, (5) employing renewed efforts at soil and water conservation, and (6) having a diversification in livestock as well as field crops on the same farm. Some would define some of these efforts as *alternative agriculture* but the semantics are irrelevant when compared to the goals.

Steve Feld

Organic produce, which is grown without pesticides. Consumer demand for such products is growing in spite of a lack of scientific evidence indicating such products are more nutritious or justifying their higher costs.

Organic farming Farming without synthetic chemical fertilizers and pesticides

Sustainable agriculture An approach to agriculture between the extremes of nonagrichemical-based organic farming and the heavy use of agrichemicals whose broad goal is to provide food for the world's population while minimizing environmental impact. A related, but not synonymous term, is *alternative agriculture*, which could include nonconventional production methods such as organic farming and the production of more nontraditional crops, livestock, and farm products.

USDA

FIGURE 19.9 Strip farming, a sustainable practice in which different crops are grown side by side. Often, some strips contain legumes that produce nitrogen and reduce the need for chemical fertilizers. Soil is preserved by contouring the strips according to the topography of the land.

The main sources of plant nutrients in alternative farming systems are animal and green manures. A green manure crop is a grass or legume that is plowed into the soil or surface-mulched at the end of a growing season to enhance soil productivity.

One broad approach to limiting use of pesticides is **integrated pest management** (IPM), which relies more on disease-resistant crop varieties and biological controls such as natural predators or parasites that control pest populations than on agrichemicals. Farmers can select tillage methods, planting times, crop rotations, and plant-residue management practices to optimize the environment for beneficial insects that control pest species. If pesticides are used as a last resort, they are applied when pests are more vulnerable or when any beneficial species and natural predators are least likely to be harmed.

IPM programs have been most effective for cotton, sorghum, peanuts, and fruit orchards, but less effective for corn and soybeans. Some of the biological controls include release of sterilized insect pests, use of insect pheromones to disrupt mating, release of natural predator pests, and use of natural insecticides.

New FDA standards are being developed for labeling organic food products.

Natural Insecticides

Many plant species contain natural protection against insects. For example, nicotine protects tobacco plants from sucking insects. However, its commercial use is limited by its high mammalian toxicity (oral LD_{50} for rats is 50 mg/kg).

One of the oldest and best-known natural insecticides is pyrethrum, a contact insecticide obtained from dried chrysanthemum flowers by extraction with hydrocarbon solvents. Pyrethrum is very effective in killing flying insects and is relatively nontoxic toward mammals (oral LD_{50} for rats is 129 mg/kg). Pyrethrum aerosol sprays are excellent home insecticides because of their safety for humans and rapid action on pests. However, pyrethrum lacks persistence against agricultural insects because of its instability to air and light, so it must be mixed with small amounts of other insecticides to kill insects that might recover from sublethal doses of pyrethrum.

Robert E. Ford/Terraphotographics

Chrysanthemum flowers being harvested in Rwanda, where they are grown as a source of pyrethrum.

Integrated pest management (IPM)
Limits the use of pesticides by relying on a combination of disease-resistant crop varieties, natural predators or parasites, and selected use of pesticides

The insecticidal properties of pyrethrum result from six esters that are collectively called *pyrethrins*. After the isolation and identification of these compounds, a number of derivatives have been synthesized that are more effective than the natural pyrethrins. For example, dimethrin is effective against mosquito larvae and is safe to use (oral LD_{50} for rats is greater than 10,000 mg/kg). The basic pyrethrin structure is shown in black, and the groups that vary in different pyrethrins are shown in yellow.

Dimethrin

Another source of effective natural pesticides is the neem tree, which grows widely in Africa and Asia. For centuries, people of India have known of the insect-fighting ability of the neem tree. The oil extracted from neem tree seeds has been found to be effective against more than 200 species of insects, including locusts, gypsy moths, cockroaches, California medflies, and aphids. *Azadirachtin,* an active ingredient of oil from neem tree seeds, interferes with insect molting, reproduction, and digestion. Tests have shown it is specific to insects without affecting pest predators. For example, use of azadirachtin on an aphid-infested field killed the aphids without harming ladybugs and lacewings, which are aphid predators.

19.7 AGRICULTURAL GENETIC ENGINEERING

Armed with the ability to insert genes into organisms (Section 15.11), it follows that introduction of genes into food plants and animals to fight pests, control diseases, and improve food quality should be of value.

Natural breeding methods require up to 10 years to produce plants suitable for field testing, but genetic engineering requires as little as 1 year. One of the first examples of a genetically engineered plant was produced by the insertion of DNA segments from the tobacco mosaic virus into the genetic code of tomato plants (Figure 19.10). The DNA segments caused the tomato plants to be strongly resistant to attack by this virus.

The first **transgenic crop** brought to market was the Flavr Savr tomato. Ordinarily, tomatoes are picked green and later ripened with ethylene gas to avoid excess softening and spoilage before the tomatoes reach the consumer. The Flavr Savr tomato was genetically modified to ripen more slowly so it could be picked at a ripened stage, enhancing its flavor and increasing its shelf life. The Flavr Savr was ultimately a financial failure. It was not always up to market standards and also was aggressively criticized by groups objecting to genetically modified foods.

The chances that today there are food products containing a genetically modified ingredient on supermarket shelves are high, however. In 2003 there were 87 transgenic crop plants approved for commercial production in the United States. The list included beets, chicory, corn, cotton, flax, melons, papaya, potatoes, rapeseed, rice, soybeans, squash, tobacco, tomatoes, and wheat, among others. Each can be assumed to have passed the scrutiny of three government agencies. The U.S. Department of Agriculture monitors experimental plantings of new crops and their potential for gene transfer to weeds or other crops. The Food and Drug Administration (FDA) evaluates the composition of the food for potential allergens, toxic agents, and nutri-

Transgenic crops Crops genetically altered by the techniques of genetic engineering

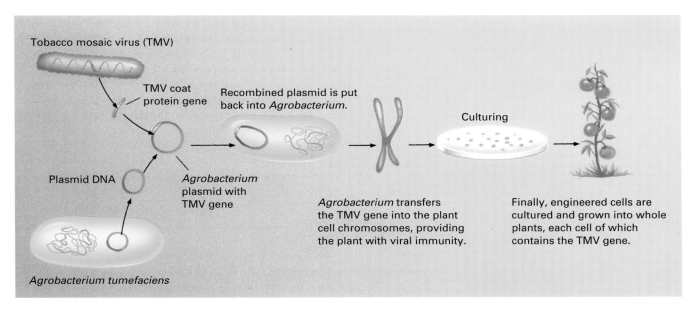

FIGURE 19.10   Steps necessary to insert a gene from tobacco mosaic virus into a tomato plant to make the plant resistant to a viral disease.

tional equivalence to the non-modified food. If there are health concerns, the FDA can forbid marketing of a food. The EPA has responsibility for the effects of pesticides produced by transgenic plants. They deal with the impact of the pesticide on human health or other plants, and must approve use of a herbicide on a plant resistant to that herbicide.

The most widely planted genetically modified food crops are soybeans, corn, rapeseed (the source of canola oil), and cotton. The percentages of these crops that were transgenic in 2002 were as follows:

- 74% of the U.S. soybean crop was transgenic. Soybeans are used in oil, flour, lecithin, and protein extracts.

- 32% of the U.S. corn crop was transgenic. The corn provides oil and a variety of other food ingredients, but very little transgenic corn was marketed as fresh corn, canned corn, or popcorn.

- 60% of the Canadian rapeseed crop, the source of canola oil marketed in the United States, was transgenic. Canola oil is incorporated into salad dressing, margarine, cheese, fried foods, and many kinds of desserts.

- 71% of the U.S. cotton crop was transgenic. Beyond its use in textiles, cotton is the source of cottonseed oil that reaches the market in cooking oils, salad dressings, and various dessert products.

By genetic engineering, a soil bacterium (*Bacillus thuringienis* or *BT*) has been modified to produce pesticides that are more toxic than natural ones. The modified genes have then been inserted into crop plants to create crops that synthesize their own pesticides. There is *Bt* corn with a toxin that destroys the European corn borer and other pests, and there is *Bt* cotton protected against the cotton bollworm and the budworm.

Genetic engineering has also been used to develop transgenic crops that are resistant to certain herbicides. This allows the control of weeds without damaging the plant. For example, cotton, corn, and soybeans have been genetically engineered with resistance to glyphosate, the active ingredient in the herbicide RoundUp, which can thus be used to destroy competing weeds.

A different type of application is to use genetic engineering to increase the amount of commercially useful substance in a plant. Transgenic rapeseed

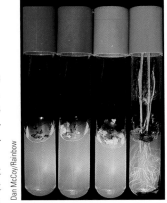

Growth of a transgenic plant. From left to right, you can see progress from a few cells that contain the altered DNA to a young plant with leaves and roots.

plants have been developed using a gene from the California bay tree that shuts off fatty acid synthesis at 12 carbons instead of 18 carbons. The resulting transgenic plant contains up to 40% lauric acid, a saturated fatty acid used to make soaps, detergents, and shampoos (Section 15.4).

The value of transgenic crops is, however, a topic that remains controversial. The potential risks and benefits to humans and the environment are not fully understood. There are different realms of uncertainty for each type of modification and each type of plant. Will overall ecological balance be disrupted? Will there be gene transfer from crops to weeds or other crops? Are beneficial insects at risk? Will insects or plant diseases become so tolerant to toxins that they cannot be stopped at all? These and other questions await further study.

CONCEPT CHECK 19B

1. Most virgin soils can support crop production for a decade or more before fertilization is needed. (**a**) True, (**b**) False.
2. What do the numbers 6–12–8 on a fertilizer mean? The 6 is the percentage of _____, the 12 is the percentage of _____, and the 8 is the percentage of _____ in the fertilizer.
3. Would the nitrogen in ammonia be considered "fixed" nitrogen?
4. Pure ammonia under ordinary conditions is (**a**) a solid, (**b**) a liquid, (**c**) a gas.
5. Approximately what percentage of the food crops of the world is lost to pests each year?
6. The first chlorinated organic insecticide was _____.
7. Which of the following is a persistent insecticide? (**a**) DDT, (**b**) Malathion, (**c**) Carbaryl, (**d**) Pirimicarb.
8. _____ is a natural insecticide extracted from chrysanthemum flowers.
9. Which is more likely to be a hormone, a selective or a nonselective herbicide?
10. The most widely used herbicide is _____.
11. Three common classes of pesticides are _____, _____, and _____.
12. What previously widely used fungicide has been implicated in destruction of the ozone layer?

 Assess your understanding of this chapterís topics with an online chapter quiz at **www.brookscole.com/chemistry/joesten4**

■ KEY TERMS

humus	nonmineral nutrients	straight fertilizers
horizons	mineral nutrients	complete (mixed) fertilizers
topsoil	primary nutrients	quick-release fertilizers
subsoil	secondary nutrients	slow-release fertilizers
loam	micronutrients	Haber process
percolation	nitrogenase	pesticides
leaching effect	chlorosis	insecticides

herbicides   nonselective herbicides   integrated pest management (IPM)

fungicides   organic farming   transgenic crops

selective herbicides   sustainable agriculture

■ THE LANGUAGE OF CHEMISTRY

1. Legume crops   a. Most common herbicide
2. Chlorosis   b. *Bacillus thuringiensis*
3. Ammonium nitrate   c. Caused by nutrient deficiency in soil
4. 2,4-D   d. First genetically engineered food on the market
5. Glyphosate   e. Causes basic soil to become acidic
6. Source of natural insecticide   f. Roots contain nitrogen-fixing bacteria
7. Pheromones   g. Solid added to neutralize acidic soil
8. Flavr Savr tomato   h. Nitrogen fertilizer also used as an explosive
9. Fixed nitrogen   i. Organophosphorus pesticide
10. Malathion   j. Herbicide inhibits amino acid synthesis in plants
11. Selective leaching   k. Sex attractants used in integrated pest management
12. Limestone   l. DDT
13. Chlorinated hydrocarbon   m. Ammonia
    pesticide

■ APPLYING YOUR KNOWLEDGE

1. What factors affect Earth's carrying capacity?
2. What are the likely consequences if the present rate of human population growth continues?
3. What is an operational definition of "soil"?
4. Give definitions and descriptions for the following:
    (**a**) Topsoil   (**b**) Subsoil
    (**c**) Loam
5. What is the structure of a typical soil?
6. Name what causes soil to be (**a**) sour, (**b**) sweet.
7. If crushed limestone is spread on soil, will it raise or lower the pH of the soil? Explain.
8. Give definitions for the following:
    (**a**) Leaching   (**b**) Selective leaching
9. What are the three nonmineral nutrients obtained from water and air required for normal plant growth?
10. Which is more easily leached from soils, nitrates or phosphates? Why?
11. (**a**) Which groups of elements are first leached from soils, the Group 1A and 2A metals or the Group 3A and transition metals? (**b**) What is the effect of this selective leaching on soil pH?
12. (**a**) What are two important roles of humus in the soil? (**b**) Do leaves turned into the soil to produce humus raise or lower the soil pH?

13. Give definitions and/or descriptions for the following:
    (**a**) Nutrients   (**b**) Nonmineral nutrients
    (**c**) Mineral nutrients
14. What are the three primary mineral plant nutrients that are considered in fertilizer formulations?
15. Which is more likely to be a problem in farming, a soil shortage of N, P, and K, or a shortage of Ca, Mg, and S? Give a reason for your answer.
16. (**a**) What does the term "nitrogen fixation" mean? (**b**) Give two natural ways that this process occurs.
17. Write the balanced equation for the Haber process. Why is this process important in the production of fertilizers?
18. What are legumes? What is nitrogenase?
19. What is chlorosis? What causes it? What are the symptoms of chlorosis?
20. An inorganic fertilizer has a grade of 10–0–0. Is this a complete fertilizer? Explain.
21. Inorganic fertilizers are graded on their content of nitrogen, phosphorus, and potassium. Are any of these actually present as the element? Are any of the reference substances actually in a bag of fertilizer?

22. Phosphate rock is treated with sulfuric acid to make superphosphate. Why isn't the phosphate rock simply used as a phosphorus-containing fertilizer?

23. Urea (NH_2CONH_2) is a common straight fertilizer supplying nitrogen. What happens to urea when it is applied to soil? What nitrogen compound is formed when urea decomposes?

24. What is a straight fertilizer? What is a complete fertilizer?

25. Give definitions or descriptions for the following:
    (a) Quick-release fertilizers
    (b) Slow-release fertilizers

26. Give definitions for the following:
    (a) Pesticide
    (b) Insecticide
    (c) Herbicide
    (d) Fungicide

27. Ammonium nitrate ($NH_4NO_3(s)$) is a common straight fertilizer. It decomposes at high temperature to produce N_2, H_2O, and O_2. Handling ammonium nitrate has led to accidental explosions. Give one reason for continued use of ammonium nitrate as a fertilizer and one reason why it should be under greater control.

28. What is the reason the use of DDT is banned in the United States? Why do you suppose that DDT is still used in other parts of the world?

29. Why is DDT fat soluble and not water soluble? What two properties make DDT a problem?

30. Trace the rise and fall of the use of DDT in agriculture. Debate whether it has been more good than bad for the human race.

31. What is the approximate percentage of food crops that is lost to pests every year? What is the estimated dollar value of these lost crops?

32. What is a persistent pesticide? Give an example.

33. What is a biodegradable pesticide? Give an example.

34. (a) What two herbicides were formulated to produce Agent Orange? (b) Which of these herbicides is currently banned in the United States for agricultural use?

35. (a) Discuss the benefits and possible harms in using pesticides in agriculture. (b) What conclusions can you draw on this controversial issue?

36. Organic farming saves energy in one area but loses it in another. Explain.

37. What is sustainable agriculture?

38. If you grew a garden, would you use synthetic fertilizers and pesticides? Why or why not?

39. What is integrated pest management?

40. Two effective natural insecticides are *Bt* toxins and neem tree seeds. Explain what these are and why they are becoming so popular.

41. (a) What are transgenic crops? (b) Give some examples.

42. What mechanism of action is followed by glyphosate and sulfonylureas? What is the relationship between "essential" amino acids and these herbicides?

43. Why are pheromones interesting compounds for insect control? How are they used?

44. Assume it is possible to use genetic engineering to improve the storage life of a ripe harvested food like oranges. Give two questions you would want answered before you would buy and eat these oranges.

45. What do you think about the connection between population growth and limitations on the amount of cultivable land? Do you see any conflict? Explain.

46. A 1992 United Nations study indicated that 4.84 billion acres of farmland soil were so degraded that it would be impossible to reclaim them. What are the main causes of this soil degradation?

47. Order the following insecticides in increasing order of toxicity based on the LD_{50} values for rats given in parentheses:
    DDT (LD_{50} = 100 mg/kg)
    Malathion (LD_{50} = 1000 mg/kg)
    Carbaryl (LD_{50} = 250 mg/kg)
    Pirimicarb (LD_{50} = 150 mg/kg)
    Dimethrin (LD_{50} = 10,000 mg/kg)

48. How many milligrams of DDT would a lab rat weighing 1.5 kg need to consume in order to reach the oral rat LD_{50} value of 100 mg/kg body weight?

49. How many milligrams of dimethrin would a 0.5 kg lab rat need to consume in order to reach the oral rat LD_{50} value of 10,000 mg/kg body weight?

50. How many acres of land are estimated to be under cultivation? What is the approximate number of acres cultivated per person? Assume Earth's population is 6.5 billion people.

51. An inorganic fertilizer has an analysis of 10–20–20. What is the percentage of N in this fertilizer? How many pounds of N would be in a 20-lb bag?

52. A fertilizer bag label displays the following grade numbers, 20–10–5. Is this a complete fertilizer? Explain. What is the percentage of N, the percentage of P as P_2O_5, and the percentage of K as K_2O in this fertilizer? How many pounds of N are in a 50-lb bag?

CHEMISTRY ON THE WEB

For up-to-date URLs, visit the text website at **www.brookscole.com/ chemistry/joesten4**

- World Population FAQ, Estimates, and Data
- Everything You Wanted to Know about Soil
- Plant Nutrients
- Fertilizers and Their Use
- Pesticides
- Herbicides
- Sustainable Agriculture
- Primer on Transgenic Crops

Appendix A: Significant Figures

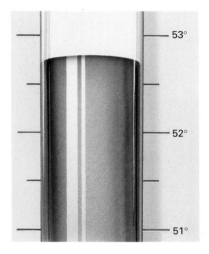

A temperature of 52.7°C can be read from the thermometer pictured in the margin by estimating the 0.7 part of the measurement. The thermometer can only measure this temperature to three **significant figures**, which are the number of digits that can be measured with certainty, plus one more that is estimated. Suppose the temperature of 52.7°C is measured during a chemical experiment and it is decided to try the next experiment at a lower temperature, one third of 52.7°C. Dividing 52.7°C by 3 on a calculator gives an answer of 17.56666667°C. But since 52.7°C has three significant figures, only three digits should be used in the answer to this calculation. The 17.56666667 on the calculator must be rounded off to give a temperature for the new experiment of 17.6°C.

In scientific calculations, it is important to report the results of measurements with only the number of digits that are significant. When these measurements are then used in calculations, the number of digits in the answer must be neither more nor less than are significant. Keeping track of significant figures in calculations requires two skills: (1) counting digits that are significant and (2) applying rules for rounding off the answers to calculations.

Counting Significant Figures

1. **All nonzero digits are significant.**

    23
    Two significant figures

    15,699
    Five significant figures

2. **Zeroes to the left of the first nonzero digit are not significant.**

    0.23
    Two significant figures

    0.00023
    Two significant figures

3. **Zeroes between nonzero digits are significant.**

    7077
    Four significant figures

    50.2
    Three significant figures

4. **Zeroes at the end of a number that includes a decimal point are significant.**

    These zeroes have been added to represent the correct number of significant figures.

    50.020
    Five significant figures

    7.00
    Three significant figures

    For a number with zeros at the end and *no* decimal point, it is impossible to know how many of the digits are significant. Does 99,000 have two, three, or four significant figures? To represent the significance of digits in numbers

Counting Significant Figures

Measured Value	Number of Significant Figures
$\overset{1}{3}$ cm	1
$\overset{12\ 34}{45.87}$ mL	4
$\overset{123}{0.223}$ g	3
$\overset{1234}{4529}$ m	4
$\overset{12}{0.70}$ kg	2
$\overset{12\ 3}{30.6}$ cm	3
$\overset{12\ 34}{65.00}$ m^3	4
$\overset{1}{0.005}$ mg	1
$\overset{1\ 2}{3.8} \times 10^3$ L	2
$\overset{1\ 23}{1.05} \times 10^{-10}$ mm	3

For calculations that have several parts, the best approach is to do the entire calculation first, and then round off the final answer. Rounding off at every separate step can cause errors.

Note that in addition and subtraction, rounding off isn't needed if all of the numbers in the calculation are whole numbers.

Rounding Off Numbers

Original Number*	Number of Significant Figures Desired	Rounded-off Number
1.6<u>7</u>	2	1.7
521.<u>1</u>	3	521
4.56<u>5</u>	3	4.57
351.<u>5</u>3	3	352
1.<u>5</u>68	1	2
16.0<u>4</u>3	3	16.0
8.2<u>3</u>5 × 10^3	2	8.2 × 10^3
7540.<u>9</u>16	4	7541

*The first number being dropped is underlined.

like this, scientists rely on scientific notation, which is discussed in Appendix B. Sometimes a decimal point is placed after a final zero to show that the zero is significant. For example, 20. has *two* significant figures.

EXAMPLE A.1 Significant Figures

How many significant figures are there in each of the following measurements?
(**a**) 0.007 m (**b**) 99.00 kg (**c**) 2.0004°C

SOLUTION
Always start counting with the first nonzero digit. In (**a**) this is the 7 and there is only one significant figure. In (**b**), you continue to count past the decimal point—the zeros have been added to show that this measurement has four significant figures. In (**c**) the zeros in the middle of the number are significant, so this measurement has five significant figures.
(**a**) $\overset{1}{0.007}$ m (**b**) $\overset{12\ 34}{99.00}$ kg (**c**) $\overset{1\ 2345}{2.0004}$°C

Exercise A.1

How many significant figures are in each of the following quantities?
(**a**) 63,332 kg (**b**) 0.06 mm (**c**) 407 L (**d**) 0.0600 s (**e**) 0.0101 g

Answers

(**a**) 5 (**b**) 1 (**c**) 3 (**d**) 3 (**e**) 3

Calculations with Significant Figures

To find the correct number of significant figures for the answer to a calculation requires three steps: (1) Do the arithmetic. (2) Decide how many significant figures the answer can have. (3) Round off the answer.

In **addition and subtraction** an answer should have no more *decimal* places than the number in the calculation with the fewest decimal places. Consider the following addition.

13.673	Three decimal places
4.0	One decimal place
[17.673]	Three decimal places (unrounded answer)

The unrounded answer has two more decimal places than 4.0 and must therefore be rounded off to include only one decimal place as in 4.0. The rules for **rounding off** the answer are simple:

If the numbers to be dropped are 5 or greater than 5, then 1 is added to the final digit that is retained.

If the numbers to be dropped are less than 5, they are dropped and the final retained digit is unchanged.

In the preceding addition, 17.673 must be rounded off to have one decimal place, which means dropping the 73 and increasing the 6 by 1 to give 17.7.

An exception to the need to round off answers occurs with numbers described as *exact*. One source of exact numbers is counting. The number of students in a class or the number of objects in one dozen are exact numbers. Such numbers do not limit the number of digits in the answer to a calculation. Another source of exact numbers is definition. For example, 1 kg is defined as *exactly* equal to 1000 g.

EXAMPLE A.2   Significant Figures in Addition

What is the total weight of three jars of chemicals weighing 14.57 g, 1835.5 g, and 0.875 g?

SOLUTION

By aligning the numbers on their decimal points for the addition and mentally drawing a vertical line after the number with the smallest number of decimal places, it is easy to see that the answer must be rounded off to one decimal place.

$$
\begin{array}{r}
14.5|7 \ \ \text{g} \\
1835.5| \ \ \ \text{g} \\
0.8|75 \ \text{g} \\
\hline
[1850.9|45 \ \text{g}]
\end{array}
$$

Rounded-off answer: 1850.9 g

The total weight of the jars is 1850.9 g.

EXAMPLE A.3   Significant Figures in Subtraction

An African Goliath beetle, one of the world's heaviest insects, weighs 99.790 g. What is the difference in weight between this beetle and a hummingbird with a weight of 10.6 g?

SOLUTION

In subtraction as in addition, align the numbers on the decimal point and use a vertical line to find the place to which the answer should be rounded.

$$
\begin{array}{r}
99.7|90 \ \text{g} \\
- \ 10.6| \ \ \ \text{g} \\
\hline
[89.1|90 \ \text{g}]
\end{array}
$$

Rounded-off answer: 89.2 g

The African Goliath beetle is 89.2 g heavier than a hummingbird.

Exercise A.2
Round off each of the following numbers to three significant figures.
(a) 187.6 (b) 5.281 (c) 1.0334 (d) 0.9265 (e) 30.099

Answers
(a) 188 (b) 5.28 (c) 1.03 (d) 0.927 (e) 30.1

Exercise A.3
Perform the following calculations and give the answers with the proper numbers of significant figures.
(a) 77.2 + 0.531 + 13.27 (b) 0.815 − 0.2346
(c) 62 + 49.1 (d) 1.386 − 1.2

Answers
(a) 91.0 (b) 0.580 (c) 111 (d) 0.2

For **multiplication and division** the answer is limited to the number of digits in the number with the fewest significant figures. When, for example, 75 is multiplied by 0.05, the answer can have only one significant figure (as in

Calculations with Significant Figures

COMPUTATION	RESULT
$18.2 + 5.35 + 20.$	44
$15.02 + 0.003 + 700.1$	715.1
$7109.3 - 40.352$	7068.9
$1.521 - 0.81$	0.71
38.2×0.95	36
17.32×1.66	28.8
$182/32.800$	5.55
$0.881/5.2$	0.17
$\dfrac{39.3 + 17.21}{190.}$	0.297
$(58.1 \times 0.82) - 1.19$	46

The number of significant figures in the answer is determined by *decimal places* in addition and subtraction, and by total *significant figures* in multiplication and division.

To use a percent in a calculation, you must move the decimal point two places to the left:

$$10\% = 0.10 \quad 59.33\% = 0.5933$$
$$230\% = 2.30$$

0.05); when 0.084 is divided by 0.00298, the answer can have only two significant figures (as in 0.084).

$$75 \times 0.05 = [3.75] \qquad \text{Rounded-off answer: } 4$$

$$\frac{0.084}{0.00298} = [28.18791946] \qquad \text{Rounded-off answer: } 28$$

EXAMPLE A.4 **Significant Figures in Multiplication**

A sample of a metal ore analyzed in the laboratory is found to contain 11.00% of titanium. How much titanium is present in 55 kg of this ore?

SOLUTION
To find the answer requires multiplying 55 kg by 0.1100

$$0.1100 \times 55 \text{ kg} = [6.05 \text{ kg}] \qquad \text{Rounded-off answer: } 6.1 \text{ kg}$$

The answer is limited to two significant figures by 55 kg. The ore contains 6.1 kg of titanium.

EXAMPLE A.5 **Significant Figures in Division**

The Pentagon building near Washington, D.C. has an area of 117,355 square meters (m^2), and a football field has an area of 5358 m^2. How many football fields would fit inside the Pentagon?

SOLUTION
To find the answer requires dividing the area of the Pentagon by the area of the football field. On a calculator the answer has ten digits. It must be rounded off to four digits as in 5358.

$$\frac{117,355 \text{ m}^2}{5358 \text{ m}^2} = [21.90276222] \qquad \text{Rounded-off answer: } 21.90$$

The Pentagon could hold 21.90, or just about 22 football fields.

Exercise A.4
Perform the following calculations and give the answers with the proper numbers of significant figures.
(**a**) 68.3×0.92 (**b**) 3.01×20.225 (**c**) $1.594/6.23$
(**d**) $\dfrac{5.0}{25.0}$ (**e**) $\dfrac{(2.151 - 1.30)}{0.591}$

Answers
(**a**) 63 (**b**) 60.9 (**c**) 0.256 (**d**) 0.20 (**e**) 1.4

Appendix B: Scientific Notation

Scientific notation, also known as **exponential notation**, is a way of representing large and small numbers as the product of two terms. The first term, the coefficient, is a number between 1 and 10. The second term, the exponential term, is 10 raised to a power—the **exponent**. For example,

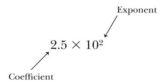

Exponent

2.5×10^2

Coefficient

The exponent indicates the number of 10's by which the coefficient is multiplied to give the number represented in scientific notation

$$2.5 \times 10^2 = 2.5 \times 10 \times 10 = 250$$

There are two great advantages to scientific notation. The first, as illustrated in Table B.1, is that the very large and very small numbers often dealt with in the sciences are much less cumbersome in scientific notation. The second is that it removes the ambiguity in the number of significant figures in a number ending with zeroes. We can, for example, write

Two significant figures	2.5×10^2
Three significant figures	2.50×10^2
Four significant figures	2.500×10^2

All the digits in the coefficient are always considered significant.

For numbers larger than 1, the exponent in scientific notation is a positive whole number, as illustrated above. For numbers less than 1, the exponent is a negative whole number that indicates the number of times the coefficient must be divided by 10 (or multiplied by 0.1) to give the number represented in scientific notation

$$1.2 \times 10^{-2} = 1.2 \times \frac{1}{10} \times \frac{1}{10} = 0.012$$

Numbers in scientific notation are read as follows:

2×10^4	"Two times ten to the fourth power"
2×10^{-4}	"Two times ten to the minus fourth power"

TABLE B.1	Large and Small Numbers in Scientific Notation	
Distance from Earth to the Sun	150,000,000,000 m	1.5×10^{11} m
Diameter of Earth	13,000,000 m	1.3×10^7 m
Height of Mt. Everest	10,000 m	1×10^4 m
Height of Empire State Building	400 m	4×10^2 m
Human height	1.7 m	1.7×10^0 m
Length of a cockroach	0.040 m	4.0×10^{-2} m
Size of a grain of sand	0.00010 m	1.0×10^{-4} m
Size of a red blood cell	0.0000065 m	6.5×10^{-6} m
Size of a polio virus	0.000000025 m	2.5×10^{-8} m

There are two points to remember in converting a number into or out of scientific notation: (1) The value of the exponent is the number of places by which the decimal is shifted. (2) The coefficient should have only one digit before the decimal point. Look through the following examples and count the number of places the decimal points have been moved.

$$10000 = 1 \times 10^4 \qquad 12345 = 1.2345 \times 10^4$$
$$1000 = 1 \times 10^3 \qquad 1234 = 1.234 \times 10^3$$
$$100 = 1 \times 10^2 \qquad 123 = 1.23 \times 10^2$$
$$10 = 1 \times 10^1 \qquad 12 = 1.2 \times 10^1$$
$$1 = 1 \times 10^0 \qquad \text{(Any number} \times 10^0 = \text{the number itself.)}$$
$$1/10 = 1 \times 10^{-1} \qquad 0.12 = 1.2 \times 10^{-1}$$
$$1/100 = 1 \times 10^{-2} \qquad 0.012 = 1.2 \times 10^{-2}$$
$$1/1000 = 1 \times 10^{-3} \qquad 0.0012 = 1.2 \times 10^{-3}$$
$$1/10000 = 1 \times 10^{-4} \qquad 0.00012 = 1.2 \times 10^{-4}$$

Some electronic calculators allow you to easily convert numbers to scientific notation. If you have such a calculator, you can change a number shown in the usual form to scientific notation simply by pressing the EE or EXP key and then the "=" key (or in some cases the "=" key and then the exponent key).

Do not use × 10 when entering exponential numbers on a calculator; this multiplies your answer by 10.

EXAMPLE B.1   Converting into and out of Scientific Notation

(a) The Sun is about 93 million miles from Earth. Express this number in scientific notation. (b) A blue whale has a weight of 1.36×10^5 kg. Convert this number from scientific notation into a conventional number.

SOLUTION
(a) Writing out the zeros in 93 million gives

93,000,000 mi

To express this distance in scientific notation requires moving the decimal point 7 places to the left. Because the number is larger than 1, the exponent is positive.

$$93,000,000 \text{ miles} = 9.3 \times 10^7 \text{ mi}$$

(b) To go the other way, the positive exponent indicates a number greater than 1, so that the decimal point must be moved to the right, in this case by 5 places.

$$1.36 \times 10^5 \text{ kg} = 136,000 \text{ kg}$$

EXAMPLE B.2   Converting into and out of Scientific Notation

(a) Convert the length of a virus from 0.00000002 m to scientific notation. (b) Convert the diameter of a protein molecule from 6.5×10^{-6} m to a decimal number.

SOLUTION

(**a**) For a number less than 1, the exponent is the negative of the number of places the decimal is moved to the *right*

$$0.00000002 \text{ m} = 2 \times 10^{-8} \text{ m}$$
$$\underset{1\,2\,3\,4\,5\,6\,7\,8}{}$$

(**b**) The negative exponent shows that the number is less than 1, so the decimal point must be moved to the left, by 6 places in this case, to convert out of scientifc notation

$$6.5 \times 10^{-6} \text{ m} = 0.0000065 \text{ m}$$
$$\underset{6\,5\,4\,3\,2\,1}{}$$

Exercise B1
Write the following numbers in scientific notation.
(**a**) 3785 (**b**) 0.093 (**c**) 19.2 (**d**) 552,000 (**e**) 0.00075

Answers
(**a**) 3.785×10^3 (**b**) 9.3×10^{-2} (**c**) 1.92×10^1
(**d**) 5.52×10^5 (**e**) 7.5×10^{-4}

Exercise B.2
Write the following numbers in conventional form.
(**a**) 4.42×10^2 (**b**) 8.9×10^{-3} (**c**) 3.01×10^{-1}
(**d**) 6.9×10^4 (**e**) 7.7×10^0

Answers
(**a**) 442 (**b**) 0.0089 (**c**) 0.301 (**d**) 69,000 (**e**) 7.7

Calculations with Numbers in Scientific Notation

For **addition and subtraction** of numbers in scientific notation, the numbers must first be converted to the same powers of 10. The coefficients are then added or subtracted as usual and the exponential term remains the same in the sum or difference.

To algebraically add a number with a + sign to a number with a − sign, drop the signs, subtract the smaller number from the larger number, and give the answer the sign of the larger number.

$$(1.234 \times 10^{-3}) + (5.623 \times 10^{-2}) = (0.1234 \times 10^{-2}) + (5.623 \times 10^{-2})$$
$$= 5.746 \times 10^{-2}$$
$$(6.52 \times 10^2) - (1.56 \times 10^3) = (6.52 \times 10^2) - (15.6 \times 10^2)$$
$$= -9.1 \times 10^2$$

To algebraically subtract numbers when they have different signs, reverse the sign of the number that is being subtracted, and then add the numbers algebraically.

For **multiplication** of numbers in scientific notation, the coefficients are multiplied in the usual manner and the exponents are added. If both exponents have the same sign, their values are added and the exponent in the answer has the same sign. If the exponents have different signs, they must be added algebraically, taking into account the + and − signs. The final answer is given with one nonzero digit to the left of the decimal and the correct number of significant figures.

$$(1.23 \times 10^3)(7.60 \times 10^2) = (1.23)(7.60) \times 10^{3+2}$$
$$= 9.348 \times 10^5$$
$$= 9.35 \times 10^5 \text{ (3 significant figures)}$$
$$(6.02 \times 10^{23})(2.32 \times 10^{-2}) = (6.02)(2.32) \times 10^{23-2}$$
$$= 13.966 \times 10^{21}$$
$$= 1.40 \times 10^{22} \text{ (3 significant figures)}$$

$$(5.2 \times 10^{-4})(6.1 \times 10^2) = (5.2)(6.1) \times 10^{-4+2}$$
$$= 31.72 \times 10^{-2}$$
$$= 3.2 \times 10^{-1} \text{ (2 significant figures)}$$

For **division** of numbers in scientific notation, the coefficients are divided in the usual fashion. The exponent of the divisor is subtracted algebraically from the exponent of the number that is being divided.

$$\frac{7.60 \times 10^3}{1.23 \times 10^2} = \frac{7.60}{1.23} \times 10^{3-2} = 6.18 \times 10^1$$

$$\frac{6.02 \times 10^{23}}{9.10 \times 10^{-2}} = \frac{6.02}{9.10} \times 10^{(23)-(-2)} = 0.662 \times 10^{25} = 6.62 \times 10^{24}$$

| EXAMPLE | B.3 | Calculations with Numbers in Scientific Notation |

The distance around the globe from New York to London is 5.6×10^3 km and the distance from London to New Zealand is 1.9×10^4 km. What is the sum of these distances? What is the difference between these distances?

SOLUTION

$$(1.9 \times 10^4 \text{ km}) + (5.6 \times 10^3 \text{ km}) = (1.9 \times 10^4 \text{ km}) + (0.56 \times 10^4)$$
$$= 2.5 \times 10^4 \text{ km}$$
$$(1.9 \times 10^4 \text{ km}) - (5.6 \times 10^3 \text{ km}) = (1.9 \times 10^4 \text{ km}) - (0.56 \times 10^4)$$
$$= 1.3 \times 10^4 \text{ km}$$

It is 2.5×10^4 km from New York to New Zealand via London, and it is 1.3×10^3 km farther from London to New Zealand than it is from New York to London.

| EXAMPLE | B.4 | Calculations with Numbers in Scientific Notation |

To compare their sizes, divide the size of a grain of sand, 1.0×10^{-4} m, by the size of a red blood cell, 6.5×10^{-6}.

SOLUTION

$$\frac{1.0 \times 10^{-4} \text{ m}}{6.5 \times 10^{-6} \text{ m}} = \frac{1.0}{6.5} \times 10^{-4-(-6)}$$
$$= 0.15 \times 10^2$$
$$= 1.5 \times 10^1 \text{ or } 15$$

A grain of sand is 15 times larger than a red blood cell.

EXAMPLE   B.5   **Calculations with Numbers
in Scientific Notation**

In 1 mol of oxygen gas there are 6.02×10^{23} O_2 molecules. How many O_2 molecules are there in 1.4×10^3 mol of oxygen gas?

$$(1.4 \times 10^3 \text{ mol } O_2) \times \left(\frac{6.02 \times 10^{23} \ O_2 \text{ molecules}}{\text{mol } O_2} \right)$$

$$= (1.4 \times 6.02) \times 10^{3+23} \text{ molecules}$$
$$= 8.4 \times 10^{26} \ O_2 \text{ molecules}$$

Exercise B.3

Perform the following calculations.

(**a**) $(8.91 \times 10^{-2}) \times (1.2 \times 10^5)$

(**b**) $(1.15 \times 10^8) + (6.2 \times 10^7)$

(**c**) $\dfrac{5.09 \times 10^5}{8.2 \times 10^{-2}}$

(**d**) $(9.82 \times 10^4) - (1.35 \times 10^3)$

(**e**) $\dfrac{(2.731 + 10^{-2}) \times (1.52 \times 10^9)}{8.3 \times 10^{14}}$

Answers

(**a**) 1.1×10^4 (**b**) 1.77×10^8 (**c**) 6.2×10^6
(**d**) 9.69×10^4 (**e**) 5.0×10^{-8}

Appendix C: Units of Measure, Unit Conversion, and Problem Solving

Units of the SI System

The metric system was begun by the French National Assembly in 1790 and has undergone many modifications. The International System of Units or *Système International* (SI), which represents an extension of the metric system, was adopted by the 11th General Conference of Weights and Measures in 1960. It is constructed from seven base units, each of which represents a particular physical quantity (Table C.1).

The first five units listed in Table C.1 are those most useful in chemistry. They are defined as follows:

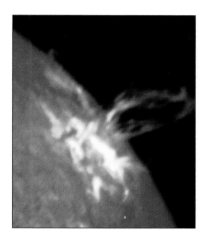

1. The *meter* is the length of the path traveled by light in vacuum during a time interval of 1/299792458 of a second.

2. The *kilogram* represents the mass of a platinum–iridium block kept at the International Bureau of Weights and Measures of Sèvres, France.

3. The *second* was redefined in 1967 as the duration of 9,192,631,770 periods of a certain line in the microwave spectrum of cesium-133.

4. The *kelvin* is 1/273.16 of the temperature interval between absolute zero and the triple point of water (the temperature at which liquid water, ice, and water vapor coexist).

5. The *mole* is the amount of substance that contains as many entities as there are atoms in exactly 0.012 kg of carbon-12 (12 g of ^{12}C atoms).

Decimal fractions and multiples of metric and SI units are designated by using the **prefixes** listed in Table C.2. The prefix *kilo,* for example, means a unit is multiplied by 10^3

$$1 \text{ kilogram} = 1 \times 10^3 \text{ grams} = 1000 \text{ grams}$$

and the prefix *centi* means the unit is multiplied by the factor 10^{-2}

$$1 \text{ centigram} = 1 \times 10^{-2} \text{ gram} = 0.01 \text{ gram}$$

TABLE C.1   SI Fundamental Units

PHYSICAL QUANTITY	NAME OF UNIT	SYMBOL
Length	Meter	m
Mass	Kilogram	kg
Time	Second	s
Temperature	Kelvin	K
Amount of substance	Mole	mol
Electric current	Ampere	A
Luminous intensity	Candela	cd

TABLE C.2	Prefixes for Metric and SI Units*				
FACTOR	**PREFIX**	**SYMBOL**	**FACTOR**	**PREFIX**	**SYMBOL**
10^{12}	tera	T	10^{-1}	*deci*	d
10^{9}	giga	G	10^{-2}	*centi*	c
10^{6}	mega	M	10^{-3}	*milli*	m
10^{3}	*kilo*	k	10^{-6}	micro	μ
10^{2}	hecto	h	10^{-9}	*nano*	n
10^{1}	deka	da	10^{-12}	*pico*	p
			10^{-15}	femto	f
			10^{-18}	atto	a

*The most common prefixes are shown in italics.

The prefixes are added to give units of a magnitude appropriate to what is being measured. The distance from New York to London (5.6×10^3 km = 5,600 km) is much easier to comprehend measured in kilometers than in meters (5.6×10^6 m = 5,600,000 m). The following is a list of units for measuring very small and very large distances:

nanometer (nm)	0.000000001 meter	$= 1 \times 10^{-9}$ meter
micrometer (μm)	0.000001 meter	$= 1 \times 10^{-6}$ meter
millimeter (mm)	0.001 meter	$= 1 \times 10^{-3}$ meter
centimeter (cm)	0.01 meter	$= 1 \times 10^{-2}$ meter
decimeter (dm)	0.1 meter	$= 1 \times 10^{-1}$ meter
meter (m)	1 meter	
dekameter (dam)	10 meters	$= 1 \times 10$ meters
hectometer (hm)	100 meters	$= 1 \times 10^{2}$ meters
kilometer (km)	1,000 meters	$= 1 \times 10^{3}$ meters
megameter (Mm)	1,000,000 meters	$= 1 \times 10^{6}$ meters

In the International System of Units, all physical quantities are represented by appropriate combinations of the base units listed in Table C.1. The result is a derived unit for each kind of measured quantity. The most common derived units are listed in Table C.3. It is easy to see that the derived unit for area is length $\times$ length = meter $\times$ meter = square meter (m^2) or that the derived unit for volume is length $\times$ length $\times$ length = meter $\times$ meter $\times$ meter = cubic meter (m^3). The more complex derivations are arrived at by the same kind of combination of units. Units such as

TABLE C.3	Derived SI Units			
PHYSICAL QUANTITY	**NAME OF UNIT**	**SYMBOL**	**DEFINITION**	**SYMBOL**
Area	Square meter	m^2	—	
Volume	Cubic meter	m^3	—	
Density	Kilogram per cubic meter	kg/m^3	—	
Force	Newton	N	(kilogram)(meter)/(second)^2	$kg\ m/s^2$
Pressure	Pascal	Pa	Newton/square meter	N/m^2
Energy	Joule	J	(kilogram)(square meter)/(second)^2	$kg\ m^2/s^2$
Electric charge	Coulomb	C	(ampere)(second)	A s
Electric potential difference	Volt	V	joule/(ampere)(second)	J/(A s)

the one for *force* have been given simple names that represent the units by which they are defined.

Conversion of Units for Physical Quantities

The result of a measurement is a **physical quantity**, which consists of a number plus a unit. To convert a physical quantity from one unit of measure to another requires a conversion factor based on equivalences between units of measure like those given in Table C.4. Each equivalence provides two

---

TABLE C.4   Common Units of Measure

Mass and Weight

1 pound = 453.59 grams = 0.45359 kilogram
1 kilogram = 1000 grams = 2.205 pounds
1 gram = 10 decigrams = 100 centigrams = 1000 milligrams
1 gram = 6.022×10^{23} atomic mass units
1 atomic mass unit = 1.6605×10^{-24} gram
1 short ton = 2000 pounds = 907.2 kilograms
1 long ton = 2240 pounds
1 metric tonne = 1000 kilograms = 2205 pounds

Length

1 inch = 2.54 centimeters (exactly)
1 mile = 5280 feet = 1.609 kilometers
1 yard = 36 inches = 0.9144 meter
1 meter = 100 centimeters = 39.37 inches = 3.281 feet = 1.094 yards
1 kilometer = 1000 meters = 1094 yards = 0.6215 mile
1 Ångstrom = 1×10^{-8} centimeter = 0.1 nanometer = 100 picometers
       = 1×10^{-10} meter = 3.937×10^{-9} inch

Volume

1 quart = 0.9463 liter
1 liter = 1.0567 quarts
1 liter = 1 cubic decimeter = 1000 cubic centimeters = 0.001 cubic meter
1 milliliter = 1 cubic centimeter = 0.001 liter = 1.056×10^{-3} quart
1 cubic foot = 28.316 liters = 29.924 quarts = 7.481 gallons

Force and Pressure

1 atmosphere = 760 millimeters of mercury (exactly) = 1.013×10^5 pascals
         = 14.70 pounds per square inch
1 bar = 10^5 pascals
1 torr = 1 millimeter of mercury
1 pascal = 1 kg/m s^2 = 1 N/m^2

Energy

1 joule = 1×10^7 ergs
1 calorie = 4.184 joules (exactly) = 4.184×10^7 ergs
       = 4.129×10^{-2} liter-atmospheres
       = 2.612×10^{19} electron volts
1 Calorie (food) = 1 kilocalorie
1 erg = 1×10^{-7} joule = 2.3901×10^{-8} calorie
1 electron volt = 1.6022×10^{-19} joule = 1.6022×10^{-12} erg
1 liter-atmosphere = 24.217 calories = 101.32 joules = 1.0132×10^9 ergs
1 British thermal unit = 1055.06 joules = 1.05506×10^{10} ergs = 252.2 calories

conversion factors that are the inverse of each other. For example, the equivalence between a quart and a liter, 1 quart = 0.9463 liter (the 1 quart is exact), gives

$$\frac{1 \text{ quart}}{0.9463 \text{ liter}} \qquad \text{There is 1 quart per 0.9463 liter.}$$

$$\frac{0.9463 \text{ liter}}{1 \text{ quart}} \qquad \text{There is 0.9463 liter per 1 quart.}$$

The cancellation of units provides the basis for choosing which conversion factor is needed: It is always the one that allows the unit being converted to be canceled and leaves the new unit uncanceled.

To convert 2 quarts to liters:

$$2 \text{ quarts} \times \frac{0.9463 \text{ liter}}{1 \text{ quart}} = 1.893 \text{ liters}$$

To convert 2 liters to quarts:

$$2 \text{ liters} \times \frac{1 \text{ quart}}{0.9463 \text{ liter}} = 2.113 \text{ quarts}$$

EXAMPLE C.1 Unit Conversion

Find the weight in (**a**) kilograms, (**b**) grams, and (**c**) decigrams of a person who weighs 115 pounds.

SOLUTION
(**a**) Table C.4 gives the equivalence between pounds and kilograms in two ways: 1 pound = 0.454 kilogram (rounded off), or 1 kilogram = 2.205 pounds. Either equivalence can be used to convert 115 pounds to kilograms by setting up the problem so that *pounds* cancels out.

$$115 \text{ pounds} \times \frac{0.454 \text{ kilogram}}{1 \text{ pound}} = 52.2 \text{ kilograms}$$

$$115 \text{ pounds} \times \frac{1 \text{ kilogram}}{2.205 \text{ pounds}} = 52.2 \text{ kilograms}$$

(**b**) Using the equivalence 1 kilogram = 1000 grams gives

$$52.2 \text{ kilograms} \times \frac{1000 \text{ grams}}{1 \text{ kilogram}} = 52,200 \text{ grams or } 5.22 \times 10^4 \text{ grams}$$

(Remember that equivalences for multiples of units, e.g., 1 kilogram = 1000 grams are exact conversion factors and do not limit significant figures.)
(**c**) Using the equivalence 1 gram = 10 decigrams gives

$$5.22 \times 10^4 \text{ grams} \times \frac{10 \text{ decigrams}}{1 \text{ gram}} = 52.2$$

$$\times 10^4 \text{ decigrams or } 5.22 \times 10^5 \text{ decigrams}$$

EXAMPLE C.2 Unit Conversion

What is the volume of a 5.00×10^2 milliliter flask in (**a**) liters, (**b**) quarts, and (**c**) ounces (1 quart = 32 ounces, exactly)?

(**a**) $\qquad 5.00 \times 10^2 \text{ milliliters} \times \dfrac{1 \text{ liter}}{1000 \text{ milliliters}} = 0.500 \text{ liter}$

(b) $0.500 \text{ liter} \times \dfrac{1.0567 \text{ quart}}{1 \text{ liter}} = 0.528 \text{ quart}$

(c) $0.528 \text{ quart} \times \dfrac{32 \text{ ounces}}{1 \text{ quart}} = 16.9 \text{ ounces}$

Exercise C.1

Use the information in Table C.4 to write the equivalences and two conversion factors for each of the following pairs of units:

(a) grams and milligrams   **(b)** kilometers and miles
(c) calories and joules

Answers

(a) 1 gram = 1000 milligrams

$$\frac{1 \text{ gram}}{1000 \text{ milligrams}} \qquad \frac{1000 \text{ milligrams}}{1 \text{ gram}}$$

(b) 1 kilometer = 0.6215 mile

$$\frac{1 \text{ kilometer}}{0.6215 \text{ mile}} \qquad \frac{0.6215 \text{ mile}}{1 \text{ kilometer}}$$

(c) 1 calorie = 4.184 joules

$$\frac{1 \text{ calorie}}{4.184 \text{ joules}} \qquad \frac{4.184 \text{ joules}}{1 \text{ calorie}}$$

Exercise C.2

Carry out the following unit conversions:

(a) 15 milligrams to grams   **(b)** 453 grams to milligrams
(c) 95 kilometers to miles   **(d)** 5 miles to kilometers
(e) 325 calories to joules   **(f)** 325 joules to calories

Answers

(a) 0.015 g **(b)** 4.53×10^5 milligrams **(c)** 59 miles
(d) 8 kilometers **(e)** 1360 joules **(f)** 77.7 calories

Sometimes it is convenient or necessary to use more than one conversion factor to convert a physical quantity into some other unit. Instead of doing one conversion at a time, a string of conversion factors can be put together and unit cancellation checked before the calculation is done. For example, atmospheric pressure is usually given in the weather report in inches of mercury. What is the atmospheric pressure in millimeters of mercury (a customary unit in scientific measurement) on a day when the weatherman reports it to be 29.2 inches of mercury? Since we have Table C.4 handy, it will be most convenient to use conversion factors based on that table. There isn't an equivalence there for millimeters and inches, but the table does give 1 inch = 2.54 centimeters. Since we know from the prefixes that 100 *centi*meters = 1 meter and 1000 *milli*meters = 1 meter, we can apply these equivalances to go from centimeters to millimeters. Combining the factors we know and checking cancellation of units gives

$$29.2 \text{ inches} \times \frac{2.54 \text{ centimeters}}{1 \text{ inch}} \times \frac{1 \text{ meter}}{100 \text{ centimeters}} \times \frac{1000 \text{ millimeters}}{1 \text{ meter}}$$

The cancellations leave only the unit we're looking for—millimeters—so the setup is correct and the answer is

$$29.2 \text{ inches} \times \frac{2.54 \text{ centimeters}}{1 \text{ inch}} \times \frac{1 \text{ meter}}{100 \text{ centimeters}}$$

$$\times \frac{1000 \text{ millimeters}}{1 \text{ meter}} = 742 \text{ millimeters}$$

Exercise C.3

(**a**) Is the following setup correct for converting milligrams to pounds?

$$225 \text{ milligrams} \times \frac{1 \text{ gram}}{1000 \text{ milligrams}} \times \frac{1 \text{ pound}}{454 \text{ grams}}$$

(**b**) Is the following setup correct for converting yards to kilometers?

$$25 \text{ yards} \times \frac{36 \text{ inches}}{1 \text{ yard}} \times \frac{1 \text{ yard}}{3 \text{ feet}} \times \frac{5280 \text{ feet}}{1 \text{ mile}} \times \frac{1 \text{ mile}}{1.609 \text{ kilometers}}$$

Answers

(**a**) Yes (**b**) No

Exercise C.4

Carry out the following calculations by using multiple conversion factors
(1 quart = 2 pints; 1 pound = 16 ounces)

(**a**) What is the volume in milliliters of 1.00 pint of milk?

(**b**) What is the mass in grams of 15.0 ounces of cereal?

(**c**) What is the distance in centimeters of the very small distance of 30.5 picometers measured by using X rays?

(**d**) What is the volume in cubic meters of a 250-gallon tank?

Answers

(**a**) 473 milliliters (**b**) 425 grams (**c**) 3.05×10^{-9} centimeters
(**d**) 0.95 cubic meter

Problem Solving by Cancellation of Units

Many kinds of numerical problems in everyday life as well as in science can be solved by extending the use of conversion factors and cancellation of units beyond unit conversion. The use of density expressed as mass per unit volume is a simple example of such an extension. Density provides the *connection* between mass and volume. Given that the density of lead is 11.4 g/cm^3, you can find the mass in grams of a piece of lead of known volume or the volume of a piece of lead of known mass. If the known information is that a piece of lead has a volume of 25.0 cm^3 and the unknown information is its mass, the problem is set up and solved as follows

$$25.0 \text{ cm}^3 \times \frac{11.4 \text{ g lead}}{1 \text{ cm}^3} = 285 \text{ g lead}$$

As an example of an everyday problem, consider assigning "units" to the number of people served by a gallon of ice cream. Suppose you know that 1 gallon of ice cream will serve 16 people—16 servings per gallon. To calculate an unknown (the number of gallons needed to serve 50 people) setting up and solving the problem gives

$$50 \text{ servings} \times \frac{1 \text{ gallon of ice cream}}{16 \text{ servings}} = 3.1 \text{ gallons of ice cream}$$

Three gallons should do the job. Applying the method again to calculate the cost, if the ice cream is $4.35 a gallon, gives

$$3 \text{ gallons} \times \frac{\$4.35}{1 \text{ gallon}} = \$13.05$$

Notice how the "conversion factors" in the two ice cream calculations are arranged so that the unwanted "units" cancel. That is the secret of what is sometimes called the *dimensional method* of problem solving.

For a more complex example, suppose a manufacturer must figure out how many people are needed to assemble 6000 Snivies each week (6000 Snivies/week). He knows that it takes a person 1 hour to assemble 8 Snivies (8 Snivies/hour) and that each person will put in 30 hours a week on the assembly line (30 hours/person). How many people must be working on the assembly line each week? Using the dimensional method, the "conversion factors" must be arranged so that the "units" cancel to give an answer of "persons per week." In single setup, the calculation can be done as follows

$$\frac{6000 \text{ Snivies}}{1 \text{ week}} \times \frac{1 \text{ hour}}{8 \text{ Snivies}} \times \frac{1 \text{ person}}{30 \text{ hours}} = 25 \text{ persons/week}$$

The manufacturer will need 25 persons on the assembly line each week.

In Section 8.2, we illustrated dimensional problem solving for calculating the numbers of moles and the masses of reactants and products in chemical reactions. The coefficients in chemical equations and the molar masses of reactants and products provide the factors that must be combined to solve problems of this kind. Consider the equation for the reaction that occurs when copper(II) sulfide is heated in oxygen

$$Cu_2S(s) + O_2(g) \longrightarrow 2 \, Cu(s) + SO_2(g)$$

The molar masses of the reactants and products are

$$159 \text{ g } Cu_2S/1 \text{ mol } Cu_2S \qquad 32 \text{ g } O_2/1 \text{ mol } O_2$$
$$63.5 \text{ g } Cu/1 \text{ mol } Cu \qquad 64 \text{ g } SO_2/1 \text{ mol } SO_2$$

How much copper can be made from 2.50 mol of Cu_2S by this reaction? The connection between Cu_2S and Cu is given by the mole ratio 1 mol Cu_2S/ 2 mol Cu. This ratio and the molar masses must be combined so that the answer has the units of grams of copper

$$2.50 \text{ mol } Cu_2S \times \frac{2 \text{ mol } Cu}{1 \text{ mol } Cu_2S} \times \frac{63.5 \text{ g } Cu}{1 \text{ mol } Cu} = 318 \text{ g } Cu$$

The mass of copper that can be produced by roasting 2.50 mol of copper sulfide is 318 g.

EXAMPLE C.4   Problem Solving

The Snivie manufacturer has one truck available to deliver his Snivies. The Snivies are packed 50 in a carton and the truck can carry 10 cartons per trip. How many trips will the truck have to make to deliver the 6000 Snivies produced each week?

SOLUTION
Expressing the known information as "conversion factors" gives 50 Snivies per carton and 10 cartons per trip. The information must be set up to yield the answer in trips per week.

$$\frac{6000 \text{ Snivies}}{1 \text{ week}} \times \frac{1 \text{ carton}}{50 \text{ Snivies}} \times \frac{1 \text{ trip}}{10 \text{ cartons}} = 12 \text{ trips/week}$$

EXAMPLE C.5   Problem Solving

Using the information before Example C.4, find the number of moles of oxygen that will be consumed in the conversion of 683 g of copper sulfide to copper metal.

SOLUTION

The connection here is supplied by the mole ratio, 1 mol O_2/1 mol Cu_2S, and the molar mass of copper sulfide.

$$683 \text{ g Cu}_2\text{S} \times \frac{1 \text{ mol Cu}_2\text{S}}{159 \text{ g Cu}_2\text{S}} \times \frac{1 \text{ mol O}_2}{1 \text{ mol Cu}_2\text{S}} = 4.30 \text{ mol O}_2$$

Exercise C.5

(**a**) If the Snivie manufacturer decides to increase his production to 20,000 Snivies per week, how many persons will he need working on the assembly line each week? (**b**) If the most trips a truck can make in a week is 20, how many trucks will the manufacturer need to deliver 20,000 Snivies per week?

Answers

(**a**) 83 persons per week (**b**) 2 trucks

Exercise C.6

Use the information about the reaction of copper sulfide with oxygen to answer the following questions:
(**a**) How many grams of copper can be made from 683 g of copper sulfide?
(**b**) How many moles of sulfur dioxide will be released during the production of the copper from 683 g of copper sulfide?

Answers

(**a**) 546 g Cu (**b**) 4.30 mol SO_2

Appendix D: Naming Organic Compounds

Chapters 12 and 14 include both common names and systematic names for organic compounds representing the various classes of hydrocarbons and functional groups. This appendix focuses on the systematic nomenclature for organic compounds, as proposed by the International Union of Pure and Applied Chemistry (IUPAC).

Hydrocarbons

The name of each of the members of the hydrocarbon classes has two parts. The first part, the prefix—*meth-, eth-, prop-, but-,* and so on—reflects the number of carbon atoms. When more than four carbons are present, the Greek or Latin number prefixes are used: *pent-, hex-, hept-, oct-, non-,* and *dec-.* The second part of the name, or the suffix, tells the class of hydrocarbon. Alkanes have carbon–carbon single bonds, alkenes have carbon–carbon double bonds, and alkynes have carbon–carbon triple bonds and are indicated by the suffixes *-ane, -ene,* and *-yne,* respectively.

UNBRANCHED ALKANES AND ALKYL GROUPS

The names of the first eight unbranched (straight-chain) alkanes are given in Table 12.5. Alkyl groups are named by dropping "-ane" from the parent alkane and adding "-yl." See Table 12.6.

BRANCHED–CHAIN ALKANES

The rules for naming branched-chain alkanes are as follows:

1. *Find the longest continuous chain of carbon atoms: this chain determines the parent name for the compound.* For example, the following compound has two methyl groups attached to a *heptane* parent.

   The longest continuous chain may not be obvious from the way the formula is written, especially for the straight-line format that is commonly used. For example, the longest continuous chain of carbon atoms in the following chain is *eight,* not *four* or *six.*

2. *Number the longest chain beginning with the end of the chain nearest the branching. Use these numbers to designate the location of the attached group. When two or more groups are attached to the parent, give each group a number corresponding to its location on the parent chain.* For example, the name of

$$\overset{7}{CH_3}\overset{6}{CH_2}\overset{5}{CH_2}\overset{4}{CH}\overset{3}{CH_2}\overset{2}{CH}\overset{1}{CH_3}$$
$$\underset{CH_3}{|}\qquad\underset{CH_3}{|}$$

is 2,4-dimethylheptane. The name of the following compound is 3-methylheptane, not 5-methylheptane or 2-ethylhexane.

$$\overset{7}{CH_3}-\overset{6}{CH_2}-\overset{5}{CH_2}-\overset{4}{CH_2}-\overset{3}{CH}-CH_3$$
$$\underset{\overset{2}{CH_2}}{|}$$
$$\underset{\overset{1}{CH_3}}{|}$$

3-Methylheptane

3. *When two or more substituents are identical, indicate this by the use of the prefixes* di-, tri-, tetra-, *and so on. Positional numbers of the substituents should have the smallest possible sum.*

$$\qquad\qquad\overset{CH_3}{|}\ \overset{CH_3}{|}$$
$$\overset{1}{CH_3}\overset{2}{CH_2}\overset{3}{C}\overset{4}{CH_2}\overset{5}{CH}\overset{6}{CH}\overset{7}{CH_2}\overset{8}{CH_3}$$
$$\qquad\quad\underset{CH_3}{|}\qquad\underset{CH_3}{|}$$

The correct name of the preceding compound is 3,3,5,6-tetramethyl-octane.

4. *If there are two or more different groups, the groups are listed alphabetically.*

$$\qquad\quad\overset{CH_3}{|}$$
$$\overset{1}{CH_3}\overset{2}{C}\overset{3}{CH_2}\overset{4}{CH}\overset{5}{CH_2}\overset{6}{CH_3}$$
$$\quad\ \underset{CH_3}{|}\ \ \underset{CH_2}{|}$$
$$\qquad\qquad\ \underset{CH_3}{|}$$

The correct name of the preceding compound is 4-ethyl-2,2-dimethyl-hexane. Note that the prefix "di" is ignored in determining alphabetical order.

ALKENES

Alkenes are named by using the prefix to indicate the number of carbon atoms and the suffix *-ene* to indicate one or more double bonds. The systematic names for the first two members of the alkene series are *ethene* and *propene*

$$CH_2{=}CH_2 \qquad CH_3CH{=}CH_2$$

When groups, such as methyl or ethyl, are attached to carbon atoms in an alkene, the longest hydrocarbon chain is numbered from the end that will give the double bond the lowest number and then numbers are assigned to the attached group. For example, the name of

$$\qquad\qquad\overset{CH_3}{|}$$
$$\overset{5}{CH_3}\overset{4}{CH}\overset{3}{CH}{=}\overset{2}{CH}\overset{1}{CH_3}$$

is 4-methyl-2-pentene. See Section 12.7 for a discussion of *cis–trans* isomers of alkenes.

ALKYNES

The naming of alkynes is similar to that of alkenes, with the lowest number possible being used to locate the triple bond. For example, the name of

is 4-methyl-2-pentyne.

BENZENE DERIVATIVES

Monosubstituted benzene derivatives are named by using a prefix for the substituent. Some examples are

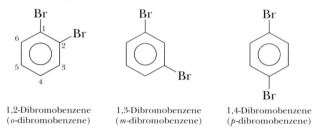

| Chlorobenzene | Methylbenzene (toluene) | Ethylbenzene |

Three isomers are possible when two groups are substituted for hydrogen atoms on the benzene ring. The relative positions of the substituents are indicated by either the prefixes *ortho-, meta-, para-* (abbreviated *o-, m-, p-*), or by numbers. For example,

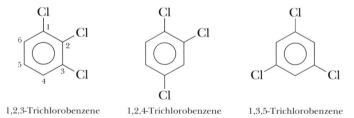

1,2-Dibromobenzene (*o*-dibromobenzene)   1,3-Dibromobenzene (*m*-dibromobenzene)   1,4-Dibromobenzene (*p*-dibromobenzene)

The dimethylbenzenes are called *xylenes* (*o*-xylene, *m*-xylene, *p*-xylene).

If more than two groups are attached to the benzene ring, numbers must be used to identify the positions. The benzene ring is numbered to give the lowest possible numbers to the substituents.

1,2,3-Trichlorobenzene   1,2,4-Trichlorobenzene   1,3,5-Trichlorobenzene

Functional Groups

An atom or group of atoms that defines the structure of a specific class of organic compounds and determines their properties is called a **functional group**. The millions of organic compounds include classes of compounds that are obtained by replacing hydrogen atoms of hydrocarbons with functional groups (Chapter 14). The important functional groups are shown in Table D.1.

The "R" attached to the functional group represents the nonfunctional, hydrocarbon framework with one hydrogen removed for each functional

TABLE D.1   Classes of Organic Compounds Based on Functional Groups*

GENERAL FORMULAS OF CLASS MEMBERS	CLASS NAME	TYPICAL COMPOUND	COMPOUND NAME	COMMON USE OF SAMPLE COMPOUND
R—X	Halide	$H-\overset{\overset{H}{\mid}}{\underset{\underset{Cl}{\mid}}{C}}-Cl$	Dichloromethane (methylene chloride)	Solvent
R—OH	Alcohol	CH_3-OH	Methanol (wood alcohol)	Solvent
$R-\overset{\overset{O}{\parallel}}{C}-H$	Aldehyde†	$H-\overset{\overset{O}{\parallel}}{C}-H$	Methanal (formaldehyde)	Preservative
$R-\overset{\overset{O}{\parallel}}{C}-OH$	Carboxylic acid†	$CH_3-\overset{\overset{O}{\parallel}}{C}-OH$	Ethanoic acid (acetic acid)	Vinegar
$R-\overset{\overset{O}{\parallel}}{C}-R'$	Ketone	$CH_3-\overset{\overset{O}{\parallel}}{C}-CH_3$	Propanone (acetone)	Solvent
R—O—R'	Ether	$C_2H_5-O-C_2H_5$	Diethyl ether (ethyl ether)	Anesthetic
$R-\overset{\overset{O}{\parallel}}{C}-O-R'$	Ester	$CH_3-\overset{\overset{O}{\parallel}}{C}-O-C_2H_5$	Ethyl ethanoate (ethyl acetate)	Solvent in fingernail polish
$R-\overset{\overset{H}{\mid}}{\underset{\underset{H}{\mid}}{N}}$	Amine‡	$CH_3-\overset{\overset{H}{\mid}}{\underset{\underset{H}{\mid}}{N}}$	Methylamine	Tanning (foul odor)
$R-\overset{\overset{O}{\parallel}}{C}-\overset{\overset{H}{\mid}}{N}-H$	Amide‡	$CH_3-\overset{\overset{O}{\parallel}}{C}-\overset{\overset{H}{\mid}}{\underset{\underset{H}{\mid}}{N}}-H$	Acetamide	Plasticizer

*R stands for a hydrocarbon group such as —CH₃ or —C₂H₅. R′ could be a different group from R.
†R can be an H atom or a hydrocarbon group.
‡The H atoms in amines and amides can also be replaced by R groups.

group added. The IUPAC system provides a systematic method for naming all members of a given class. For example, alcohols end in *-ol* (methan*ol*); aldehydes end in *-al* (methan*al*); carboxylic acids end in *-oic* (ethan*oic* acid); and ketones end in *-one* (propan*one*).

ALCOHOLS

Isomers are also possible for molecules containing functional groups. For example, three different alcohols are obtained when a hydrogen atom in pentane is replaced by —OH, depending on which hydrogen atom is replaced. The rules for naming the "R" or hydrocarbon framework are the same as those for hydrocarbon compounds.

$$CH_3CH_2CH_2CH_2CH_2OH \qquad CH_3CH_2CH_2\underset{\underset{OH}{|}}{C}HCH_3 \qquad CH_3CH_2\underset{\underset{OH}{|}}{C}HCH_2CH_3$$

1-Pentanol   2-Pentanol   3-Pentanol

Compounds with one or more functional groups and alkyl substituents are named so as to give the functional groups the lowest number. For example, the correct name of

$$\begin{array}{c} \overset{\displaystyle CH_3}{\underset{\displaystyle 1 \quad 2 \quad 3 \quad 4 | \ 5}{}} \\ CH_3\underset{\underset{\displaystyle OH}{|}}{C}HCH_2\underset{\underset{\displaystyle CH_3}{|}}{C}CH_3 \end{array}$$

is 4,4-dimethyl-2-pentanol.

ALDEHYDES AND KETONES

The systematic names of the first three aldehydes are methanal, ethanal, and propanal.

$$\overset{\displaystyle O}{\overset{\displaystyle \|}{HCH}} \qquad \overset{\displaystyle O}{\overset{\displaystyle \|}{CH_3CH}} \qquad \overset{\displaystyle O}{\overset{\displaystyle \|}{CH_3CH_2CH}}$$

Methanal   Ethanal   Propanal
(formaldehyde)   (acetaldehyde)   (propionaldehyde)

For ketones, a number is used to designate the position of the carbonyl group, and the chain is numbered in a way that gives the carbonyl carbon the smallest number.

$$\overset{\displaystyle O}{\overset{\displaystyle \|}{CH_3CCH_3}} \qquad \overset{\displaystyle O}{\overset{\displaystyle \|}{CH_3CH_2CCH_3}} \qquad \overset{\displaystyle O}{\overset{\displaystyle \|}{CH_3CCH_2CH=CH_2}}$$

2-Propanone   2-Butanone   4-Penten-2-one
(acetone)   (methyl ethyl ketone)

CARBOXYLIC ACIDS

The systematic names of carboxylic acids are obtained by dropping the final *e* of the name of the corresponding alkane and adding *oic acid*. For example, the name of

$$CH_3CH_2CH_2CH_2CH_2COOH$$

is hexanoic acid. The systematic names of the first five carboxylic acids are given in Table 14.6. Other examples are

2-Methylbutanoic acid   2-Butenoic acid

ESTERS

The systematic names of esters are derived from the names of the alcohol and the acid used to prepare the ester. The general formula for esters is

$$\overset{\displaystyle O}{\overset{\displaystyle \|}{R-C-OR'}}$$

The $R-\overset{\displaystyle O}{\overset{\|}{C}}$ comes from the acid and the R′O comes from the alcohol. The alcohol part is named first, followed by the name of the acid changed to end in -*ate*. For example,

$$CH_3CH_2\overset{\displaystyle O}{\overset{\|}{C}}-OCH_3$$

is named methyl propanoate, and

$$CH_3\overset{\displaystyle O}{\overset{\|}{C}}-OCH{=}CH_2$$

is named ethenyl ethanoate.

Appendix E: Answers to Concept Checks and The Language of Chemistry

CHAPTER 2

CONCEPT CHECK 2A

1. (a), (b)
2. an element; cannot be broken down
3. elements and chemical compounds
4. homogeneous
5. (a) True
6. chemical change; changed in identity
7. (a), (b), (c), (d), (e)
8. (a) sugar and water
9. (a) methane and oxygen   (b) hydrogen peroxide
10. (a) chemical property   (b) physical property
    (c) chemical property
11. (a) True

CONCEPT CHECK 2B

1. structure of matter
2. solids: carbon, sulfur, phosphorus, iodine, all metals except mercury; liquids; mercury and bromine; gases: hydrogen, oxygen, nitrogen, fluorine, chlorine (also neon, argon, krypton, xenon, radon)
3. C, S, P, I, metal symbols in Table 2.2 and inside front cover; Hg, Br; H, O, N, F, Cl, Ne, Ar, Kr, Xe, Rn
4. Elements: potassium, hydrogen, phosphorus, oxygen; 8 atoms
5. (a) no   (b) yes   (c) no   (d) yes
6. (a) oxygen atom (in a compound)
    (b) two molecules of oxygen
    (c) one molecule of methane   (d) yields
    (e) one water molecule   (f) two water molecules
7. nonmetal
8. (a) and (d)

CONCEPT CHECK 2C

1. a number, a unit
2. prefixes, multiples of 10
3. cubic centimeter
4. kilo-, milli-
5. mm
6. (b)
7. (a) quantitative   (b) qualitative

THE LANGUAGE OF CHEMISTRY

1. a	7. k	13. p
2. d (or o)	8. f	14. j
3. c	9. g	15. d (or o)
4. b	10. i	16. n
5. e	11. m	
6. h	12. l	

CHAPTER 3

CONCEPT CHECK 3A

1. gained, chemical
2. 50%
3. 5.88%
4. (a) the same   (b) atoms
5. (a) Yes. Although the total mass of Ag_2S may change, the percent mass of Ag in Ag_2S remains constant because the elements in a chemical compound are always present in a definite proportion by mass.
6. (a) Yes. Pure methane will have the same composition regardless of source because the percent mass of hydrogen and carbon in CH_4 is constant.
7. (b) +1
8. (b) electron
9. (a) True
10. (a) nucleus
11. 1800

CONCEPT CHECK 3B

1. 10 protons, 10 electrons, 11 neutrons
2. atomic number, mass number
3. protons, neutrons
4. electrons, protons, neutrons
5. protons and electrons
6. atomic
7. 33 electrons, 33 protons, 42 neutrons
8. 100,000
9.
10. (a) True
11. (b) False

CONCEPT CHECK 3C

1. (c) ultraviolet light
2. (a) Blue
3. farther from, closer to
4. 18
5. 2-8-8-1, 1 valence electron
6. 2-8-7, 7 valence electrons

CONCEPT CHECK 3D

1. (a) 1   (b) 2   (c) 3   (d) 3   (e) 8   (f) 7
2. (a) True
3. (b) False
4. (a) True
5. (a) True

6. (**b**) and (**d**)
7. $SrCl_2$
8. Cs
9. (**a**) oxygen　(**b**) calcium　(**c**) fluorine　(**d**) nitrogen
(**e**) sulfur　(**f**) neon
10. (**a**) True
11. (**b**) False
12. (**a**) cesium　(**b**) fluorine　(**c**) calcium　(**d**) oxygen
(**e**) sodium　(**f**) xenon

THE LANGUAGE OF CHEMISTRY I

1. d	5. c	9. j
2. a	6. h	10. e
3. i	7. b	11. g
4. k	8. f	

THE LANGUAGE OF CHEMISTRY II

1. d	5. c	9. f
2. e	6. j	10. b
3. a	7. i	
4. g	8. h	

■ CHAPTER 4

CONCEPT CHECK 4A

1. nitrogen, water vapor, carbon dioxide, methane, and ammonia
2. oxygen
3. The percentage of oxygen remains the same but the density of the air is less. This means the molecules are farther apart so a greater volume of air must be inhaled at higher altitudes to obtain the same amount of oxygen at lower altitudes.
4. The troposphere extends from Earth's surface to an altitude of roughly 10 km (~6 mi). The stratosphere is the region from the top of the troposphere to an altitude of about 50 km (~30 mi).
5. The region of highest ozone concentration (i.e., the ozone layer) lies within the stratosphere between about 15 and 30 km.
6. 300 ppm and 300,000 ppb
7. 40,000 ppm and 40,000,000 ppb
8. This is pollution that originates as a result of human activity.
9. Primary pollutants are released directly into the atmosphere from a given source. Secondary pollutants are formed by chemical reactions occurring in the atmosphere.
10. Ozone would be one example.
11. adsorb
12. absorb
13. warm; cool
14. coal-burning
15. oxides
16. In the early morning, the rush-hour traffic emits nitric oxide (NO), generated by the reaction of molecular nitrogen (N_2) and molecular oxygen (O_2) at high temperatures of the internal combustion engines, into the atmosphere. As the day progresses, the nitric oxide reacts with atmospheric oxygen (O_2) to generate nitrogen dioxide (NO_2). From mid-day to early afternoon, sunlight promotes the dissociation of nitrogen dioxide to produce oxygen atoms (O) and more nitric oxide. The oxygen atoms (O) then react with molecular oxygen to produce ozone (O_3), whose concentration usually peaks in mid- to late afternoon. (See Sections 4.6 and 4.7.)

CONCEPT CHECK 4B

1. an oxygen atom
2. The burning of sulfur-containing coal
3. $2\,SO_2 + O_2 \longrightarrow 2\,SO_3$
then
$SO_3 + H_2O \longrightarrow H_2SO_4$
sulfuric acid
4. Sulfur dioxide and carbon monoxide
5. b. lime (CaO)
6. Ruminant animals, rice growing, and decaying plants and animals all release methane into the atmosphere. Coniferous and deciduous trees release some hydrocarbons into the air.
7. (0.075 g/mi) × (1000 mi) = 750 g of non-methane hydrocarbons
8. hemoglobin
9. Remove the victim to a location where he or she can be exposed to fresh air.

THE LANGUAGE OF CHEMISTRY

1. c	6. a	11. o
2. d	7. k	12. i
3. l	8. p	13. n
4. h	9. h	14. m
5. f	10. q	

■ CHAPTER 5

CONCEPT CHECK 5A

1. ionic
2. (**a**) losing electrons
3. (**a**) +1, Li^+
(**b**) +3, Al^{3+}
(**c**) −2, S^{2-}
(**d**) −1, Br^-
4. (**a**) potassium, K^+
(**d**) sodium, Na^+
(**e**) strontium, Sr^{2+}
(**f**) silver, Ag^+
5. (**b**) gaining electrons
6. NaCl, sodium chloride
7. (**a**) Cl
(**b**) F
(**c**) S
(**d**) N
(**e**) Br

CONCEPT CHECK 5B

1. Sulfur dioxide
2. SO_3

3. Hydrogen bromide
4. Cl_2O
5. SCl_2
6. Silicon tetrachloride
7. (**a**) four electrons, (**b**) six electrons
8. (**c**) C≡C
9. (**a**) H—C≡N: (**g**) $H_2\ddot{N}$—NH_2

(**b**) H—Ö—N (with O double bond and O⁻)   (**h**) H—Ö—Ö—H

(**c**) H—C—H (with H above and below)   (**i**) H—Ö—N=Ö

(**d**) ⁻:C≡O:⁺   (**j**) H_2C=CH_2

(**e**) Ö=C=Ö   (**k**) H—C—Ö—H (with H above and below C)

(**f**) ⁻:Ö—Ö=Ö⁺

10. H—N̈—H (with H below)   :B̈r—P̈—B̈r: (with :Br: below)

Since the central atoms (N or P) are in the same group of the periodic table, they have the same number of valence electrons. Both molecules have three bonds and a lone pair on the central atom.

11. SiH_4 is tetrahedral, just like CH_4.
12. Double bonds: (**b**), (**e**), (**f**), (**i**), and (**j**); Triple bonds: (**a**) and (**d**)

CONCEPT CHECK 5C

1. (**a**) hydrogen molecule, H_2 (**b**) hydrogen chloride molecule, HCl
2. 109.5°, the tetrahedral angle
3. (**a**) H—F
   (**b**) C—O
   (**c**) H—N
4. (**a**) H_2O
5. (**d**) CCl_4

CONCEPT CHECK 5D

1. gas, liquid, solid
2. increases
3. liquid and solid
4. liquid and gas
5. Gases are compressible; solids and liquids are non-compressible.
6. Decreases
7. (**a**) True
8. hydrogen bonding

CONCEPT CHECK 5E

1. (**a**) True
2. (**a**) True
3. solute, solvent
4. (**b**) corn oil
5. (**b**) False
6. increases
7. (**a**) True
8. melting
9. sublimation

THE LANGUAGE OF CHEMISTRY

1. b   7. a   13. p
2. i   8. f   14. n
3. e   9. k   15. m
4. d   10. q   16. l
5. h   11. j   17. c
6. g   12. o

■ CHAPTER 6

CONCEPT CHECK 6A

1. The percentage of carbon dioxide in the air is about 0.035% but the percentage of carbon dioxide in an exhaled breath is about 4%.
2. UV—8%; Vis—39%; and IR—53%
3. The greenhouse effect is a phenomenon of Earth's atmosphere by which solar radiation, trapped by Earth and re-emitted from the surface as infrared radiation, is prevented from escaping by various gases (e.g., carbon dioxide and water vapor) in the atmosphere. This leads to warming of Earth.
4. infrared radiation
5. 84%
6. 15°C (59°F); −270°C
7. Wavelengths decrease in the following order: infrared, visible, and ultraviolet. The energy trend, high to low, is the opposite order: ultraviolet, visible, and infrared.
8. Neither O_2 or N_2 absorb IR radiation; CO_2 and H_2O absorb IR radiation.

CONCEPT CHECK 6B

1. Radiational cooling is Earth's loss of heat on a clear, cloudless night, caused by the absence of significant water vapor in the atmosphere, which would normally absorb much of the infrared radiation.
2. Carbon dioxide, methane, CCl_2F_2. Carbon dioxide is the least efficient of these gases at absorbing infrared radiation. Methane is 30 times more efficient than carbon dioxide and dichlorodifluoromethane is about 25,000 times more efficient than carbon dioxide.
3. Water vapor
4. Complete combustion of 1 gallon of gasoline would produce about 18 pounds of carbon dioxide and about 1 gallon of water.
5. Decaying vegetation releases carbon dioxide. Rice paddies release methane. Forest fires and volcanoes release a variety of greenhouse gases.

6. photosynthesis and dissolution in the oceans

7. Possible benefits: the area of land suitable for growing crops may increase, the treeline may move further north, some plants may grow at an increased rate due to increased carbon dioxide in the atmosphere. Possible negative effects: increased intensity of storms, flooding along low-lying coastal areas, rapid and cataclysmic climate change.

8. Global warming is a long-term climate effect and not a short-term weather phenomenon.

9.

$$6\ CO_2 + 6\ H_2O \xrightarrow{\text{Sunlight/Chlorophyll}} \underset{\text{Sugars}}{C_6H_{12}O_6} + 6\ O_2$$

Plants are able to utilize carbon dioxide from the atmosphere and water to synthesize sugars and produce oxygen. This process requires energy from the Sun.

■ CHAPTER 7

CONCEPT CHECK 7A

1. 8%

2. $\ddot{O}{=}\ddot{O}{-}\ddot{\underset{..}{O}}{:} \longleftrightarrow :\ddot{\underset{..}{O}}{-}\ddot{O}{=}\ddot{O}$

   Ozone resonance structures

3. Wavelengths between 200–400 nm

4. UVC

5. UVA

6.

$$O_2 \xrightarrow[]{1} 2\ \cdot\ddot{\underset{..}{O}}\cdot$$

$$\cdot\ddot{\underset{..}{O}}\cdot + O_2 \underset{3}{\overset{2}{\rightleftharpoons}} O_3$$

$$\cdot\ddot{\underset{..}{O}}\cdot + O_3 \xrightarrow{4} 2\ O_2$$

< 240 nm photon

< 320 nm photon

Reaction 1 shows the dissociation of molecular oxygen into two oxygen atoms triggered by UVC radiation. Reaction 1 shields us from UVC, high-energy radiation. Reaction 2 shows the combination of an oxygen atom with an oxygen molecule to produce ozone. Reaction 3 shows the dissociation of ozone, triggered by UVB radiation, to produce an oxygen atom and an oxygen molecule. Reaction 3 shields us from most of the UVB radiation. Reaction 4 shows the combination of an oxygen atom and an ozone molecule to produce two oxygen molecules.

7. Hydroxyl radicals (OH) and nitric oxide (NO) radicals can promote the destruction of ozone.

CONCEPT CHECK 7B

1. A Dobson unit (DU) is a measure of ozone concentration.

2. One DU corresponds to 1 ppb of ozone.

3. Paul Crutzen, Mario Molina, and Sherwood Roland

4. It is during this period of time that the Antarctic comes out of the darkness of winter and the destructive chemicals formed during the cold, dark winters are exposed to sunlight. This triggers the release of the ozone-destroying radicals.

5. C—Cl bonds are susceptible to cleavage and C—Br bonds are also problematic.

6. A reactive chlorine atom is formed.

7. Chlorine monoxide (ClO) and an oxygen molecule.

8. 100,000

9. CFCs have been used as refrigerants, propellants in aerosol cans, and foaming agents for blown foams.

10. There may be significant decreases in crop yields and increases in skin cancers.

■ CHAPTER 8

CONCEPT CHECK 8A

Note: When the coefficient is "1" in balanced equations, the "1" is not written. In questions that relate to balancing equations, coefficients of "1" will be placed in parentheses to provide answers for the blanks given in the question.

1. law of conservation of mass

2. mole, dozen

3. (**a**) (1) $Si + 2\ Cl_2$
   (**b**) $4\ Al + 3\ O_2 \longrightarrow 2\ Al_2O_3$
   (**c**) (1) $(NH_4)_2CO_3 + (1)\ Cu(NO_3)_2 \longrightarrow$
   (1) $CuCO_3 + 2\ NH_4NO_3$

4. *Amu* stands for atomic mass unit and is the mass $(1.661 \times 10^{-24}\ g)$ of 1/12 of a single carbon atom. *Mole* is the name used to refer to Avogadro's number of particles of something, usually molecules. *Molar mass* is the mass in grams of one mole of any substance.

5. (**a**) $6\ CO_2 + 6\ H_2O \longrightarrow (1)\ C_6H_{12}O_6 + 6\ O_2$
   (**b**) 6 molecules   (**c**) 6 mol   (**d**) 180 g   (**e**) 264 g

6. 46.01 g

7. $(186\ mol)(44\ g/mol) = 8184\ g$

CONCEPT CHECK 8B

1. reactant, product

2. energy

3. concentration, temperature

4. catalyst

5. activation energy

6. no

7. no

8. yes

9. Le Chatelier's principle

10. more $CaCO_3$ will form

CONCEPT CHECK 8C

1. released

2. absorbed

3. (**a**) True

4. energy

5. (**b**)

6. decreases

7. energy, entropy

8. the first law of thermodynamics

9. the second law of thermodynamics

10. energy, entropy

THE LANGUAGE OF CHEMISTRY

1. h
2. d
3. a
4. k
5. b
6. c
7. e
8. f
9. g
10. j
11. i

◼ CHAPTER 9

CONCEPT CHECK 9A

1. donor, acceptor
2. acid
3. a basic substance
4. $H_3O^+(aq) + OH^-(aq) \longrightarrow 2\,H_2O(\ell)$
5. (**a**) salt   (**b**) acid   (**c**) salt   (**d**) base
6. weak base, establishes an equilibrium, OH^- ions are formed
7. hydronium ion, H_3O^+
8. sulfuric acid, H_2SO_4
9. $HCl(aq) + KOH(aq) \longrightarrow KCl(aq) + H_2O(\ell)$
10. weak acids

CONCEPT CHECK 9B

1. pH of 2
2. (**b**) low hydronium ion concentration
3. (**a**) high hydronium ion concentration
4. H_2CO_3 neutralizes added base, HCO_3^- neutralizes added acid
5. 3
6. 11
7. 1.0×10^{-11} M, 1.0×10^{-3} M
8. 1.0×10^{-4} M
9. 1×10^{-2} M
10. acidic, basic
11. (**a**) sodium bicarbonate   (**c**) $MgCO_3$ (**d**) $Al(OH)_3$

THE LANGUAGE OF CHEMISTRY

1. b
2. a
3. j
4. i
5. h
6. c
7. e
8. g
9. d
10. f

◼ CHAPTER 10

CONCEPT CHECK 10A

1. (**b**)
2. reducing agent
3. (**a**)
4. (**b**)
5. (**b**)
6. oxidizing agent
7. (**a**)
8. (**a**)
9. (**a**) CH_3NH_2
   (**b**) CH_3CH_3
   (**c**) CH_3CH_2OH
10. reducing agent
11. reducing agent
12. oxidizing agent

CONCEPT CHECK 10B

1. anode, cathode
2. electrical energy is no longer produced
3. electrons to flow, charge balance is maintained
4. oxidation, anode
5. reduction, cathode
6. electrical energy is converted to chemical energy
7. electrolysis, aluminum
8. Because the rust falls away from the surface to expose a fresh surface for further corrosion.
9. oxygen, water, iron

THE LANGUAGE OF CHEMISTRY

1. a
2. j
3. i
4. c
5. g
6. d
7. e
8. f
9. k
10. b
11. h

◼ CHAPTER 11

CONCEPT CHECK 11A

1. 70%
2. the oceans
3. groundwater
4. surface water
5. 168 gal
6. 2 gal
7. pathogens, toxic metals, organic compounds
8. aquifer
9. it is returned to the sea or it evaporates
10. Only water and ammonia are capable of hydrogen bonding and, therefore, their properties are the most different when compared to the other compounds. Water has the most drastically different properties when compared with the other compounds in this table.
11. Fusion (melting) only requires that the hydrogen bonds in the crystalline ice lattice be relaxed, but vaporization requires that all intermolecular hydrogen bonding between water molecules is totally disrupted. The latter requires more energy.
12. Acid rainfall is rainfall that is more acidic than pH 5.6. Nitrogen oxides and sulfur oxides emitted by combustion processes produce nitric acid and sulfuric and sulfurous acids, which are the main contributors to acid rainfall.
13. About 2.5% of Earth's water is freshwater. Of this, about 2% is in ice caps and glaciers and about 0.5% is "readily available" for human use.
14. Thermoelectric power and agriculture/irrigation
15. Water that is fit for human consumption.

CONCEPT CHECK 11B

1. Superfund
2. (**b**) False
3.

Household waste	Harmful Chemical
automobile battery	sulfuric acid and lead
oil-based paint	organic solvents
batteries	lead, mercury, cadmium

4. plastics, unused paint, used motor oil, used auto batteries
5. biochemical oxygen demand, BOD
6. The manufacture of auto batteries can introduce lead into the groundwater. The manufacture of fluorescent lamps can introduce mercury into the groundwater.
7. The manufacture of plastics can introduce chlorinated organics into the groundwater. The manufacture of textiles can introduce chlorinated organics into the groundwater.
8. Mercury and lead

CONCEPT CHECK 11C

1. (**d**) reverse osmosis
2. (**b**) persistent
3. Aeration and filtration
4. aerobic
5. Anaerobic
6. (**b**) calcium (**c**) magnesium
7. chlorination and ozonation
8. Na^+, Mg^{2+}, Ca^{2+}, K^+
9. Osmosis

THE LANGUAGE OF CHEMISTRY

1. k	9. c	16. g
2. n	10. l	17. s
3. o	11. i	18. q
4. a	12. e	19. t
5. b	13. f	20. r
6. d	14. m	
7. j	15. h	
8. p		

CHAPTER 12

CONCEPT CHECK 12A

1. (**c**) natural gas
2. (**b**) hydrogen
3. (**a**) True
4. oxygen, water
5. distillation
6. methane
7. ethanol
8. catalytic reforming
9. catalytic cracking
10. (**c**)

CONCEPT CHECK 12B

1. tetrahedral
2. ethene or ethylene
3. ethyne or acetylene
4. —C_2H_5
5. structural
6. ethylene
7. *cis* and *trans*

CONCEPT CHECK 12C

1. hydrogen
2. 12

3. CO and H_2
4. 3
5. 3
6. (**a**) ethanol
7. R—O—R

THE LANGUAGE OF CHEMISTRY

1. d	5. j	8. b
2. f	6. a	9. e
3. i	7. c	10. g
4. h		

◼ CHAPTER 13

CONCEPT CHECK 13A

1. (**a**) True
2. (**c**) Gamma rays
3. (**b**) False
4. (**b**) False
5. $^{212}_{82}Pb$
6. (**a**) True
7. 0.5 μCi, or 18,500 dps

CONCEPT CHECK 13B

1. $^{239}_{93}Np$
2. Neptunium
3. (**a**) True
4. (**a**) True
5. Radium
6. 4 pCi/L

CONCEPT CHECK 13C

1. Food irradiation and medical imaging
2. Imaging
3. (**c**) metastable
4. (**a**) 1/8 of the original dose—(1/2)(1/2)(1/2) or $(1/2)^3$
5. The 6h half-life isotope, because it would decay more quickly producing less exposure.

CONCEPT CHECK 13D

1. (**b**) Uranium-235
2. $^{236}_{92}U$
3. $^{134}_{52}Te$
4. smaller
5. (**a**) Iron
6. (**b**) 20%
7. (**a**) True
8. Not in my backyard
9. High level—spent nuclear fuel; low level—medical waste, lab clothing, etc.

THE LANGUAGE OF CHEMISTRY

1. l	7. c	12. g
2. h	8. p	13. j
3. k	9. d	14. a
4. i	10. e	15. m
5. n	11. f	16. b
6. o		

■ CHAPTER 14

CONCEPT CHECK 14A

1. $CH_3CH_2CH_2$—
2. aromatic hydrocarbons
3. acetaldehyde
4. rubbing alcohol
5. ethylene glycol
6. 42%
7. denatured
8. 3
9. glycerol
10. CH_3OH
11. $HOCH_2CH_2OH$
12. flood the system with CH_3CH_2OH for ~3 days, under supervision

CONCEPT CHECK 14B

1. formaldehyde
2. acetic acid
3. ethanol with acetic acid
4. methyl acetate
5. 3
6. acetone
7. (**b**) esters
8. petroleum

CONCEPT CHECK 14C

1. monomers
2. free radicals, organic peroxides
3. double or triple
4. single
5. (**a**) True
6. CH_2CHCl
7. *cis*
8. (**a**) low-density polyethylene   (**b**) high-density polyethylene   (**c**) cross-linked polyethylene

CONCEPT CHECK 14D

1. adipic acid, hexamethylenediamine, nylon 6,6, water
2. condensation
3. water
4. (**b**) condensation reactions
5. (**a**) True
6. glass fibers
7. polymer
8. PET and HDPE

THE LANGUAGE OF CHEMISTRY

1. e	5. l	9. c
2. d	6. h	10. i
3. f	7. j	11. b
4. g	8. k	12. a

■ CHAPTER 15

CONCEPT CHECK 15A

1. 4
2. monosaccharides
3. glucose and fructose
4. D-glucose
5. hydrogen
6. D-glucose
7. glycerol and fatty acids
8. double bonds. A saturated fat contains only single carbon–carbon bonds while an unsaturated fat contains one or more carbon–carbon double bonds and hence fewer hydrogen atoms than a saturated fat.
9. (**a**) steroid
10. (**a**) True
11. long-chain fatty acids and long-chain alcohols
12. male sex hormones, female sex hormones

CONCEPT CHECK 15B

1. aqueous base
2. hydrophobic, hydrophilic
3. (**a**) traditional soaps
4. emollient
5. emulsion
6. shaving cream
7. smoke
8. fog
9. emulsion

CONCEPT CHECK 15C

1. glycine
2. (**a**) True
3. essential
4. 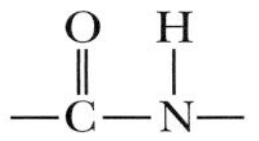   O   H
       ‖ |
     —C—N—
5. (**a**) 27   (**b**) 6
6. (**a**) amino acid sequence
   (**b**) hydrogen bonded structures to form helices or sheets
   (**c**) how the protein molecule is folded
   (**d**) shape assumed by all chains in a protein with two or more chains
7. hydrogen
8. active site
9. (**a**) catalyst
10. enzyme
11. hydrogen, ionic
12. hydrogen

CONCEPT CHECK 15D

1. CO_2 and H_2O, energy
2. ATP
3. ADP, energy
4. DNA
5. ribose, deoxyribose
6. phosphoric acid group, a sugar (ribose or deoxyribose), a nitrogen heterocyclic base (adenine, guanine, thymine, cytosine, or uracil)
7. double helix
8. hydrogen
9. T, C, A, G
10. three billion

THE LANGUAGE OF CHEMISTRY

1. d   5. g   9. f
2. c   6. i   10. h
3. a   7. j   11. k
4. l   8. e   12. b

THE LANGUAGE OF CHEMISTRY

1. d   7. o   13. l
2. i   8. k   14. c
3. n   9. p   15. j
4. g   10. h   16. e
5. a   11. m
6. b   12. f

■ CHAPTER 16

CONCEPT CHECK 16A

1. simple sugars, amino acids, glycerol plus fatty acids, energy
2. basal metabolic rate, weight in pounds
3. 1 kilocalorie
4. carbohydrates, fats, proteins
5. (**a**) True
6. (**b**) glucose
7. ATP

CONCEPT CHECK 16B

1. dietary fiber
2. triglycerides
3. high-density lipoproteins (HDLs), transport cholesterol to liver for excretion
4. low-density lipoproteins (LDLs), transport cholesterol to arteries
5. nitrogen
6. fats
7. fruits and vegetables
8. 30%

CONCEPT CHECK 16C

1. (**b**) False
2. fat-soluble, water-soluble
3. vitamin A
4. β-carotene is $C_{40}H_{56}$; vitamin A is $C_{20}H_{30}O$
5. antioxidants
6. enzymes
7. enriched
8. ions
9. electrolyte balance, acid-base balance (also fluid balance), nerve
10. anemia, osteoporosis

CONCEPT CHECK 16D

1. drying, salt or sugar
2. Generally Recognized as Safe
3. (**b**) False
4. antioxidants
5. EDTA
6. (**b**) False
7. polysaccharides
8. 100 Cal divided by 230 Cal
9. 0.30×2000 Cal

■ CHAPTER 17

CONCEPT CHECK 17A

1. generic, trade name
2. pneumonia, influenza, HIV infection
3. microorganisms
4. accidents, homicide, and suicide
5. S (sulfur) and N (nitrogen), four
6. cross-linking
7. an —NH_2 group and an —OH group
8. inside, outside
9. retrovirus
10. AIDS, DNA
11. chemotherapy
12. protease inhibitors, enzymes

CONCEPT CHECK 17B

1. hormones and neurotransmitters
2. receptors
3. proteins or steroids
4. progesterone
5. (**a**) and (**d**)
6. dopamine
7. epinephrine
8. glands
9. increasing

CONCEPT CHECK 17C

1. heroin
2. does
3. (**a**) ii   (**b**) i   (**c**) iii
4. acetylsalicylic acid
5. bleeding
6. carboxylic acid, anti-inflammatory
7. addictive (an abused drug), medical
8. depressants
9. pseudoephedrine
10. cocaine, addictive
11. (**c**) allergic reaction
12. (**b**) False
13. decongestants, antitussives, expectorants, analgesics, antihistamine
14. ultraviolet (uv) light
15. sun protection factor
16. (**a**) iii   (**b**) iii   (**c**) i   (**d**) ii
17. dose

CONCEPT CHECK 17D

1. cardiovascular disease
2. oxygen
3. (a) faster
4. enzymes
5. (a), (b), (c)
6. (a) True
7. DNA
8. kill
9. lung and bronchial; prostate
10. dose–response
11. homeopathic

THE LANGUAGE OF CHEMISTRY

1. f	7. o	13. p
2. e	8. k	14. j
3. c	9. l	15. i
4. m	10. d	16. n
5. b	11. g	
6. h	12. a	

■ CHAPTER 18

CONCEPT CHECK 18A

1. oxygen
2. aluminum
3. (a) chlorine, (i) hydrosphere
   (b) sodium chloride, (i) hydrosphere
   (c) magnesium, (i) hydrosphere
   (d) sand, (ii) earth's crust
   (e) marble, (ii) earth's crust
4. magnesium
5. reduced

CONCEPT CHECK 18B

1. iron
2. carbon
3. carbon in the form called coke
4. (a) True
5. sulfur
6. (a) True

CONCEPT CHECK 18C

1. n- and p-
2. superconductor
3. increases
4. n-type
5. p-type
6. (b) transistors
7. the one whose resistance dropped to zero at the boiling point of nitrogen

CONCEPT CHECK 18D

1. oxygen and silicon
2. (a) SiO_2, (i) silica
   (b) quartz, (i) silica
   (c) contain metal ions, (ii) silicates
   (d) clay, (ii) silicates
   (e) built up of tetrahedral units, both (i) and (ii)
3. Silica is pure SiO_2 while glass is a mixture containing metal ions, oxide ions, and SiO_2
4. clay, sand, and aluminosilicates
5. temperatures, shock
6. cement, sand, gravel

THE LANGUAGE OF CHEMISTRY

1. d	6. a	11. b
2. j	7. e	12. c
3. n	8. o	13. k
4. h	9. i	14. f
5. g	10. m	15. l

■ CHAPTER 19

CONCEPT CHECK 19A

1. 6.5 billion, 9 billion
2. clays, silts, sandy soils, loams
3. (a) acidic   (b) basic
4. particle size and chemical composition
5. a trivalent ion like Fe^{3+} (hydrolyzes more than Na^+)
6. humus
7. nitrogen, phosphorus, potassium
8. calcium, magnesium, sulfur
9. oxygen

CONCEPT CHECK 19B

1. (b) False
2. nitrogen, phosphorus (calculated as P_2O_5), potassium (calculated as K_2O)
3. Yes
4. (c) a gas
5. about 33%
6. DDT
7. (a) DDT
8. pyrethrum
9. selective herbicide
10. 2,4-D
11. insecticides, herbicides, fungicides
12. CH_3Br

THE LANGUAGE OF CHEMISTRY

1. f	6. b	11. e
2. c	7. k	12. g
3. h	8. d	13. l
4. a	9. m	
5. j	10. i	

Appendix F: Answers to Try It

CHAPTER 2

2.1. **(a)** Pb **(b)** P **(c)** HCl **(d)** $AlBr_3$ **(e)** F_2

2.2. $2H_2(g) + O_2(g) \longrightarrow 2\,H_2O(\ell)$

2.3. **(a)** Hydrogen gas reacts with chlorine gas to form hydrogen chloride gas.
(b) *Reactants:* 2 H atoms in one H_2 molecule, 2 Cl atoms in one Cl_2 molecule; *products:* 2 HCl molecules, which contain 2 H atoms and 2 Cl atoms

2.4. *mega* = 1 million, 20 megabucks = 20 million dollars ($20,000,000)

CHAPTER 3

3.1. 28 protons, 28 electrons, and 31 neutrons

3.2. $^{107}_{47}Ag$, $^{109}_{47}Ag$

3.3. **(a)** Group 2A is the only main group that has all metals. Group 1A is often regarded as having only metals since hydrogen is a nonmetal. Figure 3.11 shows hydrogen separated from the rest of the members of Group 1A for this reason.
(b) Groups 7A and 8A have only nonmetals.
(c) Groups 3A, 4A, 5A, and 6A include metalloids, nonmetals, and metals. The number of valence electrons are: **(a)** Group 1A, 1; Group 2A, 2.
(b) Group 7A, 7; Group 8A, 8. **(c)** Group 3A, 3; Group 4A, 4; Group 5A, 5; Group 6A, 6.

3.4A. **(a)** Ba **(b)** Se **(c)** Si **(d)** Ga

3.4B. **(a)** Sr **(b)** Cl **(c)** Cs

CHAPTER 5

5.1. CaF_2

5.2. **(a)** CsI **(b)** $SrCl_2$ **(c)** BaS

5.3. **(a)** Rubidium chloride **(b)** Gallium oxide **(c)** Calcium bromide **(d)** Iron(II) nitride

5.4. **(a)** CoS **(b)** MgF_2 **(c)** KI

5.5. **(a)** $MgCO_3$ **(b)** NaH_2PO_4

5.6. **(a)** H:Ï: **(b)** :B̈r:B̈r: **(c)** H:C̈:H (with H above and H below)
(d) H:Ö:C̈l:

5.7. **(a)** H:C̈:C̈:H (with H H above and H H below)
(b) H:C̈:C̈:C̈:C̈:H (with H H H H above and H H H H below) and H:C̈:C̈:C̈:H (with H H H above and H H below, with H:C̈:H branch)

5.8. Tetraphosphorus trisulfide

CHAPTER 6

6.1. 300–400 ppm; 300,000–400,000 ppb

6.2. 2.5×10^{-5} m = 2.5×10^4 nm = 25,000 nm
2.5×10^{-6} m = 2.5×10^3 nm = 2,500 nm

The IR has longer wavelengths than the UV radiation. UV radiation is more energetic.

6.3. $(5/9)(59° - 32°) = 15°$; $(9/5)450° + 32° = 842°$

6.4. $15° - 273.15° = -258.15°$

6.5.
CO_2	370 ppm	370,000 ppb
CH_4	1.8 ppm	1,800 ppb
N_2O	0.31 ppm	310 ppb
O_3	0.04 ppm	40 ppb
CCl_3F	0.00026 ppm	0.26 ppb
CCl_2F_2	0.00052 ppm	0.52 ppb

6.6. Since carbon dioxide is only 27.3% carbon, we need to solve the following equation,

6 metric tons C = 0.273 (X metric tons CO_2)

Thus, 6/0.273 = 22 metric tons of CO_2

6.7. ~132 gallons

6.8. The high standard of living enjoyed by citizens of the United States has its price, and some feel this standard of living is achieved at the expense of less fortunate countries, which struggle to attain the same standard of living.

CHAPTER 7

7.1. Molecular oxygen absorbs high-energy ultraviolet radiation (UVC) and is converted to two oxygen atoms. An oxygen atom then combines with an oxygen molecule to produce an ozone molecule.

7.2. Oxygen absorbs high energy UVC (wavelengths less than 240 nm) radiation and ozone absorbs UVB (wavelengths less than 320 nm) radiation. The bond in molecular oxygen is stronger than the bond in ozone and requires more energetic radiation to break in the protection process.

7.3. The bathtub analogy uses the level of water in the tub to illustrate the ozone concentration in the stratosphere. Naturally occurring reactions in the atmosphere produce and consume ozone at equal rates to maintain a steady-state concentration, just as the rate of water added to a bathtub could be adjusted to exactly match the rate of water draining from the tub and maintain a constant level in the tub. Additional chemical reactions of anthropogenic origin that consume ozone would correspond to the addition of extra drains to the tub so that the level of water in the tub would drop.

7.4. The weather conditions, extreme cold and little wind, coupled with long periods of darkness during the Antarctic winter, are critical to the chemical reactions leading to the production of the reactive species involved in the destruction.

7.5. The C—Cl bond is the critical bond.

7.6. The CFCs are not soluble in or reactive with water.

7.7. Decreased crop yields. Increased skin cancer rates. Increased biological mutations in organisms.

7.8. Probably not. The increased time required to get a tan in a booth that uses longer-wavelength lamps appears to offset any benefit realized from using such lamps.

■ CHAPTER 8

8.1A. $2\,Al(s) + 3\,Cl_2(g) \longrightarrow 2\,AlCl_3(s)$

8.1B. (**a**) yes (**b**) no

8.2A. A single oxygen atom has a mass of 16 amu, so if $(16\,\text{amu/O atom})(1.661 \times 10^{-24}\,\text{g/amu}) = 2.658 \times 10^{-23}\,\text{g/O atom}$, then $(2.658 \times 10^{-23}\,\text{g/O atom})(6.023 \times 10^{23}\,\text{O atoms/mole}) = 16.0\,\text{g/mole O atoms}$

8.2B. A single xenon atom has a mass of 131.3 amu, so if $(131.3\,\text{amu/Xe atom})(1.661 \times 10^{-24}\,\text{g/amu}) = 2.181 \times 10^{-22}\,\text{g/Xe atom}$, then $(2.181 \times 10^{-22}\,\text{g/Xe atom})(6.023 \times 10^{23}\,\text{Xe atoms/mole}) = 131.3\,\text{g/mole Xe atoms}$

8.3A. Na = 22.99 g/mol
Cl = 35.45 g/mol
NaCl = 58.44 g/mol

8.3B. 6 mol of C = (6 mol)(12.01 g/mol) = 72.06 g
12 mol of H = (12 mol)(1.10 g/mol) = 12.12 g
6 mol of O = (6 mol)(16.00 g/mol) = 96.00 g
So $C_6H_{12}O_6$ = 180.18 g/mol

8.4. $50.\,\text{mol Ba(NO}_3)_2 \times \dfrac{261\,\text{g Ba(NO}_3)_2}{1\,\text{mol Ba(NO}_3)_2} = 13{,}000\,\text{g Ba(NO}_3)_2$

8.5.
$$\underset{\substack{1\,\text{mol CO}\\28\,\text{g CO}}}{CO(g)} + \underset{\substack{2\,\text{mol H}_2\\2 \times 2\,\text{g/mol} = 4\,\text{g H}_2}}{2\,H_2(g)} \longrightarrow \underset{\substack{1\,\text{mol CH}_3\text{OH}\\32\,\text{g CH}_3\text{OH}}}{CH_3OH(\ell)}$$

■ CHAPTER 9

9.1. $H_2SO_4 + 2\,H_2O \rightleftharpoons 2\,H_3O^+ + SO_4^{2-}$

9.2. $\dfrac{4.0\,\text{mol NaOH}}{1\,\text{L}} \times \dfrac{40.\,\text{g NaOH}}{1\,\text{mol NaOH}} = \dfrac{160\,\text{g NaOH}}{1\,\text{L}}$

9.3. (73 g)/(36.5g/mol) = 2.0 mol
and
(2.0 mol)/(0.5 L) = 4 mol/L = 4 M

9.4. (20 g)/(40 g/mol) = 0.50 mol
and
(0.50 mol)/(2 L) = 0.25 mol/L = 0.25 M

9.5. The only way to do this would be to evaporate some of the solvent as a way of increasing the solute concentration. This is not practical in most cases.

9.6. For $[H_3O^+] = 1 \times 10^{-4}$ M, pH = 4. A pH of 4 is less acidic than a pH of 3; the tomatoes are less acidic than the cola.

9.7. 3×10^{-6} M is between 1×10^{-6} M (pH = 6) and 10×10^{-6} M (pH = 5). The exact pH is 5.523 since pH = $-\log (3 \times 10^{-6}) = -\log (3) -\log (10^{-6}) = -0.4471 + 6 = 5.5229$.

9.8. Since the product of the hydroxide ion concentration and the hydronium ion concentration is always 10^{-14}, we can calculate that the hydronium ion concentration is $(1 \times 10^{-14})/(2 \times 10^{-6}) = 5 \times 10^{-9}$ and, by bracketing, the pH must be between 8 and 9. The exact pH is 8.301.

9.9. The graph would be logarithmic in nature and would curve upward in a nonlinear fashion.

9.10. If the hydronium ion concentration changes by a factor of 4 (i.e., from 1×10^{-9} to 4×10^{-9}), the pH only changes by 0.6 pH units. An inspection shows that a tenfold change in hydronium ion concentration causes a 1-unit change in pH so an increase of 3 pH units would require a thousandfold change in hydronium ion concentration. To decrease by 5 pH units, the hydronium ion concentration would have to decrease by 100,000 (a factor of 10^5), which is to say 100,000 fold.

9.11. If pH is 10.7, this means that $-\log (H_3O^+) = 10.7$. Manipulating this, we see that
$\log (H_3O^+) = -10.7$
(H_3O^+) = antilog (-10.7)
$(H_3O^+) = 2 \times 10^{-11}$

■ CHAPTER 10

10.1. Li is oxidized (converted to Li^+); O_2 is reduced (converted to O^{2-})

10.2. (**a**) Cu is oxidized (addition of oxygen)
(**b**) $CH_3C \equiv N$ is reduced (addition of hydrogen)
(**c**) SnO is reduced (removal of oxygen)

■ CHAPTER 11

11.1. **a.** (3 g)(80 cal/g) + (3 g)(1 cal/g-°C)(70°C) = 550 cal; endothermic
b. (6 g)(1 cal/g-°C)(50°C) + (6 g)(540 cal/g) = 3540 cal; endothermic
c. (10 g)(−540 cal/g) + (10 g)(1 cal/g-°C)(−100°C) = −6400 cal; exothermic

■ CHAPTER 12

12.1. **a.** The balanced equation for the combustion of ethane is
$$2\,CH_3CH_3 + 7\,O_2 \longrightarrow 4\,CO_2 + 6\,H_2O$$
This requires the input of heat required to break the following bonds:
12 mol of C—H bonds = (12 mol)(416 kJ/mol) = 4992 kJ
2 mol of C—C bonds = (2 mol)(356 kJ/mol) = 712 kJ
7 mol of O=O bonds = (7 mol)(498 kJ/mol) = 3486 kJ
for an endothermic total of 9190 kJ

Formation of the products releases the heat associated with formation of the following bonds:

8 mol of C=O bonds = (8 mol)(−803 kJ/mol) = −6424 kJ

12 mol of O—H bonds = (12 mol)(−467 kJ/mol) = −5604 kJ

for an exothermic total of −12,028 kJ

The sum of these two is −2838 kJ for 2 moles of ethane. That is equal to −1419 kJ/mol. Since the molecular weight of ethane is equal to 30 g/mol, that equates to −47.3 kJ/g.

b. The balanced equation for the combustion of ethanol is

$$CH_3CH_2OH + 3\ O_2 \longrightarrow 2\ CO_2 + 3\ H_2O$$

This requires the input of heat required to break the following bonds:

5 mol of C—H bonds = (5 mol)(416 kJ/mol) = 2080 kJ

1 mol of C—C bonds = (1 mol)(356 kJ/mol) = 356 kJ

1 mol of C—O bonds = (1 mol)(336 kJ/mol) = 336 kJ

1 mol of O—H bonds = (1 mol)(467 kJ/mol) = 467 kJ

3 mol of O=O bonds = (3 mol)(498 kJ/mol) = 1494 kj

for an endothermic total of 4733 kJ

Formation of the products releases the heat associated with formation of the following bonds:

4 mol of C=O bonds = (4 mol)(−803 kJ/mol) = −3212 kJ

6 mol of O—H bonds = (6 mol)(−467 kJ/mol) = −2802 kJ

for an exothermic total of −6014 kJ

The sum of these two is −1281 kJ for 1 mole of ethanol. Since the molecular weight of ethanol is equal to 46 g/mol, that equates to −27.8 kJ/g.

The introduction of oxygen into a hydrocarbon reduces the amount of heat released upon combustion.

12.2. (a) 2-methylpentane (b) 3-methylpentane

(c) 2,2-dimethylbutane (d) 2,3-dimethylbutane

12.3. 3-methyl-1-butene

CHAPTER 13

13.1. $^{237}_{93}Np \longrightarrow\ ^{4}_{2}He + ^{233}_{91}Pa$

13.2. $^{234}_{91}Pa \longrightarrow\ ^{0}_{-1}e + ^{234}_{92}U$

CHAPTER 14

14.1.   (a) (b)

(c)

14.2.

CHAPTER 15

15.1.   (b) is chiral.

(d) is chiral.

15.2.

15.3.   AGGCTA

CHAPTER 16

16.1.   160. lb × 10 kcal/lb = 1600 kcal
1600 kcal × 1.3 = 2080 kcal

16.2.   Healthy Chicken:

$$15\ \text{Cal fat} \times \frac{1\ g}{9\ \text{Cal}} = 2\ g\ \text{fat}$$

$$\frac{2\ g\ \text{fat}}{65\ g\ \text{fat}} \times 100\% = 3\%\ \text{of daily fat allowance}$$

Cream of Mushroom:

$$80\ \text{Cal fat} \times \frac{1\ g}{9\ \text{Cal}} = 9\ g\ \text{fat}$$

$$\frac{9\ g\ \text{fat}}{65\ g\ \text{fat}} \times 100\% = 14\%\ \text{of daily fat allowance}$$

16.3.
$$3 \text{ g protein} \times \frac{4 \text{ kcal}}{\text{g protein}} = 12 \text{ kcal}$$

$$8 \text{ g fat} \times \frac{9 \text{ kcal}}{\text{g fat}} = 72 \text{ kcal}$$

$$40 \text{ g carbohydrate} \times \frac{4 \text{ kcal}}{\text{g carbohydrate}} = 160 \text{ kcal}$$

$$\text{Total} = 244 \text{ kcal per slice}$$

■ CHAPTER 19

19.1. Formula weight: 2 N × 14 = 28, 4 H × 1 = 4, 3 O × 16 = 48. Total = 80.

$$\text{Percentage N} = \frac{28}{80} \times 100\% = 35\%$$

There is no K or P in ammonium nitrate, so their values are zero.

Appendix G: Answers to Applying Your Knowledge

CHAPTER 1

1. zoology
3. amount of exposure
5. biochemistry or genetics
7. Air conditioning certainly contributed to the growth of these cities in a very hot climate. This growth has led to increased pollution, much of it due to greater automobile traffic in these sprawling cities. Overall, most would say the benefits have outweighed the negative effects.
9. Individual answers may vary but should be based on facts and not on ignorance, unsubstantiated assertions, or fear of chemicals.

CHAPTER 2

1. Answer will depend on each person's experience.
3. No, they would be the same substance.
5. manufacturing nitroglycerin
7. (a) physical property   (b) physical property
   (c) chemical property   (d) physical property
   (e) chemical property   (f) chemical property
9. (a) element, contains only Hg atoms
   (b) mixture of water, minerals, proteins, fats
   (c) compound, contains only one kind of molecule (H_2O)
   (d) mixture of cellulose, water. Wood changes weight when dried.
   (e) mixture of dye and solvent
   (f) mixture of water, caffeine, tea extract
   (g) compound, solid pure water containing only one kind of molecule
   (h) element, contains only C atoms
   (i) element, contains only Sb atoms
11. Properties of iron do not change because all particles in iron are atoms of iron. Steel is a mixture of iron and other atoms. The type of steel depends on what is added to the iron.
13. True
15. True

17.

	MAJOR SOURCE	COMPOUND
nitrogen	air	ammonia, NH_3
sulfur	underground deposits	sulfuric acid, H_2SO_4
chlorine	sea water	sodium chloride, NaCl
magnesium	sea water	Milk of Magnesia, $Mg(OH)_2$
cobalt	mineral deposits	cyanocobalamin, Vitamin B_{12}

19. Cytoxan, $C_7H_{15}O_2N_2PCl_2$
    (a) twenty-nine, 29 atoms total
    (b) carbon, hydrogen, oxygen, nitrogen, phosphorus, chlorine
    (c) 15 hydrogens/2 nitrogen
    (d) yes, it is organic

21.

	SOLID	LIQUID	GAS
pure substances	dry ice, CO_2 iron, Fe copper, Cu	mercury, Hg octane, C_8H_{18}	helium, He nitrogen, N_2 methane, CH_4
mixtures	butter steel 14 K "gold"	sea water homogenized milk coffee	natural gas fuel (hydrocarbons mixed with mercaptans) a person's exhaled breath (CO_2, O_2, H_2O)

23. Yes, a mixture of H_2 and O_2 can exist at room temperature. This mixture will be stable as long as no spark or activation energy is added. A reaction produces water, H_2O, which contains both elements.
25. (d) reactant; products

27. **(a)** Two sodium atoms react with 1 chlorine molecule to form two formula units of sodium chloride solid.
   (b) One nitrogen molecule reacts with three chlorine molecules to produce two molecules of nitrogen trichloride.
   (c) One molecule of carbon dioxide reacts with one molecule of water to produce one molecule of carbonic acid.
   (d) Two molecules of hydrogen peroxide react to produce one molecule of oxygen gas and two molecules of water liquid.

29. **(b)** $N_{2(g)} + 3\,Cl_{2(g)} \longrightarrow 2\,NCl_{3(g)}$
   $\underbrace{\phantom{N_{2(g)} + 3\,Cl_{2(g)}}}_{\text{reactants}} \quad \underbrace{\phantom{2\,NCl_{3(g)}}}_{\text{products}}$
   (d) $2H_2O_2(aq) \longrightarrow 2\,H_2O(\ell) + O_{2(g)}$
   $\underbrace{}_{\text{reactants}} \quad \underbrace{\phantom{2\,H_2O(\ell) + O_{2(g)}}}_{\text{products}}$

31. The tea in tea bags is a mixture. It can be partially separated by dissolving some water-soluble substances with hot water. Instant tea is a mixture of the water-soluble substances in tea.

33. **(a)** 1 gram = 1000 milligrams
   (b) 1 kilometer = 1000 meters
   (c) 1 gram = 100 centigrams

35. **(a)** 9 cal/gram; no.
   (b) 100 cm/meter; no.
   (c) 1.5 g/mL; yes, grams/milliliter is mass/volume.
   (d) 454 g/lb.; no.

37. 5.5 acres/55 cows

39. 10,000 meters; ? meters $= 10\ km \times \dfrac{1000\ m}{1\ km} =$ 10,000 meters

41. **(a)** 0.04 m   **(b)** 43 mg
   (c) 15500 mm   **(d)** 0.328 L
   (e) 980 g

43. $70\ kg \times \dfrac{1000\ g}{1\ kg} \times \dfrac{1000\ mg}{1\ g} = 70{,}000{,}000\ mg$

45. **(a)** 8.0×10^7 **(b)** 3.0×10^5
   (c) 1.6×10^{-13} **(d)** 9.7×10^1

47. **(a)** 450,000,000 watts   **(b)** 4,500,000 bulbs

49. 22,420 g

51. 0.34   The mass of the Al object is 0.34 as much as the Fe object.

■ CHAPTER 3

1. Matter is neither created nor destroyed. Examples: flash bulb before and after use, hard boiling an egg, yarn knitted into clothing, melting ice cubes in a glass. There is no change in mass during the chemical changes (first two) or the physical changes.

3. John Dalton

5. Experiments indicated that matter was conserved. Elements had been identified. Compounds had been shown to be composed of definite amounts of specific elements. The composition of a compound had been shown to always be the same regardless of its source. The composition of a combination of elements could be predicted.

7. **(b)** Law of Definite Proportions

9. **(a)** The number of protons in the nucleus of an atom.
   (b) The total count of protons *plus* neutrons in the nucleus of an atom.

(c) The mass of an average atom of an element compared to an atom of ^{12}C which is assigned a mass of exactly 12 atomic mass units.
(d) Atoms with the same number of protons but with different numbers of neutrons.
(e) On Earth, the % that this isotope is of all the atoms of that element
(f) 1 amu is one-twelfth of the mass of the carbon-12 atom.

11. See Tables 3.2 and 3.3.

13. Atoms of the same element all have the same atomic number. This means they have the same number of protons. For example these two atoms both have 10 protons $^{21}_{10}Ne$, $^{22}_{10}Ne$.

15. Lithium has 3 protons; the mass number 7 means that the sum of the numbers of protons and neutrons = 7, so it has 4 neutrons. The number of neutrons is determined by subtracting the atomic number from the mass number.
   A = mass number; Z = atomic number
   A − Z = number of neutrons; 7 − 3 = 4

17. **(e)** $_{34}Se$

19. The atomic number has the symbol Z. The number of protons equals the atomic number.
   (a) Germanium, Z = 32 or 32 protons
   (b) Silicon, Z = 14 or 14 protons
   (c) Nickel, Z = 28 or 28 protons
   (d) Cadmium, Z = 48 or 48 protons
   (e) Iridium, Z = 77 or 77 protons

21. Yes, the statement is correct. The atomic mass of 24.305 is a weighted average based on the masses of individual Mg isotopes and their abundances.

23. $v = 4 \times 10^{14}$ cycles/s

25. **(d)** energy increases

27. **(c)** IR radiation has a lower frequency than UV radiation.

29.

Bohr arrangement	Number of electrons	Element
2-4	6	Carbon
2-8	10	Neon
2-8-3	13	Aluminum
2-8-8-2	20	Calcium

31. Na, 1;   Mg, 2;   Al, 3;   Si, 4;   P, 5;   S, 6;   Cl, 7;   Ar, 8

33. When elements are arranged in order of their atomic numbers, their chemical and physical properties show repeatable trends.

35. The known elements were arranged in order and gaps between elements occurred. These gaps suggested the existence of elements not yet isolated.

37. Metals are good conductors of heat and electricity. Metals are malleable and ductile. Nonmetals are the opposite, they are not ductile, not malleable and are poor conductors of heat and electricity.

39. **(a)** There are 7.
   (b) There are 8 representative groups.
   (c) There are two representative groups that are all metals if you ignore H in 1A.
   (d) The elements in Group 7A and 8A are all nonmetals.
   (e) Yes, all the elements in period 7 are metals.

41. 10^3 or 100,000
43. (**a**) B, Al, Ga, In, Tl each have 3 valence electrons
    (**b**) C, Si, Ge, Sn, Pb all have 4 valence electrons
    (**c**) F, Cl, Br, I, At all have 7 valence electrons
    (**d**) H, Li, Na, K, Rb, Cs, Fr all have 1 valence electron
45. (**a**) Be is more metallic than B
    (**b**) Ca is more metallic than Be
    (**c**) K is more metallic than Ca
    (**d**) Ge is more metallic than As
    (**e**) Bi is more metallic than As
47. a completed outer shell or set of eight electrons in the outermost energy level

49.

ATOMIC NUMBER	ELEMENT NAME	NUMBER OF VALENCE ELECTRONS	PERIOD	METAL, M OR NONMETAL, NM
6	carbon, C	4	2	NM
12	magnesium, Mg	2	3	M
17	chlorine, Cl	7	3	NM
37	rubidium, Rb	1	5	M
42	molybdenum, Mo	6	5	M
54	xenon, Xe	8 or 0	5	NM

51. The outer or last electron in Cs is in a higher energy level, $n = 6$. Li has outer electron in $n = 2$. The higher the level the greater the average distance from the nucleus.
53. The smaller the radius of the nonmetal atom, the more reactive it is.
55. (**a**) 700 nm (**b**) 500 nm (**c**) 570 nm (**d**) 450 nm
57. K (largest), Al, P, S, Cl (smallest)
59. If element 36 is a noble gas, then element 35 is a halogen and element 37 is an alkali metal.
61. 0.017 seconds
63. (**a**) pass right though the foil or be deflected very little because atoms of Au are mostly empty space.

■ CHAPTER 4

1. (**a**) air that contains unwanted and harmful substances
    (**b**) mixtures containing water droplets and/or tiny particulates suspended in a gas
    (**c**) smog-containing compounds formed in reactions initiated by sunlight
    (**d**) substances formed by reactions of emitted compounds with components in the air
3. nitric oxide from internal combustion engines; sulfur oxides from coal combustion; carbon monoxide from inefficient burning of hydrocarbons
5. The abbreviation CAA refers to the Clean Air Act. The first act was passed in 1970 and originally controlled air pollution from cars and industry. Amendments in 1977 imposed stricter auto emission standards. The 1990 CAA extends to manufacturing and commercial activity. This version of the CAA regulates particulates, ozone, carbon monoxide, oxides of nitrogen and sulfur, carbon dioxide, and substances that would deplete stratospheric ozone. A newer version was passed in 1996.

7. Industrial smog is chemically reducing while photochemical smog is chemically oxidizing. Industrial smog contains sulfur dioxide mixed with soot, fly ash, smoke, and partially oxidized organic compounds. Photochemical smog is essentially free of SO_2, but contains ozone, ozonated hydrocarbons, organic peroxide compounds, nitrogen oxides, and unreacted hydrocarbons.
9. b, c, a, and d in order
11. Of the nitrogen oxides (NO_x) in the atmosphere, 97% come from natural sources. Lightning strikes during electrical storms produce NO. Nitric oxide, NO, is so reactive that it combines with O_2 in the atmosphere to produce NO_2. Some bacteria produce N_2O. About 3% of the atmospheric nitrogen oxides come from human activity such as combustion in automobile engines.
13. Hydrocarbons are released into the atmosphere by natural sources and human sources. Hydrocarbons are put into the atmosphere by living plants such as deciduous trees, by the decay of dead plants and animals, and by excrement from insects and animals. Human activity introduces hydrocarbons when organic solvents are used, for example in the handling of petroleum products, etc. Generally only human activities are within our control.
15. Nitrogen dioxide plays a role in the formation of ozone in the troposphere.
17. Nitrogen dioxide (NO_2) will dissociate to form an oxygen atom and nitric oxide, if it is excited by a sufficiently energetic photon of UV light. The O atom forms O_3 by reaction with O_2.

$$NO_2(g) + h\nu \longrightarrow NO(g) + O(g)$$
$$O + O_2 \longrightarrow O_3$$

19. All three are oxidizing agents. Ozone, sulfur dioxide and nitrogen dioxide cause lung damage.
21. Air is precooled below $0°C = 273°K$ to remove water vapor as ice. The temperature is decreased to less than $-78°C$ or 194°K to remove CO_2. The dry air, free of CO_2, is compressed to more than 100 atmospheres. This compression heats the air because energy is added to compress it. The air is cooled to room temperature. The room temperature air is allowed to expand and cool. This compression expansion cycle is repeated until all of the air is liquefied. The liquid air is allowed to warm and each gas in the mixture can be collected when it "boils" off at its own boiling point.
23. lung diseases, cancer and mutagenic effects
25. Oil-burning electric fuel plants generate SO_2. The amount of SO_2 emitted can be reduced by either using low sulfur fuel oil or by passing plant exhaust gas through molten sodium carbonate to form sodium sulfite.
27. (**c**) 400 ppm is equal to 0.04%

■ CHAPTER 5

1. (**a**) A cation is an ion with a positive charge.
    (**b**) An anion is an ion with a negative charge.
    (**c**) Atoms react to acquire an electron configuration with 8 electrons in the outermost shell to match the configuration of the nearest noble gas.
    (**d**) The formula unit is the simplest element ratio for an ionic compound like NaCl instead of Na_2Cl_2.

3. **(a)** a pair of electrons shared between two atoms
   (b) four electrons (2 pairs) shared by two atoms
   (c) six electrons (3 pairs) shared between two atoms
   (d) pair of valence electrons on an atom that are not shared with another atom.
   (e) a pair of electrons shared between two atoms
   (f) either a double or a triple bond.

5. **(a)** bond between two atoms that share electrons equally.
   (b) bond between two atoms that do not attract shared electrons equally.

7. **(a)** A hydrocarbon is a compound consisting only of carbon and hydrogen.
   (b) a compound consisting only of carbon and hydrogen with only single bonds
   (c) a compound consisting only of carbon and hydrogen with one or more multiple bonds between carbon atoms
   (d) Alkenes are hydrocarbons with one or more carbon–carbon double bonds.

9. **(a)** Br^{1-}
   (b) Al^{3+}
   (c) Na^{1+}
   (d) Ba^{2+}
   (e) Ca^{2+}
   (f) Ga^{3+}
   (g) I^{1-}
   (h) S^{2-}
   (i) Group 1A atoms lose one valence electron to form a +1 ion, see 9c above.
   (j) Group 7A atoms have seven valence electrons and gain one to form a 1− ion, see 9g.

11. **(a)** AlI_3 aluminum iodide
    (b) $SrCl_2$ strontium chloride
    (c) Ca_3N_2 calcium nitride
    (d) K_2S potassium sulfide
    (e) Al_2S_3 aluminum sulfide
    (f) Li_3N lithium nitride

13. **(a)** calcium sulfate
    (b) sodium phosphate
    (c) sodium bicarbonate
    (d) potassium hydrogen phosphate
    (e) sodium nitrite
    (f) copper(II) nitrate

15. **(a)** Lithium is in Group 1A so it forms the Li^{1+} ion. Tellurium is in Group 6A and forms the Te^{2-} ion. The formula for the neutral ionic combination is Li_2Te.

$$\widehat{Li^{1+} \quad Te^{2-}}$$

   (b) $MgBr_2$
   (c) Ga_2S_3

17. **(a)** ionic
    (b) covalent
    (c) ionic
    (d) covalent
    (e) covalent

19. **(a)** NO, nitrogen monoxide
    (b) SO_3, sulfur trioxide
    (c) N_2O, dinitrogen oxide
    (d) NO_2, nitrogen dioxide

21. Ionic bonding exists between oppositely-charged ions. The polar covalent bond exists between atoms that share electrons unequally. The nonpolar bonding exists between atoms that share electrons equally (atoms with similar attractions for electrons).

23. **(a)** The electron attracting power differences determine the polarity of the bonds. The farther apart two elements are in the periodic table the more polar the bond. The C—Cl bond is more polar because they are farther apart in the periodic table.
    (b) The C—F bond is more polar than the C—Cl bond.

25. eleven valence electrons

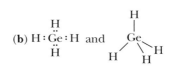

27. **(a)** $:C:::O:$ Polar; O is the negative end and C is the positive end. The arrow indicates the direction electrons shift in molecule.

    (b) H:Ge:H and [tetrahedral Ge structure]   Nonpolar; the bonds are nonpolar. The molecule is tetrahedral and symmetric.

    (c) [BCl₃ structure]   Nonpolar; the bonds are polar but the planar triangular shape puts the center of positive charge and the center of negative charge in the middle of the boron atom.

    (d) $H:\ddot{\underset{..}{F}}:$ Polar; F is the negative end and H is the positive end.

29. **(a)** The gaseous state has no definite shape or volume; a gas takes the shape and volume of its container. Gases are compressible.
    (b) A liquid has definite volume and assumes the shape of the container. The liquid molecules are free to move past each other, allowing the liquid to flow and assume the shape of its container. Liquids are relatively incompressible.
    (c) The solid state has definite shape and definite volume. The molecules or ions making up the solid are in direct contact and located at fixed positions. The solid is incompressible due to direct contact between the particles.

31. **(b)** solid

33. The number of gas molecules in a given volume of gas decreases as you move away from the earth's surface. The fewer number of molecules collide with any object less often and exert less push or pressure. Additionally, the force of Earth's gravity decreases with altitude. This results in less weight or force per unit area at greater altitudes.

35. Gas molecules are constantly moving. The gas molecules of the perfume vapor diffuse away from the person wearing the perfume.

37. Water has an unusually high normal boiling point and surface tension. Additionally it expands on freezing.

39. Evaporation of water from one's skin draws energy from the skin. This produces a resulting "chilling" sensation. We are cooled when the energy flows into the evaporating liquid.

41. Without hydrogen bonding the boiling point for ammonia would be $-100°C$. See Figure 5.9.

43. The increased pressure creates more collisions of gas molecules with the solvent surface. These more frequent collisions create more opportunities for the gas to dissolve in the solvent. Champagne, sparkling waters, Coca-Cola, and Pepsi are examples.

45. three times as much

47. solid because temperature is lower than mp

49. below 646°C at approximately 500°C.

CHAPTER 6

1. (a) The greenhouse effect is the warming of Earth by the trapping of IR radiation seeking to escape Earth by certain so-called greenhouse gases such as carbon dioxide. This keeps Earth warmer than it otherwise would be and allows life to flourish.
   (b) Global warming is the term used to describe warming of Earth above and beyond what is normal and desirable.
   (c) *Anthropogenic* means to originate from human activity.
   (d) CFC is the abbreviation for chlorofluorocarbons, a class of compounds implicated in the destruction of the ozone layer and also contributing to global warming.
   (e) Kyoto Protocol was a global warming policy developed in Kyoto, Japan in 1997 as a first response to this environmental problem.
   (f) Infrared radiation is the type of electromagnetic radiation that is sufficiently energetic to stretch and bend chemical bonds.

3. True
5. (b)
7. False
9. False
11. True
13. False
15. False
17. (d)
19. False
21. (e)
23. Each student's answer may vary here depending on his or her social and political views.
25. Each student's answer may vary here depending on his or her social and political views.
27. Each student's answer may vary here depending on his or her social and political views.
29. 2.5×10^{-3} cm = 2.5×10^{-5} m. This allows us to calculate a frequency of 1.2×10^{13} s^{-1}. This corresponds to an E$_{photon}$ = 1.9×10^{-21} cal.
   2.5×10^{-4} cm = 2.5×10^{-6} m. This corresponds to a frequency of 1.2×10^{14} s^{-1} and an E$_{photon}$ = 1.9×10^{-20} cal.

CHAPTER 7

1. (a) an area of decreased stratospheric ozone largely centered above the Antarctic.
   (b) chlorofluorocarbons
   (c) hydrochlorofluorocarbons
   (d) a first attempt at an agreement to regulate the production and use of ozone-destroying chemicals.
   (e) two or more Lewis dot structures that together represent the actual bonding in a real molecule.
   (f) ultraviolet radiation with a wavelength between 320–400 nm
   (g) ultraviolet radiation with a wavelength between 280–320 nm
   (h) ultraviolet radiation with a wavelength between 200–280 nm
   (i) a set of chemical reactions that explains how oxygen and ozone protect us from UVB and UVC.
   (j) hydrofluorocarbons
   (k) the region (from ;10–50 km) of the atmosphere in which the ozone layer lies.
   (l) a state in which the concentration of a chemical remains essentially constant even though the chemical may be undergoing numerous chemical reactions.
   (m) a measure of ozone concentration in the atmosphere equal to 1 ppb of ozone.

3. (a) true (b) false (c) true (d) true (e) true
   (f) false (g) false (h) true (i) false (j) false

5. Once in the stratosphere, CFCs react with sunlight to release chlorine atoms that react with ozone to produce chlorine monoxide and oxygen. The C—Cl bond is the critical bond.

7. About the thickness (3 mm) of two nickels stacked on top of each other.

9. UVB (<320 nm)

11. Because the chemical bonds in oxygen are stronger than the chemical bonds in ozone, oxygen absorbs higher-energy UV radiation than does ozone.

13. CH_3Br

15. CCl_2F_2 and CCl_3F

17. Weather conditions and sunlight are critical for many of the chemical reactions leading to ozone destruction.

19. Any molecule that is capable of absorbing IR radiation may act as a greenhouse gas in the atmosphere.

21. (b) oxygen

23. (d)

25. (b)

27. UVC

29. Ground-level ozone is very reactive toward many molecules it encounters and is destroyed before rising into the stratosphere.

31.

	Global Warming	Ozone Hole
(a)	troposphere	stratosphere
(b)	IR	UV
(c)	CO_2, H_2O, CH_4, etc.	CFCs primarily
(d)	Kyoto Protocol	Montreal Protocol
(e)	Mixed acceptance	Widely accepted
(f)	Little impact at present	Major impact in reducing problem

CHAPTER 8

1. (a) 6 and 6
   (b) 7 and 7
   (c) 1, 3, 2, 3
   (d) The number of atoms of each element is the same on both reactant and product side.
3. (a) $2\,Al + 3\,Cl_2 \longrightarrow 2\,AlCl_3$
   (b) $3\,Mg + N_2 \longrightarrow Mg_3N_2$
   (c) $2\,NO + O_2 \longrightarrow 2\,NO_2$
   (d) $2\,SO_2 + O_2 \longrightarrow 2\,SO_3$
   (e) $3\,H_2 + N_2 \longrightarrow 2\,NH_3$
5. (a) $Ba(s) + 2\,H_2O(\ell) \longrightarrow Ba(OH)_2(aq) + H_2(g)$
   (b) $3\,Fe(s) + 4\,H_2O(\ell) \longrightarrow Fe_3O_4(s) + 4\,H_2(g)$
   (c) $2\,Na(s) + 2\,H_2O(\ell) \longrightarrow 2\,NaOH(aq) + H_2(g)$
   (d) $2\,Li(s) + 2\,H_2O(\ell) \longrightarrow 2\,LiOH(aq) + H_2(g)$
7. (a) 10 molecules H_2 (b) 5 molecules N_2
   (c) 200 molecules NH_3
9. The carbon-12 isotope with a *defined* mass of *exactly* 12 amu is the basis.
11. Each is 0.500 mol of that element.
13. Avogadro's number is 6.022×10^{23} to 4 significant digits or 6.02×10^{23} to 3 s.d.
15. The mole is defined as the number of atoms in exactly 12 g of carbon-12, this equals roughly 6.02×10^{23} atoms.
17. Some reactions are fast because reactants have weak bonds. These reactions have low activation energies; the reactants have a "low hill to climb" to form products. Other reactions are slow because reactants have strong bonds and a high activation energy; these reactants have a high "energy hill to climb" to form products.
19. Slows reaction rates. Freezing slows reaction rates, slowing spoilage and slowing growth of mold or bacteria.
21. Reactions with high or large activation energies are slower than reactions with low or small activation energies.
23. The activation energy is high so there is no reaction at room temperature. Once started by the spark (source of activation energy), the reaction continues with production of energy because the products have lower potential energy than the reactants.
25. Faster because contact with oxygen molecules would be better and more frequent.
27. Reversible reactions occur in both forward and reverse directions.
29. (a) shift to form $CaCO_3$
   (b) shift to form CaO and CO_2
31. Reactants are consumed to make more products until a new balance between reactants and products is reached.
33. Unreacted reactants are essentially zero; all are converted to products.
35. (a) stored energy
   (b) a process that gives off energy
   (c) a process that consumes energy
   (d) a measure of disorder
   (e) a process that favors products at the expense of reactants
37. Every spontaneous process occurs with an increase in entropy for the universe. This entropy increase consumes energy that cannot be used for other purposes. Every spontaneous process has a built-in amount of this "wasted" energy.
39. (a) 18 g H_2O (b) 254 g I_2 (c) 56 g KOH
   (d) 17 g NH_3
41. (c) 12 mol H_2 = 24 g
43. (a) 500 moles O_2 (b) 16000 g O_2
   (c) 320 pounds O_2
45. (a) 0.5 mol copper   (b) 1.5 mol $CuCl_2$

CHAPTER 9

1. (a) Excess hydronium ion, $[H_3O^{1+}] > [OH^{1-}]$, pH < 7
   (b) Excess hydroxide ion, $[H_3O^{1+}] < [OH^{1-}]$, pH > 7
   (c) Equal amounts of hydronium and hydroxide, pH = 7 and $[H_3O^{1+}] = [OH^{1-}]$
3. A salt is the substance formed between the anion of an acid and the cation of a base.
5. (a) measures the fraction of molecules that ionize
   (b) ionizes 100%
   (c) ionizes 100%
   (d) only partially ionized, majority of acid molecules are still intact
   (e) partially ionized, the majority of base molecules or ions are still intact
7. pH above 7
9. (a) acidic   (b) acidic   (c) basic
   (d) basic   (e) acidic   (f) basic
11. (a) Molarity is a concentration measure equal to the moles of solute dissolved in one liter of solution.
   (b) Concentration is a measure of the relative amount of solute in a specific amount of solution.
13. An acid buffer is an aqueous mixture of both a weak acid and its anion that will maintain a stable pH when either acid or base is added. A basic buffer is a mixture of a weak base and its cation that will maintain a stable pH.
15. (a) $CH_3COOH(aq) + KOH(aq) \longrightarrow H_2O(\ell) + CH_3COOK(aq)$
   (b) $H_2SO_4(aq) + Ca(OH)_2(aq) \longrightarrow CaSO_4(aq) + 2\,H_2O(\ell)$
   (c) $H_2SO_4(aq) + 2\,NaOH(aq) \longrightarrow Na_2SO_4(aq) + 2\,H_2O(\ell)$
17. $2\,HCl(aq) + CaCO_3(s) \longrightarrow CaCl_2(s) + CO_2(g) + H_2O(\ell)$
19. black coffee with the pH = 5.0 The lower the pH the more acidic the solution.
21. lemon, pH = 2.3
23. (a) Ammonium ion reacts with base and ammonia reacts with acid.
   (b) Hydrogen fluoride reacts with base and sodium fluoride reacts with acid.
25. B with pH = 1.1 will be the stronger acid.
27. 240 g NaOH
29. 0.0200 mol HCl
31. (a) 0.137 M   (b) 1.00 M NaOH
33. 9.8 g H_2SO_4
35. (a) pH = 11   (b) pH = 3
37. 75 mL
39. No. Mixture is basic; 0.0035 mols of KOH is more than 0.0025 mols of HCl.

CHAPTER 10

1. (a) Oxidation is the loss of electrons
    (b) Reduction is the gain of electrons
    (c) Oxidation is the gain of oxygen
    (d) Reduction is the gain of hydrogen
    (e) Oxidizing agent is a substance that accepts electrons
    (f) Reducing agent is a substance that gives up electrons
3. (a) CO_2 (b) NO_2 (c) SO_3 (d) N_2
5. water, H_2O
7. carbon monoxide, CO
9. (a) oxidized   (b) oxidized
    (c) reduced   (d) reduced
    (e) reduced
11. reducing because there is a ready supply of hydrogen to act as a reducing agent
13. Zinc is oxidized.
15. Rusting is less of a problem in Arizona because there is less water in the air.
17. (a) True
19. Both produce electrical energy. The reactants are added to a fuel cell and products are removed. A battery is usually sealed or self-contained.
21. See answer to question 19. The reactants and products are still inside the battery. No material was added or removed.
23. An advantage is that emissions from the vehicle will be lower. Disadvantage is that fossil fuels used to generate electricity to recharge the car often produce atmospheric pollutants.
25. Six 2-volt cells are linked together.
27. These react with the oxygen in any air that touches the food so this oxygen cannot react with substances in the food to cause spoiling.
29. NO_2 has an odd number (17) of electrons so one of its electrons is unpaired.
31. The fuel cell is a reaction vessel. Reactants are converted to products, but total mass stays the same.
33. The concentration of Cu^{2+} ions does not increase.
35. Oxide of iron is a loose layer so corrosion of Fe continues. Al_2O_3 bonds tightly to Al, protecting Al from additional oxidation.
37. C oxidized; Al^{3+} reduced

CHAPTER 11

1. (a) water available in rivers, lakes, and streams.
    (b) water beneath the earth's surface.
    (c) fresh water contaminated by salty seawater.
    (d) any condition that causes the natural usefulness of water, air, or soil to be diminished.
    (e) pumping water into aquifers to maintain water volume
    (f) drinkable water; water that is safe to drink
3. Most of the rainwater that falls on the United States each day returns to the atmosphere by evaporation or transpiration from plants.
5. Groundwater can be contaminated with pollutants when rainwater runs over or filters through materials, dissolves the pollutants, then percolates into the ground water.
7. Agriculture is the largest single user of water in the United States.

9. Water use per day differs from person to person.
11. Groundwater recharge describes the process of pumping water into aquifers to maintain water volume. Purified recycled water from sewage effluent is used as recharge water.
13. The user was responsible for water quality prior to the 1977 Clean Water Act. The discharger is now responsible for maintaining water quality.
15. Two methods for disposal of solid wastes from industry and households are landfills and incineration. Landfills have the greater potential for water contamination.
17. A landfill can be made more secure with respect to water quality protection by controlling the materials placed in the landfill. The materials can be immobilized.
19. batteries, heavy metals
    furniture polish, organic solvents
    bathroom cleaners, acids or caustics
    paint, organic polymers
    oven cleaners, caustics
21. Some of the solder used in joints in older copper pipe plumbing contains lead. This lead will dissolve in acidic water passing through these pipes, so the lead can be picked up by the water.
23. Water is vaporized from one container and condensed in another while dissolved, nonvolatile substances remain behind.
25. Settling separates high-density suspended solids such as sand, while filtration removes suspended low-density matter such as algae.
27. The concentration of these chlorinated hydrocarbons and branched chain hydrocarbons rises because they accumulate in the environment. They are fat soluble and stored by living organisms. They accumulate in organisms at the top of the food chain.
29. Ammonia, $NH_3(aq)$, and ammonium ion, $NH_4^+(aq)$, can be removed from wastewater by using denitrifying bacteria. These bacteria convert the ammonia and ammonium ion to nitrogen, $N_2(g)$. The unbalanced reaction is:

$$NH_3(aq) \text{ or } NH_4^+(aq) \xrightarrow{\text{denitrifying bacteria}} N_2(g)$$

31. Both kill bacteria by oxidizing organic compounds. Aeration depends on O_2 as the oxidizing agent while chlorination uses Cl_2.
33. Mechanical pressure is used to force water molecules through a semi-permeable membrane from the salty aqueous side to the pure water side of the membrane.
35. 1820 cal endothermic
37. Water boils at 100°C but steam can be heated well beyond 100°C so the heat content of steam can be much larger than that of boiling water. In fact, steam can be made hot enough to ignite a match simply by passing it through a copper coil which is heated with a Bunsen burner.

CHAPTER 12

1. fuel formed by decomposition of plant and animal matter

3. compound containing only carbon and hydrogen
5. the energy released when a compound reacts completely with oxygen
7. (**a**) molecules with the same general formula but different molecular structures
   (**b**) compounds with a linear chain of CH_2 units and terminal CH_3 units
   (**c**) compounds with a backbone like a straight-chain hydrocarbon, but with some C_nH_{2n+1}, alkyl groups replacing Hs along the chain
9. n = 1   methane CH_4
   n = 2   ethane C_2H_6
   n = 3   propane C_3H_8
   n = 4   butane C_4H_{10}
11. (**a**)

$$CH_3CH_2CH_2CH_3 \text{ with } CH_3 \text{ branch}$$

(**b**)

$$CH_3CH_2CH_2CCHCH_2CH_2CH_3 \text{ with } CH_3, H_3C, CH_2CH_3 \text{ branches}$$

(**c**)

$$CH_3C{=}CHCH_2CH_2CH_3 \text{ with } CH_3 \text{ branch}$$

13. pentane   2-methylbutane

$$CH_3CH_2CH_2CH_2CH_3$$

$$CH_3CHCH_2CH_3 \text{ with } CH_3 \text{ branch}$$

2.2-dimethylpropane

$$CH_3CCH_3 \text{ with two } CH_3 \text{ branches}$$

15. para-dimethylbenzene   meta-dimethylbenzene

ortho-dimethylbenzene

17. There are *cis*- and *trans*- isomers for 2-butene because there are carbon chains on both double-bonded carbon atoms. The 1-butene has only hydrogens on the #1 carbon so there is no group that can be *cis* or *trans* to the groups on the #2 carbon.
19. Hot petroleum at about 400°C is distilled in a fractionation tower. There are trays at various levels in the tower, each at a unique temperature. Each volatile component will vaporize. These will condense at a temperature range of a specific tray. The compounds that are most volatile are collected near the top. The compounds like tars that are least volatile with the highest boiling point are collected at the bottom.
21. Aromatic hydrocarbons, branched-chain hydrocarbons, 2-methyl-2-propanol, methanol, ethanol, or methyl-tertiary-butyl ether

23. Synthesis gas, a mixture of CO and H_2, is produced by passing super-heated steam over pulverized coal.

$$C + H_2O + 31 \text{ kcal} \longrightarrow CO + H_2$$

Coal gasification using a catalyst, crushed coal and synthesis gas will produce methane.

$$2\,C + 2\,H_2O + 2 \text{ kcal} \longrightarrow CH_4 + CO_2$$

25. the percentage of isooctane in an isooctane/heptane mixture with the same knocking properties.
27. Legal requirements for enhanced oxygenated gasoline. A need for cleaner burning gasoline. Higher prices for gasoline from crude oil, petroleum.
29. Yes; the oxygen helps gasoline burn more completely.
31. Treating pulverized coal with superheated steam forms either carbon monoxide and hydrogen or, using a catalyst, carbon dioxide and methane.
33. M100 means 100% methanol.
   E100 means 100% ethanol.
   M85 means 85% methanol, 15% gasoline.
   E85 means 85% ethanol, 15% gasoline.
   FFV means flexible-fueled vehicle which can operate on gasoline or other fuels.
35. (**a**) methane, alkane
   (**b**) benzene, aromatic
   (**c**) 1-butene, alkene
   (**d**) acetylene, alkyne

■ CHAPTER 13

1. gamma rays
3. The more hazardous radioisotope is $^{222}_{86}Rn$, more ionizing radiation is emitted over a shorter time period.
5. It is a gas and can be inhaled. The high density makes it difficult to exhale and emitted alpha particles produce cell damage. Radioactive daughters can cause more damage.
7. The mass number on the phosphorus atom should be 30, not 49.
9. very slightly more than 1/16 of original
11. a sequence of radioactive decay steps that starts with radioactive U-238 and ends with stable Pb-206
13. diagnosis and treatment of diseases
15.   relative charge   relative mass
   (**a**) −1   0
   (**b**) +2   4
   (**c**) 0   0
   (**d**) +1   0
   (**e**) 0   1
17. Gamma rays can damage the DNA sequence needed for duplication of the genetic code in the rapidly multiplying cancer cells which duplicate more often than normal cells.
19. natural radiation in food, water and air
21. 17,200 years (three half-lives)
23. (**a**) discovered ionizing radiation was emitted by uranium compounds
   (**b**) demonstrated that radioactivity was an atomic property
   (**c**) proposed that radioactive decay was a natural change of an isotope of one element into an isotope of another element

25. no effect on the mass number of the daughter
27. catastrophic accident and release of radioactive material, plutonium production, disposal of radioactive wastes
29. (**a**) separation of useful plutonium and uranium from other nuclides in spent fuel
    (**b**) plutonium may be diverted for nuclear weapons, radioactive material may escape into the environment

31.

	FUELS	BENEFITS	PROBLEMS	CURRENT STATUS
Fission	Uranium Plutonium	renewable from breeder reactors	waste storage and disposal Pu toxicity	commercially available, proven method
Fusion	Deuterium Tritium	unlimited fuel supply	radioactive hardware from power plants, no containment vessel available	research only

33. (**a**) $^{64}_{29}\text{Cu} \longrightarrow {}^{64}_{30}\text{Zn} + {}^{0}_{-1}\text{e}$ (**b**) $^{69}_{30}\text{Zn} \longrightarrow {}^{69}_{31}\text{Ga} + {}^{0}_{-1}\text{e}$
    (**c**) $^{131}_{53}\text{I} \longrightarrow {}^{131}_{54}\text{Xe} + {}^{0}_{-1}\text{e}$
35. (**a**) 24,000 years.
    (**b**) It does not dissipate quickly and plutonium is chemically extremely toxic.
37. The "magnetic bottle" uses magnetic fields to contain a plasma by interacting with the magnetic fields generated by the ions in the plasma.
39. 100,000 alpha particles

■ CHAPTER 14

1. Carbon atoms can bond to other carbon atoms in almost unlimited numbers and in a variety of ways. Introducing other elements allows different molecules due to different atom sequences (functional groups and isomers).
3.
   (**a**) carboxylic acid, pentanoic acid
   (**b**) alkene, 2-pentene

   (**c**) symmetric ether, diethyl ether
   (**d**) secondary alcohol, 3-pentanol

   (**e**) tertiary alcohol, 2-methyl-2-propanol
   (**f**) aldehyde, ethanal

5. (**a**) acid, acetic acid
   (**b**) ether, dimethyl ether

   (**c**) alcohol, ethanol
   (**d**) ketone, acetone

   (**e**) aldehyde, formaldehyde
   (**f**) ester, methyl acetate

7. 90 proof $= \dfrac{90}{2}\% = 45\%$

9.

ethanol	ethylene glycol	glycerol
$\text{CH}_3\text{CH}_2\text{OH}$	$\text{CH}_2\text{OHCH}_2\text{OH}$	$\text{CH}_2\text{OHCHOHCH}_2\text{OH}$
one —OH group	two —OH groups	three —OH groups
use: solvent	use: automobile antifreeze	use: cosmetics

11. (**a**) aldehyde
    (**b**) alcohol
    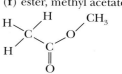

    (**c**) ketone
    (**d**) carboxylic acid

    (**e**) ester
    (**f**) ether
    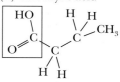

    (**g**) aldehyde

13. Primary alcohols have the —OH group bonded to a carbon atom attached to one other carbon. Secondary alcohols have the —OH group bonded to a carbon atom attached to two other carbon atoms. Tertiary alcohols have the —OH group attached to a carbon atom that is bonded to three other carbon atoms.

15. Ethanol is oxidized to acetaldehyde which is oxidized to acetic acid. This is further oxidized to CO_2 and H_2O.

17. A railroad train has a series of repeating, identical, linked units, the railroad cars. Polystyrene is a long chain of repeating, identical, linked units, the styrene molecules.

19. A macromolecule is a molecule with very high molecular mass.

21. Monomers need to contain a multiple bond in order to form addition polymers.

23. Natural rubber is poly-*cis*-isoprene. The individual monomer is isoprene.

25. Sulfur is reacted with rubber to align rubber polymers and bond the polymer chains to one another.

27. Cross linking makes the polymer more rigid.

29. (a) Yes, styrene can undergo an addition reaction because it can add to double bond in the ethene group.
    (b) Yes, propene can undergo an addition reaction because it can add to the double bond.
    (c) No, there are only single bonds so nothing can add to any multiple bonds in ethane.

31.

(a) Vinyl chloride (b) Styrene

(c) Butadiene (d) Propylene

33. (a) Polyester is a polymer built from ester monomers.
    (b) Polyamide is a polymer built from amide monomers.
    (c) Nylon 6,6 is a polymer built from 6-carbon monomers. The condensation of hexamethylene diamine and adipic acid yields a polymer with 6 carbon atoms between the nitrogen atoms.

(d) A diacid is a molecule containing two carboxylic acid groups.

(e) A diamine is a molecule containing two amine groups.

(f) A peptide linkage has a carbonyl group, $C=O$, adjacent to an amine —NH— group.

35. Condensation reactions occur between two molecules to produce a larger molecule from the smaller reactant molecules and a separate small molecule like water, which is eliminated.

37. Petroleum. Yes, this will change. Plant material sources are renewable and relatively cheaper so they will probably become the major source eventually. In addition, Earth's petroleum reserves are limited and petroleum will become more expensive and will not last indefinitely.

39. The four parts to successful recycling are: collection, sorting, reclamation, end-use. Unfortunately, collected materials will not be sorted or reclaimed if the supporting infrastructure for processing doesn't exist.

41. (a) aldehyde, alcohol
    (b) ketone, alkene, alcohol
    (c) aromatic, aldehyde, ether, alcohol
    (d) alcohol, carboxylic acid

43. Polymer composite materials are substances that have a polymer matrix with reinforcing fibers. Glass-fiber-reinforced polyesters are used for boat hulls and car body panels. Graphite fiber-epoxy composites are used in fishing rods and tennis racquets.

45. $\dfrac{100{,}000 \text{ ethylene monomers}}{1 \text{ polyethylene polymer}}$

47. 4×10^7 g/mol

■ CHAPTER 15

1. It describes objects that are nonsuperimposable mirror images of one another.

3. a 50–50 mixture of enantiomers with no net optical activity

5. a nonsugar molecule that stimulates the sweetness receptors in human taste buds.

7. a naturally occurring organic substance in living systems that is soluble in nonpolar organic solvents, but insoluble in water.

9. a fatty acid which has had some $C = C$ double bonds converted to single $C - C$ bonds.

11. The charged end of the soap and detergent molecule is hydrophilic (attracted to water). The hydrocarbon end of the molecule is attracted to nonpolar solutes but not to water (hydrophobic). A layer of soap or detergent molecules forms around the greasy solute. This envelops the solute particles so the solute particles "float away" in the water.

13. Nonionic detergents do not form insoluble calcium or magnesium salts; hence, less is needed and they are more effective.

15. The listing of water after the names of the oils indicates that the mixture is a water-in-oil emulsion.

17. The peptide bond is the bond formed between the amino group of one amino acid and the carboxyl carbon in another amino acid.

19. An amino acid contains the carboxyl, $-COOH$, and amino, $-NH_2$, functional groups

21. Order of amino acid residues in a protein or peptide

23. The protein primary structure is the sequence of amino acid residues.
The secondary structure of proteins is usually either the regular repetitive pattern of the α-helix or the β-pleated sheet. The tertiary structure is the three-dimensional structure resulting from the folding of the protein molecule.

25. (**a**) the spiral secondary protein structure
(**b**) a common secondary protein structure where protein chains form parallel strands held together by hydrogen bonds

27. An active site is a part or region of an enzyme where the substrate sits down to interact with the enzyme.

29. (**a**) 24 possible combinations.
(**b**) The 24 possible combinations for the amino acids glycine, Gly; alanine, Ala; serine, Ser; and cystine, Cys, are shown in the following table.

Gly-Ala-Ser-Cys	Ala-Ser-Cys-Gly	Ser-Cys-Gly-Ala	Cys-Gly-Ala-Ser
Gly-Ala-Cys-Ser	Ala-Ser-Gly-Cys	Ser-Cys-Ala-Gly	Cys-Gly-Ser-Ala
Gly-Ser-Cys-Ala	Ala-Cys-Gly-Ser	Ser-Gly-Ala-Cys	Cys-Ala-Ser-Gly
Gly-Ser-Ala-Cys	Ala-Cys-Ser-Gly	Ser-Gly-Cys-Ala	Cys-Ala-Gly-Ser
Gly-Cys-Ala-Ser	Ala-Gly-Ser-Cys	Ser-Ala-Gly-Cys	Cys-Ser-Ala-Gly
Gly-Cys-Ser-Ala	Ala-Gly-Cys-Ser	Ser-Ala-Cys-Gly	Cys-Ser-Gly-Ala

31. Valylcystyltyrosine.

33. Enzymes have active sites that can only fit D-glucose and not the L-glucose.

35. Protein chains in hair are held together by hydrogen bonds, ionic bonds, and disulfide bonds. Hydrogen bonds form between H atoms attached to very electronegative atoms (N and O) and the electron rich atoms with lone pairs. Ionic bonds result when carboxylate ions ($RCOO^-$) and the protonated amino groups $-NH_3^+$ are attracted to one another. Disulfide bonds form between sulfur atoms in cysteine fragments in adjacent strands.

37. The disulfide bond is a single bond between two sulfur atoms. The disulfide bonds between cysteine amino acid units hold parallel strands of hair protein in place.

39. DNA and RNA differ structurally because each contains a different sugar for the polymer chain and they contain different bases. DNA contains the sugar, α-2-deoxy-D-ribose, $C_5H_{10}O_4$, while RNA contains α-2-D-ribose, $C_5H_{10}O_5$. Both DNA and RNA contain adenine, guanine, and cytosine. The base thymine occurs only in DNA. The base uracil occurs only in RNA.

41. The monomers that polymerize to form DNA and RNA are nucleotides. The nucleotide monomers contain a phosphoric acid unit, a ribose unit, and a nitrogenous base. See Figure 15.19 in text.

43. bases like adenine-thymine and guanine-cytosine. There are two hydrogen bond sites between the A-T pair and three hydrogen bond sites between the G-C pair.

45. an animal with genes from another species spliced into its genes

47. (**a**)
DNA sequence   T G T C A G T G G G C C G C T
mRNA sequence A C A G U C A C C C G G C G A
(**b**)
mRNA sequence A C A G U C A C C C G G C G A
tRNA sequence   U G U C A G U G G G C C G C U
(**c**)
Cysteine Glutamine Tryptophan Alanine Alanine.

49. hydrogen bonding between specific bases that have complementary hydrogen bonding.

51. DNA sequence . . .   G T A G C . . .
mRNA sequence   C A U C G . . .

53. mRNA sequence C U G U

CHAPTER 16

1. Digestion is the process of breaking large molecules in food into substances small enough to be absorbed by the body from the digestive tract.

3. (**a**) 9 kcal/g
(**b**) 4 kcal/g
(**c**) 4 kcal/g

5. The basal metabolic rate, BMR, is the rate at which the body uses energy to support the normal maintenance operations of the body. It is the minimum energy required for a person to stay alive.

7. A triglyceride is a triester formed from glycerol and three fatty acids.

9. (**a**) a disease that is associated with the buildup of fatty deposits on the inner walls of arteries
(**b**) a lipid steroidal alcohol that contributes to the development of atherosclerosis.
(**c**) the yellowish deposit of cholesterol and lipid-containing material on artery walls that is a symptom of atherosclerosis
(**d**) complex assemblages of lipids, cholesterol, and proteins that serve to transport water-soluble lipids in the bloodstream through the body.
(**e**) lipoproteins richer in lipid than in protein with a corresponding lower density
(**f**) lipoproteins richer in protein than in lipid with a corresponding higher density

11. (**a**) nutrients the body needs in large amounts.
(**b**) nutrients needed by the body in small amounts.

13. (a) plants; they are found in foods like pasta, rice, potatoes, corn.
    (b) produced in the liver and limited amounts are stored in liver tissue and muscle.
    (c) plants.
15. an organic compound essential to health; it is needed in small amounts in the diet.
17. a substance that dissolves in water to produce ions. The electrolyte balance refers to a condition of proper transfer of material through osmosis, normal nerve impulse transmission, normal acid-base balance and extracellular volume.
19. phosphorus, calcium, magnesium, sodium, potassium, chlorine, and sulfur
21. to tie up metal ions to keep them from catalyzing the decomposition of food
23. 30% of daily Calories
25. Milk should be promoted as a healthy food for the general public because it is an excellent source of calcium and protein. Milk should not be promoted as a healthy food because people who are lactose intolerant may experience serious adverse reactions and become ill when they consume milk.
27. Cholesterol is water insoluble because water is a small polar molecule that will have very weak attractions to the large, organic nonpolar cholesterol molecule. High levels of cholesterol are linked to atherosclerosis or hardening of the arteries. There are 27 carbon atoms in cholesterol, $C_{27}H_{45}OH$.
29. Women produce less estrogen after menopause. Estrogen inhibits calcium loss and bone erosion. Reduced estrogen levels result in more rapid calcium and bone loss. Osteoporosis can be minimized if adequate calcium intake occurs, especially from adolescence through young adulthood, and if estrogen replacements are taken after menopause.
31. Iodine deficiency leads to an enlargement of the thyroid. This condition is called "goiter." Potassium iodide, KI, is added to table salt to make iodized salt in order to prevent iodine deficiency.
33. There is no right answer for this question. Each answer will depend on the specific product.
35. 1100 Calories (estimated daily caloric need = 1760 Calories)
37. 20 minutes
39. 77 minutes for 100 g bread
41. 14758 mg = 15000 mg
    This is 15 grams with only two significant digits.
43. 598 Calories total
45. 2750 Calories total, grams of fat = 92 g; grams of carbohydrate = 413 g; grams of protein = 69 g.

■ CHAPTER 17

1. Malaria, pneumonia, bone infections, gonorrhea, gangrene, tuberculosis, and typhoid fever are bacterial diseases. Polio, AIDS, and Rubella (German measles) are viral diseases.
3. The United States Food and Drug Administration (FDA) has the responsibility for classifying drugs as either over-the-counter drugs or prescription drugs.
5. Three major classes of antibiotics are penicillins, cephalosporins, and tetracyclines.
7. Chemotherapy is the treatment of disease with chemical agents.
9. A retrovirus is a virus that uses RNA-directed synthesis of DNA instead of the usual DNA-directed synthesis of RNA.
11. (a) Vasodilators are used to treat heart disease and asthma. The vasodilator relaxes the walls of blood vessels and creates a wider passage for blood flow. This lowers blood pressure and reduces the amount of work that the heart must do to pump blood. It also enables a person to breathe more deeply and easily by dilating the bronchial tubes.
    (b) Alkylating agents are used to treat cancer. Alkylating agents react with the nitrogen bases of DNA in cancer cells and normal cells. Alkyl groups are added to the nitrogens in the bases. This has an effect on both cancer cells and normal cells, but cancer cells are usually dividing and duplicating DNA more often, so the cancer cells are impacted more.
    (c) Beta blockers are used to treat heart disease. Propranolol (Inderol) is used to treat angina, cardiac arrhythmia, and hypertension. Beta blockers act to keep epinephrine and norepinephrine from stimulating the heart.

Inderal, Propranolol

    (d) Antimetabolites are used to treat cancer. Antimetabolites interfere with DNA synthesis, and cancer cells are more susceptible than normal cells because the cancer cells are generally replicating DNA more frequently than normal cells.
13. Histamines and neurotransmitters bind to receptor sites.
15. (a) Methotrexate is an antimetabolite cancer drug.
    (b) Chlorphenirimine (in Chlortrimeton) is an antihistamine.
17. Codeine is
    (a) an analgesic (b) an opioid
    (d) a controlled substance, a Schedule 2 drug.
19. Angina is chest pain on exertion and a symptom of heart disease.
21. A barbiturate, such as phenobarbital, is a depressant. Barbiturates bind to a receptor for GABA. This keeps channels for chloride ion, Cl^{1-}, transmission open. This inhibits transmission of nerve impulses. The physiological effects are a progression from sedation or relaxation, to sleep, to general anesthesia, to coma and death.
23. Heart muscle contracts more often and heart rate increases.
25. Estradiol has an alcohol group and a hydrogen on a carbon in the 5-membered ring while ethynyl estradiol has both an alcohol group and an ethynyl group on that carbon.

27. (a) Dopamine helps control memory, emotion, and regulate fine muscle movement.
    (b) Epinephrine is a neurotransmitter that produces increased blood pressure, dilation of blood vessels, and widening of the pupils of the eye.
29. Over-the-counter, prescription, unregulated non-medical drugs, controlled substances.
31. (a) hallucinogen (b) hallucinogen
    (c) hallucinogen (d) hallucinogen (e) depressant
    (f) antidepressant or stimulant
33. (a) Vasodilators treat angina. Vasodilators act to expand or dilate veins. This lowers resistance to blood flow and blood pressure. The heart needs to do less work to circulate the blood.
    (b) Diuretics treat hypertension. Diuretics reduce blood pressure and hypertension by stimulating excretion of sodium ion and urine production. Blood volume and pressure decrease when urine output increases.
35. Surgery, irradiation, chemotherapy
37. Hallucinogens cause a person to experience vivid illusions, fantasies, and hallucinations. Examples of hallucinogens are mescaline, PCP, and LSD.
39. (a) Morphine is a more effective pain killer than heroin and codeine.
    (b) Heroin is not a natural alkaloid.
    (c) Heroin is so addictive that it is not legal to sell or use it in the United States.
41. Nitroglycerine is a heart muscle relaxant. It is used to treat angina, which is a symptom of heart disease.
43. The blood-brain boundary is defined physically by small openings in capillaries and the astrocyte cell membranes that prevent passage of large polar molecules. The barrier keeps large toxic polar molecules from passing into the brain. Drugs must be soluble in blood and soluble in the lipid layer of the membrane in order to reach the brain.

■ CHAPTER 18

1. (a) sedimentary rock: rock formed by deposition of dissolved or suspended substances
   (b) igneous rock: rock formed by solidification of molten rock
3. (a) Annealing is the process of heating and then slowly cooling a substance to make the substance less brittle and reduce strain in the solid.
   (b) Amorphous solids have no regular crystalline order.
   (c) Ceramics are materials generally made from clays and then hardened by heat.
   (d) Cement is a substance able to bond mineral fragments into a solid mass.
5. Magnesium is extracted from seawater. It is used for alloys for auto and aircraft parts.
7. Gold, copper, and platinum all can be found as pure metallic elements in nature.
9. The slag is lower density than the molten iron.
11. Carbon is the main impurity in pig iron. Pig iron is brittle partly because it contains Fe_3C.
13. (a) $2\,Cu_2S(s) + 3\,O_2(g) \longrightarrow 2\,Cu_2O(s) + 2\,SO_2(g)$
    (b) $2\,FeS(s) + 3\,O_2(g) \longrightarrow 2\,FeO(s) + 2\,SO_2(g)$

15. ductile   malleable   electrical conductor
    copper wire   steel sheet metal gold connectors
    steel tubing   aluminum foil   mercury switches

    heat conductor   lustrous
    copper pots   silver mirrors
    aluminum pans   gold jewelry
17. The p-type of semiconductors have a shortage of electrons. The p-type of semiconductors can be made using dopants like B, boron, or Ga, gallium.
19. The structure of an SiO_4 unit is tetrahedral.

21. Superconductivity is the behavior displayed by a material at temperatures at or below the superconducting transition temperature. The material has no electrical resistance at these conditions. Superconductivity research is aimed at minimizing energy losses when transferring electricity.
23. Iron(III) compounds give glass a yellow color.
25.

Property	Metal	Ceramic
Hardness	variable	yes
Strength	yes	yes
Ductility	yes	no
Electrical Conductivity	yes	no
Brittleness	no	yes

27. Lime, CaO, reacts with CO_2 to form $CaCO_3$
29. No, the silicates will crystallize first because of their higher melting points.

■ CHAPTER 19

1. Earth's carrying capacity depends on the amount of productive land, agricultural practices, biotechnology advances, the rate of degradation of farmland, global pollution, the amount of water for irrigation, and the level of an acceptable quality of life.
3. Soil is a mixture of mineral particles, organic matter, water, and air.
5. Soil has a structure made up of a series of layers called *horizons* that are loosely packed and permeable near the surface and gradually change to impermeable solid rock.
7. Limestone will neutralize acids in the soil and raise the pH.
9. carbon, hydrogen, and oxygen
11. (a) Metal ions from Group 1A and Group 2A
    (b) The result is a more acid soil, and the pH goes down.
13. (a) substances needed by green plants for healthy growth
    (b) Nonmineral nutrients are carbon, oxygen, and hydrogen. They are available from water and atmospheric carbon dioxide.
    (c) Mineral elemental nutrients are water soluble substances that plants can only absorb as solutes through their roots.
15. A soil shortage of N, P, and K is more likely. The nutrients N, P, and K are primary elemental nutrients and are needed in greater amounts. They are more easily leached from the soil.

17. $N_2(g) + 3 H_2(g) \rightleftharpoons 2 NH_3(g)$. It is the synthetic source of ammonia fertilizer.

19. Chlorosis is a plant condition of low chlorophyll. It is caused by a deficiency of any one of the three nutrients: magnesium, nitrogen, or iron. A symptom of chlorosis is the presence of leaves that are pale yellow instead of green.

21. The fertilizer labeling system indicates the percentage by weight of nitrogen (N), phosphorus as P_2O_5, and potassium as K_2O (potash). None of the pure elements are present in the fertilizer. Each element is present in some compound. None of the reference substances are present in the fertilizer bag either.

23. Urea, H_2NCONH_2, decomposes in the soil to form ammonia, NH_3.

25. **(a)** A quick-release fertilizer dissolves readily in water. It dissolves easily and can be picked up quickly by the plant roots.
    (b) A slow-release fertilizer is not as water soluble. It dissolves very slowly in water and it is available to the plant more slowly, so it is taken up slowly by the plant.

27. Ammonium nitrate is inexpensive and a good source of water-soluble nitrogen. Farmers benefit because cheap fertilizers keep food production costs down. Consumers benefit because food prices are low. The dark side of ammonium nitrate is that it is an explosive. Handling it can lead to industrial accidents. It can be used as an explosive by terrorists as was done in 1995 when approximately a ton of ammonium nitrate was detonated next to the Federal Building in Oklahoma City.

29. DDT is fat soluble because "like dissolves like" and both are nonpolar. DDT poses problems because it is a carcinogen and does not degrade quickly. It is stored in the fat tissue of animals. This accumulation problem is exaggerated when one organism high in the food chain eats many other ones lower in the chain.

31. About 33% to 40% of food crop production is lost to pests each year worldwide. The monetary value of these losses is estimated at $20 billion per year.

33. A biodegradable pesticide is one that is quickly converted to harmless products by microorganisms and natural processes. Pyrethrins are an example of a class of biodegradable insecticides. An example of a pyrethrin is shown here.

Dimethrin

35. **(a)** Pesticides can increase crop yields and protect them when stored. They reduce the loss of crops to pests. Every bit of food saved is food that need not be replaced. Food supplies increase and food is more plentiful at lower costs. Diseases carried by pests can be controlled or even eliminated. Pesticides can cause problems of water and soil contamination if they are misused. Pesticide residues can contaminate crops and create long-term poisoning problems. Resistant strains of pests can develop and require higher levels of pesticides.
    (b) Pesticides should be used early enough to require smaller amounts. They should be used only when absolutely necessary and when alternative methods do not exist.

37. Sustainable agriculture is a set of practices intended to improve profits, limit use of agrichemicals and increase the use of environmentally friendly farming procedures.

39. Integrated pest management involves the use of disease-resistant plant varieties and biological controls such as predators or parasites to control pests.

41. **(a)** Genetically altered crops with specific genes inserted to produce plants with desirable properties.
    (b) Flavr Savr tomato, a weevil resistant garden pea and Bollgard cotton seed.

43. Pheromones are insect sex attractants. They are used to lure insects into traps so insecticide spraying is not needed.

45. Yes, there is a conflict. Population growth will decrease the number of productive acres per person. The present 0.82 acre per person will decrease and the efficiency of farming methods will need to improve in order to avoid famine and food shortages. This problem will be compounded because population centers usually are in the middle of prime agricultural land. The growth of urban areas will remove such land from production.

47.

Agrichemical	LD_{50}	
DDT	100 mg/kg	most toxic
primicarb	150 mg/kg	
carbaryl	250 mg/kg	
malathion	1,000 mg/kg	
dimethrin	10,000 mg/kg	least toxic

49. 5,000 mg

51. $\dfrac{\text{10 pounds of N}}{\text{100 pounds total}} \times 20 \text{ pound bag} = 2 \text{ pounds of N}$

Glossary

A

Achiral Can be superimposed on its mirror image

Acid Molecule or ion able to donate a hydrogen ion to a base. Acids produce H_3O^+ ions in aqueous solutions.

Acid rain Rain that has an unusually low pH (below 5.6)

Acid–base indicators Substances that change color with changes in acidity or basicity of a solution

Acid–base reaction Reaction in which a hydrogen ion is exchanged between an acid and a base

Acidic solution A solution that contains a higher concentration of H_3O^+ ions than OH^- ions

Activation energy (E_{act}) Quantity of energy needed for successful collision of reactants; influences reaction rate

Active site The region of an enzyme where the catalytic chemistry occurs. This region is usually composed of only a few critical amino acid residues.

Activity The number of radioactive nuclei that disintegrate per unit of time

Addition polymers Polymers made from monomers (usually containing carbon–carbon double bonds) joined directly with one another

Aerosols Tiny pollutant particles, so small that they remain suspended in the atmosphere for long periods

Alcohols Organic compounds containing a hydroxyl (OH) functional group

Aldehydes Organic compounds containing a —CHO functional group as in RCHO.

Alkali metals Elements in Group 1A in the periodic table

Alkaline earth metals Elements in Group 2A in the periodic table

Alkanes Hydrocarbons with carbon–carbon single bonds

Alkenes Hydrocarbons with one or more carbon-carbon double bonds

Alkyl groups Alkanes with a hydrogen atom removed

Alkynes Organic compounds containing a carbon–carbon triple bond

Allotropes Different molecular forms of the same element; in this case the two molecular forms are diatomic O_2 and triatomic O_3

Alloy A mixture of two or more metals

Alpha particles Positively charged particles identical to helium-4 nuclei; emitted by certain radioactive isotopes

Amide An organic compound containing a —$CONH_2$ (or —CONHR) functional group

Amine (primary) An organic compound containing an —NH_2 functional group

Amino acids Biomolecules containing an alpha–amino group and a carboxyl group; building blocks of proteins

Analgesics Drugs that relieve pain

Annealing Heating and cooling a metal, glass, or alloy in a manner that makes it less brittle

Anion An atom or group of bonded atoms that has a negative charge; for example, Cl^-, NO_3^-

Anode Electrode at which electrons flow out of an electrochemical cell and oxidation takes place

Antacid Base used to neutralize excess hydrochloric acid in the stomach

Anthropogenic Originating from human activity

Anthropogenic pollution Pollution that is attributable to human activity

Antibiotic A substance that inhibits the growth of microorganisms

Anti-inflammatory Reduces inflammation

Antipyretic Reduces fever

Applied science Science with a well-defined, short-term goal of solving a specific problem

Aquifer A layer of water-bearing porous rock or sediment held in place by impermeable rock

Aromatic compounds Hydrocarbons and other compounds with one or more benzene rings or rings with benzene-like chemical properties

Artificial transmutation Experimental conversion of one element into another

Atmospheric pressure The pressure exerted by the weight of the atmosphere at a given altitude

Atomic mass unit (amu) The unit for relative atomic masses of the elements; 1 amu = $1/12$ the mass of the carbon-12 isotope. 1 amu = 1.6605×10^{-24} g

Atomic number The number of protons in the nuclei of the atoms of an element

Atomic weight The number that represents the average atomic mass of the element's isotopes weighted by percentage abundance.

Avogadro's number 6.02×10^{23}; the number of items in one mole

B

Balanced chemical equations Chemical equations in which the total number of atoms of each kind is the same in reactants and products

Basal metabolic rate (BMR) The minimum energy required by an awake but resting body

Base Molecule or ion able to accept a hydrogen ion from an acid. Bases release or produce $^-OH(aq)$ in aqueous solutions.

Basic science Pursuit of scientific knowledge without the goal of a practical application

Basic solution (alkaline solution) A solution that contains a higher concentration of OH^- ions than H_3O^+ ions

Battery A series of electrochemical cells that produces an electric current; commonly refers to any electrochemical, current-producing device

Beta particles Electrons ejected at high speeds from certain radioactive isotopes

Binary compound Chemical compound composed of two elements

Binding energy The force holding neutrons and protons together in a nucleus

Biochemical oxygen demand (BOD) The amount of dissolved oxygen required by microorganisms that oxidize dissolved organic compounds to simple substances like CO_2 and H_2O

Biochemistry The chemistry of living things

Biodegradable Describes a substance that can be broken down into simple molecules by the action of microorganisms

Bleaching agent A chemical that removes unwanted color by oxidation of the colored chemical

Boiling point The temperature at which the vapor pressure of a liquid equals the atmospheric pressure and boiling occurs. If the atmospheric pressure is 1 atm, it is the **normal boiling point**.

Bond energy Amount of energy required to break one mole of bonds between a specified pair of atoms

Bonding pairs Pairs of electrons shared between two atoms, frequently represented by lines (instead of dots) for convenience in Lewis structures

Branched-chain isomers Structural isomers of hydrocarbons that have carbon–carbon bonds in side chains

Buffer Combination of an acid and a base that can control pH

Buffer solution Solution of an acid plus a base that controls pH by reacting with added base or acid

C

Carbohydrates Biomolecules composed of simple sugars (monosaccharides, $C_6H_{12}O_6$) and including disaccharides, starch, and cellulose

Carboxylic acids Organic compounds containing a —COOH functional group as in RCO_2H

Carcinogen An agent that causes cancer

Carcinogenic Cancer-causing

Cardiovascular disease Disease of the heart or blood vessels

Cast iron Molded pig iron or other carbon–iron alloy

Catalyst Substance that increases the rate of a chemical reaction without being changed in identity

Catalytic cracking process Process by which larger kerosene fractions are converted into hydrocarbons in the gasoline range

Catalytic re-forming process Process that increases octane rating of straight-run gasoline by converting straight-chain hydrocarbons to branched-chain hydrocarbons and aromatics

Cathode Electrode at which electrons flow into an electrochemical cell and reduction takes place

Cation An atom or group of bonded atoms that has a positive charge

Cation An ion with an overall positive charge; for example, Li^+, Ca^{2+}, NH_4^+

Cement A material that bonds mineral fragments into a solid mass; any material that bonds other materials together

Ceramics Materials and products made by firing mixtures of nonmetallic minerals

Chain reaction A process in which neutrons from one fission reaction can cause multiple fission reactions in nearby nuclei

Chapman Cycle The set of four reactions that represents the steady-state formation and destruction of ozone in the stratosphere.

Chemical compounds Pure substances composed of atoms of different elements combined in definite, fixed ratios

Chemical equations Representations of chemical reactions by the formulas of reactants and products

Chemical equilibrium Condition in which a chemical reaction and its reverse are occurring at equal rates

Chemical formula Written combination of element symbols that represents the atoms combined in a chemical compound

Chemical properties Properties that result in chemical reactions and changes in the identity of one or more reactants

Chemical reaction Process in which one or more pure substances are converted to one or more different pure substances

Chemistry The study of matter and the changes it can undergo.

Chiral Cannot be superimposed on its mirror image

Chlorofluorocarbons (CFCs) A term used to refer collectively to compounds containing only carbon, chlorine, fluorine, and sometimes hydrogen. CF_2Cl_2, developed initially for use as a refrigerant in air conditioning systems, is an example.

Chlorosis Yellowing of plant leaves due to nutrient deficiency

Cis **isomers** Stereoisomers with groups on the same side of a carbon–carbon double bond

Codon A three-base sequence on mRNA that codes a specific amino acid in protein synthesis

Coefficients In a chemical equation, numbers written before formulas to balance the equation

Coenzyme A nonprotein molecule needed by an enzyme to make the enzyme function possible

Coke A mostly carbon product formed by heating coal in the absence of air to drive off volatile materials

Colloids Particles larger than most molecules or ions and dispersed in a solvent-like medium

Combustion Rapid oxidation that produces heat and (usually) light

Combustion The rapid combination of a fuel with oxygen; the process of burning

Complete (mixed) fertilizers Contain the three primary nutrients (nitrogen, phosphorus, potassium)

Concentration of a solution The quantity of a solute dissolved in a specific quantity of a solvent or solution

Condensation polymers Polymers made from monomers (each with two functional groups) that combine so that small molecules are split out between their functional groups

Condensation The change of molecules from the gaseous state to the liquid state

Condensation reaction Reaction between two molecules in which a larger molecule is formed and a small molecule is eliminated

Continuous spectrum A spectrum that contains radiation distributed over all wavelengths

Corrosion The unwanted oxidation of metals during exposure to the environment

Covalent bond A bond in which two atoms share electrons

Critical mass The mass of fissionable material able to sustain a chain reaction

Crystallization (solidification) Formation of a crystalline solid from a melt or a solution

Cycloalkanes Saturated hydrocarbons with carbon atoms joined in a ring

D

Denaturation Destruction of proper functioning of a protein by changing its structure

Denatured alcohol Ethanol with small added amounts of a toxic substance, such as methanol, that cannot be removed easily by chemical and physical means

Denitrifying bacteria Bacteria that convert ammonia or ammonium ion to nitrogen

Density is mass per unit volume. The customary measurement units (Section 2.8) for density are grams (g) for mass (weight) and cubic centimeters (cm^3) for the volume of solids or milliliters (mL) for the volume of liquids.

Deoxyribonucleic acids (DNAs) Nucleic acids that function as genetic information storage molecules; contain deoxyribose

Diatomic molecules Molecules composed of two atoms

Dietary fiber Food polysaccharides that are not broken down by digestion

Digestion The breakdown of food into substances small enough to be absorbed from the digestive tract

Disinfection by-products Compounds formed by the action of disinfectants like chlorine and ozone on substances found in water

Diuretic Causes excretion of fluid from the body

Dobson unit A measure of stratospheric ozone concentration; 1 Dobson unit is 1 ppb of ozone in air.

Doping (of semiconductor) The addition of atoms of an element with extra or fewer valence electrons than the element of the semiconductor material

Double bond A bond in which two pairs of electrons are shared between two atoms

Dynamic equilibrium A state of balance between opposite changes occurring at the same rate

E

Electrochemical cell A device in which a chemical reaction generates an electric current

Electrodes Conducting materials at which electrons enter and leave an electrochemical cell

Electrolysis reaction A redox reaction driven by electrical energy from an external power supply

Electrolyte A compound that conducts electricity when melted or dissolved in water

Electronegativity A measure of the relative tendency of an atom to attract electrons to itself

Electrons Negatively charged subatomic particles found in the space around the nucleus

Emulsifying agent A compound that has a water-soluble part as well as an oil-soluble part and that stabilizes an emulsion

Emulsions Stable mixtures of water and an oily component; may have a creamy appearance

Enantiomers (optical isomers) A chiral molecule and its nonsuperimposable, mirror-image molecule. Enantiomers are one type of stereoisomer

Endergonic A chemical process in which there is a net consumption of energy

Energy The capacity for doing work or causing change

Enthalpy change A measure of the quantity of heat transferred into or out of a system as it undergoes a chemical or physical change

Entropy change A measure of the change in the disorder of a system as it undergoes chemical reaction or physical change

Enzymes Biomolecules that function as catalysts for biochemical reactions

Essential amino acids Amino acids that are not synthesized by the body and, therefore, must be obtained from the diet

Esters Organic compounds containing a —COOR functional group

Ethers Organic compounds with the general formula R—O—R′

Excited state An unstable, higher energy state of an atom

Exergonic A chemical process in which there is a net energy release

F

First law of thermodynamics Energy can be converted from one form to another but cannot be destroyed

Fission Splitting of a heavy nucleus into smaller fragments, caused by bombardment with neutrons

Fluid A state of matter that is capable of flowing; a gas or a liquid

Formula unit In ionic compounds, the simplest ratio of oppositively charged ions that gives an electrically neutral unit

Fractional distillation Separation of a mixture into fractions that differ in boiling points

Free radical A species that has one or more unpaired electrons

Functional group An atom or group of atoms that is part of a larger molecule and that has a characteristic chemical reactivity

Fusion The combination of small nuclei to make heavier nuclei

G

Gamma rays High-energy electromagnetic radiation emitted from radioactive isotopes

Gangue Unwanted substances mixed with a mineral during production of metal from ore

Generic name The generally accepted chemical name of a drug

Genes Segments of DNA that direct synthesis of proteins

Genome Total sequence of base pairs in the DNA of a plant or animal

Glass A hard, noncrystalline, usually transparent substance made by melting silicates with other substances

Global warming Warming of Earth above a level that is desirable from a human viewpoint as a result of increased concentrations of one or more greenhouse gases

Greenhouse effect A phenomenon of Earth's atmosphere by which solar radiation, trapped by Earth and re-emitted from the surface as infrared radiation, is prevented from escaping by various gases in the atmosphere. This leads to warming of Earth.

Ground state The condition of an atom in which all electrons are in their normal, lowest energy levels

Groundwater recharge Return to groundwater of water from treated sewage

Groups Vertical columns of elements with similar properties in the periodic table

H

Haber process Industrial process for the reaction of nitrogen and hydrogen gases to form ammonia

Half-life The period required for one-half of a sample of a radioactive substance to undergo radioactive decay

Halogens Elements in Group 7A in the periodic table

Hallucinogens Drugs that cause perceptions with no basis in the real world

Hazardous waste Industrial and household wastes responsible for water pollution

Heat of combustion The quantity of heat released when a fuel is burned; variously expressed as kcal/mol, kcal/g, kJ/mol, or kJ/g

Heat of fusion The heat required to melt a given quantity of a solid at its melting point

Heat of vaporization The heat required to vaporize a given quantity of liquid at its boiling point

High-density lipoproteins (HDLs) Lipoproteins that transport cholesterol from body tissues to liver (the "good" lipoproteins)

Homeopathy The treatment of a disease system by using minute doses of natural substances that, in a healthy person, would produce symptoms of the same disease

Homogeneous mixtures are uniform in composition and **heterogeneous mixtures** are not.

Horizons Layers within the soil

Hormones Chemical messengers secreted in the bloodstream by endocrine glands

Humus Decomposed organic matter in soil

Hydrocarbons Compounds that contain only carbon and hydrogen atoms, such as methane (CH_4).

Hydrogen bonding Attraction between a hydrogen atom bonded to a highly electronegative atom (O, N, F) and the lone pair on an electronegative atom in another or the same molecule

Hydronium ion A hydrated proton, H_3O^+

Lewis structure

Hydrophilic Water loving; dissolves in water

Hydrophobic Water fearing; does not dissolve in water

Hydrosphere All freshwater and saltwater that is part of planet Earth

Hypertension Elevated blood pressure

Hypertonic solution A solution with a concentration of solute higher than that found in its surroundings

I

Incomplete combustion Combustion in which there is insufficient oxygen for complete combustion

Industrial smog Smog generally caused by industrial activity such as coal burning; also known as London smog

Infectious diseases Diseases caused by microorganisms

Infrared (IR) radiation Electromagnetic radiation whose wavelength is in the range between 2.5×10^{-5} m and $2.5 \times$

10^{-6} m (see Figure 3.6); it is often referred to as heat radiation

Inner transition elements Elements 58–71 (lanthanides) and 90–103 (actinides), which lie between Groups 2A and 3A of the main group elements and are usually placed at the bottom of the periodic table

Inorganic compounds Compounds of all elements other than the organic compounds of carbon

Insoluble Describes a substance that will not dissolve in a given solvent

Insulators Poor conductors of heat and electricity

Integrated pest management (IPM) Limits the use of pesticides by relying on a combination of disease-resistant crop varieties, natural predators or parasites, and selected use of pesticides

Intermolecular forces Attractive forces that act between molecules; weaker than covalent bonds

Ionic bond The attraction between positive and negative ions

Ionic compounds Compounds composed of positive and negative ions

Ions Atoms or groups of atoms with a positive (cation) or negative (anion) charge, for example, sodium ion (Na^+), chloride ion (Cl^-), and sulfate ion (SO_4^{2-})

Isomers Different compounds with the same formula but different structures

Isomers Two or more compounds with the same molecular formula but different arrangements of atoms

Isotopes Atoms of the same element having different mass numbers

K

Ketones Organic compounds containing a C=O (carbonyl) functional group between two carbon atoms as in RCOR′

Kinetic energy The energy of objects in motion

L

Law of conservation of matter Matter is neither lost nor gained during a chemical reaction

Law of definite proportions In a compound, the constituent elements are always present in a definite proportion by weight

Leaching effect The dissolving of soil material as water moves through the soil

Le Chatelier's principle When a stress is applied to a system at equilibrium, the equilibrium shifts to relieve the stress

Lewis dot symbol A representation of a molecule, ion, or formula unit by atomic symbols with the valence electrons shown as dots

Lipids Class of biomolecules not soluble in water but soluble in organic solvents; includes fats, oils, steroids, and waxes

Lipoproteins Assemblages of lipids, cholesterol, and water-soluble proteins

Loam Soil consisting of a friable mixture of varying proportions of clay, sand, and organic matter

Lone pairs (or *non-bonding or unshared pairs*) Pairs of valence electrons that are not shared between two atoms in the Lewis structure but instead are associated with a single atom in the structure; should always be shown as dots in Lewis structures

Low-density lipoproteins (LDLs) Lipoproteins that transport cholesterol from the liver to tissues throughout the body (the "bad" lipoproteins)

M

Macroscopic Large enough to be seen, felt, and handled

Mass A measure of the quantity of matter in an object

Mass number Number of neutrons plus number of protons in the nucleus of an atom

Melting point The temperature at which a solid substance turns into a liquid

Metalloids Elements with properties intermediate between those of metals and nonmetals, and falling between them in the periodic table

Metals Elements that conduct electric current; most are malleable and ductile

Micronutrients Elements required by plants in relatively small amounts (less than 100 mg/day)

Microscopic Visible only with the aid of a microscope

Mineral (as dietary nutrient) An element other than carbon, hydrogen, nitrogen, and oxygen that is needed for good health

Minerals Naturally occurring, solid inorganic compounds

Mineral nutrients Plant nutrients absorbed through roots as solutes in water

Miscibility Ability to mix in all proportions

Molar mass Mass in grams of one mole of any substance

Molarity Number of moles of solute per liter of solution

Mole Unit for amount of substance; contains Avogadro's number of atoms, molecules, or other particles

Molecular formulas Chemical formulas that represent molecules with atom symbols and subscripts

Molecule The smallest chemical unit of a compound that retains the composition and properties of the compound and can exist independently

Monomers The small molecules that combine to form polymers.

N

Nanoscopic In the range of the nanometer (0.000000001 m); atoms and molecules are in the nanoscopic size range

Natural rubber Poly-*cis*-isoprene from the *Hevea brasiliensis* tree

Neutral solution A solution in which the pH is 7 and $[H_3O^+] = [OH^-] = 1 \times 10^{-7}$

Neutralization reaction Reaction of equivalent quantities of an acid and a base

Neutrons Electrically neutral subatomic particles found in the nucleus

Nitrogenase Bacterial enzyme that catalyzes nitrogen fixation

Nitrogen fixation Conversion of atmospheric nitrogen into soluble nitrogen compounds available as plant nutrients

Noble gases The gases of periodic table group 8A; helium, neon, argon, krypton, xenon, and radon; long thought to be completely unreactive. They are colorless and exist as single atoms at room temperature.

Nonbiodegradable Describes a substance that cannot be broken down by microorganisms and therefore persists in the environment

Nonelectrolyte A compound that does not conduct electricity when melted or dissolved in water

Nonmetals Elements that do not conduct electrical current

Nonmineral nutrients Carbon, hydrogen, and oxygen in soil

Nonpolar Describes a bond or molecule which is not polar because it has no polar bonds or because its polar bonds are oriented symmetrically so that they cancel each other

Nonselective herbicides Herbicides that kill all plant life

Nuclear reaction A process in which an isotope of one element is transformed into an isotope of another element

Nucleic acids Biomolecules that control heredity (DNA, RNA); polymers of nucleotides that contain deoxyribose or ribose, nitrogen bases, and phosphate groups

Nucleons Protons and neutrons

Nucleotide Biomolecule with a five-carbon sugar bonded to a nucleic acid base and a phosphate group; monomer in DNA

Nucleus Small central core of the atom; contains the protons and neutrons

Nutrient Substance in food that provides energy or biochemical raw material (fats, proteins, carbohydrates, vitamins, minerals)

Nutrition The science that deals with diet and health

O

Octet rule In forming bonds, main-group elements gain, lose, or share electrons to achieve a stable electron configuration with eight valence electrons

Orbital A region of three-dimensional space around an atom within which there is a significant probability (usually shown as 90%) that a given electron will be found

Organic chemistry The chemistry of carbon compounds

Organic compounds Compounds of carbon with hydrogen, and derivatives of these compounds

Organic farming Farming without synthetic chemical fertilizers and pesticides

Osmosis The flow of water molecules through a semi-permeable membrane from a less-concentrated to a more-concentrated solution (i.e., from the side of purer water to the side of saltier water to dilute it).

Osmotic pressure The external pressure required to prevent osmosis from taking place

Osteoporosis Reduction of bone mass associated with calcium loss

Over-the-counter drugs Drugs that can be purchased without a prescription

Oxidation The gain of oxygen, loss of hydrogen, or loss of electrons

Oxidation-reduction reactions (redox reactions) Reactions in which one reactant is oxidized and another reactant is reduced

Oxides Compounds of oxygen combined with another element

Oxidizing agent A reactant that causes oxidation of another reactant

Oxygenated gasolines Blends of gasoline that contain oxygenated organic compounds such as alcohols or ethers to cause the gasoline to burn more cleanly

Ozone layer The stratospheric layer of gaseous ozone (O_3) that protects life on Earth by filtering out most of the harmful ultraviolet radiation emitted by the Sun

P

Particulates Solid particles and liquid droplets suspended in the atmosphere

Parts per billion 1 part out of every billion parts (1 ppb)

Parts per million 1 part out of every million parts (1 ppm)

Pathogenic Capable of causing disease

Pathogens Disease-causing microorganisms

Peptide bond Amide group that connects two amino acids; found in proteins

Periodic table Arrangement of elements in atomic number order in rows so that elements with similar properties fall together in vertical columns

Periods Horizontal rows in the periodic table

Permeable Allows substances to pass through

Percolation Movement of water through subsoil and rock formations

Pesticides Chemicals used to kill pests

    Insecticides Pesticides that kill insects

    Herbicides Pesticides that kill weeds

    Fungicides Pesticides that kill fungi

Petroleum fractions Mixtures of hundreds of hydrocarbons with boiling points in a certain range that are obtained by fractional distillation of petroleum

pH Negative logarithm of the hydronium ion concentration of a solution

Pharmacology The branch of medicine concerned with the uses, effects, and modes of action of drugs

Photochemical smog Smog formed by the action of sunlight on photoreactive pollutants in the air

Photodissociation Decomposition of a reactant, caused by the energy of light

Physical properties Properties that can be observed without changing the identity of a substance

Plasma A high-temperature state of matter consisting of unbound nuclei and electrons

Polar Describes a bond or molecule in which charge is unevenly distributed, creating positive and negative regions

Polyatomic ion A positive or negative ion composed of two or more atoms

Polymers Very large molecules composed of repeating units derived from small molecules

Polypeptide Condensation polymer of amino acids

Positron A positively charged electron

Potable Describes water suitable for drinking

Potential energy Energy in storage by virtue of position or arrangement

Precipitation Formation of a solid product during a chemical reaction between reactants in solution

Prescription drugs Drugs that can be purchased only at the direction of a physician

Pressure Force per unit area

Primary battery A battery that cannot be recharged because its reaction is not easily reversible; a throw-away battery

Primary nutrients Major elements required by plants (nitrogen, phosphorus, potassium)

Primary pollutants Pollutants emitted directly into the atmosphere

Primary protein structure Amino acid sequence in a protein

Products Substances formed as the result of a chemical reaction

Proteins Biomolecules that are polymers of amino acids; may also include nonamino acid parts

Protons Positively charged subatomic particles found in the nucleus

Q

Qualitative Describes information or experiments that are not numerical

Quantitative Describes information or experiments that are numerical

Quantum The smallest increment of energy, for example, in an atom emitting or absorbing radiation

Quaternary protein structure Structure of a protein with more than one polypeptide

Quick-release fertilizers Water soluble

R

Racemic mixture Mixture of equal amounts of enantiomers; does not rotate polarized light

Radiational cooling The phenomenon of enhanced loss of heat from Earth on a clear, cloudless night due to the absence of significant water vapor in the atmosphere in the form of clouds

Radioactivity Spontaneous decomposition of unstable atomic nuclei; produces alpha, beta, and gamma radiation

Rainout The process by which pollutants in the air are removed by natural precipitation such as rain

Reactants Substances that undergo chemical change in a chemical reaction

Reaction rate Amount of reactant converted to product in a specific amount of time

Receptor A molecule or portion of a molecule that interacts with another molecule to cause a change in biochemical activity

Reducing agent A reactant that causes reduction of another reactant

Reduction The loss of oxygen, gain of hydrogen, or gain of electrons

Reformulated gasoline Gasoline whose composition has been changed to reduce the percentage of olefins, aromatics, and sulfur and to add oxygenated compounds

Reinforced plastics (composites) Plastics with fibers of a substance, such as graphite, imbedded in a polymer matrix

Replication The process by which DNA is copied when a cell divides

Representative or main-group elements Elements in the A groups of the periodic table

Resonance structures Two or more possible Lewis dot representations of a molecule or ion in which the only difference is the position of the valence electrons. The position of the nuclei remain constant. The actual structure is a hybrid of the structures..

Reverse osmosis The application of pressure on a solution to cause water molecules to flow through a semipermeable membrane from a more-concentrated to a less-concentrated solution

Ribonucleic acids (RNAs) Nucleic acids that transmit genetic information and direct protein synthesis; contain ribose

Roasting (of ore) Heating an ore in air, usually to produce an oxide

S

Salt Ionic compound composed of the cation from a base and the anion from an acid

Saturated hydrocarbons Hydrocarbons that are alkanes

Saturated solution Solution that has dissolved all the solute that it can dissolve at a given temperature

Second law of thermodynamics The total entropy of the universe is constantly increasing

Secondary battery A battery that can be recharged by reversing the flow of current, which reverses the current-producing reaction and regenerates the reactants

Secondary nutrients Elements required by plants in moderate amounts (calcium, magnesium, sulfur)

Secondary pollutants Pollutants formed by chemical reactions in the atmosphere

Secondary protein structure Shape of the backbone structure of a protein

Sedation Induction of a relaxed state, usually by a medication

Selective herbicides Herbicides that kill only a particular group of plants

Semiconductors Materials with electrical conductivity intermediate between that of metals and insulators

Semipermeable membrane A membrane that allows water molecules but not ions or larger molecules to pass through

Sequestrant Chemical that ties up metal atoms or ions by forming bonds with them; used as food additives

Shell A principal energy level defined by a given value of n

Silica Naturally occurring silicon dioxide

Silicates Minerals composed of metal cations and silicon—oxygen anions

Slag Mixture of nonmetallic waste products separated from a metal during its refining

Slow-release fertilizers Combinations of plant nutrients that are slow to dissolve in water

Solubility The maximum quantity of a solute that will dissolve in a given amount of a solvent at a specified temperature

Solute In a solution, the substance dissolved in the solvent

Solutions Homogeneous mixtures, which may be in the solid, liquid, or gaseous state

Solvent In a solution, the substance present in the greater amount

Specific heat capacity The amount of heat required to raise the temperature of a 1 g sample of matter of a given size by 1°C

Steady state A state in which the concentration of a chemical remains essentially constant even though the chemical may be undergoing numerous chemical reactions

Steel Malleable iron-based alloy with a relatively low percentage of carbon

Stereoisomerism Describes isomers with the same molecular formulas and the same atom-to-atom bonding sequence, but different arrangement of the atoms in space

Stereoisomers Molecules with the same molecular formula but different three-dimensional structures

Steroids Lipids with a four-ring structure; they include cholesterol and male and female sex hormones

Straight-chain isomers Structural isomers of hydrocarbons with no side chains

Straight fertilizers Contain only one nutrient

Stratosphere The region of Earth's atmosphere that extends from the troposphere to an altitude of about 50 km (~31 mi)

Strong acid Acid that is 100% ionized in aqueous solution

Strong base Base that is 100% ionized in aqueous solution

Structural formulas Chemical formulas that show the connections between atoms in molecules as straight lines

Structural isomers Isomers that differ in the order in which the atoms are bonded together

Sublimation The process whereby molecules escape from the surface of a solid into the gas phase

Subscripts In chemical formulas, numbers written below the line (for example, $_2$ in H_2O) to show numbers or ratios of atoms in a compound

Subshells Different energy levels (orbitals) within a given shell

Subsoil The layer of soil just beneath the topsoil

Superconductor A substance that offers no resistance to electrical flow

Surface tension The amount of energy required to overcome the attraction for one another of molecules at the surface of a liquid

Sustainable agriculture An approach to agriculture between the extremes of agrichemical non-based organic farming and the heavy use of agrichemicals whose broad goal is to provide food for the world's population while minimizing environmental impact. A related, but not synonymous term, is *alternative agriculture*, which could include nonconventional production methods such as organic farming and the production of more nontraditional crops, livestock, and farm products.

T

Technology Application of scientific knowledge in the context of industrial production, our economic system, and our societal goals

Teratogen A chemical or factor that causes malformation of an embryo

Tertiary protein structure Shape of an entire folded protein chain

Thermal inversion An atmospheric condition in which warmer air is on top of cooler air

Thermodynamics The science of energy and its transformations

Topsoil The layer of soil exposed to the air

Trade name The manufacturer's name for a drug

Transgenic crops Crops genetically altered by the techniques of genetic engineering

***Trans* isomers** Stereoisomers with groups on opposite sides of a carbon–carbon double bond

Transcription The process by which the information in DNA is read and used to synthesize RNA

Transistor Semiconductor device that controls electron flow in circuits

Transition elements Elements in periodic table rows 4–7 in which d or f orbitals are being filled; they lie between main-group elements

Translation Process for sequential ordering of amino acids, directed by RNA during protein synthesis

Transuranium elements Artificial elements, all with atomic numbers greater than 92

Triglycerides Triesters of glycerol and fatty acids; fats and oils

Triple bond A bond in which three pairs of electrons are shared between two atoms

Troposphere The region of Earth's atmosphere that extends from sea level up to about 10 km (1 km = 0.62 mi)

U

Unsaturated hydrocarbons Hydrocarbons that contain fewer than the maximum number of hydrogen atoms; alkenes and alkynes

Unsaturated solution Solution in which less than the maximum amount of solute is dissolved at a given temperature

Uranium series The series of steps in the naturally occurring decay of uranium-238 to lead-206

V

Valence electrons Outermost electrons in an atom

Vapor pressure The pressure above a liquid caused by molecules that have escaped from the liquid surface

Vitamin Organic compound essential to health that must be supplied in small amounts by the diet

Volatile Easily vaporized

Vulcanized rubber Rubber with short chains of sulfur atoms that bond together the polymer of natural rubber

W

Weak acid Acid that is only partially ionized in aqueous solution and establishes equilibrium with nonionized acid

Weak base Base that is only partially ionized in aqueous solution and establishes equilibrium with nonionized base

Credits

These pages constitute an extension of the copyright page. We have made every effort to trace the ownership of all copyrighted material and to secure permission from copyright holders. In the event of any question arising as to the use of any material, we will be pleased to make the necessary corrections in future printings. Thanks are due to the following authors, publishers, and agents for permission to use the material indicated.

PHOTOGRAPHS

Chapter Opener 1. 1: © Scientific American, Inc./George V. Kelvin
Chapter Opener 2. 17: Charles D. Winters
Chapter Opener 3. 39: Charles D. Winters
Chapter Opener 4. 71: Charles D. Winters
Chapter Opener 5. 93: Charles D. Winters
Chapter Opener 6. 125: © Alexei Kalmykov/Reuters/Corbis
Chapter Opener 7. 143: Eyebyte/Alamy
Chapter Opener 8. 161: Brand X Pictures/Getty Images
Chapter Opener 9. 185: Charles D. Winters
Chapter Opener 10. 203: NASA
Chapter Opener 11. 221: Marvy!/Corbis
Chapter Opener 12. 251: © Age Fotostock/Superstock
Chapter Opener 13. 285: National Solar Observatory/NOAO
Chapter Opener 14. 317: © Claudius/zeta/CORBIS
Chapter Opener 15. 351: © Digital Art/Corbis
Chapter Opener 16. 397: © Stockbyte Platinum/Getty Images
Chapter Opener 17. 425: Royalty-free/CORBIS
Chapter Opener 18. 461: Courtesy of Fraunhofer-Institut
Chapter Opener 19. 485: EyeWireCollection/Getty Images

Chapter 1. 2: center left, Charles D. Winters **3:** bottom right, Larry Cameron **5:** bottom left, Stapleton Collection/Corbis **8:** center left, Charles D. Winters **10:** center left, Charles D. Winters **11:** © Image Source/ Getty Images **13:** Grant Heilman/Grant Heilman Photography

Chapter 2. 18: top left, IBM Research, Almaden Research Center **20:** top left, (a) ©Brand X Pictures/Getty Images; top right, (b) Photodisc Green/Getty Images; bottom left, (a) Tom Pantages ; bottom right, (b) Ken Edward/Science Source/Photo Researchers **22:** top center, Charles D. Winters. **24:** bottom left, © Grant Heilman **24:** bottom center, © Scott Camazine/Photo Researchers, Inc. **24:** bottom right, © Clive Freeman/Biosym Technologies/Photo Researchers, Inc. **25:** © The Granger Collection **26:** center left, Charles D. Winters **28:** top left,Charles D. Winters **29:** top center, © Scott Camazine/ Photo Researchers, Inc. **31:** bottom right, Charles D. Winters **33:** center right, Charles D. Winters

Chapter 3. 41: top right, The Bettmann Archive/Corbis **44:** center, C.E. Wynn-Williams. Courtesy AIP Emilio Segre Visual Archives **46:** top left, IBM Corporation Research Division, Almaden Research Center **48:** top left, Charles D. Winters **49:** bottom left, Runk/Schoenberger/Grant Heilman, Inc. **51:** top right, American Institute of Physics **57:** top left, Oesper Collection in the History of Chemistry/University of Cincinnati **50:** top left, Charles D. Winters **50:** bottom center, Courtesy of Department of Athletics, Vanderbilt University **63:** bottom left, Charles D. Winters **64:** bottom left, Charles D. Winters

Chapter 4. 75: bottom right, Charles D. Winters **76:** top left, Richard T. Nowitz/Corbis; center left, (a) Dan McCoy/Rainbow; (b) center left, Jack Rosen/Photo Researchers, Inc. **78:** center, (b,c) John D. Cunningham/Visuals Unlimited **80:** center left, Alan Pitcairn/Grant Heilman Photography **81:** top right, Charles D. Winters **86:** bottom left, Charles D. Winters **87:** center right, © Yoav Levy/Phototake NYC

Chapter 5. 96: center left, Larry Cameron **97:** center right, Charles D. Winters **100:** center left, Charles D. Winters **103:** bottom right, Frances Simon, Courtesy AIP Emilio Segre Visual Archives **107:** bottom, Charles D. Winters **115:** bottom center, Charles D. Winters **116:** center left, Charles D. Winters **118:** top center, Charles D. Winters

Chapter 6. 126: center left, © Roger Ressmeyer/Corbis **127:** top right, © NASA **129:** bottomright, © Photodisc Green/Getty Images **134:** center left, Comstock Images/Getty Images **135:** center right, Stock Connection/Alamy **136:** center left, Royalty-Free/Corbis **139:** top right, Dane Andrew/ZUMA/Corbis; bottom right, Royalty-Free/Corbis

Chapter 7. 148: top right, Courtesy of S. Solomon **150:** center right, © Paul R. Kennedy/University of California, Irvine; top right, (a) Bossu Regis/Corbis Sygma; top right, (b) Brooks Craft/Corbis **151:** bottom center, Courtesy of Robinair

Chapter 8. 166: top center, Charles D. Winters **168:** top right, Bettmann/Corbis **169:** top and center right, Charles D. Winters **171:** center right, Richard Megna/Fundamental Photographs; bottom right, Charles D. Winters **172:** top, AP/Wide World Photos; top left, Larry Cameron **173:** center right, Ecosphere Associates, Ltd. Of Tucson, Arizona **176:** top right and bottom center (a, b), Charles D. Winter **178:** bottom right, Dan McCoy/Rainbow **179:** center right, Institute of Scrap Recycling Industries, Inc.

Chapter 9. 187: center right, Charles D. Winters **188:** center left, Charles D. Winters **191:** center right, © 1990 M & A Doolittle/Rainbow **192:** top left, © Leonard Lessin/Peter Arnold, Inc. **193:** center right, Charles D. Winters **195:** center right, Charles D. Winters **196:** bottom left, (a) Charles Steele; bottom right, (b) Leon Lewandowski **198:** bottom left, Charles D. Winters **199:** bottom right, Larry Cameron

Chapter 10. 205: top right, Charles D. Winters **206:** top left, Yoav Levy/Phototake NYC **206:** top right, © Bettmann/Corbis **208:** top left, Charles D. Winters **210:** bottom center, (a,b,c) Charles D. Winters **211:** bottom right, Charles D. Winters **213:** center right, © 1990 Richard Megna/Fundamental Photographs **215:** top right, American Honda Motor Co., Inc **217:** bottom right, Tom Pantages

Chapter 11. 223: center right, Charles D. Winters **224:** bottom center, © Charles D. Winters; bottom right, (d) Richard Megna/Fundamental Photographs; bottom left, (a) Runk/Schoenberger/Grant Heilman Photography **226:** center left, Simon Fraser/SPL/Photo

Researchers **227:** bottom right, Bruce F. Molnia/Terraphotographics **229:** bottom right, M. Timothy O'Keefe/Tom Stack & Associates **231:** top right, Courtesy of U.S. Geological Survey **232:** center left, Charles D. Winters **233:** center right, USDA/Soil Conservation Service **234:** bottom center, Dennis Barnes **236:** center left, Alan Pitcairn/Grant Heilman Photography **240:** top left, Grant Heilman Photography **242:** bottom left, Doug Wechsler **410:** top right, Runk/Schoenberger/Grant Heilman Photography

Chapter 12. 259: center right, © Matthia Kulka/Corbis **259:** center right, Charles D. Winters **261:** top, Clive Freeman, The Royal Institution/Photo Researchers, Inc. **263:** bottom right, © Robert Harding Picture Library Ltd./Alamy **266:** bottom left, Kjell B.Sandved/Visuals Unlimited **267:** bottom center, © Charles D. Winters **268:** top right, (c, c) Charles D. Winters **278:** center right, Charles D. Winters **272:** center left, Jan Halaska/Photo Researchers, Inc. **277:** top center, (b) Charles D. Winters **278:** center left, Hank Morgan/Rainbow **279:** bottom center, © CHAIWAT SUBRASOM/Reuters/Corbis **280:** top right, © Bettmann/CORBIS; bottom center, © David R. Frazier Photolibrary Inc., Alamy

Chapter 13. 287: bottom right, © Stock Montage **294:** top, Beverly March, Courtesy of Long Island Jewish Hospital **298:** top right, Lawrence Berkeley National Laboratory **299:** bottom right, Charles D. Winters **302:** bottom, (a, b) © CNRI/SPL/Photo Researchers, Inc **305:** top right, AIP Emilio Segre Visual Archives Herzfeld Collection; bottom right, The Bulletin of Atomic Scientists **307:** top right, © Dan McCoy/Rainbow

Chapter 14. 327: center left, Charles D. Winters **323:** bottom right, Charles D. Winters **326:** center right, Charles D. Winters **327:** bottom right, Charles D. Winters **329:** top right, Charles D. Winters **330:** center left, Charles D. Winters **331:** bottom left, © Pascal Goetghche/Photo Researchers, Inc. **332:** center left, Charles D. Winters **333:** bottom right, Charles D. Winters **334:** top left, Charles D. Winters **337:** top right, Charles D. Winters **338:** bottom left, Custom Medical Stock Photo; top left, Charles D. Winters **339:** bottom center, Charles D. Winters **340:** center right, Andrew Hollbrook/CORBIS **341:** top right, Courtesy of Dupont bottom center **342:** top center, (b) © Ann Kotowicz/Rainbow; top left, (a) © Tom Hollyman/Photo Researchers, Inc.; top right, © Tom Pantages **343:** top right, Patagonia, Inc.

Chapter 15. 352: center left, Charles D. Winters **353:** top right, Charles D. Winters **356:** center, Charles D. Winters **358:** top left, Charles D. Winters **359:** bottom right, Nigel Cattlin/Photo Researchers, Inc. **361:** top right, Charles D. **368:** center left, North Wind Picture Archives **370:** top left, Charles D. Winters **371:** bottom right, Charles D. Winters **373:** top left, Charles D. Winters **382:** top left, Charles D. Winters **383:** bottom right, Charles D. Winters **388:** center left, Charles D. Winters and Susan Young **389:** top right, Biophoto Associates/Photo Researchers, Inc. **377:** © George V. Kelvin

Chapter 16. 402: bottom left, Jose Luis Palaez, Inc./Corbis; center left, Michael Cole/Corbis **403:** top right, Charles D. Winters; bottom right, Prof. P. Motta/Dept. of Anatomy/Univeristy "La Sapienza," Rome/SPL/Photo Researchers **409:** center right and bottom right, Charles D. Winters **410:** bottom left, Charles D. Winters **412:** center left, © Michael Klein/Peter Arnold, Inc. **414:** top left, Charles D. Winters **415:** center right, Charles D. Winters **416:** top, Charles D. Winters **418:** bottom and center left, Charles D. Winters **419:** bottom right, Charles D. Winters **326:** top left, C.D. Winters

Chapter 17. 428: bottom left, (a) © Dr. Tony Brain/SPL/Photo Researchers, Inc.; bottom right, (b) " NIBSC/SPL/Photo Researchers, Inc. **429:** top right, SPL/Photo Researchers, Inc.; center right, Andrew McClenaghan/SPL/Photo Researchers, Inc. **431:** top right, © Will and Deni McIntyre/Photo Researchers, Inc. **432:** center right, Claro Cortes IV/Reuters/CORBIS **433:** top center, Courtesy of Abbott Laboratories **440:** center left, Dr. Jeremy Burgess/SPL/Photo Researchers, Inc.; bottom left, Charles D. Winters **444:** center left, © R. Konig/Jacana/Photo Researchers, Inc.; bottom left, © Carolina Biological Supply Company, Phototake NYC **445:** top center, © Scott Camazine/Photo Researchers, Inc. **447:** top left, (a) Coco McCoy/Rainbow; top right, (b) Buddy Mays/CORBIS; center right, © Bryant F. Peterson/CORBIS **449:** top and bottom right, Charles D. Winters; center right, David Scharf/Peter Arnold, Inc.

Chapter 18. 463: bottom right, Christine L. Case/Visuals Unlimited **467:** center, © 1988 Paul Silverman/Fundamental Photographs **468:** top right, © Photo Alto/ Getty Images **469:** top right, T. J. Florian/Rainbow **470:** bottom left, James Stoot/Lawrence Livermore National Laboratory; top right, Sam Forencich **471:** top right, James Cowlin/Image Enterprises; top right, Charles D. Winters **472:** top left and center, (a, b) Charles D. Winters; top right, © Sam Ogden/Photo Researchers. Inc. **473:** bottom center, (a, b) Courtesy of AT&T Archives **474:** Roger Dubruisson/CORBIS **475:** top right, Courtesy of Siemens Solar Industries, Camarillo, California **477:** center right, Charels D. Winters **478:** bottom left, (a) © James L. Amos/Peter Arnold, Inc.; bottom center and right, (b, c) Tom Pantages **479:** bottom right, Charles D. Winters **480:** top left, © Fulvio Roiter/Corbis **481:** top right, © Fulvio Roiter/Corbis

Chapter 19. 488: top left, USDA/Soil Conservation Service **491:** top right, Gordon Garrado/SPL/Photo Researchers, Inc. **495:** bottom right, © Scott T. Smith/Corbis **497:** center right, William Felger/Grant Heilman Photography **499:** center right, Charles D. Winters **501:** top right, Gilbert S. Grant/Photo Researchers, Inc. **503:** bottom right, T. McCabe/Visuals Unlimited **504:** center left, Steve Feld **505:** top, USDA; center left, Robert E. Ford/Terraphotographics **507:** bottom right, Dan McCoy/Rainbow

TABLES AND FIGURES

Chapter 12. 264: Figures 12.4a, b, International Energy Outlook 2005, www.eia.doe.gov/oiaf/ieo/index.html.

Chapter 14. 343: Figure 14.13, 2003 National Post-Consumer Plastics Recycling Report of the American Plastics Council, http://www.plasticresource,com/S_plasticsresource/docs/1700/1646.pdf. **344:** Figure 14.15, 2003 National Post-Consumer Plastics Recycling Report of the American Plastics Council, http://www.plasticresource.com/S_plasticsresource/docs/1700/1646.pdf.

Chapter 17. 454: Figure 17.6 and 17.8, From Jemal et al., *CA: A Cancer Journal for Clinicians* 55(1): 10–30, Fig. IG1. Copyright © 2005 by American Cancer Society. Reprinted by permission **456:** Figure 17.8, From Jemal et al., *CA: A Cancer Journal for Clinicians* 55(1): 10–30, Fig. IG5. Copyright © 2005 by American Cancer Society. Reprinted by permission.

Index

Page numbers followed by *f* denote figures; page numbers followed by *t* denote tables.